Techniques in Protein Chemistry IV

TECHNIQUES IN PROTEIN CHEMISTRY IV

Edited by

Ruth Hogue Angeletti
Albert Einstein College of Medicine
Bronx, New York

ACADEMIC PRESS, INC.
Harcourt Brace & Company
San Diego New York Boston
London Sydney Tokyo Toronto

Academic Press Rapid Manuscript Reproduction

This book is printed on acid-free paper. ∞

Academic Press, Inc.
1250 Sixth Avenue, San Diego, California 92101-4311

United Kingdom Edition published by
Academic Press Limited
24–28 Oval Road, London NW1 7DX

International Standard Book Number: 0-12-058757-2 (Hardcover)

International Standard Book Number: 0-12-058758-0 (Papercover)

PRINTED IN THE UNITED STATES OF AMERICA
93 94 95 96 97 98 EB 9 8 7 6 5 4 3 2 1

Contents

Section IV
Peptide Synthesis

Section V
Amino Acid Analysis

Section VIII
Protein Sequencing

Section IX
Protein Conformation, Folding, and Modeling

Section X
NMR Analysis of Protein and Peptide Structure

Foreword

For most ventures it would be premature to assert that a fourth volume of the series ''carries on a distinguished tradition,'' but a case can certainly be made for such a description of this volume. This is the fifth book based on abstracts selected from those presented at a Protein Society Symposium (the fourth to carry the title ''Techniques in Protein Chemistry''). They were selected from an all-time record of 615 abstracts submitted to this Sixth Protein Society Symposium. The task of the associate editors was a difficult one indeed.

We would have been overwhelmed by the flood of abstracts had it not been for our friends in the Association of Biomolecular Resource Facilities (ABRF) who meet with the Protein Society Symposia. They volunteered to organize the poster sessions and did a superb job. All agreed that the posters were of very high quality. I am pleased that they have once again volunteered their services for next year.

The Seventh Symposium will be held July 24–28, 1993, in the Town and Country in San Diego. This site will not only increase poster space, but also the time in which they can be examined, so we look forward to an even bigger and more interactive meeting.

My most sincere thanks on behalf of the Protein Society to Ruth Hogue Angeletti and her associate editors for their dedicated work on this volume. Their efforts and the top quality contributions by the authors of the articles do indeed ''carry on a distinguished tradition.''

Mark Hermodson
President
The Protein Society

Preface

The president of the Protein Society has declared that this series of volumes is now a "distinguished tradition." However, this tradition has a dynamic quality, for the "Techniques in Protein Chemistry" series must embody change. In the first volume, advances in the organic chemistry of proteins—sequencing, fragmentations, separations—formed the core of the book. In this volume, the rapid developments in mass spectrometry of proteins dominate. Breakthroughs in technology enable new insights into protein structure and function. It is our job as editors and associate editors to put aside our parochial interests and continue to ensure that areas of rapid development are well represented.

The associate editors are at the forefront of identifying important contributions. This year they were John Crabb, Sheenah Mische, Al Smith, John Stults, Wesley Stites, and Leonard Spicer. Once again, the staff of Academic Press has ensured that the book will be published rapidly and as error free as possible. The members of the society are responsible for the continued high quality of this volume.

Ruth Hogue Angeletti

Acknowledgments

The Protein Society acknowledges with thanks the following organizations who through their support of the Society's program goals contributed in a meaningful way to the sixth annual symposium and thus to this volume.

AAA Laboratory
Amersham
Amgen, Inc.
Applied Biosystems, Inc.
Beckman Instruments, Inc.
Biosym Technologies, Inc.
Boehringer Mannheim Corporation
Digital Equipment Corporation
Dionex Corporation
Du Pont Biotechnology Systems
EM Science
Finnigan MAT
Genentech, Inc.
Hewlett-Packard
Jasco, Inc.
Lilly Research Laboratories
Merck Sharp & Dohme Research Laboratories
Michrom BioResources, Inc.
MilliGen, Inc.
Millipore Corporation
Molecular Dynamics, Inc.
Molecular Simulations, Inc.
Monsanto Co.
National Biomedical Research Foundation
Oxford Molecular, Inc.
PerSeptive Biosystems, Inc.
Pharmacia LKB Biotechnology
Pickering Laboratories, Inc.
Promega Corporation
Rainin Instrument Co., Inc.
Savant Instruments, Inc.
Scientific Analysis Laboratories Ltd.
SCIEX
TosoHaas, Inc.
Tripos Associates, Inc.
The Upjohn Co.
Vestec Corporation
VYDAC, The Separations Group, Inc.

SECTION I

Mass Spectrometry of Proteins and Peptides

Mass Spectrometry in Protein Sequence and Structural Investigations

A. L. Burlingame
Department of Pharmaceutical Chemistry
The Mass Spectrometry Facility and the Liver Center
University of California, San Francisco, California

I. Overview

The growing realization of the importance of mass spectrometry in protein science is based upon quite recent recognition of the relative ease with which classically difficult or often intractable questions in protein biology may now be readily and successfully addressed. The progression of events which underpin the present state of the art was triggered by discoveries of revolutionary new ionization techniques, viz. LSIMS or FAB, electrospray and matrix-assisted laser desorption, able to handle polar, labile macromolecules mass spectrometrically and the subsequent development of new instruments utilizing these techniques. These methods are virtually ideally suited to deal effectively with the challenges of protein sequencing and structural analysis at the picomole level. Currently active "export" of some of this new instrumental capability to the biochemistry laboratories themselves is under way due to the burgeoning availability of more user-friendly, relatively low cost commercial instrumentation. The most recent examples are instruments designed to permit exploitation of the two newest ionization methods, namely electrospray (ES) (1) and matrix-assisted laser desorption (maLD) (2) techniques.

Over the last decade a number of mass spectrometry laboratories have been key participants in establishing the unique power and versatility of these techniques in solving previously tedious and often impossible problems. There is a growing literature of examples of questions which have been tackled successfully representing increasingly more difficult biological problems (See recent reviews 3-6). The present excitement about mass spectrometry in the protein biochemistry community is not due to any *recently* recognized need to have a better structural understanding of the machinery of cells, but rather to a growing awareness that the tools are now in hand with which to discover and dissect even minute structural details with high accuracy and fidelity through application of these mass spectrometric techniques. Indeed, the time is ripe to exploit these new methodologies, both to gain a deeper

TECHNIQUES IN PROTEIN CHEMISTRY IV

understanding of protein biochemistry with a reasonable investment of time and effort, and save labor costs incurred in pursuing outmoded, ineffective biochemical strategies in pursuit of similar goals. With the arsenal of new methods and instruments, and the ease with which definitive answers can now be obtained, it is almost to the point where laboratories cannot afford *not* to have access to mass spectrometry in order to remain competitive. This point has been well illustrated by the immediate massive investments that several of the leading biotechnology companies have made to procure the entire suite of new instruments and methods of mass spectrometry over the last several years. Already, many traditional academic protein laboratories have begun to add mass spectrometry to their capability. These trends will certainly spread rapidly as lower cost instrumentation becomes commercially available.

II. Using the Tools Now Available

How do you go about solving your problem or getting into the business? Initially the easiest way is to become a user or collaborator with an established protein mass spectrometry group. The NIH/NSF biotechnology resources (7) will provide assistance in both how to go about solving your problem and the access to necessary instrumentation and interpretative expertise without the need of making a major investment yourself. Depending on your needs, anything from obtaining a few spectra on your synthetic peptides to taking a mini-sabbatical to learn the techniques and solve your problem firsthand can be arranged.

Regardless of how you choose to proceed, it is of prime importance to discuss how mass spectrometry may eventually be used in solving your problem with an experienced protein mass spectrometrist *prior* to starting *isolation* of your protein. This dialogue can save both parties time and considerable mutual frustration by *determining at the outset* what the best stage in the problem would be to involve the use of the different capabilities of mass spectrometry (including LC/MS) and particularly to avoid use of common biochemical experimental materials which could preclude eventual successful mass spectrometric analysis. These include non-volatile buffers, detergents which are mixtures of polymers (e.g. Triton X-100, NP-40, etc.), plasticized labware, ion exchange columns which bleed polymers, and impure solvents, to mention a few. Volatile buffers such as ammonium acetate or bicarbonate and N-ethylmorpholine acetate, octyl beta glucoside and SDS can be successfully removed (8). You will recall that matrix-assisted laser desorption ionization is said to be more "tolerant" of the presence of common biochemical "cocktails" (2) used to preserve biological activity, including SDS. While that is true to some extent, one of maLD's real practical advantages will probably be felt in the discovery of how to experimentally "handle" new difficult problems, such as heterogeneous membrane-bound glycoproteins (9), hydrophobic transmembrane components (10), etc. Bear in mind, for sequence and structural analysis of components in your chemical or enzymatic digest, it is likely that it will still be necessary to make use of other ionization techniques, vibrational excitation and ion optical systems possessing different capabilities which will *not*, unfortunately, tolerate these ingredients in the sample (11).

III. The Tools

A. *Ionization: Determination of Molecular Weight, Purity and Heterogeneity*

Three ionization methods are in common use; namely (1) sputtering of surface active samples from a viscous liquid surface by irradiation with kiloelectron volt kinetic energy cesium ion (liquid secondary ion mass spectrometry, LSIMS) or xenon neutral atom(fast atom bombardment, FAB) beam; (2) nebulization of an acidified solution of a sample into an atmospheric pressure desolvation bath gas (electrospray); and (3) irradiation of a solid solution of sample with a laser frequency tuned to the chromophore of the solvent matrix.

These ionization/ejection techniques can be thought of as sampling of acid-base type ionization equilibria existing in solution, either by protonation of a basic site in a given molecule (M) under acidic conditions to form MH^+ or conversely, deprotonation of an acidic site to form $(M\text{-}H)^-$. These processes are sometimes referred to as "soft" ionization due to the fact that the ejection (desorption and desolvation) of these naked MH^+ or $(M\text{-}H)^-$ species into the gas phase occurs while maintaining their ambient vibrational energy content, thus creating a preponderance of low internal energy intact molecular species. This is an ideal situation for very sensitive detection of molecules and their mixtures, but obviously not for investigation of sequence or structural questions as discussed below. It should be mentioned in passing that the accuracy with which mass can be measured using any ionization method depends on the quality and performance as well as the proper calibration of the ion optical system used (4). It is crucial to be able to routinely measure the nominal mass of peptides up to 3000-4000 Da to $\pm$ 0.2 Da on the ^{12}C mass scale in order to be sure of a proper composition assignment (for example the mass difference of asn vs. asp or a C-terminal carboxyl vs. C-terminal amide is only 1.0 Da).

B. *Collision-Induced Dissociation of MH^+ or $(M\text{-}H)^-$ (CID)*

Sequence and structural information can be obtained readily by interpretation of the mass spectra obtained by using tandem mass spectrometers. Tandem mass spectrometers function both to separate (in MS-1) a selected component from impurities or mixtures as well as to vibrationally activate that selected component to undergo gas phase unimolecular dissociation reactions then recorded by MS-2 as its collision-induced dissociation (CID) mass spectrum. There are two ion activation energies currently in common use, namely high and low, which are carried out on four sector and triple quadrupole instruments, respectively (12). As might be expected intuitively, the high energy mode deposits more vibrational energy in the MH^+ ion selected and hence in addition to peptide backbone cleavages (also observed under low energy conditions), both amino acid side chain cleavages and immonium ions are formed as well. This class of higher energy fragmentation processes is of considerable importance in permitting the distinction between and assignment of the isobaric pair leu/ile (13), the location (14-16) and structure (15, 16) of covalent modifications and induction of saccharide ring cleavages which permit branching assignments in C-terminal glycan anchors (17) to give a few examples.

C. LC Coupled ES-MS

The combination of reverse phase liquid chromatographic systems with electrospray mass spectrometers has become a routine operation in many laboratories. It is the first LC/MS combination which is technically virtually ideally suited for the molecular weight mapping of protein and glycoprotein digests. It has replaced "flow FAB" due to superior performance in the size range accessible, and the sensitivity and faithfulness of detection of most components ranging from the very hydrophilic tri-mer sizes (but retained on the RP-HPLC column) to all but the most hydrophobic in a single gradient (18, 19). For example, detection of larger peptides from incomplete digestion and large glycopeptides may also be readily achieved. From selected reaction monitoring of monosaccharide ions formed under collisional activation conditions (20, 21), the retention times of both small and large glycopeptides may be detected.

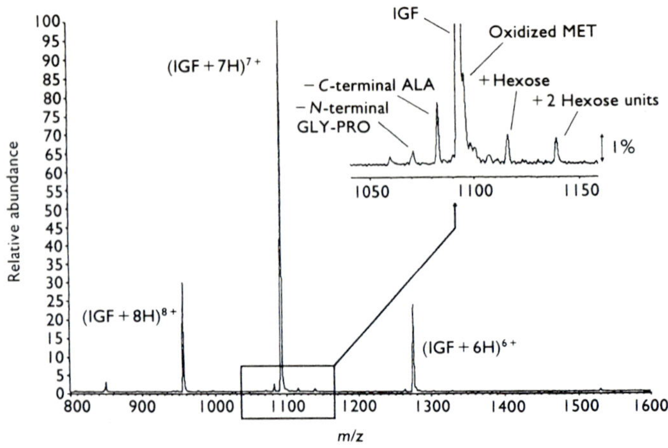

Figure 1. The ES spectrum of recombinant IGF illustrating the detection and identification of minor components at the 1-3% level. (Reproduced with permission from ref. 3.)

IV. Examples Using the Tools

A. *Molecular Weight Determination, Purity*

At this writing electrospray ionization employing either quadrupole or sector mass analyzers offers the highest accuracy of mass measurement at the highest mass resolutions (18, 19, 22). In addition, sensitivities are adequate for quadrupoles [low picomole quantities (18, 19, 23)] and are significantly better (low femtomoles) using sector instruments with multichannel array detection systems (24).

It is straightforward to obtain an accurate measurement of the molecular weight together with reasonably faithful detection, molecular weight measurement and semi-quantitation of other "similar" components present in a peptide or protein sample. For example, consider the ES-MS of recombinant IGF shown in Figure 1. As can be seen from the signals accompanying the multiply (MH_7^{7+} chosen for example) protonated molecular ion region occurring around m/z 1090, six additional mass resolved IGF-related polypeptides are present at the 1-3% level. The masses observed for these correspond to components having N- and C-terminal truncations, methionine oxidation and mono- and di-mannosylation (25).

A further example concerns analysis of a sample of commercial "ribonuclease B". The ES-MS is shown in Figure 2. Signals representing six significant molecular species are observed between 13,500 and 16,000 Daltons. These signals correspond

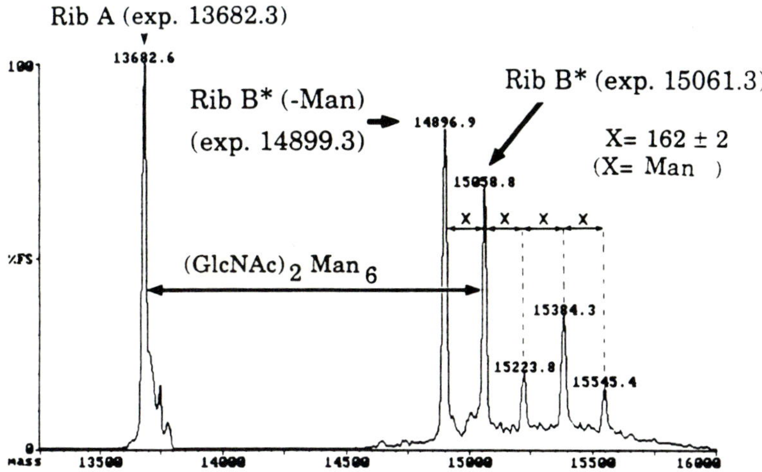

Figure 2. Electrospray mass spectrum of ribonuclease B (bovine pancreatic, Sigma R 5500) purified by reverse phase HPLC. Solvent: H_2O:CH_3OH 1:1, 2% acetic acid. (Reproduced with permission from author.)

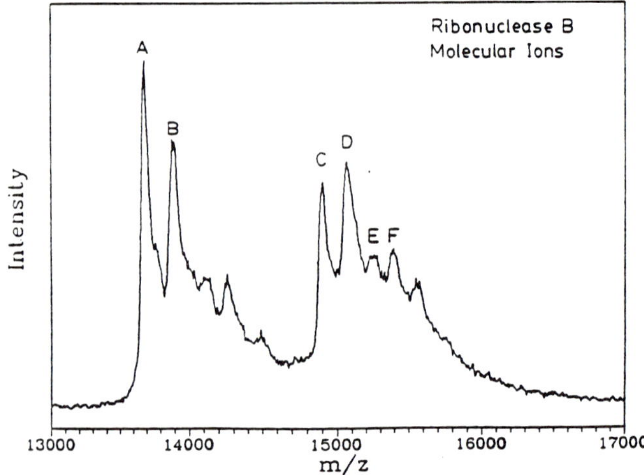

Figure 3. The protonated molecular ion region of the mass spectrum of a sample of bovine pancreatic ribonuclease B, using sinapinic acid as the matrix and $\lambda = 355$ nm. The peaks are: (A) ribonuclease A; (B) ribonuclease A + 1 N-acetylglucosamine; (C) - (F) different glycosylation states of ribonuclease B, i.e.. ribonuclease A + di-N-acetylchitobiose + 5-9 mannose residues. (Reproduced with permission.)

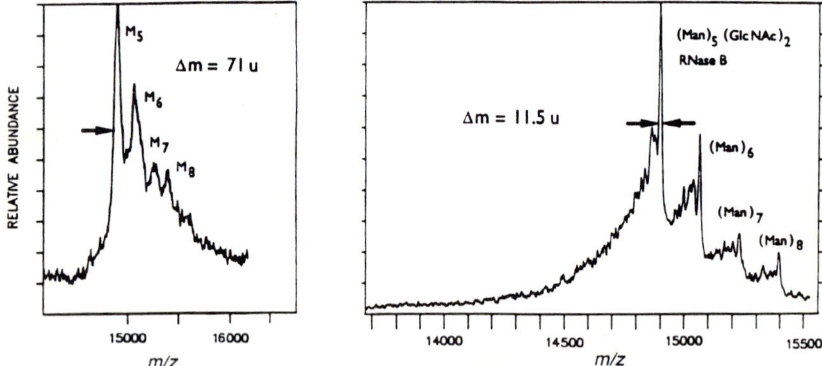

Figure 4. Matrix-assisted laser-desorption spectrum of the glycoprotein ribonuclease B. $[M+H]^+$ 14 900.4 (average). Left: TOF spectrum [obtained with a modified VESTEC 2000 instrument (VESTEC Corp., Houston, TX. USA)] showing only the molecular-ion region, acquired with nitrogen laser (337 nm) using sinapinic acid as a matrix. Right: magnetic sector spectrum acquired as a single segment using an excimer laser (353 nm) and DHB as a matrix. (Reproduced with permission from ref. 28.)

to the presence of both ribonuclease A and several glycoforms of ribonuclease B (26). The molecular weights of such samples can be measured by maLD-TOF and maLD-sector instruments as well. Spectra of ribonuclease B obtained on both TOF and double focusing instrumentation are shown in Figures 3 and 4, respectively. Similar molecular species are observed in the maLD-TOF spectrum (Figure 3) of "ribonuclease B" with consumption of less sample, but recorded with significantly lower mass resolution. It is illustrative to note that this particular preparation of "ribonuclease B" contains additional components such as peak B, corresponding to ribonuclease A with one GlcNAc attached (27). Thus, it is readily apparent from two 5 minute experiments that these two preparations are not identical, and one can clearly discern their significant differences in molecular composition. Of course, these molecular weight measurements suggest, but certainly do not prove the precise structural nature of each component. To obtain that information would require further work, probably involving CID spectral analyses of enzymic digest components (see discussion in refs 3-6). Turning to yet another maLD spectrum (Figure 4) of a third preparation of ribonuclease B, both the TOF and sector spectra illustrate the enhanced mass resolution which may be obtained from use of the double focusing instrument (28). Note that these latter spectra indicate the presence of only the high mannose glycoforms of ribonuclease B.

Finally, using ES with double focusing instrumentation higher mass resolution can be readily obtained. For example, in studying genetic variants of human hemoglobin by measurement of the molecular weights of both the intact α- and β-chains, it is easy to detect amino acid substitutions when the mass difference is large enough to be resolved by the mass analyzer. This is illustrated in Figure 5 for the hemoglobin of a patient with Hemoglobin Yamagata. The $asn^{132}\rightarrow lys$ in the β-chain differs by only 14 Da. at M_r 15,867.2 (normal β-chain) and is clearly mass resolved with the instrument resolution of 2,000 used in making this measurement (22). A mass difference this small at 16,000 Da. would have been difficult to fully resolve using a normal quadrupole mass analyzer.

It should be noted that for ES-MS the sample must be soluble in the solvent (e.g., water, acetonitrile, 0.1% TFA) being employed for introduction in to the ion source. Hydrophobic and integral membrane proteins may be run in appropriate solvents (29). At this point in time, ES-MS works well up to around 100,000 Da. (1), while maLD would be the method of choice up to 250,000 Da. (2), and also for measurement of mixtures of smaller proteins (30).

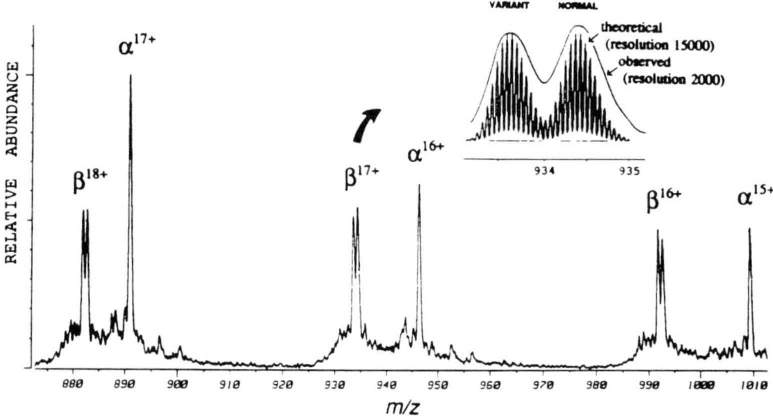

Figure 5. Partial ESI mass spectrum of hemoglobin chains from a patient with Hemoglobin Yamagata at a resolution of 2000. The shape of the doublet of 17-charged ions was compared with the peak shape for the theoretical isotopic distribution at a resolution of 15,000. (Reproduced with permission from ref. 22.)

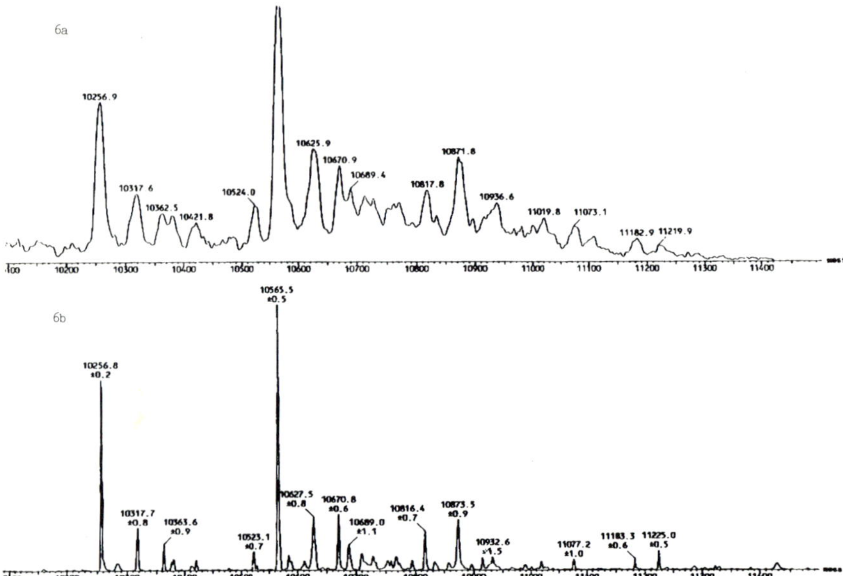

Figure 6a. Electrospray spectrum of the MH_{10}^{+10} species from lys C asn[181]-glycopeptide isolated from the scrapie glycoprotein by RP-HPLC.
Figure 6b. Maximum Entropy deconvolution result of electrospray spectrum shown in 6a.

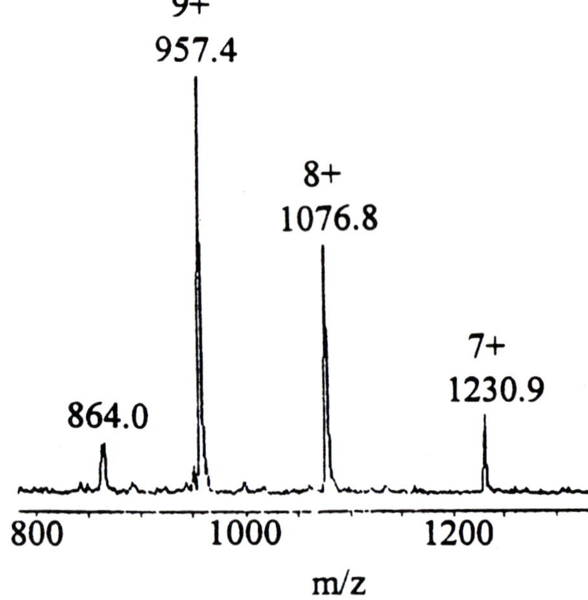

Figure 7. Electrospray spectrum of the PNGase F peptide product (asp[181]-111→185) of the scrapie N-linked site at asn[181].

B. Peptide Mixture Analysis

All of these ionization methods can be used to investigate mixtures of peptides, including those obtained from protein or glycoprotein digests. Exactly how to proceed depends upon the information sought. All of these methods work fairly well on simple mixtures, especially with components of similar hydrophilicity/hydrophobicity ratios (such as RP-HPLC fractions). More complex mixtures should be separated (at least partially) and run either in a batch mode (fractions) or by using a small aliquot on-line using LC-ES-MS (18, 19, 31). The latter has the advantage of providing retention time information as well as molecular weights for each component.

C. Glycopeptides

The degree of elaboration of glycosylation on a given glycopeptide will determine its extent of hydrophilicity as compared with the same un- or deglycosylated peptide sequence. Very hydrophilic substances are usually not sputtered with good sensitivity using LSIMS (3). However, ES-MS and LC-ES-MS work particularly well as noted earlier (see Section IIIC). Many relatively simple examples have been reported (3-5, 11, 18, 19, 23, 31).

However, N-linked glycosylation at even a single asparagine site can be very complex (extreme "microheterogeneity"). ES-MS is ideal for determination of the global qualitative and semi-quantitative degree of glycoform heterogeneity of such systems. A case in point involves the tryptic glycopeptide (residues 111-185) containing asn^{181} in the hamster scrapie glycoprotein (32). The calculated mass of this sequence is 8608.6 Da. without the incremental mass of the glycosylation. This glycopeptide was not observed in attempted LSIMS analysis. However, it was readily observed by electrospray carried out in the batch inlet mode as shown in Figure 6a. While the mass resolution and mass assignment were not optimal in the recording of the raw data from very limited quantities of this glycopeptide, the number of components and their proper mass assignments were clearly determined through use of a deconvolution (image enhancement) algorithm called Maximum Entropy (33) as shown in Figure 6b. To establish that the over 30 glycoforms detected were not artifacts, PNGase F was used to deglycosylate this glycopeptide. The ES-MS of the resulting asp^{181}-peptide is shown in Figure 7. The measured molecular weight corresponds to that of the expected deglycosylated peptide within experimental error.

D. Sequence and Structure Elucidation

As indicated above, these soft ionization methods do not usually provide sufficient fragmentation in a given mass spectrum for structure elucidation purposes without

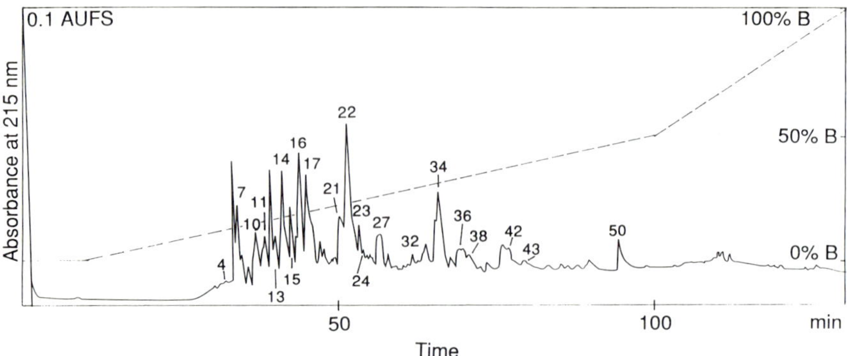

Figure 8. HPLC chromatogram of the tryptic digest of the Galβ1,3(4)GlcNAc α2,3-sialyltransferase. Sixty fractions were collected and subjected to mass spectrometric analysis. Fractions labeled yielded sequences by CID analysis, or the eluting peptides were later identified by molecular weight based on the cDNA-deduced amino acid sequence. (Reproduced with permission from *J. Biol. Chem.*)

vibrational activation. In addition, even low energy collisional activation is not usually sufficient for unambiguous sequence determination or the characterization of true unknowns (34). However, low energy activation may be adequate for certain types of problems such as the location of amino acid substitutions in protein genetic variants (35), verification of recombinant protein expression (18), quality control of batches of recombinant protein preparations (19), and partial sequence analysis in conjunction with Edman degradation on the same sample (36). We have seen that both matrix assisted laser desorption (maLD) and electrospray techniques can provide direct measurement of the molecular weight and purity or heterogeneity of a protein. Despite the fact that electrospray presently offers the highest accuracy of mass measurement available on intact proteins (for quadrupole instruments, ±.01%), even this level of accuracy is not sufficient to determine some detailed features, such as whether a single disulfide bond is oxidized or reduced (± 2 Daltons) even in an intact small protein such as shark liver fatty acid binding protein (37), or whether a single asparagine is deamidated to aspartic acid (38). The latter was shown to be the case unambiguously by high energy CID sequence analysis of a tryptic peptide in myoglobin being used as a "known" mass standard for maLD mass scale calibration. Of course, higher levels of mass measurement accuracy are readily achieved using double focusing sector instrumentation with electrospray ionization (22, 24).

Finally, we turn to a brief discussion of the power and versatility of high sensitivity four sector tandem mass spectrometry as a general tool for sequence and structure determination. The first example concerns recent collaborative work from our laboratory on determination of the primary sequence of Galβ1-3(4)GlcNAc α2,3-sialyltransferase, the membrane bound enzyme which catalyzes the final step in the biosynthesis in the sialyl Lewis X antigen, the tetrasaccharide moiety recognized by the E-selectin receptor. This enzyme was first purified from rat liver almost a decade ago by our collaborators, Paulsen and co-workers, and since then several attempts were made to obtain sufficient sequence for cloning, as well as raise antibodies to it.

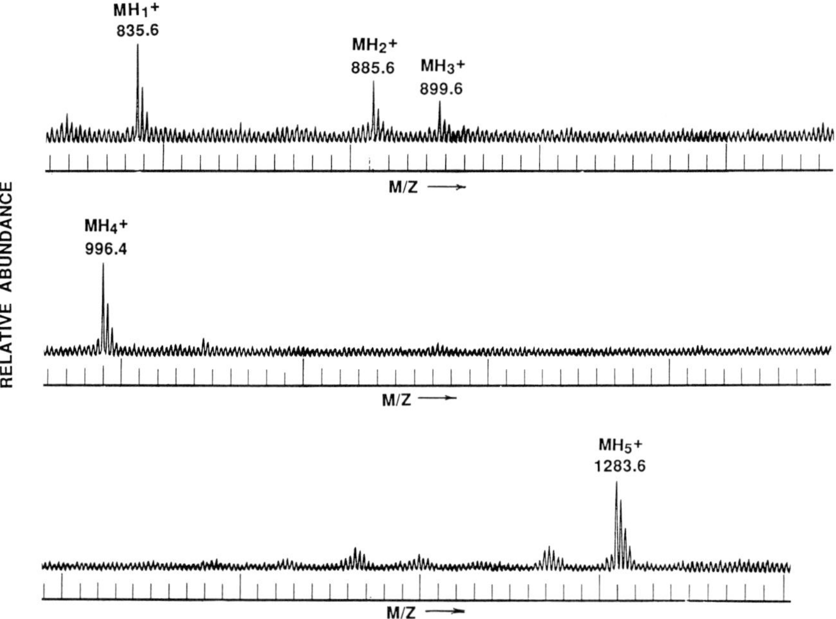

Figure 9. LSIMS spectrum of fraction #22 from the tryptic digest of the Galβ1,3(4)GlcNAc α2,3 sialyltransferase. The spectrum was recorded in the positive mode, using glycerol:thioglycerol 1:1 acidified with 1% TFA as matrix.

These attempts were unsuccessful due to the small amounts of impure protein which were readily obtainable.

Twelve micrograms (300 pmoles) of this 48 kDa protein were isolated from the membranes of 2 kg of rat liver. The active protein was stored in 350 ml of 30 mM sodium cacodylate (pH 6.5), 100 mM sodium chloride, 0.1% Triton® CF 54, and 50% glycerol, as described elsewhere (8). As discussed above, from a mass spectrometric point of view this nonvolatile buffer and detergent combination could have thwarted the successful sequence analysis of this protein preparation. To overcome these problems we pursued the strategy outlined briefly below. We chose to reduce and carboxymethylate this protein in a mixture of this buffer and additional Tris and guanidine hydrochloride prior to dialysis of the carboxymethylated protein against 4 l of 50 mM N-ethylmorpholine acetate. SDS was added to the dialysis wells late in the dialysis to make up a total of 0.1%. After dialysis, rigorous removal of SDS and remaining traces of Triton® CF 54 was carried out by acetone precipitation of the protein. The protein pellet was then taken up in a buffer of Tris, 2 M urea and calcium chloride and digested with trypsin for 18 h. Separation of the digest was carried out using a reverse phase microbore HPLC and some 50 fractions were collected. Microbore HPLC is advantageous for providing small solvent volumes for these picomole levels of peptide, since the fractions must be kept in solution (not blown down to dryness) to avoid serious adsorptive losses. Thus, microbore HPLC provides appropriate concentrations of peptides in the fractions for mass spectral analysis and sequencing. The HPLC chromatogram of this digest is shown in Figure 8. Upon

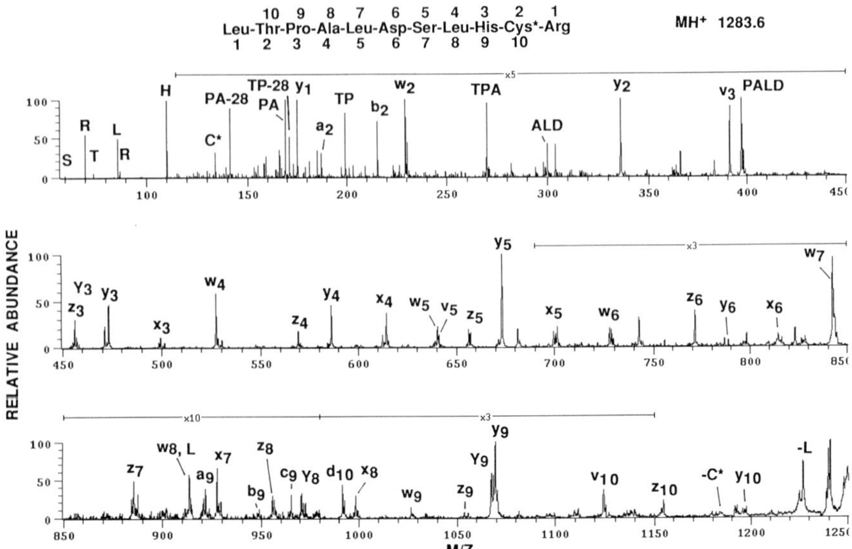

Figure 10. CID spectrum of peptide T-22B from the Galβ1,3(4)GlcNAc α2,3 sialyltransferase. The peptide sequence is Leu-Thr-Pro-Ala-Leu-Asp-Ser-Leu-His-Cys*-Arg, MH^+ = 1283.6. Cys* represents carboxymethyl cysteine. Ions with charge retention at the N-terminus are labeled as a, b, c ions and the C-terminal analogs are designated as x, y, z. The a and x ions are products of a cleavage between the α carbon and the carbonyl group. Ions y and b are formed when the peptide bond is cleaved. Ions c and z are present due to the cleavage between the amino group and the α carbon. The numbering of these fragments is always initiated at the appropriate peptide terminus. The side chain fragmentation occurs between the β and γ carbons of the amino acids, yielding the so-called d (N-terminal) and w(C-terminal ions. C-terminal v_i ions are formed via α,β bond cleavage from a number of ions, for example, from y_i ions. The immonium ions and the side chain losses from the protonated molecular ion are indicated by the one letter code of the corresponding amino acids. Internal fragment ions are labeled similarly. The nominal masses of all sequence ions observed are shown together with the sequence deduced above the spectrum.

measurement of the molecular weights of these fractions by liquid SIMS, it was clear that most of these fractions were mixtures of peptides, such as was the case for fraction 22 shown in Figure 9. This fraction contained 5 peptides, two of which were sequenced by high energy CID analysis without further separation. For example, selection of one of the protonated molecular ions at 1283.6 and its collisional activation provided the CID spectrum shown in Figure 10. Interpretation of this CID spectrum established the sequence LTTALDSLHC*R. The second component selected, protonated molecular weight 996.4, was determined to be VSASDGFWK (data not shown). From this single experiment, the sequence of 99 of a total of 374 amino acids of this protein were obtained (39). The enzyme was then cloned using PCR methodology. Finally, the knowledge of the molecular weights of other components in various HPLC fractions was used to verify the correctness of the final cDNA-deduced protein sequence.

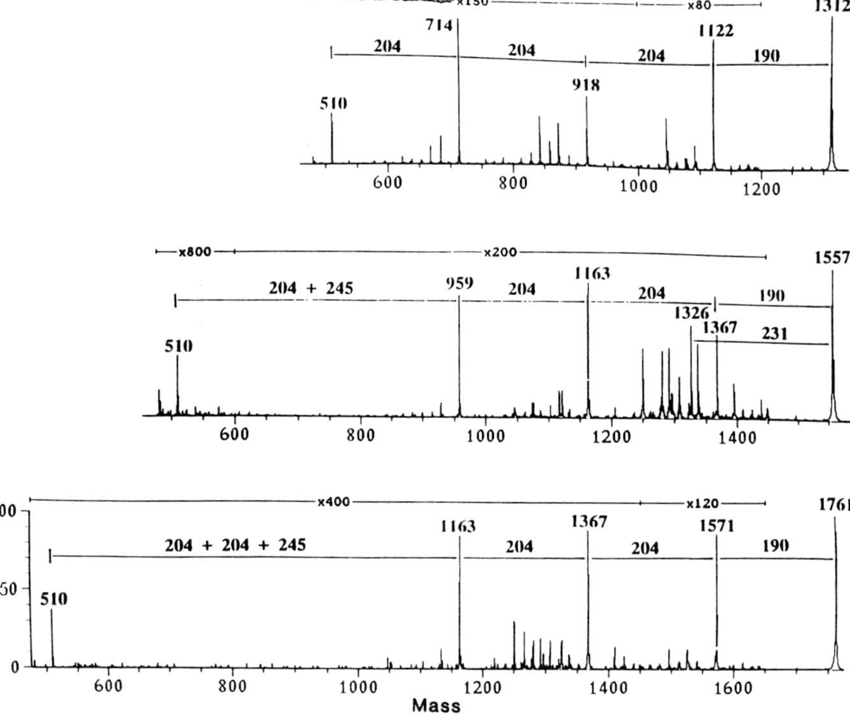

Figure 11. High energy CID spectra of the permethylated glycan species of m/z 1312, 1557, and 1761 (nominal masses rounded down to integer values), in the mass range > 500 Da. (Reproduced with permission from ref. 17.)

The last example concerns our collaborative work on the site of attachment and structural nature of the scrapie glycoprotein phosphatidylinositol glycan membrane anchor. Our investigation of the hamster scrapie glycoprotein is an interesting example of a challenging structural problem exacerbated by the availability of only low nanomole to subnanomole quantities of material at any given time. The complexity of glycosylation at asn^{181} was discussed above. This discussion will summarize briefly our work on the GPI anchor, since all techniques of current mass spectrometry were helpful in complementary ways in addressing this structural problem.

Since the protein preparation PrP 27-30 used in many of these studies is sensitive to phospholipase C, it was employed to remove the glycerol acyl ether from the phosphatidylinositol glycan attached to the protein C-terminus. Upon subsequent digestion of the resulting glycoprotein with endoproteinase lys C, the C-terminal peptide-GPI anchor was isolated by reverse phase HPLC. Treatment of this fraction with 50% HF at 4° permitted isolation of the peptido-ethanolamine as established by measurement of its protonated molecular weight by LSIMS as 1374. This figure was consistent with predicted lys C cleavage at lys^{220} in the cDNA sequence, and the

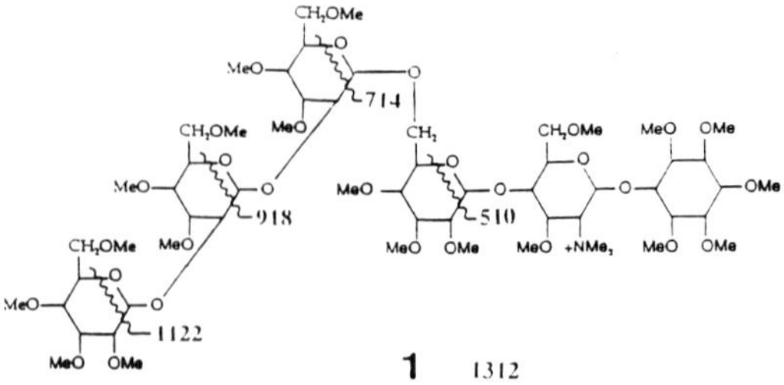

1 1312

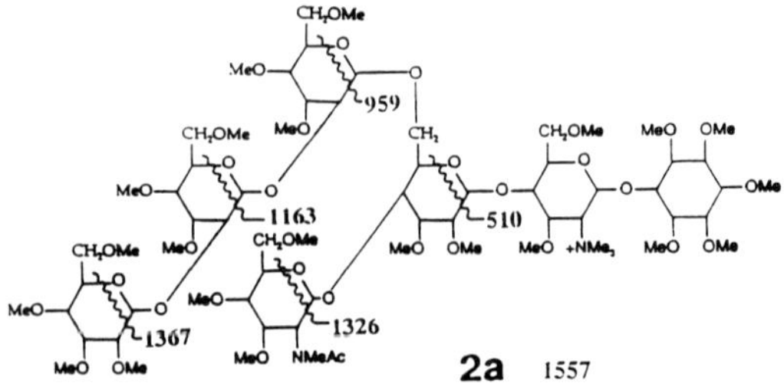

2a 1557

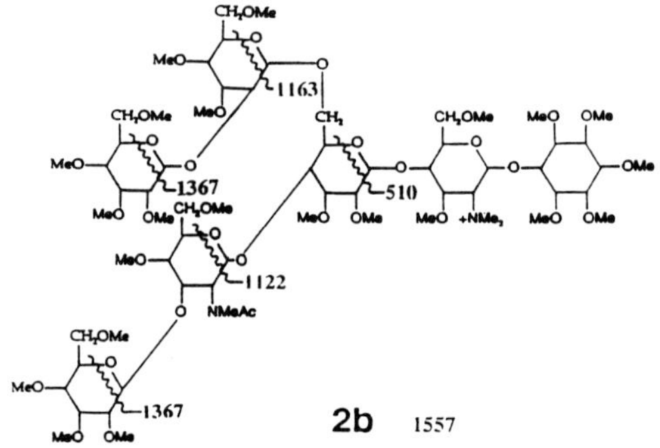

2b 1557

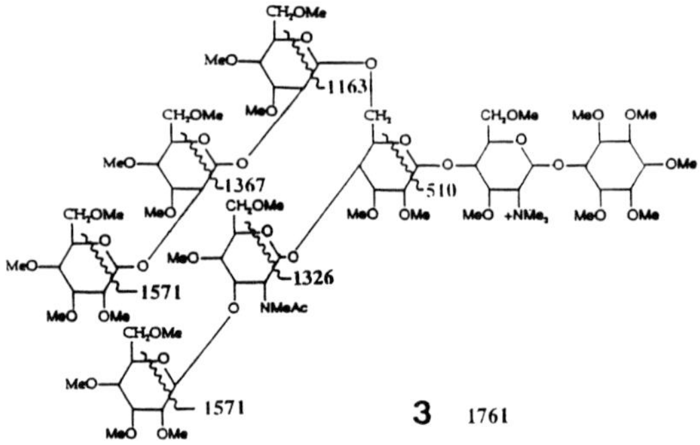

3 1761

Figure 12. Structures of the glycans corresponding to the permethylated ions of m/z 1312 (1), 1557 (2a and 2b), and 1761 (3). The formation of the $^{1,5}X$ ions is illustrated. (Reproduced with permission from ref. 17.)

known phosphodiester selective cleavage properties of cold aqueous HF. This experiment established the GPI attachment site at ser^{231} (40). Another experiment was then carried out to obtain the very hydrophilic non-retained fraction eluting in the

Figure 13. GPI anchor glycoforms. (Reproduced with permission from ref. 11.)

HPLC void volume. This fraction was subjected to permethylation to increase its hydrophobicity for LSIMS analysis. The permethylated product was extracted with chloroform and subjected to LSIMS. Three components were detected at 1312, 1557, and 1761 Da. In order to obtain structural information on these components, they were analyzed by selection of the ^{12}C isobar of each component in MS-I of a four sector tandem instrument. High energy CID spectra of each component from this mixture were recorded (41) and are shown in Figure 11. Interpretation of these CID spectra provided the sequence and branching of the glycans shown in Figure 12. However, since it is known that HF treatment might degrade unsuspected features of the native structure, further experiments using the native lys C-derived C-terminal peptido-glycan were performed upon availability of ES-MS capability in our laboratory. From the ES measurement of the molecular weights in this preparation, larger components were detected which corresponded to one additional phosphoethanolamine moiety and one sialic acid residue. In order to place the attachment of the sialic acid residue, another more careful aqueous HF, permethylation was carried out. This experiment resulted in detection of two additional higher mass components in LSIMS at 1918.8 and 2122.9 Da. High energy CID analysis was then employed to determine

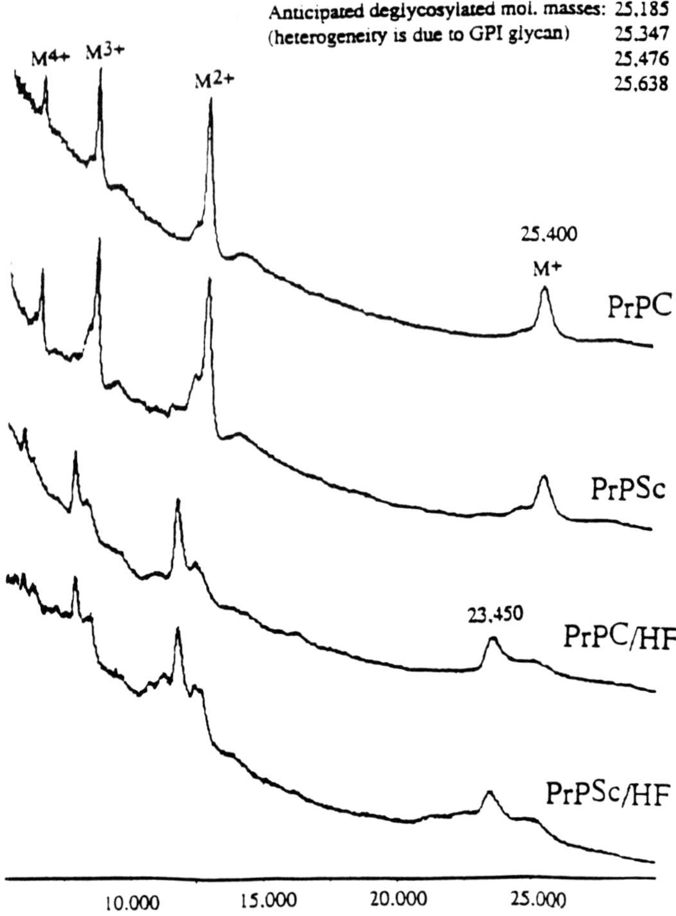

Figure 14. LDMS of PrP^{Sc} and PrP^{C}.

that the sialic acid residue was attached to the branched portion of the anchor glycan shown in Figure 13 (11). Finally, maLD experiments were performed on various preparations of the scrapie glycoprotein to obtain a global view of the complexity and heterogeneity of the protein preparation (42), after enzymatic removal of the two heterogeneous N-linked carbohydrates. These analyses (shown in Figure 14) have been challenges even for maLD, but our present data was obtained by solubilization of these membrane proteins in hexafluoroisopropanol before mixing with chloroform/methane/water containing the maLD matrix, 4-hydroxy-α-cyanocinnamic acid.

It is clear that without judicious use of microchemical manipulations and all of the new ionization and instrumental techniques in the arsenal, problems such as the GPI anchored scrapie glycoprotein would not be structurally tractable.

V. Conclusions and Outlook

Mass spectrometry is destined to become a commonplace set of tools alongside HPLC in the biomedical research laboratory. It has a variety of advantages not previously experienced by the biological research community, such as the ability to produce high quality, detailed molecular information at high absolute sensitivity (comparable to the best biological-based assays). Mass spectrometry can routinely "visualize" the entire composition of the sample (including things you don't care to know about), and easily handle heterogeneous substances (e.g. glycosylation, phosphorylation, etc.) and mixtures of substances. Its main disadvantage at present is the unusual care required to assure suitable sample preparation, particularly in connection with problems where mass spectrometry has not been previously employed.

While the investment in sufficient instrumentation and skilled practitioners of the art may be on the high side for enjoyment of the full power of the arsenal, this situation is beginning to change rapidly. Hence, it should be anticipated that growing reliance on the daily use of these versatile and powerful new tools (so clearly in evidence at this Sixth Symposium) will become widespread in the biomedical community and the biotechnology industry in the near future.

Acknowledgments

I wish to acknowledge my co-workers and collaborators who contributed to various examples discussed in this chapter. Financial support was provided by the Biomedical Research Technology Program of the National Center for Research Resources of the National Institutes of Health (Grant RR01614), the Biological Instrumentation Program of the National Science Foundation (Grant DIR-8700766), National Institute of Diabetes and Digestive and Kidney Diseases (Grant AM 27643), and the NIEHS Superfund Program (Martyn Smith, UCB) Grant ES04705.

References

1. Fenn, J. B., Mann, M., Meng, C. K., and Wong, S. F. (1989). Science **246,** 64.
2 Hillenkamp, F., Karas, M., Beavis, R. C., and Chait, B. T. (1991). Anal. Chem., **63,**1193A.
3. Burlingame, A. L., and McCloskey, J. A., eds. (1990). "Biological Mass Spectrometry" 700 pp. Elsevier Science Publishers, Amsterdam.
4. "Methods in Enzymology" (1990) (J. A. McCloskey, ed.), Vol. 193, 960 pp.
5. Carr, S. A., Hemling, M. E., Bean, M. F., and Roberts, G. D. (1991). Anal. Chem. **63**, 2802-2824.
6. Burlingame, A. L., Baillie, T. A., and Russell, D. H. (1992). Anal. Chem. **64**, 467R-502R.

7. See listing in NIH Publication No. 92-1430, Directory, "Resources for Biomedical Technology", which may be obtained by calling 301-251-4970.
8. Wen, D. X., Livingston, D. D., Medzihradszky, K., Kelm, S., Burlingame, A. L., Paulson, J. C. (1992). J. Biol. Chem., in press.
9. Baldwin, M. A.,.Wang, R., Pan, K.-M., Hecker, R., Stahl, N.,.Chait, B. T., and Prusiner, S. B. (1992). This volume.
10. Unpublished results, this laboratory.
11. Stahl, N., Baldwin, M. A., Hecker, R., Pan, K.-M., Burlingame, A. L., and Prusiner, S. B. (1992). Biochemistry, **31**, 5043.
12. For a recent discussion of these differences, see ref. 6 , pp.478R-480R and references therein.
13. Biemann, K., and Scoble, H. A. (1987). Science, **237**, 992.
14. Kaur, S., Hollander, D., Haas, R., and Burlingame, A. L. (1989). J. Biol. Chem., **264**, 16981.
15. Buechter, D. D., Medzihradszky, K. F., Burlingame, A. L., and Kenyon, G. L. (1992). J. Biol. Chem., **267**, 2173.
16. Ojingwa, J., Ding, A., McDonagh, A. F., Burlingame, A. L., and Benet, L. Z. Proc. Natl. Acad. Sci. USA, submitted.
17. Baldwin, M. A., Stahl, N., Reinders, L., Gibson, B. W., Prusiner, S. B., and Burlingame, A. L. (1990). Anal. Biochem., **191**, 174.
18. Hemling, M. E., Roberts, G. D., Johnson, W., Carr, S. A., and Covey, T. R. (1990). Biomed. Env. Mass Spectrom., **19**, 677.
19. Ling, V., Guzzetta, A. W., Canova-Davis, E., Stults, J. T., Hancock, W. S., Covey, T. R., and Shushan, B. I. (1991). Anal Chem., **63**, 2909.
20. Carr, S. A., Huddleston, M. J., and Bean, M. F. (1992). "Proceedings of the 40th Annual Conference on Mass Spectrometry and Allied Topics" pp 973-974.
21. Schindler, P. A., Hall, S. C., Collet, X., Fielding, C. J., and Burlingame, A. L. (in preparation).
22. Wada, Y., Tamura, J., Musselman, B. D., Kassel, D. B., Sakurai, T., and Matsuo, T. (1992). Rapid Commun. Mass Spectrom. **6**, 9-13.
23. Maltby, D., Hall, S. C., Medzihradszky, K., and Burlingame, A. L. (1992). "Proceedings of the 40th Annual Conference on Mass Spectrometry and Allied Topics" pp 24-25.
24. Musselman, B. D., Tamura, J., and Cody, R. B. (1992). "Abstracts, Kyoto '92 International Conference on Biological Mass Spectrometry" p 97.
25. Poulter, L., Green, B. N., and Burlingame, A. L., (1990). *In* "Biological Mass Spectrometry" (A. L. Burlingame and J. A. McCloskey, ed.), pp 119-128. Elsevier, Amsterdam.
26. Bitsch, F. (1991). "Ph. D. Thesis, University of Strasbourg".
27. Beavis, R. C., Aduru, S., Chait, B. T. (1990). "Proceedings of the 38th Annual Conference on Mass Spectrometry and Allied Topics" pp 281-282.
28. Annan, R. S., Köchling, H. S., Hill, J. A., and Biemann, K. (1992). Rapid Commun. Mass Spectrom. **6**, 298-302.
29. Falick, A. M., Schindler, P. A., and van Dorsselaer (1992). "Proceedings of

the 40th Annual Conference on Mass Spectrometry and Allied Topics" pp 1625-1626.

30. Beavis, R. C., and Chait, B. T. (1990). Proc. Natl. Acad. Sci. USA **87**, 6873-6877.
31. Medzihradszky, K. F., Settineri, C. A., Maltby, D. A., and Burlingame, A. L. (1992). This volume.
32. Stahl, N., Baldwin, M. A., Teplow, D. B., Hood, C., Gibson, B. W., Burlingame, A. L., and Prusiner, S. B. Biochemistry, submitted.
33. Max. Entropy Software, Fisons Instruments, UK.
34. See discussion and references therein in reference 6, pp 478R-480R.
35. Witkowska, H. E., Bitsch, F., and Shackleton, C. H. L. (1992). Hemoglobin, submitted.
36. Hunt, D. F., Michel, H., Dickinson, T. A., Shabanowitz, J., Cox, A. L., Sakaguchi, K., Appella, E., Grey, H. M., and Sette, A. (1992). Science **256**, 1817-1820.
37. Medzihradszky, K. F., Gibson, B. W., Kaur, S., Yu, Z., Medzihradszky, D., Burlingame, A. L., and Bass, N. (1992). Eur. J. Biochem. **203**, 327-339.
38. Zaia, J., Annan, R. S., and Biemann, K. (1992). Rapid Commun. Mass Spectrom. **6**, 32.
39. Burlingame, A. L., Medzihradszky, K. F., Wen, D. X., Livingston, B. D., Kelm, S., and Paulson, J. C. (1992). "Abstracts, LVII Cold Spring Harbor Symposium on Quantitative Biology: The Cell Surface, May 27-June 3" p. 32.
40. Stahl, N., Baldwin, M. A., Burlingame, A. L., and Prusiner, S. B. (1990). Biochemistry **29**, 8879-8884.
41. Baldwin, M. A., Stahl, N., Burlingame, A. L., and Prusiner, S. B. (1990). *In* "Methods: A Companion to Methods in Enzymology", Vol. 1, pp 306 -314.
42. Baldwin, M. A., Pan, K-M., Hecker, R., Wang, R., Beavis, R., Chait, B., Stahl, N., Burlingame. A. L., and Prusiner, S. B. (1992). "Proceedings of the 40th Annual Conference on Mass Spectrometry and Allied Topics" pp 1593-1594.

Elucidation of Covalent Modifications and Noncovalent Associations in Proteins by Electrospray Ionization Mass Spectrometry

Joseph A. Loo[1a], Rachel R. Ogorzalek Loo[2a], David R. Goodlett[a], Richard D. Smith[a], Alfred F. Fuciarelli[b], David L. Springer[b], Brian D. Thrall[b], and Charles G. Edmonds[a,b]

[a]Chemical Sciences Department
and
[b]Biology and Chemistry Department
Pacific Northwest Laboratory
Richland, WA 99352

I. Introduction

Electrospray ionization mass spectrometry (ESI-MS) (1, 2) provides powerful insight into the structure of intact proteins and their enzymatically derived peptides (3, 4). ESI affords a distribution of multiply charged molecular ions which may be analyzed by, for example, conventional quadrupole mass analyzers ($M/\Delta M \leq 1000$) and mass may be determined with precision ($\pm$ 0.01%) substantially better than available by conventional biochemical techniques. The elucidation of primary structure of unknown proteins is one important area of application. Additionally, where the primary structure of the analyte is constrained (*e.g.*, among mutant protein species) or where a known structure has been subjected to chemical or natural post-translational transformations, this technique provides an extraordinarily sensitive and selective probe of the nature and extent of these modifications. For example, acrylamide adducted hemoglobin subunits and the products of proteolytic digestion may be directly analyzed by ESI-MS with potential application as a biomarker of occupational exposure. Similarly, the radiation induced modifications and the natural post-translational modifications of histone proteins

---

[1]Present address: Parke-Davis Pharmaceutical Research Division, 2800 Plymouth Road, Ann Arbor, MI 48106-1047

[2]Present address: University of Michigan Medical School, Ann Arbor, MI 48109

TECHNIQUES IN PROTEIN CHEMISTRY IV

may be studied.

In contrast, the gentle nature of the ESI process permits noncovalent associations of such species to be probed by this technique. Proteins are folded into specific molecular conformations, maintained by noncovalent interactions. Proteins interact with other molecules to form complex structures, accounted for by noncovalent interactions within and between macromolecules. ESI-MS has demonstrated the capability for examining a range of protein noncovalent interactions. ESI-MS can be used to study protein dimers and peptide-protein associations. In order to better understand the nature of these interactions, we have used ESI-MS to analyze a variety of noncovalently bound complexes.

II. Materials and Methods

The ESI source, utilizing a liquid sheath electrode, has been previously described (2). Methanol or distilled deionized water was used as the sheath liquid, with the applied electrospray voltage at +4 kV and +5.5 kV, respectively. Sulfur hexafluoride was used as the ESI source sheath gas and is an effective electron scavenger to suppress corona discharge, especially at the higher ESI voltages necessary for spraying highly aqueous solutions (*i.e.*, water sheath) (5). The single quadrupole mass spectrometer (Extrel components, Pittsburgh, PA; *m/z* limit 2000) and triple quadrupole mass spectrometer (Sciex TAGA 6000E, Thornhill, Ontario, Canada; *m/z* limit 1400) utilize a modified atmospheric pressure inlet incorporating a differentially pumped interface, also previously reported (2). Adjustment of the voltage difference between the nozzle and skimmer elements (Δ NS) is used to control droplet desolvation in this region, and at higher levels of Δ NS, collisional dissociation of the molecule.

Reference proteins were purchased from Sigma Chemical (St. Louis, MO) and were used without further purification. Peptide and protein solutions were prepared in distilled deionized water with addition of various amounts of acetic acid (typically 5% v/v). A syringe pump controlled the analyte delivery flow rate to the ESI source at 0.35 to 0.50 $\mu L\ min^{-1}$. A separate syringe pump delivered the sheath liquid flow at a rate of 3 $\mu L\ min^{-1}$.

HPLC separations for the hemoglobin subunits and for the proteolytic digests were on C4 and C18 reversed phase columns, respectively, with aqueous trifluoroacetic acid (TFA)/acetonitrile gradient. Detection and quantification of ^{14}C acrylamide adducted species were by standard liquid scintillation techniques. V8 proteinase digestion was carried out in ammonium carbonate (pH 7.8) buffered solution with enzyme/substrate ratio of 1/50.

Histone proteins were isolated from fresh chicken erythrocytes (10 ml packed volume). The cells were lysed by freezing and thawing in lysis buffer (0.01 M MOPS, pH 7.2; 0.15 M NaCl; 0.2% Nonidet NP40), washed twice with lysis buffer without detergent, and nuclei were pelleted at 2000 x G for 10 minutes. Proteins were extracted from nuclei in 0.4 N H_2SO_4 by stirring on ice for 4 hours, and the acid insoluble material was removed by centrifugation. The supernatant was dialyzed against water overnight and concentrated in Centricon-10 cartridges (Amicon). Histones were separated and collected by reversed-phase HPLC with a Biorad RP318 column and a solvent system of acetonitrile and 0.1% TFA. Purified H2B fraction was mixed with 10x or 100x concentrations of thymine or thymidine in deionized water and irradiated in a ^{60}Co gamma source to total doses ranging from 1 to 10 Gy while saturated with nitrous oxide.

III. Results and Discussion

A. *Analysis of Covalent Modifications by ESI-MS*

1. Characterization of acrylamide adducted hemoglobin

Hemoglobin (Hb) adducts have been used to monitor occupational exposure to a number of compounds. Until recently the most common procedure to identify and characterize Hb adducts has been to cleave the adduct from the protein and derivatize and characterize it by gas chromatography-mass spectrometry (GC-MS). To extend these approaches, we examined acrylamide adducted Hb using ESI-MS in an attempt to identify the adducting moiety as well as to obtain information on the amino acid residues involved. ^{14}C acrylamide and purified human Hb (type A_0) were incubated under conditions that yielded high adduct levels. ESI-MS analysis of the intact HPLC-separated subunits indicated distinct differences in the ion profiles for the adducted and nonadducted species. Mass spectra of the intact β-subunit demonstrated three distinct mass increments of approximately 71, 102 and 136 Da. ESI-MS of HPLC purified fractions of the V8 proteinase digestion localizes the first of these on the peptide bearing Cys-93 (V91-101). The measured mass increment for this species (M_r^{Meas} 71.5 $\pm$ 1.0) is consistent with the expected thioether derivative (M_r^{calc} 71.08). The structure of the second observed adduct is unknown but it is likewise localized on cysteine bearing peptides, including V102-121 which bears Cys-112. Similarly, the structure of the third adducted species is unknown and as it does not appear among the adducted peptides we hypothesize that it is not localized on a particular amino acid residue, *e.g.*, the cysteine bearing peptides. Studies to confirm the cysteine residue as the location of the adducting moiety by tandem mass spectrometry and to further refine the picture of acrylamide adduction in this system are in progress.

2. Characterization of radiation induced modifications of histones

A basic subunit of chromatin structure is the nucleosome consisting of 146 base pairs of DNA wrapped around a central histone protein octamer with two additional histones binding the linker regions. The histones, lysine and arginine rich proteins, participate in this reversible complex with DNA called nucleohistone. Exposure of cellular constituents to free radicals generated with ionizing radiation results in modifications which can play a significant role in altered cellular function. Interaction of hydroxyl radicals with DNA results in modifications to base and deoxyribose moieties. In addition, hydroxyl radicals can result in generation of DNA-protein cross links in nucleohistone. Methodology involving acid hydrolysis of irradiated nulceohistone followed by trimethylsilylation and subsequent GC-MS has revealed the presence of DNA base-amino acid dimers (6). However, sequence information is not provided by this methodology. Application of ESI-MS to the measurement of DNA-protein cross links in nucleohistone provides a useful method to obtain information not available with other mass spectrometric techniques.

We have conducted preliminary experiments in which ESI-MS analysis of chicken erythrocyte H2B (M_r 13 789) exposed to 5 Gy of ionizing radiation revealed unique peaks with mass increments of 126.1 Da (near the expected value for thymine-H2B adduct, 125.1 Da) and 140.7 Da. Irradiation of histone H2B in the presence of excess thymine resulted in a dose-dependent increase in species with mass increments consistent with the formation of thymine adducts. The results of these experiments suggest the feasibility of analyzing radiation-induced modifications of histone proteins by ESI-MS.

3. Characterization of natural post-translational modifications of histones

The nature and function of covalent modifications of proteins occurring after translation and simple proteolytic processing is a recurring theme in structural biology. In the case of histone proteins, such post-translational modifications include acetylation, methylation, phosphorylation, mono- and poly(ADP)ribosylation and ubiquitination. ESI-MS permits the direct analysis of the intact modified protein and the evaluation of the nature and extent of modification, as illustrated in Figure 1 for the analysis of the core histones from chicken erythrocytes. Measured mass increments differ from the values predicted by the known sequence by values consistent with known acetylation and methylation post-translational modifications of histone proteins. Ongoing experiments are being conducted to further define the nature of these modifications, as well as to understand the relationship between these modifications and cellular behaviour during both normal cellular replication as well as after insult by chemicals and radiation.

B. Analysis of Noncovalently Bound Complexes by ESI-MS

1. Protein-heme and protein-oligonucleotide complexes

The early ESI-MS studies of proteins involved the analysis of heme-proteins such as cytochrome *c* (12 kDa) and myoglobin (17 kDa) (1,7,8). Such analyses were generally performed with acidic pH solutions to take advantage of the proteins' basic amino acids and the ESI multiply charging (protonation) phenomenon. ESI mass spectra of cytochrome *c* in 5% acetic acid solutions (pH 2.4) show multiply charged ions reflecting a molecular mass that includes

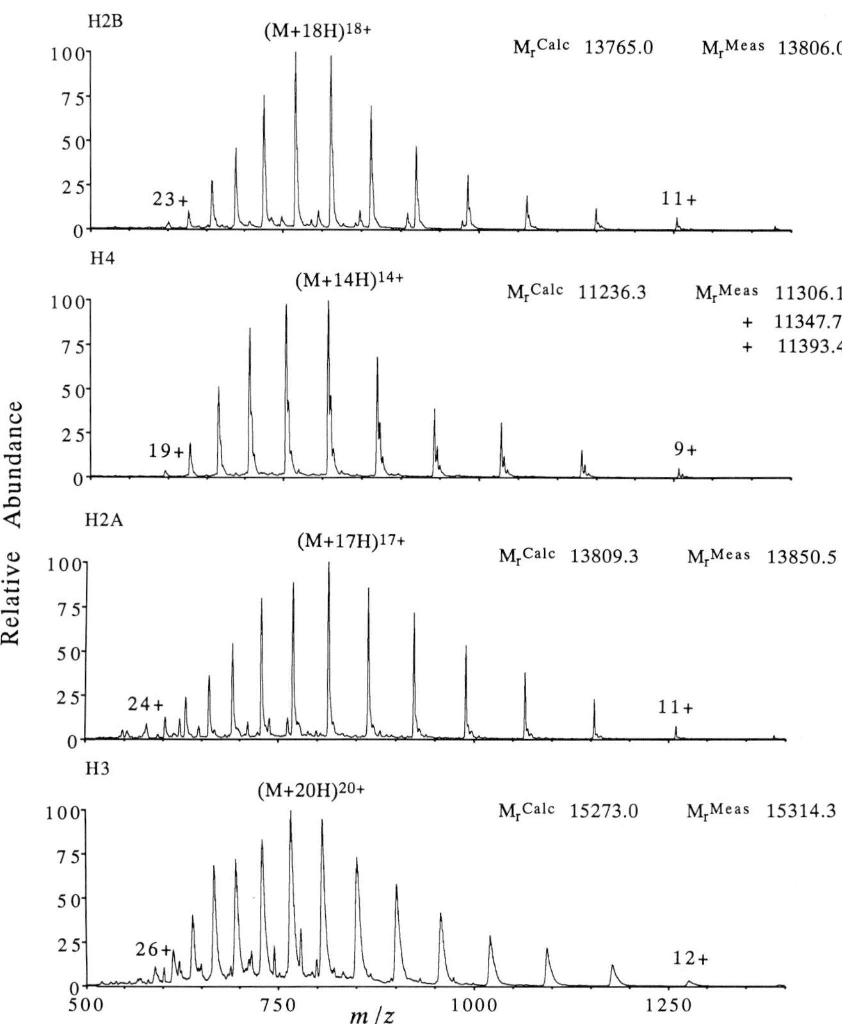

Figure 1. ESI mass spectra of HPLC purified histones from chicken erythrocyte. Measured mass increments are consistent with post-translational methylation and acetylation of these proteins.

the heme group covalently attached to the protein by thioether linkages. On the other hand, the mass spectrum of myoglobin shows only ions for the apo-protein (and in many cases, a singly charged ion for heme), as shown in Figure 2a. The heme prosthetic group is *noncovalently* bonded in a hydrophobic crevice in the myoglobin molecule. However, more recently, several groups have shown that ESI-MS has the capability to examine noncovalently bound systems (9,10), such as myoglobin (11,12), by adjustment of solution and pH conditions. For example, Figure 2b shows the ESI mass spectrum of myoglobin taken from a pH 5.5 solution, showing primarily ions for the protein-heme complex.

Complexes between proteins and nucleic acids are involved in cell processes that maintain genetic information. Many important enzymatic processes require nucleotide ligands, or coenzymes, binding to the enzyme. Multiply charged complexes between a variety of proteins and mono- and larger oligonucleotides have been observed by ESI-MS experiments. Adenylate kinase (AK) catalyzes the reaction between AMP and MgATP for the biosynthesis of ADP (13). AK from chicken muscle (194 residues, M_r 21 594) yields an ESI mass spectrum with multiply charged ions to 35+ (M_r 21 597 ± 3). Addition of AMP yields an additional set of peaks corresponding to an AK-AMP complex with M_r 21 945 ± 1. Whether or not the observation of such noncovalently bound complexes by ESI-MS reflects the solution chemistry remains an open question, the potential of this technique is certainly promising.

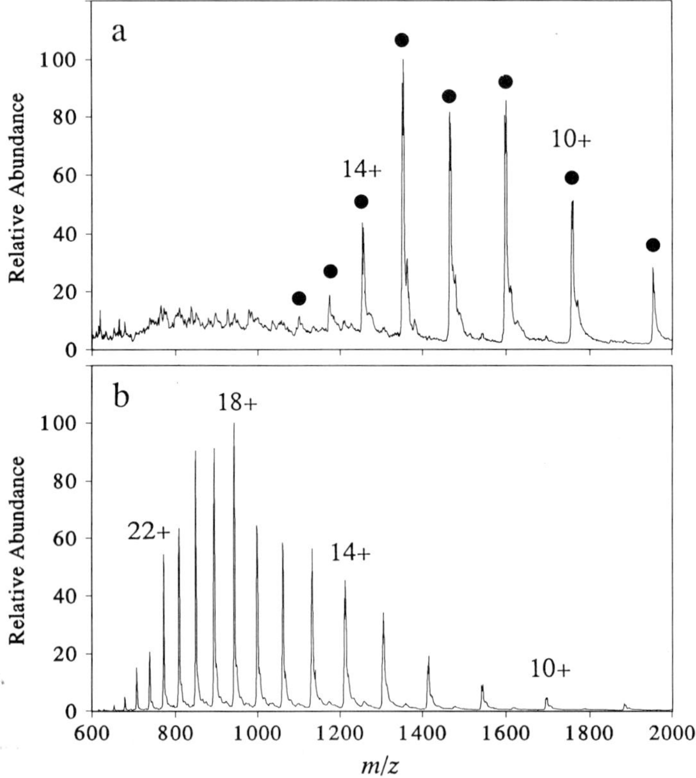

Figure 2. ESI mass spectra of horse heart myoglobin (a) in distilled deionized water (pH 5.5) and (b) 5% acetic acid (pH 2.4). Peaks labeled with (•) represent the polypeptide-heme complex (17,568 Da). Peaks in (b) represent the apo-protein species (16,951 Da).

2. Protein-protein interactions

Noncovalent interactions between proteins and other polypeptides and proteins are important in a variety of biochemical systems (*e.g.*, antibody-antigen complexes, enzyme-substrate, protein subunits). Bovine ribonuclease A (RNase A, M_r 13 682) is cleaved by subtilisin at the peptide bond between residues 20 and 21 to give S-peptide (residues 1-20, M_r 2166) and S-protein (residues 21-124, M_r 11 534). The separated fragments can reassociate noncovalently with high affinity to yield fully active ribonuclease S (RNase S) (14). The overall dissociation constant, K_D, is *ca.* 4 orders of magnitude greater at acidic pH than pH 7. With Δ NS at +0 V to minimize collisional dissociation, the ESI mass spectrum of RNase S from a pH 2.4 solution shows only multiply charged S-peptide and S-protein ions. The ESI mass spectrum from a H_2O solution shows predominantly ions for the intact RNase S complex, consistent with the trend in K_D values as a function of pH. Increasing Δ NS to +50 V produces a mass spectrum showing ions for the separated S-peptide and S-protein molecules. More studies are necessary to access the applicability for which ESI-MS can be used to monitor such noncovalent interactions.

3. Metal ion binding

Metal ion binding is an important facet of the function of many protein systems. Calmodulin and parvalbumin are among a family of proteins containing the EF-hand structural motif for calcium binding (15). Regulation of enzyme activity has generally been found to require Ca^{2+}. Calmodulin binds 4 moles of Ca^{2+} per mole with high specificity. Greater than 3 calcium ions are observed to bind to calmodulin by both positive (Figure 3) and negative ion ESI-MS under various pH conditions. Other metals have been observed to bind to a variety of polypeptides and proteins by ESI-MS, including carbonic anhydrase, a 29 kDa zinc metalloenzyme (16) containing one essential metal ion liganded by histidine residues.

To better understand the nature of the interactions with metal ions, ESI-tandem MS results for a variety of peptides with and without metal have been compared. For angiotensin peptides (*e.g.*, human angiotensin I: DRVYIHPFHL), dissociation of the His-X bond is dramatically enhanced in the presence of Zn^{2+}, suggesting that Zn is in the vicinity of histidine. Further work is ongoing to study the specificity of such peptide interactions with zinc and other metal ions.

Acknowledgments

Work is supported in part by Laboratory Directed Research and Development of the Pacific Northwest Laboratory under Contract DE-AC06-76RLO 1830 for the U.S. Department of Energy. The hemoglobin work is supported by U.S. EPA (CR-81216). Instrumentation and methods development is supported by the National Science Foundation, Instrumentation and Instrument Development Program (DIR 8908096). Pacific Northwest Laboratory is operated by Battelle Memorial Institute.

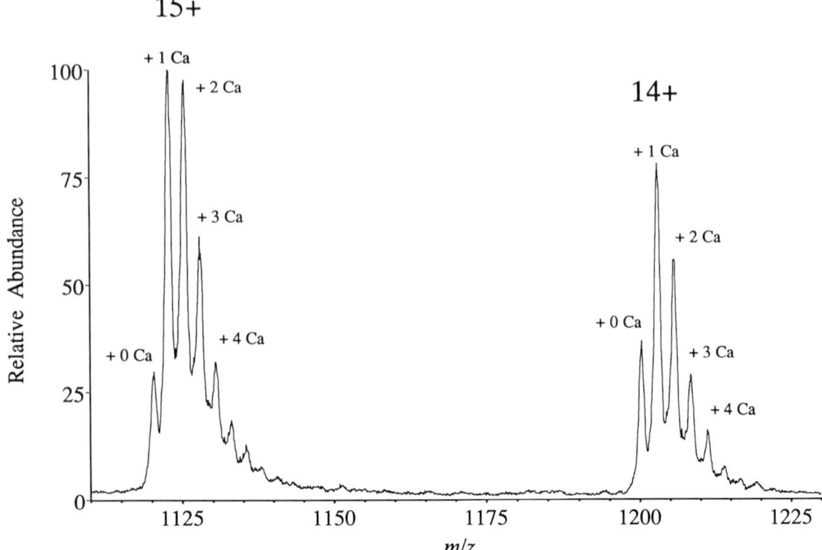

Figure 3. ESI mass spectrum of 0.1 mM bovine brain calmodulin (16,792 Da) in 5% acetic acid and in the presence of 10 mM $CaCl_2$.

References

1. Fenn, J.B., Mann, M., Meng, C.K., Wong, S.F., and Whitehouse, C.M. (1989). *Science* **246**, 64-71.
2. Smith, R.D., Loo, J.A., Edmonds, C.G., Barinaga, C.J., and Udseth, H.R. (1990). *Anal. Chem.* **62**, 882-899.
3. Covey, T.R., Huang, E.C., and Henion, J.D. (1991). *Anal. Chem.* **63**, 1193-1200.
4. Brockerhoff, S.E., Edmonds, C.G., and Davis, T.N. (1992). *Protein Sci.* **1**, 504-516.
5. Ikonomou, M.G., Blades, A.T., and Kebarle, P. (1991). *J. Am. Soc. Mass Spectrom.* **2**, 492-496.
6. Dizdaroglu, M., Gajewski, E., Reddy, P., and Margolis, S.A. (1989). *Biochemistry* **28**, 3625-3628.
7. Covey, T.R., Bonner, R.F., Shushan, B.I., and Henion, J. (1988). *Rapid Commun. Mass Spectrom.* **2**, 249-256.
8. Loo, J.A., Udseth, H.R., and Smith, R.D. (1989). *Anal. Biochem.* **179**, 404-412.
9. Ganem, B., Li, Y.-T., and Henion, J.D. (1991). *J. Am. Chem. Soc.* **113**, 6294-6296.

10. Ganem, B., Li, Y.-T., and Henion, J.D. (1991). *J. Am. Chem. Soc.* **113**, 7818-7819.
11. Katta, V. and Chait, B.T. (1991). *J. Am. Chem. Soc.* **113**, 8534-8535.
12. Loo, J.A., Ogorzalek Loo, R.R., Light, K.J., Edmonds, C.G., and Smith, R.D. (1992). *Anal. Chem.* **64**, 81-88.
13. Tsai, M.-D. and Yan, H. (1991). *Biochemistry* **30**, 6806-6818.
14. Richards, F.M. and Wyckoff, H.W. (1971). *In* "The Enzymes" (Boyer, P.D., ed.), pp 647-806, Academic Press, New York.
15. Cohen, P. and Klee, C.B. (1988). "Calmodulin," Elsevier, Amsterdam.
16. Silverman, D.N. and Lindskog, S. (1988). *Acc. Chem. Res.* **21**, 30-36.

Identification of Conserved Protein Surface Metal-Binding Sites in Related Proteins by Mass Spectrometry

T. William Hutchens and Tai-Tung Yip
Protein Structure Laboratory
USDA/ARS Children's Nutrition Research Center
Department of Pediatrics
Baylor College of Medicine
Houston, TX 77030
and
Randall W. Nelson
Vestec Corporation
Houston, Texas 77054

I. Summary

Matrix-assisted UV laser desorption/ionization (MALDI) time-of-flight (TOF) mass spectrometry has been used to identify amino acid sequences which appear to represent highly conserved metal-binding sites on the surface of human histidine-rich glycoprotein (HRG) and two newly discovered histidine-rich glycoproteins purified from bovine and porcine plasma by metal ion affinity chromatography. Cu(II)-binding peptides were identified in unfractionated plasmin digest maps of human, bovine, and porcine plasma HRG. Thus, analysis of the unfractionated proteolytic digest products of related proteins by MALDI-TOF, in the presence and absence of added transition metal ions, greatly facilitates the identification of conserved proteolytic digestion sites and conserved amino acid sequences involved in metal ion binding.

TECHNIQUES IN PROTEIN CHEMISTRY IV

II. Introduction

Transition metal ions affect a wide variety of biological regulatory events by altering protein-protein and protein-DNA recognition. Several common high affinity metal-binding structural motifs have been identified (1). In contrast, because the peptide- or protein-metal ion affinities operational for metal ion transport functions are generally relatively low, it has been more difficult to identify and evaluate those peptides and surface-exposed residues on proteins involved in plasma metal ion transport and intracellular metal ion trafficking. This work has progressed slowly due to the lack of enabling technologies.

Often, the mere identification of low- to intermediate-affinity metal-binding peptides and proteins involved in metal ion transport requires knowledge of precisely how much of the macromolecule is present in its apo (unoccupied) or holo (occupied) form under a wide variety of conditions. Accurate answers require as little physical perturbation as possible. In the case of metal binding peptides and proteins with multiple binding sites, average molar stoichiometries (e.g., results from atomic absorption) have been misleading. It is more desirable and more difficult, however, to address the *specific* molar distribution of a given protein or peptide occupied with 0, 1, or more metal ions. Unfortunately, information of this type has been nearly impossible to obtain, especially when the protein or peptide of interest is present as only one of several unrelated components in a mixture.

Matrix-assisted UV laser desorption/ionization time-of-flight (MALDI-TOF) mass spectrometry (2) and electrospray ionization (ESI) mass spectrometry (3) represent important new developments in the mass analysis of intact peptides and proteins. MALDI-TOF, in particular, appears to be well suited for the mass determination of peptides and proteins in complex, unfractionated biological fluids (e.g., 4-6). We have now demonstrated that coordinate covalent complexes resulting from the sequence-specific interaction of proteins and peptides with certain transition metal ions can remain stable to the desorption/ionization processes upon which MALDI-TOF and ESI mass spectrometry are based (6-15). Factors affecting metal ion interaction specificity and the relative stabilities of metal ion interactions with N-, S-, and O-electron donor groups, including the maintenance of secondary and tertiary structure conformations have been discussed.

One objective of this investigation was to determine if MALDI-TOF could also be used to identify highly conserved metal-binding domains in related proteins of *unknown* sequence. Human histidine-rich glycoprotein (HRG) is a 74-kDa Zn(II)- and Cu(II)-binding plasma protein with a high percentage of both His (11-13 mol%) and Pro (14-16 mol%) residues in the C-terminal region (e.g., 16,17); the cDNA sequence of human HRG is known (17). We have recently discovered two new species of histidine-and proline-rich glycoprotein in bovine and porcine plasma and milk (18). The bovine and porcine plasma HRGs were isolated by metal ion affinity chromatography (18) and were found to have molecular weights, amino acid compositions, glycosylation sites, and N-terminal amino acid sequences similar to human plasma HRG (19); the full amino acid sequence of these proteins remains unknown. The data presented here suggest the presence of highly conserved plasmin digestion sites in all 3 species of HRG. At least one peptide derived from the plasmin digestion of each of these 3 proteins is shown by MALDI-TOF to be indistinguishable in mass and to bind Cu(II); a common amino acid sequence is predicted.

III. Experimental

A. *Purification of Human, Bovine, and Porcine Plasma Histidine-Rich Glycoproteins by Zn(II) Ion Affinity Chromatography*

Intact human, bovine, and porcine plasma histidine-rich glycoproteins were purified by affinity chromatography on immobilized tris(carboxymethyl)ethylenediamine (TED) with bound Zn(II) ions exactly as described previously (18). Protein purity and mass were verified by reverse-phase HPLC and by mass spectrometric analyses (MALDI-TOF). Protein identities were established by analyses of amino acid compositions, N-terminal amino acid sequence, carbohydrate composition, trypsin and plasmin digest maps, and cross-reactivity with purified anti-human HRG antibodies (19).

B. *Matrix-Assisted UV Laser Desorption Time-of-Flight Mass Spectrometry*

Aliquots of the unfractionated HRG digest mixtures were added 1:1 (v:v) to a saturated aqueous solution of 2,5-dihydroxybenzoic acid (154.12 Da) (20). Two μL of a 10 nmol/mL peptide solution (in water) was mixed with 2 μL of 2,5-dihydroxybenzoic acid solution in the presence and absence of copper sulfate (2 μL of a 20 mM solution); 2 μL of this mixture was applied to the 2-mm diameter stainless steel probe tip and air-dried at room temperature. The dried peptide-matrix deposit on the probe tip was washed in Milli-Q water to remove residual salts, redried, and then inserted through a vacuum lock into the mass spectrometer

(Vestec Model 2000, Vestec Corp., Houston, TX). The laser desorption linear time-of-flight mass spectrometer used the frequency-tripled output from a Q-switched neodymium-yttrium aluminum garnet (Nd-YAG) pulsed laser (355 nm, 5 ns pulse). Ions desorbed by pulsed laser irradiation were accelerated to an energy of 30 keV and allowed to drift along a 2-m flight path (maintained at 30 μPa) to a 20-stage focused mesh electron multiplier. Time-of-flight spectra were recorded at 200 MHz. All spectra shown were taken in the positive ion mode. Real-time signal averages of multiple (100) laser shots were used to generate each spectrum. Data reduction (peak centroid calculations and time to m/z conversions) were performed with PC-based software.

C. *Mass-Dependent Identification of Human HRG Peptide Sequences*

Tentative sequence identities of peptides resulting from the controlled proteolytic digestion of human HRG were assigned based on the mass values determined by MALDI-TOF mass spectrometric analyses. This approach, as described previously (6), was aided by the program PROCOMP (Phil C. Andrews, University of Michigan, Ann Arbor, MI). The amino acid sequence of individual HRG digest fragments was verified after peptide purification by reverse-phase HPLC and by metal ion affinity chromatography.

IV. Results

Figure 1 shows the amino acid sequence of human plasma histidine-rich glycoprotein as derived from its cDNA sequence (17).

Figure 2 shows the MALDI-TOF mass spectra (1000 to 5000 Da) of the unfractionated digest products obtained by the plasmin digestion of native human, bovine, and porcine plasma HRG proteins. The similarity of these spectra indicate the extent to which overall structure, specifically, plasmin digestion sites have been conserved among these three species of HRG (19). Figure 3 shows more detailed MALDI-TOF mass spectra (1400 to 1800 Da) for the unfractionated plasmin digest products of human, bovine, and porcine HRG; in this region of the spectra a peak at m/z = 1624 was common to each species. Sequential analyses of the same digestion mixtures were obtained by MALDI-TOF in the absence then presence of Cu(II) ions added *in situ* (Figure 3). Table I summarizes the observed mass and metal-binding capacity of this specific metal-binding peptide fragment as identified by MALDI-TOF in Figure 3.

Human Histidine-Rich Glycoprotein (HRG)

```
        10         20         30         40         50         60
VSPTDCSAVE PEAEKALDLI NKRRRDGYLF QLLRIADAHL DRVENTTVYY LVLDVQESDC
        70         80         90        100        110        120
SVLSRKYWND CEPPDSRRPS EIVIGQCKVI ATRHSHESQD LRVIDFNCTT SSVSSALANT
       130        140        150        160        170        180
KDSPVLIDFF EDTERYRKQA NKALEKYKEE NDDFASFRVD RIERVARVRG GEGTGYFVDF
       190        200        210        220        230        240
SVRNCPRHHF PRHPNVFGFC RADLFYDVEA LDLESPKNLV INCEVFDPQE HENINGVPPH
       250        260        270        280        290        300
LGHPFHWGGH ERSSTTKPPF KPHGSRDHHH PHKPHEHGPP PPPDERDHSH GPPLPQGPPP
       310        320        330        340        350        360
LLPMSCSSCQ HATFGTNGAQ RHSHNNNSSD LHPHKHHSHE QHPHGHHPHA HHPHEHDTHR
       370        380        390        400        410        420
QHPHGHHPHG HHPHGHHPHG HHPHGHHPHC HDFQDYGPCD PPPHNQGHCC HGHGPPPGHL
       430        440        450        460        470        480
RRRGPGKGPR PFHCRQIGSV YRLPPLRKGE VLPLPEANFP SFPLPHHKHP LKPDNQPFPQ
       490        500     507
SVSESCPGKF KSGFPQVSMF FTHTFPK
```

Figure 1. Amino acid sequence of human plasma histidine-rich glycoprotein.

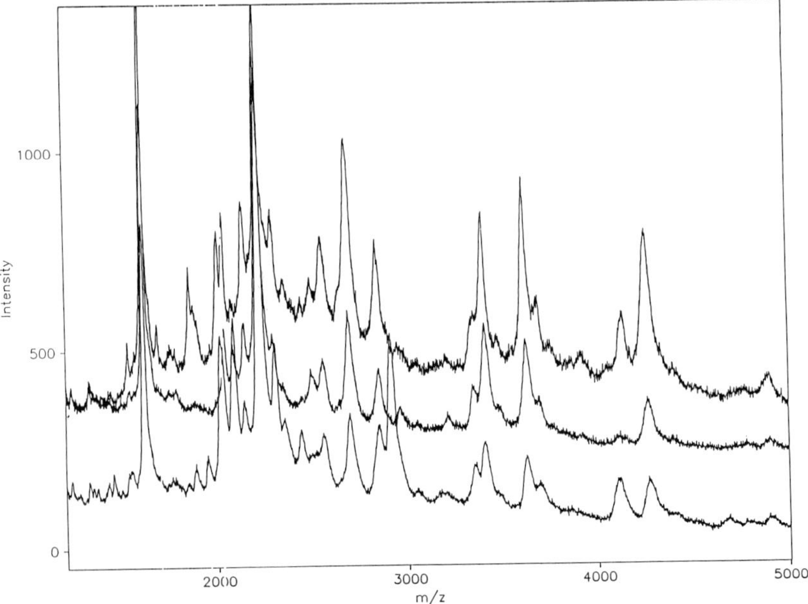

Figure 2. MALDI-TOF mass spectra (1000 to 5000 Da) of the unfractionated digest products obtained by the plasmin digestion of native human (top), bovine (middle) and porcine (bottom) plasma HRG proteins.

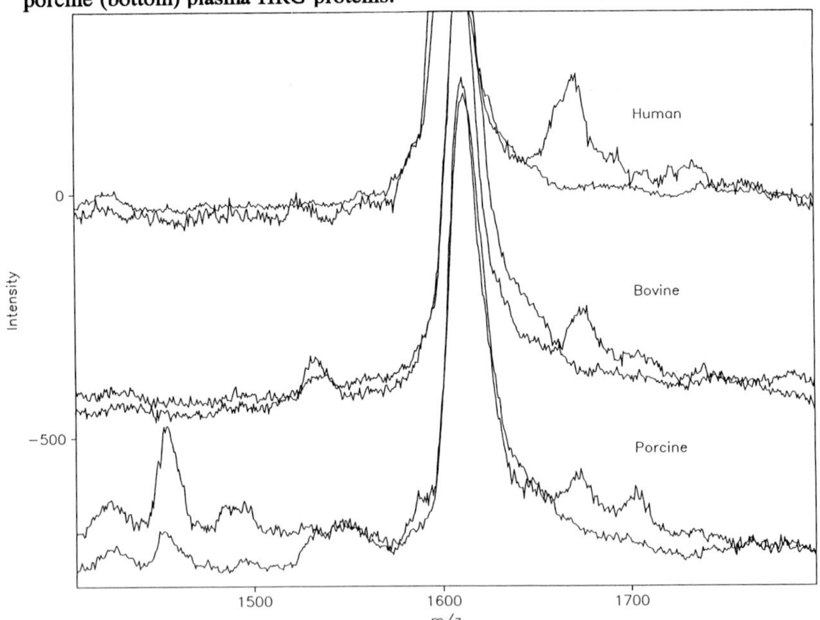

Figure 3. Detailed view of the MALDI-TOF mass spectra (1400 to 1800 Da) for the unfractionated plasmin digest products of human (top), bovine (middle), and porcine (bottom) HRG obtained in the presence and absence of added Cu(II) ions.

Table I. Mass-Dependent Identification of a Conserved Metal-Binding Amino Acid Sequence in Human, Bovine, and Porcine HRG

HRG species	Metal Binding Plasmin Digest Fragment Observed by MALDI-TOF Molecular Mass (Da)	No. of Cu-bound
Human plasma[a]	1624	1 to 2
Porcine plasma	1624	1 to 2
Bovine plasma	1624	1 to 2

[a]The human HRG plasmin digest fragment observed at 1624 Da has been identified as the sequence H322-K335 HSHNNNSSDLHPHK. The calculated chemical average mass of this peptide is 1623.68 Da.

V. Discussion

The metal ion transport protein known as histidine-rich glycoprotein has now been demonstrated to be present in human (17), rabbit (16), bovine (18,19) and porcine (18,19) plasma; a detailed presentation of their physicochemical properties and evidence of structural homologies is to be presented elsewhere (19).

Peptide mapping by MALDI-TOF revealed that human, bovine and porcine plasma HRG proteins have similar plasmin digest sites. Such analyses performed sequentially in the absence then in the presence of added metal ions demonstrated that these proteins contain at least one highly conserved metal-binding domain (1624 Da) that is released upon plasmin digestion. The sequence of this peptide, from bovine and porcine HRG, was tentatively identified based on mass identity with the the human HRG peptide (known sequence); these sequences are being confirmed by sequential Edman degradation.

Comparative analyses of unfractionated proteolytic digest fragments by MALDI-TOF mass spectrometry may greatly facilitate the identification of conserved sequences. Furthermore, such analyses performed in the presence of added transitional metal ions aids in the identification of conserved metal-binding sequences.

Acknowledgments

This project has been funded with federal funds from the U.S. Department of Agriculture, Agricultural Research Service under Cooperative Agreement number 58-6250-1-003. The contents of this publication do not necessarily reflect the views or policies of the U.S. Department of Agriculture, nor does mention of trade names, commercial products, or organizations imply endorsement by the U.S. Government.

Address correspondence to: T. William Hutchens, Department of Pediatrics (CNRC), Baylor College of Medicine, 1100 Bates, Houston, TX 77030.

References

1. Berg, J.M. (1990). J. Biol. Chem. **265,** 6513.
2. Karas, M., and Hillenkamp, F. (1988). Anal. Chem. **60,** 2299.
3. Meng, C.K., Mann, M., and Fenn, J.B. (1988). Proceedings from the 36th Annual Conference on Mass Spectrometry and Allied Topics, p. 771.
4. Beavis, R.C., and Chait, B.T. (1990). Proc. Natl. Acad. Sci. USA. *87,* 6873.
5. Nelson, R.W., and Vestal, M.L. (1991). Proceedings of the 39th Conference on Mass Spectrometry and Allied Topics, paper no. MP104, Nashville, TN, ASMS, East Lansing, MI.
6. Hutchens, T.W., Nelson, R.W., and Yip, T-T. (1992). FEBS Letters **296,** 99.
7. Hutchens, T.W., Nelson, R.W., and Yip, T-T. (1991). J. Mol. Recog. **4,** 151.
8. Nelson, R.W., and Hutchens, T.W. (1992). Rapid Commun. Mass Spectrom. **6,** 4.
9. Hutchens, T.W., Nelson, R.W., Li, C.M., and Yip, T-T. (1992). J. Chromatogr. **604,** 125.
10. Hutchens, T.W., Nelson, R.W., Allen, M.H., Li, C.M., and Yip, T-T. (1992). Biol. Mass Spectrom. **21,** 151.
11. Allen, M.H., Grindstaff, D.J., Yip, T-T., Li, C.M., and Hutchens, T.W. (1992). The 40th American Society for Mass Spectrometry (ASMS) Conference on Mass Spectrometry and Allied Topics, Washington, DC, May 31-June 5.
12. Yip, T-T., and Hutchens, T.W. (1992). FEBS Lett **308,** 149.
13. Allen, M.H., and Hutchens, T.W. (1992). Rapid Commun. Mass Spectrom. **6,** 308.
14. Hutchens, T.W., and Allen, M.H. (1992). Rapid Commun. Mass Spectrom. **6,** 469.
15. Yip, T-T., and Hutchens, T.W. (1992). Techniques in Protein Chemistry (see accompanying paper; T123).
16. Morgan, W.T. (1978). Biochim. Biophys. Acta **533,** 319.
17. Koide, T., Foster, D., Yoshitake, S., and Davie, E.W. (1986). Biochemistry **25,** 2220.
18. Yip, T-T., and Hutchens, T.W. (1991). Protein Expression and Purification **2,** 355.
19. Yip, T-T., and Hutchens, T.W. (1992). Submitted.
20. Karas, M., Bahr, U., Ingendoh, A., Overberg, A., Stahl, B., Strupet, K., Hillenkamp, F., and Giebmann, U. (1991). Proceedings of the 39th Conference on Mass Spectrometry and Allied Topics, paper no. MP117, Nashville, TN, ASMS, East Lansing, MI.

Matrix-Assisted Laser Desorption/Ionization Mass Spectrometry of Membrane Proteins: The Scrapie Prion Protein

Michael A. Baldwin, Rong Wang*, Keh-Ming Pan, Rolf Hecker, Neil Stahl, Brian T. Chait*, and Stanley B Prusiner

Department of Neurology, University of California, San Francisco, CA 94143-0518
and
* Rockefeller University, New York, NY 10021-6399

I. Introduction

The role of mass spectrometry in protein analysis has been transformed by the development of new ionization techniques such as matrix assisted laser desorption/ionization (MALDI) (1-3). This is capable of ionizing intact proteins and providing molecular masses with a high degree of accuracy, in favorable cases to within 0.01%, sufficient to identify protein mutations and post-translational modifications. Although MALDI is a solid phase technique in which the ions are ablated from a dried target, it is dependent on protein solubility as the sample must co-crystallize with a UV-absorbing matrix compound that is mixed in large excess with the analyte (10^4:1). The sample and matrix must dissolve in a common solvent that encourages crystal growth as it evaporates. Solvents drying to leave glassy, oily or amorphous solids usually result in poor quality spectra.

The first matrix employed was nicotinic acid and although other compounds have subsequently been found to give superior results, this is typical of the general class of compound in that it is a substituted aromatic acid. Sinapinic acid (SA) (4) and 4-hydroxy-α-cyanocinnamic acid (4-HCCA) (5) are more recent examples, both of which are soluble in aqueous buffers and are normally prepared in 1:2 (v/v) acetonitrile, 0.1% trifluoroacetic acid at a concentration of 10 mg/mL for SA (approx. 50 mM) or 5 mg/mL for 4-HCCA.

The scrapie prion protein (PrP^{Sc}) is a GPI-anchored membrane sialo-glycoprotein encoded by a chromosomal gene and is implicated in the development of a unique group of fatal neurodegenerative diseases including several human diseases, scrapie of sheep and bovine spongiform encephelopathy of cattle (6). PrP^{Sc} is derived from a normal cellular isoform (PrP^{C}) by an unidentified posttranslational event which causes physical changes and alters the location of PrP in the cell (7). PrP^{Sc} is more resistant to proteolysis and after limited digestion by proteinase-K, an N-terminally truncated form accumulates (PrP 27-30). We have carried out extensive peptide mapping of PrP^{Sc} using enzyme digestion and HPLC separation of the resulting peptides followed by

TECHNIQUES IN PROTEIN CHEMISTRY IV

amino acid analysis, Edman sequencing and mass spectrometry (8) (Stahl *et al*, submitted).

MALDI has the capability of revealing whether any labile posttranslational modifications have been lost or overlooked in the peptide mapping experiments. Furthermore, as a step toward determining the biochemical mechanisms involved in the transmission and development of prion diseases it is necessary to determine whether there are any covalent differences between PrP^{Sc} from different strains of the disease and also between PrP^{C} and PrP^{Sc}. MALDI requires very little material and has the potential to reveal whether the two forms differ in the nature of a posttranslational modification that is manifested by a mass difference. However, membrane proteins are insoluble in aqueous buffers in the absence of detergents such as sodium dodecylsulfate (SDS) which is incompatible with MALDI, or denaturants such as guanidinium hydrochloride (GuHCl) which inhibits ionization if used at concentrations in excess of 1 M, even if this is then diluted tenfold in the matrix. Formic acid (70-90%) has been used as a solvent and some preliminary data were obtained for both PrP 27-30 and PrP^{Sc} (8, 9), but it was found to partially suppress the signal, even for a water soluble protein such as myoglobin, and it is liable to introduce mass errors by formylation of basic sites. Furthermore, treatment with formic acid could remove acid-sensitive modifications. We therefore adapted a method that has been reported for electrospray ionization MS analysis of hydrophobic proteins, in which hexafluoroisopropanol was used as the solvent for the protein and was injected into a flowing stream of 2:5:2 chloroform, methanol, water (10).

II. Materials and Methods

Protein preparation

Bacteriorhodopsin was supplied by Sigma Chemical Co. PrP^{Sc} was isolated from the brains of scrapie-infected Syrian hamsters and was purified by ultrafiltration. After reduction and carboxymethylation, N-linked sugars were removed with PNGase F. The protein was separated by SDS PAGE, eluted from the gel with 10 mM ammonium bicarbonate/0.1% SDS, filtered, precipitated with ethanol and pelleted by ultracentrifugation. Any remaining SDS was removed by Konigsberg precipitation (11). In some experiments the lipid portion of the GPI anchor was removed by treatment with PIPLC and selected fractions were treated overnight with 50% aqueous HF at $4^{o}C$ to remove the GPI anchor.

Mass spectrometry

Mass spectra were obtained on a laser desorption time-of-flight mass spectrometer constructed at the Rockefeller University and described previously (12, 13). Pulses of 10 ns duration of 355 nm radiation from a Nd(YAG) laser were directed at the solid sample/matrix mixture. The resulting ions were accelerated through a potential difference of 30 kV and detected at the end of a 2 m long flight tube by a hybrid microchannel plate/gridded discrete dynode electron multiplier detector. The laser fluence and spot position were varied manually during data acquisition. Spectra were recorded in a LeCroy Model

8828D transient digitizer (200 MHz sampling rate) at a repetition rate of 2.5 Hz. They were subjected to a time-to-mass conversion but no smoothing, background subtraction or other data manipulation was employed.

III. Results and Discussion

Experiments with various solvent/matrix systems

In earlier MALDI experiments with PrP^{Sc} we employed SA as the matrix and 70% formic acid as the solvent. We also used 70% formic acid as the solvent with the protein and mixed this 1:1 with SA in 1:2 acetonitrile, 0.1% TFA. The results were inconsistent and we frequently obtained no protein signal. The GPI anchor for PrP^{Sc} is known to be heterogeneous (14) and it was anticipated that the MALDI peaks would be broad but when protein spectra were obtained the width of the molecular ion peak was substantially greater than calculated, probably due to, (a) multiple and variable formylation by the solvent, and (b) the well-known laser-induced covalent addition of a 206 Da photodehydration product of SA. The molecular mass of deglycosylated but not delipidated PrPSc was measured as 25,540 Da, approximately 150-200 Da higher than anticipated. It was decided to use 4-HCCA as the matrix as this does not give rise to photo-adducts although it can lead to adduction of contaminating copper and we found it to give stronger peaks. We also established that the signal due to myoglobin was significantly suppressed when using 70% formic acid as the solvent. Thus it was decided to explore the use of an organic solvent hexafluoroisopropanol (HFIP) that is known to be an effective solvent for membrane proteins. We have established that HFIP dissolves native PrP^{Sc}, which becomes denatured and loses its resistance to proteolysis (unpublished work).

Due to the limited amount of PrP available it was decided to employ bacteriorhodopsin as a model membrane protein. Although this is extremely hydrophobic its properties are more amenable to analysis than PrP as it readily forms well-defined crystals whereas PrP aggregates to give an amorphous gum. Attempts to obtain MALDI spectra either of myoglobin or bacteriorhodopsin with SA or 4-HCCA directly in HFIP were unsuccessful. Various solvents containing 4-HCCA were then mixed 1:1 with HFIP containing the protein. The following solvents were investigated:

A 1:2 Acetonitrile, 0.1% TFA/water
B 1:1:1 Acetonitrile, isopropanol, 0.1% TFA/water
C 1:2:3 Formic acid, isopropanol, 0.1% TFA/water
D 1:2:3 Formic acid, tetrahydrofuran, 0.1% TFA/water
E 2:5:2 Chloroform, methanol, 0.1% TFA/water

For completeness we also investigated bacteriorhodopsin dissolved in 70% formic acid and mixed with solvent A.

All of the above combinations gave spectra but they varied significantly in quality. The least satisfactory was solvent D which was slow to dry and gave very unstable signals. In general the mixtures containing less water were the least effective solvents for 4-HCCA but they were more readily miscible with HFIP and gave the best quality data. Thus solvent A which is the "standard" matrix solvent was immiscible with HFIP, causing the matrix and analyte to partition to different regions of the target on drying with very little mixing. This phenomenon was evident to some to degree for all of the solvents tested but solvent E which contained the lowest amount of water was the least problematical. 4-HCCA readily dissolved in solvent E at about 5 mg/mL, the concentration normally employed in MALDI. To evaluate HFIP compared with formic acid, the signal strength obtained for bacteriorhodopsin with 1:1 solvent E and HFIP was both more intense and more stable than that given by 70% formic acid 1:1 with solvent A.

Analysis of PrP^{Sc}

In an initial experiment, 10 μL 70% formic acid was added to an Eppendorf tube that was estimated to contain ~100 pmol PrP^{Sc}. Several aliquots of 0.5-1 μL were removed and mixed with various matrices but no protein signals were observed by MALDI. The formic acid was then removed on a Speedvac and HFIP was added. An aliquot was mixed 1:1 with 4-HCCA in solvent A and a stong spectrum was obtained, thus vindicating the decision to employ HFIP rather than formic acid. Furthermore, after being immersed in formic acid for two hours this sample also gave a molecular mass ~200 Da higher than anticipated. Subsequent experiments were carried out with HFIP and solvent E.

Fresh samples of PNGase-treated PrP^{Sc} were analysed. Spectra were obtained displaying ions ranging from $M+H^+$ to $M+4H^+$, multiply charged species being more prevalent with 4-HCCA than with SA (5). The peaks were substantially narrower than had been observed on previous occasions using SA and formic acid (9) and were also narrower than those in the 4-HCCA spectrum of the sample which had been treated with formic acid. In a separate experiment myoglobin was added to the matrix solution as an internal mass standard, giving a molecular mass for PrP of 25,394 Da, approximately 150 mass units less than had been measured after immersion in formic acid and close to the arithmetic mean of the predicted molecular masses for the four major species of 25,412 Da. This latter value is probably high as the calculation should be weighted to favor the two lowest mass forms which are present in higher abundance. Other minor peaks observed in the spectrum were attributable to a singly glycosylated protein resulting from incomplete PNGase digestion, a small amount of delipidated protein which the PIPLC had digested successfully, and a non-GPI-containing form truncated at Gly-228 which was identified in our earlier mapping experiments (15).

A reduction in molecular mass was observed after treatment with 50% aqueous HF, a procedure which should eliminate the GPI anchor by cleavage at the ethanolamine-phosphodiester bond which is the attachment site to the C-terminus of the protein, thereby removing the remaining source of heterogeneity and giving a single sharp peak. Despite the shift to lower mass the reaction was incomplete and resulted in broader rather than narrower peaks. It may be possible to incubate the samples with HF for a longer period but there is evidence that the amino acid chain can be cleaved by this procedure.

IV. Conclusions

We have established that MALDI mass spectra can be obtained for highly aggregated and hydrophobic membrane proteins using a new approach that gives enhanced sensitivity and does not cause the mass artefacts associated with solvation by formic acid. The data obtained to date indicate that Syrian hamster PrPSc after deglycosylation has a molecular mass very close to that predicted from a knowledge of the gene sequence and the structure of the GPI.

Acknowledgements

We wish to thank Dr A Falick for advice concerning solvents for hydrophobic proteins and Dr R Beavis and S Chaudhary for initial assistance with MALDI. This work was supported by NIH grants AG02132, NS14069, AG08967, NS22786 (SBP) and GM38274 and RR00862 (BTC).

References

1. Karas, M. and Hillenkamp F. (1988) *Anal. Chem.* **60,** 2299.
2. Karas, M., Bahr U., and Hillenkamp F. (1989) *Int. J. Mass Spectrom. Ion Processes* **92,** 231.
3. Karas, M., Bahr U., Ingedoh A., and Hillenkamp F. (1989) *Angew. Chem. Intl. Ed. Engl.* **28,** 760.
4. Beavis, R.C. and Chait B.T. (1989) *Rapid Commun. Mass Spectrom.* **3,** 432.
5. Beavis, R.C., Chaudhary T., and Chait B.T. (1992) *Org, Mass Spectrom.* **27,** 156.
6. Prusiner, S.B. (1991) *Science* **252,** 1515.
7. Borchelt, D.R., Scott M., Taraboulos A., Stahl N., and Prusiner S.B. (1990) *J. Cell Biol.* **110,** 743.
8. Stahl, N., Baldwin M.A., Teplow D., Hood L.E., Beavis R., Chait B., Gibson B., Burlingame A.L., and Prusiner S.B., in *Prion Diseases in Humans and Animals,* S.B. Prusiner, *et al.*, Editor. 1992, Ellis Horwood: London.
9. Baldwin, M.A., N. S., Hecker R., Beavis R., Chait B.T., Burlingame A.L., and Prusiner S.B. (1991) *Proc. 39th ASMS Conf. Mass Spectrom Allied Topics, Nashville, TN.* 558.
10. Falick, A.M., Schindler P.A., and Van Dorsselaer A. (1992) *Proc. 40th ASMS Conf. Mass Spectrom. Allied Topics, Washington DC*
11. Henderson, L.E., Oroszlan S., and Konigsberg W. (1979) *Anal. Biochem.* **93,** 152.
12. Beavis, R.C. and Chait B.T. (1989) *Rapid Commun. Mass Spectrom.* **3,** 233.
13. Beavis, R.C. and Chait B.T. (1990) *Anal. Chem.* **62,** 1836.
14. Stahl, N., Baldwin M.A., Hecker R., Pan K.-M., Burlingame A.L., and Prusiner S.B. (1992) *Biochemistry* **31,** 5043.
15. Stahl, N., Baldwin M.A., Burlingame A.L., and Prusiner S.B. (1990) *Biochemistry* **29,** 8879.

Characterization of the Proteins c-kit Ligand and DHFR by Electrospray Mass Spectrometry

Hanno Langen, Bernadette Sander, Francis Vilbois and Hans-Werner Lahm

F. Hoffmann-La Roche Ltd., Pharmaceutical Research – New Technologies
CH-4002 Basel, Switzerland

INTRODUCTION

In the last few years it was shown that electrospray ionisation mass spectrometry (ESI-MS) can be used for accurate mass determinations of peptides and proteins with molecular weights up to 100 kD (1). Thus the predicted molecular weight of recombinant proteins can be checked easily and in addition post-translational side chain modifications and micro-heterogeneity can be readily detected. Verifying the identity and checking the purity of recombinant proteins is very important for quality assurance and product registration. Here we show the usefulness and some limitations of ESI-MS as a technique in protein chemistry exemplified with two recombinant proteins; c-kit ligand (KL) and dihydrofolate reductase (DHFR).

The ligand of the c-kit proto-oncogene is a cell surface protein constitutively expressed on bone marrow stromal cells. The soluble form of the c-kit ligand (KL) stimulates multi-lineage hematopoietic progenitor cells from bone marrow in synergy with other colony stimulating factors (2).

DHFR (3) is a key enzyme in the biosynthesis of purines, pyrimidines, and several amino acids. DHFR is the target of a number of drugs, including the antimicrobial agent trimethoprim. In general, bacterial dihydrofolate reductases have a much higher affinity for the inhibitor trimethoprim than the mammalian enzymes, this difference in binding to the inhibitor is the basis for chemotherapeutic selectivity.

MATERIALS AND METHODS

Protein Purification. Human KL expressed in *E.coli* was purified in 8M urea/40 mM Tris-HCl, pH 7 on DEAE-Sepharose Fast Flow (Pharmacia, Uppsala, Sweden). After 100-fold dilution and dialysis against PBS followed by 40 mM Tris-HCl, the protein was passed over Q-Sepharose Fast Flow (Pharmacia). The fractions of the main peak eluting with about 150mM NaCl were pooled and applied

TECHNIQUES IN PROTEIN CHEMISTRY IV

to a Superdex 200 (Pharmacia) column. KL eluted with an apparent molecular weight of about 35 to 40 kD. The protein was desalted on an Aquapore C4 (ABI, Foster City, USA) column for ESI-MS use.

Staph. aureus DHFR was expressed in *E. coli* and purified by chromatography on Q-Sepharose Fast Flow and Blue-Sepharose (Pharmacia). Purified DHFR was dialyzed for several days against 20 mM ammonium bicarbonate followed by lyophilization. The protein content of the lyophilized powder was 87% as determined by amino acid analysis.

Electrospray Mass Spectrometry. A SCIEX API*III* electrospray mass spectrometer (ESI-MS) (Thornhill, Canada) was employed. Samples were injected into a flow (adjusted between 2 and 30 µl/min) of methanol/water (50/50). The spray capillary was at a potential of 5KV. The resolution of the MS was set to 0.8 dalton at half peak height. We used a Rheodyne (Cotati, USA) 7010 injection valve and a Harvard infusion pump (Southnatick, USA) for flow injection. The following solvent mixtures were employed to introduce the single samples into the API*III*:

0.1%, 1%, or 5% formic acid in water/acetonitrile (1:1)	A1, A4, A7
0.1%, 1%, or 5% formic acid in water/methanol (1:1)	A2, A5, A8
0.1%, 1%, or 5% formic acid in water/isopropanol (1:1)	A3, A6, A9
0.1%, 1%, or 5% acetic acid in water/acetonitrile (1:1)	B1, B4, B7
0.1%, 1%, or 5% acetic acid in water/methanol (1:1)	B2, B5, B8
0.1%, 1%, or 5% acetic acid in water/isopropanol (1:1)	B3, B6, B9
0.1% trifluoro acetic acid (TFA) in water/acetonitrile (1:1)	C1

A Hewlett Packard (Waldbronn, Germany) model 1090 HPLC with a mobile phase flow rate of 40 µl/min without postcolumn stream splitting was employed for LC/ESI-MS coupling. Separations were performed on-line on Aquapore OD-300 (7 µm, 100 x 1.0 mm; ABI, Foster City, USA) and off-line on Aquapore OD-300 (7 µm, 100 x 2.1 mm). The mobile phase for the HPLC separations consisted of 0.1% TFA/water as solvent A and 0.1% TFA/acetonitrile as solvent B.

Enzymatic Digestions. KL and DHFR were digested with TPCK-treated trypsin (Worthington, Freehold, USA) for 24 h at 37 °C with a substrate-to-enzyme ratio of 50:1 in 50 mM ammonium bicarbonate buffer, pH 7.8. The reactions were stopped by addition of 10% TFA.

In addition DHFR was digested with endoproteinase Glu-C (Boehringer Mannheim, Germany) for 6 h at 37 °C in a 25 mM ammonium bicarbonate buffer, pH 6.6 with a 50:1 substrate-to-enzyme ratio.

RESULTS AND DISCUSSION

In analysis, each additional sample handling step will expose the protein sample to new surfaces, resulting in loss and contamination of protein. Coupling of liquid chromatography (LC) to electrospray mass spectrometry (ESI-MS) (4) would therefore improve sample availability to the mass spectrometer and boost the overall throughput of analyses. Microbore columns (1mm I.D.) could be used if larger quantities (100 pmol and more) are available.

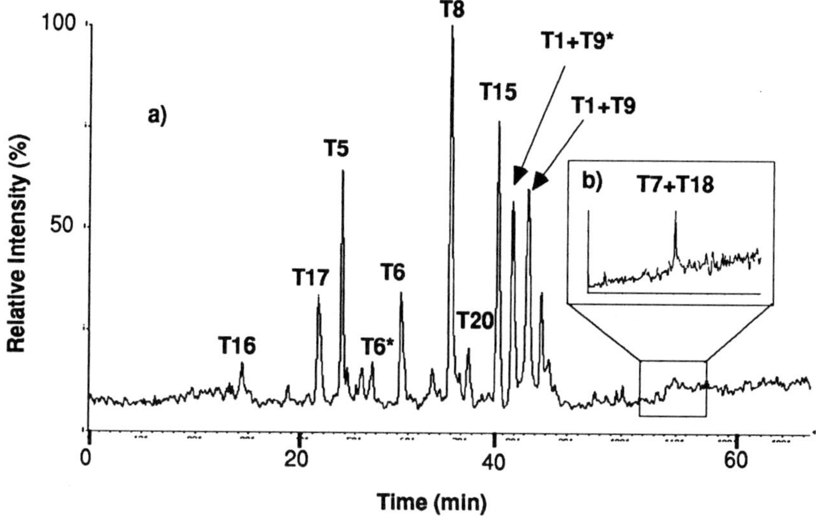

Figure 1: Total ion current of an LC/ESI-MS run. Sample: 200 pmol human c-kit ligand digested with trypsin. Flow rate: 40 µl/min. The ESI-MS was operated in positiv ion mode, dwell time: 0.5 ms, mass step 0.2 AMU, scanning: a) from 500 to 1200 AMU b) from 1400 to 2200 AMU. * Peptides containing methionine sulfoxide.

The disulfide connectivity of KL could be determined by LC/ESI-MS (see Fig.1). Cys^{4} (T1) was connected to Cys^{89} (T9) and Cys^{43} (T7) was connected to Cys^{138} (T18). This technique was very fast and reliable. Nevertheless, N-terminal sequencing and amino acid analyses were needed to interpret the molecular weight information of the unexpected peptides (with oxidized methionine) from ESI-MS. Table 1 combines the results from both experiments the LC/ESI-MS run and the direct injection of the complete tryptic digest of KL.

Peptides with short elution times were determined by flow injection of the complete digest into the ESI-MS. The results indicated that the combination of LC with ESI-MS was very useful for the interpretation of tryptic digests and for the elucidation of disulfide bridges. The determination of the small peptides proved to be difficult, due to their elution position on HPLC, and interference from salt. For a complete data set samples should be digested with several proteases. Another important aspect regarding the analysis of recombinant proteins was the ability to detect side chain modifications. The preparation of KL shown here revealed a substantial oxidation of the N-terminal methionine (40%) and Met[27] (25%). These results were confirmed by amino acid analysis of the corresponding peptides.

In a sample of human KL, analyzed by flow injection, the observed mass (M_r= 18607±2) was higher than the calculated mass (M_r = 18590.5) which could be explained by the presence of one oxidized methionine on average per molecule.

Fragment		Sequence	Calculated M_r	Observed M_r
T 1	0- 5	MEGICR	708.3	T1+T9*
T 2	6- 7	NR	289.2	289.4**
T 3	8- 13	VTNNVK	674.4	674.4**
T 4	14- 17	DVTK	462.3	462.6**
T 5	18- 24	LVANLPK	754.5	754.4
T 6	25- 31	DYMITLK	883.5	883.4
T 7	32- 62	YVPGMDVLPSHCWISEMVVQLSDSLTDLLDK	3490.7	T7+T18*
T 8	63- 78	FSNISEGLSNYSIIDK	1786.9	1787.4
T 9	79- 91	LVNIVDDLVECVK	1458.8	T1+T9*
T10	92- 96	ENSSK	564.3	564.6**
T11	97- 99	DLK	375.2	375.4**
T12	100-100	K	147.1	not found
T13	101-103	SFK	381.2	381.4**
T14	104-108	SPEPR	585.3	584.8**
T15	109-117	LFTPEEFFR	1185.6	1185.6
T16	118-121	IFNR	549.3	549.2
T17	122-127	SIDAFK	680.4	680.4
T18	128-148	DFVVASETSDCVVSSTLSPEK	2200.0	T7+T18*
T20	152-164	VSVTKPFMLPPVA	1385.8	1385.8
T19	149-151	DSR	377.2	377.6**

Table 1: Tryptic digest of KL * These peptides contain cysteines which were involved in disulfide bonds. T1+T9: M_r = 2167.0 (observed) and M_r = 2167.1 (calculated); T7+T18: M_r = 5691.3 (observed) and M_r = 5690.7 (calculated). ** These peptides were found by flow injection of the whole tryptic digest mixture.

A sample of DHFR was analyzed by flow injection (Fig.2). The observed mass of DHFR (18122±2) was in agreement with the calculated mass (18120.8). The presence of incompletely processed Met^0-DHFR was clearly visible by a second ion series. About 30% of the sample was Met^0-DHFR, which was in agreement with amino acid analysis determinations. N-terminal protein sequencing gave slightly higher values (38%).

For further investigations DHFR was cleaved with several proteases (trypsin, endoproteinase Glu-C) and analyzed on ESI-MS. The peptides were collected from an LC run, dried and flow injected into the ion source of the ESI-MS (see table 2).

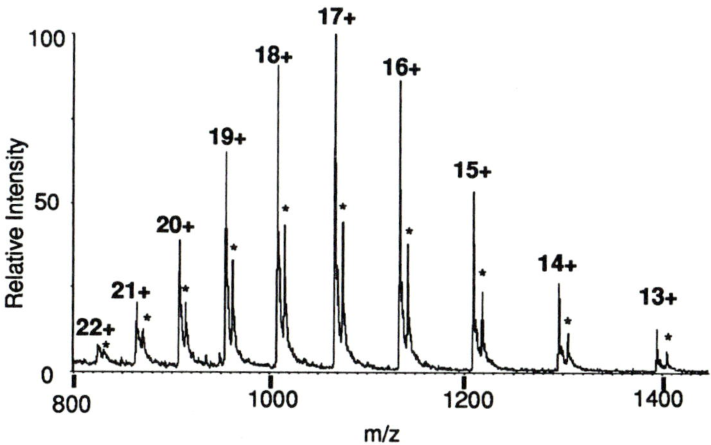

Figure 2: ESI-MS of 50 pmol of DHFR by flow injection; flow rate: 30 µl/min. ESI-MS parameters: dwell time 0.5 ms, mass step 0.2 AMU , scanning from 800 to 1450 AMU. * These peaks corresponded to the protein with Met^0. DHFR: M_r = 18122 and Met^0-DHFR: M_r = 18252.

Fragment	Calculated M_r	Observed M_r	Fragment	Calculated M_r	Observed M_r
G 1 (0- 17)*	2042.1*	2042.0*	T 1 (0- 12)*	1496.8*	1497.4*
G 1 (1- 17)	1911.1	1911.6	T 1 (1- 12)	1365.8	1366.0
G 2 (18- 48)	3625.0	3626.8	T 2 (12- 29)	2020.1	2020.3
G 3 (49- 71)	2543.4	2544.0	T 3 (30- 32)	382.2	383.6
G 4 (72- 80)	968.5	967.9	T 5 (34- 44)	1171.6	1171.8
G 5 (81- 99)	2181.1	2178.0**	T 7 (46- 57)	1358.7	1359.4
G 7 (101-114)	1684.8	1684.5	T10 (105-116)	1382.7	1382.1
G 8 (115-129)	1808.9	1808.9	T11 (117-118)	322.2	322.4
G 9 (130-132)	449.2	449.2	T12 (119-140)	2508.1	2508.8
G10 (133-138)	591.3	591.4	T13 (141-144)	504.3	504.4
G11 (139-143)	561.3	561.4	T14 (145-156)	1079.6	1079.6
G12 (144-158)	1846.1	1846.7			

Table 2: Trypsin (T) and Endoproteinase Glu-C (G) fragments of DHFR *Met^0 containing peptide ** mass lower than expected.

The spectrum of the endoproteinase Glu-C fragment G5 (81-99) was only observed by flow injection DHFR digest. The observed mass was lower than the expected mass. This peptide was part of the longer fragment T9 (59-104) (expected mass: 5152.6) from the tryptic digestion, which could probably not take up enough charges to be in the range of the ESI-MS instrument (maximum m/z: 2400). This situation occurred frequently and could be overcome by using another protease to collect a complete data set (Table 3).

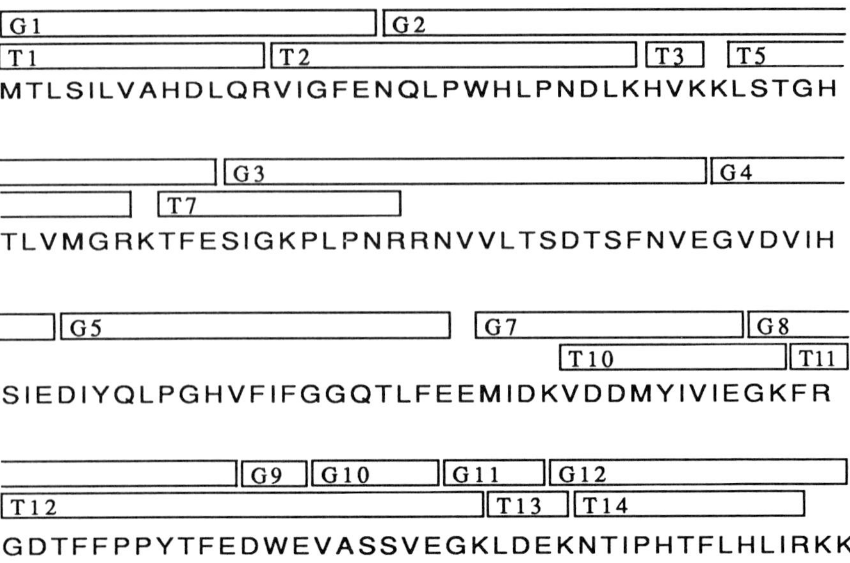

Table 3: Summary of the results from the Trypsin (T) and Endoproteinase Glu-C (G) digests of DHFR. Identified peptides were marked by bars.

To obtain ESI-MS data from minor components in a protein mixture, measurements in the low pmol range were required. Decreasing the injection flow-rate to 2 µl/min improved the signal intensity substantially (Fig. 3), while at the same time improving the signal to noise ratio.

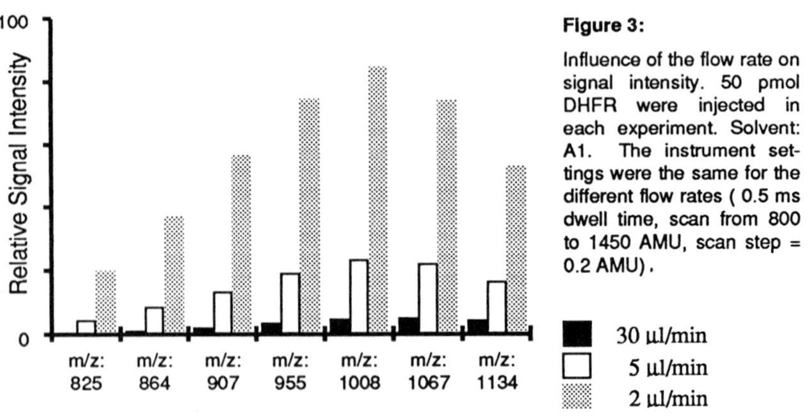

Figure 3:

Influence of the flow rate on signal intensity. 50 pmol DHFR were injected in each experiment. Solvent: A1. The instrument settings were the same for the different flow rates (0.5 ms dwell time, scan from 800 to 1450 AMU, scan step = 0.2 AMU).

Flow rates below 2 µl/min yielded a discontinuous spray, but flow injection might be possible down to 100 nl/min under different experimental conditions (5).

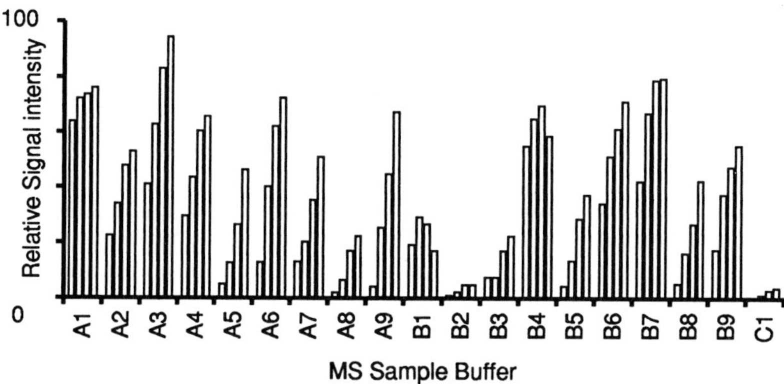

Figure 4: Influence of different solvent mixtures on signal intensity. 10 pmol DHFR were injected in each experiment with a flow rate of 10 µl/min. The instrument setting was the same for the different solvents (0.5 ms dwell time, scan from 800 to 1450 AMU, scan step 0.2 AMU). For solvent mixtures see Materials and Methods section. For each buffer m/z 825, 864, 907, and 955 were plotted (see also Fig.2).

Another important parameter for sensitivity was the sample buffer. The electrospray worked best when about 50% organic solvent was present. The highest sensitivity could be achieved with solvent mixture A3 (isopropanol and 0.1% formic acid) (Fig.4). But this solvent dissolved plastizicers from the sample tubes very well. With the solvent mixture A1 (methanol and 0.1 % formic acid) the least plastizicers were dissolved and the sensitivity was only 20 % lower than with A3. However under favourable conditions 0.5 pmol of DHFR could be detected (Fig.5).The widely used RP-HPLC buffer C1 (0.1% TFA in water/acetonitrile (1:1)) turned out not to be well suited for electrospray ESI-MS.

Electrospray-MS is very useful for verifying the identity and homogeneity of a protein with a known sequence produced by recombinant DNA techniques. Proteins and peptides must, however, carry enough charges so that their m/z ratios are within the range of the MS. One tryptic fragment of DHFR (T9: 59-104; M_r = 5152.6) could not be detected by ESI-MS for this reason. Furthermore samples must be free of salts to obtain optimal results.

On-line LC/ESI-MS improves sample handling by avoiding losses and contamination through additional manipulation steps. Microbore columns can be used if larger quantities (over 100 pmol) of protein is available. Capillary columns can handle even lower amounts of sample with the added benefit of better ESI-MS sensitivity through reduced flow rates.

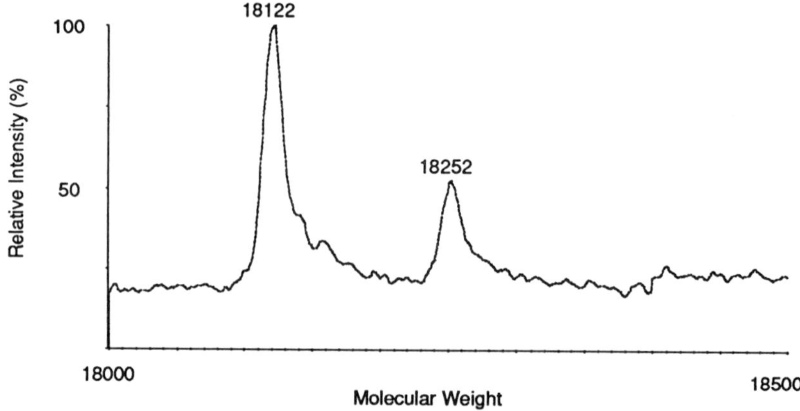

Figure 5: Deconvoluted mass spectrum of 0.5 pmol of DHFR. Flow Rate: 2 μl/min; Solvent: A1; Dwell time: 0.5 ms; Mass step: 0.2 AMU. DHFR: M_r = 18122 and Met^0 - DHFR = 18252.

ESI-MS data need careful interpretation and it is often necessary to employ different analytical techniques to arrive at a valid conclusion. In our hands ESI-MS proved to be a powerful tool in protein chemistry, allowing us to work with protein samples in the lower pmol range as long as good sample handling techniques were used.

ACKNOWLEDGEMENTS

We thank Dr. W. Vetter for helpful discussions, Mr. P. Fischlewitz and Mr. W. Meister for technical assistance with the ESI-MS, and Mr. M. Manneberg for amino acid analysis.

REFERENCES

1. Fenn, J.B., Mann, M., Meng, C.K. Wong, S.F., and Whitehouse, C.M. (1989) Science, **246**, 64 -71.
2. Huang, E., Nocka, K., Beier, D.R., Chu T.-Y., Buck, J., Lahm, H.-W., Wellner D., Leder, P., and Besmer, P. (1990) Cell, **63**, 225-233.
3. Hartmann, P.G., Stähli, M., Kocher, H.-P., and Then, R.L. (1988) FEBS Letters,**242**,157-160.
4. Covey,T.R., Huang, E.C., and Henion, J.D. (1991) Anal. Chem., **63**, 1193-1200.
5. Griffin, P.R., Furer-Jonscher, K., Hood, L.E., Yates, J.R., III, Schwartz, J., and Jardine, I. (1992) Techniques in Protein Chemistry III (Angeletti, R.H., Ed.), 467-476.

RAPID PURIFICATION, SEPARATION AND IDENTIFICATION OF PROTEINS AND ENZYME DIGESTS USING PACKED CAPILLARY PERFUSION COLUMN LC AND LC/MS

D.B. Kassel[1§], M.A. Luther[1], D.H. Willard[1], S.P. Fulton[2] and J.-P. Salzmann[3]

[1§]Glaxo Research Institute, 5 Moore Dr., Research Triangle Park, NC 27709

[2]PerSeptive Biosystems, University Park at MIT, Cambridge, MA 02139

[3]LC Packings (USA, Inc.), 80 Carolina St., San Francisco, CA 94103

I. INTRODUCTION

Flow-through, or perfusive-particle chromatography, has emerged recently as a powerful technique for the rapid separation of biomolecules (*e.g.*, peptides and proteins)[1-3]. Perfusive-particle columns are packed with derivatized porous polystyrene divinylbenzene (DVB), operate at very high column flow rates (*e.g.*, 5-10 ml/min for 4.6 mm i.d. columns), and permit very rapid separation times. The packing is unique in that it provides a tremendous speed advantage over conventional LC by enabling transport of analyte molecules through particle pores by both competitive diffusive and convective processes. Like conventional columns, sample is transported to the surface of the particles by convective flow . The particles, however, contain two types of pores, so-called throughpores, which allow convective flow, and diffusive pores, which provide high adsorptive surface area. At high flow rates, convection dominates diffusion and allows direct access to the high surface area diffusive pores. Thus, rapid perfusive transport is made possible and column resolution and capacity maintained independent of flow rate. Theoretical descriptions of perfusion column chromatography have been described elsewhere[4,5]. Regnier et al. have demonstrated the high speed, high resolution capabilities of these columns for both protein purification and peptide mapping[6,7].

Our interest has been to couple perfusion column chromatography directly to mass spectrometry in order to make use of the tremendous speed advantage these columns afford over micro LC techniques for protein and peptide molecular weight determinations. Liquid chromatography has been coupled successfully to mass spectrometers utilizing various ion source interfaces (*e.g.*, Flow- or Frit-FAB[8,9], thermospray[10], electrospray ionization[11,12]). Frit-FAB interfaces accommodate a small range of flow rates (between 1-10 μl/min). Thus, coupling can be achieved directly using capillary columns (50-320 μm i.d.) or via post-column splitting to conventional or microbore LC columns. Perfusion columns

TECHNIQUES IN PROTEIN CHEMISTRY IV

would require post-column split ratios between 1000:1 and 10000:1. Thus, only a very small fraction of analyte could be infused into the FAB ion source and consequently both the dynamic range and detection limits of the technique would be affected dramatically . The thermospray interface might be capable of accommodating flow rates of perfusion columns but, due to the nature of the ionization technique, would not be amenable to the molecular weight determination of large thermally labile compounds, such as peptides and proteins. Electrospray (or Ion Spray) ionization mass spectrometry (ESI/MS) is capable of handling both low and moderate flow rates (between 1-200 μl/min) and permits the analysis of thermally labile biomolecules. Still, the 5-10 ml/min flow rates of perfusion columns are not well-suited to ESI/MS and would require post-column split ratios of 50:1 to 200: 1 using analytical scale perfusion columns. Split ratios of this magnitude would result in a tremendous loss in sensitivity of detection.

In our laboratory, 320 μm i.d. packed capillary LC columns (HyperCarb, C4 packings) are coupled routinely to ESI/MS and operate at optimum flow rates between 3-6 μl/min. It was believed that perfusion packed capillary liquid chromatography would, in principle, permit column flow rates between 30-100 μl/min, thereby allowing packed capillary pefusion columns to be coupled directly to ESI-MS without the need for any post-column splitting. The packing of 320 μm i.d. fused silica capillary columns with derivatized polystyrene divinylbenzene perfusive particles and their coupling to electrospray ionization mass spectrometry has been achieved recently in our laboratory. Preliminary results for the purification and characterization of proteins and enzyme digests are presented. Parameters affecting analysis time and separation efficiency, such as column flow rate, gradient slope and column length have also been investigated. Presented within is the first application of packed capillary perfusion column chromatography coupled to ESI-MS for the separation of protein and enzyme digests.

II. MATERIALS AND METHODS

A. *Protein Purification*

C-src tyrosine kinase SH2 domain was placed in a pGEX-3X vector and expressed in XLI Blue cells. Glutathione-agarose, reduced glutathione, Na_2HPO_4 and EGTA were from Sigma. Factor Xa from human plasma, APMSF, aprotinin, bestatin, leupeptin and pepstatin were from Boehringer-Mannheim. Tris-HCl, NaCl and EDTA were purchased from Research Organics. $CaCl_2$ was from Aldrich. *E. coli* cells (22.3g wet weight) obtained from a 5 liter fermentation were thawed in a 1:5 dilution of lysis buffer (10 mM Na_2HPO_4, 30 mM NaCl, 10 mM EDTA, 1mM EGTA, 50 μM APMSF, 1 μM aprotinin, 10 μM bestatin, 10 μM leupeptin and 10 μM pepstatin, pH 7.0). The resulting sample was passed through a 40,000 psi piston 2X in a SLM Amicon French Press. The disrupted cells were centrifuged at 20,000x g to remove cellular debris. The supernatant was diluted two-fold with lysis buffer and dounce homogenized for two minutes before centrifugation at 48,000x g. The resulting supernatant contained the GST-SH2 fusion. A total of 30 mg of fusion product was purified using a glutathione-agarose affinity column.

Other proteins used in this study were thrombin (from Sigma) and recombinant human FKBP-FK506 binding protein. Using a BL21 (DE3) cell line, recombinant human FKBP-FK506 was placed in a pET116 plasmid and expressed and cloned in *E. coli*. (G.B. Wisely and I.R. Patel, Glaxo Research Institute, unpublished results).

B. Reduction/Alkylation of Thrombin

Thrombin was reduced using a solution containing a 500 molar excess of dithiothreitol (DTT) in 1.5 M guanidinium-HCl (GnHCl) for 30 min at 37 °C. Alkylation was achieved by incubating the reduced protein in a solution containing a 550 molar excess of iodoacetic acid in 1.5M GnHCl in the dark at room temperature for 1 hour.

C. Enzymatic and Chemical Cleavages

Factor Xa digestion of 30 mg of purified GST-SH2 was carried out in 20 mM Tris-HCl, 100 mM NaCl, 2mM $CaCl_2$, pH 8.0 at an enzyme to substrate ratio of 1:20. The digest was performed at 37 °C. Approximately 10 per cent of the fusion protein was digested over 24 hours as determined by monitoring the appearance of free GST and SH2 by polyacrylamide gel electrophoresis.

CNBr cleavage of FKBP was carried out by incubating 1 nmole of protein in 60% TFA containing an excess (50 mg) of CNBr. The reaction was carried out in the dark at at room temperature overnight. The reaction was terminated by speed-vac lyophilization (Savant, Inc.) and subsequent reconstitution in 0.1% TFA.

D. Liquid Chromatography

An ABI 140A syringe pump HPLC system was used for all analyses. Reverse-phase perfusion and C4 capillary columns were custom packed by LC Packings (San Francisco, CA). A 30 cm piece of 320 μm i.d. fused silica capillary was packed with aquapore butyl (C4) (Applied Biosystems, Foster City, CA). The silica-based C4 column was operated at a flow rate of 6 μl/min. This flow rate was achieved and maintained constant using a pre-column split as described elsewhere[13]. Both 5 cm and 10 cm pieces of 320 μm i.d. fused silica capillary were packed with Poros R/H reverse-phase perfusion packing material (PerSeptive Biosystems, Cambridge, MA) and operated between 40-100 μl/min without pre-column splitting.

Solvents were as follows: Buffer A was water containing 0.05% TFA (Pierce Chemicals, Rockville, IL) and Buffer B was 9:1 acetonitrile/water containing 0.05% TFA. Proteins and products of enzyme digestion were eluted from packed capillary perfusion columns using a gradient of 0 to 60% Buffer B in 5 min. These steep gradients did not produce reasonable chromatography using the silica-based C4 capillary column. Therefore, to maintain chromatographic integrity, proteins and enzyme digests were eluted from the packed capillary C4 column using gradients of either 5% to 65% Buffer B in 20 min or 5% to 65% Buffer B in 30 min. An ABI 740A detector equipped with a "z-shaped" capillary flow cell[14] was placed "in-line" with the mass spectrometer and operated at a wavelength of λ=210 nm.

E. *Electrospray Ionization Mass Spectrometry*

Electrospray ionization was performed using a Sciex API III triple quadrupole mass spectrometer equipped with an atmospheric pressure ionization source[15]. The outlet of the capillary LC column was coupled to a 50 µm i.d. fused silica transfer line that terminated at the electrospray needle, which was floated at 5000 volts. Q1 of the mass spectrometer was scanned from 400-1900 u in 4.5 seconds. Both Q2 and Q3 were operated in rf-only modes. Unit resolution was employed for all experiments. Nebulizer gas pressure was 35 psi and curtain gas flow was 0.8 ml/min. The orifice potential was set to 80 eV for the analysis of CNBr cleavage products and 100 eV for Factor Xa cleavage products. The electron multiplier was set at 3700 eV.

III. RESULTS AND DISCUSSION

A. *Rapid Desalting of Proteins*

We first investigated the capacity of perfusion packed capillary columns to rapidly purify high salt-containing proteins. Proteins submitted to our laboratory for mass spectral analysis are frequently solubilized in chaotropic agents, 8M urea or 6M GnHCl. These buffers are incompatible with direct infusion electrospray ionization and require a desalting step prior to their analysis. Desalting C18 cartridges are not particularly useful, especially when protein content is low, because of low protein recovery. Further, when presented with only sparing amounts of protein, it is of paramount importance to minimize the number of transfer steps prior to mass spectrometric analysis. Therefore, desalting methods which may be coupled directly to mass spectrometry are desirable. The flow rates of capillary LC are ideally suited to ESI/MS. However, conventional silica based packed capillary columns do not "clear" high molar salt solutions particularly rapidly. Low solvent strength wash times for C4 and Hypercarb packed capillaries (320 µm i.d.) are between 10-15 minutes, using 5 µl/min column flow rates. The unique feature of perfusion columns is the capacity to load samples and perform column washes at high flow rates.

Gn-HCl is a buffer often used for the solubilization of hydrophobic (*i.e.*, sticky) peptides and for reduction/alkylation reactions. Thrombin, a serine protease, has two chains, α and β, linked by one disulfide bridge. The protein contains a total of four disulfide bonds[16]. Thrombin was reduced and alkylated to provide as a model of a "typical" high salt-containing protein solution. The product of the reduction/alkylation reaction was HPLC purified using a 10 cm packed capillary perfusion column operated at 50 µl/min and monitored by LC/ESI/MS. The in-line UV chromatogram was recorded for this analysis and is shown in Figure 1a. Clearance of the salt solution was achieved in the first 30 seconds following injection. A gradient of 10% to 60% Buffer B in 5 min enabled the two subunits to be baseline resolved in only 4.2 min. The electrospray spectrum of the α subunit is shown in Figure 1b.

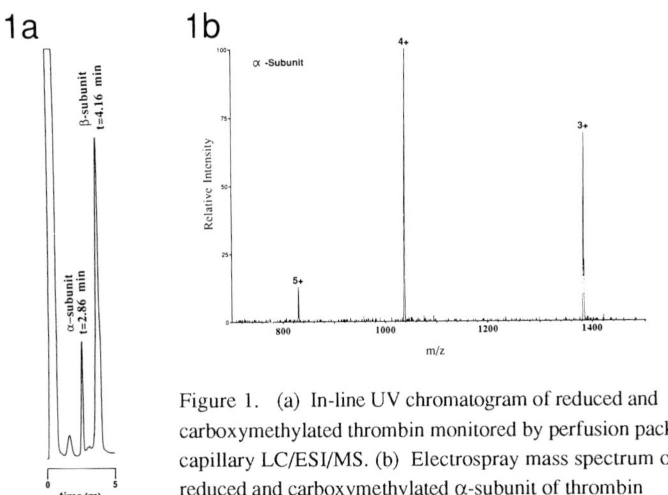

Figure 1. (a) In-line UV chromatogram of reduced and carboxymethylated thrombin monitored by perfusion packed capillary LC/ESI/MS. (b) Electrospray mass spectrum of reduced and carboxymethylated α-subunit of thrombin

To achieve similar chromatographic resolution of the α and β subunits using the silica based C4 packed capillary column required a shallower gradient and an analysis time of approximately 35 min using a 6 µl/min column flow rate (data not shown).

B. Monitoring Fusion Protein Cleavage Reactions

Affinity column chromatography is a fast and convenient method for purifying proteins. A technique employed for over-expression of novel proteins is to link the recombinant protein of choice with a fusion peptide or protein. This enables one to purify the soluble recombinant protein of interest in a single affinity chromatography step. Examples of fusion partners include Glutathione-S-Transferase (GST), Maltose Binding Protein and various peptides (*e.g.*,poly-his, Z-Z domain). In addition, protease or chain modification sites can be engineered within the fusion proteins to allow cleavage and separation of the intact recombinant protein from its fusion partner.

Elution from affinity supports often requires high concentrations of salt. These types of conditions may also be needed for efficient digestion by the protease or chemical reagent of choice. Perfusion column chromatography coupled with mass spectrometry would be an invaluable, rapid on-line monitoring method available to the protein chemist for the identification and preparative scale up of expressed fusion proteins.

The utility of perfusion column LC/ESI/MS was investigated for monitoring the products of cleavage of a GST fusion protein. The SH2 domain of c-src tyrosine kinase was expressed as a GST-fusion protein and purified using glutathione-agarose. The subsequently purified protein was then cleaved by limited proteolysis with Factor X_a to produce a free SH2 domain. The products of this digestion were analyzed by LC/ESI/MS using a 30cm silica based C4 packed capillary column operated at a flow rate of 6 µl/min. The UV chromatogram of the fusion protein enzyme digest is shown in Figure 2a. An equal amount of the digestion mixture was loaded onto a 5 cm perfusion packed capillary perfusion column and gradient eluted using a 50 µl/min flow rate (Figure 2b).

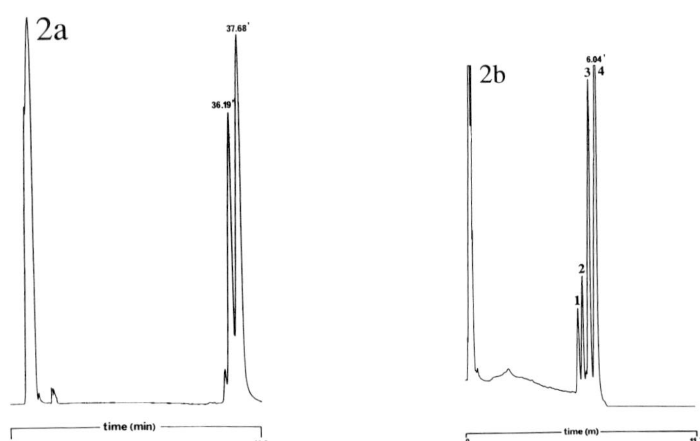

Figure 2. (a) Capillary C4 LC/ESI/MS UV chromatogram and (b) Capillary perfusion LC/ESI/MS UV chromatogram of the GST-SH2 cleavage reaction using Factor Xa.

Two striking observations could be made from the capillary perfusion chromatography experiment; analysis time was greatly reduced (from 37 min using C4 to less than 7 min using Poros™) and chromatographic resolution actually increased! We found that analysis time could be reduced further by increasing the column flow rate and/or the slope of the gradient. These changes, however, did degrade chromatographic resolution to a small extent.

The electrospray mass spectra for peaks labelled 2 and 4 are shown in Figure 3a and 3b, respectively. Peak #4 represents the cleaved GST protein and Peak #2 represents the SH2 portion of the fusion protein. Both spectra were characterized by excellent signal to noise. The molecular weight of the GST protein was found to be 27181.8 +/- 3.8 Da and matched with prediction.

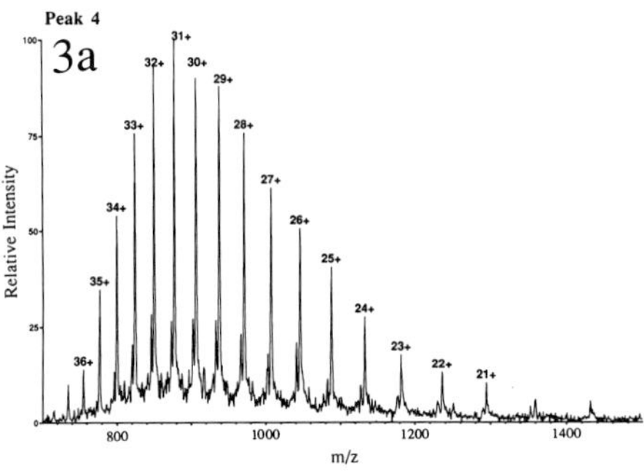

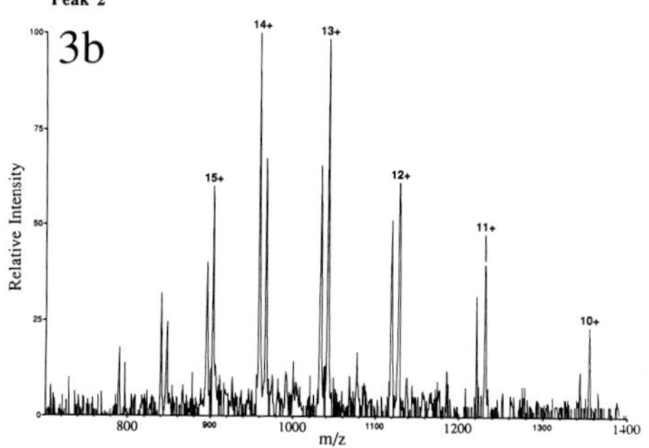

Figure 3. (a) Real-time electrospray mass spectrum of the GST domain (Peak #4) and (b) Real-time electrospray mass spectrum of the SH2 domain (Peak #2) from the perfusion packed capillary LC/ESI/MS analysis of GST-SH2 cleavage reaction.

However, the ESI spectrum of the SH2 peptide showed a species whose molecular weight was 610 Da greater than expected. This mass difference could be rationalized assuming that two (of a total of three) of the free cysteine residues of the SH2 domain became glutathionylated during purification. Indeed, this was not entirely unexpected because purification and elution of the GST-SH2 fusion is performed in the presence of high concentration of reduced glutathione (5-15 mM). Interestingly, peak 1 was also an SH2 related peptide, having a mass that would be consistent with the modification of all three SH2 domain cysteine residues. We confirmed that the SH2 domain was gluthathionylated by dialyzing the isolated SH2 peptide in excess DTT overnight at 37 °C. The reduced peptide was characterized by capillary perfusion column LC/ESI/MS and its molecular weight decreased exactly by the mass of the glutathione adducts (data not shown). Of particular interest was that the perfusion liquid chromatography experiment demonstrated that it was possible to separate the two SH2 peptides whose mass differed by only one glutathione adduct. Peak 3 corresponded to factor Xa and the molecular weight was determined to be 66730 +/- 6 from its ESI mass spectrum.

C. *Rapid Monitoring of CNBr Cleavage Reactions*

Another interest was to investigate the utility of the perfusion columns for the rapid separation and mass spectral characterization of peptide mixtures generated from both chemical and enzymatic cleavages. As a model, FK-506 binding protein (FKBP) was cleaved with CNBr producing a total of four CNBr fragments. The sequences and predicted masses of these peptides are summarized in Table 1. Using silica based microbore (1.0 mm i.d.) and capillary (0.32 mm i.d.) columns, it was possible to separate completely the 4 peptides. However, separation of the CNBr fragments could only be achieved using long analysis times (65 min) and shallow gradients (*e.g.*, 0% ---> 60% B in 60 min, data not shown).

TABLE 1

Expected CNBr Fragments for FKBP

Fragment	Sequence
Fragment 1.	GVQVEETISPGDGRTFPKRGQTCVVHYTGM Mr = 3074
Fragment 2.	LEDGKKFDSSRDRNKPFKFM, Mr = 2398
Fragment 3.	LGKQEVIRGWEEGVAQM, Mr = 1880
Fragment 4.	SVGQRAKLTISPDYAYGATGHPGIIPPHATL VFDVELLKLE, Mr = 4376

We attempted to separate the four CNBr fragments using the 10 cm packed capillary perfusion columns. Steep gradients (5% ---> 60% B in 6 min) enabled three of the four peptide fragments to be effectively separated. The UV trace corresponding to the capillary perfusion column separation is shown in Figure 4. The LC/ESI/MS analysis identified peak 1 as CNBr fragment 2, peak 2 as being a mixture of CNBr fragments 1 and 3 and peak 4 as CNBr fragment 3 (see Table 1).Thus, coupling these perfusion columns to mass spectrometry provides a major speed advantage and precludes the necessity to completely resolve (chromatographically) the cleavage fragments. Using shallower gradients (*e.g.*, 5% ---> 55% B in 25min) it was possible to chromatographically resolve the four CNBr fragments in under 15 min. This still provided a reasonable speed advantage over silica-based mcirobore and capillary column separations. The application of packed capillary perfusion column chromatography to the rapid separation of tryptic peptides has not yet been achieved. Small hydrophilic peptides have not shown to date reasonable retention on these short columns (*i.e.*, 10 cm length).

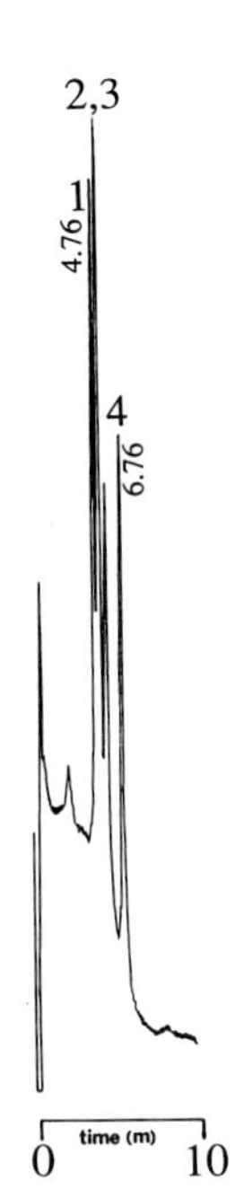

Future efforts will be directed toward separating tryptic fragments on 15cm and 30 cm packed capillary perfusion columns. Another important observation is that these capillary perfusion column have not yet approached the speed of analysis of analytical based columns for the purification and separation of protein mixtures as described previously by Regnier et al[6]. It must be emphasized that this has been the first attempt to pack 320 µm i.d. fused silica with perfusive particles and results can only be regarded as preliminary. Dramatic decreases in analysis time will undoubtedly be achieved in our future work. It is clear that at this early stage of development, these columns provide a tremendous speed advantage over silica based capillary columns.

Figure 4. UV chromatogram of the perfusion packed capillary LC/ESI/MS analysis of FKBP CNBr cleavage reaction.

IV. CONCLUSIONS

Demonstrated within was the first successful coupling of custom packed 320 µm i.d fused silica capillaries with reverse-phase perfusion packing (Poros II R/H™) to electrospray ionization mass spectrometry. These columns were found to provide tremendous speed advantage over silica based packed capillary LC columns for the analysis of proteins and enzyme digests. Perfusion packed capillary columns permitted flow rates between 50 - 100 µl/min and this enabled salt-contaminated proteins to be rapidly purified in-line with mass spectrometric detection. In addition, we demonstrated the ability to separate and mass spectrometrically identify components of enzymatic and chemical cleavage in 1/10 the time required using silica based packed capillary columns.

ACKNOWLEDGMENTS

We wish to thank Tona Gilmer, Tim Lansing and Jeff Robbins for the cloning and expression of the GST-SH2 fusion construct. The authors also wish to thank Cees Meijvogel for the packing of capillary perfusion columns.

REFERENCES

1. Afeyan, N.B., Fulton, S.P., and Regnier, F.E. (1991) J. Chromatogr. **544,** 267.
2. Regnier, F.E. (1990) Bio/Technology **8,** 203.
3. Afeyan, N.B., Gordon, N.F., Fulton, S.P. and Regnier, F.E. (1992) in Techniques in Prot. Chem. III, Academic Press, Inc. pp. 135.
4. Afeyan, N.B., Dean, R.C. and Regnier, F.E. (1989) U.S. Pat. Appl., 376.
5. Afeyan, N.B., Gordon, N.F., Mazsaroff, G.I., Varady, L., Fulton, S.P., Yang, Y.B. and Regnier, F.E. (1990) J. Chromatogr. **519,** 1.
6. Regnier, F.E. (1991) Nature **350,** 634.
7. Afeyan, N.B., Fulton, S.P. and Regnier, F.E. (1991) LC-GC **9,** 824.
8. Ito, Y., Takeuchi, T., Ishii, D. and Joto, M.J. (1985) J. Chromatogr. **346,** 161.
9. Kassel, D.B., Musselman, B.D. and Smith, J.A. (1991) Anal. Chem. **63,** 1091.
10. Harbour, G.C., Garlick, R.L., Lyle, S.B., Crow, F.W., Robins, R.H. and Hoogerheide, J.G. (1991) in Techiques in Protein Chemistry. III, Academic Press, Inc. pp. 487.
11. Griffin, P.R., Coffman, J.A., Hood, L.E. and Yates, J.R. (1991) Intl. J. Mass Spectrom.**111,** 131.
12. Huang, E.C. and Henion, J.D. (1990) J. Amer. Soc. Mass Spectrom. **1,** 158.
13. Chervet, J.P., Meijvogel, C.J., Ursem, M. and Salzmann, J.P. (1992) Liq. Chromatogr.-Gas Chromatogr. **10,** 140.
14. Chervet, J.P. (1989) High Res. Chromatogr. & Chromatogr. Commun. **12**, 278 .
15. Bruins, A.P., Covey, T.R. and Henion, J.D. (1987) Anal. Chem. **59,** 2642.
16. Butkowski, R.J., Elion, J., Downing, M.R. and Mann, K.G. (1977) J. Biol. Chem. **252,** 4942.

Sample Immobilisation Protocols for Matrix Assisted Laser Desorption Mass Spectrometry

K. K. Mock, C. W. Sutton, A. Keane, and J. S. Cottrell
Finnigan MAT Ltd., Hemel Hempstead, Herts., HP2 4TG, UK.

I. Introduction

The development of two new ionisation techniques, electrospray and matrix assisted laser desorption (LD), has greatly increased the range of biologically important molecules accessible to analysis by mass spectrometry [1,2]. The most numerous applications of both techniques have been in protein characterisation and, in many respects, their strengths and weaknesses are complementary.

One of the outstanding characteristics of LD is its unusual tolerance to contaminants such as buffer salts and surfactants. This tolerance is in marked contrast to earlier desorption techniques such as fast atom bombardment. Dramatic examples of the application of LD to the analysis of crude biological fluids can be found in the literature [3]. As an example, Figure 1 shows an LD spectrum obtained from the analysis of intact human blood. A drop of blood was diluted 50 fold in water, mixed with an equal volume of 50mM sinapinic acid matrix, and 1μl of the mixture dried in the centre of a sample target. The spectrum is dominated by the hemoglobin alpha and beta chains, and is not significantly different from the spectrum of highly purified hemoglobin. This tolerance to contaminants makes LD an excellent choice for preliminary screening experiments on crude biological extracts, the real time analysis of buffered enzymatic digests, and the analysis of complex mixtures in general.

Unfortunately, although the contamination tolerance of LD is excellent when compared with other ionisation techniques, it still falls short of coping with many common buffer systems, preservatives, and surfactants at their practical concentrations. The presence of excessive levels of contaminants can cause spectral degradation in the form of reduced signal intensity, poor persistence, and peak broadening. There are several contaminants which will cause virtually total loss of signal if present above a certain concentration, (see Table 1). The extent of suppression is both sample and matrix dependant. We have found that certain samples such as insulin, lactalbumin, and serum albumin provide spectra under conditions which would totally suppress the signal from other, apparently similar, proteins.

TECHNIQUES IN PROTEIN CHEMISTRY IV

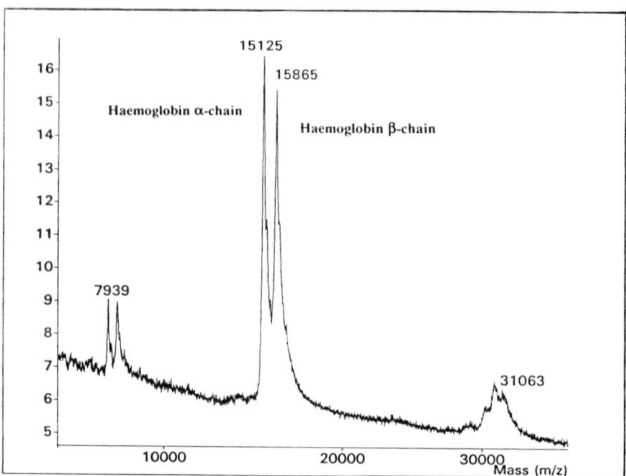

Figure 1. The spectrum of intact human blood in sinapinic acid, the sum of 10 laser shots.

If the sample is present at a reasonably high concentration ($>10\mu$M), and a suppression problem is suspected, one option is to dilute out the contaminants by the addition of clean solvent. Sometimes, this is all that is needed to obtain a spectrum of acceptable quality. If there are large quantities of sample available, HPLC or one of the many traditional purification strategies can be employed. However, there are many occasions when this is not feasible, and it would be useful to have a method of quickly and efficiently cleaning up a few picomoles of an unknown protein.

A method for sample purification *in situ* on the sample target was developed for use in ^{252}Californium Plasma Desorption (PDMS) [4]. The general approach was to immobilise the sample on a layer of nitrocellulose and then rinse away unbound contaminants such as salts and surfactants. This method of sample loading greatly improved the quality of the spectra, partly because of the purification step, but also because the nitrocellulose provided a substrate which was more compatible with the physics of desorption by fast fission fragments. We have adapted this protocol to enable *in situ* sample clean-up prior to LD analysis. Using the new procedures, spectra can be obtained from low picomole quantities of contaminated protein samples provided that the contaminants do not interfere with the binding of the sample to either a gold or nitrocellulose substrate [5].

Phosphate buffer > 20mM	Tris buffer > 50mM
Detergents > 0.1%, except for SDS, >0.01%	Alkali metal salts > 1M
Glycerol > 2%	Ammonium bicarbonate > 30mM
Guanidine > 1M	Sodium azide > 0.1%
EDTA > 20mM	Coomassie blue stain > trace

Table 1: Contaminant thresholds for the onset of spectral degradation. These are typical values for proteins in sinapinic acid

II. Experimental

The mass analyser used in this work was a Lasermat (Finnigan MAT Ltd., UK). The sample targets are formed from stainless steel sheet, and the points irradiated by the laser are confined to a 2mm diameter central spot. This spot is roughened by bead blasting and flashed with gold so as to provide a wettable surface with many nucleation sites to encourage uniform crystallisation of the evaporating sample and matrix solution.

The nitrocellulose was applied to the Lasermat sample target by electrospray deposition [6]. A fragment of nitrocellulose membrane (Millipore Immobilon-NC) was dissolved in acetone to give a concentration of approximately 2mg ml^{-1}. The solution was electrosprayed from a 26 gauge square cut hypodermic needle by an applied field of 2 to 3 kV per cm. The electrosprayed deposit was restricted to the 2mm spot in the centre of a target by means of a mask. The coverage, calculated from the average flow rate and typical spray pattern, and assuming the layer to be uniform, was around 50 μg cm^{-2}.

III. Sample Loading Protocol

The general immobilisation scheme has three stages. First, to adsorb the protein on the target surface. Second, to rinse away the contaminants while retaining as much as possible of the protein, Third, to add the matrix solution in such a way as to release some of the remaining protein back into solution.

A. Sample Adsorption The gold target surface was readily wetted by aqueous sample solution. The electrosprayed nitrocellulose was more hydrophobic, and required the presence of organic solvent or detergent in the sample solution to promote wetting. We found the addition of 15% methanol to be effective in most cases. In fact, the presence of methanol has been observed to enhance the binding of some proteins to nitrocellulose [7]. Ethanol and acetonitrile up to a concentration of 25% also worked well. Typical loading conditions for adsorption were 1 μl of sample solution containing 15 pmol of protein. Whenever possible, the sample solution was allowed to dry on the substrate so as to maximise any concentration dependent binding.

B. Rinsing Unless specified otherwise, analytical grade water was used as the rinse agent. The water was a gentle jet directed from a wash bottle for five seconds. Excess liquid was shaken off and the target allowed to dry. It was important that the target was completely dry before adding the matrix solution, otherwise the matrix droplet would spread over the surface of the target in an uncontrolled fashion. We usually accelerated the drying process by exposing the target to a moderate vacuum.

C. Addition of Matrix The final step is the addition of the matrix in a solvent system which will elute some of the protein from the surface. According to the literature, recovery of sample from nitrocellulose requires fairly vigorous conditions and extended elution times [7]. One possibility which was investigated was to use 80% acetonitrile, 20% water containing 0.1% TFA as the solvent system, since this should release the protein by physically dissolving the nitrocellulose. This proved not to be necessary, and equally good results were obtained using an acetonitrile content anywhere between 50% and 70%. Approximately 0.5 μl of the matrix solution was placed on the centre spot of the target and the target immediately covered with a microscope slide so as to retard drying of the droplet. The cover rested on the raised rim of the target to form a miniature humidity chamber. By allowing the matrix solution to soak for a couple of minutes, the elution of the protein into solution was improved, typically resulting in a two fold increase in signal intensity. After two minutes, the cover slide was removed and the droplet allowed to dry naturally.

IV. Results and Discussion

Nitrocellulose and polyvinylidene difluoride (PVDF) were obvious candidates for the immobilisation substrate because of their widespread use in Western blotting. The other surface which we investigated was the gold surface of the standard target. The immobilisation of proteins on metals has been the subject of extensive research because of its importance in relation to the biocompatibility of medical prostheses. Gold, along with silver and copper, has been reported to bind protein with exceptional efficiency [8].

Simply attaching a piece of nitrocellulose or PVDF membrane to the target is not a viable option because it will become electrically charged and interfere with mass analysis. Additionally, the bulk of the sample will be distributed throughout the interior of the membrane while only the immediate surface layers are accessible to the laser. A thin film coating of nitrocellulose or PVDF on the target could be produced by painting, but the experience gained in plasma desorption indicates that a sprayed deposit functions much more efficiently than a solid film, probably due to its larger surface area. Because of the practical difficulty of finding a solvent system for PVDF which was compatible with the electrospray process, we chose to experiment with nitrocellulose exclusively.

A typical figure for the binding capacity of a nitrocellulose membrane is 70 μg cm^{-2} for BSA on a membrane of 0.45 μm pore size [9]. This corresponds to a binding capacity of approximately 1% by weight. It is extremely difficult to quantify the surface area of either the membrane or the electrosprayed deposit, which has a granular appearance. From scanning electron micrographs [5], it looks probable that the two have similar surface areas per unit mass, but it is questionable whether the lower layers of the denser electrospray deposit are as accessible for binding as the interior of the spongy membrane. If this is so, the figure of 1% by weight would be an upper limit for the binding capacity of the electrosprayed deposit. Thus, a 2mm diameter area sprayed with 100 μg cm^{-2} of nitrocellulose has an estimated binding capacity for BSA of less than 0.5 pmol.

	1% Sod. Azide		**1% SDS**		**5% Glycerol**	
	Gold surface	**NC surface**	**Gold surface**	**NC surface**	**Gold surface**	**NC surface**
Renin substrate (1.76kDa)	81	149	28	154	92	78
Glucagon (3.48kDa)	106	93	34	6	69	106
Insulin (5.73kDa)	38	41	6	20	50	60
Lactalbumin (14.2kDa)	5	122	0	38	76	352
Chymotrypsin (25.2kDa)	0	246	262	0	523	415
Aldolase (39.2kDa)	2	46	52	34	2	52
Bov. serum albumin (66.4kDa)	0	39	14	0	0	17
Immunoglobulin G (150kDa)	0	38	0	29	0	110

Table 2: The effectiveness of the immobilisation protocol on gold and nitrocellulose substrates for a range of proteins and contaminants. Values correspond to the heights of the protonated molecular ion peaks after immobilisation and rinsing. For each sample, values are normalised to 100 for the non-contaminated protein, conventionally loaded.

The behaviour of the gold and nitrocellulose surfaces was investigated using a range of proteins and contaminants. The results are summarised in Table 2. Reference spectra were obtained from clean material, conventionally loaded by mixing sample and matrix solutions on the target. The samples contaminated with 1% sodium azide and 1% SDS gave no useful signals whatsoever when analysed without purification. Several samples gave acceptable signals in the presence of the 5% glycerol contaminant.

In general, the smaller proteins were recovered most successfully. As expected, the presence of SDS inhibited binding [9,10]. Glucagon, chymotrypsin and aldolase actually gave better results on gold than nitrocellulose in the presence of SDS. For glycerol and sodium azide contamination, nitrocellulose gave acceptable results in all cases. The presence of nitrocellulose appeared to be more important for the larger proteins.

Although these data were generated from deliberately spiked standards, we have been using the protocols on "real" problem samples for over a year. As an example, a sample of mouse IgG which gave no signal whatsoever was suspected to contain 0.1% sodium azide as a preservative. The spectrum in Figure 2 was obtained by loading an estimated 5 to 7 pmol onto a nitrocellulose coated target, rinsing, and then adding sinapinic acid matrix solution.

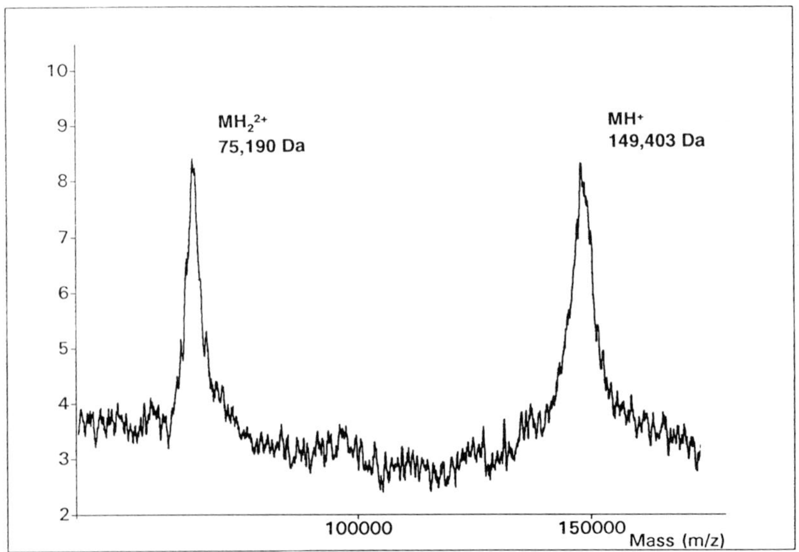

Figure 2. The molecular ion region of a sample of mouse IgG after immobilisation on nitrocellulose and rinsing away contaminants with water.

If the sample contains only a single component, or maybe a small number of similar components, then the efficiency of the immobilisation protocol is only a concern if it fails to provide a strong enough signal to measure an accurate molecular weight. In the case of a disparate mixture such as an enzymatic digest, an additional concern is that certain components may be retained while others are lost. Figure 3 compares the LD spectrum of a tryptic digest of bovine prolactin loaded conventionally with the results of immobilisation and rinsing on gold and nitrocellulose substrates. A few peaks are greatly reduced in intensity in the spectra from the rinsed samples, while others appear stronger. We have not been able to see any clear correlation between the physical properties of the peptides and their retention efficiency. For example, the peptides which are lost are not systematically different in hydrophobicity from those which are retained. It is also noticable that there is little difference between the spectra from the nitrocellulose and gold substrates.

The immobilisation on gold of three proteins of differing isoelectric point was studied as a function of the pH of the rinse solution. The contaminant in all cases was sodium azide. The results are summarised in Figure 4. It appears that the optimum pH of the rinse solution was not coincident with the isoelectric point of the sample, as might be expected, but rather some 3 to 4 pH units more acidic. One potential artefact in this experiment is the effect of the residues left by the buffers in the rinse solutions. The buffer systems were sodium hydroxide - acetic acid for the acidic pH range and Tris - HCl for the basic pH range.

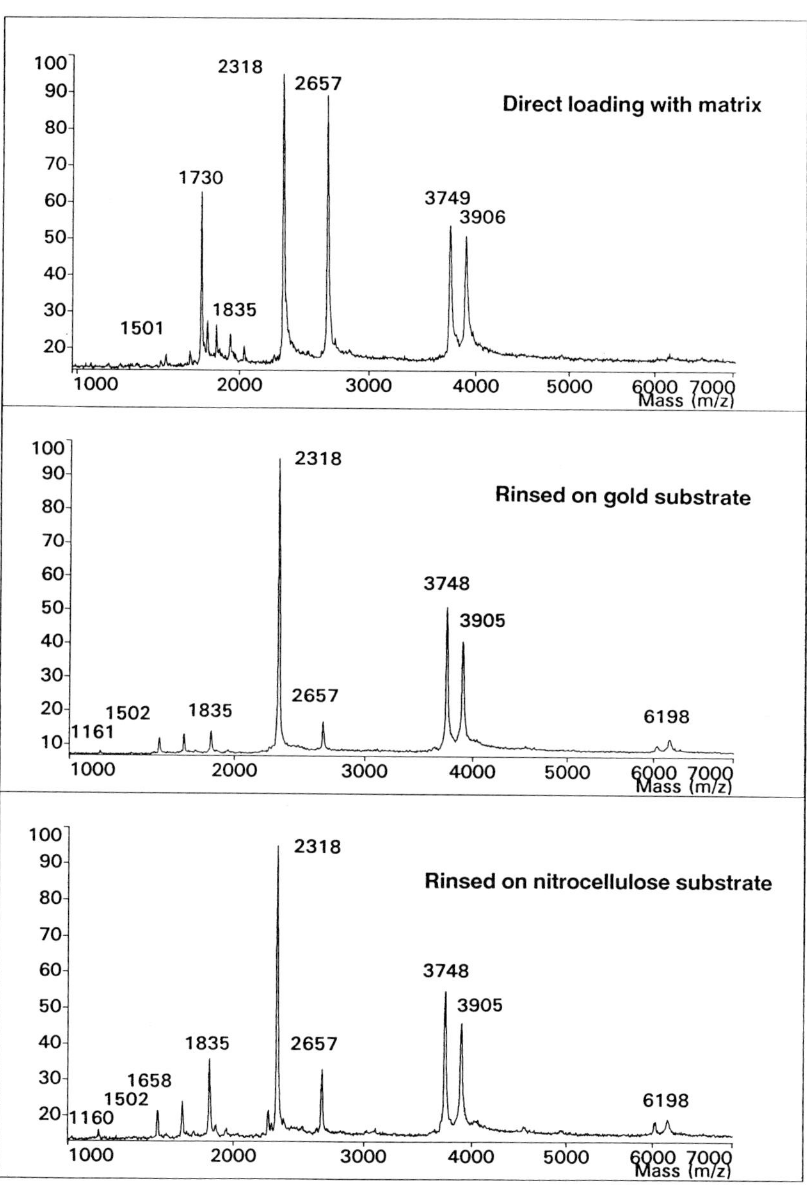

Figure 3. LD spectra of a tryptic digest of bovine prolactin: *upper trace* loaded conventionally, *middle trace* immobilised on a gold substrate and rinsed, *lower trace* immobilised on a nitrocellulose substrate and rinsed. In all three cases, the matrix was alpha-cyano-4-hydroxycinnamic acid.

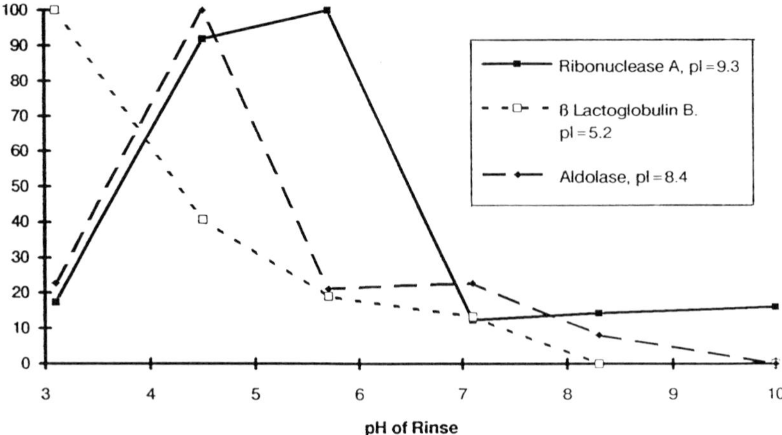

Figure 4. The effect of varying the pH of the rinse solution for three proteins dried onto a gold surface. The Y axis is the relative intensity of the molecular ion peak and the matrix was sinapinic acid.

For samples which could be successfully purified by immobilisation on nitrocellulose, the qualitative appearance of the spectrum closely resembled that of the clean sample, conventionally loaded on gold. However, it is also important to verify that molecular weight measurements are not compromised by the presence of the nitrocellulose. A small, but statistically significant shift was found between molecular weights determined using a gold target and using a nitrocellulose coated target, (e.g. 0.24% higher for bovine insulin on nitrocellulose). This shift can be fully corrected by internal calibration, i.e. using one or two peaks within the spectrum of known mass to define the mass scale. The use an internal mass standard for calibration is, in any case, advisable to obtain best mass measurement accuracy.

References

1. R. D. Smith, J. A. Loo, R. R. Ogorzalek Loo, M. Busman, H. R. Udseth, Mass Spectrom. Rev. **10**, 359-451 (1991).
2. F. Hillenkamp, M. Karas, R. C. Beavis, B. T. Chait, Anal. Chem **63**, 1193A-1203A (1991).
3. R. C. Beavis, B. T. Chait, Proc. Natl. Acad. Sci. USA **87**, 6873-6877 (1990).
4. G. P. Jonsson, A. B. Hedin, P. L. Håkansson, B. U. R. Sundqvist, B. G. S. Säve, P. F. Nielsen, P. Roepstorff, K-E. Johansson, I. Kamensky, M. S. L. Lindberg, Anal. Chem. **58** 1084-1087 (1986).
5. K. K. Mock, C. W. Sutton, J. S. Cottrell, Rapid Commun. Mass Spec. **6** 233-238 (1992).
6. C. J. McNeal, R. D. Macfarlane, E. L. Thurston, Anal. Chem. **51** 2036-2039 (1979).
7. B. A. Baldo, E. R. Tovey, Eds. Protein Blotting, Karger AG, Basel, (1989).
8. D. F. Williams, I. N. Askill, R. Smith, J. Biomed. Mater. Res. **19** 313-320 (1985).
9. H. K. Kihara, H. Kuno, Anal. Biochem. **24** 96-105 (1968).
10. R. Jahn, W. Schiebler, P. Greengard, Proc. Natl. Acad. Sci. USA **81** 1684-1687 (1984).

Assessing the Multimeric States of Proteins by Matrix-Assisted Laser Desorption Mass Spectrometry: Comparison with Other Biochemical Methods

Terry B. Farmer and Richard M. Caprioli

The Analytical Chemistry Center
and the Department of Biochemistry and Molecular Biology
University of Texas Medical School
Houston, Texas 77030

I. Introduction

Matrix-assisted laser desorption ionization mass spectrometry (MALDI MS) can be used to determine the molecular weight of monomers in protein multimeric complexes. However, unless special conditions are employed, the intact multimer complex will not be seen in the MALDI MS spectrum using existing sample analysis procedures. In order to preserve the quaternary structure as it exists in solution, chemical cross-linking agents can be used to covalently link the noncovalently bound subunits.[1-3]

Glutaraldehyde ($CHO(CH_2)_3CHO$) is a chemical cross-linking reagent that is thought to react primarily with the ϵ-amino group of lysine and the amino terminus of the polypeptide chain. When protein concentrations are kept relatively low, cross-linking will occur mostly between associated subunits, but not between multimeric forms normally unassociated in solution.[4]

Cross-linking with group directed reagents in conjunction with sodium dodecyl sulfate polyacrylamide gel electrophoresis (SDS-PAGE) has been used to determine the quaternary structure of protein complexes in solution. [5-8] Only species with covalent cross-links between chains have a significant difference in molecular weight. The accuracy of molecular weight determination by gel electrophoresis in such applications is $\pm$10-15%.[9]

TECHNIQUES IN PROTEIN CHEMISTRY IV

Electrospray mass spectrometry (ESI MS) has been used to investigate high molecular weight proteins. Generally noncovalent multimeric complexes are not observed with ESI MS although a few noncovalent ligand-protein complexes have been observed.[10]

In the work reported here, we have compared MALDI MS, ESI MS, and SDS-PAGE in the study of protein quaternary structure after chemical cross-linking with glutaraldehyde.

II. Experimental Conditions

A. Sample Preparation

Baker's yeast alcohol dehydrogenase and Baker's yeast glucose 6-phosphate dehydrogenase (both obtained from Sigma Chemical Co.) were dissolved in water or a buffer to a concentration of 10 pmol/μl (10 μM). Glutaraldehyde (25% solution by volume) (Sigma Chemical Co.) was added to a final concentration of 0.8% (% volume). The reaction mixture was incubated at room temperature up to 60 minutes and then analyzed by MALDI MS. For other analytical methods a 10-fold molar excess of glycine was used to quench the further reactivity of glutaraldehyde for the reaction mixtures used to obtain the MALDI spectra. These reaction mixtures were lyophilized and restored in gel electrophoresis loading buffer for SDS-PAGE.

B. Mass Spectrometry

The UV matrix-assisted laser desorption time-of-flight mass spectra were obtained with a Finnigan-MAT Lasermat or a Vestec VT 2000 (flight tubes = 0.5 m), both utilizing a nitrogen laser. Sinapinic acid (Aldrich Chemical Co.) was the matrix and was dissolved in 70% acetonitrile/30% water with 0.1% TFA (volume %) for the Finnigan MAT and 30% acetonitrile/70% water with 0.1% TFA (volume %) for the Vestec VT 2000. The relative amount of each multimeric species present was determined by calculating the ratios of the areas under the peaks.

Electrospray spectra were recorded on a Finnigan-MAT TSQ 70 with a Vestec source. Thirty-two scans were averaged over the mass range (600-3500 u) recorded. The samples were directly infused with a flow rate of 1.15 μl/min. The proteins were dissolved to 10 pmol/μl (10 μM) in a 47.5% methanol/47.5% water/5% acetic acid solution. Deconvolution of the resulting spectra was performed on a DECStation 2100 with Finnigan-MAT software.

C. SDS-PAGE

Cross-linked proteins and non-cross-linked proteins are loaded and separated on 5% acrylamide gels using the Weber-Osborne method.[11] The gels were run with constant current (30 mA) using a phosphate buffer, pH 7. Non-cross-linked proteins were run on the same gel to act as controls for the cross-linking reaction. Molecular weight markers were used to determine the molecular weight of the resulting bands after silver staining.[12] Scanning densitometry was performed with a BioMed Scanning Laser Densitometer to determine the relative amount of each band.

III. Results and Discussion

The addition of glutaraldehyde to the protein is accompanied by an increase in the molecular weight of that protein. The increase depends upon the number of reacting groups on the side-chains accessible to the cross-linking agent in solution and the intermolecular distance of the reacting side chains.

A. MALDI Mass Spectrometry

MALDI MS spectra were obtained for aliquots of the reaction mixture before the addition of cross-linker and at various times during the cross-linking reaction. Matrix solution (0.5 μl) was placed on the probe and 0.5 μl of the reaction was placed on top of the matrix. The solution was dried under gentle heat and the resulting spectrum obtained.

Glucose 6-phosphate dehydrogenase from Baker's yeast is a homodimer with a molecular weight of approximately 104,000.[13] As shown in Figure 1A, without the addition of glutaraldehyde the predominant peak in the MALDI MS spectrum was the monomer. After reaction with glutaraldehyde the dimer signal (56%) was significantly increased in relative intensity to that of the monomer (44%) with no significant formation of higher multimeric forms seen in the spectrum.

Baker's yeast alcohol dehydrogenase is a tetramer in solution formed from the association of identical subunits.[14-15] Before treatment with glutaraldehyde, monomer and dimer are present in almost equal amounts in the spectrum. After the addition of glutaraldehyde, some monomeric (7%) and dimeric (13%) ion peaks are visible but the main peak corresponds to the tetrameric molecular ion (80%) (Figure 1B).

The reaction of glutaraldehyde with other monomeric (lysozyme and carbonic anhydrase), dimeric (ovine lutropin), and tetrameric (hemoglobin and pyruvate kinase) proteins with subsequent analysis by MALDI MS has been shown to give the stoichiometry of the biologically active multimer.[16]

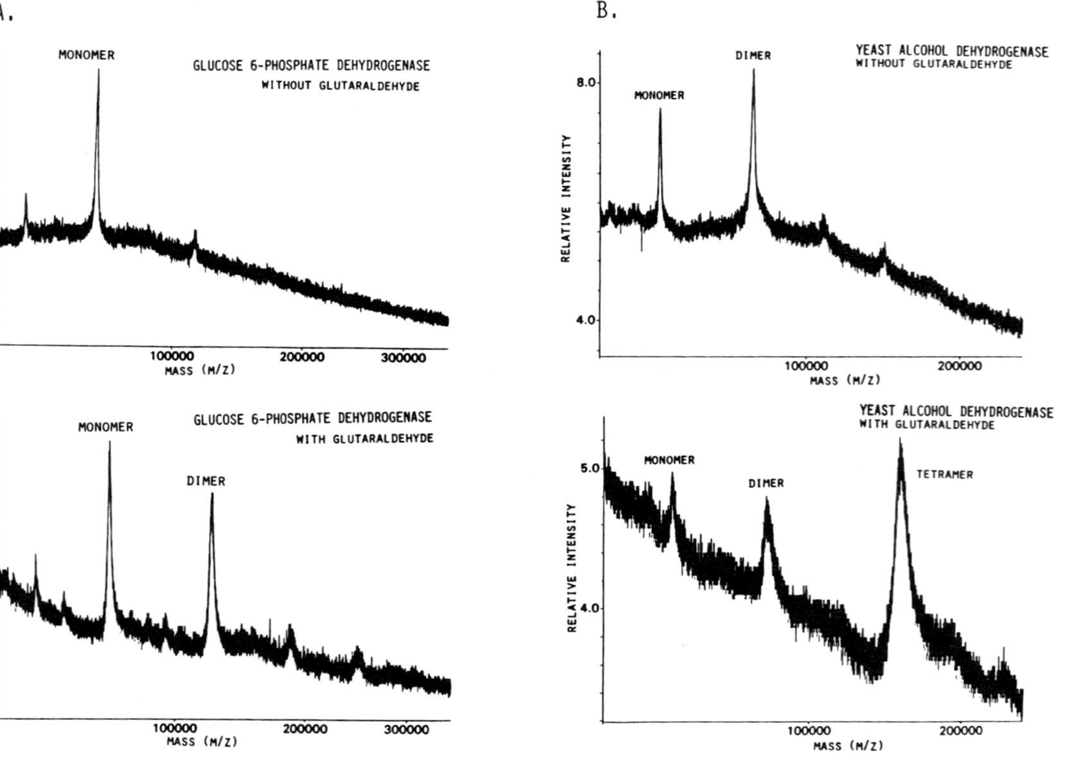

Figure 1. MALDI MS spectra of Baker's yeast glucose 6-phosphate dehydrogenase (A) and Baker's yeast alcohol dehydrogenase (B) before and after the addition of glutaraldehyde.

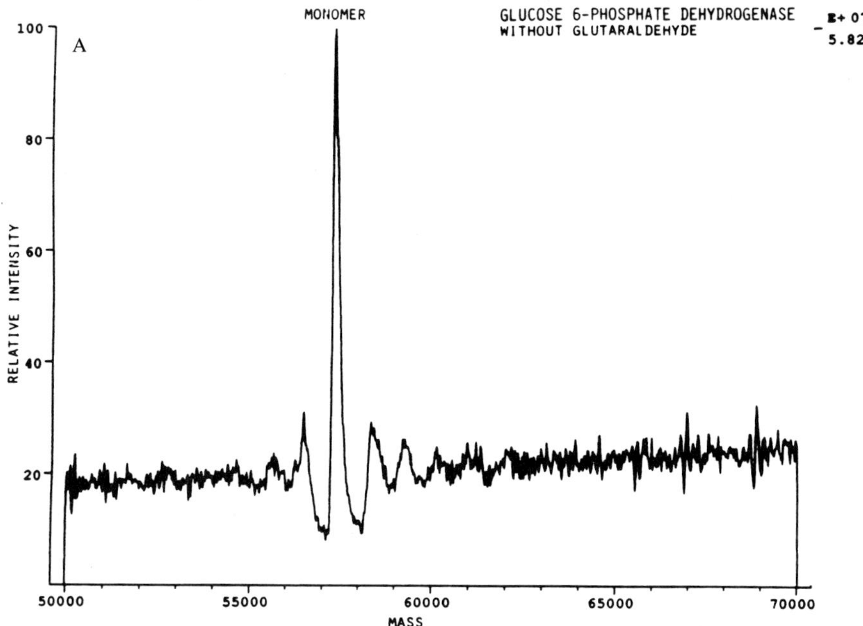

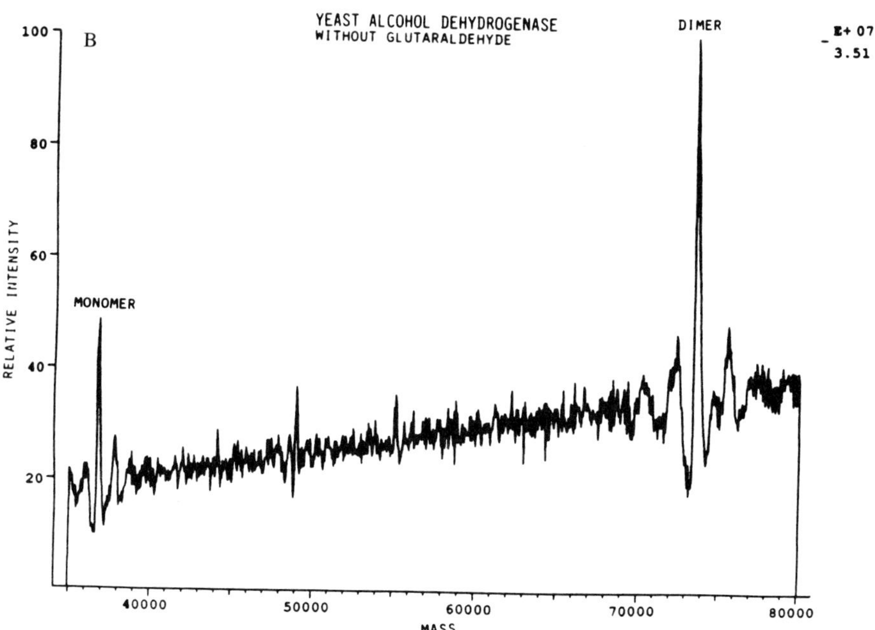

Figure 2. Deconvoluted ESI spectra of Baker's yeast glucose 6-phosphate dehydrogenase (A) and Baker's yeast alcohol dehydrogenase (B) before the addition of glutaraldehyde.

B. ESI Mass Spectrometry

The ESI mass spectra of non-cross-linked multimeric proteins typically showed signals for monomers, but not the multimeric forms (Figure 2A). Thus, as in the case of MALDI MS, noncovalent protein-protein interactions appear to be unable to survive the analysis process. An exception is the spectrum of non-cross-linked yeast alcohol dehydrogenase (Figure 2B). After cross-linking with glutaraldehyde, we were not successful in obtaining signals for monomeric or multimeric protein species even though these could be seen by SDS-PAGE and MALDI MS.

The reactivity of glutaraldehyde with proteins has been characterized by Habeeb and Hiramoto.[17] Using Edman degradation, they examined the reaction of glutaraldehyde with model compounds such as amino acids and peptides as well as with proteins. For amino acids and peptides, extensive reaction occurred with the α-amino group of glycine as well as with the α- and ϵ-amino groups of lysine. Tyrosine and histidine showed partial reaction.

The derivatization of these amino acid residues reduces the number of primary amino groups in a protein resulting in the formation of higher values of m/z on ESI analysis. These higher m/z values place the ions outside the range of most current instruments. The use of chemical cross-linkers that do not modify these residues may allow the detection of the cross-linked multimeric ions by ESI MS.

C. SDS-PAGE

Non-cross-linked glucose 6-phosphate dehydrogenase showed a band that corresponded only to the monomer after electrophoresis and silver staining. Upon cross-linking, the dimer was the major species present (Figure 3A). Non-cross-linked yeast alcohol dehydrogenase showed bands that corresponded to both monomer and dimer. After glutaraldehyde reaction, the presence of monomer, dimer and tetramer were detected by silver staining. Densitometry indicated the presence of almost 100% tetramer (Figure 3B). Chemical cross-linking reagents can also bind to the same amino acid residues to which dyes or stains bind to visualize the protein bands. Lysine, histidine, cysteine, and methionine are important amino acid side chains involved in silver staining.[18] Reaction of proteins with glutaraldehyde may lead to a decrease in silver staining and therefore greater errors in the qualitative and quantitative identification of cross-linked protein bands in the SDS-PAGE gel.

Coomassie blue and Fast green FCF staining were tried, but for several proteins the staining was minimal before glutaraldehyde cross-linking and after cross-linking the quantitation was inconsistent and sometimes difficult to obtain. Silver stain gave the most consistent results and is far more sensitive than the other stains for the small amounts of protein loaded.

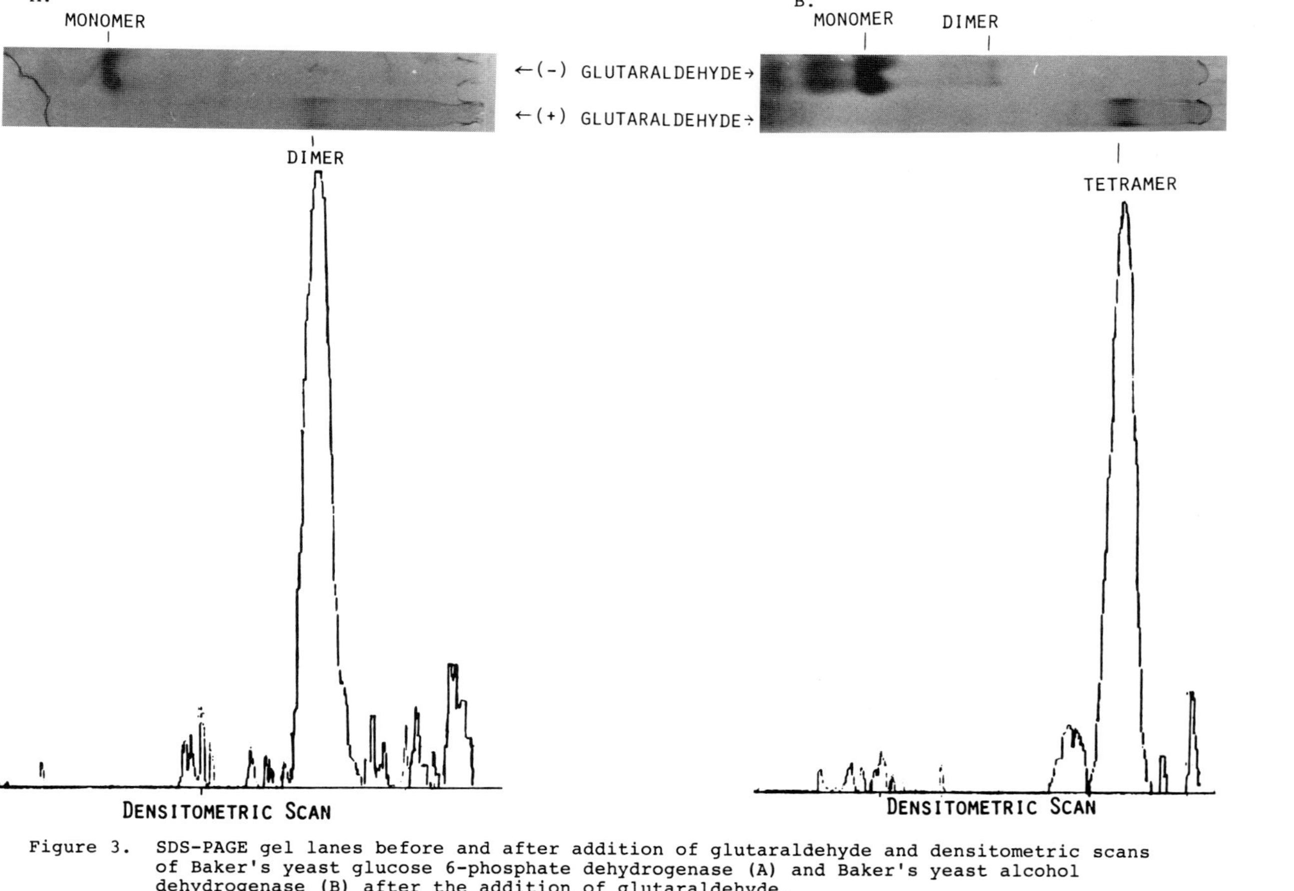

Figure 3. SDS-PAGE gel lanes before and after addition of glutaraldehyde and densitometric scans of Baker's yeast glucose 6-phosphate dehydrogenase (A) and Baker's yeast alcohol dehydrogenase (B) after the addition of glutaraldehyde.

IV. Conclusions

Preliminary results of the comparison of the SDS-PAGE and MALDI MS results obtained for glutaraldehyde cross-linked proteins are very similar qualitatively. Quantitatively, the results are more complex. The MALDI MS peak area ratios of some dimeric and tetrameric proteins agree very well with data obtained from the densitometric scans of SDS-PAGE gels. Other cross-linked proteins, however, do not correlate as well (Table 1). These differences could be the result of differences in the staining of the proteins run on the SDS-PAGE gel or they could be due to differences in ionization and detection of the proteins with MALDI MS.

Our results suggest that the glutaraldehyde cross-linked proteins are beyond the range of detection of electrospray mass spectrometry with quadrupole analyzers. It is likely that chemical cross-linkers that react with amino acid side chains other than lysine can be recorded with ESI instruments having low m/z ranges (i.e. <4000).

Table I. MALDI MS vs. SDS-PAGE for Glutaraldehyde Cross-linked Dimer and Tetramer Proteins

		Relative Peak Area %	
Proteins	Multimer	MALDI MS	SDS-PAGE
Alcohol Dehydrogenase (Baker's yeast)	1 2 3 4	7 13 - 80	2 - - 98
Lactic Dehydrogenase (Rabbit muscle)	1 2 3 4	37 29 9 24	6 15 - 79
Pyruvate Kinase (Rabbit muscle)	1 2 3 4	24 54 3 19	1 45 - 54
Glucose 6-Phosphate Dehydrogenase (Baker's yeast)	1 2	44 56	- 100
Carbamyl Phosphate Synthetase (E. Coli)	1 (low) 1 (high) 2	48 28 24	51 26 23

Further work is in progress to analyze the quaternary structure of proteins utilizing different chemical cross-linking agents. Comparisons of the results of analysis of these cross-linked multimeric complexes using MALDI MS, ESI MS, SDS-PAGE, CZE, gel filtration, and ultracentrifugation will be done.

Acknowledgements

The authors wish to thank Dr. Frank Rausch for the *E. Coli* carbamyl phosphate synthetase. We also wish to thank the National Institute of Health (Grant No. GM 43783-02).

References

1. M. Das and C.F. Fox (1979). *Ann. Rev. Biophys. Bioeng.* **8**, 165-193.
2. H. Fasold, J. Klappenberger, C. Meyer, and H. Remold (1971). *Angew. Chem. Internat. Edit.*, **10**, 795-801.
3. R.L. Lundblad (1991), Chemical Reagents for Protein Modification, CRC Press, Inc., New York, 287-304.
4. W.S. Craig (1988). "Determination of Quaternary Structure of an Active Enzyme Using Chemical Cross-linking with Guteraldehyde," Methods in Enzymology. Vol. 156, 333-350.
5. G.E. Davies and J. Gordin Kaplan (1972). *Can. J. Biochem.* **50**, 416-422.
6. R. Hermann, R. Rudolph, and R. Jaenecke (1979). *Nature* **277**, 243.
7. W.S. Craig (1988). "Determination of Quaternary Structure of an Active Enzyme Using Chemical Cross-linking with Guteraldehyde," Methods in Enzymology. Vol. 156, 333-350.
8. G.E. Davies and G.R. Stark (1970). *Proc. Nat. Acad. Sci.* **66**, 651-656.
9. J.C.H. Steele, Jr and T.B. Nielsen (1978). *Anal. Biochem.* **84**, 218-224.
10. Henion, J. and Li, Y-T (1991). *Proceedings of the 39th ASMS Conference on Mass Spectrometry and Allied Topics*, Memphis Tenn., 1155-1156.
11. K. Weber and M. Osborn (1969). *J. Biol. Chem.* **244**, 4406-4412.
12. B.D. Hames (1990). "One-dimensional Polyacrylamide Gel Electrophoresis", Gel Electrophoresis of Proteins, B.D. Hames and D. Rickwood, ed.s, IRL Press at Oxford University Press, 1-147.
13. P. Andrews (1965). *Biochem. J.* **96**, 595.
14. D.W. Russell, M. Smith, V.M. Williamson and E.T. Young (1983). *J. Biol. Chem.* **258**, 2674.
15. H. Jornvall (1977). *Eur. J. Biochem.* **72**, 425.
16. T.B. Farmer and R.M. Caprioli (1991). *Biol. Mass Spectrom.* **20**, 796-800.
17. A.F.S. Habeeb and R. Hiramoto (1986). *Arch. Biochem. Biophys.* **126**, 16-26..
18. Merril, C.R., (1986) "Recent Advances and Applications in the Art and Science of Detecting Proteins and Nucleic Acids by Silver Staining", *Electrophoresis* **86**, M.J. Dunn, ed., VCH, 273-290.

N-TERMINAL MODIFICATION OF MALARIAL ANTIGENS FROM *E. coli*

Jelle Lahnstein, Shanny L. Dyer[1], Neil H. Goss
Biotech Australia, 28 Barcoo St., East Roseville, N.S.W. 2069

Mark Duncan
Biomedical Mass Spectrometry Unit, University of N.S.W., Kensington, N.S.W., 2033

Raymond S. Norton[2]
School of Biochemistry, University of N.S.W., Kensington, N.S.W., 2033

I. Introduction

Malaria remains one of the major risks to human health in many parts of the world. To date however attempts to generate a vaccine against this disease have been unsuccessful. As part of a broad program directed at developing a vaccine against this disease we have been involved in the production of malarial antigens.

Conventional purification of recombinant products often requires the development of a new process for each protein, which then consists of several chromatographic steps. A rapid and generic purification process has been developed by Hochuli et al[1,2] and involves a histidine polymer (n=2-6) being introduced at either the N or C terminus of the r-protein. This provides a region which has a high affinity for Ni-chelate chromatography resin, allowing selective purification in a single step.

The gene encoding for a malarial merozoite surface antigen, MSA-2, was inserted into a bacterial expression vector such that the protein represented the near full length mature MSA-2 with an N-terminal extension of MRGSHHHHHH. This construct, termed Ag1624, was expressed as a soluble protein in *E. coli*.

Sequencing of purified Ag1624 gave methionine and an unknown amino acid at the N-terminus. The unknown has been identified as N-Methyl-Methionine (N-Me-Met), using mass spectrometry, reaction with ninhydrin, and NMR.

---

[1]Present address: CSIRO Division of Animal Production, North Ryde, N.S.W., 2113

[2]Present address: Biomolecular Research Institute, Parkville, Vic., 3052

TECHNIQUES IN PROTEIN CHEMISTRY IV

The proportion of N-Me-Met varied between 15% and 80% and was found to be dependent on the cell lysis conditions. The use of the denaturant guanidine hydrochloride (GnHCl) resulted in levels of over 50%, while without GnHCl the level was less than 30%.

Other soluble malarial antigens expressed by us also show this modification, while constructs of MSA-2 lacking the N-terminal extension, and lysed without GnHCl, did not. We conclude that in the cases described the modification appears to be a consequence of the presence of the extension at the N-terminus of the molecule.

II. Methods

A. *Purification of Ag1624*

Cells were suspended in 20 mM TrisHCl, pH 9 m6M GnHCl and lysed using a Manton-Gaulin MR15 homogeniser. The lysate was mixed with Ni-chelate resin (Roche) batch binding Ag1624, which was then eluted stepwise with 20 mM TrisHCl of decreasing pH containing 8M urea. Further purification was effected using a Vydac Protein C4 column with a gradient of 10% acetic acid aq. to 10% acetic acid in ethanol. Fractions containing Ag1624 were lyophilised.

B. *Sequencing*

N-terminal sequencing was carried out on an ABI 470A protein sequencer. PTH-amino acids generated were identified on an on-line 120A HPLC system using conditions as specified by ABI.

C. *Mass Spectrometry*

1. Sample preparation. The fraction containing the unknown was collected from the sequencer HPLC. Buffer salts were removed by rechromatography on a Vydac narrowbore C18 HPLC column using a water/acetonitrile gradient.
The collected fraction was lyophilised and submitted for mass analysis along with PTH-methionine and PTH-methionine sulphoxide (Pierce) as reference compounds.

2. Chemical Ionisation Mass Spectrometry. Analyses were performed on a Finnigan 3200 mass spectrometer in positive ion chemical ionisation mode using methane reagent gas. The spectrometer was interfaced with an Incos data system. The source temperature was maintained at 100°C by filament emission, the electron energy was 100 eV, and the electron multiplier was operated in the range 1.2 to 1.5 kV. Source pressure was 1 Torr. Samples were evaporated in a small glass cup and introduced into the ion source using a direct insertion probe. Thirty seconds after insertion the probe was then heated at 10°C per second.

3. Low resolution mass spectrometry and accurate mass determinations. The 70 eV probe electron impact (EI) spectra were obtained with a VG Autospec-Q mass spectrometer operating at an accelerating potential of 8 kV. Low resolution spectra were recorded after tuning to a resolution of 1,000 (10% valley definition) and calibration with perfluorokerosene (PFK) prior to analysis. Data were acquired in the centroid mode. Accurate mass results were obtained at a resolution of at least 5000 (10% valley definition). Results reported are the average of at least five magnet scans across the mass range m/z 500-50 at a scan rate of 5 seconds per decade. Data were acquired in the centroid mode and then mass measured against PFK which was introduced along with the sample.

D. Isolation and characterisation of the N-terminal dipeptides of Ag1624

1. Tryptic digest and HPLC separation of peptides.
Ag1624 (20 mg) was cleaved with trypsin (200 μg) in a solution of 1M Urea, 25mM TrisHCl, pH 8.5 (20 mL), generating N-terminal dipeptides. The digest was carried out at 37°C for 21 hours and was stopped by the addition of TFA (200 μL).

Peptides were separated on a Vydac Protein C_4 HPLC column (250X21mm) using a linear gradient of 0 to 100%(B) over 60 minutes, where eluent (A) was 0.1% TFA and (B) was 30% acetonitrile, 0.085% TFA. The flow rate was 16 mL/min and peak detection was at 220 nm.

2. Analyses of peptides. Early eluting fractions were identified using N-terminal sequencing as described above. Amino acid compositions were determined essentially as per the Waters-Millipore Pico-Tag method. The isolated N-terminal dipeptides were then reacted with ninhydrin to test for primary amines. Sample (100 μL) was mixed with 0.2 mM KCN in pyridine (8 μL, ABI) and 0.28 M ninhydrin in ethanol (4 μL, ABI) in a glass vial and heated at 100°C for 10 minutes.

Samples were methionine (6.7 nmol in 0.1% TFA), proline (8.7 nmol in 0.1% TFA), Unknown-Arg (28.6 nmol ex RP-HPLC), and Met-Arg (12.9 nmol ex RP-HPLC). Colour development was observed by eye and then recorded using colour photography.

The unknown-Arg dipeptide was further analysed by ^1H NMR. Spectra were recorded at 500.14 MHz on a Bruker AM-500 spectrometer using a 5mm (od) sample tube. The sample (180 μg) was dissolved in 90% H_2O/10% D_2O at a pH of 2.9 and spectra recorded at 300K. The water resonance was presaturated for 2s, 8192 data points were used, and the sweep width was 5495 Hz. Chemical shifts were measured using an internal standard of dioxane at 3.751ppm from DSS.

III. Results

A. *Sequencing*

The expected sequence, MRGSHHHHHH, was obtained, however the level of Met_1 was low compared with the following amino acids and another peak was observed in cycle 1 eluting between Phe and Ile (Fig. 1).

This unidentified peak was not seen when other proteins

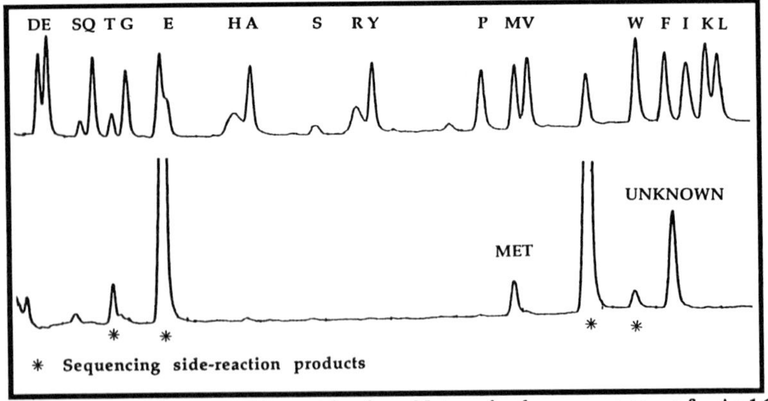

Figure 1. First residue of the N-terminal sequence of Ag1624 compared with PTH standards.

and peptides were sequenced, including those with an N-terminal methionine, and it is therefore not an artefact of the sequencing chemistry. It was considered that the unknown was likely to be a PTH derivative of a modified methionine.

Two obvious candidates, PTH-Met sulphoxide and PTH-Met sulphone, are excluded, because both are more hydrophilic than PTH-Met and elute before it. Ag1624 was sequenced through Met$_{31}$ and Met$_{37}$ and at both residues only methionine was seen. A tryptic peptide containing the only other methionine was also sequenced. Again, only methionine was seen, indicating that the modification is specific to the N-terminus.

B. Mass Spectrometry

Fragmentation patterns from electron impact spectra of PTH-methionine (MW 266) and the unknown were similar except that the major fragment ions for the unknown are shifted upwards 14 mass units (Fig. 2). The fragmentation pattern is consistent with the additional mass being associated with the PTH ring system. Chemical ionisation analyses confirm that the molecular mass of the unknown is 280. The high resolution mass spectrum gave a molecular ion consistent with the elemental formula $C_{13}H_{16}N_2OS_2$, i.e. PTH-methionine plus CH_2. Comparison with the spectrum of PTH-methionine sulphoxide confirms that the extra mass is not associated with the side chain of methionine.

The sequencing and mass spectrometry results show that the unknown is a PTH derivative of a modified methionine.

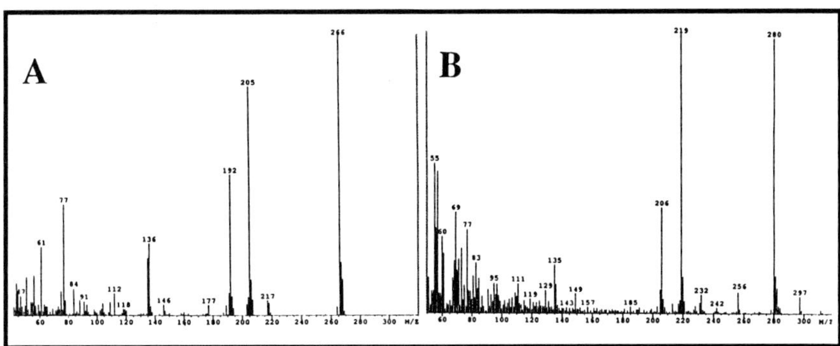

Figure 2. Low resolution EI spectra of (A) PTH-Met and (B) Unknown (CI and high resolution MS have shown peaks at m/z 256 and 297 to be contaminants).

The extra CH_2 most likely represents a substitution of a hydrogen with a methyl group. This can be on the nitrogen or the α-carbon.

C. *N-terminal dipeptides*

Reversed phase separation of the tryptic peptides resulted in the isolation of the two N-terminal dipeptides. Fractions 1 and 2 (Fig. 3) were analysed using N-terminal sequencing and found to contain Met-Arg and Unknown-Arg respectively. Amino acid analysis confirms the identity of fraction 1. Only arginine was found in fraction 2.
On reaction with ninhydrin both methionine and the dipeptide Met-Arg gave the characteristic violet colour which indicates the presence of a primary amine. The reaction with proline resulted in the expected pale yellow solution, while the solution containing the peptide Unknown-Arg remained essentially colourless, indicating the absence of a primary amine. The unknown would therefore appear to be N-methyl-methionine. The structure of this dipeptide was confirmed by the NMR results. Resonances from the Arg residue (with chemical shifts in ppm) were observed as follows: NH 8.84;

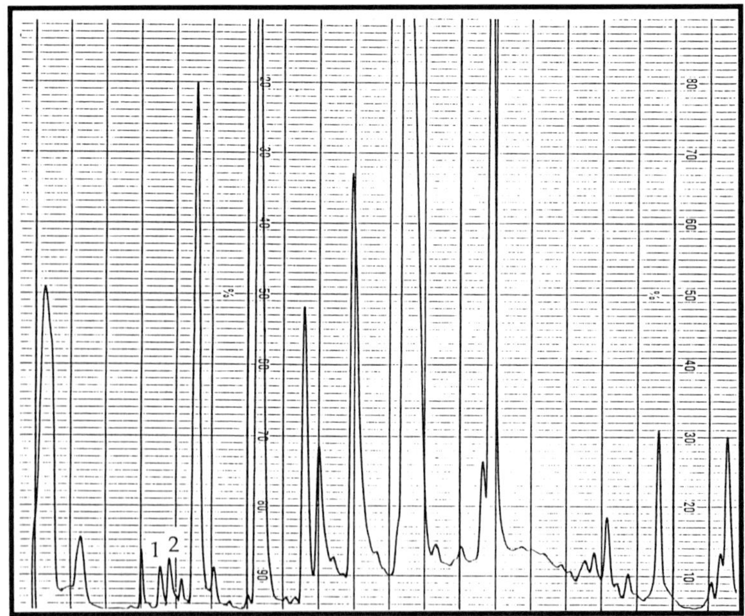

Figure 3. RP-HPLC separation of Ag1624 Tryptic peptides

$C^{\alpha}H$ 4.3 (partially overlapped with sidebands from the residual water resonance); $C^{\beta}H$ 1.79, 1.91; $C^{\gamma}H$ 1.65, 1.65; $C^{\delta}H$ 3.22, 3.22; $N^{\varepsilon}H$ 7.21; NH2 6.64. A splitting of 6.5 Hz was observed between the $C^{\delta}H_2$ and $N^{\varepsilon}H$ resonances, and 6.7 Hz between the amide and $C^{\alpha}H$ resonances. The chemical shifts for resonances of the Arg residue were very similar to those for Arg in a peptide (3), except for the amide resonance which was expected at 8.27 ppm. For the Met residue, resonances of $C^{\beta}H$ (2.20,2.20), $C^{\gamma}H$ (2.61,2.65) , and $C^{\varepsilon}H$ (2.11) were as expected for Met in a peptide (3), but the $C^{\alpha}H$ resonance at 4.03 ppm was upfield, reflecting the altered electronic distribution caused by the positive charge and the methyl group at the N-terminus. The N-methyl resonance occurred as a sharp singlet at 2.69 ppm.

D. Extent of Modification

Although Ag1624 is expressed in a soluble form, binding of contaminants to the Ni-chelating resin is significantly reduced in the presence of GnHCl. During development of the purification process GnHCl was added either before or after cell lysis. A lower level of degradation of the antigen and a higher level of N-methyl-Methionine are found if GnHCl is present during cell lysis (Table 1).

Production of other malarial antigens has also resulted in modification. Ag1623 is identical to Ag1624, except that the N-terminal extension has only four histidines. Addition of GnHCl after lysis followed by purification of this antigen resulted in an N-terminus which consisted of 20% N-Me-Met. An allelic variant of MSA-2 (from *P. falciparum* isolate CVD-1) and RESA (ring-infected erythrocyte surface antigen) were expressed with the same N-terminus as Ag1624. Both were soluble and purification after lysis in the presence of 6M GnHCl yielded N-termini with >50% N-Me-Met. Other constructs of MSA-2 which have an N-terminal methionine but do not have the hexa-His extension do not show modification of this methionine. However, another malarial

Table 1: Variation in level of N-Me-Methionine found

Lysis conditions	%N-methyl-methionine[a]
6M GnHCl	50 - 80 (n = 6)
No denaturant	15 -30 (n = 3)

a. calculated from peak heights of PTH-Met, PTH-N-Me-Met.

antigen, RAP-2, containing the N-terminal extension and expressed in inclusion bodies was found to have an unmodified N-terminal methionine.

IV. Discussion

We conclude from these results that the poly-His extension is a factor in the N-methylation of the malarial antigens described, and that the formation of inclusion bodies in the case of RAP-2 prevented methylation. Loss of the methyl group occurs on cell lysis. This appears to be inhibited by the presence of GnHCl, presumably due to denaturation of the enzyme(s) involved.

We would expect that other proteins containing this N-terminal extension, and which are expressed in a soluble form in *E. coli* would also be methylated.

N-terminal methylation of proteins is known in both prokaryotes and eukaryotes. Stock(4) divides these into several classes and suggests that N-terminal sequence similarities are significant and therefore provide a recognition site for four or five as yet unidentified methyltransferases. The sequences of *E. coli* proteins (4) are compared to the N-terminal extension of the malarial antigens in Table 2.

Similarities which can be seen between the N-termini of the malarial antigens and the *E. coli* proteins IF-3, L33, and S11 are an N-terminal methionine, a basic second residue, a small third residue, and the overall basicity of the sequence.

Table 2: Comparison of *E. coli* proteins with N-Me-Met and Malarial Antigens

Protein		N-terminal sequence						
CheZ		**MeM**	M	**Q**	**P**	S	I	K
L16		**MeM**	L	**Q**	**P**	K	E	R
IF-3		**MeM**	**K**	G	G	K	R	V
L33	(**MeM**)	**MeA**	**K**	G	I	R	E	K
S11		**MeA**	**K**	A	P	L	R	A
L11		**Me3A**	**K**	K	V	Q	A	Y
Mal. Ags.		MeM	R	G	S	H	H	H

Putative recognition residues are in bold print.

References

1. Hochuli, E et al(1987) J. Chromatogr., 411, 177.
2. Hochuli, E et al(1988) Bio/Technology, 6, 1321.
3. Wuthrich (1986) "NMR of proteins and Nucleic Acids", Wiley Interscience, New York.
4. Stock, A. (1988) Adv. Exp. Med. Biol., 231, 387 - 399.

Mass Spectrometric Characterisation of C-terminal "Ragged Ends" in Peanut Agglutinin

Pierre Thibault, David C. Watson[†], Makoto Yaguchi[†] and N. Martin Young[†]

Institute for Marine Biosciences, National Research Council of Canada, 1411 Oxford Street, Halifax, Nova Scotia, Canada B3H 3Z1
and
[†]Institute for Biological Sciences, National Research Council of Canada, 100 Sussex Drive, Ottawa, Canada K1A 0R6

I. Introduction

Cleavage of pro-proteins by proteolytic enzymes is a common form of post-translational modification that is often essential for production of active forms. In the legume lectin family two groups of lectins typified by concanavalin A and the pea lectin,respectively, are known to undergo post-translational proteolysis. An internal segment and a C-terminal segment are removed, resulting in a two chain structure (1,2), which in concanavalin A may then undergo chain fusion (1). In contrast most of the legume lectins are single-chain proteins, but sequence alignments show that sequences obtained by cDNA methods are often about fourteen residues longer than those obtained by protein sequencing (3). For example, the protein sequence of peanut agglutinin (PNA) is 236 residues long (4) but the sequence predicted from the cDNA has 250 residues (N.Sharon, personal communication). This suggests that the single-chain lectins also undergo post-translational trimming. We have used ionspray (nebuliser-assisted electrospray) mass spectrometry (5) to characterise this modification in several single-chain lectins and here report experiments on PNA that confirm evidence for C-terminal proteolytic processing to "ragged ends".

II. Materials and Methods

The PNA sample was identical to that used for the amino acid sequence determination (4). Trypsin (sequencing grade) and Lys-C were obtained from Boehringer Mannheim Canada Ltd (Laval, Canada). Buffer chemicals were purchased from Sigma Biochemicals (St. Louis, MO). HPLC-grade acetonitrile (BDH Chemicals), and distilled and deionized water (18 Mohm; Millipore Q water system, Millipore Inc.) was used in the preparation of the samples and mobile phases.

All mass spectra were acquired on a SCIEX (Thornhill, Ont., Canada) API III triple quadrupole mass spectrometer equipped with an atmospheric pressure ionization (API) source and an IonSpray® interface. Data acquisition and processing was achieved using a Macintosh IIx computer. Ionspray mass spectra of proteins were obtained by injecting 2-5 μL (1 mg/mL) into a stream of solvent (50% acetonitrile, 0.1% TFA) introduced to the mass spectrometer at a flow rate of 15 μL/min. Calibration of the mass range was achieved by injecting a solution of horse heart myoglobin (Mr: 16951). The computer program MacProMass (Terry D. Lee and Sunil Vemuri, Beckman Research Institute, City of Hope, CA) was used to support interpretation of the conventional mass spectra and the predicted fragment ions observed in the MS-MS experiments. For LC-MS analyses, an HP1090L liquid chromatograph (Hewlett Packard), equipped with a tertiary DR5 solvent delivery system, was coupled directly to the mass spectrometer via the IonSpray® interface. Separations were achieved using a 2.1 mm I.D. x 25 cm Vydac 218TP52 column (Vydac Separation Group, Hisperia, CA). Injections of 20 μL were usually made on the HPLC column. The effluent from the column was split such that a flow rate of approximately 15 μL/min was introduced to the mass spectrometer. The voltage on the ionspray needle was maintained at 5 kV. High purity air was used as nebulizing gas at an operating pressure of 40 psi. A post-column Valco submicroliter injection valve with a 0.5 μL loop was used to optimize the LC-MS interface. The LC-MS analyses were performed in full-scan mode (m/z 500-1500), using dwell times of 4 msec per Da.

Combined LC-MS-MS analyses were achieved using the same chromatographic system as described above. Tandem mass spectra were obtained following mass selection of precursor ions by the first quadrupole as they were eluting from the HPLC column. Fragment ions arising from the dissociation of the precursor ions were analyzed by scanning the third quadrupole mass analyzer, with a dwell time of 5 msec per Da. Acquisition of the mass spectral signal was made in multiple channel acquisition mode. Collisional activation was effected at energies of 100 eV (laboratory frame of reference) by introducing argon (target thickness of 4×10^{14} atoms cm^{-2}) into the rf-only quadrupole.

III. Results

The ionspray mass spectrum of PNA is presented in Figure 1. This spectrum was obtained through flow injection analysis using a total of only 3 μg of protein. Intense multiply-protonated ions extending from 11 to 29 charges are clearly observed in Figure 1. A series of closely associated peaks are also noted in each individual charge state cluster. The pattern of this microheterogeneity is better visualized from the reconstructed molecular weight profile of the mass spectrum (insert of Figure 1) as obtained from a

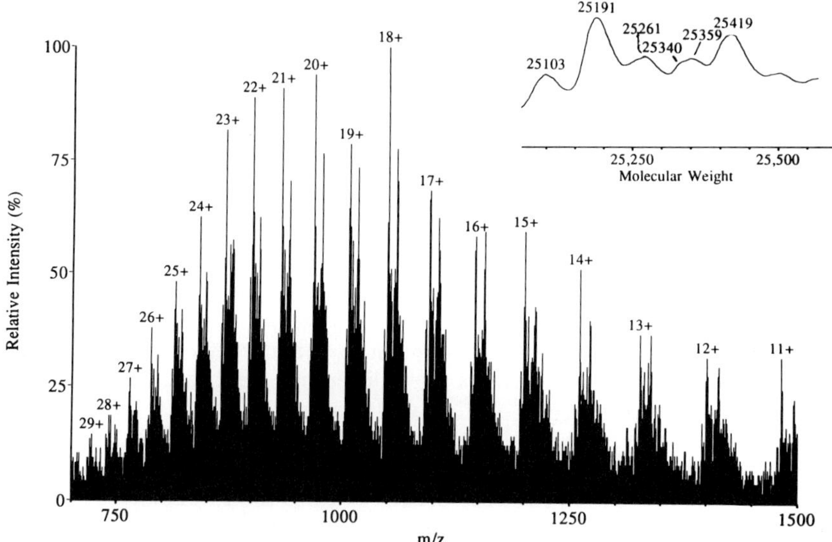

Figure 1. Ionspray mass spectrum of a purified sample of PNA. The spectrum corresponds to the injection of a total of 3 μg of protein.

Table I. Assignment of major components in the mass spectrum of PNA

Protein-derived C-terminal Sequence:- -TTTRRS[236]
cDNA-derived C-terminal Sequence:- -TTTRRSIDNNEKKIMNMASA[250]

Mass observed (Da)	Mass Calculated (Da)	Assignment
25103	25102.1	1-235
25191	25189.2	1-236
25419	25417.5	1-238

deconvolution algorithm program (6). The observed masses for this PNA preparation are 25103, 25191, 25261, 25340, 25359 and 25419 Da. The three most abundant peaks at 25103, 25191 and 25419 Da were in good agreement with masses deduced from the reported amino acid sequence (4), taken with the C-terminal extension found in cDNA sequencing (Table 1). The three assigned species represent products from cleavages at residues 235, 236 and 238, the most abundant being the 236 form in agreement with the protein sequencing. The isoform of greater mass is assigned to a species with an additional Ile-Asp, while the smaller one lacks the terminal Ser. PNA is not glycosylated, and the mass differences cannot be explained by attachment of a single hexose or hexosamine possibly overlooked in earlier sequencing work. The three minor species of intermediate masses may arise from genetic variants.

In order to confirm the reported sequence (4) and to rule out any other modification that might have given a mass difference identical to an Ile-Asp combination, the tryptic digest of PNA was analyzed by LC-MS (Figure 2). The total ion current (m/z 500-1500) profile is presented in Figure 2A along with the 2-dimensional contour profile of ion intensities (m/z vs time) in Figure 2B. All the tryptic peptides predicted from the PNA sequence (4) were observed as indicated in Figure 2B and Table 2, except the two below the mass limit, 75-77 and 202-203. It is noteworthy that the LC-MS analysis shown in Figure 2 corresponds to the injection of a total of 20 μg of the protein digest, and only 10% of this sample loading was introduced to the

Table II. Assignment of tryptic peptides of PNA analyzed by LC/MS

Retention time (min)	Mass Observed (Da)	Fragment Assignment
4.9	630	40-45
6.2	712	196-201
14.8	779	216-221
15.7	948	46-53
12.5	1112	163-172
16.0	1142	204-215
16.0	1343	150-162
18.6	1500	222-234
22.6	2310	54-74
18.9	2387	173-195
21.3	3537	78-112
17.9	3967	113-149
20.7	4201	1-39

Peanut agglutinin sequence (4) with cDNA extension (Italics):-

1 AETVSFNFNSFSEGNPAINFQGDVTVLSNGNIQLTNLNKVNSVGRVLYAM
51 PVRIWSSATGNVASFLTSFSFEMKDIKDYDPADGIIFFIAPEDTQIPAGS
101 IGGGTLGVSDTKGAGHFVGVEFDTYSNSEYNDPPTDHVGIDVNSVDSVKT
151 VPWNSVSGAVVKVTVIYDSSTKTLSVAVTNDNGDITTIAQVVDLKAKLPE
201 RVKFGFSASGSLGGRQIHLIRSWSFTSTLITTTRRS*IDNNEKKIMNMASA*

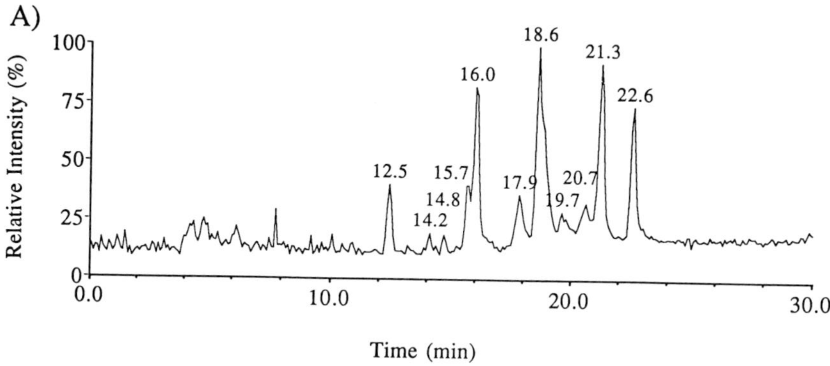

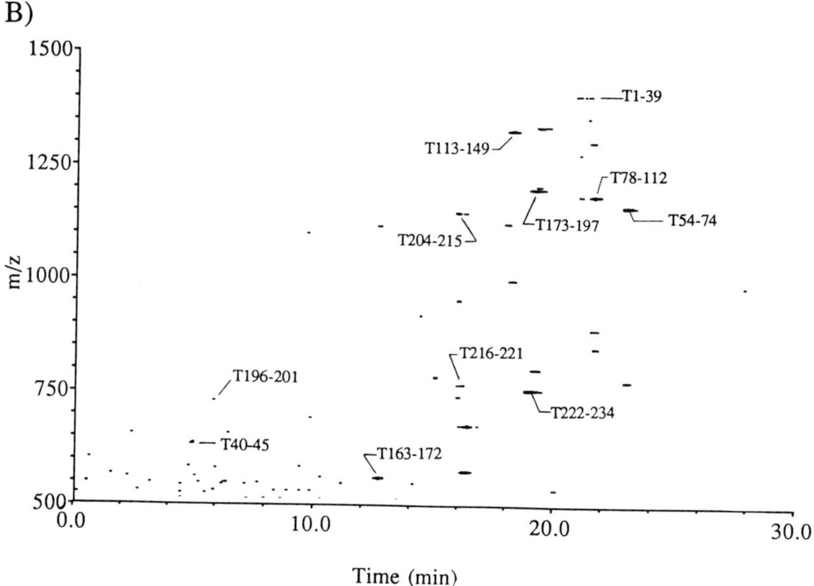

Figure 2. **LC-MS analysis of peptides arising from the tryptic digestion of PNA.**

(A) Total ion chromatogram (m/z 500-1500).

(B) Contour profile of ion intensities (m/z vs retention time) for the corresponding analysis.

Conditions: Vydac 218TP52 HPLC column, 0.2 mL/min flow rate, 1:15 split to send 1.5μg of digest to the mass spectrometer, linear gradient 10-90% acetonitrile (0.1% TFA) in 30 min.

mass spectrometer. The remainder of the sample was diverted to a waste line, but could alternatively have been directed to the UV detector or to a fraction collector for subsequent analyses.

Further confirmation of the sequence assignments was obtained by conducting tandem mass spectral analyses of selected peptides in combined LC-MS-MS experiments. An example of the application of LC-MS-MS for peptide sequencing is presented in Figures 3A and 3B for the two C-terminus tryptic peptides 222-235 and 236-238, respectively. The observed fragment ions are annotated according to a nomenclature proposed by Biemann for known dissociation products of peptides (7). The MS-MS spectrum of the tripeptide Ser-Ile-Asp (Figure 3B) confirms the expected sequence of the C-terminus ragged end, as predicted for the larger isoform of 25419 Da. Since residue 235 is arginine, the tryptic cleavage of the 236 residue isoform would give free serine, whose mass is below the range scanned by the mass spectrometer. Peptides arising from proteolytic cleavage with Lys-C were therefore also submitted to LC-MS analysis. Ions at m/z 1248 and 1287, corresponding to the triply-protonated forms of Lys-C fragments 204-236 and 204-238, were observed at 18.6 min in the LC-MS analysis of this digest (data not shown), further substantiating the above assignments.

IV. Discussion

The ionspray mass spectrometric results provide strong evidence that PNA has multiple forms resulting from cleavage at three close sites at residues 235, 236, and 238. Since an Asp residue is involved, some of the isoforms seen in iso-electric focussing probably originate from this heterogeneity (8). Difficulties encountered with C-terminal methods during the sequencing work (4) are also explained. We have found that the phenomenon of "ragged ends" also occurs in other single chain and two-chain lectins of the legume family (Young *et al.*, ms in preparation) though the presence of glycosylation in many of them rendered their analysis more difficult.

The biological significance of these findings is not clear. Some years ago it was shown that the two isoforms of the *Dolichos biflorus* lectin differed only in their C-termini, with one form being longer than the other (9). It was postulated that this difference was related to differences in the carbohydrate-binding properties of the two forms, but the C-terminal region should be at the opposite side of the subunit opposite to that of the carbohydrate-binding site, judging from the lectin crystal structures so far obtained. A role in targetting the protein to vacuoles is possible (10). It has been found that the C-terminal region of the barley and rice lectins (which are from the *Gramineae*, a totally different plant family) is used to target these proteins to vacuoles and, during or after this process, it undergoes C-terminal processing (11). A similar process may be occurring in the legume lectins, though in the case of *Phaseolus vulgaris* hemagglutinin, it appears that an internal signal is used (12).

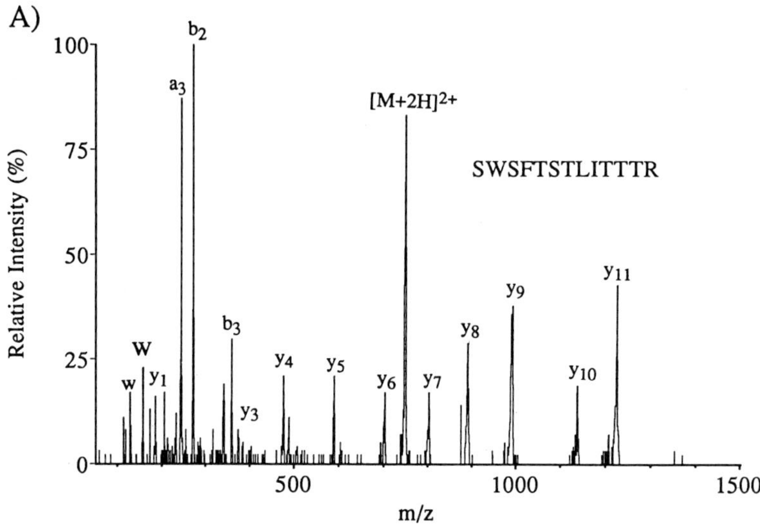

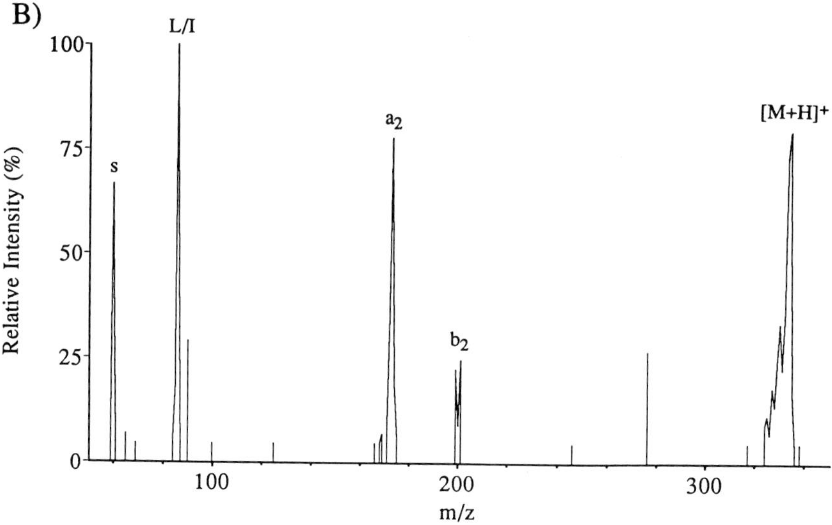

Figure 3. Tandem mass spectra of tryptic fragments 222-235 (A) and 236-238 (B) obtained from a combined LC-MS-MS experiment. Chromatographic conditions were identical to those of Figure 2; Ar target gas thickness, 4 x 10^{14} atoms cm^{-2}.

The data also shed some light on two other aspects of the PNA sequence. The sequence obtained from the cDNA differed at two internal positions from the protein sequence, Lys196→Thr and Leu197→Val. The mass spectrometry confirmed the protein sequence, suggesting that either different plant varieties were used in the two studies or multiple genes are present. Evidence was obtained during the protein sequencing for genetic variation since peptides with alternate residues at one or more positions were obtained. One variation was confirmed by the LC/MS, that introduces a new tryptic cleavage, Leu212→Arg. Two of the other changes lead to loss of a cleavage site, which would result in large peptides that were not detected. The intermediate forms seen in the mass profile probably are due to these genetic variations but assignments could not be made on the basis of the presently known ones, hence other variations must occur.

Acknowledgements

We thank Dr. Nathan Sharon for providing the PNA sequence obtained from its cDNA, and Mrs. P. Blay and Mrs. R. MacArthur for assistance in the preparation of the manuscript.

References

1. Carrington, D.M., Auffret, A. and Hanks, D.E. (1985) *Nature* **313**, 64-67.
2. Higgins, T.J.V., Chandler, P.M., Zurawski, E., Button, S.C. and Spencer, D. (1983) *J.Biol.Chem.* **258**, 9544-9549.
3. Sharon, N. and Lis, H. (1990). *FASEB J.* **4**, 3198-3208.
4. Young, N.M., Johnston, R.A.Z and Watson, D.C. (1991) *Eur. J. Biochem.* **196**, 631-637
5. Covey, T.R., Bonner, R.F., Shushan, B.I. and Henion, J.D. (1988) *Rapid Commun. Mass Spectrom.* **2**, 249-256.
6. Mann, M., Meng, C.K. and Fenn, J.B. (1989) *Anal. Chem.* **61**, 1702-1708.
7. Biemann, K., (1990) *Methods Enzymol.*, **193**, 886-887.
8. Pueppke, S.G. (1991) *Arch. Biochem. Biophys.* **212**, 254-261.
9. Roberts, D.M., Walker, J. and Etzler, M.E. (1982) *Arch. Biochem. Biophys.* **218**, 213-219.
10. Chrispeels, M.J. and Raikhel, N.V. (1992) *Cell* **68**, 613-616.
11. Bednarek, S.Y. and Raikhel, N.V. (1991) *Plant Cell* **3**, 1195-1206.
12. Tague, B.W., Dickinson, C.D. and Chrispeels, M.J. (1990) *Plant Cell* **2**, 533-546.

ACCURATE DETERMINATION OF MOLECULAR MASS FOR PROTEINS ISOLATED FROM POLYACRYLAMIDE GELS

T. N. Asquith[1], T. W. Keough[2], R. Takigiku[1], M. P. Lacey[2], M. P. Purdon[1] and D. L. Gauggel[1]

(1) Human and Environmental Safety Division and (2) Corporate Research Division, Miami Valley Labs, the Procter and Gamble Company, Cincinnati, OH, 45239-8707

I. Introduction

The molecular mass (MM) of a protein provides important information for identification and purification. SDS-PAGE is commonly used to determine protein MM values because of its simplicity and wide applicability. However, the accuracy of MM values determined from SDS-PAGE is limited because they are a function of relative mobility, which is subject to many experimental variables. This is especially true for small (<10 kD) proteins (1). Because SDS-PAGE is still an excellent method for separating mixtures of proteins, it would be very useful to combine the high resolution of SDS-PAGE with a more accurate method of MM determination.

Applying mass spectrometry (MS) to determine MM values of proteins isolated from native gels has been described with FABMS (2) and with PDMS (3). However, the important question of whether the high resolving power of SDS-PAGE could be combined with MS was left unanswered. Further, optimization of a method for recovering proteins specifically for MS was not explored. The recovery method is an important consideration since it is unlikely that any single method can recover all proteins, and because sample preparation requirements differ according to the type of MS used.

This paper describes a method to remove proteins from SDS gels for analysis by time-of-flight MS. Initial efforts were directed toward sample preparation for PDMS. More recently, we extended our sample preparation method to matrix-assisted

TECHNIQUES IN PROTEIN CHEMISTRY IV

laser desorption MS (MALDI). PDMS gave accurate MM values (errors were <0.1 %) for peptides and small proteins (<14 kD), and was reasonably insensitive to gel components such as SDS and dyes. Recent efforts were extended to MALDI, which indicated that it was more sensitive and accurate than PDMS for small proteins (<7 kD). However, MALDI analysis of larger proteins (> 14 kD) often resulted in poor quality spectra. Current attempts to reproducibly analyze larger proteins by MALDI are described.

II. Experimental

A. Materials and Equipment

Proteins were separated under reducing conditions in Daiichi 10-20 %, 8 x 10 x 0.1 cm precast gels (Integrated Separation Systems, Hyde Park, MA) as recommended by ISS. Ultra-low MW standards (3.5-29 kD) and tricine/SDS buffers were from ISS. Staining was done in 0.1 % (w/v) Coomassie Blue dye (CB) in ethanol:acetic acid:water (50:5:45) for twenty minutes, then destained in 30 % ethanol (v/v) for twenty minutes. Bands of interest were excised with a clean razor blade. Sonication of gel slices was done in an ultrasonic bath at 37° (Fisher Scientific). Electroelution of proteins was done with an electroelution device (Amicon, Danvers, MA), and the eluates concentrated with Centricon 3 filters (Amicon). PD mass spectra were obtained with an Applied Biosystems (Foster City, CA) linear time-of-flight mass spectrometer equipped with a 10 µCi ^{252}Cf ion source. Matrix-assisted LD mass spectra were obtained on a Vestec VT2000 instrument (Vestec Incorporated) from sinapinic acid or 2,5-dihydroxybenzoic acid matrices.

B. Methods

Proteins were extracted from gel slices with 200 µl of 50% (v/v) formic acid or 50% (v/v) trifluoroacetic acid by crushing 1-2 gel slices in a 1.5 ml capped centrifuge tube with a Kel-F Kimble pellet pestle (Vineland, NJ). Gel slices were then sonicated at a high setting for one hour. The homogenate was centrifuged (ultrafree-MC centrifugal filters, Millipore Inc.,

Marlborough, MA) prior to analysis to remove gel fragments. Electroelution was done as recommended by Amicon except that 0.01% (v/v) thioglycerol was added to the anode buffer prior to electrophoresis. Smaller proteins were extracted because they would not be retained by the Centricon membranes. Percent recovery was quantitated by reverse phase HPLC. Mass spectral analyses were done as previously described for PDMS (4) and MALDI (5).

III. Results and Discussion

A. PDMS

As previously reported (2, 7 and 8), minimal staining of proteins in gels maximized recovery and subsequent analysis. In our hands, only sonication plus concentrated acid (50% formic (8) or 50% TFA), could consistently remove enough of the larger proteins (5-14 kD) from polyacrylamide gels for PDMS analyses. Average percent recovery values for formic acid extractions ranged from 70-80% for bradykinin , 40-50% for oxidized insulin B chain to 15-20% for ubiquitin. Low recovery of the larger analytes (6-14 kD) was overcome by extraction with 50% TFA. Other solvents such as acetic acid (2) or formic acid/acetonitrile/ 2-propanol/water (7) yielded much poorer signal-to-noise ratios, which were interpreted as resulting from lower extraction efficiencies.

A series of control experiments were run to determine the effects of SDS-PAGE reagents on PD mass spectra of α-lactalbumin. Figure 1 shows that while spectral quality was unaffected by boiling the protein in SDS at a 100:1 molar ratio, little or no protein was detected after incubating α-lactalbumin with high levels of CB (100:1 molar ratio). Extracts of blank gel had no affect on PDMS spectra (data not shown). Similar effects were documented for the smaller proteins.

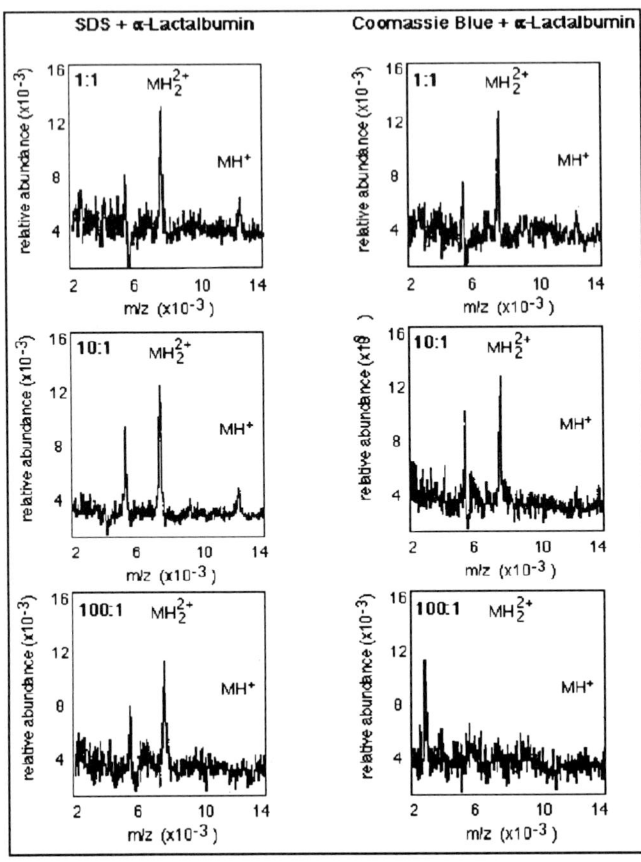

Figure 1. Effects of SDS and Coomassie Blue R250 on PD mass spectra.

Figure 2 shows high quality PD mass spectra obtained for peptides and proteins isolated by extraction. Note the high signal-to-noise ratios for substance P, glucagon, aprotinin and α-lactalbumin. The MM measurements obtained from gel isolates are given in Table 1. For peptides and small proteins, mass measurements are typically accurate to within ±0.1 %. In some cases like aprotinin, the accuracy achieved was poorer, possibly due to the presence of unresolved species. Ubiquitin and α–lactalbumin were also recovered from gel slices by electroelution for analysis by PDMS. The ubiquitin MM values were equivalent

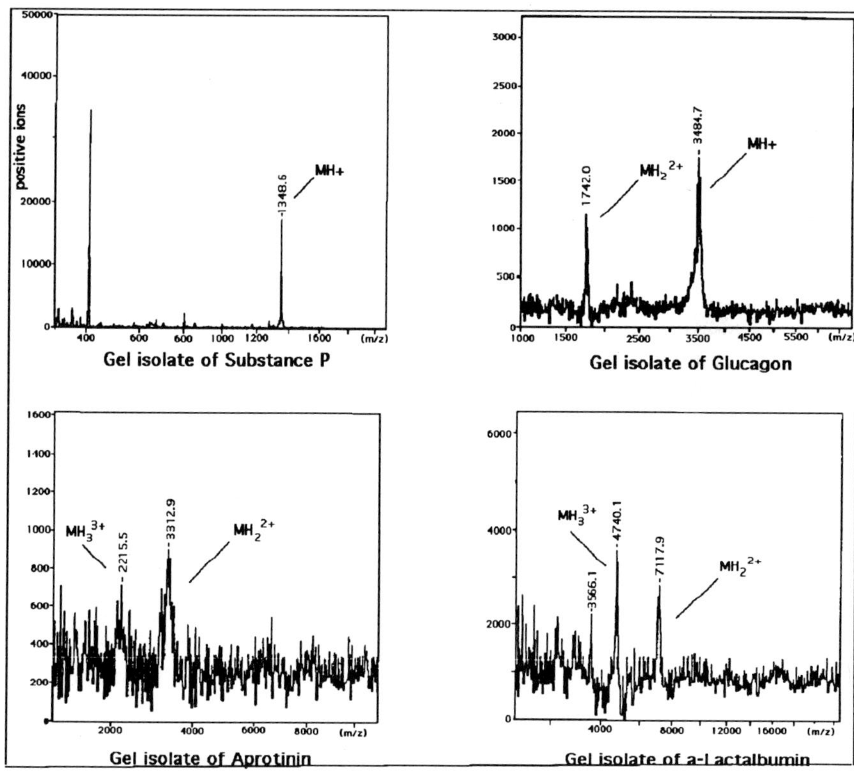

Figure 2. PD mass spectra of substance P, glucagon, aprotinin and α-lactalbumin after extraction from gel slices.

to those from extracted samples whereas α-lactalbumin values were closer to the calculated values.

B. PDMS Accuracy versus SDS-PAGE

Mass values derived from SDS-PAGE measurements (Table 1) were significantly different from the calculated masses. Generally, the smaller the protein the greater the deviation between calculated and observed mass. Only ubiquitin had an observed mass value reasonably close to the calculated value. But even in that case, the MM determined by gels was one order of magnitude less accurate than the PDMS result.

Table 1. Calculated and Observed Mass Values determined by PDMS and by SDS-PAGE

Protein	mass calculated	mass measured	Δ %	gel mass	Δ %
Bradykinin	1060.2	1060.2	0.00	8016	755
Substance P	1347.7	1347.6	-0.01	7010	520
Glucagon	3482.8	3482.9	0.01	--------	------
Ox. Ins B Chain	3496.9	3496.5	-0.02	7328	210
Aprotinin	6518.6	6634.7	0.78	10,162	56
Ubiquitin	8451.8	8460.8	0.11	8655	2
α-lactalbumin	14187.0	14228.1	0.28	11,292	-21
α-lactalbumin[1]	14187.0	14175.5	-0.08		

1. Electroeluted α-lactalbumin.

C. MALDI

Figure 3 shows MALDI mass spectra of substance P, glucagon, aprotinin and α-lactalbumin obtained from 100-fold dilutions of the gel extracts used for PDMS. Results for the smaller proteins were easily obtained by mixing diluted aliquots of gel extract into sinapinic acid matrix. However, spectral quality for α-lactalbumin varied widely between different preparations of analyte in 2,5-dihydroxybenzoic acid matrix. No response was seen when α-lactalbumin isolates were analyzed in a sinapinic acid matrix. Gel isolates of other "larger" proteins often did not yield observable spectra even in 2,5-dihydroxybenzoic acid. This may be due to an inhibitory effect of dyes. Mixing as little as 0.001 % (w/w) of CB or amido black with larger proteins (> 14 kD) often yielded no MALDI signals. Addition of soluble gel components also resulted in significantly inferior LD mass spectra, thus negating the possibility of using indirect stains. Low amounts of SDS did not adversely affect these spectra because they were analyzed in a dihydroxybenzoic acid matrix (9). We are currently evaluating the mass measurement accuracy of the MALDI method when applied to peptides and proteins isolated from SDS gels.

In an attempt to remove dyes and other contaminants,proteins were first blotted onto membranes (Problott®, Immobillon® or nitrocellulose) and then

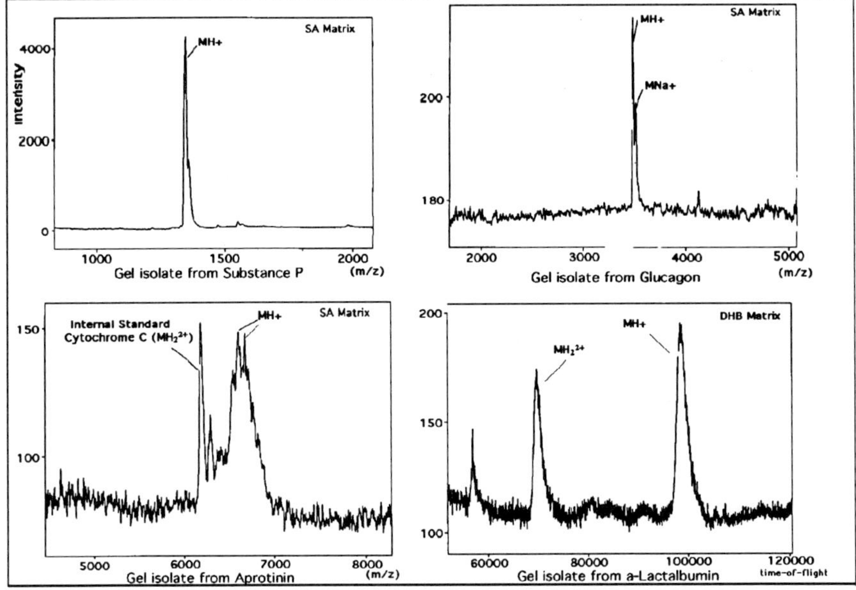

Figure 3. MALDI spectra of substance P, glucagon, aprotinin and α-lactalbumin after extraction from gel slices.

extensively washed. This approach was also unsuccessful, presumably due to soluble materials from the membranes which inhibited MS response (data not shown). Current efforts are focused on separating soluble SDS-PAGE components from extracted proteins by either capillary electrophoresis or microbore HPLC.

IV. Summary

Microgram amounts of proteins, isolated from CB stained and fixed SDS gels, can be mass analyzed by PDMS and MALDI. Because PDMS is not adversely affected by small amounts of SDS or CB dye, it is a useful method to analyze peptide and small protein gel isolates. For the proteins shown here mass measurements by PDMS were generally within expected accuracies (+/- 0.1 %). Next steps are to extend this technique to analyze larger proteins by improving the method for MALDI and for pneumatically-assisted electrospray MS.

References

1. Huang, J. and Matthews, H. R. (1990) Anal. Biochem. **188**, 114-117.
2. Camilleri, P., Haskins, N. J. and Hill, A. J. (1989) Rapid Commun. Mass Spectrom. **3**, 346-347.
3. Klarskov, K. and Roepstorff, P. (1990) In "Ion Formation from Organic Solids", A. Hedin, B. U. R. Sundquist and A. Benninghoven (eds.), John Wiley & Sons, Chichester, U. K., 67-71.
4. Lacey, M. P. and Keough, T. W. (1989) Rap. Commun. Mass Spectrom. **3**, 323-328.
5. Hillenkamp, F. and Karas, M. (1990) Meth. Enzym. **183**, 280-294.
6. DeWald, D. B. and Pearson, J. D. (1989) Anal. Biochem. **180**, 340-348.
7. Feick, R. G. and Shiozawa, J. A (1990) Anal. Biochem. **187**, 205-211.
8. Vanfleteren, J. R. (1989) Anal. Biochem. **178**, 385-390.
9. Strupat, K., Karas, M. and Hillenkamp, F. (1991) Int. J. Mass Spectrom. Ion Processes **111**, 89-102.

SECTION II

Analysis of Glycoproteins

Carbohydrate Characterisation of a Glycoprotein by Matrix Assisted Laser Desorption Mass Spectrometry

C. W. Sutton, A. C. Poole, and J. S. Cottrell
Finnigan MAT Ltd., Hemel Hempstead, Herts., HP2 4TG, UK

I. Introduction

The complete elucidation of the oligosaccharide structures attached to a glycoprotein is a complex task, generally involving a variety of analytical techniques [1,2]. Once the characterisation has reached a certain point, and if sufficient amounts of pure oligosaccharides can be isolated, definitive structures can be determined by two dimensional NMR. While chromatography or mass spectrometry based methods do not in themselves provide such complete structural information, they are more applicable to the analysis of mixtures, or samples which are scarce or impure.

Digestion with a series of highly specific exo-glycosidases can provide extensive structural information for naturally occurring oligosaccharides, [3], but the technique most commonly used to analyse the digestion products, P4 gel chromatography, has certain limitations:

- Each chromatography run takes several hours
- Radio-labelling is required to achieve good sensitivity
- Chromatographic resolution is poor, so analysis of mixtures can be problematic

We have investigated the use of matrix assisted laser desorption mass spectrometry (LDMS) as an alternative "read-out" for glycosidase digestion. LDMS has characteristics which make it well suited to this application [4]. Typical sample loading is 1 to 10 pmol, and analysis time is around 5 minutes. Through the use of an internal standard, mass measurements are generally accurate to better than 0.1%. The spectra are simple, showing no fragmentation, so that complex mixtures may be analysed. The practical mass range is in excess of 150 kDa, and the ionisation process is relatively insensitive to many common contaminants such as the buffer systems used for glycosidase digestion.

Since LDMS offers the capability of measuring the molecular weights of either the intact glycoprotein, the glycopeptides, or the free oligosaccharides, it is useful to consider the differing degrees of structural information provided by analysing these three species. It can be seen from

TECHNIQUES IN PROTEIN CHEMISTRY IV

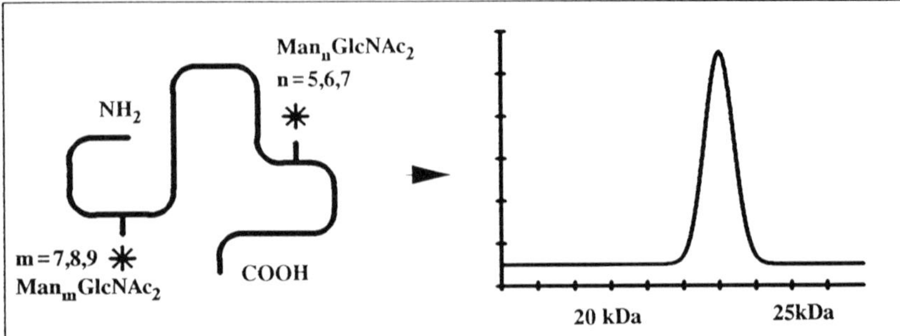

INTACT GLYCOPROTEIN

* Glycoforms may be mass degenerate or too close in mass to be resolved.
* Overall carbohydrate content can be determined by measuring shifts in average molecular weight caused by action of specific glycosidases.

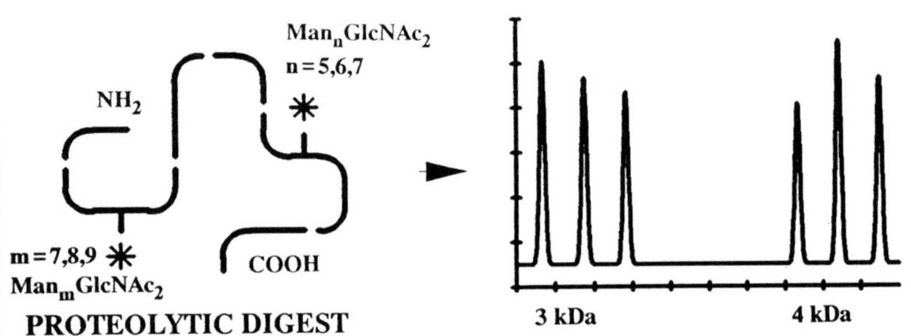

PROTEOLYTIC DIGEST

* Glycan distributions at each glycosylation site are separated on the mass scale by the masses of the attached peptides. Overlaps can be resolved by HPLC.
* Mass degeneracy is lifted. Structural information is site specific. Only site to site correlations are lost.

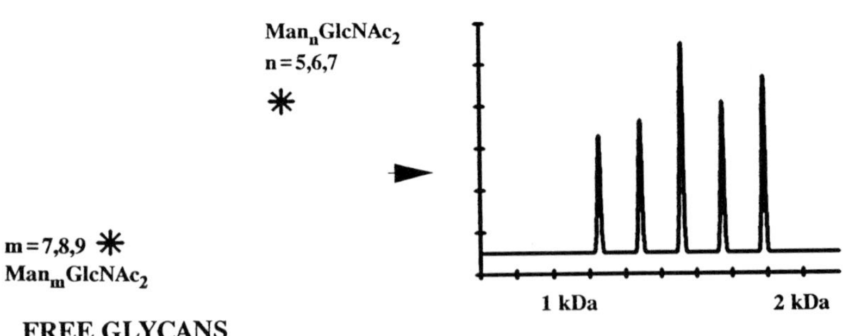

FREE GLYCANS

* All site specific information lost
* Peaks collapse into limited mass region causing overlap problems.
* Principal value is as glycan fingerprint

Figure 1. A schematic representation of the information provided by the LDMS analysis of glycoproteins, glycopeptides, and free sugars.

Figure 1 that maximum information results from analysing the glycopeptides, with only site to site correlation data being lost in this approach.

While LDMS analysis is intrinsically fast and sensitive, a further "multiplex" advantage comes from the ability to degrade and analyse the complete oligosaccharide population from a specific site as a group, rather than having to repeat the entire process for each individual species. The tolerance to contaminants of LDMS is also important since it would handicap both speed and sensitivity if HPLC clean-up was required prior to each analysis.

As a model glycoprotein, we have chosen tissue inhibitor of metalloproteinases (TIMP), an extensively disulphide bridged protein with two N-linked glycosylation sites (Asn30 and Asn78). Although the amino acid sequence and disulphide bridging of TIMP are known [5], we have been unable to find any published details of the oligosaccharide structures.

II. Experimental

The experimental strategy was:

1. Digest the protein with a protease chosen such that each potential glycosylation site is separated onto a different glycopeptide
2. Fractionate the digest by RP-HPLC, and identify glycopeptides by their characteristic mass spectra (distributions of peaks separated by mass differences of 146, 162, 203, & 291 Da)
3. Confirm the identities of glycopeptides by treatment with PNGase F
4. Assign probable oligosaccharide compositions by comparing measured masses with a database of naturally occurring structures
5. Treat glycopeptides with a sequence of specific exo-glycosidases chosen to confirm the composition and elucidate the linkage and anomeric configuration to the extent required or possible

General. Specific exo- and endo- glycosidases were purchased from Oxford Glycosystems Ltd., Abingdon, UK. Human recombinant TIMP, expressed in Chinese Hamster Ovary cells and purified from cell culture supernatants by immuno-affinity chromatography, was the gift of Celltech Ltd., Slough, UK.

Tryptic Digest. 120 μg of lyophilised TIMP was dissolved in 50μl of 5mM ammonium bicarbonate pH 7.8. To 44μl of this solution, 1μl of trypsin (TPCK treated, 2mg/ml) was added, and the mixture incubated at 35°C for 24 hours.

HPLC. An aliquot of the TIMP tryptic digest (10μl, 0.98nmol) was reduced with DTT and the peptides separated by RP-HPLC on an Applied Biosystems 130A separation system using a Brownlee RP-300 column (C_8,

100mm x 2.1mm). Fractions of 200 μl were collected, lyophilized and stored at -20°C until required.

PNGase F digestion. A 1 μl aliquot (50 pmol) of each glycopeptide containing tryptic fraction was incubated with 0.4μl of 50% (v/v) acetonitrile, 50mM ammonium bicarbonate pH7.8, 0.1μl 100mM DTT and 0.5 μl of peptide-N-glycosidase F (*Flavobacterium meningosepticum*) at room temperature for 16 hours.

Exo-glycosidase digestion. A 6 μl aliquot (300 pmol) of each glycopeptide containing tryptic fraction was digested with a series of exo-glycosidases for 24 hours at room temperature. After each digest, an aliquot (0.2 μl to 0.5μl) was removed for LDMS analysis before the addition of the next enzyme.

1. Neuraminidase (2μl, 0.02U, *Arthrobacter ureafaciens*) in 20 mM sodium acetate pH 5.0, 250 μg/ml BSA.

2. β-Galactosidase (1 μl, 1mU, *Streptococcus pneumoniae*) in 20 mM sodium cacodylate pH 6.0, 10 mM sodium azide, 1mg/ml BSA.

1 μl of the incubation mixture was then removed and treated with:

3. N-Acetyl-β-D-hexosaminidase (0.2μl, 100mU, jack bean) in 20mM sodium citrate pH 5.0, 0.1% sodium azide, 250 μg/ml BSA

The remainder was treated with:

4. N-Acetyl-β-D-hexosaminidase (1μl, 1mU, *Streptococcus pneumoniae*) in 50mM sodium phosphate pH 7.0, 10mM sodium azide, 0.5 mg/ml BSA.

Mass spectrometry. Aliquots of each digest of the TIMP glycopeptides (0.5μl, approx 12.5 pmol) were analysed by LDMS using a Lasermat (Finnigan MAT Ltd., Hemel Hempstead, UK) in the presence of an internal calibrant of substance P (1347.6 Da), or porcine renin substrate tetradecapeptide (1759.0 Da) in a matrix of α-cyano-4-hydroxycinnamic acid. Intact TIMP was analysed in the presence of horse skeletal apomyoglobin (16951.5 Da) in a matrix of sinapinic acid.

III. Results and Discussion

The measured molecular mass of the intact TIMP was 24.3 kDa, which compares with 20.7 kDa calculated from the amino acid sequence. Thus, the overall carbohydrate content was 15% by weight.

A sample of TIMP was then digested with trypsin and fractionated by HPLC. By screening the fractions using LDMS, the glycopeptides were immediately recognisable. For example, Figure 2 shows the spectrum of fraction 12, which proved to contain the T3 tryptic glycopeptide, (residues 23-37). There are many sharp peaks on the leading edge of the distribution, but the higher mass peaks, later shown to contain varying amounts of sialic acid, are weaker and broader. (At present, the detection of sialylated oligosaccharides, as free sugars or glycopeptides, is poor relative to neutral carbohydrates. We are searching for either better matrix materials or some simple derivatisation procedure to rectify this problem). The characteristic of this spectrum which immediately indicates that it is probably a glycopeptide is that all of the mass differences between the peaks correspond to the masses of monosaccharide residues, such as 162 Da for a galactose or a mannose.

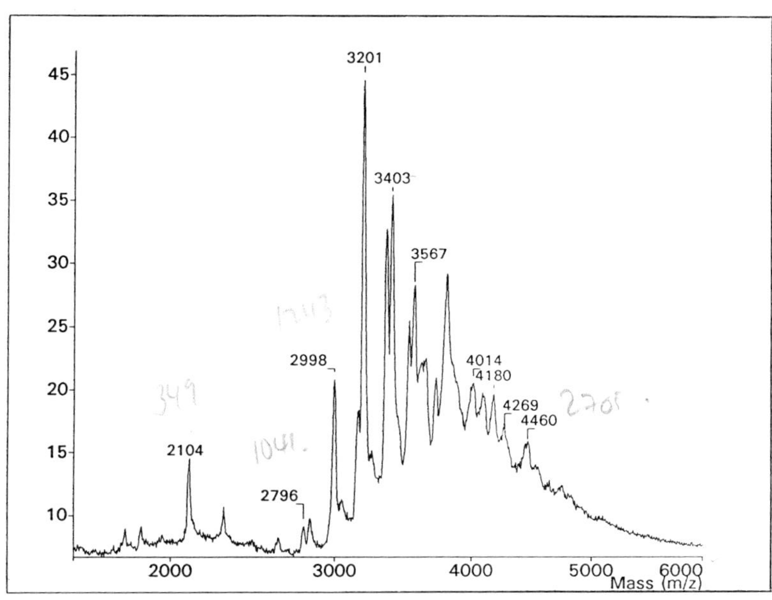

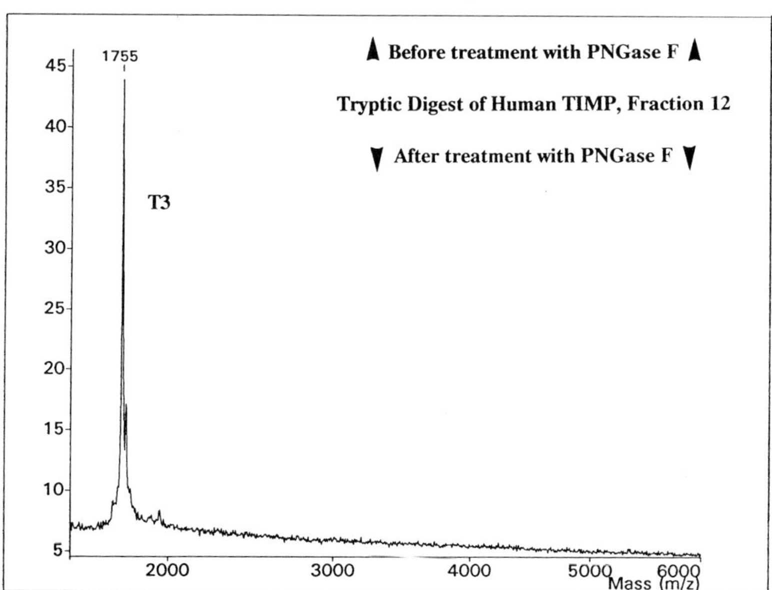

Figure 2. HPLC fraction 12 of the TIMP tryptic digest, containing the T3 glycopeptide, before and after treatment with PNGase F.

It has been our experience, screening digests by LDMS, that proteolysis rarely goes to completion. As expected, initial screening of the TIMP digest revealed more than just two glycopeptides, so glycopeptide containing fractions were treated with PNGase F and re-analysed. Each glycopeptide distribution was observed to collapse down to a single peak which both confirmed the initial identification and also provided the mass of the peptide component (Figure 2). In the case of TIMP, the predominant glycopeptide containing the glycosylation site at Asn78 proved to be a partial (T9,10; residues 76-88) indicating that cleavage after Arg79 may be inhibited by the close proximity of the oligosaccharide structures.

Having determined the mass of the core peptide, the oligosaccharide masses can be calculated for each observed glycopeptide peak. It is then possible to make a reasonable guess as to the composition of each species from the molecular weight alone. This may seem surprising at first sight, since there are mass degeneracies even at the monosaccharide level; mannose and galactose residues both weigh 162 Da. Nevertheless, if we take a representative sample of naturally occurring N-linked oligosaccharide structures, there are very few mass degeneracies which give rise to ambiguities of composition. For example, out of the 30 oligosaccharides listed in the current Oxford Glycosystems catalog there is only a single case of mass degeneracy (Table 1).

Cat. #	Mass	Cat. #	Mass	Cat. #	Mass
R-001001	390.3	C-004301	1486.3	M-002900	1906.6
R-002000	447.4	C-005300	1543.4	C-124300	1955.7
R-002001	593.5	M-002700	1582.4	C-024311	2013.8
R-002100	609.5	***H-003510***	***1664.5***	C-035300	2029.8
R-002101	755.7	***C-024300***	***1664.5***	C-035301	2175.9
R-002300	933.8	C-004311	1689.5	C-224300	2247.0
R-002301	1079.9	M-002800	1744.5	C-224301	2393.1
M-002500	1258.1	C-006300	1746.6	C-046300	2395.1
C-004300	1340.2	C-024301	1810.6	C-335300	2903.6
M-002600	1420.2	C-024310	1867.7	C-446300	3560.2

Table 1. Mass values for a sample of N-linked oligosaccharide structures. The OGS catalog number indicates the class by the prefix letter (R=root, M=high mannose, H=hybrid, C=complex). The six digit number, reading from left to right, indicates the number of residues of sialic acid, galactose, GlcNAc, mannose, bisecting GlcNAc, and core fucose. The mass values are for the free sugar, cationised with sodium.

In practice, even this ambiguity could be resolved by looking at the compositions of the surrounding peaks. If the peak with a carbohydrate mass of 1664 Da forms part of a series of hybrid sugars, then it is likely that it too has the hybrid composition 003510. If,on the other hand, there are accompanying peaks whose compositions can only be those of complex oligosaccharides, e.g. 004300 and 014300, then it becomes more likely that

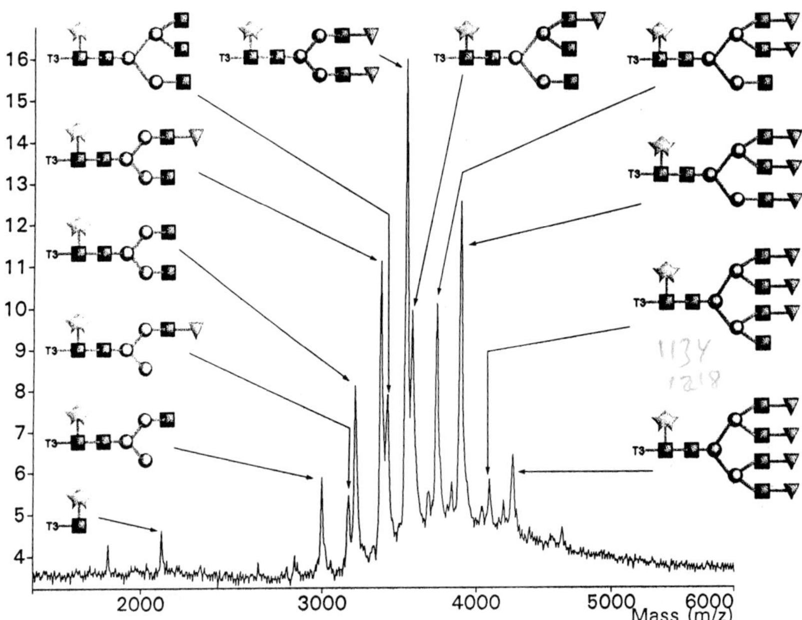

Figure 3. The T3 tryptic glycopeptide after treatment with neuraminidase.

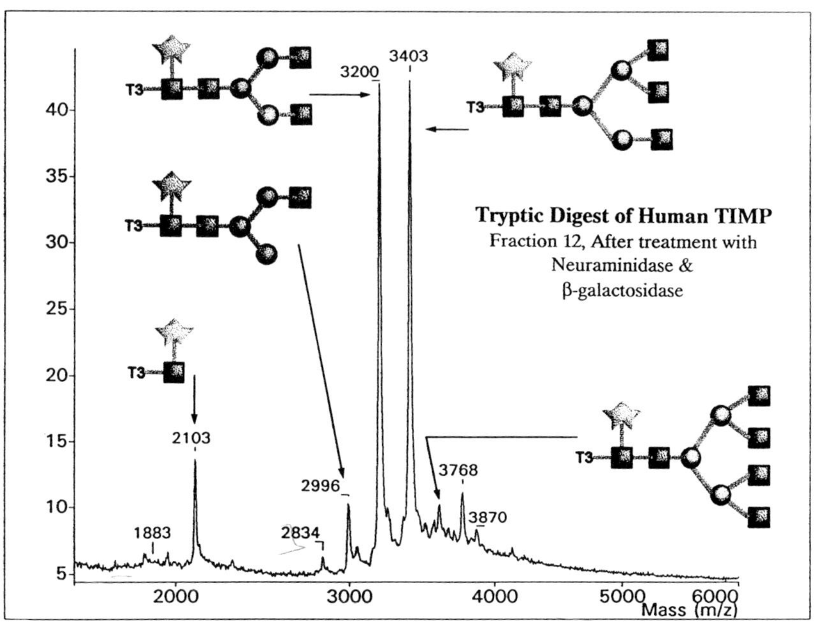

Figure 4. The T3 tryptic glycopeptide after treatment with neuraminidase and β-galactosidase.

the ambiguous peak is also a complex sugar of composition 024300. A similar incidence of ambiguity is shown in the compilation of Kobata et. al. where there are only 3 pairs of mass degenerate compositions out of 86 oligosaccharide structures [6].

The next stage of the process was to treat the glycopeptides with a series of specific exo-glycosidases. By following the changes in molecular weight resulting from glycosidase action, the composition assignment can be confirmed and the sequence and linkage determined to the extent permitted by the specificity of the glycosidases.

Figure 3 shows the result of treating the T3 glycopeptide with neuraminidase from *Arthrobacter ureafaciens*. All sialic acids were lost which simplifies and sharpens the distribution so that probable compositions can be assigned with greater confidence.

This was followed by treatment with β-galactosidase from *Streptococcus pneumoniae* which is specific to galactose linked β1->4 to GlcNAc or GalNAc. This was found to remove all of the putative galactoses, specifying their linkage and confirming that they were indeed galactoses and not mannoses (Figure 4).

The next stage was to divide the sample and treat with two N-acetyl hexosaminidases of differing specificity. Under the right conditions, that from *Streptococcus pneumoniae* cleaves only GlcNAc linked β1->2 to mannose whereas the enzyme from jack bean cleaves β1->2,3,4, & 6 linkages. By continuing this process with mannosidases and fucosidases, it was possible to deduce rigorously the structures which are illustrated in Figure 2. This work is still in progress, and a full description, including the analysis of the structures found on the T9,10 glycopeptide, is in preparation.

IV. Conclusions

The combination of sequential glycosidase digestion and LDMS provides a uniquely sensitive and rapid methodology for the site specific characterisation of oligosaccharides. All structures at a given site may be analysed in parallel. Sequence and linkage can be determined to the limits permitted by the specificities of available glycosidases.

References

1. M. W. Spellman, Anal. Chem. **62** 1714-1722 (1990)
2. R. A. Laine, Methods in Enzymology **193** 539-553 (1990)
3. T. Taniguchi, T. Mizuochi, M. Beale, R. A. Dwek, T. W. Rademacher, A. Kobata, Biochemistry **24** 5551-5557 (1985)
4. K. K. Mock, M. Davey, M. P. Stevenson, J. S. Cottrell, Biochem. Soc. Trans. **19** 948-953 (1991)
5. R. A. Williamson, F. A. O. Marston, S. Angal, P. Koklitis, M. Panico, H. R. Morris, A. F. Carne, B. J. Smith, T. J. R. Harris, R. B. Freedman, Biochem. J. **268** 267-274 (1990)
6. A. Kobata, K. Yamashita, S. Takasaki, Methods in Enzymology **138** 84-94 (1987)

Post-translational Modifications of Recombinant Proteins Determined by LC/Electrospray Mass Spectrometry and High Performance Tandem Mass Spectrometry

Katalin F. Medzihradszky
Christine A. Settineri
David A. Maltby
A. L. Burlingame
Department of Pharmaceutical Chemistry and the Mass Spectrometry Facility
University of California
San Francisco, California

I. Introduction

Mass spectrometry has played a significant role in structural identification of covalent modifications to proteins at the modified amino acid level for many years (1). Examples indicating the type and diversity among commonly occurring post-translational modifications are N-terminal acetylation (2), phosphorylation (3), glycosylation (4-6), determination of the site of attachment and glycan structure in phosphotidylinositol glycan membrane anchoring (7), palmitoylation (8), and lysine methylation (9). The emergence of liquid matrix sputtering ion sources (liquid SIMS and FAB) has enabled the use of a mass spectrometric based strategy for the identification of proteins through sequence analysis of the intact components of proteolytic digests. This strategy involves collision-induced dissociation (CID) of the molecular ion of each peptide using a tandem sector instrument (10). The tandem method provides a stage of separation for mixtures based upon mass selection of the ionized molecular weights of those components of interest. It also provides for the vibronic activation of each selected molecular ion, which triggers unimolecular dissociation reactions. The sequence of the peptide and structural information on

TECHNIQUES IN PROTEIN CHEMISTRY IV

modifying moieties may be deduced from the mass differences of these dissociation reactions. There are practical limitations to liquid matrix sputtering ionization in terms of molecular size with maintenance of good sensitivity. These limitations have been largely eliminated by the recently developed electrospray ionization technique (11). It offers a major extension in mass range, equivalent or better sensitivity, and ease of coupling to microbore liquid chromatographic inlet systems (12, 13). Used as an LC-quadrupole MS method, it is well suited for the molecular weight determination of virtually all components in a protein digest including glycopeptides, with a mass accuracy on the order of 1.5 Da. Having in hand these mass measurements for the entire digest facilitates the ready detection of covalently modified peptides if the cDNA sequence is known. The electrospray technique is also able to measure the molecular weights of intact proteins with relatively high accuracy (± 0.01%), and hence detect directly mass differences which represent post-translational or chemical modifications (14, 15).

In this report we wish to illustrate the utility of microbore liquid chromatography coupled to electrospray quadrupole mass spectrometry for the rapid location of putative post-translationally modified peptides in proteolytic digests of recombinant proteins. In addition, we will demonstrate the importance of high energy CID analysis for the unambiguous identification of the modified peptide and site of covalent modification.

II. Experimental

A. *Materials*

Human platelet-derived growth factor B-chain (hPDGF-B) and human low-affinity nerve growth factor receptor (hNGF-R) were expressed and purified according to the procedures described elsewhere (4, 16). The purified proteins were reduced, carboxymethylated and digested with either trypsin or trypsin/chymotrypsin according to a protocol reported previously (4, 16).

B. *Mass Spectrometry*

LC/ESMS experiments were performed using a Carlo Erba Phoenix 20 dual syringe pump system (Fisons) interfaced to a VG Biotech BIO-Q electrospray mass spectrometer. Peptide separation was carried out using an Aquapore 300 column (C18, 1x100 mm, Applied Biosystems Inc). Solvent A was 0.1% TFA in water, solvent B 0.08% TFA and 70% acetonitrile in water. Aliquots of protein digests were injected at 0% B followed by a linear increase of solvent B to 80% at 0.5 or 1%/min at a flow rate of 4 µl/min. The eluant was monitored at 215 nm using a UV-detector (Linear, model 204) equipped with a U-Z View capillary flow cell (LC-Packings) in-line between the end of the column and the entrance into the ion source. In order to improve the spraying properties of the eluant a post column addition of 1:1 acetonitrile/water containing 0.1% TFA was accomplished at 2 μl/min. The mass spectrometer was scanned over a range of m/z 300-2000 at 7 sec/scan.

High energy CID experiments were carried out on a Kratos Concept IIHH high performance tandem mass spectrometer equipped with electro-optical multichannel array detection as described in detail elsewhere (4).

III. Results and Discussion

Many recombinant proteins have been analyzed in our laboratory for the purposes of the confirmation of cDNA deduced sequences and the investigation of post-translational modifications. Prior to the advent of liquid chromatography-electrospray mass spectrometry, it was common practice to analyze fractions collected manually from an HPLC separation of a protein digest batchwise by LSIMS. Using this methodology one could never be sure that all of the components in such a digest would be detected, especially very hydrophilic peptides and glycopeptides (17). However, the recent emergence of several configurations with direct coupling of microbore HPLC

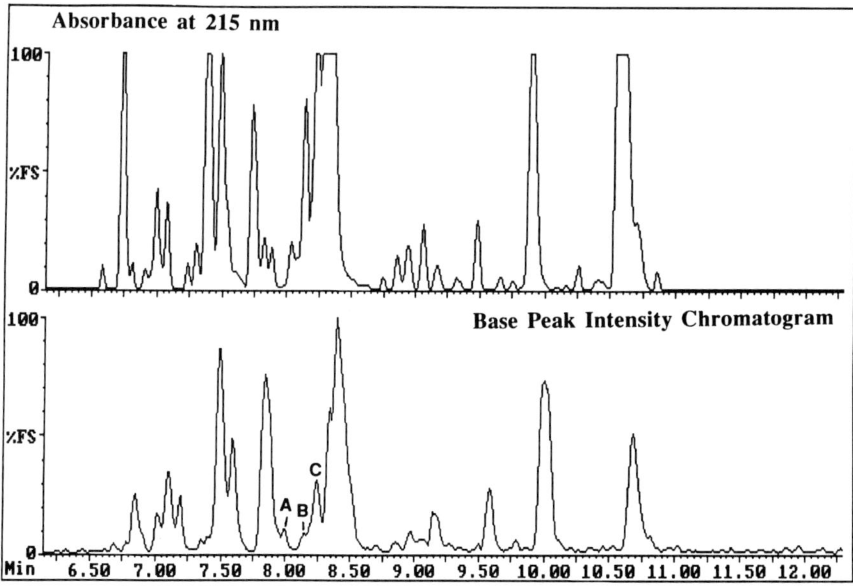

Figure 1. HPLC chromatogram of 200 pmoles of the tryptic digest of recombinant human PDGF-B. Peptides were detected a) by the UV absorbance at 215 nm and b) by scanning the mass spectrometer from m/z 300 to m/z 2000. This trace represents the base peak intensities measured. The earlier eluting di-mannosyl tryptic peptide [86-98] (peak A) was base line separated from the corresponding mono-mannosyl and non-glycosylated peptides (Peak B and C, respectively).

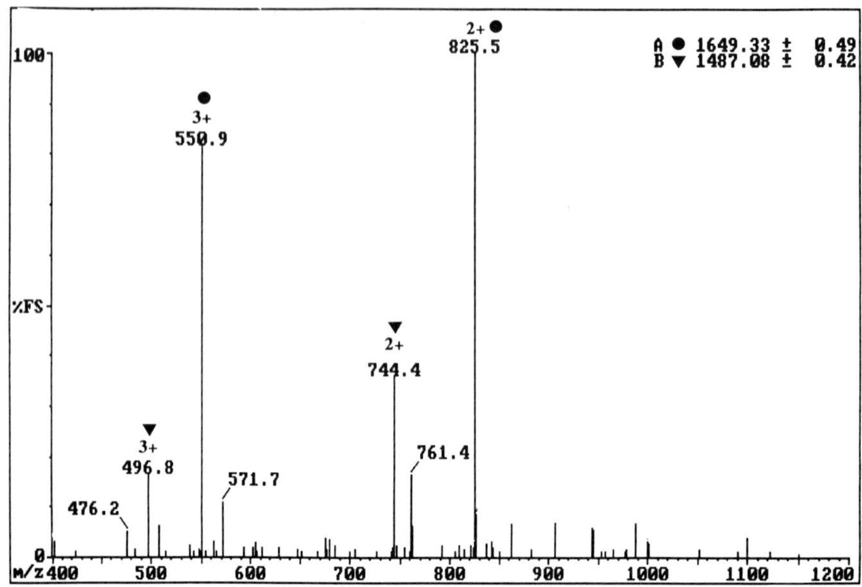

Figure 2. Electrospray mass spectrum of peak B from the tryptic digest of PDGF-B in Figure 1. The major component is the tryptic glycopeptide [86-98] bearing one mannose residue with a calculated molecular mass of 1648.8. The glycosylated and non-glycosylated peptides were not completely separated under these chromatographic conditions, thus ions corresponding to the latter were also detected in this fraction. The calculated molecular mass of the non-glycosylated peptide is 1486.8. (Chemical average value is used for all the masses determined by ESMS.)

systems with electrospray mass spectrometers has not only obviated most of these difficulties, but has also significantly reduced the sample amount required and the analysis time.

Since recombinant human PDGF expressed in yeast has been studied previously in our laboratory in some detail (4), we used this protein to evaluate the relative merits of the new LC/ESMS methodology. Thus, a 200 pmoles aliquot of a tryptic digest was analyzed and the U.V. (panel a) and base peak intensity (panel b) chromatograms are shown in Figure 1. Earlier studies revealed different levels of *O*-glycosylation in this recombinant protein. Mannosyl-glycopeptides can be identified based on the 162 Da or 324 Da mass differences (corresponding to one or two mannose residues) between the expected and observed molecular masses in the protein digest. A di-mannosyl tryptic glycopeptide [86-98] was observed in the fraction labeled A. It was base line separated from a corresponding mono-mannosyl glycopeptide, which eluted in the fraction labeled B (Figure 2). Ions corresponding to the analogous non-glycosylated tryptic peptide were also detected in this mass spectrum, although the non-glycosylated peptide eluted mainly in fraction C. The glycopeptides were isolated by analytical HPLC and subjected to high energy CID analysis. Figure 3 shows the CID spectrum of the mono-mannosyl glycopeptide. Since the diagnostic C-terminal fragment ions

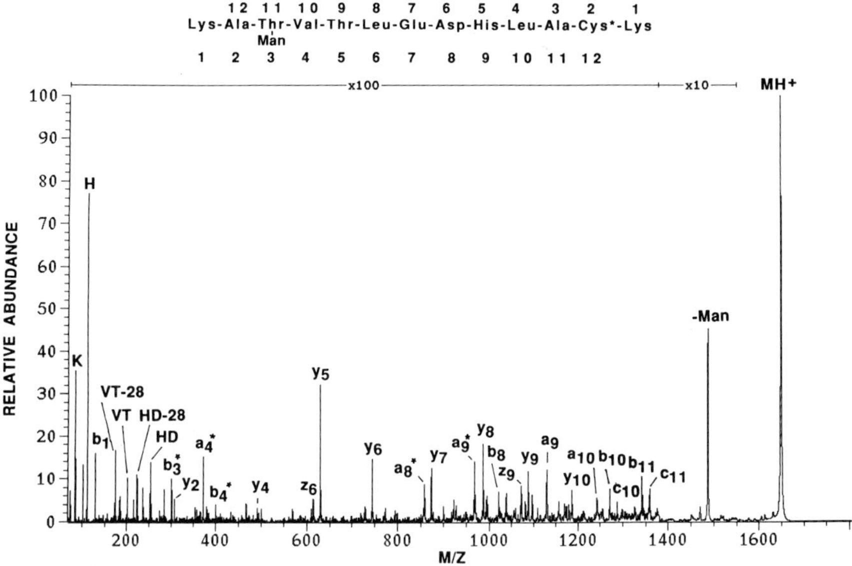

Figure 3. High energy CID spectrum of a tryptic glycopeptide [86-98] from recombinant human PDGF-B (MH^+=1648.8). Cys* stands for carboxymethyl cysteine. C-terminal fragment ions $\mathbf{z_9}$, $\mathbf{y_9}$, $\mathbf{y_{10}}$ do not show the expected 162 Da shift suggesting that Thr^{88} is glycosylated with one mannose unit. Ions labeled with asterisks are gas-phase de-glycosylated peptide fragments. The nomenclature used for peptide fragmentation was defined by K. Biemann (21).

$\mathbf{z_9}$, $\mathbf{y_9}$, $\mathbf{y_{10}}$ do not show the 162 Da shift which would have been expected if Thr^{90} had been mannosylated, Thr^{88} must be glycosylated with one mannose unit. This observation is in agreement with earlier data (5). LC/ESMS analysis could provide a quick, reliable means for the analysis of new sample lots expressed under different conditions.

We have investigated a construct of recombinant human NGF-receptor which contains the N-terminal 222 amino acid extracellular domain terminating just before the transmembrane region of the native full length receptor (16). We wished to determine the nature of the *N*-linked glycosylation of this extracellular domain expressed in CHO-cells. This protein contains one consensus site for *N*-linked glycosylation at Asn^{32}. A search of the LC/ESMS run of 200 pmoles of a tryptic/chymotryptic digest revealed a spectrum which contained three coeluting glycopeptides as shown in Figure 4. Deconvolution of the ions observed in this spectrum yielded the three molecular masses 6373.0, 6081.7, and 5790.6 Da, each differing by 291 Da, indicating structural heterogeneity in the oligosaccharide structure due to attachment of different numbers of sialic acid residues.

These masses correspond to di-, mono-, and asialo fucosylated complex biantennary sugar structures attached to a peptide of mass 4022.3. This value matches the

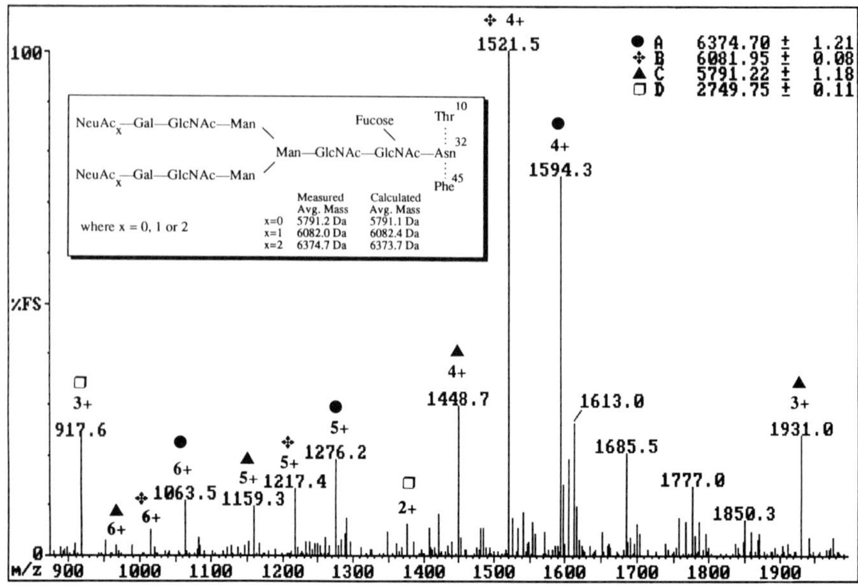

Figure 4. LC/ESMS spectrum of three coeluting glycopeptides from 200 pmoles of a recombinant human NGF-receptor tryptic/chymotryptic digest. Calculated molecular masses are 6373.5, 6082.3 and 5791.0, for the di-, mono-, and asialo glycopeptides, respectively. From the relative peak heights in the spectrum, the glycopeptide containing a monosialylated sugar structure appears to be the major species, followed by the disialylated and then the asialo species. Component D is NGF-R residues [127-149].

anticipated mass of residues 10-45 in the recombinant human NGF-receptor and has the following sequence: THSGECCKACNLGEGVAQPCGA**N**QTVCEPCLDSVTF, where the cysteine residues have been carboxymethylated. From the relative ion abundances observed in the electrospray spectrum, the glycopeptide containing a monosialylated carbohydrate structure appears to be the major species, followed by the disialylated species and the asialylated species. These results are in agreement with our previous observation on the PNGase-cleaved ABOE (*p*-amino benzoic acid octyl ester) derivatized oligosaccharide mixture (16).

The identification of glycopeptides in a protein digest is not trivial even when the amino acid sequence of the protein is known and some preliminary information is available on the carbohydrate structures. One reason for this is the presence of various sugar structures at a given glycosylation site. Glycopeptides of identical peptide sequence but bearing different carbohydrate structures divide the detectable signal into several different molecular masses. In other mass spectrometric methods linked with chromatography, screening for certain characteristic ions at given mass values has increased the sensitivity dramatically. This approach can be applied in searching an enzyme digest for glycopeptides. Glycopeptides containing N-acetylhexosamine residues feature an abundant ion at m/z 204 in their CID spectra. If fragmentation is induced as the eluant enters the mass spectrometer, fragment ions can be monitored and glycopeptide-containing fractions can be readily identified at the low picomole level (18, 19). With the appropriate computer program, selected ion monitoring and molecular weight measurement can be carried out in a single run, as recently presented by S. A. Carr (20).

IV. Conclusions

LC/ESMS can be used to readily identify peptides in a protein digest which show anomalies to the predicted primary structure at the low picomole level and at high speed. However, further verification is usually necessary to confirm the suspected structures. Isolation of the modified peptides followed by high energy CID analysis provides one approach for resolution of this problem.

Glycopeptides continue to present major challenges for structure elucidation, due to their size, complexity and heterogeneity. LC/ESMS has made it possible to identify glycopeptides in a digest at the low picomole level as well as to address glycosylation structural heterogeneity, since the peptide and carbohydrate structures can be detected and analyzed together. In addition, LC/ESMS coupled with selected ion monitoring of diagnostic carbohydrate ions is the most sensitive method to date for identification of glycopeptide-containing fractions in a glycoprotein digest.

Acknowledgement

The authors thank F. R. Masiarz, B. Chapman and L. S. Cousens from Chiron Corporation for providing the protein samples. We gratefully acknowledge F. C. Walls for his technical assistance. This work was supported by NIH National Center for Research Resources Grant RR01614 , NSF Biological Instrumentation Program Grant DIR-8700766, and the NIEHS Superfund Grant ES04705.

References

1. Carr, S. A., and Biemann, K. (1984). Meth. Enzymol. **106**, 29-58.

2. Hopper, S., Johnson, R. S., Vath, J. E., and Biemann, K. (1989). J. Biol. Chem. **264**, 20438-20447.

3. Sanders, D. A., Gillece-Castro, B. L., Burlingame, A. L., and Koshland, D. E., Jr. (1992). J. Bacteriol. **174**, 5117-5122.

4. Settineri, C. A., Medzihradszky, K. F., Chu, C., George-Nascimento, C., Masiarz, F. R., and Burlingame, A. L. (1990). Biomed. Environ. Mass Spectrom. **19**, 665-676.

5. Medzihradszky, K. F., Gillece-Castro, B. L., Settineri, C. A., Townsend, R. R., Masiarz, F. R., and Burlingame, A. L. (1990). Biomed. Environ. Mass Spectrom. **19**, 777-781.

6. Carr, S. A., Hemling, M. E., Folena-Wassermann, G., Sweet, R. W., Anumula, K., Barr, J. R., Huddleston, M. J., and Taylor, P. (1989). J. Biol. Chem. **264**, 21286-21295.

7. Stahl, N., Baldwin, M. A., Hecker, R., Pan, K.-M., Burlingame, A. L., and Prusiner, S. B. (1992). Biochemistry **31,** 5043-5053.

8. Stults, J. T., Griffin, P. R., Lesikar, D. D., Naidu, A., Moffat, B., and Benson, B. J. (1991). Am. J. Physiol. **261**, L118-L125.

9. Dever, T. E., Costello, C. E., Owens, C. L., Rosenberry, T. L., and Merrick, W. C. (1989). J. Biol. Chem. **264**, 20518-20515.

10. Bieman, K., and Scoble, H. A. (1987). Science **237**, 992-998.

11. Fenn, J. B., Mann, M., Meng, C. K., Wong, S. F., and Whitehouse, C. M. (1989). Science **246**, 64-71.

12. Hemling, M. E., Roberts, G. D., Johnson, W., Carr, S. A., and Covey, T.R. (1990). Biomed. Environ. Mass. Spectrom. **19**, 677-691.

13. Ling, V., Guzetta, A. W., Canova-Davis, E., Stults, J. T., Hancock, W. S., Covey, T. R., and Sushan, B. I. (1991). Anal. Chem. **63**, 2909-15.

14. Poulter, L., Green, B. N., Kaur, S., and Burlingame, A. L. (1990). *In* "Biological Mass Spectrometry" (A. L. Burlingame and J. A. McCloskey, eds.), pp 119-128. Elsevier, Amsterdam.

15. Smith, J. B., Thevenon-Emeric, G., Smith, D. L., and Green, B. (1991). Anal. Chem. **193**, 118-124.

16. Settineri, C. A., Leung, I., Cousens, L. S., Chapman, B., Masiarz, F. R., and Burlingame, A. L. (1992). *In* "Techniques in Protein Chemistry III" (R. H. Angeletti, ed.), pp 295-301. Academic Press, Inc., San Diego.

17. Naylor, S., Findeis, A. F., Gibson, B. W., and Williams, D. H. (1986). J. Am. Chem. Soc. **108**, 6359.

18. Huddleston, M. J., Bean, M. F., Barr, J. R., and Carr, S. A. (1991). *In* "The Proceedings of the 39th ASMS Conference on Mass Spectrometry and Allied Topics, Nashville, TN.", pp 280-281.

19. Unpublished results from this laboratory

20. Carr, S. A., Huddleston, M. J., and Bean, M. F. (1992). *In* "The Proceedings of the 40th ASMS Conference on Mass Spectrometry and Allied Topics, Washington DC", pp 973-974.

21. Biemann, K. (1990). Methods Enzymol. **193**, 886-887.

IDENTIFICATION AND CHARACTERIZATION OF GLYCOSYLATION SITES IN CARCINOEMBRYONIC ANTIGEN (CEA) BY MASS SPECTROMETRY

Kristine M. Swiderek, Constance S. Pearson, John E. Shively

Division of Immunology, Beckman Research Institute of the City of Hope, Duarte, CA 91010

1. Introduction

Carcinoembryonic antigen (CEA) is one of the most thoroughly characterized proteins in terms of biochemical and structural properties, tissue distribution and usefulness for clinical diagnostics. However little is known about the function of this protein. It is thought today that posttranslational modifications like the glycosylation of proteins, are important to the final functionality of a protein. CEA contains 28 asparagine-linked glycosylation sites (1) and is about 50 % carbohydrate by weight. The characterization of these carbohydrate structures may give a better understanding of the role of CEA in the cell. Here we demonstrate as an example the isolation and identification of two glycopeptides from CEA and the characterization of their carbohydrate structure by mass spectrometry.

2. Materials and Methods

CEA was isolated from liver metastases of colon tumors and purified as described (2). Before enzymatic digestion 10 nmoles CEA were reduced and alkylated with 4-vinylpyridine according to (3) and after dialysis treated with neuraminidase

TECHNIQUES IN PROTEIN CHEMISTRY IV

overnight at room temperature. CEA was digested with purified trypsin and chymotrypsin (Worthington) in 50 mM Tris-HCl, pH 8.5 at a protein/enzyme ratio of 40:1 (wt/wt) for 36 hr at 37°C. The tryptic/chymotryptic peptides were purified by using different HPLC techniques. The first step was separation on reverse-phase HPLC with a Vydac C4 (4.6 x 250 mm) column using a linear gradient of acetonitrile in 0.1% (vol/vol) trifluoroacetic acid. The fractions were rechromatographed on a Vydac C18 (4.6 x 250 mm) under similar conditions and then applied to hydrophilic HPLC using a Polyhydroxyethyl Aspartamide column (4.6 x 250) (PolyLC, Columbia, MD). Samples, dissolved in 70 % acetonitrile, were eluted with a linear gradient from 80 % acetonitrile to H_2O in 10 mM triethylamine phosphate pH 3.0. Fractions were rechromatographed again on reverse-phase HPLC with a Vydac C18 (2.1 x 250) using a fast, linear gradient of acetonitrile in 0.1% (vol/vol) trifluoroacetic acid to remove salt.

For sequence analysis peptides were spotted on a poly-vinylidene difluoride (PVDF) membrane (Millipore) and subjected to automated Edman degradation on a gas-phase microsequencer built at the City of Hope (4). The phenylthiohydantoin amino acid derivatives were identified by on-line reverse-phase HPLC.

Mass spectra were recorded in the positive ion mode using a TSQ-700 triple quadrupole instrument (Finnigan-MAT, San Jose, CA) equipped with a cesium ion gun (Phrasor Scientific, Inc, Duarte, CA) for SIMS. 1µl of sample was applied on the sample stage using thioglycerol (3-mercapto-1,2-propanediol) as matrix. Cesium ions were bombarded onto the sample stage with an energy of 10 keV.

3. Results and Discussion

3.1. Purification and Identification of Glycopeptides

To prepare CEA for characterization the protein was purified as described, treated with neuraminidase and digested with trypsin and chymotrypsin to insure complete digestion. Removal of sialic acid and the enzymatic digestion of CEA in the presence of two proteases with different specificity resulted in a more sufficient degradation. The tryptic/chymotryptic map for CEA is

A
1.0
0.5
A214
0
15
30
45
60
75
Time (min.)
Site 26
Site 1
40
20
0
% Solvent B
B
0.16
0.08
A214
Site 26
100
50
0
% Solvent B
20
30
40
Time (min.)
C
0.64
0.32
A214
100
50
0
% Solvent B
Site 1
0
10
20
30
Time (min.)
D
0.64
0.32
A214
100
50
0
% Solvent B
Site 26
10
20
30
Time (min.)
E
0.32
0.16
A214
CH
LSGALNLS
(Site 26)
100
50
0
% Solvent B
0
10
20
Time (min.)
F
0.16
0.08
A214
CH
SGREIIPNASL
LIQ (Site 1)
100
50
0
% Solvent B
0
10
20
Time (min.)

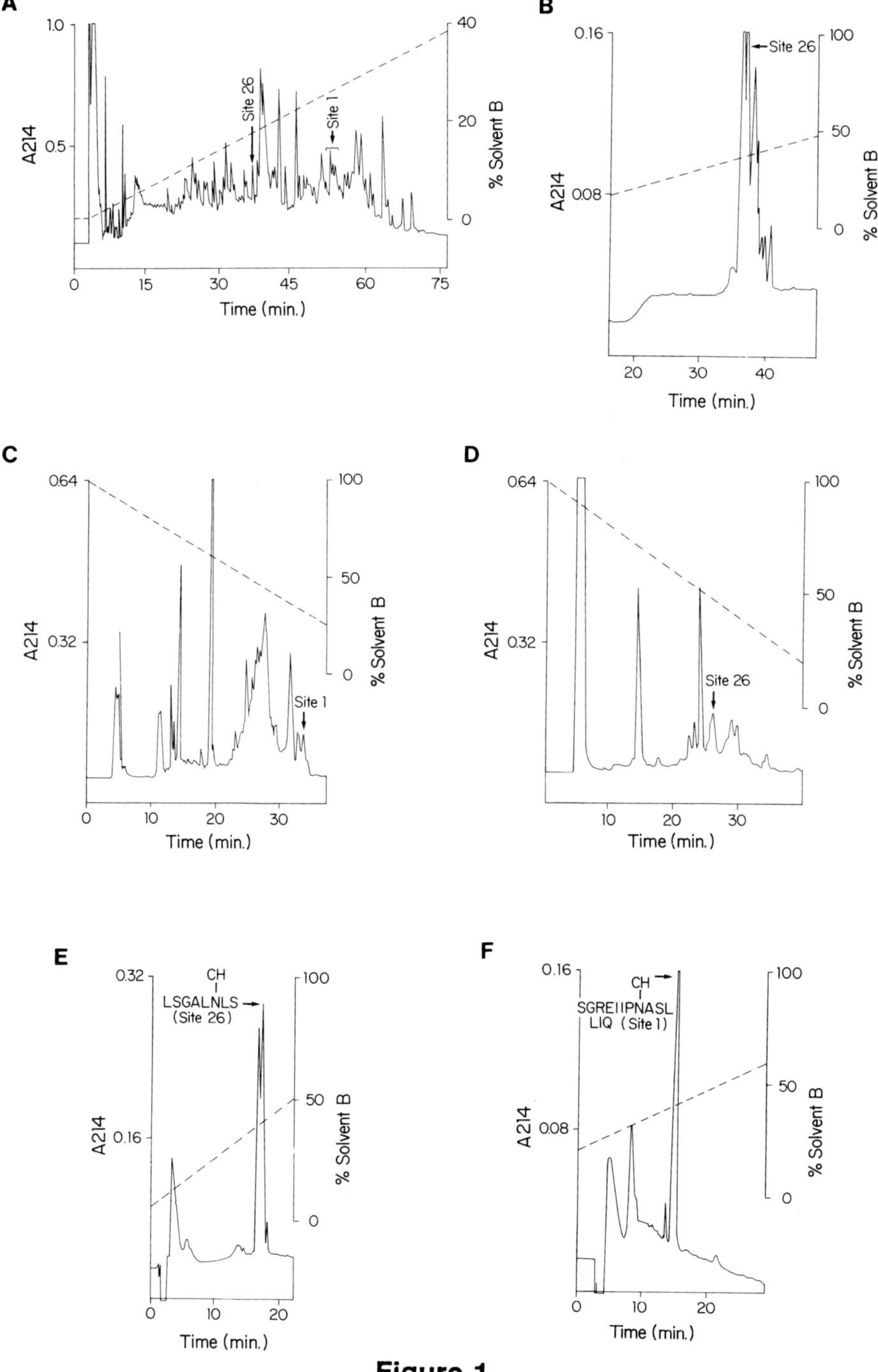

Figure 1

Figure 1: Purification of glycopeptides from CEA
As an example the purification of site 1 and 26 by different HPLC steps is shown.

Figure 1a: Tryptic/chymotryptic map
The tryptic/chymotryptic digest of CEA was separated by reverse-phase HPLC on a C4 column. Peptides were eluted with a linear gradient of 100 % buffer A (0.1 % trifluoroacetic acid (vol/vol)) to 40 % buffer A/60 % buffer B (vol/vol) in 120 min. Buffer B is 0.1 % trifluoroacetic acid/9.9 % H_2O/90 % acetonitrile (vol/vol).

Figure 1b: Rechromatography of the fraction containing glycosylation site 26
The rechromatography was carried out by reverse-phase HPLC using a C18 column with a linear gradient from 100 % buffer A to 40 % buffer A/60 % buffer B in 60 min. Buffer A and B are the same as described in Fig. 1a.

Figure 1c/d: Rechromatography by hydrophilic HPLC
Fractions were applied to hydrophilic HPLC after reverse-phase HPLC using a Polyhydroxyethyl Aspartamide column. Samples were injected in about 70 % acetonitrile and peptides were eluted with a decreasing gradient from 100 % buffer B (80 % acetonitrile/20 % 50 mM triethylamine phosphate pH 3.0 in H_2O (vol/vol)) to 100 % buffer A (10 mM triethylamine phosphate pH 3.0) in 50 min. The UV absorbance profile at 214 nm is shown for two fractions containing site 26 (**Fig. 1c**) and site 1 (**Fig. 1d**).

Figure 1e/f: Desalting of fractions after hydrophilic HPLC
The rechromatography was carried out by reverse-phase HPLC using a C18 column with a linear gradient from 95 % buffer A/ 5 % buffer B to 45 % buffer A/55 % buffer B in 25 min (**Fig. 1e**) and a linear gradient from 80 % buffer A/20 % buffer B to 40 % buffer A/60 % buffer B in 30 min (**Fig. 1f**). Buffer A and B are the same as described in Fig. 1a.

shown in Fig.1. Since the amino acid sequence of CEA and the position of the glycosylation sites is known all generated fractions were submitted to sequence analysis to screen for glycopeptides. Glycopeptide containing fractions were identified by reading the consensus sequence of glycopeptides even in a mixture of peptides. Our experience is that mass spectrometric analysis of poorly separated glycopeptides is mostly impossible. Therefore all glycopeptide containing fractions then had to be rechromatographed several times by reverse phase and hydrophilic HPLC to isolate the glycopeptides to purity. Fig. 1 shows an example of the purification of two glycopeptides. The collected fractions labeled site 26 and site 1 (Fig. 1a) were rechromatographed by reverse-phase HPLC (shown for site 26 in Fig. 1b) and hydrophilic HPLC (Fig. 1c/d). As demonstrated in Fig. 1b the resolution obtained on a C18 column in comparison to the C4 column was increased; however an acceptable separation of glycopeptides from all other impurities after chromatography on C18 was achieved only by application of the fraction to hydrophilic HPLC. Fig. 1c and 1d show the final purification of two glycopeptides carrying site 26 and site 1. Both examples demonstrate clearly that fractions after reverse-phase HPLC contained still a considerable amount of other peptides. That demonstrates the importance of the separation of the glycopeptides from CEA by hydrophilic HPLC. Typically glycopeptides we isolated from CEA by hydrophilic HPLC did not give sharp peaks in the UV absorbance but broad, later eluting peaks due to heterogeneity in their carbohydrate content. Contaminating peptides carrying no carbohydrate structures usually eluted early in the gradient and gave sharp peaks in the UV absorbance profile. As a last step before sequence and mass spectrometric analyses, samples were desalted after hydrophilic HPLC by reverse-phase HPLC using C18 columns and a fast, shallow gradient (Fig. 1e/f). This step was necessary especially for the preparation for mass spectrometric analysis to remove triethylamine phosphate from the samples. Through sequence analysis of the fractions we could identify the following sequences and the position of glycosylation:

Site 26: L S G A L <u>N</u> L S

Site 1: S G R E I I P <u>N</u> A S L L I Q

The typical consensus sequence for N-linked glycosylation is N-X-S/T, site 26 and site 1 peptides both contain this sequence. Additionally, sequence analysis of these peptides revealed a "blank cycle" in the position where we would expect an asparagine residue. We have been able to isolate 16 glycopeptides to purity out of 28 known glycosylation sites using the strategy described above (results not shown).

% Relative Intensity
m/z
1675.2
1877
2025
2228
2552
2714
2902
2917
3064
3079
3226
3444
3576
3591
3738
Peptide
GlcNAc
Fucose
Mannose
Galactose
(-Fuc=1877)
(-Fuc=2917)
3226 (-Fuc=3079)
3738 (-1Fuc=3591 -2Fuc=3444)
889
1092
1295
1442
2131
2278
2293
2496
2643
2805
3008
3155
3170
3317
3519
3666
3813
(-Fuc=1295)
3316
(Fucose)
GlcNAc - Galactose
(-Fuc=3008)
(-Fuc=2496)
(-Fuc=3170)
Gal
GlcNac
3666 (+Fuc=3813 -Fuc=3519)
2131 (+Fuc=2278)

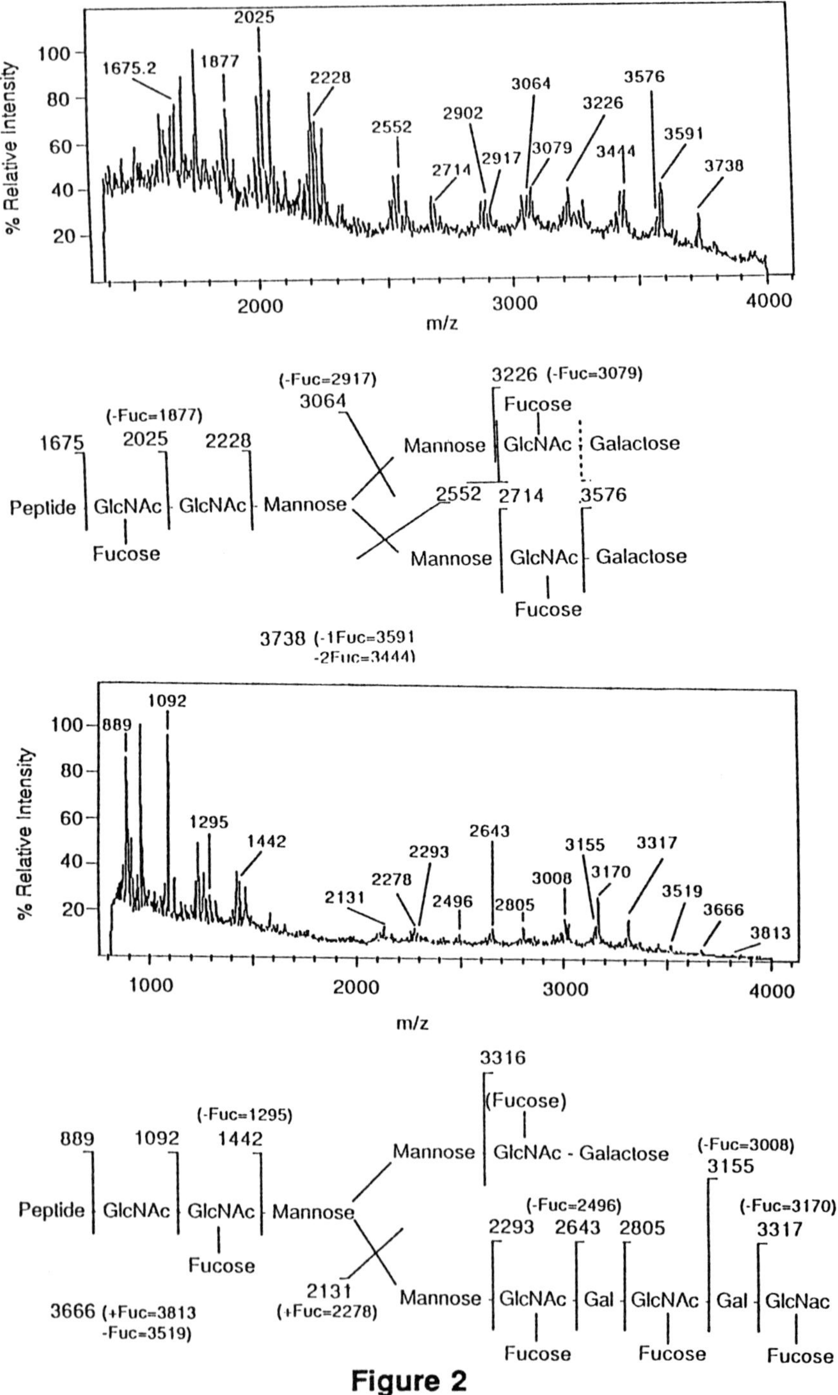

Figure 2

Figure 2: Mass spectrometric analysis of glycopeptides Samples were analyzed by secondary ionization mass spectrometry (SIMS) as described in methods. Scans were taken in the mass range from 400 to 4000. The indicated masses represent the major fragments. The different types of ring cleavages typically occurring in carbohydrates during mass spectrometric analysis are not spefically labeled.

(—) bound where fragmentation occurs

(- - -) alternative break resulting in the same mass

Figure 2a: SIMS spectrum of the glycopeptide carrying site 1 and the proposed structure deduced from the mass spectrum.

Figure 2b: SIMS spectrum of the glycopeptide carrying site 26 and the proposed structure deduced from the mass spectrum.

3.2. Mass Spectrometric Analysis of Glycopeptides

The combination of sequence and mass spectrometric analysis made it possible to suggest a carbohydrate structure for glycopeptides. The mass of the amino acid sequence of glycopeptides was calculated after sequence analysis, the mass of the glycopeptide was determined by mass analysis. Also the analysis of glycopeptides by mass spectrometry shows a characteristic pattern due to the fragmentation of the carbohydrate structure during the ionization process. This fragmentation pattern can be used for the interpretation of the structure. Sequence analysis of the glycopeptide carrying site 1 revealed a peptide sequence with a calculated mass of 1675 (MH^+). Mass spectrometric analysis of this glycopeptide gave a parent mass of at least 3738 (MH^+). Therefore the mass difference of 2061.1 would correspond to the attached carbohydrate structure. The proposed structure is illustrated in Fig. 2a. Mass spectrometric analysis of the glycopeptide with site 26 as well as its proposed structure is shown in Fig. 2b.

4. Conclusion

We demonstrated a purification protocol for the isolation of glycopeptides out of a complex digestion mixture for a protein with a mass of about 100,000. The purity of glycopeptides was proved by automated sequence analysis as well as by secondary ion mass spectrometry. As an example we presented the carbohydrate structures of two glycopeptides from CEA. Our goal is to isolate and identify all glycosylation sites from this protein to gain some more understanding about its structure and its possible function.

References

1. Paxton, R. J., Mooser, G., Pande, H., Lee, T. D. & Shively, J. E. (1987) Proc. Natl. Acad. Sci. 84, 920-924
2. Pritchard, D. G. & Todd, C. W. (1976) Cancer Res. 36, 4699-4701
3. Hawke, D. H. (1987) Protein Society Meeting
4. Hawke, D. H., Harris, D. C. & Shively, J. E. (1985) Anal. Biochem. 147, 315-330

This work was supported by NIH grant CA 37808

Monosaccharide and Oligosaccharide Analysis of Recombinant Erythropoietin Electro-transferred onto Polyvinylidene Fluoride Membranes

Michael Weitzhandler, Douglas Kadlecek, Nebojsa Avdalovic, and R. Reid Townsend*

Dionex Corporation, Sunnyvale, CA 94088
and
*Dept. of Pharmaceutical Chemistry, University of California, San Francisco, CA 94143-0446

I. Introduction

Erythropoietin (EPO) is a heavily glycosylated hormone which circulates in picogram amounts and stimulates the differentiation of erythroid precursors in bone marrow (1). Recombinant EPO (rEPO) has an apparent M_r $\simeq$ 38,000, $\simeq$40% carbohydrate by weight (2), and has been shown to have a dramatic therapeutic benefit for the anemia of chronic renal disease. Its use in other disorders is under investigation (1). N-glycosylation is crucial to the efficacy of *in vivo* administrated rEPO (3-6). With the established importance of glycosylation in rEPO function, efficient methods are needed to analyze its carbohydrates during expression, purification from tissue culture media, and storage. Substances which interfere with carbohydrate analyses (e.g., salts, glucose, other glycoproteins) are usually present at each of these stages. We report herein a sensitive, facile strategy for identifying the monosaccharide composition and oligosaccharides of rEPO after purification by SDS-PAGE and electro-transfer onto PVDF membranes.

II. Materials and Methods

Reagents

Glass distilled deionized water was used for the preparation of all buffers, eluents, and electrophoresis solutions. The storage flask and spigot were glass throughout to prevent contact with tubings which might support microbial growth and yield artifactual sugars during monosaccharide analysis. Sodium hydroxide (50% solution) and acetic acid were from Fisher, methanol from VWR, powderless gloves from Tagg Industries, and PVDF sheets (Immobilon PSQ) from Millipore. Microfuge tubes were

TECHNIQUES IN PROTEIN CHEMISTRY IV

from Sarstedt, autosampler vials from Sun Brokers, TFA from Pierce, and reduced Triton X-100 from Calbiochem. All reagents for electrophoresis were from BDH, molecular weight standards from Pharmacia, and monosaccharide and oligosaccharide standards from Dionex Corporation. rEPO, expressed in CHO cells, and the amidase, PNGase F, were supplied by Boehringer Mannheim. Endo-β-galactosidase from *Escherichia freundii* was kindly provided by Drs. Minoru and Michiko Fukuda, LaJolla Cancer Research Foundation.

Electrophoresis and transfer to PVDF membranes

SDS-PAGE (12.5%) was performed (7) using mini-gels run in a Tall Mighty Small unit (Hoefer Scientific) at constant amperage (20 mA) using an LKB Multidrive 3.5-kV power supply. The glycoproteins were transferred to a single sheet of PVDF membrane using a constant current (400 mA for 5 h) in a Hoefer TE 42 transfer unit cooled to 15°C by an LKB Multitemp II Recirculator. The transfer buffer consisted of 20% methanol in tris(25 mM)-glycine(190 mM) buffer, pH 8.8. The proteins were stained on the PVDF membrane using 0.1% Coomassie brilliant blue R-250 in water/ methanol/acetic acid (5/4/1) for 10 min, de-stained with 90% methanol, and finally washed in distilled water. The PVDF membrane was allowed to dry at room temperature and was stored at 4°C until further analysis.

Monosaccharide composition analysis

For analysis of neutral sugars, stained bands were excised from PVDF membranes, wetted with methanol, placed into 1.5-mL microfuge tubes, and submerged in 400 μL of 2 M TFA. To analyze amino sugar, stained, excised bands were submerged in 400 μL of 6 M HCl. Both samples were briefly vortexed and centrifuged. All tubes were capped and subjected to hydrolysis at 100°C. At the indicated times, the hydrolysates were centrifuged for 2 min in a bench-top microcentrifuge (Fisher) to unite the condensate with the bulk fluid. The hydrolysates were then dried in a Speed Vac centrifuge (Savant). The dried samples were reconstituted in 200 μL of water and transferred to autosampler vials. Sialic acids were released in a separate, milder hydrolysis (8). Stained, excised bands were pre-wetted in methanol and hydrolyzed in 400 μl of 0.1 M HCl (80°C) for 30 min.

Endoglycosidase and amidase digestion

To release and further analyze oligosaccharides, a rEPO band was treated sequentially with PNGase F and endo-β-galactosidase. Stained bands on PVDF, after excising and wetting in methanol, were immersed in 150 μL of sodium phosphate buffer (5 mM, pH 7.6) containing 0.1% reduced Triton X100 in a 1.5- mL microfuge tube. PNGase F (1 mU) was added, followed by incubation at 37°C for 16 h. An aliquot (75 μL) was removed for oligosaccharide analysis by HPAEC/PAD. To detect poly-lactosamine oligosaccharides, endo-β-galactosidase (1 mU) was added to PNGase F digests after the addition of 8 μL of 200 mM sodium acetate (pH 5.0). Detergent (reduced, 0.1% Triton X-100) was required to release oligosaccharides in all amidase and endoglycosidase digestions.

Monosaccharide and oligosaccharide analyses by high pH anion exchange chromatography with pulsed amperometric detection (HPAEC/PAD)

Neutral and amino monosaccharides were separated isocratically (16 mM NaOH) (9) on a CarboPac PA1 column. The column was regenerated after 25 min with 200 mM NaOH for 10 min, followed by return to 16 mM NaOH. Sample injections were separated by 50 min. Sialic acids were separated using a 50-180 mM sodium acetate gradient over 20 min.

Oligosaccharides were separated on a CarboPac PA100 column. The column was equilibrated in 100 mM NaOH and 20 mM sodium acetate. After 5 min, an acetate gradient was developed over 60 min to a limit of 200 mM while the concentration of sodium hydroxide remained at 100 mM.

Mono- and oligosaccharides were detected without the addition of post-column base using the PAD. The pulse sequence used for monosaccharides (300 nA, full scale) and oligosaccharides (100 nA, full scale) was $E1 = +0.05$ V, $t_1 = 480$ ms; $E2 = +0.6$ V, $t_2 = 120$ ms; $E3 = -0.6$V, $t_3 = 60$ ms.

Calculation and interpretation of monosaccharide and oligosaccharide chromatographic data

Monosaccharides were identified from a library of retention times as detailed in the *Dionex GlycoStation Users' Manual* (Fuc, 5.87 min; GalN, 12.0 min; GlcN, 14.3; Gal 15.7; Glc, 17.1; Man, 18.6). The identity and quantity of each monosaccharide in unknown samples was estimated by comparing them with monosaccharide standards retention times and peak areas that were run in the same schedule.

Oligosaccharide mapping was performed as described (10) using disialylated biantennary oligosaccharides from a standard mixture of oligosaccharides from bovine fetuin (fetuin alditols, Dionex) as an internal standard. Its retention time, using the above acetate gradient for the analysis of oligosaccharides, was 35.7 ± 0.6 min (s_{n-1}, $n = 8$) over a period of 24 h. The oligosaccharide designation codes are those previously described(10).

III. Results and Discussion

Direct carbohydrate analysis of glycoproteins immobilized onto PVDF membranes requires that the released sugars (mono- and oligosaccharides) do not re-bind to PVDF after cleavage from the peptide backbone. We have shown (11) that the four major classes of monosaccharides found on glycoproteins (neutral, amino, deoxyhexose, and anionic sugars) and the major oligosaccharide classes remained completely in solution in the presence of PVDF membrane strips. We compared the kinetics of monosaccharide release from soluble rEPO and rEPO electro-transferred onto PVDF. Figure 1 shows that the time-dependent release of neutral (Gal and Man), amino (GlcN), and the 6-deoxy sugars (Fuc) from rEPO in solution (1A) was similar to that observed for electro-transferred rEPO (1B). The percent recovery of monosaccharides upon comparing electro-transferred and soluble rEPO hydrolysates was Gal, 80%; GlcN, 91%; Man, 70%; Fuc, 83%; and GalN, 89%. The above incomplete recoveries are partially due to incomplete electro-transfer; the gel stained faintly positive for rEPO after the transfer had been performed. We concluded from the above studies that both mono- and oligosaccharides, once released by either acid or enzymatic hydrolysis from glycoproteins on PVDF membranes, could be recovered non-selectively and in high yields.

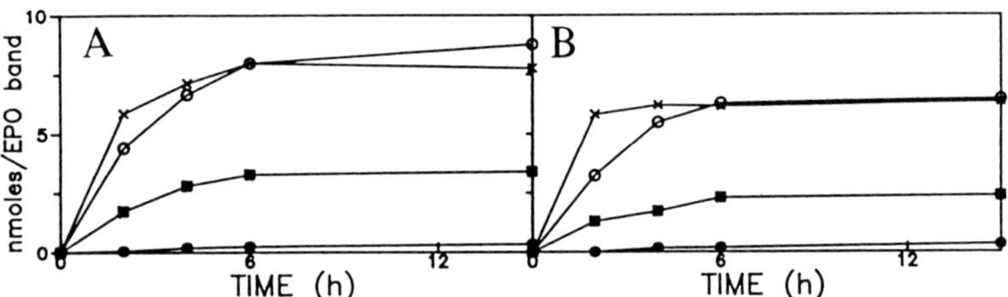

Figure 1. Release of monosaccharides from hydrolysis of soluble rEPO (A) or rEPO on PVDF (B). rEPO ($\simeq$10 μg) either soluble (A) or on PVDF (B) was hydrolyzed for the indicated times with 2 M TFA at 100°C. GlcN(×), Gal(○), Man(■), and Fuc(●) were quantified as described under "Materials and Methods."

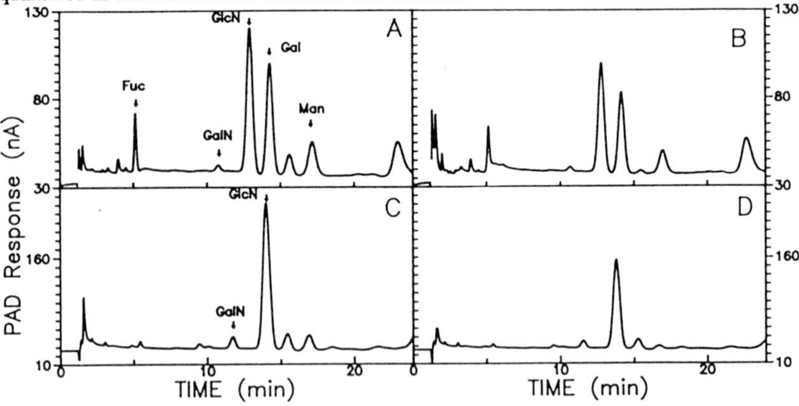

Figure 2. Monosaccharide analysis of soluble rEPO and rEPO on PVDF. Soluble rEPO ($\simeq$10 μg) was hydrolyzed with either 2 M TFA (A) or 6 M HCl (C) for 4 h at 100°C. The major stained band from two individual lanes ($\simeq$10 μg rEPO loaded/lane) was excised from PVDF, wetted with methanol, and similarly hydrolyzed with 2 M TFA (B) or 6 M HCl (D). The elution positions of the standard monosaccharides are indicated by the labeled arrows.

Table I. Monosaccharide analysis of rEPO

Investigation	Monosaccharides[1]					
	Man	Gal	GlcN	Fuc	Neu5Ac	GalN
Present study						
Direct hydrolysis	8.3	20	23	5.2	--	0.9
Hydrolysis of rEPO on PVDF	5.8	16	21	4.3	14	0.8
Sasaski et al. (2)[2]	8.9	14	18	3.2	11	1.1
Higuchi et al. (6)	8.5	13	17	3.6	11	1.0
Imai et al. (5)	8.6	15	18	3.2	11	0.8

[1]Expressed as moles of monosaccharide per mole of protein. Hydrolyses were performed as described under "Materials and Methods." In the present study, identical amounts (≃ 10 μg) were directly hydrolyzed or loaded on the gel lane.
[2]Expressed as the average of 4 batches.
-- = not performed.

The neutral and amino sugars released from soluble rEPO after hydrolysis with 2 M TFA or 6 M HCl are shown in Figs 2A and 2C, respectively. The acid-released neutral and amino monosaccharides from rEPO on PVDF are shown in Figs 2B and 2D. The absence of any interfering peaks from SDS-PAGE and electro-transfer is also apparent.

Quantification of individual sugars from hydrolysis of soluble rEPO or rEPO on PVDF and comparison with literature values are given in Table I. Approximately equal ratios of Gal to Man suggests the presence of lactosamine-type oligosaccharides [bi-(Gal:Man = 2:3), tri- (3:3) and tetra-antennary (4:3)], while larger ratios (Gal:Man) indicate poly-lactosamine oligosaccharides. rEPO had a Gal:Man ratio of 20:8.3 and a Gal:GlcN ratio of 20:23, consistent with the presence of Galβ(1→4)GlcNAc repeats (poly-lactosamine structures) attached to the tri-mannosyl core of the N-linked chains (2).

The acidic sugar, N-acetylneuraminic acid (Neu5Ac) was analyzed using the same HPLC method with an acetate gradient. The sialic acid (Neu5Ac) analysis of a single electro-transferred band of rEPO is shown in Fig. 3. The moles of Neu5Ac per mole of protein are given in Table I. The ratio of sialic acids to Gal provides information of the degree to which chains are terminated with sialic acid, provided that the oligosaccharides do not contain polysialic acids (12). Monosaccharide analysis revealed that rEPO on PVDF remained 91% sialylated (Table I), the same value obtained from the stock solution of rEPO that was applied to the gel (data not shown). Significant differences were found in the monosaccharide composition of the rEPO used in this study and in three other investigations (Table I). We found a ratio of Gal:Man of 2.4, whereas literature values were 1.5-1.7. These results suggest that either this rEPO contains longer poly-lactosamine chains, more branches have lactosamine repeats, or a greater proportion of tetra-antennary structures. The ratio of Neu5Ac to Gal supports the last possibility.

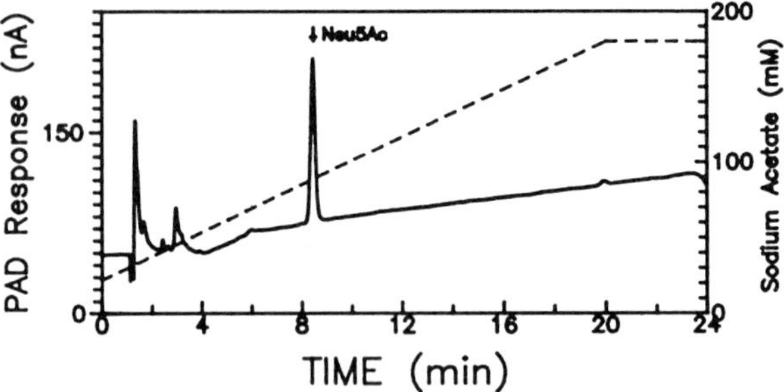

Figure 3. Sialic acid analysis of rEPO on PVDF. A stained band (≃10 μg) was excised, treated with 0.1N HCl, and analyzed for sialic acid. The elution position of standard Neu5Ac is shown by the labeled arrow.

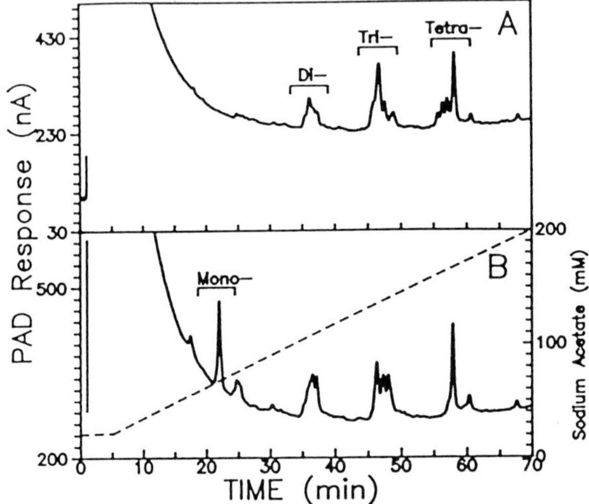

Figure 4. Oligosaccharide mapping of PNGase F-released and endo-β-galactosidase-treated oligosaccharides from rEPO on PVDF (≃ 10 μg). The excised band was incubated with 1 mU of PNGase in 150 μL of buffer (A). After 16 h, 75 μL were removed and analyzed. The pH value was adjusted to approximately 5.0 using 8 μL of 200 mM sodium acetate buffer (pH 5.0). 1 mU of endo-β-galactosidase was then added. After 16 h, the remaining volume was analyzed (B).

Figure 4A shows the profile of oligosaccharides which were released with the PNGase from the electro-transferred rEPO. The proportions of tetra-, tri-, and bi-sialylated species and their isomeric forms released from the electro-transferred rEPO were identical to those observed after digestion of soluble rEPO. The large peak detected from 1 to 10 min of elution time is from the glycerol present in this PNGase preparation. Mono- (4%), di- (16.1%), tri- (43.7 %), and tetra-sialylated (33%) species were observed.

Endo-β-galactosidase (13) is invaluable for identifying and analyzing poly-lactosamine oligosaccharides on glycoproteins (14). PNGase F released oligosaccharides from rEPO on PVDF were treated with endo-β-galactosidase in the same tube, and analyzed using HPAEC/PAD (Fig. 4B). After PNGase F treatment, four major tetra-sialylated oligosaccharides were seen. HPAEC/PAD of an endo-β-galactosidase digest of the PNGase F-treated rEPO band revealed the disappearance of the first three eluting tetra-antennary peaks (compare Figs 4A and 4B). The relative area of the unchanged peak (R_t = 58.3 min) was 17.5% and 16.6% before and after endo-β-galactosidase treatment, respectively. From previous studies of the oligosaccharides from rEPO (2), we assigned this peak as a tetra-sialylated tetra-antennary structure (C4-457300.40.00). We concluded that three of the four major tetrasialylated species contained poly-lactosamine structures (C4-457300.40.1R), which likely differ only in the branch location of the sialylated lactosamine repeats. The profile of the tri-sialylated complex was also significantly changed after endo-β-galactosidase treatment (Fig. 4B). The four incompletely resolved peaks seen in the PNGase F-released tri-sialylated oligosaccharide region appeared as three sharp peaks of approximately equal areas (R_t = 46.5-48.3 min) after endo-β-galactosidase treatment. These newly appearing oligosaccharide peaks in the tri-sialylated region are either derived from the tetra-sialylated forms with one sialylated poly-lactosamine structure or represent tri-sialylated oligosaccharides without such units. The overall reduction of the area of the tri-sialylated complex from 43.7% to 30.4% and the known predominance of tetra-antennary oligosaccharides with one sialylated lactosamine repeat (2) strongly suggest the presence of tri-sialylated poly-lactosamine structures in this rEPO preparation. A new peak (R_t = 21.3), which comprised 19.6% of the total electrochemical response, appeared near the monosialylated complex (Fig 4B). From the described specificity of endo-β-galactosidase and previous structural studies of rEPO (2, 13), we concluded that this peak (R_t = 21.3 min) was Neu5Acα(2→3) Galβ(1→4) GlcNAcβ(1→3)Gal. The smaller amount of sialylated poly-lactosamine structures with two and three repeats, which have been found in rEPO, likely comprise the mono-sialylated complex which elutes after the above peak (R_t = 24 min). New peaks were not apparent in the region of the chromatogram where neutral species elute (5-18 min), suggesting that all the poly-lactosamine oligosaccharides released by endo-β-galactosidase are sialylated. This observation held true when the digestion was repeated with glycerol-free PNGase and the magnitude of the breakthrough peak was sharply reduced (11).

rEPO from different sources varies in the proportion of sialylated species (2). The methods described here enabled the identification of such features from a single, stained band on PVDF. A comparison of the chromatographic profiles of rEPO oligosaccharides, before and after endo-β-galactosidase treatment, showed a significant change in the profile of the sialylated complexes and the appearance of a diagnostic enzyme cleavage product. Such an approach should prove particularly useful for the analysis of poly-lactosamine structures on natural EPO and rEPO from different expression systems. Using only Coomassie blue stained rEPO electro-transferred onto PVDF membranes, we report a strategy for i) obtaining a complete monosaccharide composition, ii) obtaining an oligosaccharide map which separates most forms according to size, charge and isomerity, and iii) identifying poly-lactosamine oligosaccharides. This approach should prove useful for further understanding the glycobiology of rEPO and also providing a much needed tool to monitor the consistency of rEPO as a therapeutic drug.

Acknowledgments

The authors thank Richard Lahti and Sylvia Morris for their assistance in preparing this manuscript and Dr. Jeffrey Rohrer for his critical comments.

References

1. Graber, S. E. and Krantz, S. B. (1989) Hematology/Oncology Clinics of N. America. **3**, 369-400.
2. Sasaki, H., Bothner, B., Dell, A., and Fukuda, M. (1987) *J. Biol. Chem.***262**, 12059-12076
3. Fukuda, M., Sasaki. H., Lopez, L., and Fukuda, M. (1989) *Blood* **73**, 84-89
4. Takeuchi, M., Inoue, N., Strickland, T. W., Kubota, M., Wada, M., Shimizu, T., Hoshi, S., Kozutsumi, H., Seiichi, T., and Kobata, A. (1989) *Proc. Natl. Acad. Sci.* USA. **86**, 7819-7822.
5. Imai, N., Higuchi, M., Kawamura, A., Tomonoh, K., Oh-Eda, M., Fujiwara, M., Shimonaka, Y., and Ochi, N. (1990) *Eur. J. BIochem*. **194**, 457-62.
6. Higuchi, M., Oh-eda, M., Kuboniwa, H., Tomonoh, K., Shimonaka, Y., and Ochi, N. (1992) *J. Biol. Chem*. **267**, 7703-9.
7. Laemmli, U. K. (1970) *Nature* (London) **227**, 680-685.
8. Rosenberg, A. and Schengrund, C-L. eds. (1975) Roles of Sialic Acid, Plenum Press (New York) p. 43.
9. Hardy, M. R., Townsend, R. R., and Lee, Y. C. (1988) *Anal. Biochem.* **170**, 54-62.
10. Hermentin, P., Witzel, R., Vliegenthart, J. F. G., Kamerling, J. P., Nimtz, M., and Conradt, H. S. *Anal. Biochem*. **202**, 281-289
11. Weitzhandler, M., Kadlecek, D., Avdalovic, N., Forte, G. F., Chow, D., and Townsend, R. R. accepted for publication in *J. Biol. Chem.*
12. Troy II, F. A. (1992) *Glycobiology* **2**, 5-24.
13. Fukuda, M.N. and Matsumura, G. (1976) *J. Biol. Chem.* **251**, 6218,6225.
14. Fukuda, M. (1965) *Biochim. Biophys. Acta*. **780**, 119-150.

A Practical Approach for Isolation and Characterization of Glycosylation Sites of Glycoproteins Bearing N- and/or O-Linked Carbohydrate Chains

Kuo-Liang Hsi, Ling Chen and Pau-Miau Yuan
Applied Biosystems Inc., Foster City, CA 94404

I. Introduction

We (1,2) have reported the isolation and characterization of glycosylation site(s) of high mannose type glycoproteins using a dot-blot staining assay combined with a ConA-Sepharose affinity method to selectively isolate the glycopeptides. Our overall strategy has been as follows: 1) screen for glycoproteins and glycosylation types by lectin staining of dot-blotted sample on membrane; 2) selective isolation of glycopeptides from proteolytic digests by micro-batch affinity chromatography using various lectin-agaroses; 3) sequencing and mass analyses of isolated glycopeptides; and 4) deglycosylation of the glycopeptides and structure determination of the separated carbohydrate chains.

In this paper, we expand the technique to those glycoproteins that bear different types of N-linked carbohydrate chains and/or O-linked oligosaccharide chains so that both N-linked and O-linked glycopeptides can be selectively isolated from glycoproteins.

Two glycoproteins were chosen as model glycoproteins for this study, namely tissue plasminogen activator (t-PA) and human chorionic gonadotropin (HCG). t-PA (3) bears three different N-linked carbohydrate chains (i.e. high mannose, hybrid and complex types) and hCG (4,5) has several N-linked and O-linked carbohydrate chains attached.

II. Materials and Methods

A. Chemicals and Reagents. Chemicals were either reagent or HPLC grade, as appropriate. ConA-Sepharose 4B was purchased from Sigma and Jacalin-Agarose was from Vector Laboratory. Human chorionic gonadotropin was from Sigma and t-PA was a commercially available product from Genentech. Trypsin was purchased from Cal

TECHNIQUES IN PROTEIN CHEMISTRY IV

Biochem. Galactose and α-methyl mannoside were from Sigma. Solvents for HPLC and other reagents were from Applied Biosystems, Inc.

B. Instrumentation. Sequence analysis was performed either on the ABI Model 477A or 473A Sequencer using standard cycles. Mass analysis was carried out on a prototype Matrix-Assisted Laser Desorption Time-of-Flight Mass Spectrometer. Peptides were separated on an ABI Model 172 HPLC System that consisted of a 140B Solvent Delivery System, a 785A Programmable Absorbance Detector and a 112A Injector, using a solid particle, non-porous column (Sphere C_{18}, particle sizes:1.5 μ, column size: 4.6 x 30 mm),

C. Proteolytic Digestions. Glycoproteins were subjected to reduction and alkylation as described in [6] prior to protease digestion. Briefly, protein (5-10 nanomoles) was treated by 10 μl of 10% ß-mercaptoethanol and followed by 10 μl of 4-vinyl pyridine in 300 μl of 250 mM Tris-HCl and 6M guanidine hydrochloride, pH7.5, at room temperature for one hr each. Samples were desalted by dialysis against d.i. water and dried in a SpeedVac. Dried samples were then re-dissolved in 200 μl digestion buffer (200 mM NH_4HCO_3, pH 8) for digestion. Trypsin was added at a E/S ratio of 1:50. The digestion was carried out at 37°C for 6-10 hr , and then terminated by adding 10% TFA to adjust to pH 2.

D. Micro-batch Affinity Binding. A 0.5 ml of ConA-Sepharose 4B resin in an Eppendorf tube was washed three times with 200 μl of binding buffer (50 mM Tris-HCl, containing 100 mM NaCl, and 1 mM of Ca^{2+} and Mn^{2+}, pH 7.5). 200 μl of protein digestion mixture (sample from 0.1 to 10 nanomoles) in binding buffer was transferred into the resin-contained tube and allowed to stand for 20 min with occasional shaking. The supernatant was removed and the resin was washed again with binding buffer (5x with 200 μl). Bound peptides were released by suspending the resin in 200 μl of elution buffer (0.5 M of α-methylmannoside in binding buffer, pH adjusted to 5 by adding 10% of acetic acid) three times. The pooled supernatant was dried, reconstituted in 0.1% of TFA and further desalted with fractionation by RP-HPLC.

Jacalin-Agarose was used instead of ConA-Sepharose 4B for the isolation of O-linked glycopeptides using the same procedures. The binding buffer used for Jacalin was 175 mM Tris-HCl, pH 7.5, while the elution buffer was 0.5 M galactose in the binding buffer.

III. Results and Discussion

The classical approach for glycoprotein analysis of complete deglycosylation of the glycoprotein and then isolation of the resulting oligosaccharide is not ideal due to the total loss of the glycosylation site information and the destruction of the peptide portion of the molecule. The approach presented here offers a useful protocol to solve this problem. In addition, since all experiments were performed in an Eppendorf tube and the reagents and solvents were scaled down for micro preparation, this approach is suitable for sub-nanomole analysis.

A. Application to t-PA

t-PA, a recombinant glycoprotein applied clinically in the

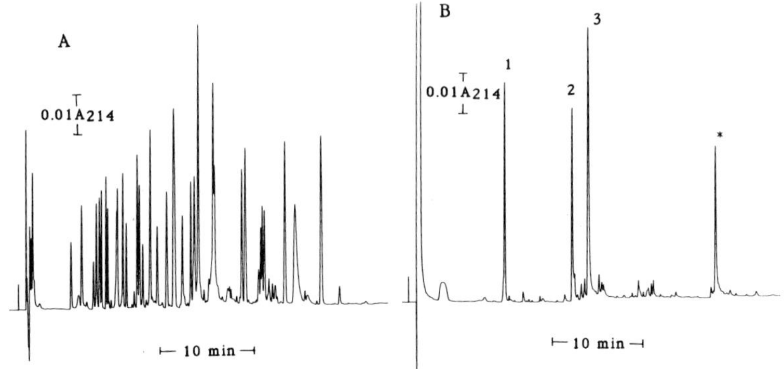

Figure 1A. RP-HPLC separation of human tissure plasminogen activator (tPA)-tryptic peptides. Column: solid particle C18 (4.6×30 mm, particle size: 1.5μ); Buffer: A: 0.1% TFA/water, B: 0.09% TFA/50% acetonitrile; Elution gradient: B%=2-70%/45min; Flow rate: 250μl/min. Sample amount loaded was 80 picomole protein digest. 1B. RP-HPLC separation of tPA glycopeptides prepared by micro-batch affinity binding using Con A-sepharose 4B. HPLC conditions were same as in 1A. Sample loaded was 300 picomole based on the protein. *Breakdown product from ConA.

treatment of myocardial infraction, consists of 523 amino acid residues and three different types of N-linked carbohydrate chains: high mannose, hybrid and complex (3). A RP-HPLC tryptic peptide map was produced before and after affinity purification of glycopeptides on ConA-Sepharose 4B (Fig.1) Following the micro-batch affinity preparation, the bound glycopeptides were eluted out from the resin with α-methyl mannoside and then fractionated by HPLC (Fig. 1B). Peaks 1, 2 and 3 were found homogeneous and represent residues 441-449, 163-189 and 102-129 of intact t-PA, respectively (Table I). Peak 4, however, was found to be derived from the breakdown products of ConA by a control experiment (data not shown). All three peptides possess a consensus sequence of (?)-X-S/T, a typical feature of N-linked glycosylation.

Mass analysis showed that peak 3 has three signals of MH^+4275, 4438, and 4600 (Fig. 2). The mass difference between each two adjacent components was 162, suggesting a difference of one mannose. The observed mass data matches the following three glycopeptide structures (3): $Man_5GlcNAc_2$-Asn-peptide, $Man_6GlcNAc_2$-Asn-peptide and $Man_7GlcNAc_2$-Asn-peptide. Peak 2 showed three mass signals (MH^+4989, 5280 and 5571) with a mass difference among them of 291, suggesting a content difference in NeuAc. This is typical for a hybrid or complex type of glycopeptide microheterogeneity. Based on the core structure and attached

Table 1. Amino Acid Sequences and Carbohydrate Structure of tPA Glycopeptides Prepared by Micro-Batch Affinity Binding

Sample	Determined Sequence	Corresponding Domain in tPA	Predicted Carbohydrate	Mass Values Obs./Calc.
	CHO			
tPA-1	CTSQHLLNR	441-449	GlcNAc4Hex5Fuc	2947/2947*
			GlcNAc4Hex5FucNeuAc	3237/3238
			GlcNAc4Hex5FucNeuAc2	3528/3530
tPA-2	YSSEFCSTPA	163-189	GlcNAc4Hex5Fuc	4989/4989
	CSEGNSDCYF		GlcNAc4Hex5FucNeuAc	5280/5280*
	CHO			
	GNGSAYR		GlcNAc4Hex5FucNeuAc2	5571/5572*
tPA-3	GTWSTAESGA	102-109	GlcNAc2Man5	4275/4275*
	CHO			
	ECTNWNSSAL		GlcNAc2Man6	4438/4437
	AQKPYSGR		GlcNAc2Man7	4600/4958

*Most abundant Peak in LD-TOF mass spectrum

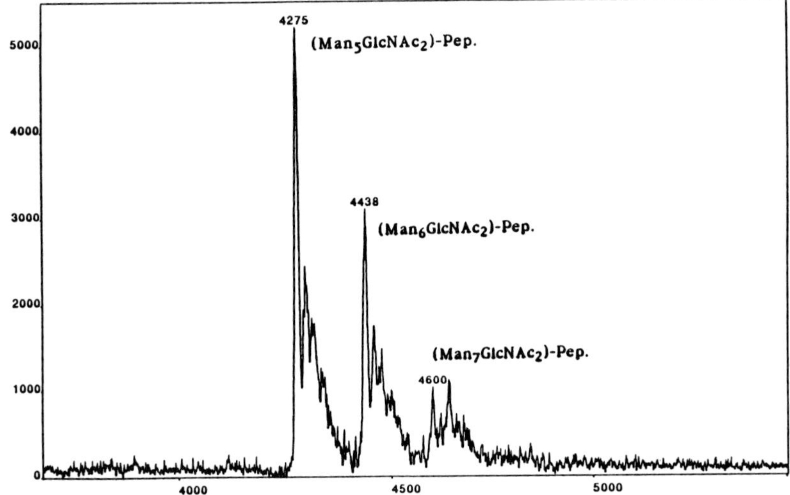

Figure 2. Matrix-Assisted laser desorption mass spectrum of tPA-3. The 162 amu mass difference between two adjacent peaks corresponds to the mass of one mannose.

building blocks (3), these three masses may represent the following structures: GlcNAc4Man3Gal2Fuc-, GlcNAc4Man3Gal2FucNeuAc-, and GlcNAc4Man3Gal2FucNeuAc2-. Similarly, three signals of MH^+2947, 3237 and 3528 from peak 1 match the predicted masses of the following three structures: GlcNAc4Man3Gal2Fuc-, GlcNAc4Man3Gal2FucNeuAc-, and GlcNAc4Man3Gal2FucNeuAc2- (Table I).

B. Application to hCG

hCG is a heterodimer glycoprotein consisting of non-covalently associated α- and ß-subunits. Asn 52 and 78 of α-chain and Asn 13 and 28 of ß-chain contain N-linked carbohydrates, whereas Ser 121, 127 , 132 and 138 of ß-chain were O-linked (4,5). After pyridylethylation and digestion by trypsin, hCG tryptic peptides were fractionated by HPLC (Fig. 3A). Lectins ConA-Sepharose 4B and Jacalin-Agarose were used for affinity isolation of N-linked and O-linked glycopeptides, respectively. The HPLC separation of the resulting glycopeptides is presented in Fig. 3B and 3C. Sequence analysis (Table II) of hCG glycopeptides demonstrated that peaks hCG-1, hCG-2 , hCG-3 and hCG-4 represented residues 52-63 (α-chain), 76-91 (α-chain), 9-20 (β-chain) and 21-43 (β-chain) of intact hCG, respectively. hCG-5 was a long form of hCG-1 derived from

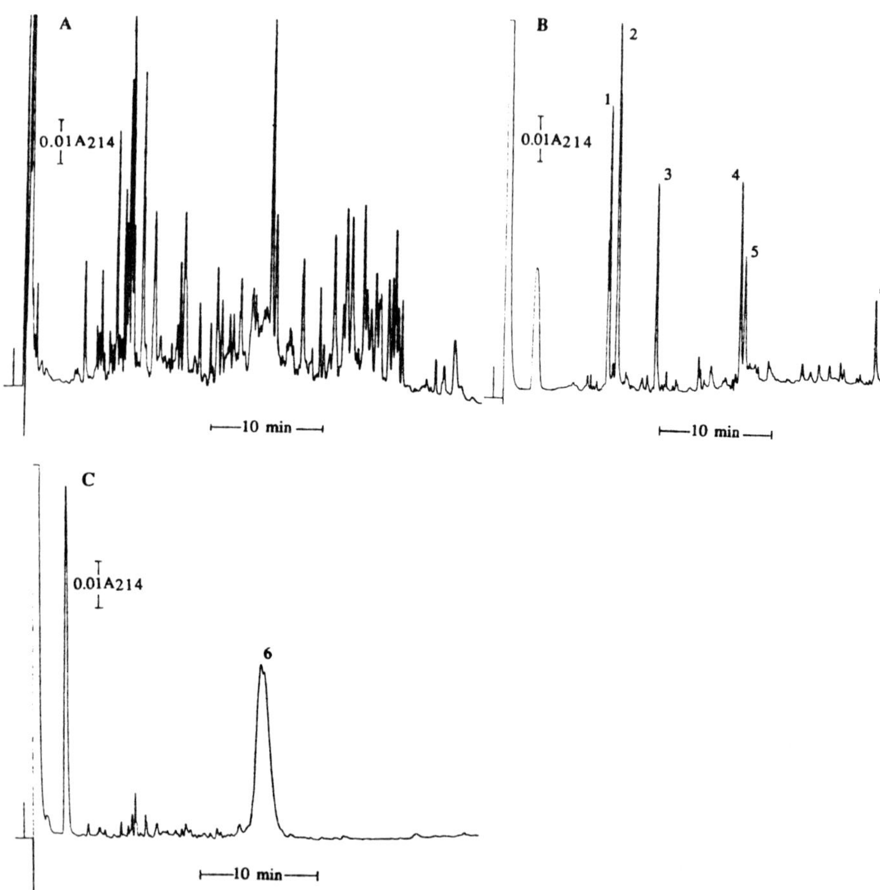

Figure 3A. RP-HPLC mapping of 100 picomole of hCG-tryptic peptides. HPLC column and run conditions were same as described in Figure 1A. **3B**. RP-HPLC profile of hCG N-linked glycopeptides by Con A-sepharose 4B affinity binding with 300 picomole of digest. HPLC conditions were same as described in Figure1A. * Breakdown product from Con A. **3C**. RP-HPLC separation of hCG O-linked glycopeptides isolated by Jacalin-agarose affinity binding with 600 picomole of protein digest. HPLC conditions were same as in Figure 1A.

Table II. Amino Acid Sequences and Carbohydrate Structures of hCG Glycopeptides Prepared by Micro-Batch Lectin Affinity Binding

Sample	Determined Sequence	Corresponding Domains in hCG	Predicted Carbohydrate Structures	Mass Values Obs./Calc.
	CHO			
hCG-1	NVTSESTCCV	52-63	GlcNAc3Hex4	2710/2711*
	AK	(α-chain)	GlcNAc3Hex4NeuAc	3002/3002*
			GlcNAc3Hex4Fuc	2863/2857
			GlcNAc4Hex5	3075/3076
			GlcNAc4Hex5Fuc	3225/3222
			GlcNAc4Hex5NeuAc	3373/3368
			GlcNAc4Hex5NeuAcFuc	3516/3514
	CHO			
hCG-2	VENHTACHCS	76-91	GlcNAc3Hex4	3472/3472*
	TCYYHK	(α-chain)	GlcNAc3Hex4NeuAc	3762/3762*
			GlcNAc4Hex5	3835/3836
			GlcNAc4Hex5NeuAc	4127/4127
			GlcNAc4Hex5NeuAc2	4417/4419
	CHO			
hCG-3	CRPINATLAV	9-20	GlcNAc3Hex4	2680/2679
	EK	(β-chain)	GlcNAc4Hex5	3045/3044
			GlcNAc4Hex5Fuc	3192/3190
			GlcNAc4Hex5NeuAc	3335/3335*
			GlcNAc4Hex5FucNeuAc	3483/3482
			GlcNAc4Hex5Fuc2NeuAc	3623/3627
			GlcNAc4Hex5Fuc2NeuAc2	3768/3773
	CHO			
hCG-4	EGCPVCITVN	21-43	GlcNAc4Hex5	4503/4501**
	TTICAGYCPT	(β-chain)	GlcNAc4Hex5Fuc	4647/4647*
	MTR		GlcNAc4Hex5NeuAc	4795/4792
			GlcNAc4Hex5NeuAcFuc	4938/4938
			GlcNAc4Hex5NeuAc2Fuc	5226/5229

*Most abundant peak in LD-TOF mass spectrum
**Mass data listed here for all hCG-4 heterogeneity forms were MNa^+

partial digestion. Mass analyses of glycopeptides hCG peaks 1-4 are listed in Table II. These results confirmed the previously published reports (4,5). hCG-6 was eluted from Jacalin-Agarose and contained two O-linked glycopeptides, a major peptide of FQDSSSSKAPPPSLP-SPSRLPGPSDTPILPQ (residues 115-145) and a minor peptide of APPPSLPSPSRLPGPSDTPILPQ (residues 123-145) of hCG ß-chain. The glycosylation sites deduced by sequence analysis of all these glycopeptides were consistent with reported data (4,5). Fig. 4 illustrates the multiple forms of O-linked glycopeptides by mass

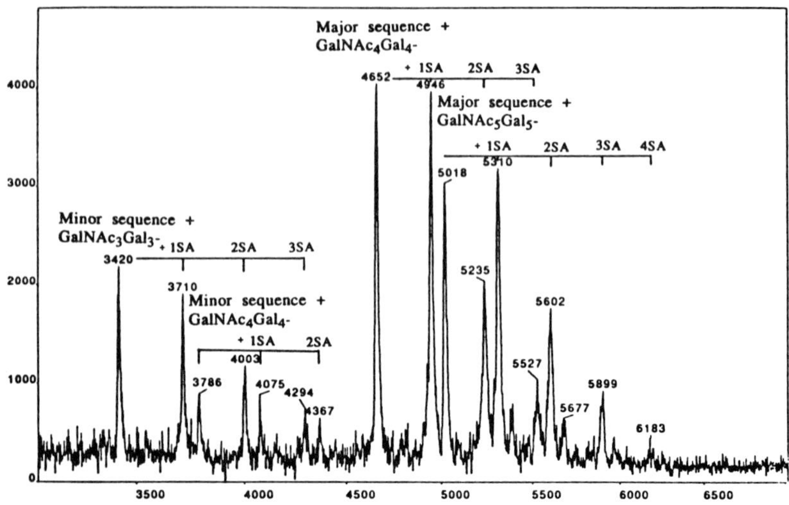

Figure 4. Matrix-assisted laser desorption mass spectrum of O-linked glycopeptides, hCG-6 in Figure 3C. Micro-heterogeneities of carbohydrate chains of two glycopeptides were illustrated.

analysis. The observed mass values can account for the corresponding structures.

IV. Summary

A micro-batch affinity preparation approach has been developed to selectively isolate N- or O-linked glycopeptides from proteolytic digests by using ConA- or Jacalin-Agarose. Sequence and mass analyses of the glycopeptides allow the characterization of glycosylation site(s) of N- or O-linked glycoproteins. This work along with our previous reports (1,2) constitute a practical strategy for the glycoprotein analysis at a sub-nanomole sample level.

References

1. Hsi, K-L., Chen, L., Hawke, D.H., Zieske, L.R., and Yuan, P-M. (1991) Anal. Biochem. **198,** 238-245
2. Hawke, D.H., Hsi, K-L., Zieske, L.R., Chen, L., and Yuan, P-M. (1992) in Techniques in Protein Chemistry III (Angeletti, R.H. Ed.) pp. 315-326, Academic Press

3. Spellman, M.W., Basa, L.J., Leonald, C.K., Chake, J.A., and O'Connor, J.V. (1989) J. Biol. Chem. **264**, 14100-14111
4. Baenziger, J.U., and Green, E.D. (1988) Biochem. Biophys. Acta **947**, 287-306
5. Kassler, M.J., Mise, T., Ghai, R.D., and Bahl, O.P. (1979) J. Biol. Chem. **254**, 7909-7914
6. Hawke, D.H. and Yuan, P-M. (1978) User Bulletin (Applied Biosystems Inc.) **28**, 1-8

Location of the Keratan Sulfate Attachment Sites in the Hyaluronate Binding Region of the Proteoglycan, Aggrecan

Peter J. Neame and Frank P. Barry

Shriners Hospital for Crippled Children,
Tampa, Florida 33612
and
Department of Biochemistry and Molecular Biology
University of South Florida, Tampa, Florida 33612

I. Introduction

The major part of the bulk of cartilage consists of proteoglycan aggregates. These are complexes of link protein (a ~45 kDa glycoprotein), hyaluronic acid (100 - 1000 kDa) and the large, aggregating chondroitin sulfate proteoglycan, aggrecan (2500 - 5000 kDa, core protein ~240 kDa depending on alternate splicing). Attached to the core protein are *N*- and *O*- linked oligosaccharides. The majority of the latter are chondroitin sulfate and keratan sulfate (KS) glycosaminoglycan (GAG) chains.

The exact locations of the GAG chains and the structures of individual chains at a given site have been difficult to determine. It is known that these substituents are very heterogeneous in size, in degree of sulfation, and in degree of epimerization. To date, characterization of these structures has relied on analysis of partially purified fragments of the proteoglycan and has averaged the structures of GAG chains at several locations. The attachment sites of chondroitin sulfate chains have been analyzed by sequencing peptides before and after β-elimination (1) and by sequencing proteins that have only one, well defined, GAG, such as decorin (2) or collagen type IX (3). It is possible to be reasonably certain that the first enzyme in chondroitin sulfate GAG attachment (xylosyl transferase) requires a consensus sequence AcidicAA-X-(Ser + Gly)-HydrophobicAA. Serine and glycine both seem to be required and are usually in the order Ser-Gly. Collagen type IX has a Gly-Ser attachment site. There may be 1-2 amino acids between the Ser-Gly and the other two required amino acids. In contrast, little is known about the attachment sites of KS chains.

TECHNIQUES IN PROTEIN CHEMISTRY IV

Part of the core protein is rich in KS (4). KS is a variable length repeating disaccharide of galactose and *N*-acetyl glucosamine. KS is variably sulfated and attached to proteins through an *O*- or an *N*-linkage. The KS domain is between amino acids 674-830 (in human aggrecan) (5). A second region that contains keratan sulfate is known as the hyaluronate binding region (HABR). The HABR is the N-terminal globular domain in aggrecan and binds to both hyaluronic acid and link protein. This domain is readily isolated by tryptic digestion of proteoglycan, isolation of the high molecular weight HA-associated complex and dissociation and separation of the HABR, link protein and hyaluronic acid by gel permeation chromatography. On a Coomassie blue-stained SDS-PAGE gel (8) this domain is a diffuse, weakly staining band of 70-130 kDa. Immunolocalization of blotted material with anti-KS monoclonal antibody stains this domain strongly. There are no regions of the HABR that show significant sequence similarity to the keratan sulfate domain, indicating that the KS attachment site(s) must have a different consensus sequence.

Keratan sulfate can be detected by monosaccharide analysis, based on the characteristic equimolar galactose and N-acetyl glucosamine. This type of analysis requires large amounts of material (> 1 nmol). A monoclonal antibody, 5D4 (6), binds to highly sulfated keratan sulfate and provides a more sensitive test for KS.

During sequence analysis of the HABR of pig aggrecan, we speculated that two threonines that were in very low yield in Edman degradation of a heterogeneous, high molecular weight, *N*-acetylglucosamine-rich glycopeptide might be substituted with KS. As GAGs are insoluble in organic solvents, such as butyl chloride or ethyl acetate, we modified the method described by Abernethy *et al* (7), and used aqueous solvents to extract derivatized, GAG-substituted amino acids from the PVDF membrane. The peptide was covalently immobilized to prevent it being washed away. To test for specific *O*-glycosylation sites, the sequentially removed amino acid derivatives were tested for immunoreactivity with 5D4.

II. Materials and Methods

HABR from pig laryngeal cartilage was isolated, digested, and the peptides separated as described elsewhere (8). The KS-containing peptide was isolated by virtue of its high molecular weight relative to other glycosylated and non-glycosylated peptides. It was detected by its positive reaction in ELISA assays using monoclonal antibody 5D4 (6). The antibody was a generous gift from Dr. Bruce Caterson, UNC-Chapel Hill and is commercially available from Seikagaku America, Rockville, MD.

Peptide was coupled to arylamine-PVDF (AA-Sequelon™) using EDC (1-ethyl-3-(3-dimethylaminopropyl)carbodiimide) as described by the manufacturer (Millipore). The membrane was washed with 0.1% aqueous TFA to remove uncoupled material. The membrane was cut into small pieces (2 x 2 mm) and placed directly into the standard reaction cartridge, without a glass fiber filter.

Table I. Modified sequencer (GAG) cycles[a]

	Reaction (GAG) cycle			Conversion (GAG) cycle			
Step	Function	Fxn#	Time	Function	Fxn#	Time	
1	Prep R2	4	6	Block Flush	23	6	
2	Deliver R2	5	20	Argon Dry	22	40	
3	Prep R1	1	6	Ready to Receive	13	151	
4	Deliver R1	2	2	Block Flush	23	6	
5	Argon Dry	29	40	Argon Dry	22	160	
6	Deliver R2	5	250	Load R4	3	9	b
7	Prep R1	1	6	Argon Dry	22	4	b
8	Deliver R1	2	2	Pause	25	600	b
9	Argon Dry	29	40	Argon Dry	22	700	
10	Deliver R2	5	250	Block Flush	23	6	
11	Prep R1	1	6	Load S4	12	13	
12	Deliver R1	2	2	Argon Dry	22	4	
13	Argon Dry	29	40	Load S4	12	13	
14	Deliver R2	5	250	Argon Dry	22	4	
15	Reaction Heater	32	50	Pause	25	70	
16	Argon Dry	29	120	Argon Dry	22	4	
17	Deliver S1	14	60	Pause	25	70	
18	Deliver S2	17	200	Argon Dry	22	4	
19	Argon Dry	29	120	Pause	25	90	
20	Load R3	9	5	Argon Dry	22	4	
21	Argon Dry	29	4	Pause	25	40	
22	Pause	33	300	Load S4	12	13	
23	Load S2	18	6	Argon Dry	22	4	
24	Block Flush	30	6	Collect Direct	18	50	
25	Argon Dry	29	120	Inject	16	1	c
26	Prep Transfer	22	30	FC Advance	19	1	
27	Deliver X1	11	20	Deliver S4	11	50	
28	Transfer w/Argon	24	50	Argon Dry	22	4	
29	Deliver X1	11	20	Empty	20	30	
30	Transfer w/Argon	24	40				
31	Deliver X1	11	20				
32	Transfer w/Argon	24	50				
33	End Transfer	25	1				
34	Reaction Heater	32	55				
35	Argon Dry	29	1800				
36	Deliver S3	20	60				
37	Argon Dry	29	120				

[a]Cycles are for an ABI model 477A protein sequencer with on-line ABI model 120A PTH-amino acid analyzer. Reagents were as supplied by ABI. Bottle X1 was S4 (10% acetonitrile in water).
[b]None of these steps are essential. If additional chromatography or mass spectrometry is performed on the removed carbohydrate, then it is useful to have a stable PTH derivative.
[c]This is only necessary if on-line analysis is used during the run. If all cycles are collected, it can be omitted.

Alternatively, a Blott™ cartridge can be used; this improves the fluid flow and transfer efficiency.

Sequence analysis on the KS-containing peptide was performed on an Applied Biosystems (ABI) 477A protein sequencer, using the manufacturer's FAST cycles and an internal standard (Nle). PTH-amino acids were separated and detected with an on-line ABI 120A HPLC.

To determine the site of KS attachment, we used a modified sequencer cycle that included a water elution of the TFA-released N-terminal derivative. The sequencing cycles used for PTH-amino acid-GAG collection were as shown in table I. The reaction cycle is a modification of the FAST cycle recommended by ABI. The modified cycle uses aqueous 10% acetonitrile (from bottle X1) to transfer the released product of TFA cleavage to the conversion flask. Three transfers are used, with argon transfers at the end of each (steps 27-32 in table 1). The cartridge is then dried for 30 minutes. Similar modifications can be made to the BLOT cycle. The lengthy argon dry at step 35 can then be reduced to 20 minutes or less, but needs to be followed by a brief pulse of TFA and a further argon dry to remove traces of water from the corners of the cartridge slot.

To test whether the modified cycle permitted sequencing, in spite of the water washes, we initiated analysis of the KS-positive peptide from HABR with 5 FAST cycles, followed by a modified (GAG) cycle, followed by 3 FAST cycles (figure 1). While a single cycle is not enough to assess the cumulative effect of changes to the Edman chemistry, it was clear that any effects of the water wash were small.

GAG cycles were used for cycles 10-19. The transferred material was converted to the PTH-derivative and sent to the fraction collector without HPLC analysis. The fractions were transferred to a 96 well microtitration plate and dried in a Speedvac (Savant model A290). For the ELISA assay, 100 µl 0.5 M sodium chloride, 50 mM sodium bicarbonate, 0.02% sodium azide was added to each well. The plate was left overnight at 37 C with shaking to enable the GAG to bind to the plastic. The plate was washed, and incubated with monoclonal antibody 5D4 for 1 hour at 37 C in PBS/0.5% Tween 20. After further washing, bound antibody was detected with 3,3′, 5, 5′- tetramethyl benzidine and horseradish peroxidase-rabbit anti-mouse antibodies (figure 2, lower frame).

III. Results

Tryptic digestion of the aggrecan HABR followed by gel permeation chromatography and reversed-phase HPLC indicated that all the highly sulfated KS in the HABR was in a single tryptic peptide from the C-terminal of the domain. The 27-residue peptide had a high molecular weight (8-15 kDa) as a result of the KS GAG chain (s). We isolated the peptide, coupled it to arylamine-substituted PVDF membranes (AA-Sequelon™) and eluted the substituted amino acid in water, together with its GAG chain. ELISA assay of the eluted amino acid enabled us to associate the GAG with a specific threonine.

By coupling the C-terminal peptide to AA-Sequelon™, it was possible to obtain clear sequence to the penultimate amino acid (figures 1 and 2, open circles, show the yields for the first 19 cycles). Yields of PTH-glu and PTH-asp were low, as they were used for carbodiimide coupling, but other PTH-amino acids

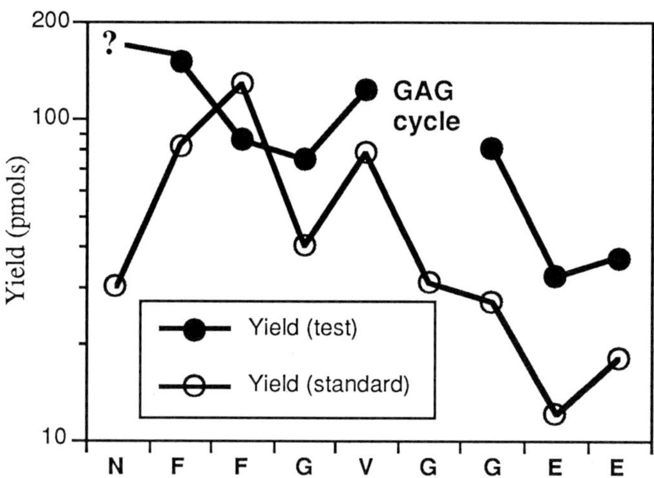

Figure 1. Yields of PTH amino acids, in pmol, obtained after sequencing covalently coupled peptide. Open circles show the first nine cycles using standard FAST cycles (150 pmol initial yield). Closed circles show a test run that was assayed for 5D4 reactivity after cycle 9 (~300 pmol initial yield). A GAG cycle was inserted at cycle 6 to check cycle yield. The first cycle in the test run was lost due to a transfer failure.

were in nominal yields, with the exception of two residues, at cycles 12 and 17, where low levels of PTH-threonine were observed. The threonine residues were confirmed by amino acid analysis (8). Another threonine, at cycle 15, was in good yield and provided a reference threonine level.

The sequence was NFFGVGGEEDI<u>T</u>IQTV<u>T</u>WPDVELPLP. The threonines that were in low yield are underlined. This peptide contained KS, on the basis of hexosamine content (approximately 25% of the amino acid plus hexosamine content was glucosamine) and 5D4 reactivity, so we speculated that KS was attached at one or both of these sites.

The standard sequencing cycle was modified to include a water elution of the TFA-released amino acid derivative. To detect KS chains, the released amino acids were converted to their PTH-form and directed to the fraction collector on the 477A. The eluted amino acid was then transferred to a 96 well microtitration plate, dried and analyzed by ELISA. High 5D4 reactivity was observed at the location of the threonine at cycle 12 (figure 2, lower frame). Another increase in 5D4 reactivity was seen at cycle 18; this was one cycle after the next low-yield threonine and may be a result of bad lag.

We have recently analyzed a different, glycosaminoglycan-substituted peptide using a modified BLOT cycle. This run consisted of 18 GAG cycles followed by 10 standard BLOT cycles. The sequence data obtained after 18 GAG cycles was still readily interpreted.

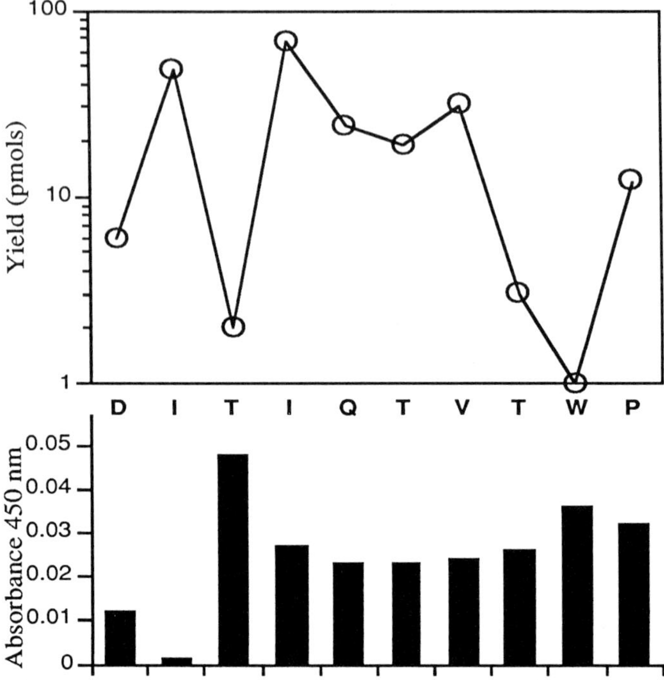

Figure 2. Yields of PTH amino acids and 5D4 reactivity in the KS-positive peptide from HABR. Amino acid yields obtained when sequencing covalently coupled NFFG: Cycles 10 - 19. Upper frame: Amino acid yields obtained when sequencing using standard FAST cycles. Lower frame: Absorbance at 450 nm from GAG cycle fractions assayed for reactivity with the anti-keratan sulfate monoclonal antibody, 5D4.

IV. Conclusions

The extracellular matrix and the cell surface have large numbers of glycosylated proteins. The carbohydrates have the potential for extraordinary structural diversity and may confer specific properties on their parent proteins. It is therefore important to determine their structure and site of attachment to proteins. Proteoglycans are abundant, heavily glycosylated molecules, found at the cell surface and in the extracellular matrix. They have proved to be rather intractable to modern, high sensitivity, analytical techniques. It is often possible to isolate peptides that are *N*- and *O*-substituted in a variety of places, but it is difficult to define the *O*-linkage sites. It is even more difficult, in cases where multiple substitution occurs, to determine which sites are substituted with particular carbohydrates.

Aggrecan contains several domains that are variably substituted with chondroitin sulfate, KS and *O*-linked oligosaccharides. To date it has not been possible to define the exact location of KS except in the broadest sense. For example, there is a domain in the core protein that apparently contains only KS; within this domain, it is reasonable to assume that *O*-linked KS is attached to the hydroxylated amino acids that occur every 6 residues. Another domain contains primarily chondroitin sulfate. It is known, from HF cleavage of the GAG chains, that chondroitin sulfate is attached at many, if not all, Ser-Gly sequences in the non-globular domain of the proteoglycan (1). The chondroitin sulfate-domain also contains KS, although the location of the KS is not known. A consensus sequence for KS attachment sites is not known.

We have covalently coupled a KS-containing peptide to derivatized PVDF. TFA-released anilinothiazolinone-amino acids can then be transferred to the conversion flask using aqueous solvents. This has permitted us to analyze each cycle of a sequencing run by ELISA assay. Using this methodology, we have identified at least one of the sites of KS substitution on the HABR domain of the proteoglycan, aggrecan.

Necessary precursors to sequence analysis using this method are (a) to have an appropriate probe, such as an antibody or a lectin, for the modified amino acid, and (b) to sequence the immobilized peptide of interest using standard cycles to obtain an estimate of covalent coupling efficiency and yield at each cycle.

Using this methodology, it is possible to isolate enough of individual substituents on heavily glycosylated proteins to characterize each site. Eluted, derivatized, amino acids may be analyzed by the methods we have described and/or by mass spectrometry to build a detailed structural map of substitutions on complex glycoproteins.

Further optimization of the reaction cycle may be performed; for example, additional washing of the immobilized peptide with water might improve the trailing 5D4 reactivity. We have not attempted to optimize the PITC coupling chemistry or the TFA cleavage chemistry for sequencing through GAG-substituted amino acids.

Acknowledgments

This work was supported by NIH grant AR 35322 and a grant from the Shriners of North America. We are grateful for the assistance of Carmen Young (protein sequencing) and Janette Gaw (peptide isolation and immunoassays).

References

1. Krueger, R. C. J., Fields, T. A., Hildreth, J. I., and Schwartz, N. B. (1990). *J. Biol. Chem.*, **265,** 12075-12087.
2. Day, A. A., Ramis, C. I., Fisher, L. W., Gehron-Robey, P., Termine, J. D., and Young, M. F. (1986). *Nuc. Ac. Res.*, **14,** 9861-9876.
3. McCormick, D., van der Rest, M., Goodship, J., Lozano, G., Ninomiya, Y., and Olsen, B. (1987). *Proc. Natl. Acad. Sci. USA*, **84,** 4044-4048.
4. Heinegård, D., and Hascall, V. C. (1974). *J Biol Chem*, **249,** 4250-4256.
5. Doege, K. J., Sasaki, M., Kimura, T., and Yamada, Y. (1991). *J. Biol. Chem.*, **266,** 894-902.
6. Caterson, B., Christner, J. E., and Baker, J. R. (1983). *J. Biol. Chem.*, **258,** 8848-8854.
7. Abernethy, J. L., Wang, Y., Eckhardt, A. E., and Hill, R. L. (1992). *In* "Techniques in Protein Chemistry III", (R. H. Angeletti, ed.) pp. 277-286, Academic Press, San Diego.
8. Barry, F. P., Gaw, J., Young, C. N., and Neame, P. J. (1992). *Biochem. J.*, **286,** 761-769.

Site Specific Heterogeneity of N-Linked Oligosaccharides on Recombinant Human Erythropoietin

Patricia L. Derby, Thomas W. Strickland, and Michael F. Rohde
Amgen, Inc., Thousand Oaks, CA

Introduction

Erythropoietin is a glycoprotein that stimulates proliferation and differentiation of erythroid precursor cells to mature erythrocytes (1). It is produced mainly in adult kidney cells. Impaired renal function, such as that seen in patients with chronic renal disease, leads to a decrease in erythropoietin production and anemia (2). Recombinant human erythropoietin (rHuEPO), 165 residues long and a molecular weight of 30 kD, is produced in Chinese hamster ovary (CHO) cells and is used to reverse the anemia observed in patients with renal disease (3).

Carbohydrate makes up approximately 40% of the mass of rHuEPO (4) and is essential for the *in vivo* activity (5-7). The carbohydrate of rHuEPO is made up of three N-linked chains at asparagine residues 24, 38, and 83 and one O-linked chain at serine 126 (8-11). The site specific glycosylation of the N-linked oligosaccharides of rHuEPO has been studied by FAB-MS (12) and more recently by reversed phase chromatography of asialo pyridylaminated derivatives (7). To further analyze the site specific heterogeneity of carbohydrate structures, glycopeptides containing the individual N-sites were isolated and characterized by N-terminal sequencing. After N-glycanase and sialidase treatment, the oligosaccharides from each N-glycosylation site were analyzed by high pH anion exchange chromatography.

Materials and Methods

Enzymes used for peptide cleavages were Sequencing Grade Glu-C from ProMega and trypsin from Boehringer Mannheim. Enzyme digests were done at 37°C for 18 hours using ratios of enzyme to protein of 1:25 and 1:50, respectively. Separation of peptides was performed on a Hewlett-Packard 1090M HPLC using a Vydac C4 column (4.6 X 250 mm). The gradient was 0-50% acetonitrile in 0.06% triflouroacetic acid in 80 minutes. N-terminal sequencing of peptides was done on an ABI

TECHNIQUES IN PROTEIN CHEMISTRY IV

473A Protein Sequencer. Cleavage of the oligosaccharides from the peptides was done with recombinant N-glycanase from Genzyme (0.25 units glycanase/20 µg peptide) in the presence of 1mM Sodium Azide in water and at 37°C for 72 hours. Sialidase (*Arthrobacter ureafaciens* neuraminidase) from CalBioChem (0.005 units sialidase/50 µg peptide) was used to remove the sialic acid from the oligosaccharides. The sialidase digest was done for 18 hours at 37°C in water. High pH anion exchange chromatography of the oligosaccharides was performed on a Dionex system using a Dionex Carbopac PA1 column (4 X 250 mm) and a Dionex Pulsed Amperometric Detector. The oligosaccharide gradient was from 100 mM NaOH to 300 mM Sodium Acetate/100 mM NaOH in 70 minutes: the asialo gradient was from 150 mM NaOH to 100 mM Sodium Acetate/150 mM NaOH in 65 minutes and 100 mM Sodium Acetate/150 mM NaOH to 350 mM Sodium Acetate/150 mM NaOH in 5 minutes for a total run time of 70 minutes.

Results and Discussion

The peptide maps of rHuEPO obtained by Glu-C and trypsin digestion are seen in Fig. 1 and 2, respectively. The three N-glycosylation sites of rHuEPO have been identified by the presence of glucosamine in the amino acid composition of the peptides containing the putative N-glycosylation sequence Asn-X-Ser/Thr (8). Table I shows the N-terminal sequence of the peptides containing those residues.

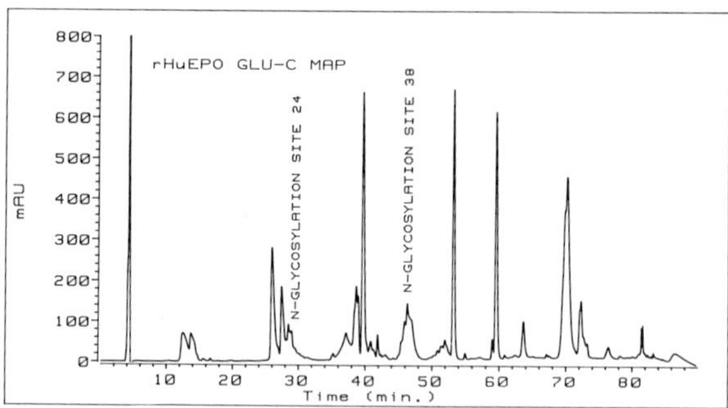

Figure 1. Glu-C peptide map of 1.6 nmoles rHuEPO . Peptides containing N-glycosylation sites 24 and 38 elute at 28 and 46 minutes respectively.

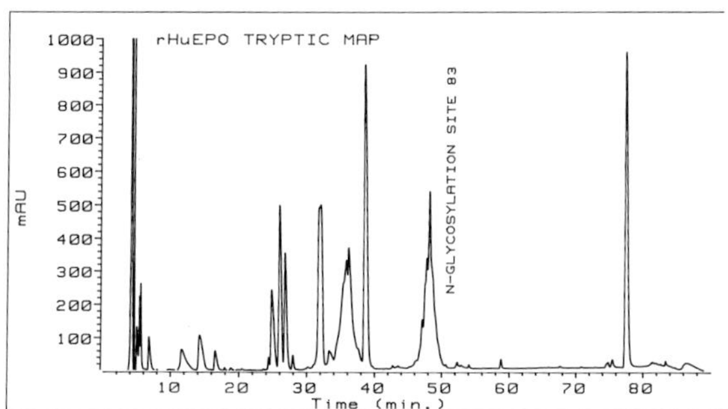

Figure 2. Tryptic peptide map of 1.6 nmoles rHuEPO . The peptide containing N-glycosylation site 83 elutes at 48 minutes.

TABLE I

N-Glycosylation Site	Peptide Sequence
Asn 24	21-A E (*) I T T G (X) S L N E-37
Asn 38	37-(*) I T V P D T K L V N F Y A W K R M E-56
Asn 83	76-G Q A L L V (*) S S Q P W E P L Q L H V D K-98

* Glycosylation Site
X Cysteine Residue

Table I. N-terminal sequence of Glu-C and tryptic peptides

Each N-linked oligosaccharide-containing peptide, as well as a sample of intact rHuEPO, was digested with N-glycanase in order to remove the oligosaccharides. Figure 3a. shows the results of high pH anion exchange chromatography of unfractionated rHuEPO. There are three major groups of peaks: at 20 - 30 minutes, 30 - 40 minutes, and 40 - 50 minutes. These groups correspond to the oligosaccharides containing two, three, and four N-acetylneuraminic acid (NANA) residues respectively (13). The resolution of several components within each group containing the same number of NANA residues is a reflection of the different branched structures such as biantennary, triantennary and tetraantennary forms present on rHuEPO (10, 11, Figure 5). The group of peaks eluting around seventy minutes have been identified as sulfate containing oligosaccharides (14). Figures 3b, 3c, and 3d show the oligosaccharide maps of N-glycosylation sites 24, 38, and 83. Site 24 contains bisialyated, trisialyated, and tetrasialyated structures; sites 38 and 83 have mainly trisialyated and tetrasialyated forms, although site 38 does contain some bisialyated structures.

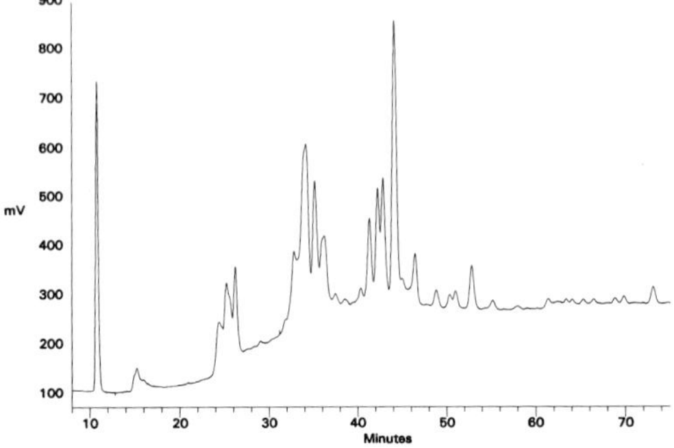

Figure 3a. Oligosaccharide map of rHuEPO (330 pmoles) from high pH anion exchange chromatography.

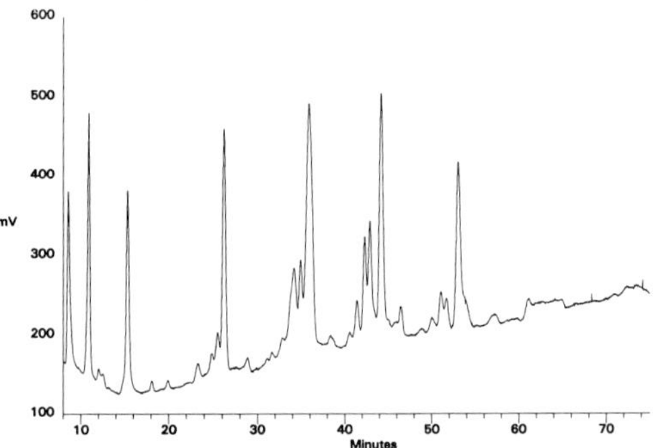

Figure 3b. Oligosaccharide map of N-glycosylation site 24 (165 pmoles).

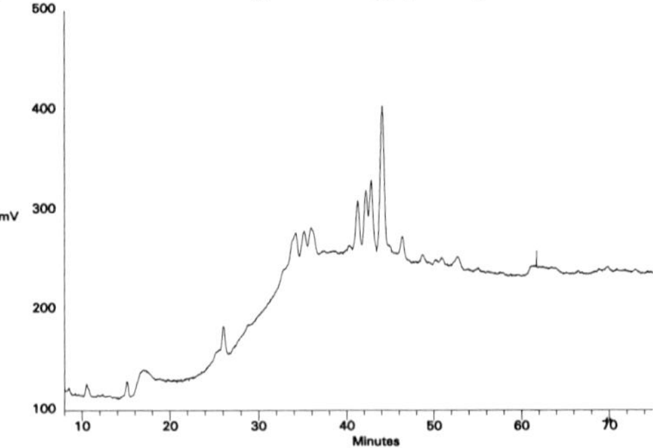

Figure 3c. Oligosaccharide map of N-glycosylation site 38 (110 pmoles).

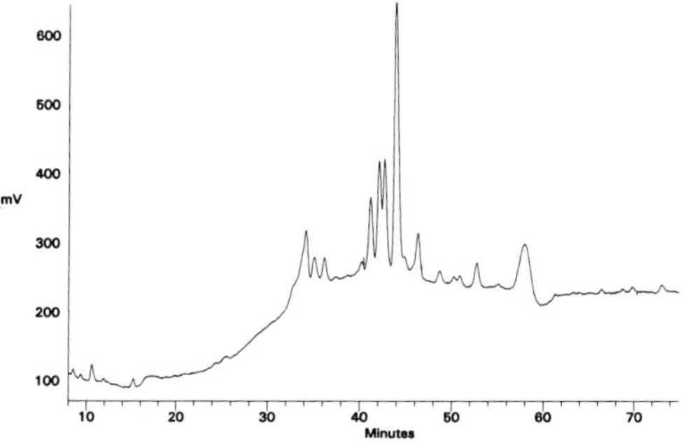

Figure 3d. Oligosaccharide map of N-glycosylation site 83 (165 pM).

To determine the core structures present at each site, NANA was removed with sialidase. The oligosaccharide maps of these asialo structures from rHuEPO and N-sites 24, 38, and 83 are seen in Figures 4a - 4d. All of the chromatograms have at least seven peaks in common.. These peaks, identified by comparison to commercial standards and from structural analysis performed by Oxford Glycosystems, are

1. biantennnary;
2. triantennary (1-3 branch);
3. triantennary (1-6 branch);
4. tetraantennary;
5. tetraantennary with 1 N-acetyllactosamine repeat;
6. tetraantennary with 2 N-acetyllactosamine repeats;
7. tetraantennary with 3 N-acetyllactosamine repeats.

These structures are seen schematically in Figure 5.

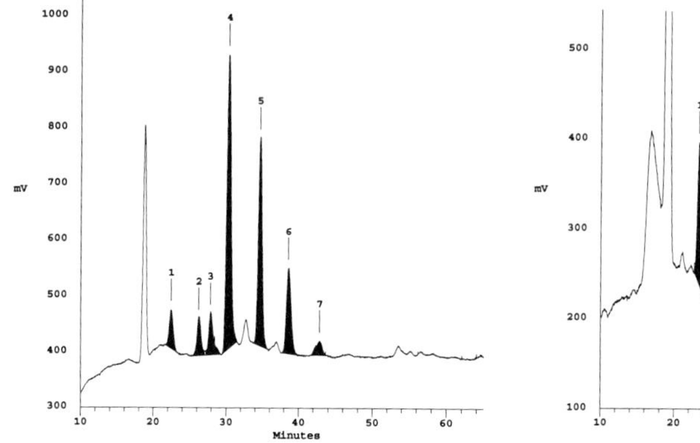

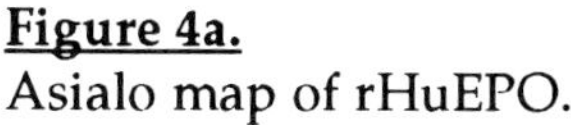

Figure 4a.
Asialo map of rHuEPO.

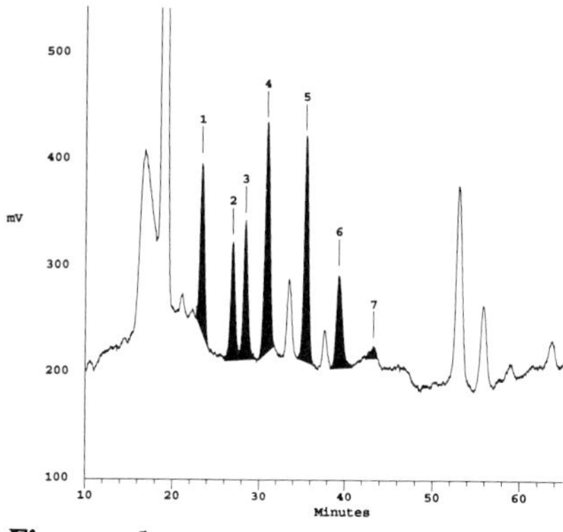

Figure 4b.
Asialo map of N-site 24

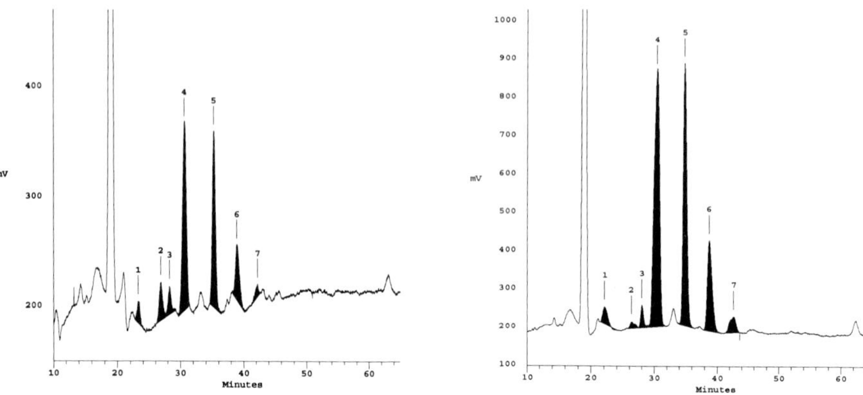

Figure 4c.
Asialo map of N-site 38

Figure 4d.
Asialo map of N-site 83

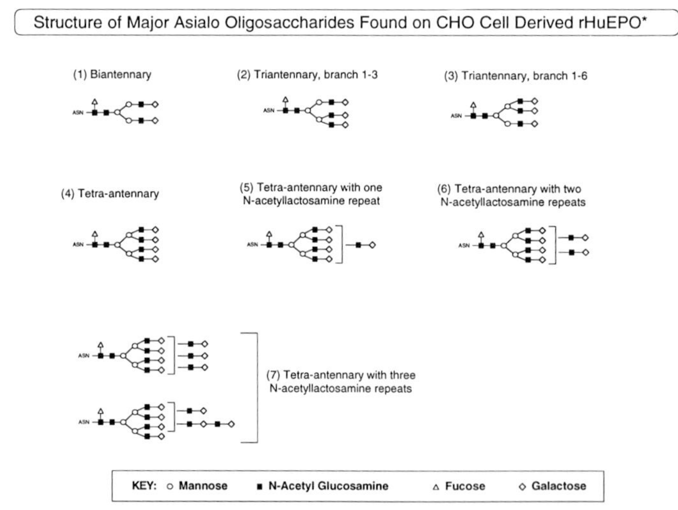

Figure 5. Schematic structures 1 - 7 on rHuEPO asialo maps.

TABLE II

Site Specific N-Glycosylation of rHuEPO

Type of N-linked Sugar Chain	Relative Amount (%) N24	N38	N83	ALL
1. Biantennary	15.61	3.58	2.51	5.04
2. Triantennary (1-3)	11.25	7.40	0.99	5.09
3. Triantennary (1-6)	12.93	4.89	2.27	6.49
4. Tetraantennary	24.83	39.08	46.14	40.89
5. Tetraantennary- 1 lactosamine repeat	22.01	30.50	30.79	26.46
6. Tetraantennary - 2 lactosamine repeats	11.86	12.15	13.96	13.14
7. Tetraantennary- 3 lactosamine repeats	1.43	2.45	3.35	2.89

Relative amounts (%) of asialo structures at rHuEPO N-glycosylation sites 24, 38, and 83 and in unfractionated rHuEPO.

From these results, it is clear that N-site 24 contains the majority of the biantennary structures in rHuEPO; it also contains more of the triantennary forms than either site 38 or 83. N-sites 38 and 83 structures are mainly tetraantennary with zero, one, two, or three N-acetyllactosamine repeats. The presence of less branched forms at the N-site 24 indicates that it undergoes less oligosaccharide processing in the Golgi of CHO cells than sites 38 or 83. This lower level of processing at Asn-24 could be caused by several different mechanisms. One possibility is steric interference with N-acetylglucosaminyl transferases, T-IV or T-V, which are the branch points on the pathway to more highly branched structures. Such steric interference could result from the closeness of the carbohydrate group at Asn-38, the folding of the protein chain in the vicinity of Asn-24, the possible interaction of the oligosaccharide side chain with the peptide sequence in this area, or a combination of these effects. Alternatatively, there may be information contained in the polypeptide backbone which, by interacting with T-IV or T-V, would make highly processed oligosaccharide chains more likely at Asn-38 and 83 than at Asn-24.

The main difficulty in this project was the separation of the N-sites 24 and 38. Currently, variations in time, temperature, and enzyme to substrate ratios are being investigated to determine the best conditions for the Glu-C digest. Another thing to consider is that this method requires sufficient material for all of the oligosaccharide digests and subsequent anion exchange chromatography. At least 500 µg of each peptide is required. Mass Spec and NMR analysis of these glycopeptides would certainly be helpful and both are planned for the future.

References

1. Krantz, S.B and Goldwasser, E. (1984). Proc. Natl. Acad. Sci. U.S.A. **81,** 7574-7578
2. Jacobson L.O., Goldwasser, E., Fried, W., and Plazk, L.F. (1957). Nature **179,** 633-634.
3. Winearls C.G., Oliver, D.O., Pippard M.J., Reid, C., Downing, M.R. and Cotes, P.M. (1986). Lancet **ii,** 1175-1178.
4. Davis, J.M. , Arakawa T.., Strickland T.W., and Yphantis, D.A. (1987). Biochem. **26,** 2633-2638.
5. Tsuda, E., Kawanishi, G., Ueda, M., Masuda, S., and Sasaki, R. (1990). Eur. J. Biochem. **188,** 405-411.
6. Yamaguchi, K., Akai, K., Kawanishi, G., Ueda, M., Masuda, S. and Sasaki,, R.(1991). J. Biol. Chem. **266,** 20434-20439.
7. Higuchi, M. , Oh-eda, M., Kuboniwa, H., Tomonah, K., Shimonaka, Y. and Ochi, N. (1992). J. Biol. Chem. **267,** 7703-7709.
8 Lai, P-H., Everett, R., Wang, F-F., Arakawa, T., and Goldwasser, E.. (1986). J. Biol Chem. **261,** 3116-3121.
9. Imai, N., Kawamura, A., Higuchi, M., Oh-eda, M., Orita, T., Kawaguchi, T., and Ochi N. (1990). J. Biochem. **107,** 352-359.
10. Sasaki, H., Brothner, B., Dell, A., and Fukuda, M.(1987). J. Biol. Chem. **262,** 12059-12076.
11. Takeuchi, M., Takasaki, S., Miyazaki, H., Kato, T., Hoshi, S., Kochibe, N., and Kobata, A. (1988). J. Biol. Chem. **263,** 3657-3663.
12. Sasaki, H., Ochi, N., Dell, A.,and Fukuda, M. (1988). Biochem. **27,** 8618-8626.
13. Rogers, G. Personal Communication.
14. Strickland, T.W. , Adler, B., Aoki, K., Asher, S., Derby, P.L., Goldwasser, E. and Rogers, G. (1992). J. Cell. Biochem. Supp 16D, 324.

SECTION III

Phosphorylated Proteins

Analysis of Serine, Threonine and Tyrosine Phosphorylation Sites with Mass Spectrometry

John W. Crabb[1], Charles Johnson[1], Karen West[1], Janina Buczylko[2], Krzysztof Palczewski[2], Jinzhao Hou[1], Kerstin McKeehan[1], Mikio Kan[1], Wallace L. McKeehan[1], Michael J. Huddleston[3] and Steven A. Carr[3]

[1]W. Alton Jones Cell Science Center, Lake Placid, NY 12946
[2] University of Washington Medical School, Seattle, WA 98195
[3]SmithKline Beecham Pharmaceuticals, King of Prussia, PA 19406

I. Introduction

Over the past several years mass spectrometry has become an indispensible tool for peptide and protein structure analysis, particularly for locating and characterizing posttranslational modifications (1). Integrated with other technologies such as Edman degradation, amino acid analysis and HPLC, mass spectrometry offers one of the most powerful approaches for quickly and accurately identifying phosphorylation sites. This year we have demonstrated the efficacy of this combined approach in two studies, namely the identification of serine and threonine autophosphorylation sites in rhodopsin kinase (2) and tyrosine phosphorylation sites in the heparin-binding fibroblast growth factor receptor *flg* (3). This report describes the methods used to identify a total of seven phosphorylation sites in these two proteins.

II. Materials and Methods

Preparation of Rhodopsin Kinase (RK). An improved purification procedure was used to isolate 100 μg of RK from 75 bovine retinas (2). Soluble contaminating proteins were washed from bleached rod outer segment membranes with a low ionic strength buffer, then RK was extracted from the

TECHNIQUES IN PROTEIN CHEMISTRY IV

membranes with 0.25% Tween 80 and purified to homogeneity by Heparin-Sepharose chromatography. Dephosphorylated RK binds tightly to Heparin-Sepharose and during elution with [γ-^{32}P] ATP the protein undergoes autophosphorylation and its affinity for heparin decreases. ^{32}P-labelled peptides were isolated by RP-HPLC following limited proteolysis of 100 μg (1.5 nmol) RK with endoproteinase Asp-N or complete tryptic digestion of about 2 nmol RK.

Preparation of Heparin-Binding Fibroblast Growth Factor Receptor *flg* (FGF-R1). Recombinant FGF-R1, a tyrosine kinase isoform of the FGF receptor (4), was purified from baculoviral-infected Sf9 insect cells (5). For phosphopeptide analysis, lysates from 1-2 x 10^8 infected insect cells were mixed with 1-2 μg of immunopurified ^{32}P-labeled receptor (2 x 10^6 cpm total), the mixture absorbed onto heparin-agarose beads, the immobilized protein treated with unlabeled ATP and then fragmented with trypsin (5 μg) while still on the beads (3). Tryptic peptides were eluted with water, concentrated and further digested with 25 μg trypsin/37°C/4h. ^{32}P-labeled peptides were purified to homogeneity by affinity chromatography using an antiphosphotyrosine antibody (Upstate Biotechnology, Inc.) followed by RP-HPLC under acidic and neutral conditions.

HPLC, Sequence/Amino Acid Analysis. RP-HPLC was performed with Vydac C8 and C18 columns at acidic pH using aqueous trifluoroacetic acid/acetonitrile solvents and at neutral pH with aqueous hexafluoroacetone/NH_3/acetonitrile solvents (6). Automatic Edman analysis (7) was performed with a gas phase protein sequencer (Applied Biosystems model 470/120/900) using 10-250 pmol of peptide. Phosphoamino acid analysis was performed by thin layer electrophoresis at pH 3.5/700V/60 min following liquid HCl hydrolysis/110°C/60 min (8). Peptide concentrations were determined by PTC amino acid analysis (9) using an Applied Biosystems automated system (models 420H/130/920).

Mass Spectrometry. Electrospray and tandem mass spectra were obtained with a Sciex API-III triple quadrupole mass spectrometer equipped with a fully articulated ion spray probe and an atmospheric pressure ionization source (1,2). HPLC fractions containing 50-100 pmol phosphopeptide each were dried and reconstituted in 20 μl of mobile phase (methanol:water (1:1) containing 0.2% formic acid). About 25% of each sample (5 μl) was injected into the mass spectrometer at a flow rate of 5 μl/min. In normal MS mode resolution was adjusted to about 20% valley between adjacent isotope peaks in a singly charged cluster, allowing singly and doubly charged ions to be identified by apparent spacing between peaks and doubly charged ions to be distinguished from those with higher charge states. In the triple quadrupole system, argon was used as the collision gas with a calculated collision energy of about 50eV. Parent ion transmission was maximized by reducing the resolution of Q1 to transmit about a 2-3 *m/z* window around the selected ion.

III. Results and Discussion

Ser and Thr Phosphorylation sites in RK

RK is a G protein-coupled receptor kinase that phosphorylates photo-activated rhodopsin (Rho*) on Ser and Thr residues in the deactivation pathway of phototransduction (10). The 66 kDa enzyme also undergoes autophosphorylation at multiple sites with a concurrent decrease in affinity for phosphorylated Rho* (11). Major and minor autophosphorylation sites have been identified in RK using a combination of methods (2).

From SDS-PAGE, sequence and audoradiographic analysis of PVDF electroblots, limited proteolysis of RK with endoproteinase Asp-N was shown to remove N- and C-terminal peptides with no ^{32}P remaining associated with 50-55 kDa undigested fragments. One major and two minor radioactive fractions were purified by RP-HPLC. The major radioactive aspartyl peptide yielded the sequence $D^{483}VGAFX^{488}X^{489}VKGVAFEKA$ by Edman analysis of about 39 pmol [numbering based upon the cDNA deduced RK sequence (12)]. Direct analysis of this peptide by electrospray mass spectrometry was not successful because of residual Tween 80 contamination that suppressed ionization. For productive ESMS analysis the peptide was subsequently cleaved with endoproteinase Lys-C and repurified by RP-HPLC (removing the C-terminal Ala and lowering the Tween contamination). Two peptide signals (Fig 1A) were observed in the mass spectra. The $(M+2H)^{2+}$ signal at *m/z* 858.2 suggests a bisphosphorylated peptide (M_r determined = 1714.4; M_r calculated = 1713.8). No signal at *m/z* 818 indicates the

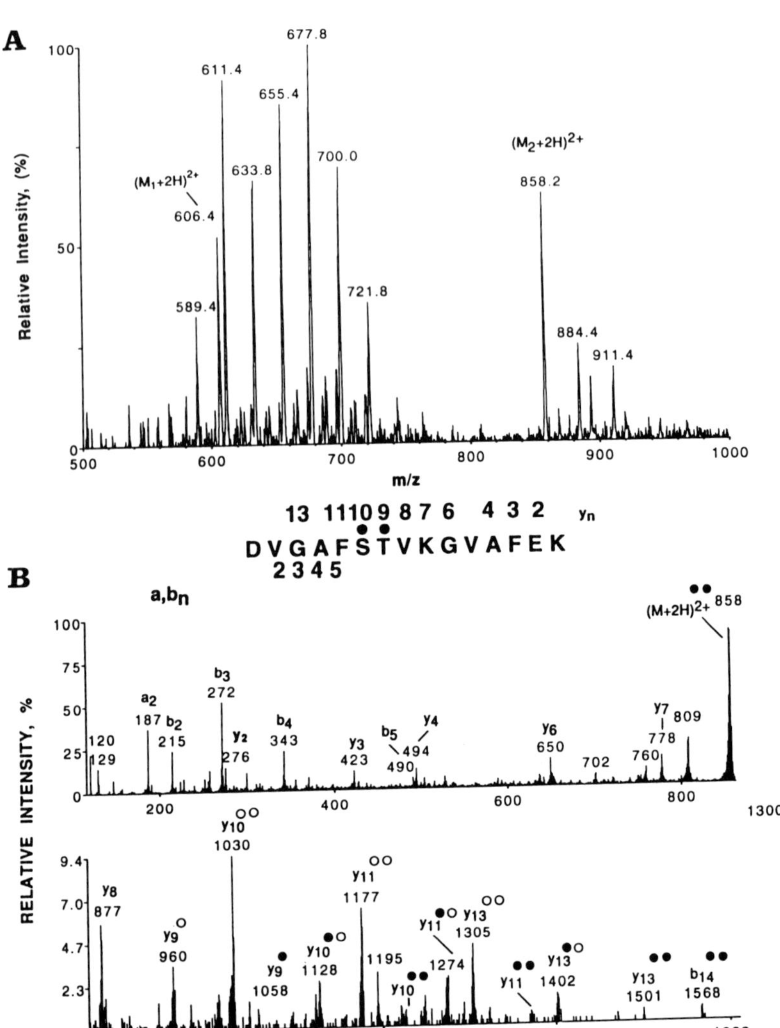

Figure 1. Mass spectrum of the major autophosphorylation site in rhodopsin kinase. A. Electrospray mass spectometry of ~20 pmol of purified peptide. B. Electrospray tandem mass spectrometry of the $(M+2H)^{2+}$ (m/z 858). See reference 1 for a discussion of fragment ion nomenclature.

absence of monophosphorylated peptide (M_r, 1633.8). The signal at m/z 606.4 corresponds to the $(M+2H)^{2+}$ for the RK peptide DGLRTHPLFR which was identified by Edman analysis as a component of this HPLC fraction. The series of signals separated by 22 m/z units from about m/z 589.4 to 721.8 presumably are $(M+2H)^{2+}$ ions for Tween 80 with varying $-C_2H_4O-$ units (44 Da). Tandem mass spectrometry of the $(M+2H)^{2+}$ parent ion provided the partial sequence -G-A-F-S^{488}-T^{489}-V-K-X-X-F-A-. The mass increments between y_8, y_9 and y_{10} along with the satellite signals from loss of either H_3PO_4 (98 Da, indicated by open circles) or HPO_3 (80 Da, indicated by closed circles) demonstrate that the phosphate groups are located at Ser^{488} and Thr^{489}. For a discussion of fragment ion nomenclature, see reference 1.

Edman analysis of minor radioactive HPLC fractions from limited aspartyl proteolysis revealed a peptide with a sequence of $I^{16}AARGX^{21}F$. ^{32}P analysis after each Edman cycle, low PTH amino acid recovery at cycle X^{21} and the deduced RK sequence indicates that Ser^{21} is a potential phosphorylation site. Quantitation of ^{32}P and results from Edman analysis of purified ^{32}P-labeled tryptic peptides supported those obtained with limited proteolysis, namely Ser^{488} and Thr^{489} are major autophosphorylation sites and Ser^{21} is a minor site.

To eliminate the possibility that the minor site at Ser^{21} had been phosphorylated with cold label prior to kinase purification, RK was treated with protein phosphatase 2A before purification; subsequent RP-HPLC profiles of limited proteolysis ^{32}P-peptides were unchanged. To eliminate the possibility that the lower level phosphorylation of Ser^{21} was due to rapid dephosphorylation due to contaminating phosphatases, thio-autophosphorylated RK was prepared, subjected to limited proteolysis and RP-HPLC and found to yield ^{35}S-peptide maps essentially identical to the ^{32}P-peptide maps (2).

Tyr Phosphorylation Sites in FGF-R1

The FGF receptor interacts with a family of at least seven fibroblast growth factors that control cell growth and differentiation. Due to alternate splicing in four distinct receptor genes, multiple isoforms of the FGF receptor exist that differ in the extracellular ligand-binding domain, the intracellular juxtamembrane domain and the intracellular tyrosine kinase domain (4). Tyrosine phosphorylation sites in the recombinant glycosylated tyrosine kinase isoform FGF-R1 have been determined using a combination of methodologies (3). For phosphopeptide analysis, baculoviral-infected insect cell lysates were mixed with

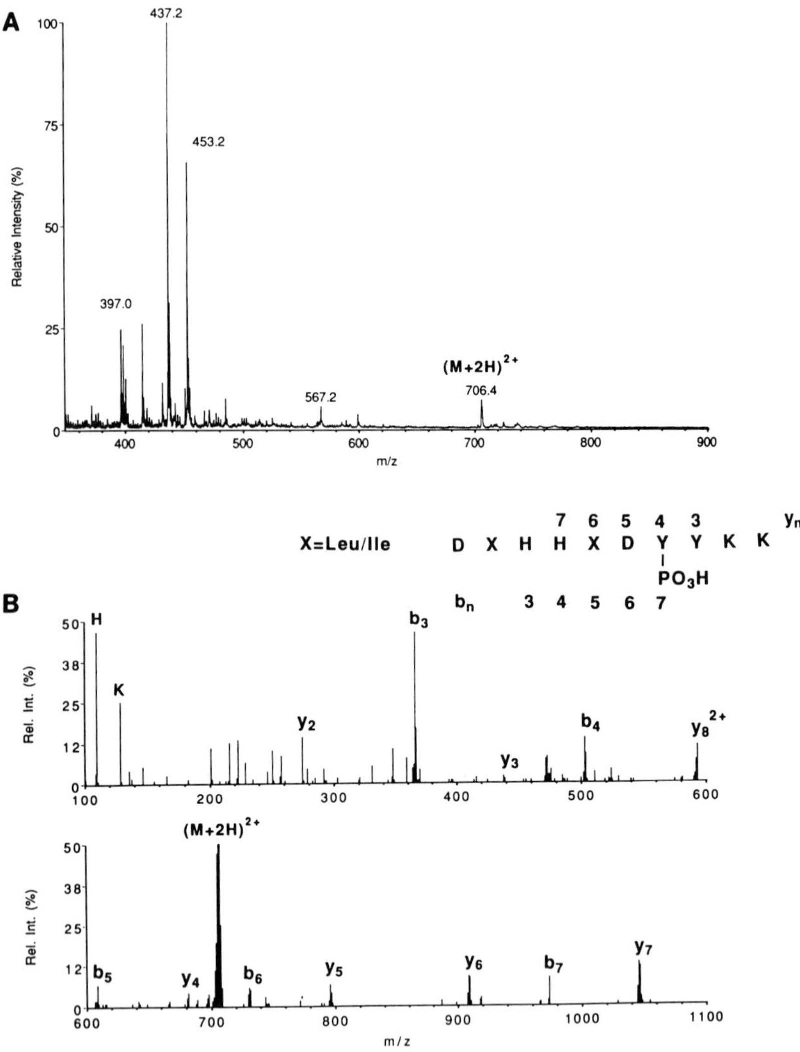

Figure 2. Mass spectrum of a major phosphorylation site in the tyrosine kinase isoform of the fibroblast growth factor receptor *flg*. A. Electrospray mass spectrometry of ~20 pmol of peptide. B. Electrospray tandem mass spectrometry of the $(M+2H)^{2+}$ (*m/z* 706.4). See reference 1 for a discussion of fragment ion nomenclature.

immunopurified ^{32}P-labeled FGF-R1 (5) and absorbed onto heparin-agarose beads The mixture was incubated with cold ATP then digested with trypsin while immobilized on the beads. Over 90% of the immobilized ^{32}P radioactivity was recovered as tryptic phosphopeptides which were first purified on a column of antiphosphotyrosine antibody then fractionated by acidic RP-HPLC. Six radioactive RP-HPLC peaks were obtained, each contained phosphotyrosine based upon phosphoamino acid analysis and all were purified further by additional RP-HPLC at either acidic or neutral pH. Edman analysis revealed one unidentifiable peptide and five peptides from tyrosine containing regions of the FGF-R1 sequence (4). The two most abundant phosphopeptides (P1 and P2) had the same N-terminal sequence, are located within the intracellular kinase catalytic consensus domain of FGF-R1 and also contain adjacent tyrosine residues Y^{653} and Y^{654}. Electrospray mass spectrometry of P1 revealed an $(M+2H)^{2+}$ signal at *m/z* 706.4 (Fig 2A) suggesting a monophosphorylated peptide (M_r determined = 1411.8; M_r calculated = 1411.6). Absence of a signal at *m/z* 746.8 (expected for a bisphosphorylated peptide) indicated that there is only one phosphate in the peptide. Tandem mass spectrometry (Fig 2B) of the $(M+2H)^{2+}$ parent ion established the location of the phosphotyrosine and provided the partial sequence -H-H-L/I-D-Y^{653}-Y- (L/I = Leu or Ile, indistinguishable by this method). The mass increments from b_6 to b_7 and from y_3 to y_4 (Fig 2B) demonstrate that the phosphate group is located at position 653; the absence of a fragment ion 80 Da greater than the observed y_3 corroborates that only Y^{653} is phosphorylated.

Edman analysis indicated phosphopeptides P3 and P4 were from the extracellular domain of FGF-R1 and gaps in the Edman analysis at Y^{154} and Y^{307} implicated these residues as potential phosphorylation sites. Further analysis of P4 (~20 pmol) by electrospray tandem mass spectrometry demonstrated that Y^{307} was phosphorylated (3). Edman analysis of P5, the least abundant of the phosphopeptides in this study, indicated that it is from FGF-R1 residues 757-784. Others (13) have identified Y^{766} as a phosphorylation site and contend it is the single, major site in FGF-R1; the absence of Y^{766} in our Edman analysis support it as a phosphorylation site. Other tyrosine residues in the deduced sequence are in a favorable context for phosphorylation and the significance of the N-terminal extracellular phosphorylation sites remains unclear. Further structure/function studies are required to determine the role of phosphorylation in activities of the FGF receptor.

IV. Conclusions

The peptide mass analyses in the present studies and partial sequence information derived from electrospray and tandem mass spectrometry were essential for locating and discriminating between mono- and bisphosphorylation sites. Notably, the FGF receptor analysis represents one of the first applications of mass spectrometry in the identification of tyrosine phosphorylation sites. These studies evolved in a timely fashion through synergistic use of a variey of methods for sample preparation and characterization, together with knowledge of gene deduced sequences.

Acknowledgements

Supported in part by USPHS grants EY08061, EY06603, DK35310, DK40739, DK38639, and CA37589.

References

1. Carr, S.A., Hemling, M.E., Bean, M.F. and Roberts, G.D. (1991) Anal. Chem. **63**, 2802-2842.
2. Palczewski, K., Buczylko, J., Van Hooser, P., Carr, S.A., Huddleston, M.J. and J.W. Crabb (1992) J. Biol. Chem. **267**, 18991-18998.
3. Hou, J., McKeehan, K., Kan, M., Carr, S.A., Huddleston, M.J., Crabb, J.W. and McKeehan, W.L. (1993) Protein Science **2**, 86-92.
4. Hou, J., Kan, M., McKeehan, K., McBride, G., Adams, P. and McKeehan, W.L. (1991) Science **251**, 665-668.
5. Xu, G., Nakahara, M., Crabb, J.W., Shi, E., Matuo, Y., Fraser, M., Kan, M., Hou, J. and McKeehan, W.L. (1992) J. Biol. Chem. **267**, 17792-17803.
6. Tarr, G.E. and Crabb, J.W. (1983) Anal. Biochem. **131**, 99-107.
7. Crabb, J.W., Johnson, C.M., Carr, S.A., Armes, L.G. and Saari, J.C. (1988) J. Biol. Chem. **263**, 18678-18687.
8. Cooper, J., Sefton, B. and Hunter, T. (1985) Methods Enzymol. **99**, 387-402.
9. West, K.A. and Crabb, J.W. (1992) *In* Techniques in Protein Chemistry III (R.H. Angeletti, ed), pp 233-242, Academic Press.
10. Palczewski, K. and Benovic, J.L. (1991) Trends Biol. Sci. **16**, 387-391.
11. Buczylko, J., Gutman, C. and Palczewski, K. (1991) Proc. Natl. Acad. Sci. U.S.A. **88**, 2568-2572.
12. Lorenz, W., Inglese, J., Palczewski, K., Onorato, J.J., Caron, M.G. and Lefkowitz, R.J. (1991) Proc. Natl. Acad. Sci. U.S.A. **88**, 8715-8719.
13. Mohammadi, M., Honegger, A.M., Rotin, D., Fisher,R., Bellot, F., Li, W., Dionne, C.A., Jaye, M., Rubinstein, M., and Schessinger, J. (1991) Mol. Cell Biol. **11**, 5068-5078.

The Identification of Phosphorylation Sites in Large Membrane Proteins Following Their Isolation by SDS-PAGE

Karen S. De Jongh, Eric I. Rotman, and Brian J. Murphy
Department of Pharmacology, University of Washington,
Seattle, WA 98195

I. Introduction

Protein phosphorylation plays an important role in the regulation of a wide range of metabolic processes (1). In order to investigate the functional effects of phosphorylation, it is desirable to identify the sites at which a protein is phosphorylated. One approach to identifying sites of phosphorylation has involved enzymatic or chemical digestion of the purified protein that has been phosphorylated in the presence of [γ-^{32}P]ATP, followed by isolation of labeled phosphopeptides using reverse phase HPLC and amino acid sequence analysis. While this approach has proven useful for a wide range of soluble proteins, a number of constraints make it technically difficult for many proteins, and alternative micropreparative procedures for peptide sequencing (reviewed in 2) must be employed. Since unambiguous internal sequence information can only be obtained from extremely pure samples, the complexity of the HPLC maps subsequent to digestion of large proteins can preclude obtaining useful information. In these cases it is desirable to simplify the HPLC traces by separating the phosphopeptides from non-phosphorylated peptides prior to HPLC mapping. In addition, since the cellular targets of phosphorylation-dependent events are often membrane proteins which are present in cells at low levels, the isolation of the purified protein in a form suitable for digestion has proven quite challenging. Generally these proteins are soluble only in the presence of detergents, which interfere with both digestion protocols and reverse phase HPLC peptide mapping.

Here we describe the protocol we used to determine the sites at which the 260 kDa rat brain sodium channel α subunit (properties reviewed in 3) and the 190 kDa rabbit skeletal muscle L-type calcium channel α1 subunit (properties reviewed in 4) are phosphorylated by cAMP-dependent protein kinase. Since a major mechanism of regulation of the ion flux through voltage-dependent sodium and calcium channels is phosphorylation (5), we were interested in determining the sites at which these proteins are phosphorylated in order to study the functional significance of this modification. Both are large, hydrophobic membrane proteins which are difficult to purify in large quantities, and thus we developed a scheme to enable the isolation of phosphopeptides derived from these proteins in reasonable yields. We employed SDS-gel electrophoresis as the final step in our purification scheme and carried out tryptic digestion directly from gel slices in order to avoid the solubility and detergent problems mentioned above. Muskynska et al. (6) have previously shown that ferric ions adsorbed to iminodiacetic acid-substituted agarose act as an affinity adsorbent for phosphoproteins, and thus ferric chelate chromatography was employed to isolate the resulting phosphopeptides prior to HPLC mapping and amino acid sequence analysis.

II. Materials

Sequencing grade porcine trypsin that had been modified by reductive methylation to reduce autolysis was obtained from Promega. Iminodiacetic acid-Sepharose was from Sigma and sterile Titer tubes used for collecting fractions prior to amino acid sequencing were from Bio-Rad.

III. Methods

A. *Phosphorylation of sodium and calcium channels*

Rat brain sodium channels were partially purified through the step of wheat germ agglutinin chromatography (≈ 30% pure) as outlined previously (7). A preparation containing 1200 pmol sodium channels (assessed by ^{3}H-saxitoxin binding) was phosphorylated by the catalytic subunit of cAMP-dependent protein kinase in the presence of 0.5 mM [γ-^{32}P]ATP (0.6 Ci/mmol) for 30 minutes at 37°C as outlined previously (8). The reaction was stopped by the addition of SDS-PAGE sample buffer and incubation at 55°C for 15 minutes.

Rabbit skeletal muscle L-type calcium channels were purified as outlined previously (9) and a preparation containing 500 pmol was phosphorylated as outlined above for the sodium channel.

B. *Generation of sodium and calcium channel tryptic peptides*

Electrophoresis of phosphorylated protein samples was carried out on 1.5 mm, 6% polyacrylamide gels (10) in a Bio-Rad mini protean apparatus. Gels were poured without sample wells to allow loading of up to 2 ml sample on each. Following electrophoresis at 200 volts the wet gels were exposed to X-ray film for localization of the ^{32}P-labeled 260 kDa sodium channel α subunit or the 190 kDa calcium channel α1 subunit . Excised gel slices containing these subunits were washed for 3 hours and then 16 hours in 10% acetic acid/10% 2-propanol followed by two 1.5 hour washes in 50% methanol (50 ml each wash). Slices were cut into 4 x 4 mm pieces, the slices from each gel were placed in a 5 ml tube and dried in a vacuum concentrator. 2.0 ml of 50 mM ammonium bicarbonate, pH 8.3 containing 2 µg sequencing grade modified porcine trypsin was added to each tube and incubated at 37°C for 16 hours. The supernatants were removed and the gel slices were washed by rotating with 1 ml 50 mM ammonium bicarbonate for 40 minutes at room temperature. Supernatants were removed and the wash repeated twice more. All supernatants from each sample were combined and concentrated to 50 µl. The amount of salt was reduced by adding 500 µl water and concentrating to 50 µl, after which 450 µl 0.1 M acetic acid, pH 3.0 was added.

C. Ferric chelate chromatography

Ferric chelate chromatography was based on the method described by Muszynska et al.(6). For each sample, 700 µl iminodiacetic acid-Sepharose was packed in a 1 cm diameter column and washed with 5 ml of each of the following solutions: water, 50 mM ferric chloride, water, 0.1 M acetic acid, pH 3.0. The tryptic peptides were bound to the ferric-substituted Sepharose by rotation for 1 hour at room temperature, and unbound material was collected. The columns were washed with the following solutions : 0.1 M acetic acid, pH 3.0; 0.1 M acetic acid, pH 5.0; 0.1 M ammonium acetate, pH 7.4, 0.1 M ammonium acetate, pH 8.9; 0.1 M ammonium acetate, pH 10.0; and 0.1 M sodium phosphate, pH 8.0. 1.0 ml fractions were collected and the radioactivity in each measured by Cerenkov counting. Fractions containing the highest radioactivity were combined and concentrated to 50 µl. The amount of salt was reduced by adding 500 µl water and re-concentrating to 50 µl.

D. Reverse phase HPLC

1% TFA was added to the isolated phosphopeptides which were separated on Aquapore BU-400 C4 or Spheri-5 RP-18 C18 columns (2.1 x 30 mm) with a flow rate of 300 µl/min. Gradient elution from 0.1 % trifluoroacetic acid to 48% acetonitrile/0.085% trifluoroacetic acid over 60 min was begun when the absorbance after injection of the sample had returned to baseline. Fractions were collected based on absorbance at 206 nm into sterile Titer tubes. Fractions containing radioactivity that were unbound to the C4 column were concentrated and rechromatographed on a C18 column under the same conditions. Amino acid sequences of fractions containing ^{32}P were determined with an Applied Biosystems model 477A protein sequencer with an on-line model 120A phenylthiohydantoin analyzer for amino acid identification. A sequence assignment of S' indicates an increase of PTH-DTT-dehydroalanine detected in a particular position compared to the previous step of Edman degradation. Amino acid recovery at each cycle were obtained with the Applied Biosystems 900A Data Analysis software.

E. Two-dimensional phosphopeptide mapping

Tryptic phosphopeptides were lyophilized, resuspended in 1% pyridine / 10 % acetic acid (pH 3.5) containing a trace of phenol red and subjected to electrophoresis for 1 hour at 400 volts, followed by ascending chromatography in 1-butanol/pyridine/acetic acid /water (15:10:3:12, v/v) on thin layer cellulose plates. The plates were dried and subjected to autoradiography.

Figure 1: The rat brain sodium channel α subunit. (A) The amino acid sequence of the rat brain type 2 sodium channel α subunit from residues 554 to 687 is shown with cAMP-dependent phosphorylation consensus sequences indicated by *. Tryptic peptides that were sequenced (table 1) are underlined. (B) 5 pmol partially purified sodium channels were phosphorylated by cAMP-dependent protein kinase and [^{32}P]ATP as outlined under Methods and electrophoresed on a 6% polyacrylamide gel which was dried and exposed to X-ray film. The arrow indicates the sodium channel α subunit.

IV. Results

A. *Sodium channel phosphopeptides*

The primary structure of the rat brain type II sodium channel α subunit predicts six potential cAMP-dependent phosphorylation sites (RRXS/T, KRXXS/T, Ref. 11) between residues 554 and 687, and our previous results have localized the *in vitro* and *in vivo* cAMP-dependent phosphorylation sites to this region (12) whose amino acid sequence is shown in figure 1A. Figure 1B shows the *in vitro* cAMP-dependent phosphorylation of the partially purified sodium channel, revealing only one phosphorylated protein corresponding to the intact α subunit. Tryptic digestion of this protein from dried polyacrylamide gel slices and extraction of the tryptic peptides yielded 73% of the radioactivity present in the gel slices containing the α subunit.

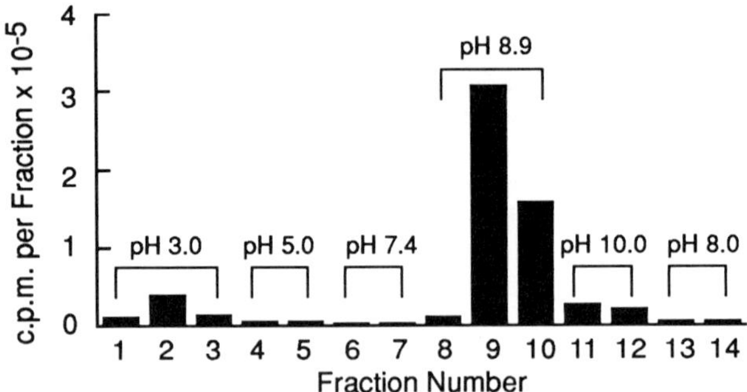

Figure 2: Ferric chelate chromatography of sodium channel tryptic phosphopeptides. Sodium channel α subunit tryptic phosphopeptides were eluted from the ferric chelate column as outlined under Methods.

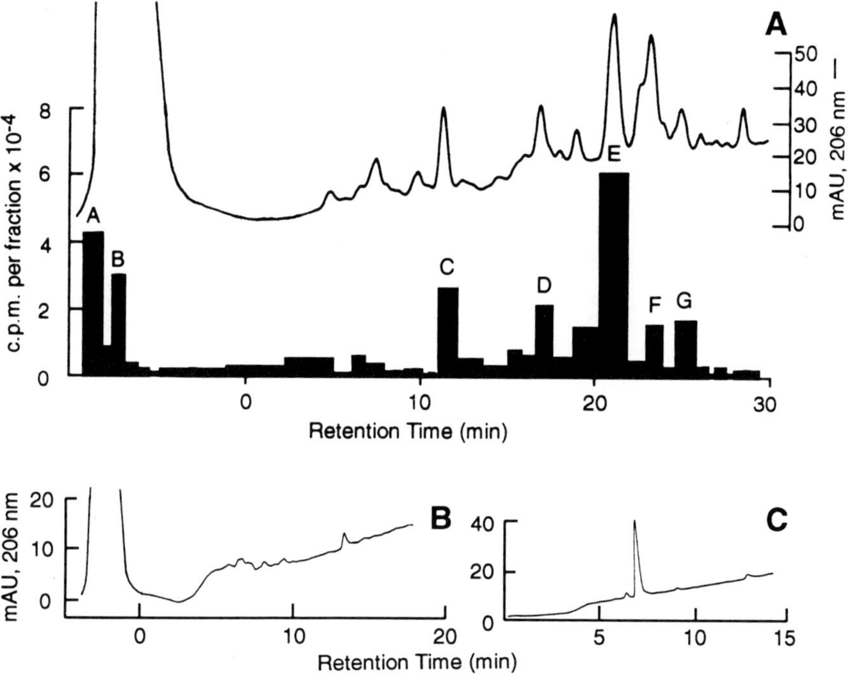

Figure 3: Reverse phase HPLC of sodium channel α subunit tryptic phosphopeptides. (A) Fractions 8-10 eluted from the ferric chelate column (figure 2) were combined, concentrated and chromatographed on a C4 column as outlined under Methods. Fractions were collected based on their absorbance at 206 nm (—) and the radioactivity in each determined by Cerenkov counting which is indicated by the histogram. 7 peaks of radioactivity were recovered (A-G). (B) Peak A eluted from the C4 column was rechromatographed on a C18 column as outlined under Methods. Fractions were collected based on their absorbance at 206 nm (—) and the radioactivity in each determined by Cerenkov counting which is indicated by the hatched area. (C) Peak B eluted from the C4 column was rechromatographed on a C18 column as outlined under Methods. Fractions were collected based on their absorbance at 206 nm (—) and the radioactivity in each determined by Cerenkov counting which is indicated by the hatched area.

In order to isolate the phosphopeptides in sufficient amount and purity to carry out sequence analysis, ferric chelate chromatography was employed on the tryptic digest. The phosphopeptides were tightly bound to the ferric chelate column, as evidenced by 1.5 % of the applied radioactivity remaining in the unbound fraction. Upon batch elution 69 % of the bound radioactivity was eluted at pH 8.9 (figure 2). Chromatography of this radioactive fraction on reverse phase HPLC yielded 7 radioactive peaks (figure 3A, peaks A-G), including two that eluted before commencement of the acetonitrile gradient. These unbound peaks were rechromatographed on a C18 column (figures 3B and 3C). The total radioactivity recovered in the seven peaks eluted from the C4 column was 41% of that applied. Four of the seven radioactive peaks yielded clean amino acid sequence when subjected to N-terminal sequence analysis, which corresponded to amino acid sequences in the sodium channel α subunit (table 1). The presence of the PTH-DTT-dehydroalanine adduct, concomitant with the absence of PTH-serine, at one

Table I. Amino acid sequence analysis of sodium channel α subunit tryptic phosphopeptides

HPLC Peak	Amino Acid Sequence	Residue Numbers	pMol in cycle 1	pMol in cycle 2
A	RNS'R	571-574	2.9	7.5
B	RPS'NVSQASR	621-630	5.4	36.2
C	RDS'LFVPHR	608-616	2.3	6.2
E	RSS'SYHVSM	685-693	4.4	12.3

NH_2-terminal sequence analysis of sodium channel α subunit tryptic phosphopeptides that were purified by reverse phase chromatography (figure 3) was carried out. Peaks yielding unambiguous amino acid sequence are shown together with their corresponding positions in the sodium channel α subunit (figure 1A). S' indicates PTH-DTT-dehyroalanine. No serine was seen at these cycles. The pmol amino acid observed in the first two cycles of sequence analysis for each peptide are indicated.

cycle of Edman degradation in each sequence suggested a phosphorylated serine residue at that position. In all cases that amino acid position in the sodium channel α subunit correspondent to a serine residue in a cAMP-dependent phosphorylation consensus sequence. The recovery of radioactivity at each stage of the procedure is outlined in table 2.

B. Calcium channel phosphopeptides

Two-Dimensional tryptic phosphopeptide mapping of the 190-kDa calcium channel α1 subunit which had been phosphorylated by cAMP-dependent protein kinase revealed two predominant phosphopeptides and a third of intermediate intensity (figure 4, T1-T3). Following isolation of the α1 subunit by SDS-PAGE, tryptic digestion and purification of the tryptic phosphopeptides by ferric chelate chromatography, reverse phase HPLC resulted in two peaks of radioactivity eluting from a C18 column. These two fractions contained 29% of the radioactivity that was present in the dried polyacrylamide gel slices. The amount of radioactivity recovered at each step of the isolation is indicated in table 2. Two-dimensional phosphopeptide mapping of the first peak eluting in the column flow-through revealed it co-migrated with one of the major phosphopeptides (figure 4, spot T2). Its NH_2-terminal sequence was KMS'R, corresponding to residues 685-688 of the calcium channel α1 sequence (not shown). The second phosphopeptide, which also eluted in 0.1% trifluoroacetic

Table II. Recovery of radioactivity at each step in the purification

Purification Step	Sodium Channel	Calcium Channel
Intact protein in dried polyacrylamide gel slices	100	100
Tryptic peptides eluted from gel slices	73	80
Ferric chelate column eluate	46	53
HPLC column eluates	19	29

The radioactivity recovered at each stage in the purification of sodium channel α subunit and calcium channel 190-kDa α1 subunit phosphopeptides was determined by Cerenkov counting and is expressed as a percentage of the radioactivity in washed, dried gel slices containing the intact proteins.

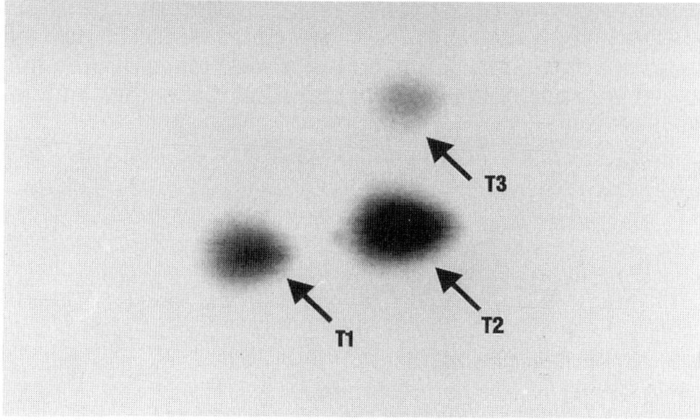

Figure 4: Two-dimensional tryptic phosphopeptide map of the 190-kDa calcium channel α1 subunit. Calcium channel α1 subunit which had been phosphorylated by cAMP-dependent protein kinase and [^{32}P]ATP and digested with trypsin was applied to thin layer cellulose plates which were run as outlined under Methods. The plates were dried and autoradiography revealed three tryptic phosphopeptides (T1-T3).

acid but was slightly retarded by the column, co-migrated with the other major phosphopeptide (figure 4, spot T1). Its NH_2-terminal sequence was XRKMS'R, corresponding to residues 684-688 of the calcium channel α1 sequence (not shown). The yield of lysine at cycle 1 of the sequence analysis of phosphopeptide T2 and cycle 3 of the amino acid sequence analysis of phosphopeptide T1 were 13.3 pmol and 2.8 pmol respectively. Thus the two major phosphopeptides on the two-dimensional phosphopeptide map both represent tryptic peptides containing serine 687 which is in a cAMP-dependent protein kinase phosphorylation consensus sequence.

V. Discussion

The approach outlined has allowed the identification of phosphorylation sites in large membrane proteins which are present in tissues at relatively low levels. The use of SDS-PAGE as the first step in the procedure allowed the use of sample preparations that were only partially pure. Tryptic digestion directly from gel slices avoided the problems associated with the presence of detergents when generating fragments from membrane proteins. In this study we inferred the presence of phosphorylated residues based on their location in phosphorylation consensus sequences and the relative amounts of PTH-serine and PTH-DTT-dehyroalanine seen during Edman degradation. Actual identification of phosphorylated residues may be confirmed by analysis of [^{32}P]phosphate release during Edman degradation (13) or mass spectrometry (14).

Starting from 500 pmol rabbit skeletal muscle calcium channel we isolated two tryptic phosphopeptides which contain the same phosphorylated residue and constitute the major site of *in vitro* cAMP-dependent phosphorylation on the 190 kDa form of the α1 subunit of this channel. Our results also delineated four major *in vitro* cAMP-dependent phosphorylation sites in the rat brain sodium channel α subunit. We were able to correlate particular phosphorylated peptides with spots on two-dimensional tryptic phosphopeptide maps. This will make the use of two-dimensional phosphopeptide mapping feasible for identifying physiologically relevant phosphorylation sites *in vivo*. The method should be readily applicable to the identification of phosphorylation sites in a wide range of proteins that are present in low abundance or that present technical difficulties with regard to size or solubility.

References

1. Taylor, S.S. (1989) *J. Biol. Chem.* **264**, 8443-8446.
2. Simpson, R.J., Moritz, R.L., Begg. G. S., Rubira, M.R. and Nice, E.C. (1989) *Anal. Biochem.* **177**, 221-236.
3. Catterall, W.A. (1988) *Science* **242**, 50-61
4. Catterall, W.A. (1991) *Science* **253**, 1499-1500.
5. Rossie, S. and Catterall, W.A. (1987) *The Enzymes* **18**, 335-357.
6. Muszymska, G., Andersson, L. and Porath, J. (1986) *Biochemistry* **25**, 6850-6853.
7. Hartshorne, R.P. and Catterall, W.A. (1984) *J. Biol. Chem.* **259**, 1667-1675.
8. Rossie, S., Gordon, D. and Catterall, W.A. (1987) *J. Biol. Chem.* **262**, 17530-17535.
9. De Jongh, K.S., Warner, C., Colvin, A.A. and Catterall, W.A. (1991) *Proc. Natl. Acad. Sci.* **88**, 10778-10782.
10. Laemmli, U.K. (1970) *Nature* **227**, 680-685.
11. Krebs, E.G. and Beavo, J.A. (1979) *Annu. Rev. Biochem.* **48**, 923-959.
12. Rossie, S. and Catterall, W.A. (1989) *J. Biol. Chem.* **264**, 14220-14224.
13. Wang, Y., Fiol, C.J., DePaoli-Roach, A.A., Bell, A.W., Hermodson, M.A. and Roach, P.J. (1988) Anal. Biochem. **174**, 537-547.
14. Moseley, M.A., Jorgenson, J.W. and Tomer, K.B. (1991) Techniques in Protein Chemistry II, 441-454.

Acknowledgements

We thank Carl Baker for providing purified sodium and calcium channels, Anita Colvin of the University of Washington Molecular Pharmacology Facility for protein sequencing and Dr. James Stull (Department of Physiology, University of Texas Southwestern Medical Center, Dallas) for introducing us to iminodiacetic acid-substituted agarose. This work was supported by Grants P01 HL44948-01A1, NS15751 and NS22625 from the National Institutes of Health and postdoctoral research fellowships from the National Institutes of Health (to E. I. R.) and the Medical Research Council of Canada (to B. J. M.).

Modification of Thiophosphorylated Proteins with Extrinsic Probes[1]

Kevin C. Facemyer
Mark R. Tibeau
Christine R. Cremo

Department of Biochemistry and Biophysics
Washington State University
Pullman, Washington

I. Introduction

The introduction of extrinsic probes onto proteins in a specific manner is a difficult process. Often the protein chemist relies upon the reactive sulfhydryls of cysteines which are usually the strongest nucleophiles in a protein. We have devised a new highly specific protein labeling technique which does not rely upon cysteine chemistry. The technique is useful to label the wide variety of proteins that can be specifically thiophosphorylated by protein kinases and ATPγS[2]. These thiophosphorylated proteins contain a phosphorothioate group with a nucleophilic sulfur (Eckstein, 1985) which can be reacted with haloacetates. This approach to labeling of proteins promises to be generally useful because of the ubiquity of protein phosphorylation, the site specificity afforded by the kinases, and the stability of the linkage between the extrinsic probe and the protein.

We have previously used the thiophosphate labeling method to modify Ser-19 on the RLC20 of smooth muscle myosin with a fluorescent probe (Facemyer & Cremo, 1992). We have demonstrated (1) the labeling is highly specific for the thiophosphate group, (2) the label is stably attached to the protein between pH 3 and 10 and under common reducing conditions for SDS gel electrophoresis, and (3) the labeled light chains can be exchanged for unlabeled ones in smooth muscle myosin. Here we extend the utility of the labeling method by further characterizing some potentially useful chemical aspects of the labeling approach.

II. Results and Discussion

Our strategy for specifically labeling thiophosphorylated proteins is shown in Figure 1. Each of the modified protein structures is marked with an underlined number which will be used throughout the text. The native RCL20 (structure 1 in Figure 1) is shown as a vertical line with

---

[1]This work is supported by the National Institutes of Health (Grant AR40917 to C.R.C.) and the American Heart Association (Established Investigatorship to C.R.C.).

[2]Abbreviations: ATPγS, adenosine 5'-O-(3-Thiotriphosphate); IAA, iodoacetic acid; RLC20, 20 kDa regulatory light chain of chicken gizzard myosin; LC17, 17 kDa light chain of chicken gizzard myosin; DTE, dithioerythritol; [^{3}H]IAEDANS, tritiated form of 5-[[2-[(iodoacetyl)-amino]ethyl]amino]naphthalene-1sulfonic acid; NEM; N-ethylmaleimide; PPDM, p-phenylenediamine.

TECHNIQUES IN PROTEIN CHEMISTRY IV

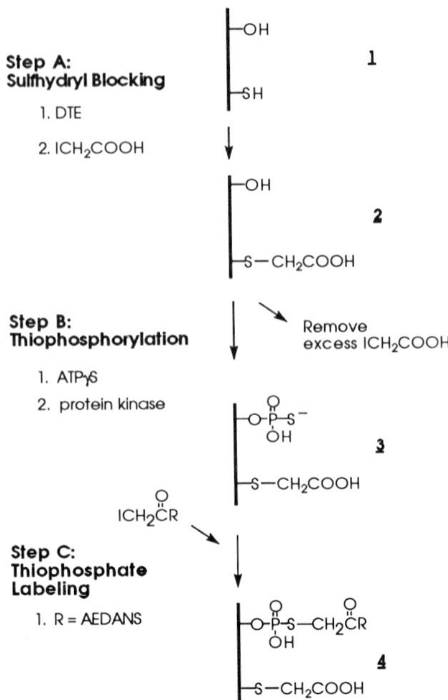

Figure 1. Outline of method for specifically labeling thiophosphorylated proteins. For specific labeling protocols see (Facemyer & Cremo, 1992). See text for explanation of structures.

the thiophosphorylatable Ser-19 at the top and the only cysteine, Cys-108, at the bottom. In most situations, cysteine sulfhydryls are the only reactive protein side chains under conditions in which the thiophosphate group is labeled. Therefore, in Step A Cys-108 is blocked by reaction

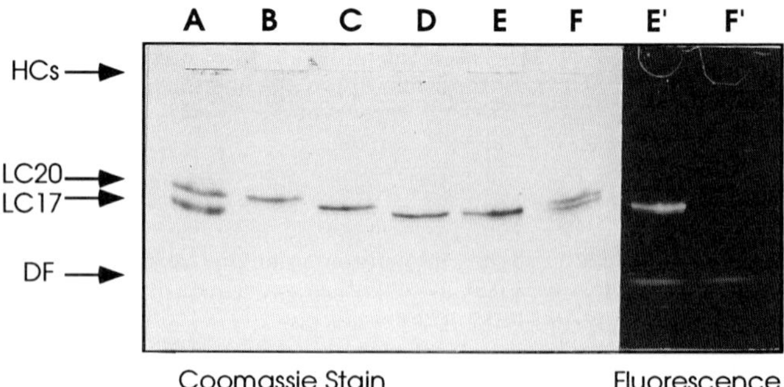

Figure 2. Urea gel assay for RLC20 modification. Gels were prepared as previously described (Facemyer & Cremo, 1992) and stained with coomassie blue. Lane A, 30 μg smooth muscle myosin; Lane B, 1.5 μg isolated RLC20; Lane C, isolated RLC20 which has been reacted at Cys-108 with IAA; Lane D, sample in Lane C that has been 100% thiophosphorylated at Ser-19; Lane E, sample in lane D that has been reacted with IAEDANS to 100%; Lane F, RLC20 which has been reacted at Cys-108 with IAA, 50% phosphorylated at Ser-19, and treated with IAEDANS exactly as sample in lane E (upper band is not phosphorylated, lower band is phosphorylated). Lanes E' and F' are lanes E and F detected by long-wave UV irradiation before coomassie blue staining to show fluorescence of IAEDANS.

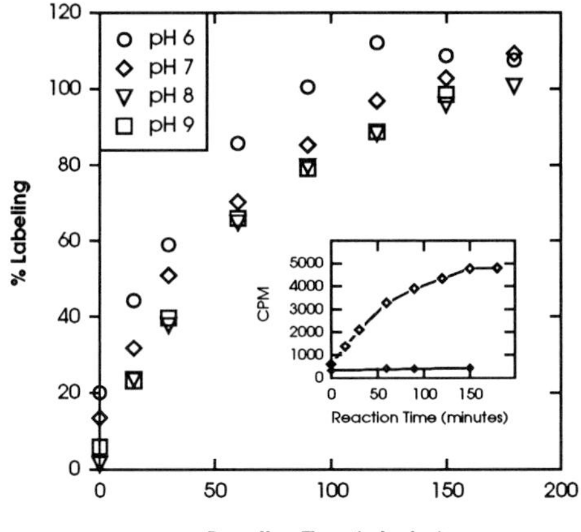

Figure 3. Time course and pH dependence of specific labeling of thiophosphorylated RLC20 at 25° C. RLC20 (0.49 mg/ml (25 μM) in 75 mM ammonium bicarbonate pH 7.9, 5 mM $MgCl_2$) was blocked at Cys-108 with IAA and then either phosphorylated or thiophosphorylated on serine-19. At time = 0, [^{3}H]IAEDANS (5756 counts per min/nmol) was added to 250 μM. At the indicated times, the triplicate samples were processed as previously described (Facemyer & Cremo, 1992). For each time point, the data are the averaged counts for the thiophosphorylated minus those for the phosphorylated RLC20. The inset shows the raw data for the time course at pH 7 (filled symbols. phosphorylated; open symbols, thiophosphorylated).

with a haloacetyl derivative, in our case iodoacetic acid (IAA) to produce structure 2. Reversible methods for blocking cysteines may also be used (Facemyer & Cremo, 1991; Facemyer & Cremo, 1992). The protein is thiophosphorylated in Step B on Ser-19 by myosin light chain kinase, using ATPγS as a substrate, to yield structure 3. Following removal of excess ATPγS, the haloacetate derivative [^{3}H]IAEDANS is added in Step C. The phosphorothioate group on the protein displaces halogen from the haloacetate to produce the stable structure 4.

The above reactions with RLC20 can be followed by urea gel electrophoresis (Figure 2). An increase of a single negative charge causes proteins to migrate faster in this gel system. Reactions corresponding to Step A and B from Figure 1 can be followed with these gels (see lanes B,C and D in Figure 2). However, structures 3 and 4 (from Figure 1) comigrate in this gel because charge is both added and removed during the labeling of the thiophosphate with [^{3}H]IAEDANS (compare lanes D and E, Figure 2). Fluorescence detection of lanes E and F indicate that the fluorescent derivative only labels the thiophosphorylated protein (E'), not an identically treated phosphorylated control (F').

The time course and pH dependence of the reaction of thiophosphorylated Ser-19 with [^{3}H]IAEDANS was examined by precipitating the protein with acetone at various times after addition of [^{3}H]IAEDANS (Figure 3). The counts per minute in each thiophosphorylated protein pellet were corrected for background by subtracting the counts which were incorporated into an otherwise identical but phosphorylated protein sample. The averaged raw data for the pH 7 time course is shown in the inset. These data show that the reaction is highly specific and stoichiometric, and does not have a strong pH dependence in the range of pH 6-9. This suggests that it may be possible in some cases to avoid pre-blocking reactive cysteines if the thiophosphate reaction is carried out at low pH for a short period of time.

It was of interest to compare the rates of reaction of maleimides and haloacetates with thiophosphates, because many extrinsic probes are commercially available in maleimide form. As a model reaction, we used ATPγS as a thiophosphate containing compound to react with either NEM or IAEDANS. The rate of the reaction was followed by capillary zone

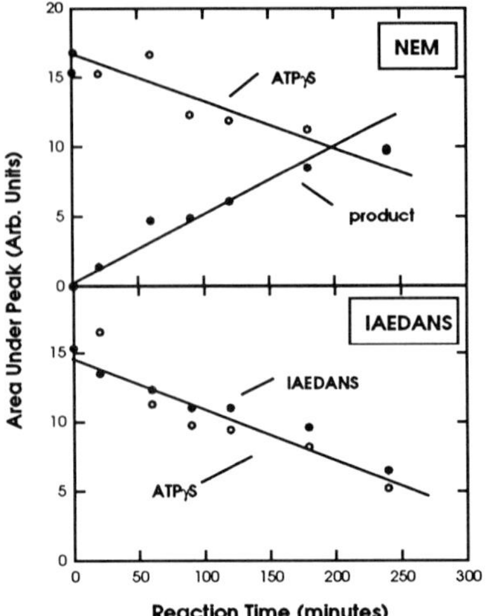

Figure 4. Time course of reaction of ATPγS with NEM (top) and IAEDANS (bottom). ATPγS (2 mM) and either NEM or IAEDANS (2 mM) were reacted on ice in 100 mM Tris (pH 8.0 at 4° C). At the times shown, 5 μl of the reaction was quenched with 10 μl of 100 mM sodium phosphate pH 2.5. The 15 μl sample was analyzed with a Biorad HPE capillary zone electrophoresis apparatus with a "Microsampler 100" cartridge (20 cm x 25 μm) using positive polarity in 100 mM sodium phosphate pH 2.5. The load time was 8 sec at 8 KV and the run was at 6 KV for 12 min. The areas under the peaks for ATPγS (3.9 min), NEM-ATPγS adduct (4.5 min), and IAEDANS (7.8 min) were calculated relative to an internal ADP (6.6 min) standard.

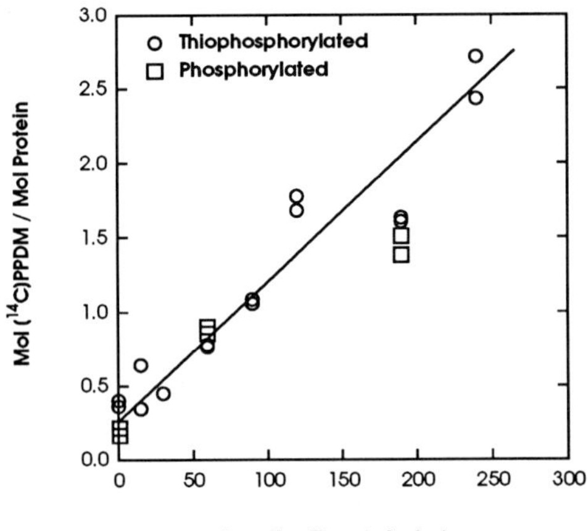

Figure 5. Reaction time course of [^{14}C]PPDM labeling of thiophosphorylated and phosphorylated light chains. Light chains (0.5 mg/ml; 25 μM) were reacted with a 20-fold molar excess of [^{14}C]PPDM in 75 mM ammonium bicarbonate pH 7.9, 5 mM $MgCl_2$ at 25° C. At various times, 30 μl aliquots were precipitated with acetone and analyzed for counts.

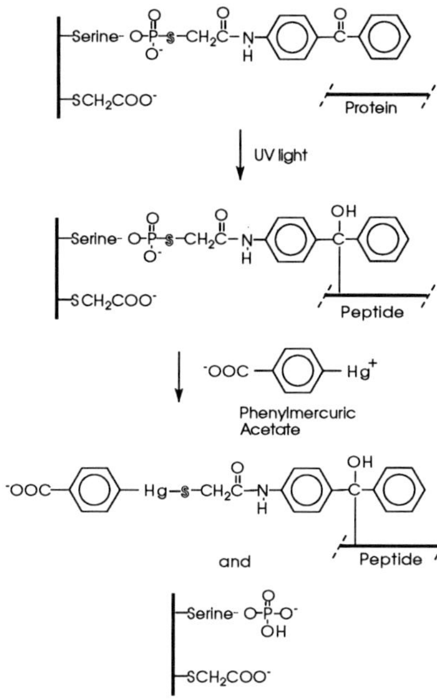

Figure 6. Use of phenylmercuric acetate in photocrosslinking experiments. The sulfur, derived from [^{35}S]ATPγS, is shown in a different font to indicate that it is radioactive. See text for details.

electrophoresis. As can be seen in Figure 4, NEM (top graph) reacts at about 2/3 the rate of IAEDANS (bottom graph). This data suggests that maleimides may be generally useful thiophosphate labeling reagents.

We then examined the rate of reaction of the maleimide, [^{3}H]PPDM, with the thiophosphate on the light chain. Figure 5 shows that the reaction is non-specific for the thiophosphate, as the control (which is phosphorylated) is labeling at about the same rate as the thiophosphorylated protein. We also found that the NEM reaction with the protein showed the same behavior using a spectrophotometric assay (data not shown). Therefore, in our system, maleimides do not appear to be useful reagents for specific thiophosphate labeling. It is likely that the maleimides are reacting with reactive amines in the light chains.

We want to use the thiophosphate labeling procedure (Figure 1) for photo-cross linking experiments. A haloacetate photo-cross linking derivative would be reacted with a thiophosphate and then irradiated to cross link with a protein. The cross linking site would then be determined after proteolytic digestion of the protein and isolation of the cross linked peptide. Typically, this kind of experiment would have two drawbacks. First, a radioactive photocrosslinking reagent would have to be synthesized. Second, sequencing of the purified radioactive peptide would give two sequences, one from the original photo-cross linker attachment site, and one from the site with which the photoreactive derivative reacts. We propose that both of these problems can be avoided using the thiophosphate labeling method as outlined in Figure 6. A benzophenone derivative is used as an example, which is shown after reaction with [^{35}S]-thiophosphorylated protein (Figure 6, top). After irradiation, the benzophenone is shown covalently attached to the

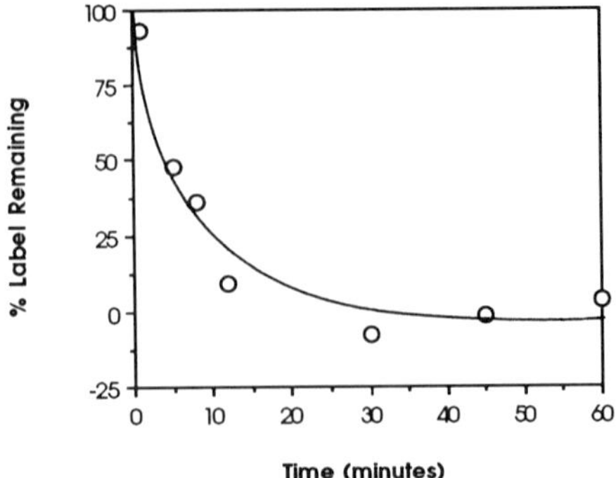

Figure 7. Reaction time course of phenylmercuric acetate with [^{3}H]IAEDANS-labeled thiophosphorylated light chain (structure 4, Figure 1). 5 µM of [^{3}H]IAEDANS-labeled thiophosphorylated RLC20 was reacted with 15 µM of phenylmercuric acetate in 200 mM sodium acetate pH 4.5, 5 mM $MgCl_2$, at 25° C for various times. The reaction was stopped by precipitating the protein with acetone. The acetone supernatants were counted. The data plotted are the averages of triplicate time points, expressed as the % of the original label incorporated.

protein which has been proteolytically digested to peptides (Figure 6, middle). Phenylmercuric acetate is then used to cleave the P-S bond (Neumann & Smith, 1967) to give the mercury substituted derivative in which the radioactive sulfur is now covalently attached to the peptide of interest (Figure 6, bottom). This approach solves both problems stated above, because the peptide of interest is radioactive without synthesis, and it has been detached from the original phosphorylated protein sequence so that only one sequence will now be obtained.

Although the reaction of phenylmercuric acetate with S-substituted thiophosphates has been previously described (Neumann & Smith, 1967), it was important to determine appropriate conditions for the reaction with the labeled protein. Figure 7 shows the time course of the reaction of phenylmercuric acetate with [^{3}H]IAEDANS-labeled thiophosphorylated light chain. All of the tritium is removed from the protein after about 30 min at pH 4.5.

In summary, we have demonstrated a new technique to specifically label thiophosphorylatable amino acids on proteins with haloacetyl derivatives which is potentially useful for many proteins. Specifically, we have applied the technique to label a thiophosphorylatable serine residue of the RLC20 of smooth muscle myosin. The thiophosphate labeling reaction rate is independent of pH between pH 6 and 9. The reaction is complete within a few hours at 25 °C and is highly specific for the thiophosphate moiety. The product is stable to a wide pH range and to the presence of high concentrations of reductant. Although we found that maleimides also react efficiently with thiophosphates, neither NEM or PPDM gave specific reaction with our thiophosphorylated light chains. We have shown that phenylmercuric acetate is a useful reagent to cleave the P-S bond and propose its use in photo-cross linking experiments.

REFERENCES

Eckstein, F. (1985) *Ann. Rev. Biochem.*. **54**, 367-402.
Facemyer, K. C., & Cremo, C. R. (1991) *Biophys. J.* **59**, 436a.
Facemyer, K. C., & Cremo, C. R. (1992) *Bioconj. Chem.*, in press.
Neumann, H., & Smith, R. A. (1967) *Arch. Biochem. Biophys.* **122**, 354-361.

A Novel Method of Identifying Phosphorylation Sites Using a Thiophosphorylated Peptide and ESI-MS

David R. Stover and Kenneth A. Walsh
Dept. of Biochemistry, University of Washington
Seattle, WA 98195

I. Introduction

The ever-expanding list of cellular functions that are controlled by phosphorylation has focussed attention on the identification of phosphorylation sites in specific proteins. Recent advances in coupled liquid chromatography-electrospray mass spectrometry(LC-MS) now facilitate detection of phosphorylation sites without the use of radioactive phosphate(For review of MS techniques see reference 1). This technique is both sensitive and effective in specifying the stoichiometry of phosphorylation in multiply phosphorylated proteins of known sequence. In essence, we compare tryptic peptides of exogenously or endogenously phosphorylated proteins with predictions from the DNA sequence. A peptide containing an extra 80 mass units (or multiple thereof) is a prime candidate for a phosphorylation site. Even if the suspect peptide is in a mixture it can be sequenced by standard MS/MS techniques. However,this protocol becomes problematic when subnanomole amounts of large proteins are analyzed. We have found that the inclusion of a thiophosphorylated synthetic carrier peptide (Fig. 1) permitted recovery of a phosphorylated peptide from 250 pmoles of a 200 kDa protein.

Many sites of phosphorylation can be predicted by consensus motifs or homology to other phosphorylated proteins(2, 3). By thiophosphorylating (with ATPγS) a synthetic peptide that corresponds to the predicted site, a carrier peptide is produced that protects the putative native phosphopeptide from nonspecific adsorption during isolation and can replace ^{32}P as a marker. This carrier thiophosphopeptide can be distinguished from the native phosphopeptide by its additional 16 mass units per attached (thio)phosphate. This technique does not require ^{32}P-labelling; moreover, endogenous phosphorylation events can be detected in isolated proteins.

In the case of CD45, we noticed several putative phosphorylation sites for casein kinase II(CKII) within a single 22 residue tryptic peptide of this transmembrane protein-tyrosine phosphatase. We describe a technique that enabled us to demonstrate that this peptide is phosphorylated in vivo. In addition, three methods of determining which residues within the peptide are modified are outlined.

TECHNIQUES IN PROTEIN CHEMISTRY IV

II. Materials and Methods

A. *Isolation of Phosphorylated CD45*

The human acute T-cell leukemia cell line Jurkat (clone E6-1, ATCC TIB 152) was grown at 37°C in RPMI 1640 with 10% newborn calf serum. Jurkat cells were loaded with ^{32}P by first placing them in phosphate-free media for 30 minutes and then transferring them to ^{32}P-containing media (100 μCi/ml) for 4 hours before activation. The cells were then activated with phytohemagglutinin-L (PHA, 5 μg/ml; Sigma) for five hours, pelleted at 500 x g, washed in PBS, and repelleted at 1000 x g. The cell pellets were resuspended in hypotonic buffer (10 mM Tris pH 7.5, 10 mM EDTA pH 8.0, 1 mM Benzamidine, 10 μg/ml Aprotinin, Leupeptin, and TLCK, 50 ng/ml PMSF, and 80 mM ß-glycerophosphate) and disrupted by three 10 second bursts of sonication. Cell particulates were pelleted by centrifugation at 20,000 x g, and resuspended by homogenization in extraction buffer (20 mM Tris pH 7.5, 10 mM EDTA, 100 mM NaCl, 1% Nonidet P-40 (NP-40), 1 mM Benzamidine, 10 μg/ml Aprotinin, Leupeptin, and TLCK, and 50 ng/ml PMSF, 80 mM ß-glycerophosphate). After extraction, membranes were removed by ultracentrifugation at 35,000 rpm in a Beckman SW-40ti rotor.

To the remaining membrane proteins was added three volumes of DE buffer (10 mM Tris pH 7.5, 5 mM EDTA, 0.1% NP-40, and 1 mM Benzamidine). The extract was applied to a 10 ml DEAE-52 (Whatman) column equilibrated in DE buffer. After washing with ~4 column volumes of DE buffer, bound proteins including CD45 were eluted with DE buffer + 400 mM NaCl. Eluted CD45 was detected by Western analysis (anti-CD45 mAB 9.4). CD45 was further purified on an anti-CD45 antibody (mAB 9.4) column as previously described(4).

B. *In Vitro Phosphorylation*

A synthetic peptide(Sα-T100), corresponding to the 100th tryptic fragment of CD45, was phosphorylated overnight (10-16 hours) at 30°C by CKII (~3 μg/ml) and 0.1 mM ATP (or ATPγS) in 50 mM Tris pH 7.5, 150 mM NaCl, 10 mM $MgCl_2$, 10 mM DTT. The products were purified by reverse phase HPLC and the extent of phosphorylation was assessed by MS.

C. *Proteolysis of CD45*

Phosphorylated CD45 (50 μg/250 pmoles) was partially digested with EndoLys-C at a ratio of 1:100 (wt/wt) for 30 minutes at 30°C, and precipitated by addition of TCA to 15%. The lyophilized pellet was mixed with 1 nmole of thiophosphorylated Sα-T100 peptide and completely digested by 5 μg of trypsin at 37°C for 24 hours in 100 mM NH_4HCO_3. The mixture of thiophosphorylated Sα-T100 and tryptic fragments of phosphorylated CD45 were separated by HPLC on a 4.6 mm x 10 cm RP-18 column (Pierce) using a 0-60% acetonitrile gradient over 60 minutes.

D. Identification of Phosphorylation Sites

HPLC fractions were collected and analyzed using a Sciex API-III electrospray triple quadrupole mass spectrometer. Approximately 10% of the fraction was used to determine the mass of putative phosphopeptides and the number of attached phosphates; another 45% provided MS/MS fragmentation data of the phosphopeptide ion. The remaining material was lyophilized and digested with 100 ng of EndoAsp-N protease (Boehringer Mannheim) in 100 mM NH_4HCO_3 at 37°C for 18 hrs. The resulting fragments were separated at 200 µl/min on a 2 mm x 3 cm RP-18 column (Hewlett-Packard); 10% of the eluate was electrosprayed directly into the mass spectrometer(LC-MS). The sites of phosphate attachment within the synthetic peptide were also examined after acid cleavage in 5% formic acid at 110°C for 2 hours under Argon. After lyophilization and resuspension in 50% acetonitrile and 0.5 % formic acid, fragments were analyzed as a mixture by MS.

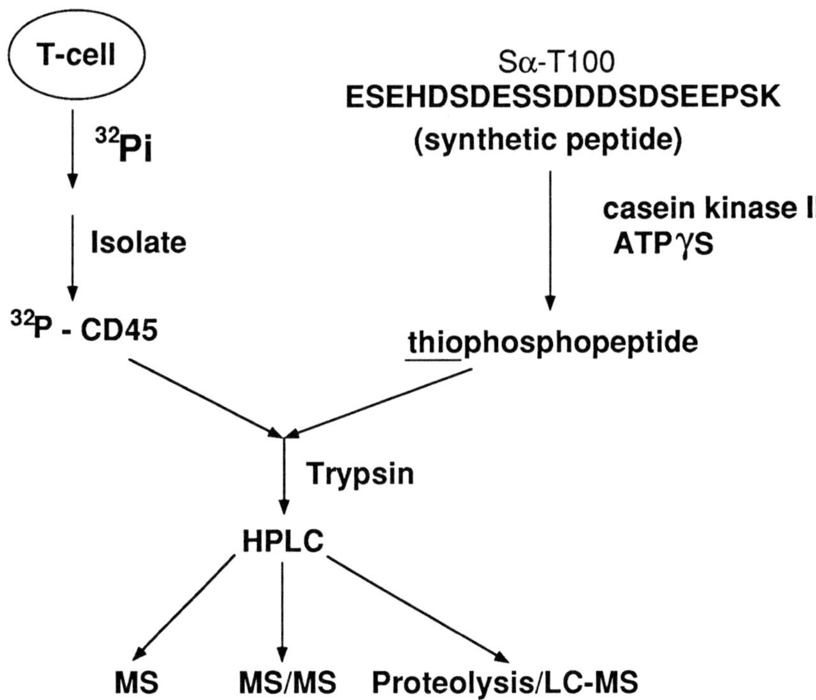

Figure 1. General Outline of the synthetic thiophosphopeptide carrier technique for identifying sites of Phosphorylation. Addition of the carrier peptide can be used as a marker for the endogenously phosphorylated peptide and protects it from losses during isolation.

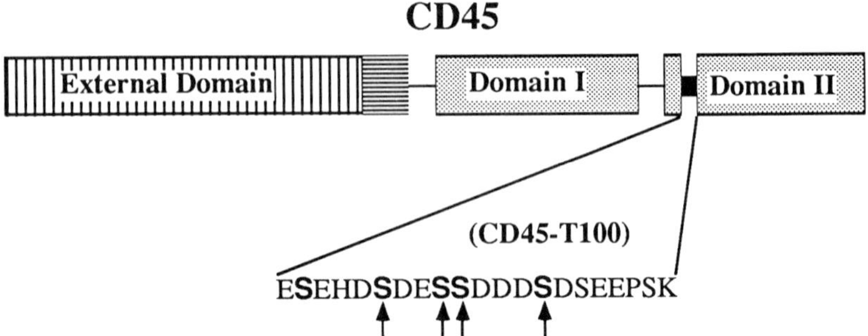

Figure 2. Illustration of the CD45 tryptic peptide that contains the five predicted sites of CKII phosphorylation (bold type) and the four serines actually observed to be phosphorylated in vivo (indicated by arrows). The corresponding synthetic peptide (Sα-T100) was used as a carrier for this endogenous peptide.

III. Results and Discussion

CD45 is a protein-tyrosine phosphatase that has been reported to be phosphorylated in vivo on both serine and tyrosine residues in response to mitogenic agents(5, 6). In vitro it is also multiply phosphorylated by CKII(4). Using conventional techniques, two separate attempts to locate the sites of in vivo phosphorylation failed using a tryptic digest of approximately 1 nmole of CD45 from ^{32}P-labelled T-cells. During RP-HPLC, only the breakthrough contained significant levels of radioactivity. Subsequent MS analysis of small radioactive peaks did not reveal any identifiable phosphopeptides. Significant losses of phosphopeptides appeared to occur during HPLC separation, indicating that addition of a carrier peptide with similar elution properties as the native phosphopeptide might specifically protect it from losses.

Comparison of the sequence of CD45 to data banks of motifs for consensus phosphorylation sites (e.g. Prosite, KeyBank, and KeyTool) yielded a list of 38 consensus serine/threonine phosphorylation sequences, 23 of which are specific for CKII. Twelve of the 23 CKII sites are either on the extracellular domain or are threonine rather than serine residues. Four of the remaining predicted CKII phosphorylation sites are found in a highly acidic region not reported in any other known PTPase domain (Fig. 2). Based on this information we synthesized a peptide corresponding to the tryptic peptide containing the four predicted CKII sites. Since this peptide is the 100th tryptic peptide of CD45 (starting at the N-terminus) we call the native peptide CD45-T100 and the corresponding synthetic peptide Sα-T100.

To test whether the carrier technique (Fig. 1) would be feasible we first compared the elution behavior of phosphorylated and thiophosphorylated synthetic peptide and found that they co-eluted from our reverse-phase C-18 HPLC column. Furthermore, MS readily distinguished the two species by the additional 16 Da per (thio)phosphate attached.

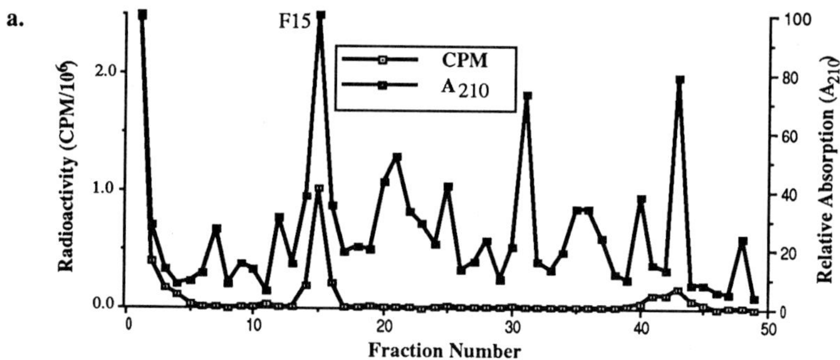

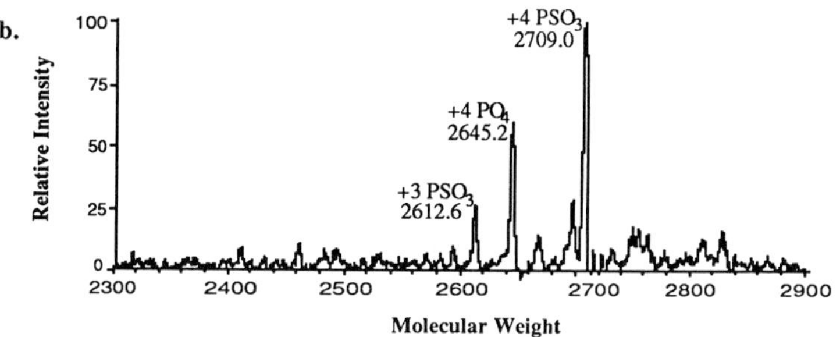

Figure 3. a. HPLC profile of CD45 tryptic peptides (250 pmol) mixed with 1 nmole Sα-T100. The peak of radioactivity (open squares) coelutes with the most prominent A_{210} peak (closed squares) that is caused by the addition of the synthetic peptide. **b.** MS analysis of fraction #15 from A. The two peaks at 2612.6 and 2709.0 have the molecular weight of Sα-T100 with 3 and 4 thiophosphates attached, respectively. The middle peak (2645.2) has the molecular weight of CD45-T100 with 4 phosphates. It is distinguished from tetra-thiophosphorylated Sα-T100 by a difference of 64 mass units (16 per (thio)phosphate).

To 250 pmoles of CD45, isolated from ^{32}P-loaded, activated T-cells, one nmole of thiophosphorylated Sα-T100 was added to act as a carrier for the endogenous peptide that we predicted would be phosphorylated. After tryptic digestion the peptides were separated by HPLC. Most of the radioactivity coeluted with the thiophosphorylated Sα-T100 (Fig. 3a), indicating that CD45-T100 might be phosphorylated in vivo. While it was encouraging to see the radioactivity in the same fraction as the Sα-T100, it should be noted that the UV absorption of the synthetic peptide was sufficient to locate the corresponding native peptide.

To establish the identity of the radiolabelled peptide and to observe the number of attached phosphates, 10% of fraction #15 was analyzed by MS (Fig. 3b). Three forms of the peptide were observed. Two match the molecular weight of Sα-T100 carrying 3 or 4 thiophosphates. The third, a peptide of mass 2545.2, corresponds to tetraphosphorylated CD45-T100. Because it contains phosphates rather than thiophosphates it must have originated from native CD45, not from the

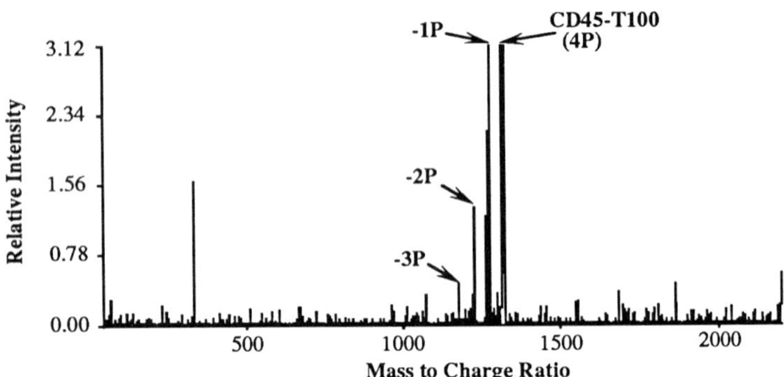

Figure 4. The ion peak with the mass/charge ratio of doubly charged CD45-T100 with 4 attached phosphates was selected for MS/MS(See Figure 3B). The parent ion and daughter ions with one, two and three phosphates removed by collisional activation are observed (loss of 98/2 per phosphate). Some sequence information can be obtained from this spectrum; however, the ion abundances are small and agree with the more convincing data in Table I.

thiophosphorylated synthetic peptide. Although many other phosphopeptides yield enough sequence information by MS/MS analysis to pinpoint the specific residues modified, the size and composition of this particular peptide limited the amount of sequence data we obtained. MS/MS analysis of 45% of fraction #15 demonstrated that this peptide contained at least 3 phosphates (Fig. 4), but little sequence information was derived.

We digested the remaining material in fraction #15 (<100 pmol) with Asp-N protease. LC-MS analysis (Table I) revealed that the digest was incomplete; however, this was fortuitous as the longer overlapping fragments were better retained and separated by the C-18 column and therefore yielded more information. Three of the peptides, ESEH, DSEEPSK, and DXDEXXDDDX (X = phosphoserine) were sufficient to identify the positions of the phosphoserines within CD45-T100 (Fig. 2). The other peptides listed in Table I and minor ions observed in the MS/MS analysis (Fig. 4) are all consistent with this assignment of the four phosphoserine positions.

Table I. LC-MS of Fraction #15 AspN Digest

M_r	Ion Intensity	Fragment*	#P's
2646	3,745,131	EZEHDZDEZZDDDZDZEEPZK	4
501	30,737,000	ESEH	0
792	9,526,000	DSEEPSK	0
1074	1,582,000	DZDZEEPZK	1
1189	607,000	DDZDZEEPZK	1
879	1,085,000	DXDEXX	3
994	1,854,000	DXDEXXD	3
1109	1,328,961	DXDEXXDD	3
1391	364,269	DXDEXXDDDX	4
2166	4,431,000	DZDEZZDDDZDZEEPZK	4

* X = Confirmed Phosphoserine
Z = Possible Phosphorylation Site
S = Confirmed Nonphosphorylated Serine

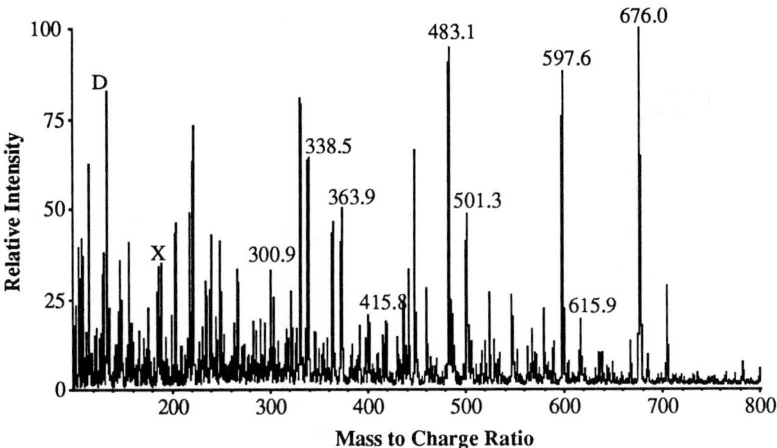

Figure 5. Phosphorylated Sα-T100 was cleaved in 5% Formic acid at 110°C for two hours. The fragments were analyzed by MS as a mixture.

In an alternative experiment, phosphorylated synthetic Sα-T100 was cleaved with acid to determine the sites of phosphorylation. Acid cleavage (5% Formic acid at 110°C for 2 hours) results in cleavage of peptides at either side of Asp. Since many of the predicted cleavage products were too hydrophilic for separation by HPLC, they were analyzed as a mixture by MS (Fig. 5). The phosphopeptides detected (Table II) indicate that CKII phosphorylated the synthetic peptide at the same sites as were found in CD45 in vivo.

Table II. Acid Cleave Phosphorylated Sα-T100

Mass/Charge	Sα-T100 Fragments*	#P's
501.0	ESEH	0
615.9	ESEHD	0
483.1	EXX	2
597.9	EXXD	2
676.0	SEEPSK	0
300.9	(ZD DZ)	1
415.8	(DZD DZD)	1
363.9	DDD	0

* X = Confirmed Phosphoserine
Z = Possible Phosphorylation Site
S = Confirmed Phosphoserine

IV. Conclusion

A thiophosphorylated, synthetic peptide coeluted with, and served as a carrier of, the corresponding endogenously phosphorylated peptide from a C-18 reverse-phase HPLC column. Mass spectrometry distinguished the endogenously phosphorylated peptides from the thiophosphorylated carrier, allowing determination of stoichiometry and identification of sites of phosphorylation on the native peptide. The isolation and analyses involving MS, MS/MS, and subdigestion used only 250 pmoles of the Mr 200,000 protein, CD45. Using this technique, four sites of in vivo phosphorylation were identified in this protein. The corresponding four serines were phosphorylated in vitro by CKII.

References

1. McCloskey, J. A. (ed.) (1990) "Methods in Enzymology" **193**. Academic Press Inc., San Diego, California.
2. Gonzalez, G. A., Yamamoto, K. K., Fischer, W. H., Karr, D., Menzel, P., Biggs, W., Vale, W. W., and Montminy, M. R. (1989) *Nature* 337(6209). 749-752.
3. Cala, S. E. and Jones, L. R. (1991). *J. Biol. Chem.* **266(1)**. 391-398.
4. Tonks, N. K., Charbonneau, H., Diltz, C. D., Fischer, E. H., and Walsh, K. A. (1988) *Biochem.* **27**: 8695-8701.
5. Autero, M. and Gahmberg, C. G. (1987) *Eur. J. Immunol.* **17** (1503-1506).
6. Stover, D. R., Charbonneau, H., Tonks, N. K., Walsh, K. A. (1991)_ *Proc. Natl. Acad. Sci. USA* **88** (7704-7707).

Protein Phosphorylation: Sequence-Specific Identification of *In Vivo* Phosphorylation Sites by MALDI-TOF Mass Spectrometry

Tai-Tung Yip and T. William Hutchens

Protein Structure Laboratory
USDA/ARS Children's Nutrition Research Center
Department of Pediatrics
Baylor College of Medicine
Houston, Texas 77030

I. Abstract

We have developed procedures that allow the *sequence*-specific identification of phosphopeptides in *unfractionated* digests of proteins phosphorylated *in vivo.* This is accomplished by sequential analyses of proteolytic digest maps generated by matrix-assisted UV laser desorption/ionization (MALDI) time-of-flight (TOF) mass spectrometry before and after *in situ* dephosphorylation. This approach was also used to reveal 1) the molar distribution of a given phosphopeptide with multiple and varying numbers of phosphate groups, and 2) upon addition of metal ions, the phosphorylation-dependent binding of iron.

II. Introduction

A variety of important biochemical reactions are regulated by the activities of specific kinases and phosphatases. To facilitate our understanding of phosphorylation-dependent alterations in the structure and function of proteins and peptides, simplified and more sensitive procedures are required to monitor the kinetics and extent of *sequence*-specific phosphorylation and dephosphorylation reactions in complex

TECHNIQUES IN PROTEIN CHEMISTRY IV

unfractionated mixtures. Detailed investigations of sequence-specific phosphorylation reaction mechanisms and higher order structural alterations associated with either phosphorylation or dephosphorylation (e.g., 1-3) frequently involve the preparation of suitable model peptides as substrates (e.g., 4). A desirable strategy for the identification and/or design of such model substrates requires knowledge of the minimum primary structure necessary to maintain phosphorylation of the selected protein target residues and domains; these efforts presently necessitate the use of labor intensive biochemical isolation and screening procedures.

We have recently reported the development of mass spectrometric techniques that enable the detailed evaluation of posttranslational modifications in peptide structure, including intramolecular disulfide bond formation (6,7), coordinate covalent interactions with specific transition metal ions (6-12), and metal ion-induced conformational changes (7). Although most of these developments involved the evaluation of purified synthetic peptides, one procedure employs the use of matrix-assisted UV laser desorption/ionization time-of-flight mass spectrometry (MALDI-TOF) to map and identify *sequence*-specific transition metal-binding protein surface domains and peptides in *unfractionated* protein digest mixtures (10). We report here procedures developed for the rapid (min), sensitive (1 to 10 pmol), and sequence-specific identification of phosphopeptides in unfractionated digests of phosphoproteins using MALDI-TOF. Human β-casein was chosen as a model for these investigations because phosphopeptides derived from the proteolytic digestion of this milk protein remain the subject of intense investigations (13), particularly with respect to the regulation of mineral and metal ion absorption (e.g., 14) and cell growth (e.g., 15).

III. Materials and Methods

A. *Purification of Human β-Casein from Milk*

Human milk was collected by breast pump and stored at -20°C until use. Thawed milk was dialyzed first against 50 mM EDTA, 0.5 M NaCl (pH 6.5) and then 20 mM phosphoserine in 50 mM 2-(N-morpholino)ethanesulfonic acid (MES) and 1 M NaCl (pH 6.5). An aliquot (25 ml) of this sample was pumped into a 50-ml column (2.2 cm x 13 cm) of immobilized iminodiacetate(IDA)-Fe(III) (Chelating Sepharose Fast Flow, Pharmacia) equilibrated with 20 mM phosphoserine in 50 mM MES containing 1 M NaCl (pH 6.5). After extensive washing with the same buffer to remove all unadsorbed proteins, casein was eluted with 100 mM EDTA, pH 7.0. High-performance reverse phase (phenyl, Vydac) and size-exclusion (Superose 12, Pharmacia) chromatography were used to separate the intact β-casein from casein fragments. The final product was detected as a single silver-stained band after polyacrylamide gradient gel electrophoresis and appeared as a single peak by MALDI-TOF with a determined molecular mass of 24,430 Da; the chemical average molecular mass of nonphosphorylated human β-casein calculated from the cDNA sequence is 23,835 Da (16). The identity of the purified β-casein was confirmed by amino acid sequence analysis.

B. Proteolytic Digestion, Enzymatic Phosphorylation and Dephosphorylation

TPCK-treated trypsin (Sigma) digestion of the purified human β-casein was initiated at an enzyme-to-substrate ratio of 1:50 at 37°C in 10 mM Tris-HCl at pH 7.8. Phosphorylation was initiated by incubating the sample in 20 mM Tris-HCl (pH 7.2) containing 9 mM magnesium acetate, 5 mM ATP with 1 μCi ^{33}P-ATP (New England Nuclear, DuPont) and 10 U of cAMP-dependent protein kinase, catalytic subunit (Promega) at 37°C. Dephosphorylation was performed by incubating the sample with 0.25 U of bovine intestine mucosal alkaline phosphatase (Sigma) in 5 mM Tris-HCl (pH 7.8) at 37°C. For *in situ* dephosphorylation, 0.01 U alkaline phosphatase in 2 μl of 5 mM Tris-HCl (pH 7.8) was added to the probe tip, allowed to air-dry and then rinsed briefly in water.

C. Synthesis of the Model Phosphopeptide

The 18-residue human β-casein N-terminal peptide (R1-K18) was synthesized on an Applied Biosystems Model 430A automated peptide synthesizer using Fmoc/NMP(9-fluorenylmethyloxycarbonyl/N-methyl-pyrrolidone) chemistry (FastMoc, Applied Biosystems, Foster City, CA) and purified using methods described previously (9). Mass and purity were evaluated by electrospray ionization mass spectrometry and by MALDI-TOF as described previously (9); sequence was verified with an Applied Biosystems Model 473A automated amino acid sequence analyzer.

D. Matrix-Assisted UV Laser Desorption/Ionization Time-of-Flight Mass Spectrometry

Peptides in the unfractionated β-casein digest were mixed 1:1 (v:v) with matrix; either a saturated aqueous solution of 2,5-dihydroxybenzoic acid (154.12 Da) or a saturated solution of 3,5-dimethoxy-4-hydroxycinnamic acid (224.21 Da) in 30% acetonitrile. One μL of this mixture was analyzed on a Vestec Model 2000 mass spectrometer (Vestec Corp., Houston, TX) exactly as described previously (8,9). All spectra shown were taken in the positive ion mode. Real-time signal averages of multiple (100) laser shots were used to generate each spectrum. Calibration was performed before and verified after each sample analysis using bovine insulin (5734.5 Da).

IV. Results

Figure 1 illustrates MALDI-TOF mass spectra (1250 to 2750 Da) showing the trypsin digest products of human β-casein (10 pmol) analyzed before (bottom) and after (top) kinase treatment *in vitro*. A single prominent shift in peptide mass (80 Da) observed indicated a single phosphorylation event involving a peptide with an observed mass of 1486.9 Da before phosphorylation (peak O) and 1566.9 Da after phosphorylation (peak P). The subsequent isolation of peak P by reverse phase (C_8) HPLC indicated that it was associated with prominant ^{33}P incorporation. By analysis of all peptide masses of 153 possible trypsin digest sites (theoretical fragments from the partial and complete digestion at 11 Lys residues and 3 Arg residues) with PROCOMP (Phil Andrews, U. of Michigan), the 1486.9 Da peptide was identified as the S105-K115 sequence (Table I). This was confirmed by complete sequencing of the purified Peak O and Peak P (17).

Figure 2 shows the same proteolytic digest maps (1800-2700 Da) of unfractionated human β-casein fragments obtained by MALDI-TOF before subsequent enzymatic treatment (middle), after phosphorylation *in vitro* by incubation with cAMP-dependent protein kinase (top), and after incubation of the digest mixture with alkaline phosphatase (bottom). Note the complete disappearance of a group of peptides observed initially at 2168.4 Da, 2248.4 Da, 2328.4 Da and 2408.4 Da after treatment with phosphatase and the increase in intensity of a peptide with an observed mass of 2088.4 Da. The 2088.4 Da peptide was identified as the R1-K18 sequence (Table I) by computer analysis of peptide mass. The peaks labeled as 1P, 2P, 3P and 4P represent the naturally occurring 1, 2, 3 and 4 phosphorylation states of the R1-K18 sequence *in vivo*.

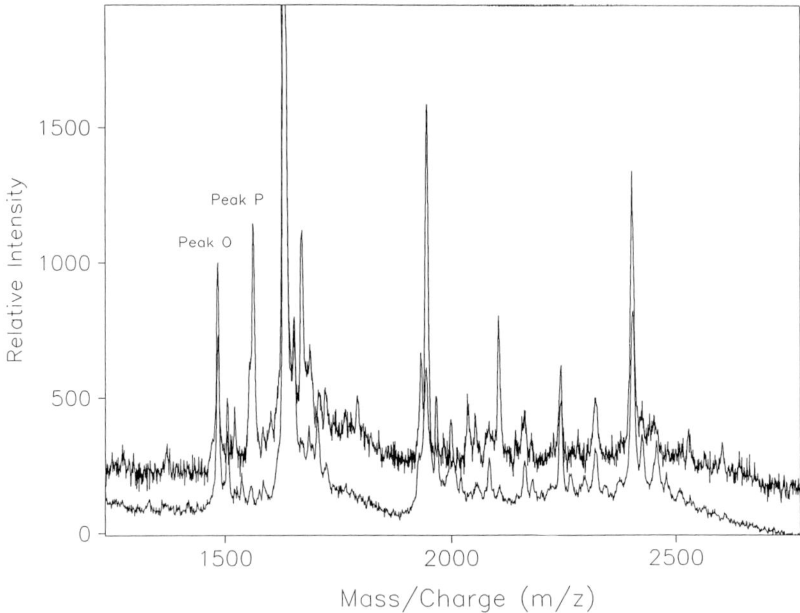

Figure 1. MALDI-TOF mass spectra (1250 to 2750 Da) of the trypsin digest (10 pmol) products of human β-casein analyzed before (bottom) and after (top) kinase treatment *in vivo*.

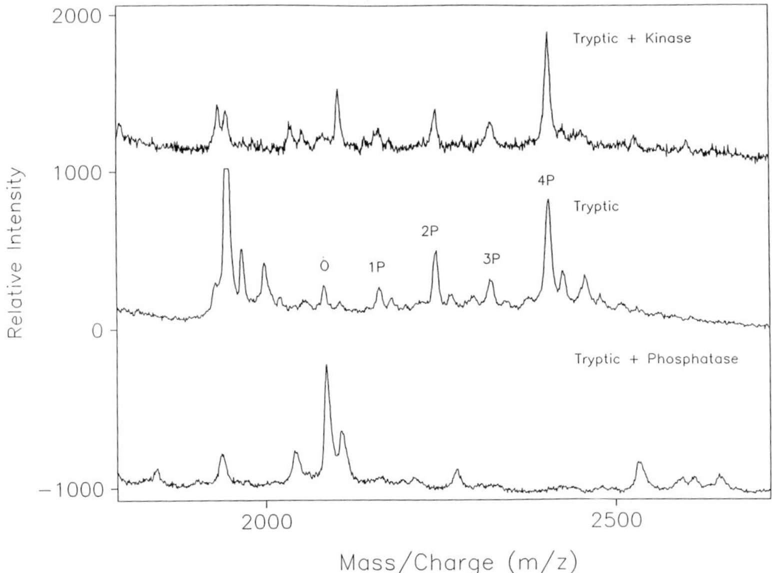

Figure 2. The proteolytic digest maps (1800-2700 Da) of unfractionated human β-casein fragments (10 pmol) obtained by MALDI-TOF before enzymatic treatment (middle), after phosphorylation *in vitro* by incubation with kinase (top), and after incubation of the digest mixture with phosphatase (bottom).

Because casein is one of the highest affinity immobilized Fe(III)-binding protein in human milk (see METHODS), we next proceeded to map the iron-binding site(s) in casein. Figure 3 shows the Fe(II) binding properties of the 2408.4-Da β-casein fragment as determined by sequential MALDI-TOF analyses before (bottom) and after (top) addition of $FeSO_4$ to the unfractionated digest fragments on the probe tip. Fe(II)-binding specificity was readily apparent. Other peptides in the unfractionated β-casein digest mixture (i.e., peptides not containing the R1-K18 sequence) showed little or no capacity for Fe(II) binding.

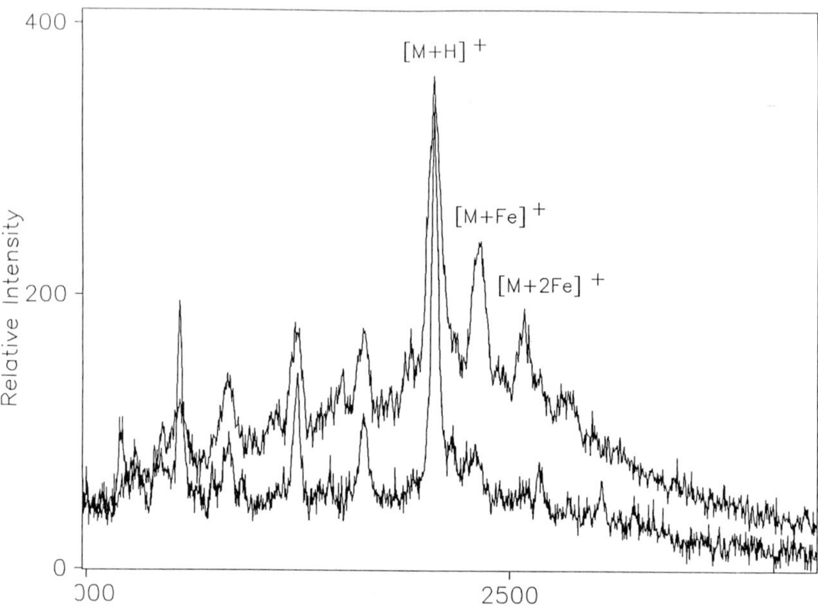

Figure 3. Sequential MALDI-TOF analyses before (bottom) and after (top) addition of $FeSO_4$ to the unfractionated β-casein tryptic fragments (5 pmol) on the probe tip.

MALDI-TOF is also a very efficient method to monitor *in vitro* phosphorylation kinetics. Figure 4 shows the sequential phosphorylation catalyzed by cAMP-dependent protein kinase of a synthetic R1-K18 β-casein peptide *in vitro*. The relatively slow phsophorylation rate corroborated with the observation that the phosphorylation state of this peptide was unchanged after only 3 hr of kinase treatment (Figure 1 and 2, tryptic $\pm$ kinase). The acidic nature (5 glu residues) of this peptide, in contrast to the S105-K117 sequence, probably affects the cAMP-dependent protein kinase activity.

Figure 5 shows again the phosphorylation-dependent Fe(II) binding properties of the synthetic β-casein peptide R1-K18 containing one

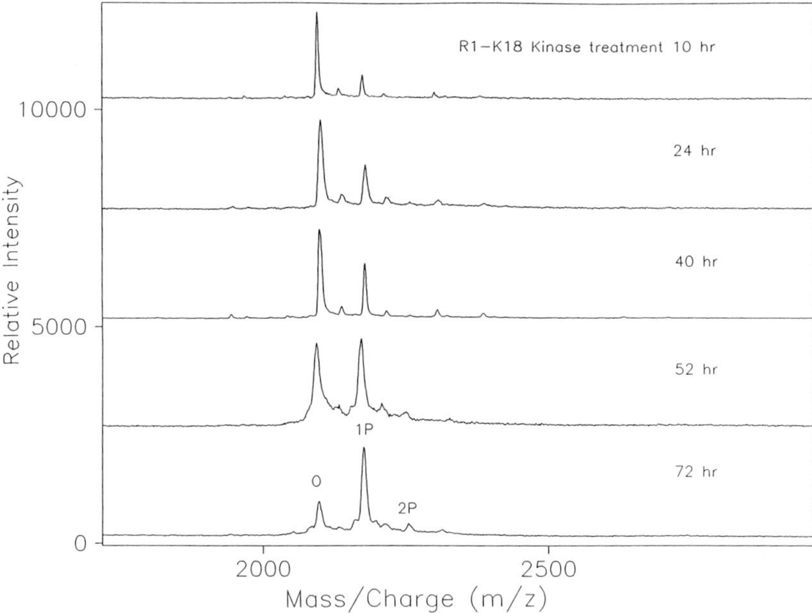

Figure 4. Sequential phosphorylation of a synthetic R1-K18 β-casein peptide (2 pmol) *in vitro.*

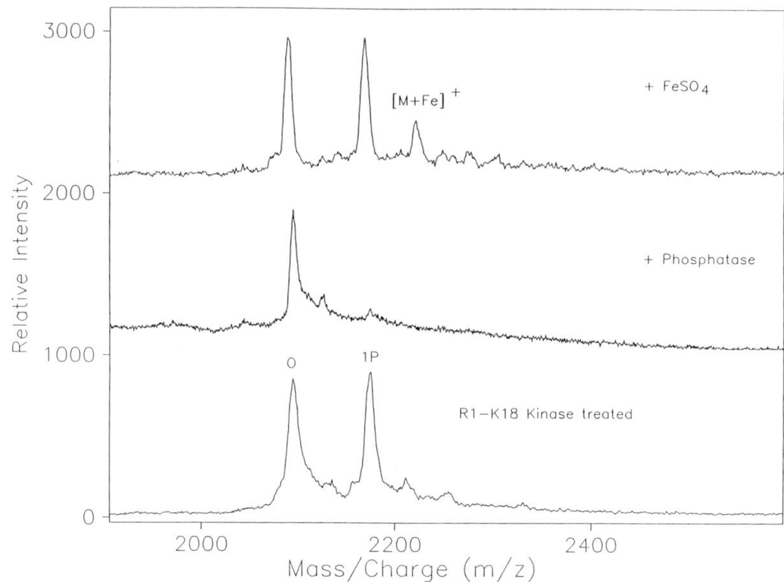

Figure 5. Sequential MALDI-TOF analyses of a phosphorylated R1-K18 β-casein peptide (2 pmol) before (bottom), after *in situ* alkaline phosphatase treatment on probe tip (middle), and after addition of $FeSO_4$ on probe tip (top).

phosphate group (designated peak 1P at 2168.3 Da). The presence of a single phosphate group on the R1-K18 peptide was confirmed by its disappearance after treatment with phosphatase (Figure 5, middle). Although the R1-K18 peptide with one phosphate was able to bind at least one Fe(II) (Figure 5, top), the nonphosphorylated form of the R1-K18 peptide was not able to bind detectable amounts of Fe(II) (Figure 5, top).

Figure 6 illustrates the phosphate-independent Ca(II) binding properties of the synthetic R1-K18 β-casein peptide. Even in the absence of phosphate groups, the acidic nature of this peptide sequence is apparently enough for Ca(II)-binding. The effect of added calcium on the iron-binding properties of the phosphorylated casein peptides is currently under investigation and is of primary importance in the understanding of iron and calcium bioavailability in the milk.

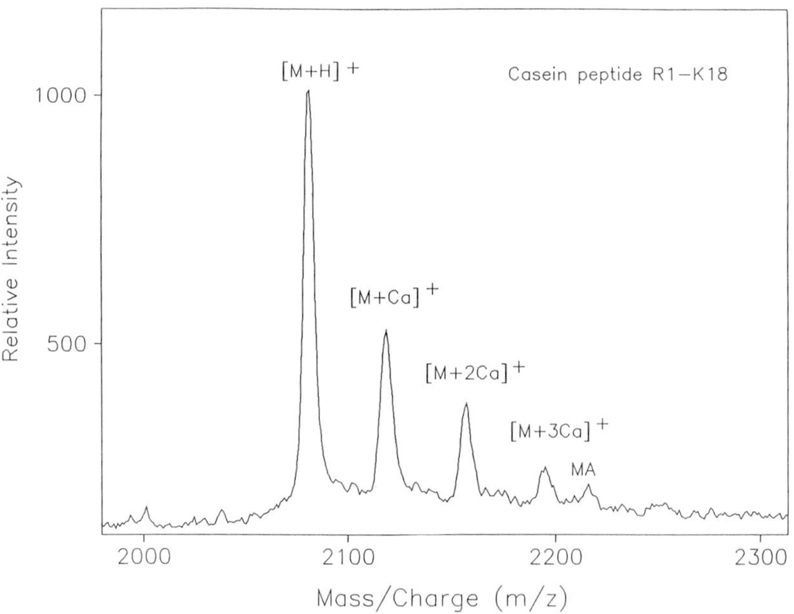

Figure 6. Ca(II) binding properties of the synthetic R1-K18 β-casein peptide (2 pmol).

Table I. Determined Mass Values for the Human β-Casein Peptides R1-K18 and S105-K117 Observed by MALDI-TOF Before and After Phosphorylation

	Protonated molecular mass $M+H^+$ (Da)	
	calculated[a]	observed[b]
RETIESLSSSEESITEYK		
Before phosphorylation	2089.23	2089.4 ± 0.6
After phosphorylation	2169.21	2169.3 ± 0.9
SPTIPFFDPQIPK		
Before phosphorylation	1487.74	1487.9 ± 0.4
After phosphorylation	1567.72	1567.9 ± 0.4

[a]Calculated chemical average mass values for the 18-residue (R1-K18) and the 13-residue (S105-K117) human β-casein peptides were derived from the published cDNA sequence (16).

[b]Average mass observed (± S.D.) from a minimum of 6-8 separate determinations.

V. Conclusions

MALDI-TOF was effective as a rapid (minutes) and efficient (<10 pmol) means to monitor *in vivo* protein phosphorylation state.

MALDI-TOF may also be used 1) to facilitate the identification of *sequence*-specific sites of protein phosphorylation and dephosphorylation, 2) to monitor protein and peptide phosphorylation and dephosphorylation reaction rates, even in complex unfractionated mixtures, 3) to determine the minimum primary structure necessary for the phosphorylation of specific protein surface domains, and 4) to evaluate the effects of intact protein phosphorylation and dephosphorylation on susceptibility to subsequent proteolytic events.

Although phosphopeptide detection by MALDI-TOF can be favored in the negative ion mode, for the purpose of these investigations, we found no significant or consistent advantages to the evaluation of spectra collected in this mode. In fact, a diminished signal intensity for the corresponding nonphosphorylated (or dephosphorylated) peptides was sometimes observed in the negative ion mode thus creating a disadvantage.

Acknowledgments

We thank Chee Ming Li and Gary L. Cook in our laboratory for peptide synthesis and amino acid sequencing. Funded, in part, with federal funds from the U.S. Department of Agriculture, Agricultural Research

Service under Cooperative Agreementnumber 58-6250-1-003. The contents of this publication do not necessarily reflect the views or policies of the U.S. Department of Agriculture, nor does mention of trade names, commercial products, or organizations imply endorsement by the U.S. Government.

References

1. Taylor, S.S., Beuchler, J.A., and Yonemoto, W. (1990). Ann. Rev. Biochem. **59**, 971.
2. Sood, S.M., Chang, P., and Slattery, C.W. (1990). Arch. Biochem. Biophys. **277**, 415.
3. Hollosi, M., Urge, L., Perczel, A., Kajtar, J., Teplan, I., Otvos, L., and Fasman, G.D. (1992). J. Mol. Biol. **223**, 673.
4. Meggio, F., Perich, J.W., Meyer, H.E., Hoffman-Posorske, E., Lennon, D.P.W., Johns, R.B., and Pinna, L.A. (1989). Eur. J. Biochem. **186**, 459.
5. Meisel, H., and Frister, H. (1988). *In* "Milk Proteins. Nutritional, Clinical, Functional and Technological Aspects" (C.A. Barth, E. Schlimme, eds), p. 150. Springer-Verlag, Darmstadt, New York.
6. Allen, M.H., and Hutchens, T.W. (1992). Rapid Commun. Mass Spectrom. **6**, 308.
7. Hutchens, T.W., and Allen, M.H. (1992). Rapid Commun. Mass Spectrom. **6**, 469.
8. Hutchens, T.W., Nelson, R.W., and Yip, T.-T. (1991). J. Mol. Recog. **4**, 151.
9. Hutchens, T.W., Nelson, R.W., Allen, M.H., Li, C.M., and Yip, T.-T. (1992). Biol. Mass Spectrom. **21**, 151.
10. Hutchens, T.W., Nelson, R.W., and Yip, T.-T. (1992). FEBS Lett. **296**, 99.
11. Nelson, R.W., and Hutchens, T.W. (1992). Rapid Commun. Mass Spectrom. **6**, 4.
12. Hutchens, T.W., Nelson, R.W., Li, C.M., and Yip, T.-T. (1992). J. Chromatogr. **604**, 125.
13. Fiat, A.-M., and Jolles, P. (1989). Molec. Cell Biochem. **87**, 5.
14. Yuan, Y.V., and Kitts, D.D. (1991). Nutr. Res. **11**, 1257.
15. Shimizu, M., Kato, M., Miwa, K., and Kaminogawa, S. (1991). Biochem. Internat. **24**, 1127.
16. Menon, R.S., and Ham, R.G. (1989). Nucleic Acid Res. **17**, 2869.
17. Yip, T.T., and Hutchens, T.W. (1992). FEBS Lett. **308**, 149.

Synthetic Phosphopeptide Epitope Mapping with Mass Spectrometric Verification

William T. Moore*, Wan-Kyng Liu[2], Shu-Hui Yen[2], Frederick L. Hall[3], and Richard M. Caprioli*

*The Analytical Chemistry Center,
the Department of Biochemistry and Molecular Biology,
the University of Texas Medical School, Houston, TX 77030
[2]Department of Pathology,
Albert Einstein College of Medicine, Bronx NY 10461,
and [3]the Division of Orthopaedic Surgery,
University of Southern California School of Medicine,
Children's Hospital, Los Angeles CA 90054

I. Introduction

A. Background

Paired helical filaments (PHF) are major constituents of the neurofibrillary tangle characteristic of Alzheimer's disease. Immunocytochemical and biochemical studies on PHF-enriched samples indicate the predominance of modified tau proteins (60-68 kD) in the PHF[1,2,3,4]. Normally tau proteins are associated with microtubules. The tau proteins in PHF (PHF-tau) however have different chemical characteristics[1]. A major biochemical difference seen in PHF-tau and normal tau (N-tau) appears to be an abnormally high phosphorylation state[5]. This difference is dramatically demonstrated by the differential reactivity of N-tau and PHF-tau to a monoclonal antibody, Tau-1. Whereas N-tau readily reacts with Tau-1 antibody, PHF-tau shows negative reactivity. If PHF-tau is treated with alkaline phosphatase however it becomes reactive, suggesting that the Tau-1 epitope is phosphorylated on the PHF-tau masking the site for recognition by the Tau-1 antibody[6]. A fine-structure map of the epitope and the site(s) of phosphorylation may suggest protein kinase candidates.

TECHNIQUES IN PROTEIN CHEMISTRY IV

B. Objective and Strategy

Since starting materials are limited and isolation yields are low for PHF-tau[4,7,8], we chose a synthetic peptide approach to map the phosphorylation dependent epitope of PHF-tau with a strategy that was described previously to map a rubella virus capsid epitope[9,10]. In the case of tau proteins, the area of interest for our synthetic focus could be limited to a 19 residue sequence corresponding to residues 189-207 of an isoform of human tau: PKSGDRSGYSSPGSPGTPG. This limited sequence was deduced from cDNA clones expressing recombinant tau segments[11].

In the present study we describe the preparation and characterization of the tau residue #189-207 synthetic peptide and its assembly derivatives that permit a more concise definition of the Tau-1 epitope. We also describe the preparation and characterization of synthetic phosphopeptide derivatives of the tau residue #189-207 region. Representative data derived from a more extensive study to be described separately[12] involving immunoanalysis and in vitro enzymatic phosphorylation by purified preparations of $P34^{cdc2}/P58^{cyclin\ A}$ proline directed protein kinase (PDPK) are presented that indicate a primary epitope-masking phosphorylation site.

II. Experimental

A. Chemical Synthesis of Peptides

All peptides were synthesized with an Applied Biosystems Inc. (ABI, Foster City, CA) Model 430A automated peptide synthesizer using Merrifield solid phase t-BOC(benzyloxycarbonyl) methodology. The starting material was a BOC-C-terminal amino acid-OCH2-Pam-copoly(styrene-1%-divinylbenzene) resin obtained from ABI. All BOC amino acids were either obtained from Peninsula Laboratories Inc.(Belmont, CA) or Advanced Chemtech (Louisville, KY). The side chain protecting groups for the non-phosphopeptide syntheses were Bzl(Ser & Thr), BrZ(Tyr), Cl-Z(Lys), OBzl(Asp), Mts(Arg). All the peptide synthesizer reagent and solvent chemicals were from ABI except trifluoroacetic acid (Halocarbon Products Corp., Hakensack, NJ), DCM, and DMF (Burdick and Jackson).

For the synthesis of O-phosphoserine and O-phosphothreonine containing peptides the approach involving "methyl" phosphate protected groups described primarily for the synthesis of O-phosphotyrosine peptides[13,14] was chosen for several reasons; 1. N-Boc-O-Dimethylphosphono-L-serine (code 15422) and threonine (code 15442) were commercially available from Peninsula Laboratories (Belmont, CA). 2. Peptides could be deprotected with trifluoromethanesulfonic acid (TFMSA) treatment with success (personal communication by Dr. Pak Ho of Peninsula Labs.). 3. The immediate availability of on-site fast atom bombardment mass spectrometry (FABMS) permitted critical assessment of the synthetic chemistry. We have previously used FABMS to complement analysis of phosphopeptides prepared by Arendt et al.[15] using the "phenyl" phosphate-protected residues and 4. An ABI Model 430A peptide synthesizer with a peptide-resin sampling feature was available to permit stepwise mass spectrometric analysis of the synthesis[16].

The synthetic peptides were prepared using small-scale (0.1mmole) rapid cycles (system software version 1.40). The run file opt 21 r vessel cycles for N-Boc-O-dimethylphosphono-L-serine were the standard ones recommended for Ser(Bzl). The same cycles were used for the N-Boc-O-dimethylphosphono-L-threonine except that the activator cycle was changed to the aboc22 version that incorporates a 0.3 mL DMF delivery in addition to a 2.5 mL DCM delivery and an extended dissolution time for a 1 mmole amount of the Boc amino acid derivative.

B. TFMSA Treatment/Peptide-Resin Cleavage and Deprotection

Micro-TFMSA cleavages performed on peptide-resin aliquots removed after coupling reaction were carried out as previously described[9,16]. Small scale preparative TFMSA cleavages used to prepare the phosphopeptides were performed as described previously with minor modifications. TFMSA cleavage was performed on 50 mg amounts of peptide-resin with appropriate scale-up of the reagent volumes. TFMSA treatment was allowed to proceed for 3 h at room temperature to promote maximal demethylation.

C. Reverse Phase HPLC

Synthetic peptides were separated by preparative HPLC using a reverse phase 1 X 10 cm C-8 Aquapore RP300 column (ABI). The column was eluted with a linear acetonitrile gradient of 0% B to 100% B developed over 30 minutes. Solvent A was 0.1% TFA and solvent B was 70% acetonitrile, 30% water and 0.1% TFA. An ABI model 151A HPLC system was used at a flow rate of 5 mL/min with UV detection at 230 nm. The peaks containing phosphopeptide derivatives were mass identified by FABMS and fractions were pooled and lyophilized.

D. Fast Atom Bombardment Mass Spectrometry (FABMS)

One μL aliquots of the RP-HPLC effluent were mixed with 1.5 μL of glycerol/thioglycerol on the stainless steel target of a sample probe. The probe was introduced into a Kratos MS50RF mass spectrometer equipped with an Ion Tech B11NF saddle field gun operated at 8 kV and 40 μa of current using Xe to create energetic atoms. The mass spectrometer was generally operated at a resolving power of 1000 to 1200 and an accelerating voltage of 8 kV. The instrument was set to scan an appropriate mass range usually from 2000 to 700 amu and data were collected with multichannel analyzer programs available on the DS90 data system. CsI was used to calibrate the instrument.

E. Immunoreagents

Tau Protein: Tau protein preparations were isolated from bovine brain extracts by heat and perchloric acid treatments and trichloroacetic acid precipitation by a procedure previously described[1]. The final protein concentration of the preparation was 10 μg/mL.
Tau-1 monoclonal antibody: Tau-1 is a mouse monoclonal anti-tau antibody provided by and originally characterized by Dr. L. I. Binder of the University of Alabama[6].

F. Competitive Enzyme-Linked Immunoadsorbent Assay (ELISA)

A competitive ELISA was selected over standard ELISA in order to avoid possible "false" negative results that may arise[17]. For the competitive ELISA, the antigens selected for adsorption to 96-well ELISA plates were either purified preparations of bovine tau protein or HF-derived Tau-1 peptide. Tau protein was preferred because of the considerations mentioned[17]. The primary antibody was Tau-1 monoclonal antibody that had been preadsorbed with or without varying amounts of synthetic peptide derivatives by incubation at 4°C overnight. The amount of primary Tau-1 antibody bound to the bovine tau protein coated wells was determined colorimetrically using a biotinylated secondary antibody and avidin-conjugated horse radish peroxidase /2,2'-azino-di(3-ethyl-benzthiazoline sulfonate) as the chromogen generating system. Attenuated color development was an indication of immunoreactivity of a synthetic peptide preparation with Tau-1 monoclonal antibody. Each assay point was determined in duplicate and two or more assays were carried out per peptide. The details of the assay will be described in another publication[12].

G. Enzymatic Phosphorylation

The phosphorylation studies involving the synthetic peptides were performed under the auspices of Dr. Frederick L. Hall of the University of Southern California School of Medicine and will be described in detail elsewhere[12]. Cytoplasmic $p34^{cdc2}/p58^{cyclin\ A}$ PDPK was isolated from FM3A Mouse mammary carcinoma cells according to methods described previously[18,19]. Synthetic peptides were phosphorylated by incubating peptides with the enzyme preparation for 30 min. at 30° C in a reaction medium containing 100 mM Tris acetate buffer (pH7.5), 10 mM magnesium acetate and 100 μM of [gamma-^{32}P]ATP (specific activity 1000-1500 cpm/pmol). Negative control reactions were carried out in peptide-minus media. The reaction was stopped by the addition of perchloric acid (7% final concentration). The labelled peptides were harvested from supernatants obtained after centrifugation at 13,000xg by adsorption to phosphocellulose papers. The papers were thoroughly washed in phosphoric acid, dehydrated in acetone and air dried before subjecting the papers to liquid scintillation counting for measurement of ^{32}P incorporation.

III. Results

TAU-1 EPITOPE MAPPING USING THE MICRO-TFMSA CLEAVAGE METHOD TO GENERATE SYNTHETIC PEPTIDE DERIVATIVES WITH FABMS STRUCTURE VERIFICATION.

A summary of all the synthetic peptide derivatives analyzed by FABMS for structural status and immunological reactivity to Tau-1 monoclonal antibody is shown in Table 1. Tau-1 peptide comprises residues 189-207 of the largest isoform of tau and is the sequence of a recombinant expression product that was found to have immunoreactivity within this region[11]. The upper portion of Table 1 summarizes the assembly product sequences, the theoretical masses, FABMS derived masses and immunological analysis results for the dephospho-synthetic peptides. In all eight dephospho-peptides were extensively analyzed. There are two entries for Tau-1 peptide. The first sequence is for the aspartyl form of the peptide and the second, distinguished by the bar over the DR sequence, represents the aspartimide form of the peptide. The peptides designated A through F in the Table 1 are micro-TFMSA derived assembly derivatives that were used to epitope map the dephospho Tau-1 peptide. Figure 1 shows the FABMS spectra obtained for three representative micro-TFMSA generated assembly products and constitutes a mass spectrometric verification of the synthetic run. The major signals observed either have assigned masses very close to theoretical (peptide B) or have masses that are -18 amu of the theoretical (peptide D and F). Since the -18 does not appear until the Asp #5 is encountered (Table 1), this difference has been attributed to aspartimide formation. Figure 2 compares the FABMS spectra for full length Tau-1 peptide. The upper panel for Figure 2 indicates a major signal that has a mass assignment that is correct. There is however a -18 amu satellite peak also present. Note in the lower panel that the -18 peak is the major signal. These two preparations of Tau-1 peptide were derived from two different micro-TFMSA treatments performed at different days on the same peptide resin and indicate that the aspartimide formation is a result of the cleavage chemistry conditions. Additional peaks are also observed in the mass spectra shown in Figures 1 and 2. The additional peaks seen in Figures 1 and 2 having masses +150 or multiples of

TABLE I

PEPTIDE SUMMARY: MASS SPECTROMETRIC ANALYSIS, IMMUNOREACTIVITY & IN VITRO PHOSPHORYLATION

		$(M+H)^+$		Relative Immunoreactivities	Relative Phosphorylation[1]
		theor.	obs.		
Tau-1 peptide	PKSGDRSGYSSPGSPGTPG	1790.8	1790.9	ND	-
Tau-1 peptide	PKSGDRSGYSSPGSPGTPG	1772.8	1773.0	100	-
Peptide A	PGSPGTPG	669.3	669.4	0	-
Peptide B	RSGYSSPGSPGTPG	1306.6	1306.6	2	-
Peptide C	DRSGYSSPGSPGTPG	1421.6	1403.8	27	-
Peptide D	GDRSGYSSPGSPGTPG	1478.6	1461.1	164	100
Peptide E	SGDRSGYSSPGSPGTPG	1565.7	1548.0	124	-
Peptide F	KSGDRSGYSSPGSPGTPG	1693.8	1676.1	100	-
PT1	GDRS*GYSSPGSPGTPG	1558.6	1540.8	12	0
PT2	GDRSGYS*SPGSPGTPG	1558.6	1541.1	12	0
PT3	GDRSGYSS*PGSPGTPG	1558.6	1540.4	5	0
PT4	PKS*GDRSGYSSPGSPGTPG	1870.8	1852.8	200	244
PT5	PKSGDRSGYSSPGS*PGT*PG	1950.8	1933.5	45	16
Bovine Tau		-	ND	28	-

ND = not determined
* denotes the position of a phosphoryl group
1. Enzymatic phosphorylation. See text for explanation.

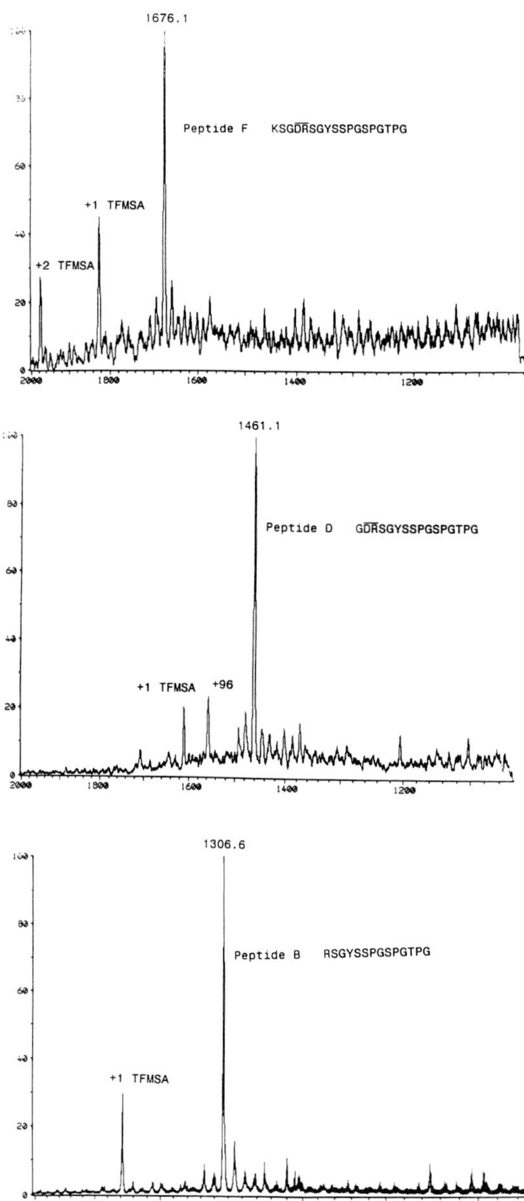

Figure 1. FABMS analysis of synthetic Tau-1 peptide assembly derivatives (Peptides B,D and F).

150 above the mass of the major signal represent noncovalent TFMSA adducts. Note the number of TFMSA adducts corresponds to the number of basic residues in the peptide. In the middle panel of Figure 1 another ion is noted having a mass +96 over that of the major peptide signal. This mass probably represents a trifluoroacetyl group that has been most likely acquired from TFA contaminated with trifluoroaceticanhydride that was used in the micro-TFMSA cleavage chemistry.

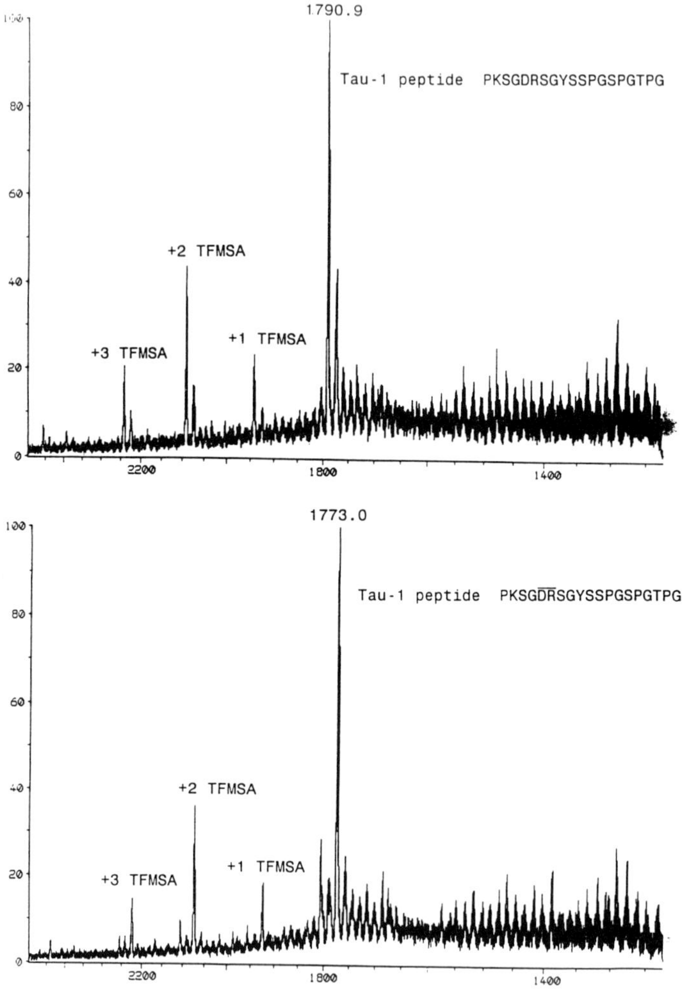

Figure 2. FABMS analyses of full length Tau-1 peptides prepared by two different cleavage conditions. Upper panel: aspartyl peptide $[(M+H)^+ = 1790.9]$. Lower panel: aspartimidyl peptide $[(M+H)^+ = 1773.0]$. The bar over the DR sequence designates the aspartimidyl form of the peptide.

Figure 3 indicates the results of a competitive ELISA analysis of the full length aspartyl and aspartimidyl forms of the Tau-1 peptide evaluated by FABMS (Figure 2) and assembly derivative peptide B evaluated by FABMS (Figure 1). From this set of competitive inhibition curves several characteristics of the epitope can be defined. First aspartimidyl-Tau-1 is 70% less immunoreactive than aspartyl-Tau-1 suggesting that the succinimide ring that has formed has disrupted the epitope. However it should be noted that the aspartimidyl-Tau-1 is 18-fold more reactive than the C-terminal 14-mer (peptide B) comprising 74% of the sequence. This immunoanalysis (Figure 3) suggests some limits for the Tau-1 epitope. These data suggest that the either the N-terminal or central portion of the sequence may be more important than the C-terminal portion of the molecule for epitope expression.

To further define or map the epitope, additional assembly derivatives spanning the N-terminal region were examined. The results for the immunoanalysis for these products are shown in Figure 4. This set of inhibition curves representing aspartimidyl form assembly derivatives show that peptide D (see Fig. 1 for FABMS evaluation) is the most immunoreactive peptide of the series. Peptide D is 1.6 times as active as the full length

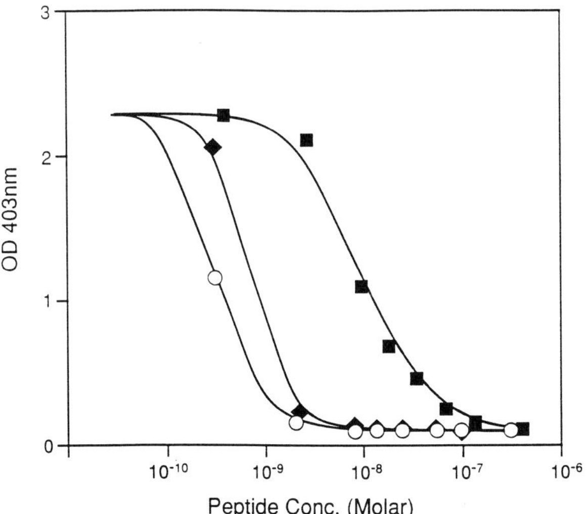

		Relative Activities (%)
○ Tau-1 peptide	PKSGDRSGYSSPGSPGTPG	100
◆ Tau-1 peptide	PKSG$\overline{\text{DR}}$SGYSSPGSPGTPG	30
■ Peptide B	RSGYSSPGSPGTPG	1.7

Figure 3. Immunoanalysis of micro-TFMSA derived Tau-1 peptide forms and the C-terminal region assembly derivative. The bar over the DR sequence designates the aspartimidyl form of the peptide.

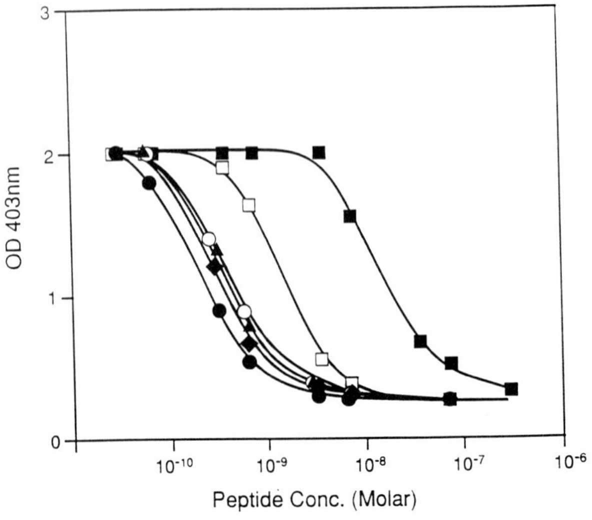

		Relative Activities (%)
○ Tau-1 peptide	PKSG$\overline{DR}$SGYSSPGSPGTPG	100
■ Peptide B	RSGYSSPGSPGTPG	1.6
□ Peptide C	$\overline{DR}$SGYSSPGSPGTPG	27.6
● Peptide D	G$\overline{DR}$SGYSSPGSPGTPG	164
◆ Peptide E	SG$\overline{DR}$SGYSSPGSPGTPG	124
▲ Peptide F	KSG$\overline{DR}$SGYSSPGSPGTPG	100

Figure 4. Immunoanalysis of the stepwise assembly derivatives of the Tau-1 peptide. The bar over the DR sequence designates the aspartimidyl form of the peptide.

aspartimidyl-form of the Tau-1 peptide. The whole set of inhibition curves suggests a further refinement of the epitope map for Tau-1 and indicates that the residues comprising the central portion of the sequence extending to the N-terminal side of the peptide are most important. We would give the epitope at this stage of analysis the following definition:

GDRSGYSS--->?

This region corresponds to residues 192-199 of the longest human tau isoform.

IDENTIFICATION OF EPITOPE MASKING PHOSPHORYLATION SITE(S) USING SYNTHETIC PHOSPHOPEPTIDE DERIVATIVES

The epitope region suggested above contains three possible phosphorylation sites and the whole Tau-1 peptide contains an additional three. In order to map the Tau-1 epitope further and to address the phosphorylation state epitope masking site(s), we prepared several synthetic phosphopeptide derivatives. The sequences of the phosphopeptide derivatives (PT1 through 5), the FABMS analysis data and immunoreactivity data are summarized in Table 1. Figure 5 indicates the quality of the phosphopeptide synthesis for one of the more important phosphopeptide derivatives and is reflective of the success achieved with the other syntheses (data not shown). Fifty mg amounts of peptide-resins yielded several mg of phosphopeptide after purification. Immediately following the small scale preparative TFMSA treatment, deblocked phosphopeptide was subjected to reverse phase HPLC. The chromatographic profiles indicated a major single peak eluting early from a preparative RP-HPLC column. Elution profiles for all the phosphopeptides were very similar to that of PT3 shown in Figure 5 (upper panel). The middle panel of Figure 5 is the FABMS analysis of the crude TFMSA-derived material. The mass spectrum indicates that phosphopeptide synthesis proceeded successfully. The spectrum indicates that the peptide is +80 that of the dephospho-relative (peptide D) shown in the middle panel of Figure 1. Other peaks that should be noted are those for noncovalent TFMSA adducts and satellite peaks representing incompletely deprotected mono and dimethyl protected phosphoryl groups. The structures of the dimethylphosphono group and derivatives are shown as an inset in the middle panel of Figure 5. The FABMS analysis of the peak fraction pool is shown in the lower panel of Figure 5 and indicates that the material is efficiently purified by the preparative C8 column. Only a minor contamination by monomethyl derivative is observed. All synthetic peptides, even the PT5 derivative containing both the phospho-threonine and serine moieties, behaved similarly (data not shown). It should be noted that crude PT5 (phosho-Ser plus phospho-Thr derivative) contained the greatest amount of monomethyl derivative that was subsequently removed by C8 runs (data not shown).

Phosphopeptide derivatives structurally verified by FABMS were subjected to competitive ELISA. As summarized in Table 1 the presence of a phosphoryl-group had a profound effect on the immunoreactivity of the Tau-1 peptide reactivity. The phosphopeptide analogs chosen, PT1 through 3 (lower portion of Table 1), were analogs of the most immunoreactive dephospho analog, peptide-D (upper portion Table 1). The competitive ELISA data indicate that immunoreactivities of the phosphopeptide derivatives PT1 through 3 are reduced 92 to 97%, the PT3 derivative being the least immunoreactive derivative. These data suggest that Ser 199 might be the site that is phosphorylated in PHF-tau. It is of interest to note that derivatives phosphorylated outside of the epitope region still retained significant immunoreactivities. Phospho-derivative PT4 having a phosphoryl serine near the N-terminus was two-fold more immunoreative than dephospho Tau-1

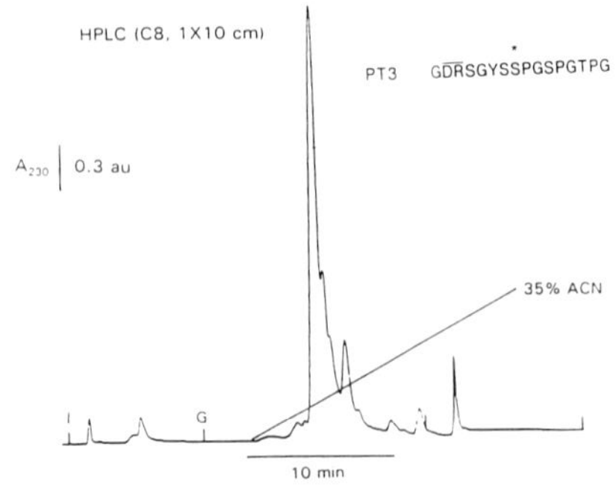

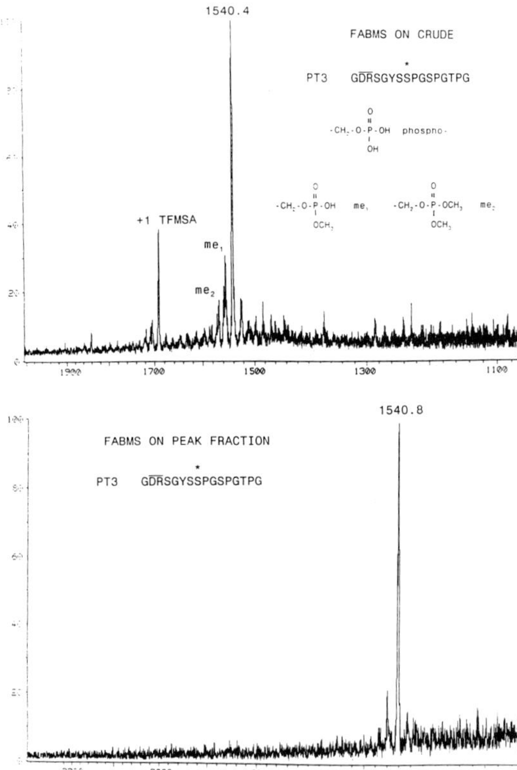

Figure 5. Characterization of synthetic phosphopeptide PT3. Upper panel: Reverse phase-HPLC profile. Middle panel: FABMS analysis of crude PT3 derived from TFMSA treatment. Lower panel: FABMS analysis of HPLC purified PT3. The bar over the DR equence designates the aspartimidyl form of the peptide.

peptide. This difference raises the possibility that these peptides may be good models for studies probing phosphorylation state induced secondary structure in small peptides.

ENZYMATIC PHOSPHORYLATION OF TAU-1 PEPTIDE DERIVATIVES BY PROLINE DIRECTED PHOSPHOKINASES (PDPK)

Another test of the structural integrity and epitope definition of the synthetic peptides and phosphopeptide analogs prepared for this study was to determine whether these peptides could serve as kinase substrates. Since $p34^{cdc2}/p58^{cyclin\ A}$ PDPK[19] catalyzes the phosphorylation of bovine tau[20] as well as other neural substrates [18,21], we examined whether the Tau-1 synthetic peptides could serve as substrates for this enzyme. Of the dephospho-peptide analogs four peptides were examined all in the aspartimidyl-form if they contained Asp residues. Peptide A the shortest C-terminal peptide containing two possible consensus PDPK phosphorylation site sequences (XSPX) did not phosphorylate upon radiometric assay suggesting that the C-terminal phosphorylation sites are not recognized by the enzyme (data not shown). Some caution must be exercised in drawing conclusions from the results of this derivative. The peptide may be too short to serve as a high affinity substrate for the enzyme or permit capture by the phosphocellulose papers used in the radiometric ^{32}P assay. Full length Tau-1 peptide and N-terminal truncated versions of the peptide did undergo phosphorylation catalyzed by the enzyme. It is of interest to note that Peptide B and full length Tau-1 peptide did undergo similar levels of phosphorylation (7.0 and 8.7 pmol/30 min). These results suggest that the N-terminal region located Ser in the Tau-1 peptide is an unlikely phosphoryl acceptor site and that the serines occupying the central region of Tau-1 peptide might be probable sites.

The availability of the well-characterized phosphopeptide analogs permitted a further evaluation of the putative PDPK phosphorylation site(s). Table 1 (Relative Phosphorylation column) shows results obtained with the most immunoreactive dephospho-analog, peptide D, and three phosphorylated derivatives of this peptide (PT1-PT3). The dephosphopeptide peptide D incorporated levels of ^{32}P (12.1 pmol/30 min) comparable to the synthetic peptides discussed above. No ^{32}P incorporation was seen in the phosphopeptide analogs PT1 through PT3 supporting the prediction that the X-S-P-X motif present in the Tau-1 peptide at residue #11 and representing Ser #199 in human tau is a likely target for this enzyme[19,22].

Two longer full-length Tau-1 peptide phosphopeptide derivatives were also examined. In these synthetic analogs the phosphoryl groups resided at hydroxyl aminoacids distanced from the PDPK motif area. PT4 having a phosphoryl group present at the N-terminally located Ser was a good substrate in the radiometric assay and served as a good positive control indicating peptide viability even though it had gone through a prolonged TFMSA treatment to fully demethylate the product. It should be noted (see mass spectral data in Table 1) that all these peptides are aspartimidyl-peptides because of the extended TFMSA treatment. The PT5 phosphopeptide having phosphoryl-groups at two possible PDPK consensus sites did serve as a substrate for the enzyme however the level of phosphorylation was low (Table 1) suggesting that these sites might be secondarily phosphorylated or that phosphoryl groups in this region negatively influence the efficacy of phosphorylation at the nearby favored site (Tau-1 peptide Ser #11 or human tau Ser position #199).

Conclusions

1. FABMS analysis provided a critical structural analysis of the synthetic peptide chemistry performed in this study for both the dephospho and phosphopeptides used in the ELISA assay to map the Tau-1 antibody. Aspartimide formation was noted and this information was taken into account for definition of the Tau-1 epitope. The mass spectrometry was essential in establishing the TFMSA cleavage conditions necessary to insure complete demethylation of the dimethylphosphono protection groups and also for the interpretation of reverse phase HPLC profiles generated by the purification runs.
2. The study demonstrates that phosphopeptides can be easily prepared with automated peptide synthesis instrumentation using commercially available dimethylphosphono tBOC-serine and threonine precursors. If aspartic acid is present, however, one should expect aspartimide formation arising from the harsher TFMSA conditions required for complete demethylation. In this particular case the aspartimide formation did not obviate the epitope mapping study.
3. The micro-TFMSA generated assembly products in combination with the competitive ELISA analysis permitted the definition of a minimal epitope for Tau-1 monoclonal antibody. The epitope corresponds to residue # 192-199 of the longest human tau isoform constituting residues 4-11 of the Tau-1 synthetic peptide shown underlined below:

```
       4           11
P K S G D R S G Y S S P G S P G T P G
      ---------------
                    ^
                    |
                   PO4
```

4. Analysis of the immunoreactivity of synthetic phosphopeptides by competitive ELISA suggested that the primary site of phosphorylation that results in the masking of the epitope in PHF-tau is Ser #199 (Tau-1 peptide #11, see above). The synthetic phosphopeptide having the phosphoryl group at this position was the least immunoreactive product. The presence of a phosphoryl group at any of the serine portions within the epitope dramatically diminished immunoreactivity. The presence of phosphoryl groups outside the designated epitope area either enhanced (PT4) or less extensively lowered (PT5) immunoreactivity.
5. Enzymatic phosphorylation studies using $p34^{cdc2}/p58^{cyclin\ A}$ PDPK and the synthetic peptides as in vitro substrates supported the epitope conclusions derived from the ELISA immunoanalysis. The presence of phosphoryl groups within the proposed epitope region blocked

incorporation of ^{32}P. Interestingly, PT4 having a phosphoryl group at the position that enhanced immunoreactivity appeared to improve substrate selectivity. These observations suggest that this peptide might be an interesting candidate for examining the effects of phosphoryl group attachment on secondary structure.

6. The approach of the present study should permit a means of analysis for addressing the identification of the specific members of the proline-directed protein kinase family that may be deregulated or ectopically expressed in Alzheimer's disease.

References

1. Ksiezak-Reding, H., Binder, L.I. and Yen, S-H (1990). *J. Neurosci. Res.* **25**, 420-430.
2. Greenberg, S.C. and Davies, P. (1990). *PNAS* **85**, 4506-4510.
3. Lee, V.M-Y., Balin, B.J., Otvos, L. and Trolanowski, J.Q. (1991). *Science* **251**, 675-678.
4. Liu, W-K., Ksiezak-Reding, H. and Yen, S-H (1991). *JBC* **266**, 21723-21727.
5. Ksiezak-Reding, H., Liu, W-K. and Yen, S-H (1992). *Brain Res.* In Press.
6. Papasozomenos, S.C. and Binder, L.I. (1987). *Cell Motil. Cytoskel.* **8**, 210-216.
7. Ihara, Y., Abraham, C., Selkoe, D.J. (1983). *Nature* **304**, 727-730.
8. Wischik, C., Novak, M., Edwards, P.C., Klug, A., Tichelaar, W. and Crowther, R.A. (1988). *PNAS* **85**, 4885-4888.
9. Moore, W.T., Wolinsky, J.S., Suter, M. J-F., Farmer, T.B. and Caprioli, R.M. (1992). In "Techniques in Protein Chemistry III" (ed. R. Hogue Angeletti) 183-197. Academic Press Inc., San Diego, CA.
10. Wolinsky, J.S., McCarthy, M., Allen-Cannady, O., Moore, W.T., Jin, R., Cao, S-N, Lovett, A. and Simmons, D. (1991). *J. Virol.* **65**, 3986-3994.
11. Kosik, K.S., Orecchio, L.D., Binder, L.I., Trojanowski, J.Q., Lee, V. M-Y. and Lee, G. (1988). *Neuron.* **1**, 817-825.
12. Liu, W-K., Moore, W.T., Williams, R.T., Hall, F.L. and Yen, S-H. (1992). Manuscript submitted for publication.
13. Kitas, E.A., Perich, J.W., Johns, R.B. and Tregear, G.W. (1988). *Tetrahedron Lett.* **29**, 3591-3592.
14. Perich, J.W., (1991). Synthesis of O-phosphotyrosine-containing peptides in "Methods in Enzymology", (Hunter, T. and Sefton, B.M., eds.) Vol. 201, pp. 234-245 Academic Press, Inc., San Diego, CA.
15. Arendt, A., Palczewski, K., Moore, W.T., Caprioli, R.M., McDowell, J.H. and Hargrave, P.A. (1989). *Int. J. Pept. Protein Res.* **33**, 468-476.
16. Moore, W.T. and Caprioli, R.M. (1991). In "Techniques in Protein Chemistry II", (J. Villafranca, ed.) p. 511-527, Academic Press, Inc., San Diego, CA.
17. Geysen, H.M., Rodda, S.J. Mason, T.J., Tribbick, G. and Schoofs, P.G. (1987). *J. Immunol. Meth.* **102**, 259-274.
18. Hall, F.L., Braun, R.K., Mitchell, J.P. and Vulliet, P.R. (1990). *Proc. West. Pharmacol. Soc.* **3**, 213-217.
19. Hall, F.L., Braun, R.K., Mihara, K., Fung, Y-K. T., Berndt, N., Carbonaro-Hall, D.A. and Vulliet, P.R. (1991). *JBC* **266**, 17430-17440.
20. Hall, F.L., Mitchell, J.P. and Vulliet, P.R. (1990). *JBC* **265**, 6944-6948.
21. Guan, R.J., Hall, F.L. and Cohlberg, J.A. (1992). *J. Neurochem.* In Press.
22. Vulliet, P.R., Hall, F.L., Mitchell, J.P. and Hardie, G.D. (1989). *JBC* **264**, 16292-16298.

SECTION IV

Peptide Synthesis

Evaluation of Peptide Synthesis As Practiced in 53 Different Laboratories

Gregg B. Fields,[1] Steven A. Carr,[2] Daniel R. Marshak,[3] Alan J. Smith,[4] John T. Stults,[5] Lynn C. Williams,[6] Ken R. Williams,[7] and Janis D. Young[8]

[1]Dept. of Lab Medicine & Pathology, University of Minnesota, Minneapolis, MN 55455
[2]SmithKline Beecham Pharmaceuticals, King of Prussia, PA 19406
[3]Cold Spring Harbor Laboratory, Cold Spring Harbor, NY 11724
[4]Beckman Center, Stanford University Medical Center, Stanford, CA 94305
[5]Genentech, Inc., South San Francisco, CA 94080
[6]Norris Cancer Research Institute, University of Southern California, Los Angeles, CA 90033
[7]Howard Hughes Medical Institute, Yale University, New Haven, CT 06536
[8]Dept. of Biological Chemistry, University of California Medical School, Los Angeles, CA 90024

I. Introduction

Peptide synthesis has become one of the most important methodologies in bioorganic chemistry. Synthetic peptides serve as probes of biological structure and function and act as important intermediates for the development of enzyme inhibitors and peptidomimetics as therapeutic agents. The automation of solid-phase synthesis has permitted a great variety of researchers to utilize synthetic peptides. The Association of Biomolecular Resource Facilities (ABRF) has ~130 member laboratories that are engaged in the synthesis of peptides and structural analysis of peptides and proteins as a service in academic, government, and research institutions and in private industry. The ABRF Committee on Peptide Synthesis and Mass Spectrometry was formed to evaluate the quality of the synthetic methods utilized in its member laboratories for peptide synthesis. Peptide synthesis, as defined by this committee, includes the chemistries used for peptide assembly, cleavage, purification, and characterization of the final product. An initial study, launched in 1991, requested the synthesis of a test peptide by ABRF member laboratories. This ABRF committee characterized products from these syntheses by amino acid analysis (AAA), high-performance liquid chromatography (HPLC), capillary electrophoresis (CE), mass spectrometry, and biological activity. Some of the results were surprising; 17% of the 36 independent ABRF laboratory crude samples did not contain any of the desired peptide product (1). The respective cleavage conditions of Fmoc and Boc solid-phase peptide synthesis were believed to be the primary source of synthetic difficulties, since non-desired products were the result of covalent adducts, not deletions. This year a similar study was designed whereby problems in peptide assembly versus cleavage were evaluated by an array of analytical techniques, including AAA, HPLC, CE, electrospray mass spectrometry (ESMS), LC-ESMS, plasma-desorption mass spectrometry (PDMS), fast-atom-bombardment mass spectrometry (FABMS), and matrix-assisted laser desorption/ionization mass spectrometry (MALDI-MS). A total of 53 ABRF laboratories participated in the study by supplying 58 crude and 33 purified samples of a peptide whose sequence was designed by this ABRF committee. In addition, these laboratories supplied 42 peptide-resin samples for Edman degradation preview sequence analysis. The second year of a high response rate of ABRF member laboratories (40%) allowed for the continued evaluation of "state-of-the-art" peptide synthesis, as well as the critical examination of the strengths and weaknesses of a variety of analytical techniques used to characterize synthetic peptides (2).

TECHNIQUES IN PROTEIN CHEMISTRY IV

II. Materials and Methods

Participating ABRF laboratories were asked to synthesize the following peptide by the methodology most commonly used in their facility:

H-Gly-Val-Arg-Gly-Asp-Lys-Gly-Asn-Pro-Gly-Trp-Pro-Gly-Ala-Pro-Tyr-OH

This particular sequence was chosen based on the potential for side reactions during assembly, side-chain deprotection, and cleavage (3). Participants were asked to provide 5 mg of crude, salt free product and were given the option of providing 60 mg of peptide-resin and 2 mg of purified product. The samples were supplied to the committee in coded form via a third party to maintain participant anonymity but allow the participants to identify data sets resulting from their samples. Samples were prepared by dissolving 2-5 mg/mL in 0.05% aqueous TFA for AAA and HPLC, 2-5 mg/mL in water for CE, and 1-2 μg in 25% acetonitrile–0.05% TFA for ESMS, PDMS, FABMS, and MALDI-MS.

A. Amino Acid Analysis

Crude and purified samples (0.5 μg) were hydrolyzed for 24 h at 112 °C in 100 μL 6 N HCl, 0.2% phenol. Analysis was performed on a Beckman 6300 with a sulfated polystyrene cation-exchange column (0.4 x 25 cm). Quantitation was by a Nelson Analytical Model 4400 Data System.

B. Analytical HPLC/Capillary Electrophoresis

Samples were analyzed on a Perkin Elmer series 4 HPLC using a BioRad rp304 column (300 Å pore size, 4.6 x 250 mm). The linear gradient extended from 0.1% aqueous trifluoroacetic acid (TFA) to 70% acetonitrile (containing 0.09% TFA) over 33 min. The flow rate was 2 mL/min and the absorbance monitored at 210 nm using a Perkin Elmer LC 95 detector. Crude samples were injected by a Perkin Elmer SS 100 autosampler, while purified samples were injected manually. CE was performed on an Applied Biosystems 270HT using the 600 Data Analysis System. The buffer was 50 mM sodium biphosphate, pH 4.4. Voltage was 20 kV at 30 °C, the program was 27 min, and electrophoresis monitored at 220 nm after a 1 to 5 sec sample injection.

C. Mass Spectrometry

ESMS and LC-ESMS methods have been described (1,4). FABMS was carried out on MS-1 of a JEOL HX110HF tandem double-focusing mass spectrometer. Samples were prepared by addition of a 1 μL aliquot of the peptide solution to 1 μL of *m*-nitrobenzyl alcohol matrix on the sample target. Ions were generated by FAB with 6 keV Xe atoms. Resolution was set at 3000, and the mass axis was scanned from 375-2500 Da in 20 sec. PDMS was performed on a BioIon 20 (Applied Biosystems, Inc.) plasma desorption, time-of-flight mass spectrometer using a ^{252}Cf fission fragment source in the positive ion mode at an accelerating potential of 16 kV for 1 h at 8000 nsec intervals. Aliquots (1-10 μL) of the peptide solution were applied to an aluminized Mylar target pre-electrosprayed with nitrocellulose. Targets were dried under a stream of N_2 and not washed. Initial PDMS analysis indicated that some Fmoc samples contained a +28 Da peak which was not present on reanalysis. As a result, only the latter PDMS analyses were used in calculating the % desired product. MALDI-MS was accomplished with a Vestec LaserTech ResearcH laser desorption time-of-flight mass spectrometer, operated at 20 keV. Samples were prepared by addition of 1 μL of the diluted peptide solution (~10 pmol) to 1 μL of matrix solution [50 mM 2,5-dihydroxybenzoic acid (DHBA) or sinapinic acid (SA) in water] on the sample target, which were then allowed to dry at ambient temperature. The N_2 laser (337 nm) was triggered at 20 Hz. Spectra were acquired at 5 nsec resolution and were a summation of ion intensity for 64 laser pulses.

D. Sequence Analysis

Edman degradation sequence analysis was performed on an Applied Biosystems 477A Protein Sequencer/120A Analyzer. Protocols for solid-phase sequencing of Boc-synthesized peptide-resins have been described (5). Protocols for sequencing Fmoc-synthesized peptide-resins were identical to standard protein sequencing protocols (5).

III. Results and Discussion

A total of 58 crude and 33 purified peptide samples and 42 peptide-resins were supplied for inclusion in this study. Sixteen of the crude peptides were synthesized by Boc chemistry (28%) and 42 by Fmoc chemistry (72%) (Table I). For the prior ABRF study, 18 of the crude peptides were synthesized by Boc chemistry and 18 by Fmoc chemistry (1). Hence, the fraction of peptides synthesized by Fmoc chemistry is significantly increased in this year's as compared to last year's study. For peptide assembly, HBTU/HOBt (6) proved to be most popular, followed by carbodiimide-mediated coupling and PyBOP/HOBt (Table I). The predominant side-chain protecting group strategy for Fmoc-based syntheses was Arg(Pmc), Lys(Boc), Asp(O*t*Bu), Asn(Trt), and Tyr(*t*Bu). Exceptions were Arg(Mtr) and Trp(Boc), used in 4 and 2 of the syntheses respectively, and no side-chain protection of Asn, used in 7 of the syntheses. The predominant side-chain protecting group strategy for Boc-based syntheses was Arg(Tos), Lys(ClZ), Trp(For), and Tyr(BrZ), with approximately equal use of Asp(OBzl) or Asp(OcHex) and 3 syntheses incorporating Arg(Mts). Of the 42 peptide-resins assembled by Fmoc chemistry, 28 were cleaved by Reagent K and 5 by Reagent R (3) (2 laboratories did not indicate cleavage conditions). Of the 16 peptide-resins assembled by Boc chemistry, 10 were cleaved by HF containing anisole alone or anisole plus other scavengers (7). Three laboratories elected to deprotect Trp(For) by treatment of the peptide-resin with ~10% piperidine–DMF prior to HF cleavage. Peptides were purified by reversed-phase HPLC, with 23 laboratories using C-18 columns, 5 using C-8 columns, and 1 using a C-4 column. The majority of laboratories used gradient elutions, where eluent A was 0.05-0.1% aqueous TFA and eluent B was 60-100% acetonitrile containing 0.08-0.1% TFA.

AAA showed 51 of 58 crude products and 31 of 33 purified products to be compositionally correct (Trp was not quantitated). Peptide deletions included 1 des(Asn), 2 des(Pro), 1 partial des(Asn,Gly), and 1 des(2Gly) crude products and 2 des(Pro) purified products. The 2 other crude products with improper compositions showed correct compositions for the subsequent purified products. Based on mass spectrometric results, one of these crude products was indeed a complex mixture with no detectable desired peptide while the other crude peptide contained >80% of the desired peptide.

The HPLC retention time of the apparent desired peptide was 10.50 ± 0.19 min for crude samples (Figure 1A) and 10.25 ± 0.16 min for purified samples, the difference attributable to automated versus manual injection. HPLC analyses indicated a high percentage of successful syntheses, as 51 of 58 crude products had >25% of the apparent desired peptide and 28 of 33 purified products had >75% of the apparent desired peptide (Table II). In general, CE indicated a higher degree of purity (85%) of the submitted samples than HPLC (75%). However, with 41 samples run consecutively by CE, the main peak migration times advanced progressively by more than 4 min. This variability in CE migration time made it impossible to determine the relative % content of the *correct* peptide, whether by comparison to a known standard or by trying to achieve reproducibility of the individual peptide itself, and thus probably accounts for the higher estimated degree of purity as compared to HPLC. These CE results are consistent with the prior study's findings of varied reproducibility (1), even though different buffer pH was used in an attempt to improve reproducibility.

Table I. Chemistry and Instrumentation Used to Synthesize ABRF Test Peptide

	#		#
Samples received		Linker-resin	
Crude	58	PAM-polystyrene	16
Purified	33	Cl-polystyrene[a]	1
Peptide-resin	42	HMP-polystyrene[b]	30
		PAC-polystyrene[b]	4
General chemistry		Pepsyn KA	6
Boc/Benzyl	16	TentaGel S-AM	1
Fmoc/tButyl	42		
		Coupling chemistry	
Instrumentation[c]		Carbodiimide	8
Applied Biosystems		Carbodiimide/HOBt	12
430A	26	OPfp/HOBt	5
431A	15	HBTU/HOBt	25
Millipore		TBTU/HOBt	2
9050	7	PyBOP/HOBt	5
9500	1	BOP/HOBt	1[d]
9600	5		
Advanced ChemTech		Cleavage chemistry	
350	1	HF	15
Vega		TFMSA	1
250	1	TFA	42
BioTech			
7600	1		

[a]This linker-resin is commonly referred to as Merrifield resin.
[b]HMP- and PAC-polystyrene provide similar linkages but are prepared differently (3).
[c]One laboratory did not supply their instrument manufacturer.
[d]OPfp was used for Asn.

Thirteen peptide-resins representing each of the various coupling chemistries were subjected to Edman degradation sequence analysis. Nine showed highly efficient peptide assembly, resulting in desired sequence purities of >88%. The other 4 peptide-resins had <10% of the desired sequence. Deletions in these sequences were Asn^8 for one sample, Pro^{12} in one sample, Gly^1, Gly^7, and Asn^8 in one sample, and Gly^1, Gly^4, and Trp^{11} in one sample. These deletions were consistent with results from AAA and mass spectrometric analyses (see below), although, in the absence of tandem MS analysis, only sequence analysis could assign the location of deleted residues.

Table II. Characterization of ABRF Test Peptide

	Desired Product[a] (%)	HPLC[b] (%) Fmoc	HPLC[b] (%) Boc	ESMS[b] (%) Fmoc	ESMS[b] (%) Boc	FABMS[b] (%) Fmoc	FABMS[b] (%) Boc	PDMS[b] (%) Fmoc	PDMS[b] (%) Boc
Crude	0	7.1	6.2	9.5	13	7.1	6.2	9.5	13
	<25	2.4	13	2.4	0	4.8	6.2	2.4	0
	25-75	40	25	31	50	31	38	40	63
	>75	50	56	57	38	57	50	48	25
Purified	0	14	0	18	0	18	0	14	0
	<25	4.5	0	0	0	0	0	4.5	0
	25-75	4.5	0	4.5	18	4.5	9	14	27
	>75	77	100	77	82	77	91	68	73

[a]A Trp(For)-containing peptide was the desired crude product for 1 Boc synthesis.
[b]The total number of samples was 42 crude Fmoc, 16 crude Boc, 22 purified Fmoc, and 11 purified Boc.

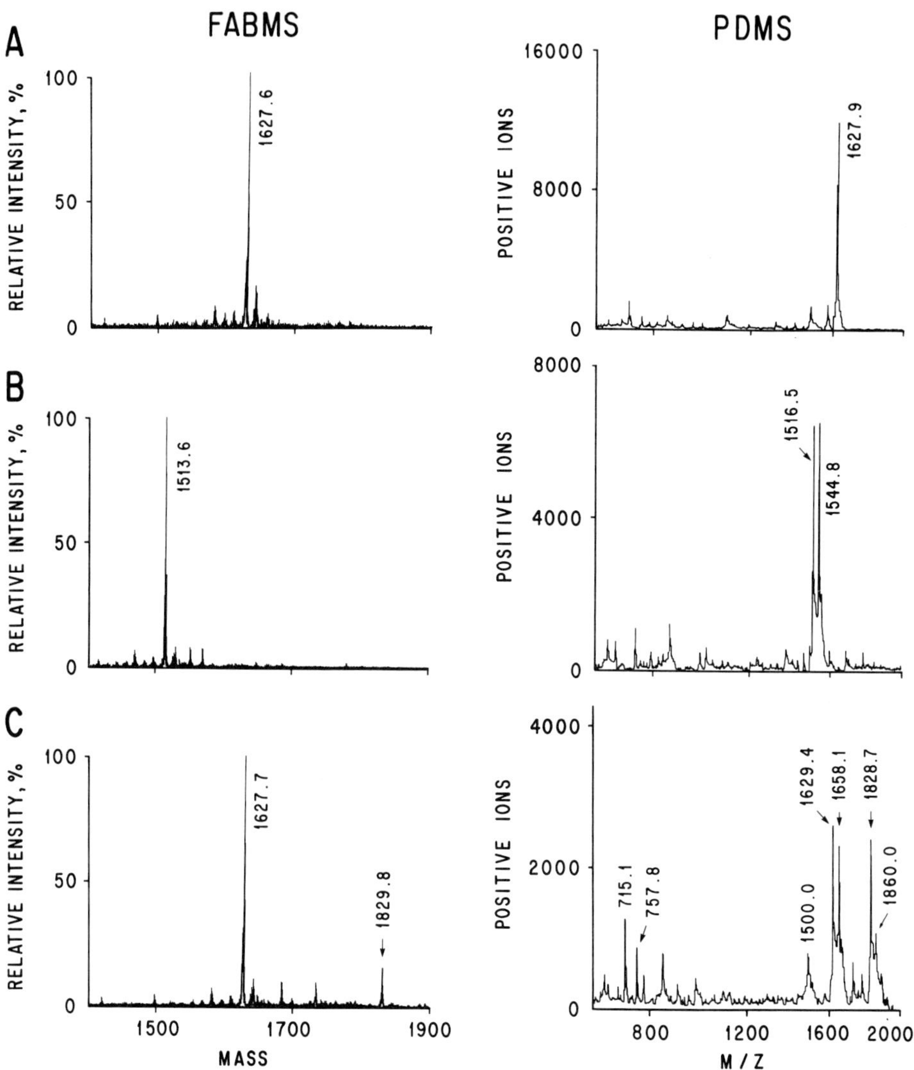

Figure 1. Analysis of selected ABRF crude peptide samples.

A. The relative abundance of the correct product was >90% by HPLC (retention time = 10.49 min), ESMS ($[M + 3H]^{3+}$ = 543.4 and $[M + 2H]^{2+}$ = 814.6), FABMS ($[M + H]^{+}$ =1627.6), and PDMS ($[M + H]^{+}$ = 1627.9).

B. The relative abundance of the correct product was 89% by HPLC (retention time = 10.51 min) and 0% by ESMS ($[M + 3H]^{3+}$ = 505.6 and $[M + 2H]^{2+}$ = 757.8), FABMS($[M + H]^{+}$ = 1513.6), and PDMS ($[M + H]^{+}$ = 1516.5).

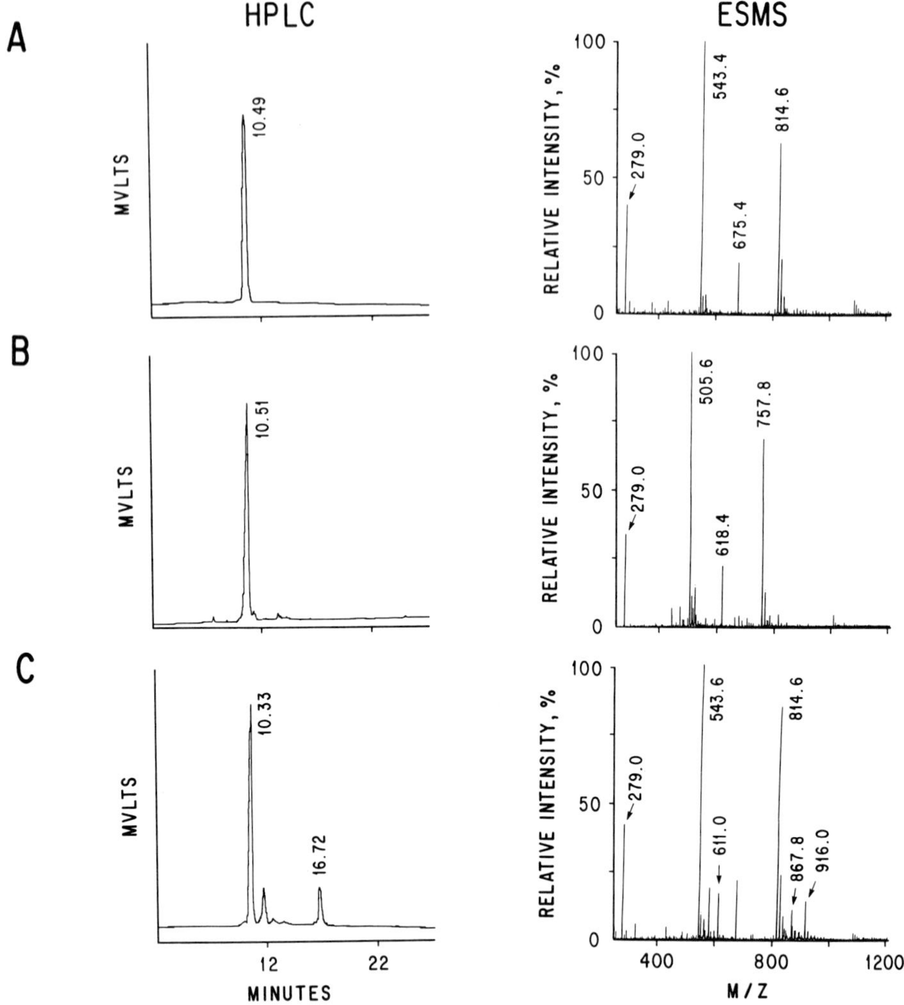

(Continuation of legend to Figure 1)

C. The relative abundance of the correct product was 68% by HPLC (retention time = 10.33 min), 69% by ESMS ($[M + 3H]^{3+}$ = 543.4 and $[M + 2H]^{2+}$ = 814.6), 75% by FABMS ($[M + H]^{+}$ = 1627.7, and 56% by PDMS ($[M + H]^{+}$ = 1629.4), while the relative abundance of the +202 Da adduct was 12% by ESMS ($[M + 3H]^{+3}$ = 611.0 and $[M + 2H]^{2+}$ = 916.0), 11% by FABMS $[M + H]^{+}$ = 1829.8), and 30% by PDMS $[M + H]^{+}$ = 1828.7).

Assessment of product purity by ESMS, FABMS, and PDMS showed *semi-quantitative* agreement with HPLC analyses (Table II). Figure 1A shows the analyses of a crude peptide containing >90% of the desired product as evaluated by HPLC, ESMS, FABMS, and PDMS. The HPLC retention time of this product was 10.49 min, while molecular ions were $[M + 3H]^{3+} = 543.4$ and $[M + 2H]^{2+} = 814.6$ by ESMS, $[M + H]^{+} = 1627.6$ by FABMS, and $[M + H]^{+} = 1627.9$ by PDMS. By comparison of the relative abundance of these desired molecular ions for all samples, the combined mass spectrometric techniques assigned 5 (11.9%) Fmoc-synthesized crude products and 2 (12.5%) Boc-based crude products as "poor" quality (<25% desired product) (Table II). One of these Fmoc-synthesized products had a molecular mass 114 Da below that of the desired peptide (Figure 1B), which could have been a deletion of 1 Asn or 2 Gly residues (Table III). Comparison with AAA and preview sequence analysis indicated an Asn deletion. Interestingly, the HPLC elution time of this des(Asn^8) peptide was the same as the desired peptide (Figure 1B). Another Fmoc synthesized crude product was a mixture of peptides, 57, 171, and 228 Da below the desired peptide. Comparison with AAA and preview sequence analysis indicated deletions at Gly^1, Gly^7, and Asn^8. Two Fmoc synthesized crude products (and their subsequent purified products) had molecular masses indicative of a Pro deletion, although one product was 97 Da and the other was 98 Da below the desired peptide. Examination of the peptide-linkers (Table I) revealed a peptide *amide* linker-resin (TentaGel S-AM) used by one of the laboratories generating a des(Pro) product, accounting for the 1 Da difference. The final Fmoc synthesized "poor" quality crude product contained numerous peptides of higher molecular masses than the desired peptide, most predominantly a +42 Da species. The source of this modification is unknown. Of the 2 "poor" quality Boc-synthesized crude products, one was primarily a deletion peptide, 300 Da lower than the desired peptide. AAA and preview sequencing showed this to be a des(Gly^1,Gly^4,Trp^{11}) deletion peptide. The other Boc-synthesized poor product was 28 Da higher than the desired product, indicative of a Trp(For) peptide. In this particular case the crude formylated product was the desired product, as the For group was removed in a separate step post-cleavage. Overall, incomplete (<90%) deprotection of Trp(For) was detected in Boc-synthesized products 4 times (out of a possible 12) by the combined mass spectrometric analyses. For Fmoc-based syntheses, the predominant side reaction resulted in products of mass 202 Da higher than the desired peptide, equivalent to either an EDT-thioanisyl or anisyl-trifluoroacetyl adduct (Table III) generated during peptide-resin cleavage. The characterization of a peptide modified by one of these adducts is shown in Figure 1C. LC-ESMS indicated that the HPLC peaks eluting at 10.3, 11.6, and 16.7 min correspond to the desired peptide, the desired peptide plus 106 Da, and the desired peptide plus 202 Da, respectively (data not shown). LC-ESMS is particularly useful for characterizing complex mixtures of peptides, enabling direct assignment of HPLC peaks as desired peptide or side products based on the observed molecular weights.

A general comparison of all mass spectrometric techniques showed PDMS to be more sensitive to covalent modifications of peptide products than ESMS or FABMS, and hence fewer crude and purified products were classified as containing >75% desired product by PDMS compared with other techniques (Table II). For FABMS, spectra of all samples showed clusters at -17 and +16 Da. Because the isotope abundances of these peaks were not identical to the desired product, they were treated as artifacts of the ionization. Thus, dehydration (-18 Da) and oxidation (+16 Da) may be underestimated by FABMS, except in cases where those peaks were significantly more intense than normal.

Nine samples containing <50% desired product (as evaluated by FABMS) were analyzed by MALDI-MS. In general, the results qualitatively agreed with those obtained by the other mass spectrometric techniques, although the abundances of the ions in a mixture varied as the laser spot was moved to different crystals on the target, and mass variance increased with higher laser power. Samples ionized from an SA matrix showed higher percentages of modified species than those ionized from a DHBA matrix, but the DHBA matrix gave a lower background signal. While these preliminary results with MALDI-MS looked promising, additional studies are clearly required.

Table III. Common By-Products Observed in ABRF Peptide by Mass Spectrometry

Mass Difference From Desired Peptide (Da)	Possible Contaminant	Crude Peptides With >10% Contaminant (%)[a]	
		Fmoc	Boc
-398	des(2Gly,Trp,Val)[b]	-	6.2
-357	des(2Gly,Asn,Lys)[b]	2.4	-
-300	des(2Gly,Trp)[b]	-	6.2
-171	des(Gly,Asn)[b]	2.4	-
-114	des(Asn) or des(2Gly)	4.8	-
-97	des(Pro)	4.8	-
-57	des(Gly)	2.4	-
-18	$-H_2O$	2.4	19
+28	formyl	-	25[c]
+42	acetyl	2.4	-
+56	*t*Bu	4.8	25
+96	trifluoroacetyl	-	6.2
+106	anisyl	2.4	-
+151	Tos	-	6.2
+172	adduct from EDT + TFA (8)	2.4	-
+202	[EDT + thioanisyl] or [anisyl + trifluoroacetyl]	19	-
+212	Mtr	2.4	-
+266	Pmc	2.4	-

[a]The number of samples analyzed was 42 Fmoc, 16 Boc. All contaminants were seen in at least 2 out of 3 different mass spectrometric analyses.

[b]The molecular mass of 1 Asn deletion is equivalent to 2 Gly deletions; thus, these possible products may be 1 Asn or 2 Gly deletions.

[c]A Trp(For)-containing peptide was the desired crude product for 1 Boc synthesis.

The combination of AAA, HPLC, ESMS, FABMS, and PDMS showed 5 (11.9%) of the crude Fmoc samples, 2 (12.5%) of the crude Boc samples, and 4 (18.2%) of the purified Fmoc samples to contain poor yields (<25%) of the desired product. Five of the 7 poor yield crude products were the result of deletion peptides (4 from Fmoc syntheses, 1 from Boc syntheses). Two of the Fmoc deletion peptides were synthesized by HBTU/HOBt (2 of 25, 8%), while the other 2 were OPfp/HOBt syntheses (2 of 6, 33%). The high percentage of failure syntheses by OPfp/HOBt suggests additional care may be needed for this methodology, such as extended or repeated couplings. Syntheses incorporating Fmoc-Asn without side-chain protection also appear problematic. Of the 4 Fmoc-synthesized crude peptides that contained >10% des(Asn), 2 were assembled without Asn side-chain protection. The only Fmoc-synthesized peptide for which all three mass spectrometric techniques detected >10% dehydration was assembled without Asn side-chain protection by DCC/HOBt. This dehydration was probably the result of a side reaction (nitrile formation) attributable to the use of unprotected Asn during coupling (3). Thus, of the 7 peptides synthesized using Fmoc-Asn, 3 (43%) had >10% of a side reaction directly attributable to Asn incorporation. The other problem detected in Fmoc syntheses was the generation of EDT-thioanisyl or anisyl-trifluoroacetyl adducts during peptide-resin cleavage. The modification of peptides by scavengers (usually via the indole side-chain of Trp) is highly dependent upon the work-up of the crude peptide following cleavage. Products that are recovered by rotovaping are more susceptible to modification than those recovered by ether precipitation (3), which in turn are modified more readily than those recovered by product dilution, extraction, and direct HPLC purification. The protocol for work-up of crude products was not requested from each participating laboratory, and therefore definitive causes of this modification cannot be determined.

Impurities in Boc-synthesized peptides were mostly due to cleavage problems. Of the 12 crude samples synthesized with Trp(For), 4 had >10% of the For group still attached. Variable success was seen when deprotection of Trp(For) was attempted during HF cleavage in the presence of anisole alone or anisole plus either butanedithiol or dimethylsulfide. Deprotection of Trp(For) was complete by treatment of the peptide-resin with ~10% piperidine-DMF for 2 h prior to HF cleavage. Dehydration and *t*-butylation were also significant problems in Boc-synthesized peptides, as was observed in the prior study (1).

IV. Conclusions

As estimated by ESMS, 30 of the 58 crude samples (52%) and 26 of the 33 purified samples (79%) contained >75% of the desired product. This represents a considerable improvement over the prior study, where the percentage of crude and purified products containing >75% of the desired material were 28 and 65%, respectively (1). Possible reasons for these improved results are any combination of (i) a synthetic peptide sequence less susceptible to side-reactions induced during peptide-resin cleavage (no Cys and/or Met present), (ii) the greater fraction of peptides synthesized by Fmoc chemistry, where cleavage conditions are less harsh, and (iii) more rigor and care in laboratory techniques following last year's results. The vast majority of poor quality peptides in the present study were due to peptide assembly difficulties, while most problems in the prior study were side-reactions induced during peptide-resin cleavage.

The wide range of peptide quality allowed us to evaluate critically the strengths and weaknesses of the analytical techniques used here. HPLC appeared to provide an accurate estimate of the complexity of the peptide samples and, in this particular situation in which a sample of the desired product was available, an accurate estimate of the content of the desired product. The only exception was the failure to detect an Asn deletion peptide which co-eluted with the desired product. Since a sample of the desired product is not usually available, additional analyses are required. CE provided little useful information in this study. AAA was the best technique for absolute amino acid quantitation, but was not helpful for detecting modifications of amino acid residues. Preview sequence analysis had the same strengths and weaknesses as AAA and the advantage of determining residue position. The mass spectrometric techniques examined here (ESMS, FABMS, PDMS, and MALDI-MS) allowed for identification of peptide modifications, including residual protecting groups. MS as used in this study does not distinguish between products of the same mass, nor allow for assignment of the positions of residue deletions and/or modifications. HPLC, ESMS, FABMS, and PDMS were usually in agreement in estimating the % desired product, with all four estimating that the mean % desired product in 56 crude samples was ~65-70%. Of the 51 ABRF laboratories responding to our survey, only 40% had MS available at their institutions and used it regularly for peptide characterization. Based on the 25 respondents who replied to this question, the major reason (80%) for not using mass spectrometry was lack of on-site availability. Our experience in the present and prior study suggests that efficient characterization of synthetic peptides is best obtained by a combination of HPLC, AAA, and MS, with sequencing by either Edman degradation or tandem MS being used to identify the positions of modifications and deletions.

Acknowledgments

We thank all of the ABRF core facilities that participated in this study and the National Science Foundation for financial support (grant DIR 9003100).

References and Notes

1. Smith, A.J., Young, J.D., Carr, S.A., Marshak, D.R., Williams, L.C., and Williams, K.R. (1992). *In* "Techniques in Protein Chemistry III" (Angeletti, R.H., ed.) pp. 219-229, Academic Press, Orlando, FL.
2. The potential pitfalls of synthetic peptide characterization has been reviewed recently: Grant, G. (1992). *In* "Synthetic Peptides: A User's Guide" (Grant, G.A., ed.) pp. 185-258, W.H. Freeman and Co., New York, NY.
3. Review: Fields, G.B., Tian, Z., and Barany, G. (1992). *In* "Synthetic Peptides: A User's Guide" (Grant, G.A., ed.) pp. 77-183, W.H. Freeman and Co., New York, NY. The reader is referred to this most recent review for specific references and abbreviations.
4. Hemling, M.E., Roberts, G.D., Johnson, W., and Carr, S.A. (1990) *Biomed. Environ. Mass Spectrom.* **19**, 677-691.
5. Kochersperger, M.L., Blacher, R., Kelly, P., Pierce, L., and Hawke, D.H. (1989) *Am. Biotech. Lab.* **7**(3), 26-37.
6. Fields, C.G., Lloyd, D.H., Macdonald, R.L., Otteson, K.M., and Noble, R.L. (1991) *Peptide Res.* **4**, 95-101.
7. Review: Tam, J.P. and Merrifield, R.B. (1987). *In* "The Peptides, Vol. 9" (Udenfriend, S. and Meienhofer, J., eds.) pp. 185-248, Academic Press, Inc., New York, NY.
8. Sieber, P. (1987) *Tetrahedron Lett.* **28**, 1637-1640.

SYNTHESIS, PURIFICATION AND INTERACTIONS OF CASEIN SIGNAL PEPTIDES

C. Creuzenet, G. Korshunova, H. Naharisoa, H. Nedev and T. Haertlé

LEIMA, Institut National de la Recherche Agronomique. BP 527, 44026 NANTES CEDEX 03, FRANCE

I. Introduction

Signal peptides 15 to 30 amino acid long are involved in protein translocation through different membranes. They play an important role addressing the proteins to the endoplasmic reticulum, mitochondria, chloroplasts in Eucaryotes, and to the cytoplasmic membrane in Procaryotes. They initiate transport of the nascent protein before translation is ended and they are classified in the family of targeting sequences (Gierasch, 1989).

Even though all signal peptides have related functions, amino acid sequence analysis has shown that they share little homology (Perlman and Halvorson, 1983). Yet, the variation of signal peptides involved in protein secretion is limited as they are always organized in three distinct domains (Von Heijne, 1985). The polar c-domain (5 to 7 amino acid long) is the most conserved. It contains the peptidase cleavage site, which has a universally conserved "(-3, -1) design" (Von Heijne, 1983) often of the type Ala-X-Ala (Perlman and Halvorson, 1983). The central h-domain contains essentially Val, Ile, Leu, Ala, Phe, Met and Trp residues and has a mean length of 12 amino acids. The high hydrophobicity of this domain seems to be essential for the function of a signal sequence. The N-terminal domain (1 to 2 amino acid long) contains Lys or Arg residues and has a mean charge of +1.7. The precise mechanism of action of signal peptides in the initiation of the transmembrane translocation is not yet well understood. Structural studies and monolayer experiments performed on chemically synthesized signal peptides, in presence of varying phospholipids, demonstrate that signal peptides have a high affinity for phospholipids, and adopt preferentially a helical conformation in a lipidic environment (Rosenblatt *et al.* 1980; Briggs *et al.* 1986; Batenburg *et al.* 1988).

Casein signal peptides are particularly interesting since they show a very high degree of sequence conservation (Mercier *et al.* 1990), as opposed to the vast majority of signal peptides, which show a great variability even among closely related proteins. The high degree of conservation of signal

TECHNIQUES IN PROTEIN CHEMISTRY IV

peptides from caseins suggests that since they have undergone powerful selection pressure they have probably approached an optimal degree of efficiency.

Hence, a series of consensus sequences of casein signal peptides has been synthesized and their interactions with two phospholipids (dimyristoylphosphatidyl- glycerol and -choline) have been studied by monolayer techniques and fluorescence anisotropy measurements. The studies, performed on simple lipid model systems, may help to understand signal peptide/lipid interactions. In addition, after simple chemical derivatization (esterification, alkylation) with compounds of interest (medicines, aroma), casein signal peptides can be useful as chemical tools to insert or transport these compounds in lipoproteic structures.

In this report the chemical synthesis and purification of casein signal peptides are presented. A comparative study of their interactions with pure phospholipids is also presented.

II. Materials and methods

Materials

Fmoc - amino acids were purchased from Applied Biosystems (France). Dichloromethane (DCM), N, N-dimethylformamide (DMF), N-methyl-pyrrolidone (NMP) were all peptide synthesis grade and purchased from Solvents Documentation Syntheses (France). Dimyristoylphosphatidylcholine (DMPC), dimyristoylphosphatydilglycerol (DMPG), trifluoracetic acid (TFA), diisopropylethylamine (DIEA), thioanisole, DMAP, DCC, ethanedithiol (EDT), phenol were purchased from Sigma Chimie (France), 2-(1H-Benzotriazol-1-yl)-1,1,3,3-tetramethyluronium hexafluorophosphate (HBTU), N-Hydroxybenzotriazole (HOBT) were purchased from Propeptide (France).

Peptide synthesis

The synthesis was carried out on an Applied Biosystems Inc (ABI) 430A peptide synthesizer using standard scale (0.25 mmol) procedure. Peptides were synthesized by solid-phase methods on HMP-resin 4-(Hydroxymethyl)phenoxymethyl-copolystyrene-1% divinylbenzene-resin using Fmoc strategy. The following protecting groups were used: Lys(Boc), Ser(t-butyl), Thr(t-butyl), Cys(trityl), Cys(t-butyl), Glu(t-butyl). The first amino acid was coupled to the resin by DCC/DMAP dicylohexylcarbodiimide/dimethylamino-4 pyridine method and the activation was performed either with DCC/HOBT (P1, P2) or HBTU/HOBT (P3 to P5). The coupling efficiency was monitored by a ninhydrin test. Resin-peptide was washed with dichloromethane and dried under vacuum overnight prior deprotection. Deprotection and cleavage from the support was performed using TFA/Phenol/EDT/thioanisole/H_2O (82.5 / 5 / 2.5 / 5 / 5) at room temperature during 2 h according to King *et al.* (1990). Cys(t-butyl) was cleaved by treatment with TFA/TFMSA. Synthesized peptides were precipitated with anhydrous diethyl ether which helped to remove scavengers.

Peptide precipitate was washed several times, centrifuged, separated and dried under vacuum.

Peptide purification

Peptides were purified by RP-HPLC (Waters) using linear gradients of water/acetonitrile, containing TFA or TFMSA. Precise conditions are given in Figure 1.

Surface tensiometry

Tensiometric measurements were performed on a KRUSS tensiometer K12, according to the Wilhelmy plate method (Demel 1974). 80 ml of sodium phosphate buffer (100 mM, pH 6.4) were poured into the thermostated sample vessel (6.65 cm diameter), then the platinum plate was automatically dipped (2 mm) into the buffer. 0.1 g/l solutions of phospholipids (DMPC or DMPG) in $CHCl_3/CH_3OH$ 1/1 or 1/2 (v/v), respectively, were used. During the study of the phospholipid/peptide interactions, aliquots of a peptide solution (0.1 mg/ml in water/methanol/TFA, 40/60/0.11, v/v) were spread over the well stabilized and equilibrated lipid monolayer. The lipid/peptide molar ratios varied from 300 to 10. All the measurements were made at 30 °C, above the transition temperature of all of the used lipids. The obtained results were expressed as the relative variation in surface tension (T) according to equation: $T = (\sigma-\sigma_0)/\sigma_0$. Where σ_0 (σ) is the surface tension of the phospholipid (phospholipid plus peptide) film at equilibrium. σ and σ_0 values are expressed in mN/m units. In the presented Figures, all the control curves obtained by spreading the solvents on the phospholipid film were subtracted.

Preparation of large multilamellar vesicles

0.4 µmol of phospholipids, 0.8 nmol of a fluorescent probe, 1, 6-diphenylhexatriene (DPH) and variable amounts of peptide were dried together under nitrogen. Subsequently, they were hydrated overnight at 30 °C with 2 ml of sodium phosphate buffer (100 mM, pH 6.4), and vigorously mixed during 10 min before use. The lipid/peptide molar ratio varied between 150 and 10.

Fluorescence measurements

Fluorescence measurements were made on a SLM 4800C AMINCO spectroflurometer. The emission spectra were recorded in the ratio mode. The excitation wavelength was 352 nm, and the emission was recorded between 360 and 600 nm. During polarization experiments, the spectrofluorometer was equipped with three polarizers, disposed in a T-format. The sample was excited at 352 nm with filtered ($\lambda < 625$ nm) and either vertically or horizontally polarized light. The filtered ($\lambda > 435$ nm) emission intensities were measured after vertical or horizontal polarization in order to deduce the parallel ($I_{//}$) and perpendicular ($I_{\perp}$) components of the emission. The fluorescence anisotropy (A) was calculated according to the following formula: $A = (I_{//}-I_{\perp}) / (I_{//}+2I_{\perp})$. During the experiments, the samples were gently stirred. The temperature was varied between 50 °C and 10 °C.

III. Results and discussion

Peptide purification

High degree of sequence conservation observed for casein signal peptides of all mammalian species and important evolutionary selection pressure exerted on these peptides suggest that they may have reached significant functional efficiency. Therefore, the following signal peptides of bovine α_1, α_2, β and κ caseins were synthesized :

P1 (P2) : M K V(F) L I L A C L V A L A L A
P3 (P4) : M K F F I F T C(S) L L A V(A) A L A
P5 : M M K S F F L V V T I L A L T L.

C-termini of all studied peptides contain clusters of hydrophobic amino acids, so different chromatographic phases have been tried in order to achieve their purification. P1, P2 were solubilized in ACN/propanol-2 (1/1) + TFMSA 0.025%; P3, P4 were solubilized in ACN/propanol-2/H_2O (4/1/5) + TFA 0.1%, P5 in ACN/H_2O (1/1) + TFA 0.1% . Chaotropic agent such as perchloric acid 0.1% increased the solubility and resolution but its use was limited since it might oxidize Met and Cys residues. In result P1, P2, P4 were purified on Nucleosil phenyl column 7 µm, 250 x 10 mm, P5 on C-18 reverse-phase column Delta Pack 300 Å, 10 µm, 300 x 7.8 mm, P3 on Nucleosil CN 10 µm, 250 x 4.6 mm at room temperature using linear gradient H_2O /ACN. Conditions were given in Figure 1.

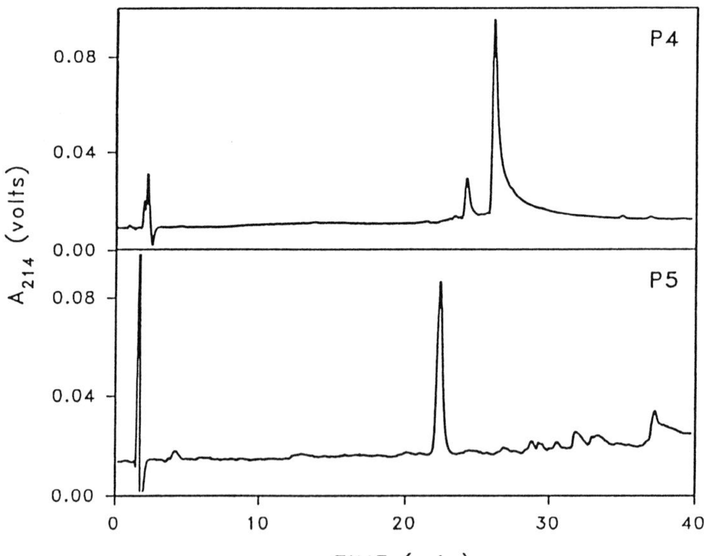

Figure 1: HPLC patterns of P4 and P5 solvent A : 1,1% TFA in H_2O, solvent B : 0,9% TFA in ACN, linear gradient from 20% to 70% B in 40mn, flow rate 1ml/min, wavelength 214nm, AUFS 0,10

Peptide characterization

The purity of peptides was verified by capillary electrophoresis completed by amino acid analysis and sequencing.

After hydrolysis of the peptides with 6N HCl for 24 h under vacuum at 110 °C in a Pico-Tag station (Waters) amino acids were derivatized with PITC according to Bidlingmeyer *et al.* (1984) and quantified by RP-HPLC on a Pico-Tag C-18 column 150 x 3.9 mm. In the case of Met and Cys residue oxidation with performic acid was accomplished before hydrolysis.

Capillary electrophoresis was carried out on P/ACE system 2000 (Beckman Instruments) using system Gold chromatography software. Separation was achieved with fused silica capillary 57 cm x 75 μm eluted with 100 mM sodium borate pH 8.4 - P3, P5, 100 mM sodium phosphate pH 2.5 - P2, P4 at 30 °C; voltage: 20 Kv to 30 Kv; detection wavelength 214 nm. Profiles demonstrated a high homogeneity of obtained product.

Sequencing (data not shown) of all peptides agreed well with their amino acids composition.

HPLC of crude peptides (P4, P5) demonstrated significant degree of purity (Figure 1) and in the case of other peptides the desired peptide was the major product, although sometimes it was accompanied by small amounts of other material probably due to oxidation of Cys residues. Retention times varied with the hydrophobicity of separated peptides.

Interaction studies

The interactions between the synthesized signal peptides and DMPC or DMPG were studied by two distinct methodologies: tensiometric assays and fluorescence polarization measurements. DMPC and DMPG were chosen as simple phospholipid models presenting following advantages. First, the saturation of the aliphatic chain excludes possible oxidation problems. Second, their charge differences at the pH of the study (pH = 6.4) could characterize electrostatic contributions to the interactions.

During tensiometric assays of interactions of both phospholipids studied, the quantity leading to a continuous monolayer organization was found to be 25 μg. The initial surface tensions were about 30 mN/m and 25 mN/m for DMPG and DMPC, respectively. The addition of peptide P5 on these films lead to a small increase in surface tension, slightly higher with DMPG than with DMPC (Figure 2a). The increase upon addition of peptide might indicate that the peptide remains at the interface and possibly intercalates between the lipid molecules. Demel *et al.* (1973) demonstrated that agents interacting only with the phospholipid polar headgroups do not induce any surface pressure increase in monolayers. The difference in surface tension increase upon addition of P5 on DMPG and DMPC films reveals that this peptide has a greater affinity for DMPG than for DMPC. The difference of interaction between the two lipids was enhanced when the amount of lipid spread on the buffer was higher (200 μg), the initial tension being the same as in the previous experiment (Figure 2b). In this case, the film is no longer a monolayer, but apparently has collapsed and might have formed superstructures characterized by a high negative charge density in the case of DMPG.

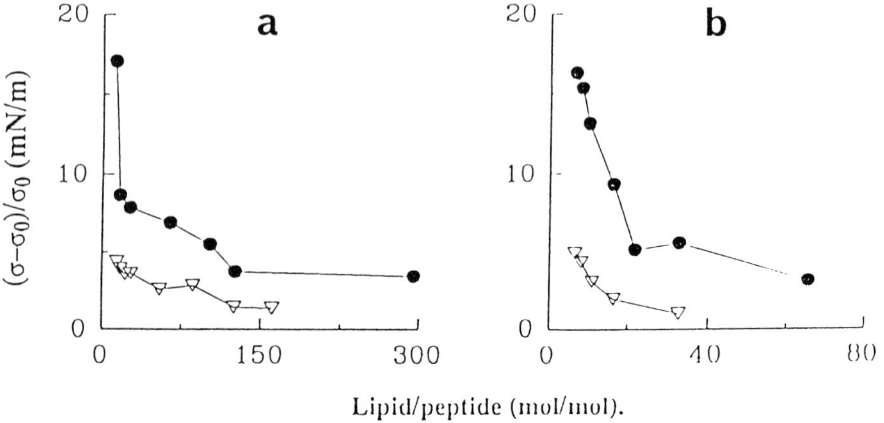

Figure 2: Relative variation in surface tension ($(\sigma-\sigma_0)/\sigma_0$) as a function of the molar lipid/peptide (P5) ratio, for 25 µg (a) or 200 µg (b) of lipid. ●: DMPG; ▽: DMPC.

The enhanced difference in affinity of P5 for the two lipids supports the idea that the electrostatics play an important part in the interaction.

The interactions between all synthesized casein signal peptides and the two phospholipids were also studied by fluorescence polarization. Large multilamellar vesicles (MLV) containing a fluorescent probe and one of the peptides at different phospholipid/peptide ratios were prepared. All the peptides were incorporated in the preparation mixture before hydration of the MLV because they were not soluble enough in aqueous solutions to be directly added to MLV suspensions. The buffer applied to hydrate the MLV was the same as the one used during tensiometric studies. 1,6-diphenylhexatriene (DPH) was used as the fluorescent probe since it is known to insert in the middle of the bilayer, near the ends of the acyl chains (Lentz *et al.*, 1976). Consequently, this probe is highly sensitive to the motions of the acyl chains. As anisotropy measurements depend on the rotational diffusion of the fluorophore (Lakowicz, 1983), any restriction in the motions of the acyl chains induced by the insertion of a peptide in the vicinity of the probe, will induce an increase in anisotropy.

Before starting the polarization experiments, the incorporation of DPH into the vesicles was checked by recording the emission spectra of the samples. The lipid/DPH molar ratio was 500. Typically, the fluorescence emission spectra (Exc. 352 nm) present three emission maxima at 404, 430 and 455 nm, plus a shoulder at 485 nm. The spectral features remained unchanged at all studied peptide/lipid ratios (data not shown).

Figure 3 shows the results obtained for peptides P2 to P5, either with DMPC or DMPG. The fluorescence polarization curves were obtained by decreasing the temperature from 50 °C to 10 °C. Examining first in detail the case of P2/DMPG interaction (Figure $3a_1$), one observes that at a lipid to peptide molar ratio equal to 109, a significant anisotropy increase occurs above the transition temperature of the lipid. This is suggesting that the studied peptide inserts readily between the acyl chains of the DMPG bilayer. No change in the transition temperature is observed (23.8 °C). The anisotropy further increases with the amount of peptide added, and the transition becomes broader, until a certain limit is achieved. At a lipid to peptide ratio of 17, the anisotropy versus temperature curve shows no transition anymore. At this ratio, the insertion of P2 into the bilayer must have stabilized the structure so much that the motions of the acyl chains are restrained as if the lipids were fixed in the gel state; phenomena were observed in the case of other peptides studied but they occurred at different lipid to peptide ratios. Hence, the obtained casein signal peptides could be classified in terms of their affinity for the lipids according to the amplitude of the anisotropy increase and to the lipid/peptide ratio inducing the disappearance of the transition temperature. The affinity of the peptides for DMPG increases in the following order : P5, P3, P2, P4. The differences in affinity of P2, P3 and P4 for DMPG are small. All these three peptides insert much more readily in DMPG bilayers than P5. When the interactions of the peptides with DMPC are analysed, the order of their affinity is the same as observed in the case of DMPG. The interaction of P3 and P5 with DMPC, however, show less differences than in the case of DMPG.

It seems interesting to compare the differences in affinity of a studied peptide for these two lipids. P2 and P3 interact with the same affinity with both lipids and P4 interacts slightly stronger with DMPC. P5/DMPC interactions seem to be weaker than P5/DMPG. Measures of polarization during P5 interactions with lipids agree well with those obtained during tensiometric studies. The interactions of this peptide with both phospholipids are weak being only a little bit stronger with the negatively charged DMPG than with the zwitterionic DMPC. These results demonstrate the importance of coulombic forces during P5 interactions. All the other peptides interact strongly with lipids independently of their charge. Hence, the hydrophobicity of the peptide, rather than its charge, may be the main determinant of the P2, P3 and P4 interactions. Hoyt and Gierasch (1991) suggest that certain hydrophobicity threshold is required for a signal sequence to be functional. The mean hydrophobicities of the synthesized peptides were calculated using the Kyte and Doolittle scale (1982). They were 2.55, 2.25, 1.87, 2.08 for P2 to P5, respectively. If only the hydrophobic character of the peptides was taken into account, the affinity for the acyl chains of the lipids would increase in the following order : P4, P5, P3 and P2. Actually, experimental data on P4 interactions do not respect this order. The main difference between P4 and the

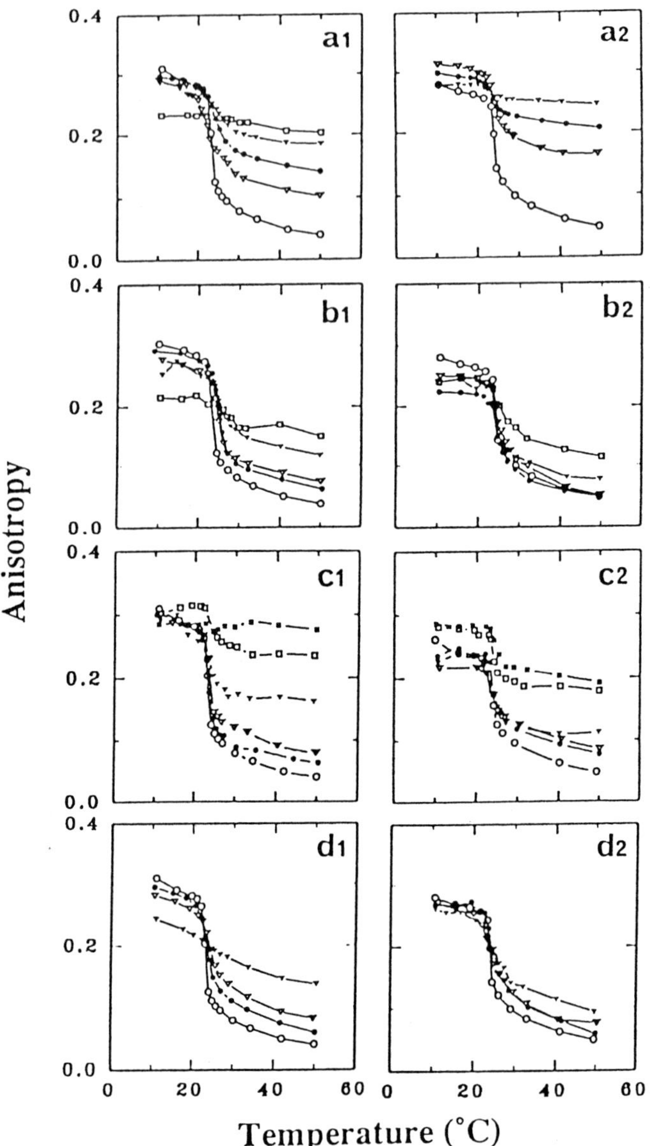

Figure 3: Steady-state fluorescence anisotropy of DPH as a function of temperature and signal peptide concentration for DMPG (a_1 to d_1) or DMPC (a_2 to d_2) MLVs. All the experiments were made in a 100 mM sodium phosphate buffer, pH 6.4. Lipid concentration was 2.10^{-10} M. Lipid to DPH molar ratio was 500. Peptide to lipid molar ratios were: P2 (a_1, a_2) : o: 0, ▿: 109, •: 54, ▾: 27, ▫: 17; P3 (b_1, b_2) : o: 0, •: 120, ▿: 100, ▾: 50, ▫: 25; P4 (c_1, c_2) : o: 0, • : 100, ▿: 75, ▾: 50, ▫: 20, ▪: 10; P5 (d_1, d_2) : o: O, •: 66, ▿: 33, ▾: 10.

other peptides is the replacement of Cys[8] in P2 and P3 by Ser in P4. This punctual change may explain the higher activity of P4. Possible dimerisation of P2 and P3, by cysteine oxidation, may create steric hindrance, so even if P4 is less hydrophobic, its insertion in the phospholipid bilayer may be easier. P2 and P3 are the naturally occurring sequences of the signal peptides of caseins ß and α, respectively. The high levels of synthesis and secretion of these caseins suggest that probably, *in vivo*, dimerisation might be avoided, thanks to the cotraductional transport of the synthesized sequence in the endoplasmic reticulum.

The difference between P5 and the other peptides also could be predicted from the calculations of the hydrophobic moments (μ) of these peptides according to the method of Eisenberg *et al.* (1984). This method evaluates the potential of a given amino acid sequence to act as a surface active or rather as a transmembranary peptide. It is based on the assumption that the analyzed sequences adopt a α-helical conformation, a general feature of transmembranary sequences in proteins. The obtained results clearly demonstrate, however, that P5 may be classified between surface active and transmembranary peptides when all the others have clear transmembranary character. This is consistent with the pattern of obtained results: the electrostatic effects are important when the peptide remains close to the polar headgroups of the lipids, as it might be observed in the case of P5. The electrostatic effect is less important when the peptide penetrates the bilayer, like presumably P2, P3 and P4 do. In the latter case, hydrophobicity effects prevail. As P5 is the signal peptide of casein κ, it must enter the lipid bilayer *in vivo*, starting translocation of casein κ through the membrane. The results of this study emphasize that even if all the signal sequences studied here are involved in casein secretion, the mechanisms of their interactions with phospholipids seem to be different (at least between P5 and the others).

The obtained results can be interpreted from an evolutionary point of view. All the peptides studied are consensus sequences of casein signal peptides. The high degree of conservation of casein signal sequences between all the mammalian species suggests that they must have evolved from common ancestor genes. The gene encoding for the P5 sequence has probably different origin, explaining its differences of sequence and observed functional properties.

References

Batenburg, A. M., Brasseur, R., Ruysschaert, J.M., Van Scharrenburg, G. J M., Slotboom, A. J., Demel, R. A. and De Kruijff, B. (1988) J. Biol. Chem. **263**, 4202-4207.

Bidlingmeyer, B. A., Cohen, S. A. and Tarvin, T. L. (1984) J. Chromatogr. **336**, 93-104.

Briggs, M. S., Cornell, D. G., Dluhy, R. A. and Gierasch, L. M. (1986) Science **233**, 206-208.

Demel, R.A., London, Y., Geurst van Kessel, W. S. M., Vossenberg, F. G. A. and van Deenen, L. L. M. (1973) Biochim. Biophys. Acta **311**, 507-519.

Demel, R. A. (1974) *Methods in Enzymology* **32**, 539-545.

Eisenberg, D., Schwarz, E., Komaromy, M. and Wall, R. (1984) J. Mol. Biol. **179**, 125-142.
Gierasch, L. M. (1989) Biochemistry **28**, 923-930.
Hoyt D. W. and Gierasch, L. M. (1991) Biochemistry **30**, 10155-10163.
King, D. S., Fields, C. G. and Fields, G. B. (1990) Int. J. Pept. Prot. Res. **36**, 255-266.
Kyte, J. and Doolittle, R.F. (1982) J. Mol. Biol. **157**, 105-132.
Lakowicz, J. R. (1983) *Principles of fluorescence spectroscopy*, 111-150, Plenum Press, New-York, N.Y.
Lentz, B. R., Barenholz, Y. and Thompson, T. E. (1976) Biochemistry **15**, 4529-4537.
Mercier, J. C., Vilotte, J. L. and Provot, C. (1990) *Genome analysis in domestic animals* 233-258 (Gedelman, H., ed) Verlag VCH, Weinheim.
Perlman, ,D. and Halvorson, H. O. (1983) J. Mol. Biol. **167**, 391-409.
Rosenblatt, M;, Beaudette, N. V. and Fasman, G. D. (1980) Proc. Natl. Acad. Sci. USA **77**, 3983-3987.
Von Heijne, G. (1983) Eur. J. Biochem. **133**, 17-21.
Von Heijne, G. (1985) J. Mol. Biol. **184**, 90-105.

The Systematic Development of Peptides Having Potent Antimicrobial Activity Against *E. coli* Through the Use of Synthetic Peptide Combinatorial Libraries

Richard A. Houghten
Kim Thy Dinh
David E. Burcin
Sylvie E. Blondelle
Torrey Pines Institute for Molecular Studies
San Diego, CA 92121

It is now possible to systematically screen tens of millions of peptides in existing pharmacologically relevant *in vitro* assay systems using synthetic peptide combinatorial libraries (SPCLs). An example of this procedure is presented here, in which an SPCL composed in total of 52,128,400 acetylated hexa-peptides was screened in a standard microdilution assay to identify peptide sequences which had potent antimicrobial activity against *E. coli*.

Introduction

Synthesizing and screening all combinations of the 20 natural L-amino acids in any assay system, for even short peptides, is impractical [there are 64,000,000 six-residue peptides (20^6), 1,280,000,000 seven-residue peptides (20^7), etc.]. The use of heterogeneous equimolar mixtures of synthetic peptides (1-5) permits the rapid screening of extremely large numbers of nonsupport-bound peptides. These can be used directly in virtually any existing *in vitro* assay system. When combined with an iterative selection process, this process permits the systematic identification of optimal peptide sequences. An example illustrating the usefulness of this approach is described here for the development of peptides which have activity against *Escherichia coli*.

TECHNIQUES IN PROTEIN CHEMISTRY IV

Materials and Methods

Preparation of Synthetic Peptide Combinatorial Libraries

The methods for preparing each of the individual peptide mixtures making up SPCLs have been described earlier (1-4). Briefly, standard t-Boc chemistry in combination with simultaneous multiple peptide synthesis [SMPS, (6)] was used to prepare peptide mixture resins using a process termed divide, couple, and recombine [DCR, (4,7)]. This process generates a single peptide on each individual resin bead. The amount of resin (200-400gm; approximately 5,000 resin beads/mg) was used which ensures that enough couples of each individual peptide in the mixture diversity are available after the addition of the final two amino acids as described below. Cysteine was excluded since its presence in the mixture positions would either require the constant presence of a reducing agent in the assay being used, or would result in difficult to define disulfide aggregates in the library. The DCR process was repeated until a final mixture of 130,321 (19^4) protected tetra-peptide resins was prepared. This peptide resin mixture (XXXX-resin) was divided into 400 aliquots and placed in numbered, porous, polypropylene packets. Synthesis of the next two defined positions and N-acetylation were carried out using SMPS (6). This library consists of all of the possible combinations of the amino acids used for a six amino acid peptide (thus the term "combinatorial" in SPCL). The peptide mixtures were deprotected and cleaved from their respective resins using low-high hydrogen fluoride method (8,9). Peptide mixtures were extracted with water or dilute acetic acid and the solutions lyophilized.

Determination of Antimicrobial Activity

The antimicrobial activity of each peptide mixture was determined as described earlier (10). In brief, *Staphylococcus aureus* ATCC 29213 was grown overnight at 37°C in Mueller-Hinton broth, then reinoculated and incubated at 37°C to reach the exponential phase of bacterial growth (i.e., a final bacterial suspension containing 10^5 to $5x10^5$ colony-forming units/ml). The concentration of cells was established by plating 100μl of the culture solution using serial dilutions (e.g., 10^{-2}, 10^{-3} and 10^{-4}) onto solid agar plates. In 96-well tissue culture plates, peptide mixtures were added to the bacterial suspension at concentrations derived from serial two-fold dilutions ranging from 1500μg/ml to 2.9μg/ml. The plates were incubated overnight at 37°C and the growth determined at each concentration by the optical density (OD) at 620nm. Although the growth values differed $\pm$10% from one assay to another, the relative percentage of growth found for each peptide mixture was consistent in three separate determinations. The IC_{50}s (the concentration necessary to inhibit 50% of the growth of the bacteria) were then calculated. The minimum inhibitory concentration (MIC) was defined as the lowest concentration of peptide at which there was no change in OD between time 0 and 21h.

Results

A wide range of different SPCLs have been prepared, which vary in length from three to seven residues, and are made up of between 20 to 60 L-, D-, and/or unnatural amino acids. The SPCL used for this study consisted of a six-residue peptide mixture set having an acetylated (Ac) N-terminal and an amidated C-terminal. The first two amino acids in each peptide chain were individually and specifically defined, while the last four amino acids consisted of equimolar mixtures of 19 of the 20 natural L-amino acids. Cysteine was omitted from the mixture positions of this SPCL, but included in the defined positions. The library can be generally represented by the sequence Ac-O_1O_2XXXX-NH_2. O_1 and O_2 are specific individual amino acids [i.e., O_1O_2 = AA, AC, AD, etc., through YV, YW, YY, for a total of 400 combinations (20^2)]. Each X position represents an equimolar mixture of the 19 amino acids; four positions result in a total of 130,321 combinations (19^4). Each of the 400 different peptide mixtures that make up this library consists of 130,321 individual hexamers, which in total represent 52,321,400 peptides. Preparation of this library involved minimal manipulation following the final cleavage from the resin. Only extraction and lyophilization are necessary prior to use. Each peptide mixture was typically used at a final assay concentration of 1.0mg/ml. Assuming that the average molecular weight of Ac-O_1O_2XXXX-NH_2 is 785, then a mixture of 130,321 peptides at a total final concentration of 1.0 mg/ml yields a concentration of every peptide within each mixture of 7.67ng/ml (9.8nM).

Equimolarity of each individual peptide making up the mixtures is the foundation of the utility of any SPCL. We have used the DCR approach in which *a priori* equimolarity is ensured via the separation and compartmentalization of starting resin, the addition of a single amino acid to each compartmentalized resin, followed by a recombination and mixing of the resulting individual resins. This is accomplished by the exactitude of the physical process of weighing the separate resins, mixing them, subdividing and weighing them again, along with testing each resin or mixture of resins for coupling completion as described earlier (4). Amino acid analysis confirmed the expected equimolarity within the limits of such analysis (± 10%).

Development of New Antimicrobial Peptides

In microdilution assays (10), the 400 peptide mixtures of the Ac-O_1O_2XXXX-NH_2 peptide library used were screened against *Staphylococcus aureus* and *Streptococcus sanguis* (Gram-positive bacteria), *Escherichia coli* and *Pseudomonas aeruginosa* (Gram-negative bacteria), and *Candida albicans* (yeast). While many antimicrobial and antifungal lead peptide mixtures were derived from this SPCL for the above organisms (1,4), a single example (Ac-FRXXXX-NH_2) found effective against *E. coli* will be used here to illustrate this approach. The initial mixture had an

IC_{50} = 1078μg/ml (Table IA). A set of 20 new synthetic peptide mixtures (Ac-FROXXX-NH_2) were prepared in which the third position of Ac-FRXXXX-NH_2 was defined. The peptide mixture Ac-FRWXXX-NH_2 (IC_{50} = 222μg/ml; Table IA) was the most effective of this set, and was more than twice as potent as the next best case in this series (Ac-FRCXXX-NH_2; IC_{50} = 507μg/ml). The activities of the 20 peptide mixtures at each iterative step are shown, along with the peptide mixture from the previous step, in Tables IA-ID. The fourth through the sixth positions were defined in a similar iterative manner (Tables IB-ID). Upon defining the fourth position, the two most active peptide mixtures were Ac-FRWWXX-NH_2 (IC_{50} = 36μg/ml) and Ac-FRWRXX-NH_2 (IC_{50} = 36μg/ml). This is approximately a four-fold increase relative to Ac-FRWXXX-NH_2. Only the iterative process for Ac-FRWWXX-NH_2 is shown here. In the next iterative step, Ac-FRWWRX-NH_2, Ac-FRWWVX-NH_2, and Ac-FRWWHX-NH_2 were found to have IC_{50} values equal to 24,44,and 52 μg/ml, respectively. Upon defining the sixth position of Ac-FRWWHX-NH_2, the hexa-peptide Ac-FRWWHR-NH_2 was the most active, with an IC_{50} = 13μg/ml and a MIC between 16 to 32μg/ml. The antimicrobial activity of this sequence against *E. coli* is equal to or higher than those found for naturally occurring peptides such as the magainins (MIC = 125μg/ml) and cecropins (MIC = 4 - 8μg/ml).

Discussion

The fact that individual, biologically active compounds can be detected in the presence of very large numbers of other inactive substances has long been used to isolate bioactive substances of pharmacological relevance. A variety of methods is now able to avoid the immense task of synthesizing and screening millions of individual peptides (Reviewed in 11). All of these rely upon the inherent selectivity of biological receptor systems. The recent developments in this area are at the forefront of a true revolution in drug discovery. Each of these methods are by many orders of magnitude more rapid than traditional drug discovery methods.

The use of synthetic peptide combinatorial libraries (SPCLs) circumvents a number of the limitations inherent in approaches that involve the use of peptides attached to an insoluble support [i.e., resin beads (12), phage particles (13-15), or glass surface (16)]. The SPCL developed and described here consists of 400 different mixtures of hexa-peptides, each mixture made up of 130,321 different hexa-peptides. At a total concentration of 1.0mg/ml, the concentration of each of the individual nonsupport-bound peptides making up these peptide mixtures is approximately 9.8nM. Thus, there is a sufficient concentration of each individual peptide present in solution to detect activity in existing assays routinely used in the study of antigen/antibody interactions, receptor/ligand interactions, enzyme/substrate interactions, etc. Thus, SPCLs can be used in functional assays such as microdilution assays to develop new peptide antibiotics; in plaque assays to develop new antivirals, etc. SPCLs ranging

Table I. Antimicrobial activity against E.coli

A	IC_{50} μg/ml	B	IC_{50} μg/ml
Ac–FRWXXX–NH_2	222	Ac–FRWWXX–NH_2	36
Ac–FRCXXX–NH_2	507	Ac–FRWRXX–NH_2	36
Ac–FRRXXX–NH_2	588	Ac–FRWKXX–NH_2	50
Ac–FRLXXX–NH_2	753	Ac–FRWLXX–NH_2	62
Ac–FRMXXX–NH_2	822	Ac–FRWCXX–NH_2	68
Ac–FRFXXX–NH_2	832	Ac–FRWFXX–NH_2	77
Ac–FRYXXX–NH_2	878	Ac–FRWIXX–NH_2	106
Ac–FRVXXX–NH_2	884	Ac–FRWMXX–NH_2	120
Ac–FRKXXX–NH_2	923	Ac–FRWTXX–NH_2	120
Ac–FRHXXX–NH_2	933	Ac–FRWHXX–NH_2	121
Ac–FRIXXX–NH_2	947	Ac–FRWYXX–NH_2	128
Ac–FRXXXX–NH_2	1078	Ac–FRWVXX–NH_2	129
Ac–FRTXXX–NH_2	1080	Ac–FRWXXX–NH_2	145
Ac–FRPXXX–NH_2	1175	Ac–FRWSXX–NH_2	180
Ac–FRAXXX–NH_2	1201	Ac–FRWGXX–NH_2	290
Ac–FRSXXX–NH_2	1387	Ac–FRWAXX–NH_2	331
Ac–FRDXXX–NH_2	> 1500	Ac–FRWNXX–NH_2	357
Ac–FRGXXX–NH_2	> 1500	Ac–FRWQXX–NH_2	403
Ac–FRQXXX–NH_2	> 1500	Ac–FRWPXX–NH_2	468
Ac–FREXXX–NH_2	> 1500	Ac–FRWDXX–NH_2	662
Ac–FRNXXX–NH_2	> 1500	Ac–FRWEXX–NH_2	> 1000

C	IC_{50} μg/ml	D	IC_{50} μg/ml	MIC μg/ml
Ac–FRWWRX–NH_2	24	Ac–FRWWHR–NH_2	13	16–32
Ac–FRWWXX–NH_2	40	Ac–FRWWHK–NH_2	15	16–32
Ac–FRWWVX–NH_2	44	Ac–FRWWHW–NH_2	27	62–125
Ac–FRWWHX–NH_2	52	Ac–FRWWHF–NH_2	30	31–62
Ac–FRWWAX–NH_2	52	Ac–FRWWHH–NH_2	31	34–62
Ac–FRWWMX–NH_2	55	Ac–FRWWHL–NH_2	31	62–125
Ac–FRWWLX–NH_2	65	Ac–FRWWHI–NH_2	34	36–62
Ac–FRWWKX–NH_2	68	Ac–FRWWHV–NH_2	37	62–125
Ac–FRWWYX–NH_2	74	Ac–FRWWHG–NH_2	38	62–125
Ac–FRWWGX–NH_2	83	Ac–FRWWHY–NH_2	39	62–125
Ac–FRWWTX–NH_2	86	Ac–FRWWHM–NH_2	42	62–125
Ac–FRWWWX–NH_2	88	Ac–FRWWHA–NH_2	43	62–125
Ac–FRWWQX–NH_2	91	Ac–FRWWHS–NH_2	48	62–125
Ac–FRWWCX–NH_2	92	Ac–FRWWHX–NH_2	52	
Ac–FRWWPX–NH_2	96	Ac–FRWWHC–NH_2	52	62–125
Ac–FRWWFX–NH_2	103	Ac–FRWWHT–NH_2	56	62–125
Ac–FRWWNX–NH_2	105	Ac–FRWWHN–NH_2	83	250
Ac–FRWWSX–NH_2	115	Ac–FRWWHP–NH_2	90	250
Ac–FRWWIX–NH_2	126	Ac–FRWWHQ–NH_2	120	>250
Ac–FRWWDX–NH_2	379	Ac–FRWWHE–NH_2	>250	>250
Ac–FRWWEX–NH_2	> 1000	Ac–FRWWHD–NH_2	>250	>250

in length from three to seven residues have been prepared which incorporate L-, D-, and/or 20 different non-natural amino acids. The SPCL concept can also be incorporated into an existing peptide of any length (i.e., incorporating two defined positions and four different mixture positions into a 10-residue peptide sequence to generate tens of millions of analogs of a known 10-residue peptide). The limitations of the SPCL approach appear to be only those currently inherent in available synthetic peptide methodologies (i.e., one cannot routinely chemically synthesize sequences greater than approximately 50-75 residues). The timeframe to define a specific peptide sequence is four to six weeks.

The utility of SPCLs is exemplified here for microdilution assays to develop new, highly effective antimicrobial peptides. The activities found are greater than a number of naturally occurring antimicrobial peptides. In other studies we have precisely identified the antigenic determinant recognized by an antibody (mAb 3E7) raised against β-endorphin (manuscript in preparation). This antigenic determinant was not fully identified by either the phage or support-bound peptide approaches (12,13). It should be noted that, whereas the libraries used in the phage library approach are often termed "random peptide libraries" (14,15), this is a misnomer as applied to SPCLs since they are prepared and screened in a nonrandom, highly systematic and ordered manner.

None of the library screening approaches, other than the SPCL method, as yet appear capable of functioning in microdilution assays or plaque assays for the development of novel antimicrobial and antiviral peptides. This illustrates one of the key advantages of working with nonsupport-bound peptides at concentrations high enough to affect activity in such functional assays.

Conclusion

Synthetic peptide combinatorial libraries as used in this laboratory (2,3,17) permit the systematic screening of tens to hundreds of millions of different peptides. A fundamental feature of SPCLs is that free peptides are generated and screened in solution at a concentration of each peptide appropriate for virtually any existing *in vitro* assay system. This, as well as all of the recently described library screening techniques (reviewed in 11 and 17), complement existing rational drug design methods currently used to study peptide/receptor interactions such as X-ray crystallography, nuclear magnetic resonance, and computer modeling, since large numbers of optimal peptide sequences can be rapidly identified for a particular receptor system of interest. We believe that the use of SPCLs as described here, which can be considered a form of "rational drug discovery" (versus what is often referred to as "rational drug design"), greatly facilitates the drug discovery process while significantly shortening its timeframe, and at the same time greatly increases the number of leads generated as potential therapeutic and diagnostic agents.

References

1. Houghten, R.A., Blondelle, S.E., and Cuervo, J.H. (1992). *In* "Innovation and Perspectives in Solid Phase Synthesis" (R. Epton, ed.), in press. Solid Phase Conference Coordination, Ltd., Canterbury.

2. Houghten, R.A., Cuervo, J.H., Pinilla, C., Appel, J.R., Dooley, C.T., and Blondelle, S.E. (1992). *In* "Peptides, Proceedings of the Twelfth American Peptide Symposium" (J.A. Smith and J.E. Rivier, eds.), p. 560. ESCOM, Leiden.

3. Houghten, R.A. and Dooley, C.T. (1992). BioOrganic & Medicin. Chem. Lett. In press.

4. Houghten, R.A., C. Pinilla, S.E. Blondelle, J.R. Appel, C.T. Dooley and J.H. Cuervo. 1991. Generation and use of synthetic peptide combinatorial libraries for basic research and drug discovery. Nature *354*:84-86.

5. Pinilla, C., J. Appel and R.A. Houghten. 1992. Synthetic Peptide Combinatorial Libraries: The Screening of Tens of Millions of Peptides for Basic Research and Drug Discovery, p.25. *In* Lerner R.A., H. Ginsberg, R.M. Chanock and F. Brown, eds. Vaccines 92. Cold Spring Harbor Laboratory, Cold Spring Harbor.

6. Houghten, R.A. 1985. General method for the rapid solid-phase synthesis of large numbers of peptides: specificity of antigen-antibody interaction at the level of individual amino acids. Proc. Natl. Acad. Sci. USA *82*:5131-5135.

7. Furka, A., F. Sebestyen, M. Asgedom and G. Dibo. 1991. General method for rapid synthesis of multicomponent peptide mixtures. Int. J. Pept. Protein Res. *37*:487-493.

8. Houghten, R.A., M.K. Bray, S.T. De Graw and C.J. Kirby. 1986. Simplified procedure for carrying out simultaneous multiple hydrogen fluoride cleavages of protected peptide resins. Int. J. Pept. Protein Res. *27*:673-678.

9. Tam, J.P., W.F. Heath and R.B. Merrifield. 1983. S_N2 deprotection of synthetic peptides with a low concentration of HF in dimethyl sulfide: evidence and application in peptide synthesis. J. Am. Chem. Soc. *105*:6442-6455.

10. Blondelle, S.E. and R.A. Houghten. 1991. Hemolytic and antimicrobial activities of the 24 individual omission analogues of melittin. Biochemistry *30*:4671-4678.

11. Birnbaum, S. and K. Mosbach. 1992. Peptide Screening. Curr. Opin. Biotechnol. *3(1)*:49-54.

12. Lam, K.S., S.E. Salmon, E.M. Hersh, V.J. Hruby, W.M. Kazmierski and R.J. Knapp. 1991. A new type of synthetic peptide library for identifying ligand-binding activity. Nature *354*:82-84.

13. Cwirla, S.E., E.A. Peters, R.W. Barrett and W.J. Dower. 1990. Peptides on phage: A vast library of peptides for identifying ligands. Proc. Natl. Acad. Sci. USA *87*:6378-6382.

14. Devlin, J.J., L.C. Panganiban and P.E. Devlin. 1990. Random Peptide Libraries: A Source of Specific Protein Binding Molecules. Science *249*:404-406.

15. Scott, J.K. and G.P. Smith. 1990. Searching for Peptide Ligands with an Epitope Library. Science *249*:386-390.

16. Fodor, S.P.A., J.L. Read, M.C. Pirrung, L. Stryer, A.T. Lu and D. Solas. 1991. Light-Directed, Spatially Addressable Parallel Chemical Synthesis. Science *251*:767-773.

17. Houghten, R.A., J.R. Appel, S.E. Blondelle, J.H. Cuervo, C.T. Dooley, and C. Pinilla. The use of synthetic peptide combinatorial libraries for the identification of bioactive peptides. *Biotechniques*, in press, 1992.

Acknowledgements

The authors wish to thank Eileen Silva for assistance in preparing this manuscript. This work was funded by Houghten Pharmaceuticals, Inc., San Diego, California.

Synthesis of Proteins by Chemical Ligation of Unprotected Peptide Segments: Mirror-Image Enzyme Molecules, D- & L-HIV Protease Analogs

R.C. de Lisle Milton, Saskia C.F. Milton, Martina Schnölzer

& Stephen B.H. Kent

The Scripps Research Institute, La Jolla CA 92037

I. Introduction

Enzymes invariably act on only one enantiomer of a chiral substrate, or generate only one diastereomer from a prochiral substrate (1). This specificity originates in the chiral structure of the enzyme molecule, including the three-dimensional folding of its polypeptide backbone and the orientation of the amino acid side chains in the folded protein molecule (1-3). Until recently only naturally occurring L-enzymes could be described, and this left the presumed properties of D-enzymes, including their folded structures, enzymatic activity and chiral specificity, as unexplored questions. However, the advances in the chemical synthesis of proteins (4-6) which made the production of homogenous crystalline L-[Aba^{67,95,167,195}]HIV-1 protease possible, have recently allowed the total synthesis of the D-enantiomer of this enzyme in our laboratory (7). This synthetic D-enzyme molecule had an identical covalent structure to its L-enantiomer. However, it had an inverted CD spectrum, and showed reciprocal chiral specificity towards peptide substrates and inhibitors. That is, the D-enzyme cut only the D-substrate and was blocked by only the D-inhibitor, while the L-enzyme recognized only the L-substrate and the L-inhibitor.

Co-crystallization of D- & L- HIV protease preparations should generate centrosymmetric crystals for high-accuracy X-ray diffraction studies (8) which would provide data that would find a wide usefulness in drug design studies for AIDS therapeutics. For this purpose, we wanted to produce each enantiomorph of the enzyme in hundred milligram quantities

SCALE: 0.2 mmole

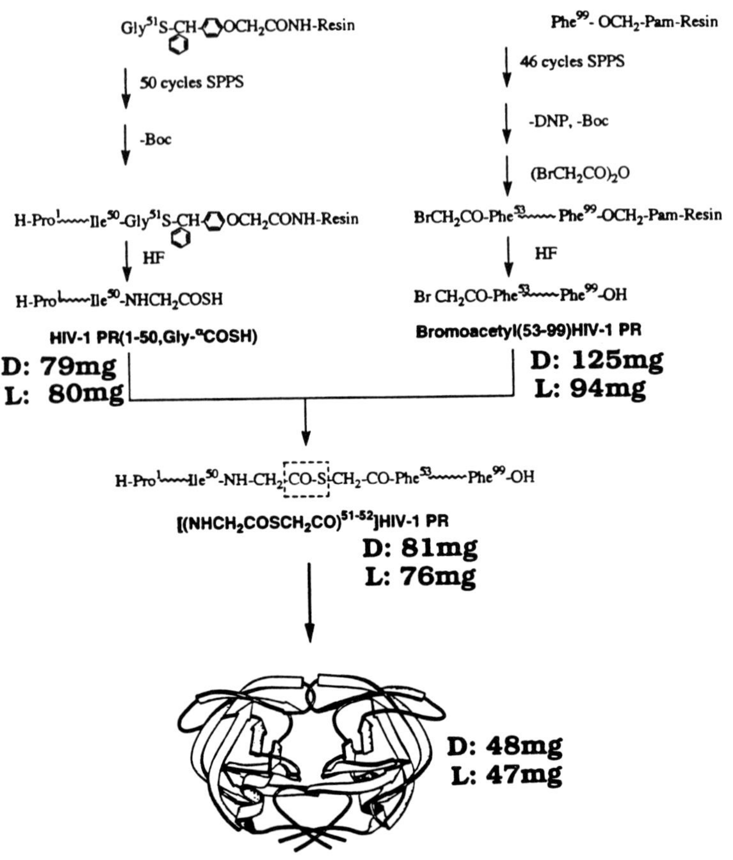

FOLDED $[Aba^{67,95}(CO\text{-}S)^{51\text{-}52}]_2$HIV-1 PR

D: 3.0%, L: 2.5% overall yield based on starting substition

Figure 1. Schematic representation of the chemical segment ligation strategy employed for the total synthesis of the D- and L- $[Aba^{67,95}(CO\text{-}S)^{51\text{-}52}]_2$HIV-1 protease analogs. HPLC purified, functionalized, unprotected peptide segments, assembled by stepwise solid-phase synthesis, were reacted in the presence of 6M GuHCl to form the ligated D- & L-$[(NHCH_2COSCH_2CO)^{51\text{-}52}]$ HIV-1 PR products. The boxed area represents the structure of the thioester analogue of the peptide bond Gly^{51}-Gly^{52} at the site where the ligation occurred. After purification, the monomers were separately folded by dialysis in the presence of D- & L- [MVT-101] inhibitor, respectively, to yield the homodimeric enzymes. Yields of each product (in mg) are given next to its name.

without the complication of autolysis during the extraction, purification and folding of the synthetic products. We recently showed that an HIV-1 protease analog, prepared by the directed chemical ligation of unprotected peptide segments and containing a thioester replacement for the natural peptide bond between Gly^{51}-Gly^{52}, has full activity (9). This segment condensation strategy also largely avoids the possibility of autodigestion by the enzymes during their preparation. We here describe the application of the ligation strategy to the preparation of 10-100mg amounts of the D- and L-HIV-1 protease enzymes.

II. Synthesis and Ligation

Protected D- and L-amino acids were obtained from the Peptide Institute (Osaka, Japan), Peptides International (Louisville, KY), Bachem Bioscience (Philadelphia, PA) and Bachem California (Torrance, CA) and had <0.03% of the opposite enantiomer. The 99-residue monomer of the HIV-1 protease molecule was prepared by directed ligation of functionalized unprotected segments (9) (Figure 1). These segments, [αCOSH]HIV-1 PR(1-51) and [N^{α}-$BrCH_2CO$]HIV-1 PR(53-99), were assembled in separate syntheses by a highly optimized machine-assisted SPPS protocol using Boc-chemistry (5,6) (Figure 2) performed on a modified ABI 430A synthesizer. The protocol comprised removal of the N^{α}-Boc group with undiluted TFA (2min total) followed by a DMF flow wash to give the TFA·peptide-resin salt, and a single 10min coupling step using HBTU activated Boc-amino acids and *in situ* neutralization with DIEA in DMF.

[αCOSH]HIV-1 PR(1-51) peptide was assembled on 4-[α-(Boc-Gly-S)benzyl]phenoxyacetamidomethyl-resin (10), and [N^{α}-$BrCH_2CO$]HIV-1 PR(53-99) was prepared by bromoacetylation of [$Aba^{67,95}$]HIV-1 PR(53-99)-OCH_2 Pam peptide-

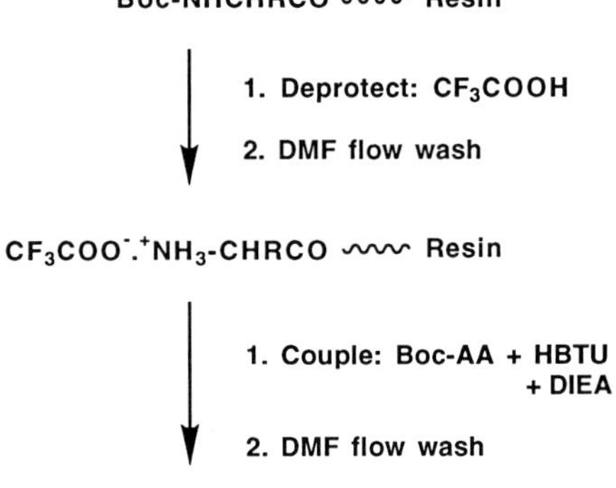

Figure 2. Schematic representation of the optimized solid-phase chain assembly tactics used in the synthesis of the functionalized peptide segments. Deprotection and coupling reactions were separated by a single flow wash step.

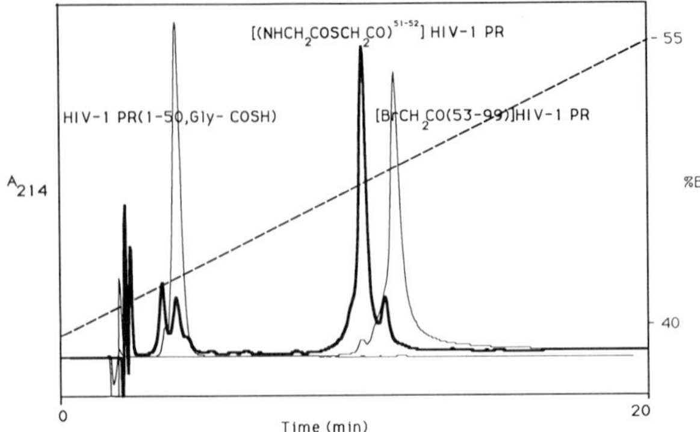

Figure 3. A composite chromatogram showing the two purified functionalized unprotected D-[αCOSH]HIV-1 PR(1-51) and D- [N^{α}-$BrCH_2CO$]HIV-1 PR(53-99) segments and the final (48 hr) D-[$Aba^{67,95}$$(CO\text{-}S)^{51\text{-}52}$]$_2$ HIV-1 PR ligation product (bold) run on a Vydac 218TP5415 column eluted by gradient (40-55%B, 20min) with 0.1% TFA (buffer A) and 0.09% TFA/CH_3CN, 1:9 (buffer B), at a flow rate of 1 ml/min.

resin. All peptides were cleaved and deprotected by high HF treatment (6).

After preparative reverse phase HPLC purification (Vydac 218TP101550 - 5x25cm, 0.1% TFA/CH_3CN & 30-50ml/min) the functionalized peptide segments were reacted in the presence of 6M GuHCl (in 0.1M phosphate buffer, pH 5.3) for 48hr to form the ligated D- and L-[$(NHCH_2COSCH_2CO)^{51\text{-}52}$]HIV-1 PR monomers (Figure 3). The products were purified by reverse phase HPLC and separately folded by dialysis in 25mM phosphate buffer, pH 5.5, in the presence of a 10-fold excess of either D- or L-[MVT-101]inhibitor (Ac-Thr-Ile-Nle-ψ-

Table I. Step reaction yields for the synthesis of the D- and L-[$Aba^{67,95}$$(CO\text{-}S)^{51\text{-}52}$]$_2$HIV-1 protease analogs.

	L		D	
	mg	%	mg	%
HIV-1 PR (1-50, Gly$^{\alpha}$ -COSH)	80.1	5.0	79.0	4.4
Bromoacetyl (53-99) HIV-1 PR	93.6	7.6	124.6	9.3
[$(NHCH_2COSCH_2CO)^{51\text{-}52}$] HIV-1 PR	75.8	67.4	80.8	44.0
Folded [$Aba^{67,95}$, $(CO\text{-}S)^{51\text{-}52}$]$_2$HIV-1 PR	47.4*	53.3	48.2*	60.0
OVERALL YIELD		3.0		2.5

*Yield extrapolated from a folding experiment with a small amount of material

[CH2NH]-Nle-Gln-Arg.NH2) (11) to yield the homodimeric enzymes. Table I gives the yields in a typical preparation of the D- and L-enzymes.

III. Product Characterization

Ion spray mass spectrometry of the HPLC purified ligated products (Figure 4) revealed single molecular species in each case with observed molecular masses of 10768.9±1.4 daltons (D-enantiomer) and 10769.4±0.9 daltons (L-enantiomer) [Calculated: 10763.9 daltons (monoisotopic) and 10770.8 daltons (average)]. Minor amounts of dehydration byproducts were also detected. The sequences of the monomers were also examined by a new protein ladder sequencing technique utilizing a one step laser desorption mass spectrometric readout (Wang, Kent & Chait - *this symposium*).

HPLC of the ligated products from 6M GuHCl revealed a single peak in each case (Figure 5), but, after folding, a number of minor autolysis products (as well as the MVT-101 inhibitor

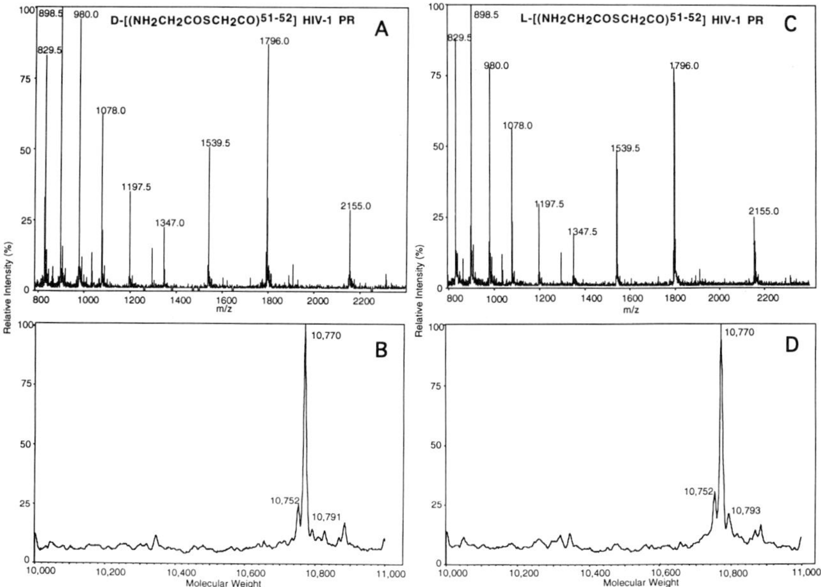

Figure 4. Ion spray mass spectra of the HPLC purified $[(NHCH_2COSCH_2CO)^{51-52}]$HIV-1 PR monomers. (**A & C**) The labelled peaks arise from a single molecular species differing in the number of excess protons. The observed molecular masses of the ligated products were 10770 daltons for both enantiomers [Calculated: 10763.9 daltons (monoisotopic) and 10770.8 daltons (average)]. (**B & D**) Reconstructed mass spectra in which the raw data shown in A & C have been reduced to a single charge state. All data points in A & C were included in the calculations and no mathematical filtering was performed. The mass regions from 10 to 11 kD are shown for clarity.

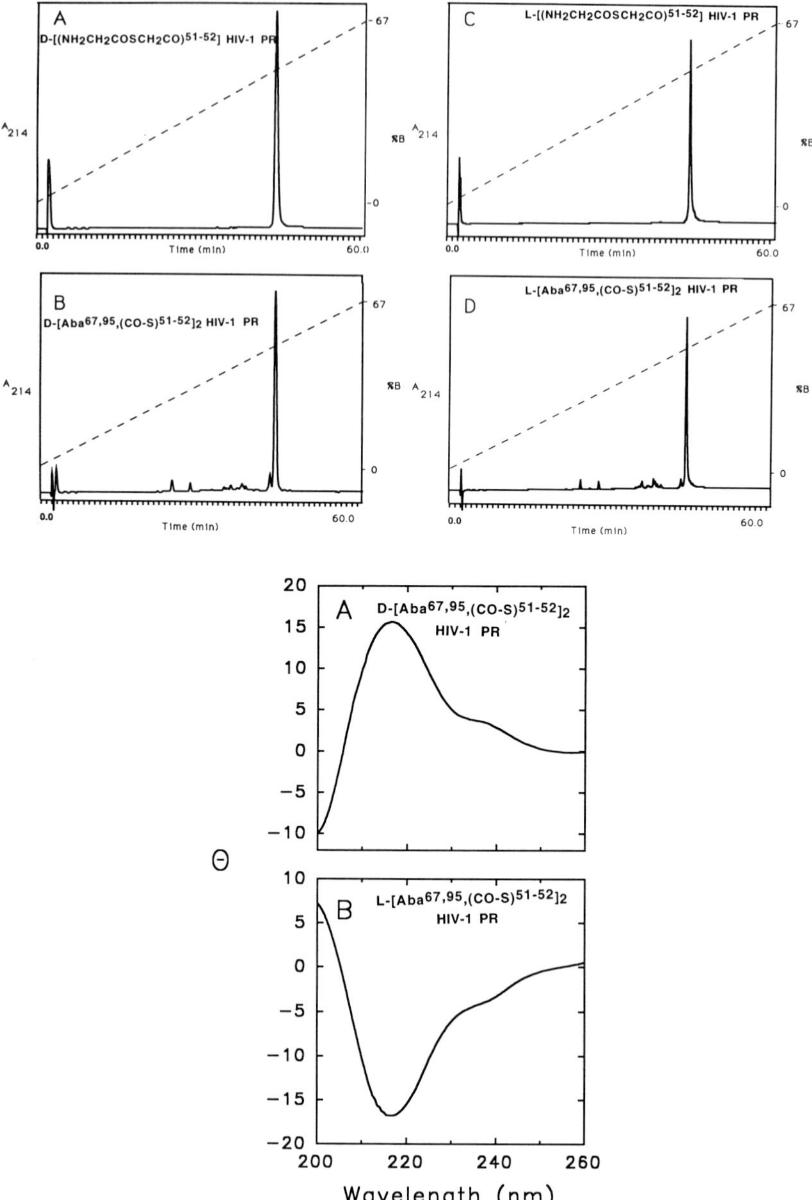

Figure 6. Far-ultraviolet CD showing (**A**) the D-$[Aba^{67,95}(CO\text{-}S)^{51\text{-}52}]_2$ HIV-1 PR spectrum and (**B**) the L-$[Aba^{67,95}(CO\text{-}S)^{51\text{-}52}]_2$ HIV-1 PR spectrum. The spectra were recorded in 25mM sodium phosphate buffer, pH 5.5 (0.4 mg/ml protease in the presence of inhibitor) at 25°C in a 1mm pathlength quartz cell.

Figure 5. (Opposite) RP HPLC characterization of the D- & L-[$Aba^{67,95}$(CO-S)$^{51-52}$] HIV-1 PR ligation products. (**A** & **C**) The purified ligated [($NHCH_2COSCH_2CO$)$^{51-52}$]HIV-1 PR monomers in 6M GuHCl. (**B** & **D**) The homodimeric enzymes folded in the presence of D- or L-[MVT-101] inhibitor, respectively, in 25mM sodium phosphate buffer, pH 5.5. The samples were run on a Vydac 218TP5415 column eluted by gradient (0-67%B, 60min) with 0.1% TFA (buffer A) and 0.09% TFA/CH_3CN, 1:9 (buffer B), at flow rate of 1 ml/min.

peptide at approx. 27min) were seen in both the D- and L-[$Aba^{67,95}$(CO-S)$^{51-52}$]$_2$HIV-1 protease preparations. It would seem that even in the presence of a large excess of inhibitor, the enzyme is still subject to a minor degree of autolysis.

CD spectra of the folded D- and L-protease preparations were of equal magnitude, but opposite sign (Figure 6), as expected for mirror image proteins (12,13).

IV. Enzymatic Activity

The enzymatic activity of the D- & L-[$Aba^{67,95}$(CO-S)$^{51-52}$]$_2$ HIV-1 PR enantiomers was determined by their action on D- and L-isomers of the hexapeptide substrate Ac-Thr-Ile-Nle-Nle-Gln-Arg.amide (an analog of the p24/p15 GAG viral processing site) (9) (Figure 7). The D-enzyme cut only the D-substrate and was inactive on the L-substrate, while the L-enzyme showed full activity towards the L-substrate, but was inactive towards the D-substrate. This reciprocal chiral specificity was also evident in the effect of chiral inhibitors. D- and L-[MVT-101] inhibited the cleavage of chiral fluorogenic substrates by the D- and L-HIV-1 PR analogs, respectively, but had no effect on the action of the opposite enantiomer (Table II). Interestingly, the achiral inhibitor, Evans Blue, which shows mixed inhibition kinetics, inhibited both enantiomers of the enzyme (Table II).

Table II. Chiral inhibitors show reciprocal chiral specificity against D- and L-[$Aba^{67,95}$(CO-S)$^{51-52}$]$_2$HIV-1 PR *

	L-MVT101	D-MVT101	Evans Blue§
L-[$Aba^{67,95}$(CO-S)$^{51-52}$]$_2$HIV-1 PR	+	-	+
D-[$Aba^{67,95}$(CO-S)$^{51-52}$]$_2$HIV-1 PR	-	+	+

* The D- and L-enzymes were seperately assayed by the fluorogenic assay method using the corresponding chiral substrate, in the presence of 5 x IC_{50} concentration of inhibitor. The fluorogenic assays were performed with 15 ul aliquots (corresponding to 1.75 (±10%) mg protein) of each enzyme enantiomer in 100mM MES buffer pH6.5 added to a solution of 50mM D- or L-fluorogenic substrate in the MES buffer. The substrate sequence was 2-aminobenzoyl-Thr-Ile-Nle-Phe(p-NO_2)-Gln-Arg.amide; it was synthesized with either D- or L-amino acid derivatives to provide the appropriate enantiomeric forms.

§ The inhibitor Evans Blue is a non-peptide, achiral mixed competitive-uncompetitive inhibitor of the HIV-1 PR enzyme.

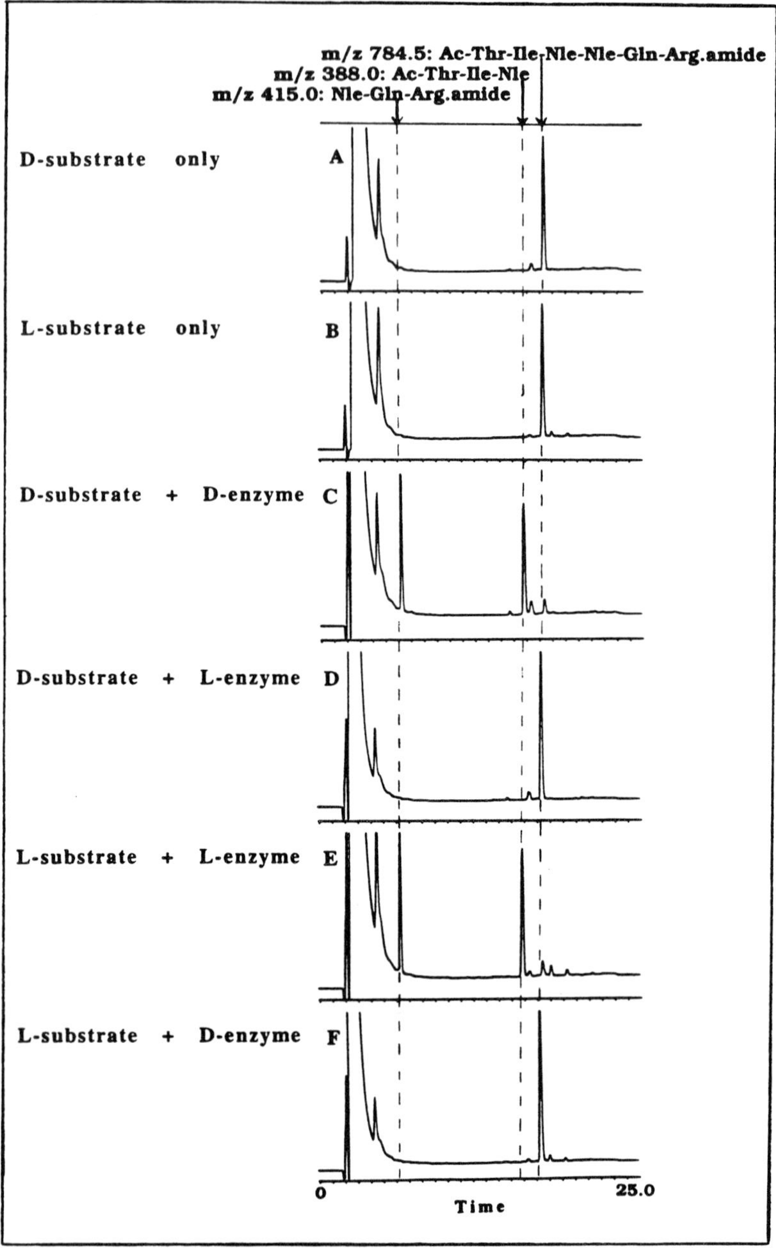
m/z 784.5: Ac-Thr-Ile-Nle-Nle-Gln-Arg.amide
m/z 388.0: Ac-Thr-Ile-Nle
m/z 415.0: Nle-Gln-Arg.amide
D-substrate only
A
L-substrate only
B
D-substrate + D-enzyme
C
D-substrate + L-enzyme
D
L-substrate + L-enzyme
E
L-substrate + D-enzyme
F
0
25.0
Time

Figure 7. (Opposite) The enzymatic activity of the D- & L-[$Aba^{67,95}(CO\text{-}S)^{51\text{-}52}]_2$ HIV-1 PR enantiomers on D- and L-isomers of the substrate Ac-Thr-Ile-Nle-Nle-Gln-Arg.amide. 50µl substrate solution (1mg/ml) was treated with 5µg enzyme (50µl of a 0.1mg/ml solution) at pH6.5 (MES buffer, 100mM) for 30min at 37°C after which an aliquot of the reaction mixture was chromatographed (Vydac 218TP5415 RP HPLC column) with a linear gradient, 0-40%, of buffer B (0.09% TFA/CH_3CN, 1:9) in buffer A (0.1% TFA) over 20min (flow rate 1ml/min, A_{214nm}). The peptide products were identified by ion spray MS as (H)-Nle-Gln-Arg.amide (m/z: 415.0 - early eluting) and Ac-Thr-Ile-Nle-(OH) (m/z: 388.0 -late eluting). Minor impurities (crude peptide substrates were used) present in the substrate preparations were not cleaved.

V. Conclusions

The HIV-1 protease enzyme exists as a homodimeric structure (11,14). It is a highly specific enzyme (15) and this specificity and its catalytic activity depend on a precise 3-D structure being formed between the folded dimer and six residues of the substrate molecule (11). The observed reciprocal specificities, therefore, show that the folded forms of the D- and L-enzyme molecules are mirror images of each other in all elements of the 3-D structure responsible for their enzymatic activity. This is consistent with their observed CD spectra (Figure 6).

The 3-D structure of a folded enzyme molecule contains numerous chiral elements in secondary and supersecondary structure, in tertiary structure and in quarternary structure (Figure 8). Since the only difference between the synthetic D-

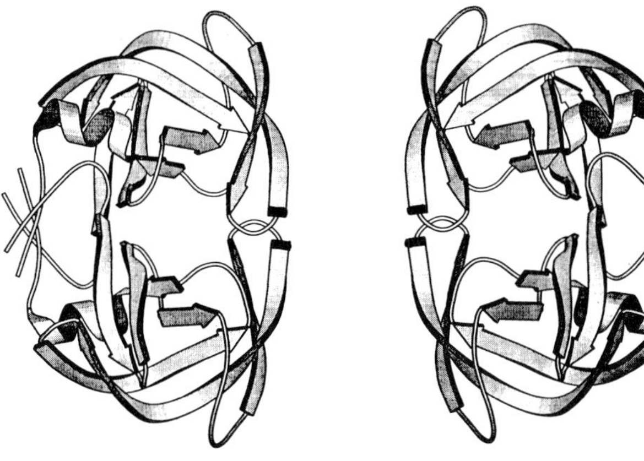

L-HIV PROTEASE D-HIV PROTEASE

Figure 8. Ribbon representations of the polypeptide backbone of the folded homodimeric L- & D-HIV-1 PR analog molecules. [Left] L-[$Aba^{67,95}(CO\text{-}S)^{51\text{-}52}]_2$ HIV-1 PR based on the X-ray crystallographic coordinates of the chemically synthesized [$Aba^{67,95,167,195}$] HIV-1 PR enzyme complexed with a substrate-derived peptide inhibitor (the inhibitor is not shown); [Right] the folding of the D-[$Aba^{67,95}(CO\text{-}S)^{51\text{-}52}]_2$ HIV-1 PR polypeptide backbone suggested by the data presented here. This model was generated by performing a mirror-image transformation of the L-enzyme data.

and L-polypeptide chains is the stereochemistry of the α-carbon atoms (and the Cβ atoms of Ile and Thr) of the amino acids, it is concluded that the stereochemistry of the backbone determines all aspects of higher structure in this protein.

The observations of reciprocal chiral specificity in the enzymatic activity of the D- and L-HIV-1 proteases here and previously (7) reported, serve to generalize and emphasize the chiral nature of the biochemical interactions of proteins. The large amounts of high purity D- and L-enzyme enantiomorphs prepared using the chemical ligation method will allow a thorough experimental evaluation of the use of D-, L-proteins in X-ray crystallography (8).

Acknowledgements

A. Aguon and G. Verduzco are acknowledged for their assistance in the preparation of the D-substrate, G. Merutka for CD, and V. Roberts for assistance in preparing Figure 8. This work was supported by the Markey Foundation.

References

1. Fersht, A. (1977) *In* "Enzyme Structure and Mechanism", p.75-81. Freeman, San Francisco.
2. Fruton, J.S. (1972) *In* "Molecules and Life", p. 79-80. Wiley, New York.
3. Mirsky, A.E. and Pauling, L. (1936) Proc. Natl. Acad. Sci. U.S.A. **22**, 438.
4. Kent, S.B.H. (1988) Annu. Rev. Biochem. **57**, 957.
5. Kent, S.B.H., Alewood, D., Alewood, P., Baca, M., Jones, A. and Schnölzer, M. (1992) In "Innovation & Perspectives in Solid-Phase Synthesis", Ed. Epton, R. SPPC Ltd. Birmingham, U.K. In press.
6. Schnölzer, M., Alewood, P., Jones, A., Alewood, D. and Kent, S.B.H. (1992) Int. J. Peptide Protein Res. In press.
7. Milton, R.C. de L., Milton, S.C.F. and Kent, S.B.H. (1992) Science **256**, 1445-1447.
8. Mackay, A.L. (1989) Nature **342**, 133.
9. Schnölzer, M. and Kent, S.B.H. (1992) Science **256**, 221-225.
10. Yamashiro, D. and Li, C.H. (1988) Int. J. Peptide Protein Res. **31**,322-334.
11. Miller, M., Schneider, J., Sathyanarayana, B.K., Toth, M.V., Marshall, G.R., Clawson, L., Selk, L., Kent, S.B.H. and Wlodawer, A. (1989) Science **246**, 1149.
12. Corigliano-Murphy, M.A., Liang, X., Ponnamperuma, C., Dalzoppo, D., Fontana, A., Kanmera, T. and Chaiken, I.M. (1985) Int. J. Peptide Protein Res. **25**, 225.

13. Zawadzke, L.E. and Berg, J.M. (1992) J. Am. Chem. Soc. **114**, 4002.
14. Wlodawer, A., Miller, M., Jaskólski, M., Sathyanarayana, B.K., Baldwin, E., Weber, I., Selk, L., Clawson, L., Schneider, J. and Kent, S.B.H. (1989) Science **245**, 616.
15. Kent, S.B.H., Schneider, J., Clawson, L., Selk, L., Delahunty, C. and Chen, Q. (1989) *In* "Viral Proteinases as Targets for Chemotherapy" (H-G. Krauslik, S. Oroszlan and E. Wimmer, eds.), p. 223. Cold Spring Harbor Laboratory Press, Cold Spring Harbor, NY.

DIFFICULT SEQUENCES IN THE SYNTHESIS OF G-PROTEIN FRAGMENTS THAT HARBOR A PUTATIVE ADHESION MOTIF

Assaf Steinschneider

University of Illinois at Chicago, Research Resources Center
Chicago, IL 60612

I. INTRODUCTION

Sequence dependent difficulties in aminoacylation have plagued peptide synthesis throughout its development (1,2, and references in both). The underlying factors are only partially understood. A substantial body of information agrees with notions that a major source for problems is the formation of beta-sheet, and/or related structures, leading to the aggregation of nascent peptide chains. This presumably reduces the availability of the N-terminal amino group to the incoming amino acid (1-4). So far, efforts to improve the molecular environment (5,6) and to develop algorithms for the prediction of difficult sequences relied heavily on conformational considerations (5,4).

Several problematic sequences were encountered in this facility in preparing a series of fragments derived from the alpha subunits of mammalian G-proteins that are central to transmembrane signal transduction. Inspection of the primary structure revealed that some of the sluggish stretches harbored tetrapeptide sequences sharing a structural motif that suggested adhesive function. These were Glu-Gln-Asp-Val, bGt1(167-170); Asp-Gln-Asp-Leu, hGs1(194-197); Lys-Lys-Asp-Val, bGt1(266-269); and Arg-Arg-Asp-Val, bGt1(309-312) (7). Together with other G-protein difficult sequences (unpublished data) they are highly conserved among known members of this family of proteins, suggesting a biological function. Representative observations will be described below.

Abbreviations. NMP, N-methylpyrrolidinone; DMF, dimethylformamide; DCC, dicyclohexylcarbodiimide; HOBt, hydroxybenzotriazole; HBTU, [2-(1H-benzotriazol-1-yl)-1,1,3,3,- hehexafluorophosphate]; t-but, t-butyl; Trt, triphenylmethyl; PMC, N^G-(2,2,5,7,8-pentamethylchroman-6-sulfonyl);
Guanyl-nucleotide binding protein alpha subunits: bGt1, bovine rod transducin; bGt2, bovine cone transducin; hGs1, human stimulatory subunit; hGi(1,2,3), human inhibitory subunit.

TECHNIQUES IN PROTEIN CHEMISTRY IV

II. MATERIALS AND METHODS

Unless specified otherwise, peptides were prepared with a carboxamide C-terminus employing an Fmoc strategy. Solid phase synthesis was performed using an ABI Model 431A Peptide Synthesizer with programs and chemicals provided by the manufacturer. Some syntheses were interrupted to secure partial fragment(s), without noticeable effect on coupling efficiencies. The main solvent was NMP. DCC/HOBt were employed as coupling agents and 4-(2,4,-dimethoxyphenyl-Fmoc-aminomethyl)-phenoxy co(polystyrene-1% divinylbenzene) peptide amide resin (Calbiochem catalog #757731) (8) served as the solid support. Double couplings were applied selectively depending on previous information from this and other laboratories but will not be indicated in the text. Side chain protecting groups were t-But (Asp, Glu, Ser, Thr, Tyr); t-Boc (Lys); Trt(His, Cys) and PMC(Arg). Resin samples were collected into 0.5% AcOH/MeOH, washed with DCM/MeOH (3:2, V/V) and then assayed with ninhydrin (9). Sequence homologies cited in the text are based on the data reviewed by Kaziro et al (10). The Chou and Fasman algorithm (11, 12), as executed by the PRONUC package, was employed for conformational predictions.

III. RESULTS

Peptide bGt1 Glu(305)-Lys(329) (ELNMRRDVKEIYSHMTCATDTQNVK-NH_2). This peptide, with its partial and homologous fragments represents the single system with most available information. Fig 1 illustrates the progress of the synthesis. In characteristic fashion (1,4), double couplings were only of limited effect. Sluggish aminoacylations were consistently encountered starting with residue Lys(313), step 17 in the synthesis, and persisted throughout the tetrapeptide (in the order of residue addition) Val-Asp-Arg-Arg. Viewed as a protein sequence this is equivalent to Arg-Arg-Asp-Val. Two straight side chain, polar, hydrogen-bonding residues are followed by aspartic acid and then by an aliphatic residue. As a structural motif, this also describes the Arg-Glu-Asp-Val fibronectin adhesion sequence identified first in B16-F10 melanoma cells, which is related to the better known Arg-Gly-Asp-Ser adhesion signal (13). Among major,known G-proteins the sequence is preserved in bGt2 with one conservative substitution to Arg-Lys-Asp-Val, and with an additional nonconservative substitution in hGi(1-3) to Arg-Lys-Asp-Thr. It is not recognizable in other members of this protein family (Table 1 and ref. 10).

When the coupling agents DCC/HOBt were replaced with 2-(1H-benzotriazol-1-yl)-1,1,3,3-tetramethyluronium hexafluorophosphate(HBTU)/HOBt (14), aminoacylations were normal throughout the synthesis of the partial sequence bGt1(311-329), which includes the first three difficult residues, and throughout the desVal(312) analog (both not shown). Since the same ninhydrin assay was used to monitor the disappearance of the terminal amino group using either combination of coupling agents, it is apparent that aminoacylation difficulties described above were genuine and not due to a problem with the assay.

Beta ('sheet') structure was predicted for residues Ser(317)-Thr(323) of bGt1(305-329), the partial sequence bGt1(311-329) and intermediate fragments. The stretch downstream in the synthesis was devoid of predicted secondary structure until adding Glu(305) when the segment

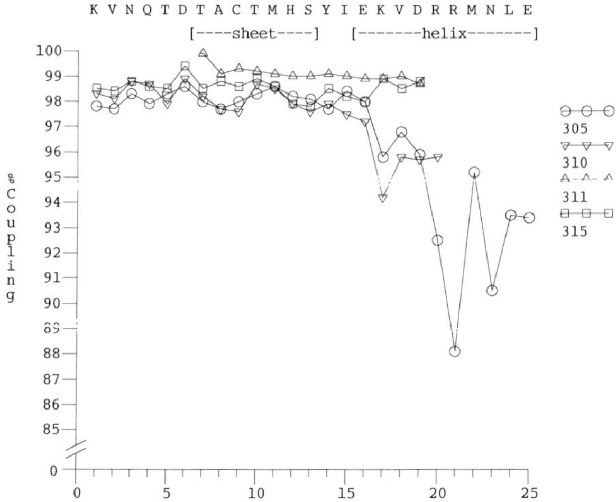

Fig. 1. Coupling efficiencies in the synthesis of peptides bGt1(305-329), bGt1(310-329), bGt1(311-323) and hGi1(315-333), labled 305, 310, 311, and 315, respectively. Conditions as in Materials and Methods. The sequence and predicted conformation apply to peptide bGt1(305-329). Peptides bGt1(311-323) and hGi1(315-333) were aligned with the homologous positions of bGt1(305-329). Fragments bGt1(305-329) and hGi1(315-333) were synthesized with interruptions after residues Asp(311) and Cys(325), respectively, to secure partial sequences.

Table I. Conservation of the Putative Adhesion Motif*

Protein	bGt1 Sequence		
	(167-170)	(266-269)	(309-312)
bGt1	**EQDV****	**KKDV**	**RRDV**
bGt2	EQDV	KKDL	RKDV
hGi(1-3)	QQDV	KKDL	**RKDT**
hGs1	**DQDL**	KQDL	-------
rGs2	DQDL	KQDM	-------

* adapted from ref. 10.

** boldface-difficult aminoacylations encountered so far.

starting at Glu(314) would assume alpha helical conformation. No beta turns were indicated by the program.

Peptide bGt1 Asp(311)-Thr(323) (DVKEIYSHMTCAT). As an exception in this survey, this peptide was prepared with a C-terminal carboxyl group, using (4-(hydroxymethyl)-phenoxymethyl-copoly (styrene-1% divinylbenzene)(HMP) resin. The stretch starting with Lys(313) is sufficiently distant from the origin to present problems (4) expected from the behavior of peptide bGt1(305-329) and its fragments, with whom the peptide also shares predicted conformational features. As seen in Fig 1 (curve 311), couplings were, however, completely normal. While one may argue that this was due to the change in resin, another interpretation is in better agreement with other observations, described in part below. Accordingly, the presence of a potentially difficult sequence may not by itself be sufficient cause for sluggishness.

Peptide hGi1 Asp(315)-Gln(333) (DTKEIYTHFTCATDTKNVQ-NH_2). This fragment is homologous to the sequence bGt1(329-311) with the following substitutions: K(329)Q, Q(326)K, M(319)F, S(317)T, and V(312)T. Except for the last, all are outside the problem area. Aminoacylations were normal throughout (Fig 1, curve 315). While it cannot be ruled out that the V(312)K replacement was critical, this is considered less likely and would not account for the improvement in the previous step of the synthesis when Lys(313) was added. Again, the unexpectedly efficient couplings may be rationalized as they were for the bGt1 (311-323) fragment described above.

Essentially the same conformational features were predicted as for the bGt1(311-329) homolog, in accord with the conservative nature of most amino acid replacements.

Peptide bGt1 Gly(162)-Ile(181) (GYVPTEQDVLRSRVKTTGII-NH_2). Coupling efficiencies were essentially constant up to Arg(172), dropping substantially for the next six residues and recovering with residue Pro(165) (Fig 2). This pattern is characteristic of many difficult sequences (1,4). The main problems involved the stretch Val(170)-Glu(173), corresponding to another putative adhesion signal Glu-Gln-Asp-Val. This sequence is the same in bGt2 and is present with one and two conservative substitutions in hGi(1-3),(Gln-Gln-Asp-Val), and hGs(1,2),(Asp-Gln-Asp-Leu), respectively (see Table 1).

Conformational parameters predicted beta structure between residues Tyr(163)-Thr(166) and Val(170)-Ile(181). Aminoacylations thus became sluggish in association with predicted beta conformation, to continue through a stretch expected to be devoid of secondary structure. Interestingly, they recovered subsequently even though the synthesis now prsumably proceeded again through a beta conformation, considered the least favorable environment (1,4,5).

Peptide bGt1 Asn(265)-Ser(280) (NKKDVFSEKIKKAHLS-NH_2). As shown in Fig 3, serious problems ensued with Val(269) and persisted to the end of the synthesis. Again, the main difficulties were encountered with members of the putative adhesion motif Lys-Lys-Asp-Glu, residues 266-269. The sequence is conserved as Lys-Lys-Asp-Glu in bGt2 and hGi(1-3), and as Lys-Gln-Asp-Leu in hGs1 (Table 1).

Alpha helical conformation was predicted for residues Glu(272)-Leu(279), upstream of the problem area.

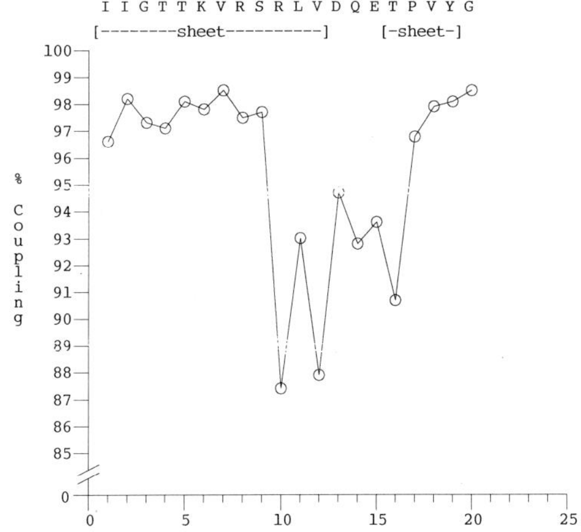

Fig. 2. Coupling efficiencies in the synthesis of peptide bGt1(162-181). Conditions as in Materials and Methods.

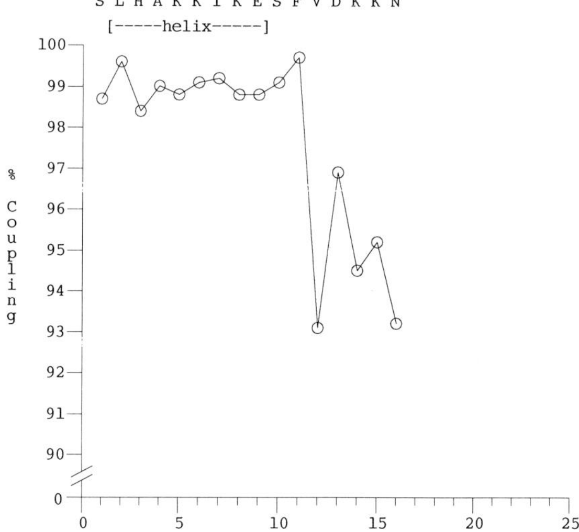

Fig. 3. Coupling efficiencies in the synthesis of peptide bGt1(265-280). Conditions as in Materials and Methods.

IV. DISCUSSION

Taken together, the results strongly suggest that residue specific interactions akin to those between adhesive protein molecules were implicated in creating conditions for sluggish aminoacylations. These occurred even though the dipolar aprotic solvent NMP was used, expecting to optimize the conditions with respect to chemical reactivity, low internal H-bonding and resin swelling. Problems may arise due to the interactions of as few as only one residue, as suggested by the abrupt difficulties with Arg(172) or Lys(313) in the syntheses of peptides bGt1(162-181) and bGt1(305-329), respectively. On the other hand the normal behavior of two fragments that host potentially difficult sequences, bGt1(311-312) and hGi1(315-333), is consistent with notions that an appropriate counterpart elsewhere in the molecule may be needed. Possibly, its role in promoting adhesive interactions is analogous to that of helper sequences implicated in biochemical studies (16,17 and references therein).

Present observations also exemplify the failure to detect a meaningful relationship between coupling efficiencies and predicted conformational attributes of the peptide and related predictors (18,4). Thus, low reactivity did not appear to correlate with predicted beta conformations, where problems might have been anticipated (1,3,4), or any other structure predicted for the problematic stretch or for the residues preceding in the synthesis. Occasionally, difficult couplings occurred within a sequence essentially devoid of predicted structure, expected to provide a favorable environment (4,5). Similar conclusions also apply to the other peptides under scrutiny (unpublished data). Whether or not the present observations constitute evidence regarding the role of actual conformations in coupling difficulties is an open question. Obviously, more data will be needed to determine whether the present study typifies a general pattern or only pertains to residues noticed previously as exceptions (4). Clearly, however, a new element has been added to the understanding and hopefully prediction of sluggish aminoacylations.

The putative adhesion motif and often the adjacent residues are highly conserved among the better known signal transducing G-proteins (10). Presumably, homologous regions in the parent proteins could engage in the stabilization of alpha subunit architecture, in intermolecular communication with neighboring proteins and diverse other effectors, as well as in interactions with unusual partners, e. g. the complexation of antibodies (16). The three putative adhesion sequences of bGt1(transducin), the full complement for this protein (unpublished data), may be localized to broadly defined regions of the protein associated with particular biological functions (19). More specific information is only starting to emerge. Thus, the sequence Arg-Arg-Asp-Val, bGt1(309-312), is present, at least in part, in fragments implicated in the interactions of rod transducin alpha subunit with its upstream receptor rhodopsin (16), and with its downstream effector, cGMP phosphodiesterase (20). Residue Asp(311) has been specifically implicated in binding to the receptor (16). Indeed, a key role for Asp is suggested both by its nonvarying appearance in the putative adhesion motif and by its ubiquitous functional role in other proteins.

REFERENCES

1) Kent, S. B. H. (1988), Ann. Rev. Biochem. **57**, 957-989
2) Fields, G. B.,and Noble,R. L. (1990) Intl. J. Peptide Protein Res. **35**, 161-214
3) Pillai, V. N. R., and Mutter, M. (1981) Accts. Chem. Res. **14**, 122-130
4) Milton, R. C. deL., Milton, S. C. F., and Adams, P. A.(1990) J. Am. Chem. Soc. **112**, 6039-6046
5) Narita, M., Ishikawa,K., Chen, J.-Y., and Kim, Y. (1984), Intl. J. Peptide Protein Res. **24**, 580-587
6) Stevens, R. L., Stevens, K. L., and Young, J. D. (1991), in Techniques in Protein Chemistry **II** (J. Villafranca ed.) Academic press Inc. San Diego, pp 209-220.
7) Steinschneider, A. (1992) The Protein Society sixth symposium, program and abstracts, San Diego, CA
8) Rink, H., (1987) Tetrahed. Lett. **28**, 3787-3790
9) Sarin V. K., Kent, S. B. H., Tam, J., and Merrifield, R. B. (1981) Analyt. Biochem. **117**, 147-157
10) Kaziro, Y., Itoh, H., Kozasa, T., Nakafuku, M., and Satoh, T. (1991) Ann. Rev. Biochem. **60**, 349-400
11) Chou, P. Y., and Fasman, G. D. (1974) Biochemistry **13**, 211-222
12) Chou, P. Y., and Fasman, G. D. (1979) Biophys. J. **26**, 367-384
13) Humphries, M. J., Akiyama, S. K., Komoriya, A., Olden, K., and Yamada, K. M. (1986) J. Cell Biol. **103**, 2637-2647
14) Knorr, R., Trzeciak, A., Bannwarth, W., and Gillessen, D. (1989) Tetrahed. Lett. **30**, 1927-1930
15) Wang, S.-S. (1973) J. Amer. Chem. Soc. **95**, 1328-1333
16) Hamm, H. E., Deretic, D., Arendt, A., Hargrave, P. A., Koenig, B., and Hofmann, K. P. (1988) Science **241**, 832-835
17) Yamada, K., M.,(1991) J. Biol. Chem. **266**, 12809-12812
18) Deber, C. M., Lutek, M. K., Heimer, E. P., and Felix, A. M. (1989) Peptide Res. **2**, 184-188
19) Ho, Y.-K., Hingorani, V. N., Navon, S. E., and Fung, B. (1989) in Current Topics Cellular Regulation **30**, pp 171-202
20) Rarick, H. M., Artemyev, N. O. and Hamm, H. E. (1992) Science **256**, 1031-1033

ACKNOWLEDGEMENT

The advice of Dr. Gordon L. Humphrey in the preparation of this manuscript is gratefully acknowledged.

SECTION V

Amino Acid Analysis

Cysteine and Tryptophan Amino Acid Analysis of ABRF92-AAA

Daniel J. Strydom[a], Thomas T. Andersen[b], Izydor Apostol[c], Jay W. Fox[d], Raymond J. Paxton[e] and John W. Crabb[f]

[a]Center for Biochem. Biophys. Sciences and Medicine, Harvard Medical School, Boston, MA 02115
[b]Department of Biochemistry and Molecular Biology, Albany Medical College, Albany, NY 12208
[c]Department of Chemistry, Purdue University, West Lafayette, IN 47907
[d]Department of Microbiology, University of Virginia Medical School, Charlottesville, VA 22908
[e]Department of Protein Chemistry, Immunex Corp., Seattle, WA 98101
[f]Protein Chemistry Facility, W. Alton Jones Cell Science Center, Lake Placid, NY 12946

I. Introduction

Amino acid analysis is one of the major analytical techniques used in biochemical and biotechnological environments. It remains the method of choice for reliable protein/peptide quantitation and in many instances provides a useful parameter in the characterization of peptides and proteins. In addition it complements other structural analysis methods, particularly Edman degradation and mass-spectrometry. As such it is useful to assess the quality of the results obtained by various investigations.

As part of an annual continuation of collaborative trial studies by the Association of Biomolecular Resource Facilities (ABRF) (1-5), this study concentrated on the determination of two problem residues, cystine/cysteine and tryptophan. Previous trials (1-5) have shown that the average accuracy for determination of cystine/cysteine and tryptophan are less than for other residues, although some laboratories reported excellent results. The aim of this study was to evaluate methods for cystine/cysteine and tryptophan analysis and to identify those methods that can be used reliably to quantitate these residues.

II. Materials and Methods

A. *Sample Preparation and Distribution*

The 1992 ABRF amino acid analysis test sample was bovine pancreatic chymotrypsin (Sigma Chemical Company, Product No. C-4129, Lot No. 63F803) and was chosen because of its relatively high Cys and Trp content (10 and 8 mole/mole protein (241 residues), respectively (6)). The Sigma preparation was dissolved in 0.1% trifluoroacetic acid, dialyzed against the same solvent to remove salts, the concentration determined by amino acid analysis and 56 μg (2.2 nmol) aliquots of chymotrypsin dried in small plastic tubes using a Savant Speed Vac. The sample was sent as an unknown by US Mail to 231 ABRF facility directors for analysis using amino acid analysis methods of their choice. Each participating laboratory was asked to report pmol of amino acids found using their standard analysis method as well as to quantify Cys and Trp using additional methods and analyses. Participants were asked to respond to specific questions regarding their methodology,

TECHNIQUES IN PROTEIN CHEMISTRY IV

calibration, and quantification procedures and patterns of usage. In addition, an abbreviated list of references was provided with the sample to encourage participation in the study.

B. Calculations

Raw data were received by an independent collaborator, identifying marks removed and the anonymous results forwarded to the 1992 ABRF Amino Acid Analysis Committee. Data reduction was performed with a personal computer as previously described in detail (5). The accuracy of each residue relative to the true value was calculated as % Error, where % Error = 100 × (experimental - true residue value)/ true residue value. The overall accuracy of composition was calculated as the Average % Error, where Average % Error = Σ absolute % Error for 16 amino acids/16. Cys and Trp errors were not included in the average percent error calculation, nor was Pro for two data sets using fluorescent based methodologies, which did not analyze for Pro.

The standard analysis results formed the major part of most facilities' composite data, and absolute yield of protein and average error was calculated from these data. In a few instances the 'standard data set' was patently inferior to that obtained by the Cys or Trp method and the better data set was used to calculate yield of protein and average error. The Cys and Trp values were calculated from their individual data sets, normalized to the amount analyzed in the standard analyses, and residues per mole calculated. Some facilities only quantified Arg and His in their special analyses for Cys and Trp (presumably because of the relatively low amount of Arg and His in chymotrypsin) and in these instances the Arg and His data were also normalized into the composite data set.

III. Results

A. Instrumentation

A total of 59 facilities participated in this study, the largest number yet in an ABRF amino acid analysis study. The employed instrumentation and methodology is summarized in Table I. Nearly all sites used instruments dedicated to amino acid analysis. Pre-column analysis was slightly more popular (58% of the sites), mostly employing PITC-derivatization. Automated derivatization has markedly increased from previous years; 16 sites used the ABI derivatizer, with 11 also doing automated

Table I. Instrumentation and Methodology used in the Analysis of ABRF-92AAA

Instrumentation				Methodology			
Post-column		Pre-column		Post-column		Pre-column	
Beckman 6/7300	19	ABI 420H	11	Ninhydrin	21	PITC	32
Waters	2	ABI 420A	5	OPA	3	OPA/FMOC	2
St. Johns	1	ABI 130A	2	Fluram	1		
Dionex	1	Waters	13				
Pharmacia	1	H/P 1090	3				
Home-made	1						
Total	25		34		25		34

hydrolyses, as compared to a total of 8 automated sites in 1991 (5). Post-column analysis sites (42%) predominantly used ninhydrin for detection.

B. Overall Accuracy and Recovery

The average % error obtained by each participating laboratory in the analysis of ABRF-92AAA is plotted in Figure 1 for all residues except Cys and Trp. Overall the accuracy of the standard analyses was good, with an average for all sites of 10.5% error (range 2.6 to 37.3%), with almost two thirds of the participants (63%) obtaining ≤ 10% average error. This level of accuracy (obtained by a median of 2 analyses of an average of 5.5 μg protein) is substantially better than that obtained with the 4 μg ABRF sample of myoglobin in 1990 (13.5% error with triplicate analyses of 1.3 μg aliqouts) but slightly lower than the 9.5% error obtained with 150 μg of BSA in the 1991 study (doing multiple or single analyses on 0.6 to 45 μg protein). A statistical comparison of the results obtained with post-column and pre-column methodologies in this year's study is presented in Table II, including amounts hydrolyzed/analyzed and errors obtained in the standard analyses as well as those for cysteine and tryptophan. Laboratories obtaining ≤ 10% error in the standard analysis were almost evenly represented by post-column and pre-column methods, namely 18 and 19 laboratories, respectively. Notably, 12 of the 21 sites with ≤ 10% Cys error and 6 of the 10 sites with ≤ 10% Trp error utilized post-column instrumentation. Predictably, however, the post-column methodologies generally used more sample than did the pre-column methods (Table II) - on average two-fold more.

Table II. Summary of 1992 ABRF Amino Acid Analysis Results[a]

Type of Analysis	All	Postcolumn Methods		Precolumn Methods	
	Responses	Ninhydrin	Fluorescence	PTC	Fluor.
Standard Analysis					
Total Sites	59	21	4	32	2
Sites with ≤ 10% Error	37	15	3	17	2
Avg amount analyzed (μg)	1.9 ± 1.8	2.6 ± 1.9	1.9 ± 1.2	1.4 ± 1.6	0.17
Avg amount hydrolyzed (μg)	5.5 ± 6.4	8.2 ± 9.2	6.7 ± 1.6	3.9 ± 3.2	1.3
Avg % Error	10.5 ± 6.6	10.7 ± 8.7	8.2 ± 13.7	10.7 ± 5.0	10.1
Cys Analysis					
Total Sites	53	21	4	27	1
Sites with ≤ 10% Error	21	10	2	9	-
Avg amount analyzed (μg)	1.9 ± 1.7	2.6 ± 1.9	2.3 ± 0.8	1.4 ± 1.5	0.16
Avg amount hydrolyzed(μg)	6.2 ± 8.3	9.3 ± 11.9	6.3 ± 3.3	3.8 ± 3.2	0.81
Average % Error	24.3 ± 25.7	23.7 ± 30.3	28.9 ± 33.2	23.9 ± 20.3	20.8
Trp Analysis					
Total Sites	40	16	4	19	1
Sites with ≤ 10% Error	10	3	3	4	-
Avg amount analyzed (μg)	2.1 ± 1.9	2.8 ± 1.9	2.6 ± 0.7	1.4 ± 1.8	0.18
Avg amount hydrolyzed (μg)	7.1 ± 9.0	10.0 ± 10.8	7.4 ± 2.8	4.9 ± 7.4	0.88
Average % Error	36.0 ± 38.2	37.7 ± 31.9	8.6 ± 12.9	40.0 ± 44.5	53.3

[a] Individual analyses were from single (39% of sites) to 10 replicates, averaging 2.4 (median 2). Average values are reported as average ± standard deviation.

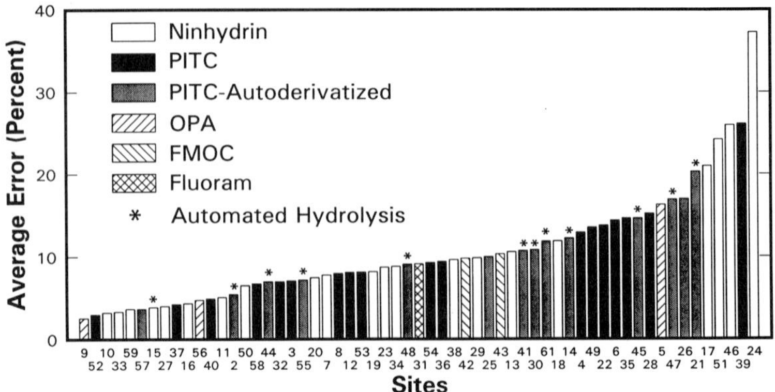

Figure 1 Average % errors of amino acid compositions determined by each site, excluding Cys and Trp. Automated hydrolyses (*) were done on one Savant and 11 ABI instruments.

The amount of chymotrypsin recovered by each laboratory was calculated relative to the 56 μg (2.2 nmol) originally dried in the sample tube. The average amount of protein quantified (43.4 $\pm$ 17.4 μg) represents a 77 $\pm$ 40% yield. The lower yield and larger standard deviation in this year's study compared to that obtained by participants in the 1990 and 1991 ABRF studies, suggests that the cause may be sample dependent and perhaps due to resolubilization, transfer and hydrolysis.

C. Cystine/Cysteine Analyses

Cystine/cysteine was determined by 90% of the participating laboratories (53 sites). The distribution of Cys % error per site and method is shown in Figure 2 and a summary of the data is presented in Table III categorized according to methodology. As expected, the average error for Cys (24.1%) was higher than that realized from the standard analyses. However, 21 sites (40% of those attempting to quantify Cys) achieved $\leq$ 10% error. A majority (76%) of those achieving $\leq$ 10% Cys error also achieved $\leq$ 10% error in the standard analysis. In contrast to last year's study (5), where two methods (alkylation and disulfide exchange) were apparently superior,

Table III. Summary of Cystine/Cysteine Analyses

Cys Method	Total Sites	Average % Cys Error	Std Dev	Sites with ≤ 10% Cys Error
a. Performic acid oxidation	24	15.8	11.5	10 (42%)
b. Dimethylsulfoxide oxidation	9	19.4	16.3	4 (44%)
c. Pyridylethylation	3	5.1	3.3	3 (100%)
d. Carboxymethylation	2	55.4	-	0
e. Dithiodipropionic acid	6	41.1	48.7	3 (50%)
f. Dithiodiglycolic acid	3	33.9	26.6	1 (33%)
g. Direct Analysis	6	42.1	20.2	0
Summary				
Oxidation (a,b)	33	16.8	13.1	14 (42%)
Alkylation (c,d)	5	25.2	30.6	3 (60%)
Disulfide Exchange (e,f)	9	38.7	42.7	4 (44%)
Total	53	24.1	25.6	21 (40%)

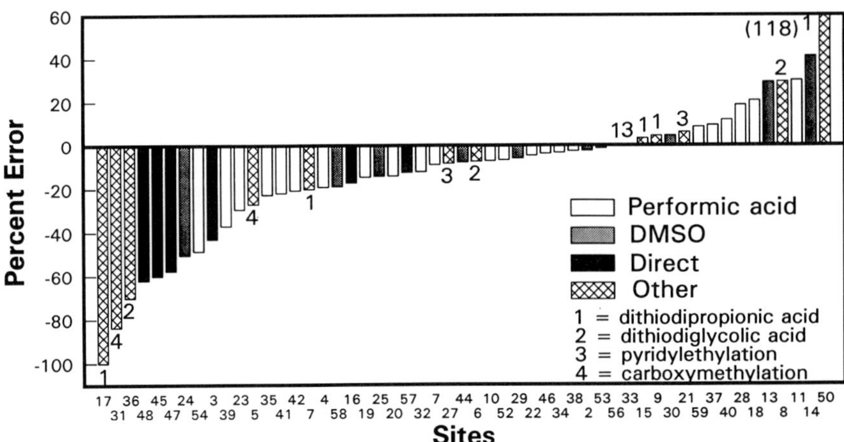

Figure 2 Cystine/cysteine analyses: Percent errors by each site.

three methods (oxidation, alkylation and disulfide exchange) yielded excellent results with no approach clearly superior to another. As in the 1991 ABRF study, oxidation was again the most popular method for Cys analysis, with 33 sites pursuing this approach. Performic acid oxidation prior to hydrolysis (7) or oxidative hydrolysis in the presence of dimethylsulfoxide in liquid HCl (8) or in HCl vapor (9) worked equally well with 42-44% of the sites using these techniques achieving ≤ 10% Cys error. Forty-four percent of the 9 sites utilizing disulfide exchange for Cys analysis achieved ≤ 10% error; dithiodipropionic acid and dithiodiglycolic acid additives were used essentially according to Barkholt and Jensen (10) and Hoogerheide and Campbell (11). Of the five laboratories that alkylated the sample, the three using pyridylethylation obtained the best results. These data indicate that more than one method can be used to quantify Cys accurately. Consistent with the experience of many others, direct analysis without prior Cys modification yielded inferior results (average Cys error 42%).

The combined Cys results from ABRF-91AAA and -92AAA suggest that alkylation and disulfide exchange methods may provide superior performance to oxidation (Table IV). However, the large difference in the number of sites utilizing these methods statistically warrants caution in such an interpretation.

Table IV. Performance of Cysteine Methodology, ABRF-91AAA and -92AAA Combined

Method	Total Sites	Sites with ≤ 10% Cys Error
Oxidation	56	22 (39%)
Alkylation	9	6 (67%)
Disulfide Exchange	13	7 (54%)
Total	78	35 (45%)

D. Tryptophan Analyses

Trp analyses by 40 sites generated the largest ABRF database to date for comparing and evaluating Trp quantification methodology. The distribution of Trp percent error by site and method is shown in Figure 3 and the data summarized according

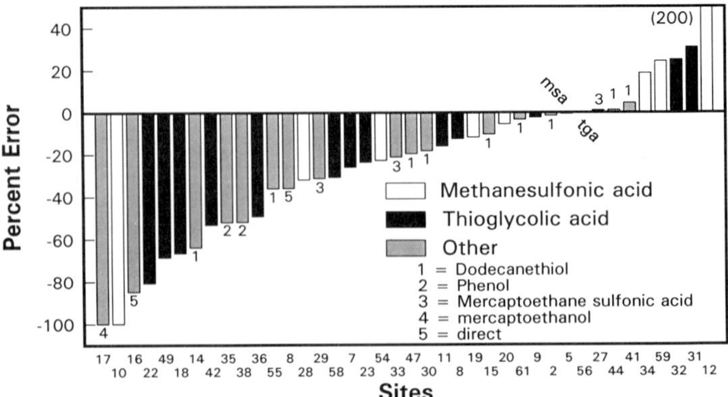

Figure 3 Tryptophan analyses: Percent errors by each site.

to Trp methodology in Table V. The results exemplify the difficulty of Trp analysis; the Trp error (36.1%) was high and only 10/40 sites achieved ≤ 10% error. Notably however, five of the sites with ≤ 10% error used the same methodology, namely HCl hydrolysis in the presence of dodecanethiol. This appears to be the superior approach to Trp analysis. Three other hydrolysis methods yielded low Trp error in conjunction with post-column analyzers, but with less statistical confidence; 2/14 sites using thioglycolic acid, 2/9 using methanesulfonic acid and 1/3 sites using mercaptoethane sulfonic acid obtained ≤ 10% Trp error. Four of the five sites that successfully used dodecanethiol carried out an automatic vapor phase HCl hydrolysis according to Bozzini *et al.* (12) and West and Crabb (9) and analyzed PTC amino acids. Site 15 did a vapor phase HCl hydrolysis (2h/165 °C) using a Savant autohydrolyzer in the presence of 0.1-1% dodecanethiol and performed post-column analysis with ninhydrin detection. Overall this year's Trp analyses are of higher quality and statistically more reliable than those of last year when 16 sites achieved an average Trp error of 60.9% with only 4 sites obtaining ≤ 10% error (5). Combination of the results for ABRF-91AAA and -92AAA (Table VI) underscores the superiority of dodecanethiol over all the other methodologies.

The overestimation of Trp in some instances implies that hydrolysis for tryptophan may not be the only factor in successful analyses, but that the chromatography and perhaps derivatization artifacts would be points where attention is also needed.

Table V. Summary of Trp Analyses

Trp Method	Total Sites	Average % Trp Error	Std Dev	Sites with ≤ 10% Trp Error
Dodecanethiol/HCl	9	17.7	19.6	5 (56%)
Mercaptoethane sulfonic acid	3	13.4	13.3	1 (33%)
Methane sulfonic acid	9	38.0	58.6	2 (22%)
Thioglycolic acid	14	39.9	22.7	2 (14%)
Phenol/HCl	2	52.0	-	0
β-Mercaptoethanol/HCl	1	100.0	-	0
HCl	2	48.7	-	0
Total	40	36.1	38.2	10 (25%)

Table VI. Performance of Tryptophan Methodology, ABRF-91AAA and -92AAA Combined

Method	Total Sites	Sites with ≤ 10% Trp Error
Dodecanethiol	11	6 (55%)
Methane Sulfonic Acid	17	4 (24%)
Mercaptoethane Sulfonic Acid	5	1 (20%)
Thioglycolic Acid	16	3 (19%)
Other	7	0
Total	56	14 (25%)

E. Hydrolysis, Calibration, Identifying the Test Protein and Survey

Hydrolysis conditions: Nearly all sites used 6*N* HCl as digestion acid for 'routine' amino acids. A variety of additives were employed for hydrolysis when quantitating 'routine' amino acids, with 15 sites using no additives. Of the latter, only three were in the upper third with regard to accuracy (overall error < 7%). Phenol was used by most sites (33), with 0.1% being the most common concentration. Other additives were β-mercaptoethanol (4 sites; 0.02-1%), thioglycolic acid (2 sites; 4-5%), or sodium sulfite (1 site). Vapor phase hydrolysis was preferred by a 2:1 margin over liquid phase hydrolysis for routine and Cys analysis (39:18), an increase over 1991 (1:1). Conditions of temperature and time of hydrolysis were distributed bimodally, as seen in previous years, were similar for all three types of analysis and did not correlate with quality of analysis. These conditions were 111 ± 2.7 °C for 22 ± 2.4h at 34 sites, or 160 ± 15 °C for 1.4 ± 0.4h at 19 sites. The top third of participants with the least error (≤ 7.1%) comprised 12 sites that used lower temperature/longer times and 6 sites that used higher temperature/shorter times while the lower third with the highest error (10.7-40%) had a similar ratio of 12 to 8 sites.

Calibration. The majority (42/59) of the respondents used the traditional unhydrolyzed free amino acids for their calibration standards; 19 used hydrolyzed free amino acids, and 12 used either hydrolyzed peptide or protein standards. Free amino acid standards analyzed per year ranged: 25-85 standards (13 sites); 100-300 (26 sites); and 400-1200 (9 sites). Standard peptides and proteins analyzed annually ranged: 1-12 samples (8 sites); 20-61 (17 sites); and 100-500 (10 sites). Participants indicated that response factors necessary for calibration were acquired from a single standard run, an average of several runs, or a standard concentration curve. Most typical was the use of either the single or averaged run methods. The concentration range of standards for the single run method was from 100 pmol to 5 nmol, averaging at about 1 nmol. Sites using the averaged run method had three standards as the mean, with an average sample amount of 250 pmol. Only 5 participants used multiple concentration level standards to generate a calibration curve.

Sites which used precolumn methods were more likely to use hydrolyzed free amino acids, peptides or proteins for their calibration standards than those participants who used post-column methods. Similarly, the precolumn method users were more likely to perform multiple single level runs to produce average response values than the post-column users. The frequency at which participants performed calibrations were widely varied, but most sites calibrated daily or with each batch.

Internal standards were used by 30 sites (mostly Nle), 20 adding it before hydrolysis, and 4 before chromatography. The accuracy of the absolute amount determined did not correlate with the use of internal standards.

The calibration methods for Trp or Cys determination were similar to the typical calibration methods for standard runs performed at the sites. For Trp the majority (26/37) used free amino acids as standards, and for Cys similarly 31 of 54 sites used free amino acids, and 16 of 54 used hydrolyzed amino acids. The remaining sites used hydrolyzed peptide or protein for standards. Response factors for both Trp and Cys were determined in equal frequency from single runs and averaged multiple runs.

Identification of the test protein. Only 16 sites attempted identification of the test protein using their amino acid analyses, and 14 were correct (as chymotrypsin or chymotrypsinogen). A few of these sites supported and initiated identification by electrophoresis or sequencing. The majority of correct identifications appear to be due to prior experience with chymotrypsin and recognition of the composition, however 5 sites successfully utilized a computer database of amino acid compositions and search software in the identification of the protein. The latter were "Scrutineer" of Sibbald *et al.* (13)(2 sites), "Finder", by G.Shaw, University of Florida, GENEPRO, by Riverside Scientific Enterprises, Seattle, WA, and PC Gene Scan.

Survey. A total of 56 sites responded to the AAA survey; 52 perform AAA as a service and 53 have dedicated instruments. Twenty six sites gave cost estimates for providing this service, but the wide range of responses precluded any conclusions. Twenty eight sites reported the amount charged per sample; the range was \$7 to \$80 with an average of \$36 ± \$19. However, it should be noted that since the "average" facility recovers only 47% of the actual cost of carrying out an amino acid analysis from user charges (14), the total cost of completing an amino acid analysis (based on these 28 sites) is probably ~\$77.

The number of experimental samples analyzed per year can be broken into two broad ranges: ≤ 400 samples (37 sites); and 744-3000 samples (13 sites). For synthetic peptides the ranges were: ≤ 250 peptides (33 sites); and 400-1000 peptides (10 sites). There is no correlation between the number of samples analyzed per year and the analytical quality found in this study. Most analyses are performed to obtain composition (42%) or composition and quantitation data (43%), with 15% of analyses for quantitation only.

Survey respondents indicated that most protein analyses are done at the 1-3 μg or 3-10 μg level. Thirty eight sites use one of these levels for ≥50% of their analyses. Only 9 sites each use <1 μg or >10 μg of protein for ≥50% of their analyses. For experimental peptides, most analyses are done at the 50-500 pmol level or >500 pmol level. Thirty eight sites use one of these levels for ≥50% of their analyses. Only 2 sites use <50 pmol of peptide for ≥50% of their analyses.

Most sites do not regularly perform cysteine and tryptophan analyses because (1) the data is not required and (2) additional work is required. Forty four sites indicated that they would do more cysteine and tryptophan analyses if the methods were more reliable and easier.

Table VII. The Best Analyses of ABRF-92AAA by Method[a]

	Post-column Methods		Pre-column Methods		
	Fluorescence	Ninhydrin	Fluorescence	PTC	Sequence[b]
Site #	9	33	42	2	
Ala	22.20	21.45	21.87	23.18	22
Arg	3.06	3.25	3.19	3.36	3
Asp	22.56	22.00	21.72	21.49	22
Cys	10.45	10.06	7.92	9.75	10
Glu	15.75	15.27	15.19	15.76	15
Gly	23.12	23.60	27.49	24.06	23
His	1.94	2.10	2.64	1.97	2
Ile	9.60	10.20	9.26	8.59	10
Leu	19.44	18.85	19.89	20.79	19
Lys	13.98	13.44	11.85	13.82	14
Met	1.96	2.04	2.43	2.00	2
Phe	6.03	5.37	6.23	6.24	6
Pro	9.45	8.55	9.42	9.26	9
Ser	25.26	27.77	23.08	25.44	27
Thr	22.10	21.94	20.50	22.37	22
Trp	7.80	6.30	3.74	7.88	8
Tyr	3.79	4.08	4.25	3.82	4
Val	22.92	22.13	20.50	20.29	23
Amt analyzed (μg)[c]	3.2	0.8	0.2	1.7	
Amt hydrolyzed (μg)[c]	6.0	3.3	0.8	1.7	
Total yield (μg)	48.7	34.4	28.1	54.1	
Error (%)	2.7	4.2	12.8	5.0	
Trp method[d]	tga	mes	tga/phenol	ddt	
Cys method[e]	dtdpa	PEC	pao	dmso	

[a]Values are reported as residues per mole of chymotrypsin. [b]Ref 6. [c]for standard analyses [d]tga = thioglycolic acid; mes = mercaptoethanesulfonic acid; ddt = dodecanethiol [e]dtdpa = dithiodipropionic acid; PEC = pyridylethylcysteine; pao = performic acid; dmso = dimethyl sulfoxide/ HCl

IV. Summary and Conclusions

The 1992 ABRF Amino Acid Analysis Study was supported by more sites (59) than in any previous year and importantly the Cys (53 sites) and Trp (40 sites) results provide the largest and most significant ABRF database to date for comparison and evaluation of methods for quantitation of these problematic amino acids. The best analyses of ABRF-92AAA from both post-column and pre-column methods are presented in Table VII. Pre-column instrumentation was used more in this study than post-column, while both methods yielded 18-19 sites achieving $\leq$10% average compositional error. Post-column analyses consumed about twice as much sample in this study as did pre-column analyses. Vapor phase hydrolysis has become decidedly more popular than the traditional liquid phase method, with two-thirds of the sites using it, as compared to half the sites a year previously (5).

Three cystine analysis methods yielded excellent results, namely oxidation, alkylation and disulfide exchange. A majority of those achieving $\leq$10% Cys error also achieved high accuracy in the standard analysis, suggesting that bench skills in amino acid analysis, rather than specific methodology, is the determining factor in high quality Cys analyses.

Tryptophan analyses, although done well by some participants using a variety of methods, were especially successful by utilizing automatic HCl vapor phase hydrolysis in the presence of dodecanethiol. This was the method of choice for Trp analysis in this collaborative study.

The present study indicates that accurate results can be achieved by the easy to use disulfide exchange reagents or DMSO/HCl reagents for cysteine, and dodecanethiol for tryptophan. Amino acid analyses, already a basic tool in the quantitative biochemical study of proteins and peptides, can be enhanced by a more complete analysis including cysteine/cystine and tryptophan. This study identifies reliable methodologies and realistic expectations for amino acid analyses in resource facilities; practical applications of these methods are up to the individual scientists and facilities.

Acknowledgements

We thank all participants in this study, K. West and C. Johnson (W Alton Jones Cell Science Center) for expert assistance in the preparation and distribution of the sample and Dr B. Holmquist (Harvard Medical School) for receiving raw data and helping us maintain the anonymity of collaborating facilities. This work was supported in part by NSF grant DIR 9003100 (to JWC) on behalf of the ABRF.

References

1. Niece, R.L., Williams, K.R., Wadsworth, C.L., Elliot, J., Stone, K.L., McMurray, W.J., Fowler, A., Atherton, A., Kutny, R. and Smith, A. (1989) *in* "Techniques in Protein Chemistry (T.E. Hugli, ed), Academic Press, San Diego, pp 89-101.
2. Crabb, J.W., Ericsson, L.H., Atherton, D., Smith, A.J. and Kutny, R. (1990) *in* "Current Research in Protein Chemistry (J.J. Villafranca, ed) Academic Press, San Diego, pp 49-61.
3. Ericsson, L.H., Atherton, D., Kutny, R., Smith, A.J. and Crabb, J.W. (1991) *in* "Methods of Protein Sequence Analysis 1990 (H.Jörnvall and J.-O.Höög, eds), Birkhauser Verlag, Basel.
4. Tarr, G.E., Paxton, R.J., Pan, Y.-C.E., Ericsson, L.H., & Crabb, J.W. (1991) *in* "Techniques in Protein Chemistry II", Academic Press, pp 139-150.
5. Strydom, D.J., Tarr, G.E., Pan, Y-C.E. and Paxton, R.J. (1992) *in* Techniques in Protein Chemistry III (R.H.Angeletti, ed), Academic Press, San Diego, pp 261-274.
6. Hartley, B.S. and Kauffman, D.L. (1966) *Biochem. J.* **101**, 229-231.
7. Hirs, C.H.W. (1956) *J.Biol.Chem.* 219, 611-621.
8. Spencer, R.L. and Wold, F. (1969) *Anal. Biochem.* 32, 185-190.
9. West, K.A. and Crabb, J.W. (1992) *in* Techniques in Protein Chemistry III (R.H.Angeletti, ed.), Academic Press, San Diego, pp 233-242.
10. Barkholt, V. and Jensen, A.L. (1989) *Anal. Biochem.* 177, 318-322.
11. Hoogerheide, J.G. and Campbell, C.M. (1992) *Anal. Biochem.* 201, 146-151.
12. Bozzini, M., Bello,R., Cagle,N., Yamane, D. and Dupont, D. (1991) Applied Biosystem Research News, February 1991.
13. Sibbald, P.R., Sommerfeldt, H. and P. Argos (1991) *Anal.Biochem.* 198, 330-333.
14. Niece, R.L., Beach, C.M., Cook, R.F., Hathaway, G.M. and Williams, K.R. (1991) *FASEB J.* 5, 2756-2760.

Compositional Protein Analysis Using 6-Aminoquinolyl-N-Hydroxysuccinimidyl Carbamate, a Novel Derivatization Reagent

Steven A. Cohen, Kathryn DeAntonis and Dennis P. Michaud

Millipore Corporation, Milford MA, 01757

I. Introduction

Today many standard protein chemistry protocols such as Edman degradation and peptide mapping can be accomplished with low or even sub-picomole sample amounts through the use of modern liquid chromatography (1). Amino acid analysis has been assisted in this drive for higher sensitivity by the use of an array of pre-column derivatization reagents that yield easily detected labels. Fluorescent tags like ortho-phthalaldehyde (OPA) and fluorenyl methyl chloroformate (FMOC-Cl) provide the most sensitive tags, but the UV absorbing Edman reagent phenylisothiocyanate (PITC) remains the most widely used tag for compositional analysis of peptide and protein hydrolyzates, perhaps because of its familiarity to protein chemists (2, 3). Despite the wide choice in labelling chemistry, the desire for improved high sensitivity amino acid analysis continues to spur much active research into new reagents. In addition many existing reagents exhibit less than ideal characteristics such as insufficient derivative stability, the formation of multiple products from a single amino acid, large interfering peaks from excess reagent or poor reproducibility or response linearity. We have recently described (4) the synthesis of a novel activated carbamate, 6-aminoquinolyl-N-hydroxysuccinimidyl carbamate (AQC) which forms the basis for a pre-column derivatization method and provides very accurate amino acid analysis of hydrolyzed peptides and proteins. A rapid reaction with amino acids forms highly stable asymmetric urea compounds with good fluorescence characteristics. A dramatic shift in fluorescence emission maximum for the amino acid adducts compared to the major reagent peak allows the direct injection of the sample reaction mixture with no interference from excess reagent. We herein report the application of AQC to the analysis of protein and peptide hydrolyzate samples, with an emphasis on the accuracy and reproducibility of the compositional data.

TECHNIQUES IN PROTEIN CHEMISTRY IV

II. Materials and Methods

A. Chemicals

Di(N-succinimidyl) carbonate (DSC) was purchased from Fluka Chemical, sodium acetate trihydrate (HPLC grade) and disodium ethylenediamine tetraacetic acid were from Baker Chemical (Phillipsburg, PA), triethylamine and 6-aminoquinoline (AMQ) were purchased from Aldrich Chemical Co. (Milwaukee, WI). Amino acid standards were from Pierce Chemical Co. (Rockford, IL); peptides and ß-Lactoglobulin A were obtained from Sigma Chemical Co. (St. Louis, MO). Bovine serum albumin (BSA) was supplied by the Amino Acid Subcommittee of the Association for Biomolecular Resource Facilities.

Synthesis of AQC: Details of the synthesis are given in Reference 4. Briefly, DSC (3g, 12mmol) and AMQ (1.5g, 10mmol) were dissolved in dry acetonitrile, and the AMQ was added dropwise to the carbonate solution. After 30 minutes the reaction mixture was concentrated by rotary evaporation to about half its volume and after cooling, the resulting crystals were filtered and washed with acetonitrile to yield 1.95 g (66% of crude product). The product was recrystallized from acetonitrile as off-white crystals, mp 210-215°C (dec) .

B. Amino Acid Derivatization

Purified AQC was dissolved in dry acetonitrile at a concentration 10m**M**. Standard solutions of amino acids were made from dilutions of commercial standard mixtures. Aliquots of standards in 5-50 µL of water or dilute HCl were pipetted into glass tubes (6 x 50 or 12 x 75 mm) and 0.2**M** sodium borate buffer, pH 8.8 was added to bring the volume up to 40-400 µL. Derivatization was carried out by adding one volume of AQC solution to four volumes of buffered sample.

C. Fluorescence Studies

The AQC derivative of Ala (AQC-Ala) was synthesized by adding 1mL of 1**M** Ala in borate buffer to 1mL of a 10 m**M** AQC solution in acetonitrile. Fluorescence spectra of AQC, AMQ and AQC-Ala were recorded on a Shimadzu RF-5000 fluorescence spectrophotometer. Stock solutions at 3m**M** were diluted in acetonitrile to give a final concentration of 0.3m**M**.

D. Chromatography

Two HPLC systems were used routinely. System I consisted of 2 - 510 pumps, a M712 autosampler, a M470 scanning fluorescence detector and a TCM column heater (all from Millipore Corp., Milford, MA). Mobile phase A was 140m**M** sodium acetate, with 17m**M** TEA titrated to pH 5.05 with phosphoric acid containing 1mg/L disodium ethylenediamine tetraacetic acid (EDTA). This eluent can be made as a concentrate using 190 g of acetate, 17.2 g of TEA, 10 mg of EDTA and 1 liter of water, titrated to pH 5.02. Working eluent consists of 100 mL of concentrate mixed with 1000mL of water. Mobile phase B was 60% acetonitrile in water (v/v). System II consisted of a 625 LC System, a M715 or M717 autosampler (Millipore Corp.), a M470 scanning fluorescence detector and a TCM column heater. Mobile phase A was as described above, Eluent B was water and Eluent C was acetonitrile. Both systems used multi-step linear gradients with the column thermostatted at 37°C. The flow rate was 1.0 mL/min.

E. Preparation of Samples

Proteins and peptides (25-10,000ng) were hydrolyzed with gaseous HCl as previously described (5). After removing the excess HCl under vacuum, the amino acids were resolubilized with 20 µL of 10-20m**M** HCl, the volume brought to 80 µL with 0.2**M** borate buffer, pH 8.8, and the derivatives formed via the addition of 20 µL of 10 m**M** AQC in acetonitrile. In one study, 50µL of a salt (0.1**M**) or detergent solution (0.2% w/v) was added to the sample and dried prior to hydrolysis.

F. Data Analysis

Compositional accuracy was calculated according to the method described in Reference 3:

%Error = 100 x (Experimental - True Residue Value)/True
Average % Error = Sum (Absolute % Error)/16

III. Results

A. Reagent Chemistry

The reaction of amino acids with AQC occurs very rapidly and is essentially complete within seconds. In a reaction with a half-life of approximately 15 seconds, the excess reagent is subsequently hydrolyzed to yield AMQ, N-hydroxysuccinimide and carbon dioxide (Figure 1). A three-fold molar excess of reagent over total amine content is sufficient to quantitatively convert the amino acids to their AQC derivatives, although a 10-fold excess is desirable. Amino acid derivative yields were essentially unaffected as the pH was varied in the range 8.2 to 10.0 (4). Below pH 8.2, Asp, Glu and Lys were the most sensitive to the decrease in pH, with yields at pH 7.5 ranging 45-65% of that at pH 8.8.

Only Tyr shows any evidence of multiple derivative formation. At room temperature a minor component eluting after Phe slowly converts to the major product eluting just prior to Val. It is hypothesized that this corresponds to the hydrolysis of a diderivatized Tyr molecule to the stable mono-derivatized asymmetric urea (4). This reaction can be driven to completion in 10 minutes by heating to 50°C. Routine derivatization thus occurs as follows (a) buffer sample to pH 8.8 with 80 µL of 0.2**M** borate buffer (b) add 20 µL of 10m**M** AQC in acetonitrile (c) heat at 50°C for 10 minutes, and (d) inject the sample.

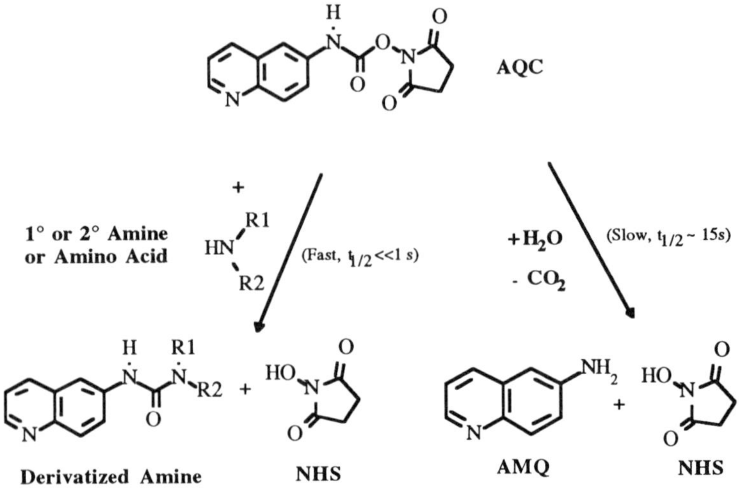

Figure 1. Chemistry of amino acid derivatization with AQC.

B. Fluorescence Studies

Fluorescence emission spectra (excitation at 245nm, in acetonitrile) recorded for AMQ and AQC-Ala (the derivatization product of Ala) showed a 60 nm blue shift in the wavelength maximum for the amino acid product compared to the free amine. At 395nm, the approximate maximum of AQC-Ala, the relative fluorescence of AMQ is less than 1% of that of the amino acid, which allows for the direct injection of the derivatization mixture without prior removal of the excess reagent. Only a small, easily resolved AMQ peak is contributed by the excess, which simplifies chromatographic separation. NHS produced by the reaction does not interfere at all as it does not fluoresce.

C. Chromatography of Derivatized Amino Acids

HPLC separations development was accomplished through the optimization of several key chromatographic variables including buffer pH and ionic strength, column temperature and gradient slope. Final optimized conditions, shown in Figure 2A, gave nearly complete baseline resolution for all components in approximately 35 minutes. Total analysis time was 50 minutes. The low pH mobile phase positions the small AMQ peak at the front of the chromatogram, and results in a reversal of elution order for Glu and Ser relative to separations performed at pH > 5.2. High sensitivity is demonstrated by the separation of a 1 pmol injection in Figure 2B. Detection limits range from a low of ~40 fmol (Phe) to ~310 fmol (Asp) except for Cys (~800 fmol) which most likely suffers from internal fluorescence quenching. Excluding Cys, the average detection limit is approximately 160fmol. Table I gives relative area response for the hydrolyzate amino acids as well as reproducibility for a series of 50 pmol analyses.

Table I: Reproducibility for Peak Response and Relative Fluorescence Response for Amino Acid Standards

Amino Acid	Area %CV [a]	Relative Response Factors [b]
Asp	0.82	0.20
Ser	0.58	0.28
Glu	1.32	0.22
Gly	1.36	0.23
His	1.60	0.35
Arg	1.17	0.36
Thr	1.05	0.35
Ale	0.76	0.39
Pro	0.47	0.18
Cys	1.34	0.05
Tyr	0.99	0.38
Val	0.59	0.66
Met	0.79	0.53
Lys	0.54	0.31
Ile	0.37	0.84
Leu	0.38	0.83
Phe	0.49	1.00
Average %CV	0.86	

[a]Data from 10 injections at the 50pmol level.

[b]Values are calculated from the area counts recovered relative to that of Phe, averaged from three analyses.

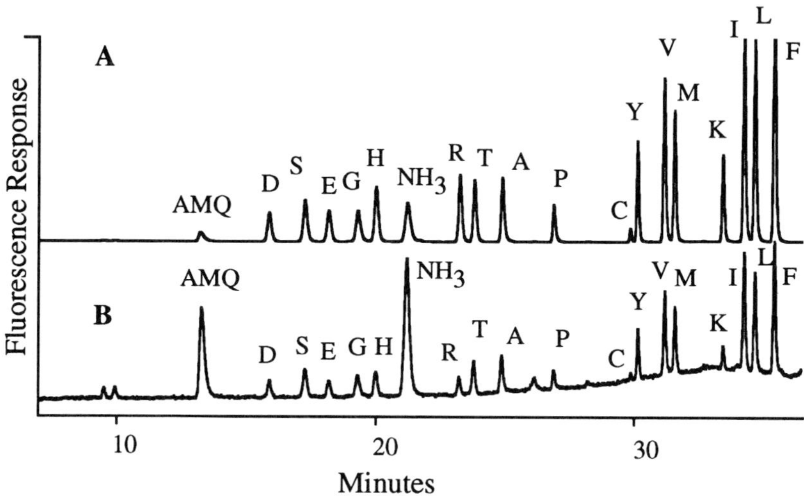

Figure 2. Chromatography of amino acid standards at the (A) 50pmol and (B) 1 pmol levels. Detector full scale responses were (A) 90 and (B) 4.5 mV.

D. Compositional Analysis of Proteins

Amino acid compositional data were generated from the hydrolysis of purified bovine serum albumin and ß-Lactoglobulin A using a wide dynamic range in sample amount. Data for ß-Lactoglobulin were compiled from a total of eight separate hydrolysis batches done over a period of two months, while the BSA data were from three hydrolysis batches done in a one month period. Figure 3 shows typical chromatograms for protein hydrolyzates. Table II summarizes compositional data from the entire data set while Table III shows selected individual analyses. Samples were divided into groups according to the amount hydrolyzed as calculated from the amino acid analysis: trace analysis (<100ng), high sensitivity (100-500ng), low intermediate (500-1000ng), high intermediate (1000-2000ng) and high (>2000ng) range analyses. Figure 4 plots the average error data for each range for the individual amino acids.

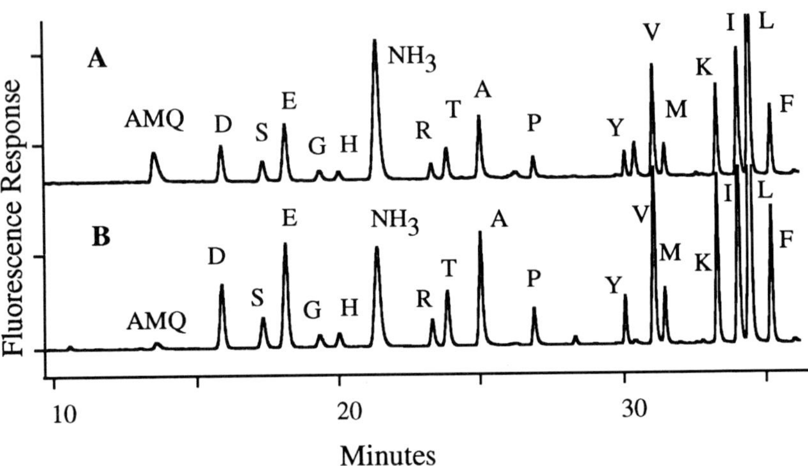

Figure 3. Chromatography of hydrolyzed, derivatized ß-Lactoglobulin. (A) 160ng hydrolyzed, 10% injected, full scale = 6mV. (B) 1500ng hydrolyzed, 5% injected, full scale = 20mV.

Table II: Summary of Compositional Error Data

	ß-Lactoglobulin				Serum Albumin		
n	11	5	12	2	2	2	7
Range (ng)[a]	90-389	587-790	1038-1915	9400-9770	30-33	175-185	1305-1932
Average (ng)	212	684	1665	9587	31	180	1739
Range of Average Errors	5.5-17.3	4.9-14.6	4.1-12.2	4.3-4.7	10.5-25.9	4.9-6.8	3.4-7.7
Average of Average Errors	10.7	8.6	6.4	4.5	18.2	5.8	5.5

[a] Amounts are calculated from the amino acid recovery data.

Table III: Compositional Analyses for Selected Samples

	ß-Lactoglobulin			Bovine Serum Albumin			
	Res/mole			Res/mole			
	Expected	Calculated	Calculated	Expected	Calculated	Calculated	Calculated
Asp	16	17.6	17.1	54	53.7	59.5	58.2
Ser	7	6.5	6.3	28	29.6	26.9	24.7
Glu	25	26.8	26.7	79	73.7	85.4	83.8
Gly	3	3.4	3.0	16	22.9	16.8	15.8
His	2	2.0	1.9	17	13.7	17.7	16.6
Arg	3	3.2	3.1	23	23.0	23.6	23.5
Thr	8	7.9	7.9	34	31.4	33.5	33.0
Ala	14	14.4	14.5	46	56.1	47.9	47.4
Pro	8	7.7	8.1	28	28.2	25.4	27.8
Tyr	4	3.6	3.7	19	16.2	17.7	17.8
Val	10	9.8	10.1	36	34.5	33.9	35.9
Met	4	3.5	3.8	4	4.6	4.1	4.1
Lys	15	15.5	16.0	59	52.0	61.9	60.7
Ile	10	9.5	9.6	14	14.3	13.4	13.8
Leu	22	21.8	22.7	61	54.8	57.8	61.5
Phe	4	4.0	4.0	27	25.8	26.3	27.6
Amount[a]		316	1801		30	185	1305
Average % Error		5.6	4.1		10.5	4.9	3.4

[a] Values are nanograms of sample hydrolyzed and derivatized

Some significant trends concerning the accuracy of amino acid compositional data as a function of sample amount hydrolyzed can be observed through analysis of the overall average data (Table II). Although there is a steady increase in the average error as sample amount decreases, even with 100-500ng of sample, compositional analyses are very good with average errors consistently below 10%. Above 500ng this is typically reduced to 5-6%. Above 1µg there is only a slight improvement in analysis accuracy. Although this trend has been observed before, the magnitude of the errors routinely observed with AQC derivatized samples is significantly smaller than that previously reported for manual hydrolysis procedures used in conjunction with pre-column derivatization (6). This is particularly evident with samples <1µg.

Analysis of specific amino acid trends (Figure 4) is also of interest. The largest errors are for Gly which, due to background, increase as sample amount is reduced. Met and Tyr exhibit reduced yields with lower sample amounts, presumably due to increased oxidative losses, despite the inclusion of phenol in the hydrolysis step. Arg errors also increase somewhat with reduced sample amount; the cause may be an increase in methionine sulfoxide (generated from Met oxidation) which is eluted close to Arg. This may be the cause of the 13 and 16% positive errors in Arg at the two lowest levels of ß-Lactoglobulin, which has only 3 Arg residues, but BSA with 23 Arg residues shows increased Arg errors only with <100ng samples. With >1µg, Arg errors are consistently <%5, regardless of the amount of Arg in the protein.

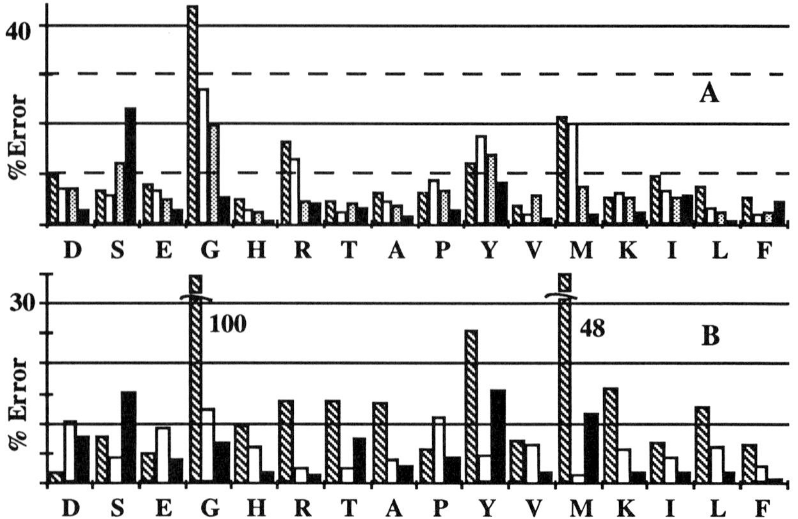

Figure 4. A) Error data for ß-Lactoglobulin as a function of hydrolysis amount
<500ng 500-1000ng 1000-3000ng >3000ng
B) Error data for bovine serum albumin as a function of hydrolysis amount
25-35ng 160-190ng 1300-2000ng

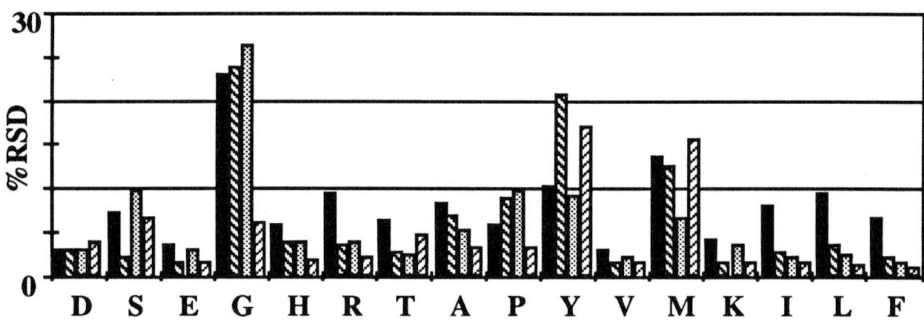

Figure 5. Reproducibility for compositional analysis as a function of hydrolysis amount.
<500ng LG 500-1000ng LG 1000-3000ng LG 1000-2000ng BSA

Analysis of compositional reproducibility (Figure 5) also reveals some important insights into the quality of the entire AQC-based method. Nearly all amino acids have relative standard deviations <10%, and most analyses with >500ng samples have %RSD<5%. In general, residues with higher % errors (e.g. Gly, Tyr, and Met) are those residues showing greater variability. However, this is not true for Ser, for which reproducibility is significantly better than the accuracy. Variability for samples >1μg is consistently under 5% except for Ser, Gly, Tyr and Met. Of the three most variable amino acids, oxidation of Met and Tyr would apparently be the cause of increased variability and less than quantitative yields (errors for these are all negative) at all levels, especially with <0.5μg of sample. Gly errors are positive presumably due to a relative increase in the level of contamination, and the data clearly demonstrate that this contamination is extremely variable in trace analysis.

Samples hydrolyzed in the presence of many common biochemical salts and detergents can be analyzed without cleanup. Table IV shows error calculations for hydrolyzed ß-Lactoglobulin in the presence sodium chloride, sodium dodecyl sulfate (SDS), sodium phosphate and sodium acetate. Good compositional analyses are obtained in all cases, although there is a slight decrease in overall accuracy. Some of this may be the result of less sample used for some of the analyses (e.g acetate), but there is evidence of increased breakdown of labile residues such as Met, Thr, and Ser in the presence of salts. All of the experiments used 50μL of additive except for the phosphate addition which was limited to 10 μL; using the standard derivatization protocol with larger volumes of phosphate gave poor results due to the reaction pH <8.0. It is very possible that increased buffering of the reaction medium would result in significantly better quantitation.

Table IV: Effect of Salt and Detergents on %Compositional Errors for ß- Lactoglobulin[a]

	Control	SDS	Sodium Chloride	Sodium Acetate	Sodium Phosphate
Asp	6.7	9.1	10.2	15.0	2.3
Ser	11.7	20.3	12.1	30.4	43.4
Glu	10.6	13.0	13.4	11.3	4.8
Gly	9.2	17.3	20.9	22.7	40.4
His	1.3	2.6	3.5	5.9	3.5
Arg	3.6	6.5	5.6	3.9	17.1
Thr	7.4	10.7	8.4	31.9	24.2
Ala	1.9	5.0	1.4	3.6	0.6
Pro	2.3	5.0	2.5	3.1	2.7
Tyr	1.2	10.2	2.3	3.8	1.2
Val	3.0	1.3	3.2	1.0	0.8
Met	4.5	13.3	4.3	12.7	18.5
Lys	2.8	3.1	3.3	4.2	1.0
Ile	6.2	4.7	6.8	4.6	2.5
Leu	1.4	1.9	1.5	0.8	1.0
Phe	2.5	1.9	1.4	6.9	3.6
Average Errors	4.8	7.9	6.3	10.1	10.5
Average ng hydrolyzed	1390	1133	1189	564	845

[a] Values given are for % errors in the calculated composition for the individual amino acids and are the average of three analyses.

IV. Summary

Pre-column derivatization with AQC forms the basis of a new method for amino acid analysis with increased accuracy and improved capabilities. Derivatized amino acids are formed rapidly and quantitatively in seconds and the sample can be analyzed without removal of excess reagent. Compositional analysis of hydrolyzed proteins is extremely accurate and reproducible, with typical average errors in the 4-7% range without correction for losses due to hydrolysis or incomplete cleavage of peptide bonds. Samples hydrolyzed in the presence of salt or detergent are amenable to analysis with only a small change in overall compositional accuracy.

References

1. Young P., Astephen, N. A., and Wheat, T. W. (1992) LC/GC, **10**, 26-32.
2. Cohen, S. A. and Strydom, D. J. (1988), Anal. Biochem., **174**, 1-16.
3. Strydom, D. J., Tarr, G. E., Pan, Y.-C. E., and Paxton, R. J. (1992) *in* "Techniques in Protein Chemistry III", R. H. Angeletti, Editor, Academic Press, Inc., pp 261-274.
4. Cohen, S. A. and Michaud, D. P., submitted to Anal. Biochem, August 1992.
5. Bidlingmeyer, B. A., Cohen, S. A., and Tarvin, T. L. (1984), J. Chromatogr., **336**, 93-104.
6. West, K. A., and Crabb, J. W. (1990) *in* "Current Research in Protein Chemistry: Techniques , Structure, and Function", J. J. Villafranca, Editor, Academic Press, Inc., pp 37-47.

Sensitive Analysis of Cystine/Cysteine using 6-Aminoquinolyl-N-Hydroxysuccinimidyl Carbamate (AQC) Derivatives

Daniel J. Strydom
Center for Biochemical and Biophysical Sciences and Medicine,
and Department of Pathology, Harvard Medical School, Boston, Massachusetts

Steven A. Cohen
Millipore Corporation, Milford, Massachusetts

I. Introduction

The analysis of cystine or cysteine in proteins continues to be a troublesome and difficult aspect of amino acid analyses despite many decades of development, and even though many possible approaches are available. This is due to the instability of these residues under normal amino acid hydrolysis conditions.

Various approaches to the modification of Cys before hydrolysis of the parent protein have therefore been considered over time. Two major approaches are oxidation to cysteic acid, and reduction to cysteine followed by alkylation with an alkylating reagent. Thus reduction of the protein, followed by alkylation with a variety of reagents, leads to stable derivatives, each perhaps necessitating special chromatographic procedures for identification and quantitation. Oxidation, with performic acid (1), or with DMSO/HCl (2,3), yields cysteic acid, which is an early-eluting peak in ion-exchange and reversed phase chromatography. The oxidation process may give low yields, while the early elution of cysteic acid may coincide with break-through peaks or other baseline disturbances. All of these methods, in the hands of expert workers, can yield excellent results, but are technically demanding and when amino acid analyses for cystine are not routinely done, they are many times irreproducible - see *e.g.* the results of collaborative and other amino acid analysis studies (3-5).

The recent development of the concept of disulfide interchange during conventional hydrolysis (6) offers a third approach to these analyses. While very few laboratories have published studies employing these techniques, the 1991 Association of Biomolecular Resource Facilities amino acid analysis survey (5) showed that all four of the laboratories that used such reactions performed accurate analyses of cysteine. More recently such derivatives were employed in combination with the widely used PITC-methodology (7).

A need for high sensitivity (low pmol/ high fmol) amino acid analyses exists as protein recognition, isolation and sequencing techniques are becoming more sensitive. Difficulties in cystine analysis are compounded by this general need to analyze sub-nanomole quantities of amino acids, and the more specialized need to analyze low pmol and sub-pmol amounts of protein. The recent development of the highly sensitive and stable AQC-derivatives for fmol

TECHNIQUES IN PROTEIN CHEMISTRY IV

analyses of amino acids (8,9) provides the framework for the generation of workable analyses of cysteine at such low levels.

We here describe the combined use of the three approaches to cystine analysis with derivatization to AQC derivatives and demonstrate the success of disulfide interchange in providing analyses of cysteine to fmol sensitivity. The cleanliness of the chromatographic environment surrounding the elution position of one such derivative and the stability of the derivatives combine to provide an opportunity to quantify low nanogram quantities of proteins in the presence of the ubiquitous (and at present inescapable) background of amino acids.

II. Materials and Methods

A. Materials

Aminoquinolyl-N-hydroxysuccinimidyl carbamate (AQC) was synthesized as described (8,9). Most proteins were commercially available preparations. Angiogenin was a recombinant product with Met-30 changed to Leu, provided by Hoechst, AG.

B. Hydrolysis and AQC-Derivatization Procedure

Dithiodiglycolic acid (Sigma Chemical Co.), dithiodipropionic acid (DTDPA) and dithiodibutyric acid (carboxypropyl disulfide) (both from Aldrich Chemical Co.) were dissolved either in 0.2*M* NaOH or in methanol to a concentration of 1 or 10 mg/ml. This solution (10 μl) was added to the sample in a small test tube and dried under vacuum. The dried mixtures were hydrolysed with 6*N* HCl (containing 0.06% phenol) in the vapor phase at 110°C for 18 hr and briefly dried.

Dried hydrolysates were dissolved in 10 μl of 10 mM HCl containing appropriate amounts of Nle as internal standard. Borate buffer (30 μl) and 10 μl reagent solution were added, the tubes were covered with parafilm and heated at 55°C for 15 min. Borate Buffer : 2.47 g boric acid, 372 mg disodium EDTA, 200 ml water, 1*N* NaOH to a pH of 9.3. Aliquots are kept at 4°C and fresh samples are opened every week. AQC Reagent : 3 mg/ml dry acetonitrile. Kept over dry silica gel at room temperature, in the dark.

D. Chromatography

The AQC-derivatives were chromatographed on a C18 column (Novapak, 15 x 0.39 cm, Millipore) under the following gradient conditions: 100%A to 98%A in 0.5 min; to 93%A in 14.5 min; to 90%A in 4 min; to 68%A in 13 min; hold for 1 min; wash column with 100%B for 4 min; equilibrate with 100%A for the next injection at 55 min. Linear gradient segments were used at a flow rate of 1 ml/min. The column temperature was 35°C. Solvent A is a sodium acetate-triethylamine-phosphate buffer at pH 5.05, and solvent B 60% acetonitrile. The eluates were monitored with a Waters Model 470 Scanning Fluorescence

Detector with excitation wavelength of 245 nm, emission wavelength of 395 nm, and a gain setting of 100. Data were collected on a Nelson Analytical Data system at 2 points per second. Aliquots of up to 20 μl were injected into the HPLC.

III. Results

A. Oxidation with Performic Acid

Amino acid standards were oxidized with freshly prepared performic acid (1) and derivatized. Chromatography of the ensuing AQC-cysteic acid showed excellent separation from aminoquinoline, the major reagent related peak, (Figure 1). Quantitation of cystine in lysozyme was good, with 7 of 8 residues being detected in 18 hr hydrolysates. The fluorescent color yield is similar to that of aspartic acid, with the consequence that sensitive analyses at sub-pmol levels are theoretically feasible. Methionine sulfone separates and quantifies adequately. The required chemical yields at low pmol levels are, however, difficult to achieve reproducibly.

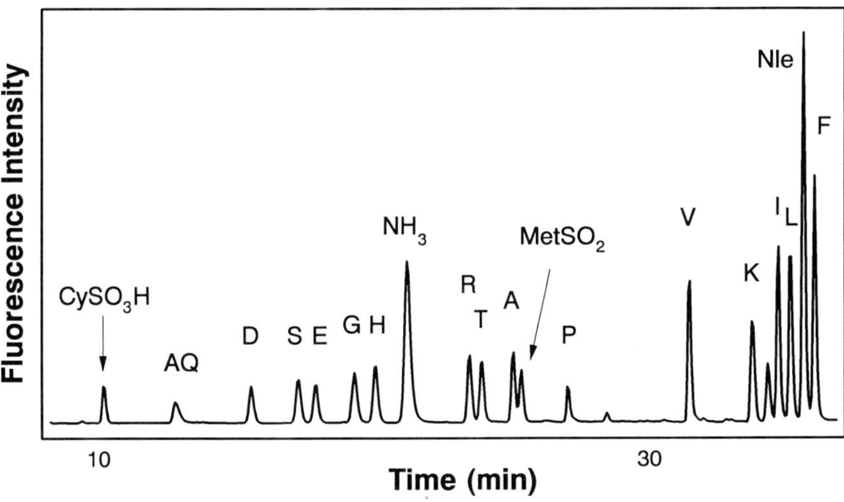

Figure 1 Chromatography of AQC-derivatives of amino acid standards oxidized with performic acid.

B. Disulfide Interchange

The experimentally simple disulfide interchange methodology (6,7) was first assessed for ease of chromatographic separation of the ensuing AQC-derivatives. The derivatives of dithiodiglycolic acid and dithiodipropionic acid (DTDPA) separated well in the standard chromatographic system, while the dithiodibutyric acid derivative needed gradient changes to separate from valine

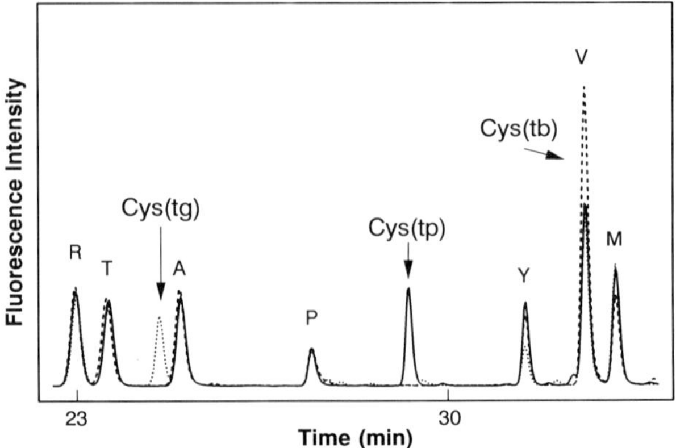

Figure 2 Chromatography of derivatives of cystine formed by hydrolysis of amino acid standards in the presence of dithiodiglycolic acid (Cys(tg)), dithiodipropionic acid (Cys(tp)) or dithiodibutyric acid (Cys(tb)) , and their subsequent reaction with 6-aminoquinolyl-N-hydroxysuccinimidyl carbamate.

(Figure 2). The DTDPA product with cystine (Cys(tp)) separates exceptionally well from all common amino acids in a very clear part of the chromatogram and was therefore chosen to investigate the utility of the methodology. Standard amino acid mixtures (250 pmol) were hydrolyzed in the presence of DTDPA, appropriate dilutions of the AQC-derivatives chromatographed, and Cys(tp) quantified. Figure 3 demonstrates that the response of the analytical system is linear over a wide range.

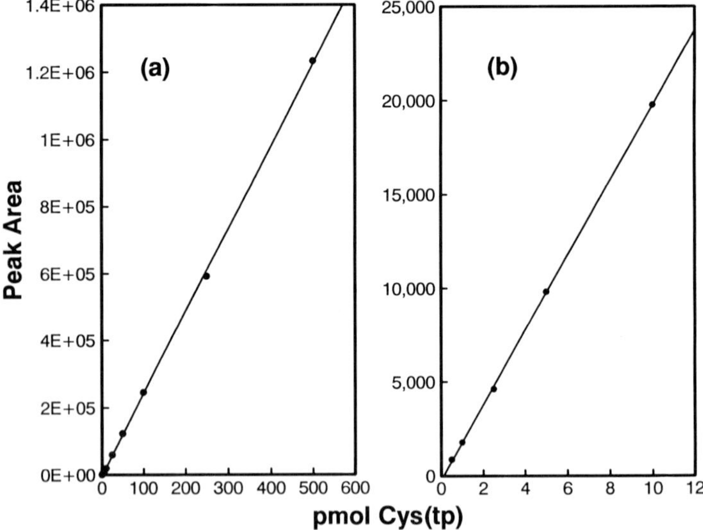

Figure 3 Linearity of quantitation of Cys(tp). The average area of duplicate determinations are shown for each amount loaded on the HPLC column. (a) The range from 0.5 to 500 pmol, expanded in (b) to show 0.5 to 10 pmol.

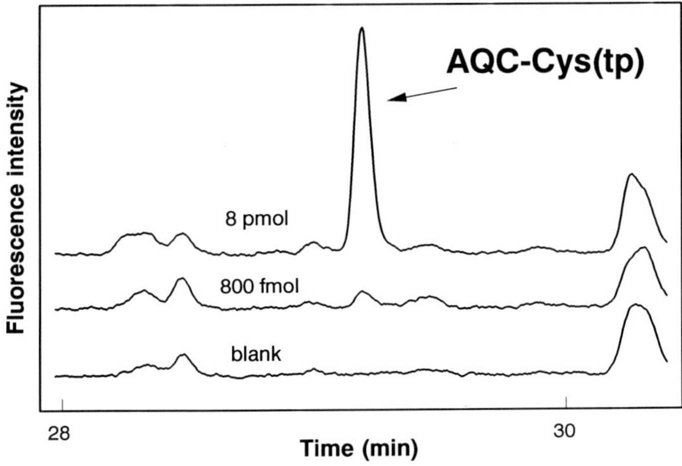

Figure 4 Analysis of cystine after hydrolysis in the presence of DTDPA. Cystine (10 and 1 pmol) was hydrolysed, and 40% of the derivatized sample injected onto the HPLC column.

Analysis of Cystine using hydrolysis by DTDPA/HCl: The disulfide exchange reaction of cystine with DTDPA at very low levels was examined by hydrolyzing 1, 5 and 10 pmol amounts of cystine in the presence of DTDPA as described above. Figure 4 shows the chromatography of 40% of such samples. The 800 fmol Cys(tp) injected out of the nominally 2 pmols of Cys(tp) derived from 1 pmol of cystine, resolves well from the general background, and can be quantified.

Table I: Amino Acid Analyses of Human Lysozyme[a]

	Performic acid[b]	DTDPA[c]	Sequence
Cys/2	7.17	6.90	8
Asp	18.85	18.20	18
Glu	9.17	11.26	9
Ser	5.13	7.59	6
Gly	14.18	16.52	11
His	0.96	1.45	1
Arg	13.30	12.32	14
Thr	6.31	6.61	5
Ala	13.69	13.22	14
Pro	3.06	3.01	2
Tyr	0.00	5.20	6
Val	9.35	8.93	9
Met	1.55	1.62	2
Ile	3.92	4.62	5
Leu	7.37	7.81	8
Phe	2.34	2.34	2
Lys	nd	6.65	5
% Compositional Error[d]	20.2	20.9	

[a]Presented as residues per mole. [b]Average of duplicate analyses. [c]Single analysis. [d]Average of the absolute % errors of each amino acid.

Table II. Cystine/Cysteine Compositions[a]

Protein	Cys (Sequence)	Cysteic acid 20 hr	Cys(tp) 18 hr	Cys(tp) 48 hr
Angiogenin	6	4.86	5.60	6.14
Serum Albumin	35	30.28	33.30	32.67
Chymotrypsin	10	9.00	8.94	9.88
Ovalbumin	6	6.62	5.82	6.34
Actin	5	5.23	4.95	4.69

[a]Presented as residues per mole, based on the theoretical amount of Arg, Leu and Phe in each protein and are the average of duplicate analyses. The protein was hydrolyzed at 110°C for the times indicated.

Analyses of proteins: Table I compares the analysis of lysozyme by performic acid oxidation and the DTDPA method. Similar values are found for the cystine content. Cysteic acid was calibrated by analysis of an amino acid standard that had been oxidized, while Cys(tp) was calibrated by analysis of amino acid standards which had been hydrolyzed in the presence of DTDPA for 18 hr. The oxidized sample shows the loss of tyrosine, lower yield of $MetSO_2$ and loss of information on lysine, while the DTDPA-reacted sample shows slightly low yields of Tyr and Met, and higher amounts of Glu, Ser, His and Gly.

Table II summarizes the results for cysteine obtained from analyses of various proteins, hydrolyzed at 0.5 to 1.6 μg amounts in the presence of DTDPA, and compares them to the values for the same amounts analyzed after performic acid oxidation. The cystine values for angiogenin and chymotrypsin are low at 18 hr of hydrolysis, most probably due to slow hydrolysis of the

Table III : Amino Acid Analyses of Angiogenin[a] using DTDPA/AQC

	20 ng	40 ng	100 ng	Sequence
Cys(tp)	6.06	6.63	6.27	6
Asp	15.79	15.08	15.08	15
Glu	12.33	12.37	10.52	10
Ser	13.25	11.89	9.61	9
Gly	24.91	17.01	11.99	8
His	5.86	6.91	5.68	6
Arg	12.21	12.92	12.92	13
Thr	9.58	9.41	9.06	8
Ala	8.72	7.49	6.78	5
Pro	10.23	10.20	8.93	8
Tyr	9.49	5.46	4.19	4
Val	10.44	8.00	6.23	5
Met	0.00	0.00	0.00	0
Ile	13.15	10.72	8.84	7
Leu	9.06	7.42	7.38	7
Phe	6.37	5.98	5.72	5
Lys	8.02	7.76	7.38	7
% Compositional Error	51.5	29.8	13.4	
pmol protein	1.42	2.91	6.92	

[a]Reported as residues per mole, based on Asp + Arg = 28; single analyses.

relatively large number of refractory Val-Cys and Ile-Cys sequences in these proteins, and appropriately increased to values closer to the theoretical on extended hydrolyses. The values for actin and serum albumin show a slight decrease on extended hydrolysis, suggestive of some hydrolytic destruction, such as generally seen to a larger extent with serine and threonine.

Analyses at nanogram levels of protein are illustrated by that of angiogenin in Table III. The cystine values are assessed within 10% of theory, while many of the other amino acids show large deviations, due to the general background.

C. Reduction and Alkylation

We determined the chromatographic properties of various additional alkyl cysteine derivatives, summarized in Table IV. A difference of more than 0.01 in relative retention time signals full separation of the two compounds. Chromatographic separation of most were adequate, with only carboxymethylcysteine not resolving from glutamic acid. Reduction and alkylation usually involves additional experimental procedures, such as desalting, and as a consequence is not effective for low amounts of protein, although in principle these derivatives can be analyzed with the same sensitivity as the common amino acids.

Table IV: Elution Positions for Cysteine Derivatives

AQC-Amino Acid	Retention Time Relative to Phe
Cysteic Acid	**0.275**
Glutamic Acid	*0.493*
Cys, carboxymethyl	**0.494**
Threonine	*0.661*
Cys, sulfopropyl	**0.670**
Cys, NNN-trimethylamino-ethyl	**0.670**
Cys, thioglycolic	0.685
Alanine	*0.695*
Cys, NNN-trimethylamino-propyl	**0.708**
Proline	*0.751*
Cys, thiopropionic	**0.794**
Cys, NNN-triethylamino-propyl	**0.822**
Cystine	**0.838**
Tyrosine	*0.848*
Cys, thiobutyric	**0.873**
Valine	*0.873*

IV. Discussion

The sensitive analysis of cystine in the company of most of the naturally occurring protein amino acids is possible by a variety of methods employing the fluorescent AQC derivatives, as summarized above in Table IV. The hydrolysis of proteins with hydrochloric acid in the presence of dithiodipropionic acid (6) followed by derivatization with AQC (8,9) provides for the most sensitive analyses of cystine/cysteine, valid even when amounts of cystine and protein are limited. The derivative of cysteine can be detected and quantified to fmol levels and analysis is linear over a wide range. The methodology cannot yet replace standard amino acid analyses, since some losses of Tyr and Met are seen and some background is carried into samples with the disulfide reagents.

When proteins are conventionally analyzed after hydrolyzing amounts lower than 1 μg, and especially amounts lower than 100 ng, the general laboratory background of amino acids makes accurate compositional determinations next to impossible. Some categorization is still possible, but many adventitious amino acids such as Glu, Gly, Ser, Ala, Leu, etc. are superimposed. Quantitation of the protein is only possible in such instances on consideration of the experimentally determined amounts of e.g. Phe, Pro, His, etc., residues which occur less abundantly in the background. The clean background seen in the chromatograms of AQC derivatives from 28 to 31 min, *i.e.* around the elution time of the thiopropionic acid derivative of cysteine, make estimation of cystine feasible to below the 500 fmol range (200 fmol can be determined above the electronic noise), *with no contribution from the background.* Experiments on analyses of various proteins hydrolysed from nominally low nanogram levels show that the cystine estimations are sufficiently accurate to allow quantitation of the proteins, even though the variable background makes it difficult to use most other residues for this purpose. Thus angiogenin was quantified at the 20 nanogram level. The quantitative manipulation of such low amounts of protein remains very difficult and since such amounts are generally obtained from very dilute solutions, the quantitative transfer of protein to the hydrolysis system is crucial in allowing accurate assessment of protein amounts.

Acknowledgements

We thank Wynford V. Brome, Dennis P. Michaud and John Petersen for excellent technical assistance with these studies. This work was supported in part by funds from Hoechst, A.G. under agreements with Harvard University.

References

1. Hirs, C.H.W. (1956) *J.Biol.Chem.* **219**, 611-621.
2. Spencer, R.L. and Wold, F. (1969) *Anal. Biochem.* **32**, 185-190.
3. West, K.A. and Crabb, J.W. (1992) *in* Techniques in Protein Chemistry III (R.H.Angeletti, ed.), Academic Press, San Diego, pp 233-242.
4. Crabb, J.W., Ericsson, L.H., Atherton, D., Smith, A.J. and Kutny, R. (1990) *in* "Current Research in Protein Chemistry (J.J. Villafranca, ed) Academic Press, San Diego, pp 49-61.
5. Strydom, D.J., Tarr, G.E., Pan, Y-C.E. and Paxton, R.J. (1992) *in* Techniques in Protein Chemistry III (R.H.Angeletti, ed), Academic Press, San Diego, pp 261-274.
6. Barkholt, V. and Jensen, A.L. (1989) *Anal. Biochem.* **177**, 318-322.
7. Hoogerheide, J.G. and Campbell, C.M. (1992) *Anal. Biochem.* 201, 146-151.
8. Cohen, S.A. and Michaud, D.P. (1992) *Anal. Biochem.* submitted
9. Cohen, S.A., De Antonis, K. and Michaud, D.P. (1993) *in* Techniques in Protein Chemistry IV (R.H.Angeletti, ed), Academic Press, this volume.

AMINO ACID ANALYSIS AND MULTIPLE METHYLATION OF LYSINE RESIDUES IN THE SURFACE PROTEIN ANTIGEN OF *RICKETTSIA PROWAZEKI*

W.-M. Ching, H. Wang, J. Davis[a], G. A. Dasch

Rickettsial Diseases Program, Naval Medical Research Institute,
Bethesda, MD 20889
and
[a]Laboratory of Biochemistry National Heart and Lung and Blood Institute
NIH, Bethesda, MD 20892

I. Introduction

Typhus and spotted fever group rickettsiae possess organized paracrystalline arrays comprised of protein (S-layers) on the surfaces of their cell envelopes. The surface protein antigens (SPAs) of both typhus and spotted fever rickettsiae are highly immunogenic in humans and animals and elicit good protective immune responses; they contain strain-specific, species-specific, and both group-specific and group-cross-reactive epitopes recognized by human and animal antisera and immune lymphocytes (1-5). Due to its importance as a subunit vaccine candidate, we have characterized the biochemical properties of SPA purified from the etiologic agent of epidemic typhus, *Rickettsia prowazeki* (1, 6). Amino acid composition analysis of purified SPA by conventional methods indicated that more than half of the lysine residues predicted from the gene sequence were not detected. ε-N-monomethyl-, dimethyl- and trimethyllysine have been identified as natural components of proteins (7). Modified amino acids have been detected in the S-layer protein of *R. prowazeki* (5, 8, 9). SPA from the avirulent live vaccine strain Madrid E had more ε-N-methyllysine and no detectable ε-N-trimethyllysine when compared with the SPAs from two virulent strains, revertant Madrid EVir and Breinl from Poland (8-10). To further investigate the possible relationship of level of methylation and strain virulence, we isolated SPAs from two virulent strains, Madrid 1 and Madrid 2, which had been isolated in guinea pigs on the same day from the blood of two separate patients ill during the Spanish epidemic of 1941. Madrid 1 is the parent isolate from which the Madrid E strain was selected by guinea pig passage (11). For comparison, we examined SPAs purified from strain Madrid E and cloned strain Madrid E^r (selected for erythromycin resistance from strain Madrid E) (12).

II. Materials and Methods

A. Reagents

H-Lys (Me)-OH.HCl and H-Lys $(Me)_2$-OH·HCl·0.5·H2O were obtained from Bachem Bioscience Inc. and ε-N-trimethyl-L-lysine bis (p-hydroxyazo-benzene-p-sulfonate) was obtained from Calbiochem. 6M HCl, amino acid standard and fluoraldehyde (OPA) were purchased from Pierce, and Tris was purchased from BioRad Laboratories. Sodium eluents for post-column derivatization analysis were obtained from Pickering Laboratories and Pico-Tag eluents for pre-column derivatization were obtained from Waters. All the chemicals and solvents used for the Applied Biosystems Inc. (ABI) Model 420 automated amino acid analyzer were purchased from ABI. Crude SPA extract obtained from gradient purified rickettsiae was purified by Mono Q anion exchange column chromatography (1, 6).

B. HPLC amino acid analysis systems

Our Waters system includes a Waters Maxima 820 controller, a Waters 490E detector, two Waters model 510 pumps, a Waters 712 WISP injector, and a temperature controller and heating box combined with a post-column derivatization unit which includes a Waters 420 fluorescence detector and two external pumps. Our ABI automated amino acid analyzer consists of a 420H hydrolyzer and derivatizer, a 130A separation system, and a 920A data analysis module.

C. Post-column derivatization analysis of dimethyllysine and trimethyllysine

ε-N-dimethyllysine and ε-N-trimethyllysine were analyzed on a Waters™ Amino Acid Analysis HPLC column (4.6 x 250 mm, 9 $\pm$ 0.5 μm particle size) packed with styrene-divinyl benzene copolymer with sulfonic acid in Na^+ form. 7-10 μg of samples were hydrolyzed in HCl vapor phase at 150 °C for 1.50 hr. Neutralized hydrolysates were diluted in 0.1 M HCl and about 1-5 μg of sample was loaded onto the column equilibrated at 64 °C. Eluent A was 0.2 N sodium citrate (pH 3.15) and eluent B was 1.0 N sodium citrate (pH 7.4). The linear gradients for analysis of methylated lysines increased eluent B from 0% to 40% in 10 min and then to 90% after another 50 min with a flow rate of 0.5 ml/min. The linear gradients for standard amino acid analysis increased eluent B from 0% to 80% in 45 min and then to 100% after another 15 min. OPA solution (0.175%) was prepared in 0.125 M potassium borate buffer pH 10.4. The post-column derivatization solution was pumped into the HPLC eluent at a flow rate of 0.4 ml/min. The fluorescence detector was set at λ ex = 338 μm and λ em = 425 μm.

D. Pre-column derivatization analysis of monomethyllysine

Waters HPLC system and Pico-Tag column (3.9 x 300 mm I.D., and 4 μm particle size, silicon based C18) were used to analyze ε-N-monomethyllysine. Following protein hydrolysis of samples as described above, derivatization of the hydrolysate by PITC and column chromatography were carried out according to the procedures provided by Waters. Samples were also hydrolyzed and derivatized automatically on an ABI 420H system. Brownlee^{TM} PTC-C18 column (2.1x 220 mm, 5 μm spherical silica sorbents) was maintained at 34°C for separation of derivatized amino acids. Elution buffers and gradient were as recommended by ABI.

III. Results

Precise quantitation of the amount of methylated amino acids in proteins is usually carried out by HPLC although many techniques have been used (7). The major difficulties in HPLC analyses are the small quantities of modified amino acids present in a sample and their coelution with other amino acids. Initial separation of basic amino acids may simplify elution profiles but can lead to loss in sample and precision in quantitation. Since the amounts of each sample available to us were quite limited, we developed a method for the quantitation of ε-N-mono-, di-, and trimethyllysine present in the protein hydrolysate without initial separation of basic amino acids. We first tried isocratic elution with 0.2 M sodium citrate, pH 6.48 on a Dionex DC-5A resin packed column (4 mm x 150 mm). In this system, ε-N-di- and trimethyllysine could not be resolved completely. With a Waters amino acid analysis HPLC ion exchange column, we were able to separate ε-N-mono-, di-, trimethyllysine, but were unable to separate monomethyllysine from unmodified lysine (compare Fig. 1A and 1B). In the hydrolysate of SPA from the virulent strain Madrid 1, ε–N-di- and trimethyllysine were detected (Fig. 1C). The elution position of each compound was confirmed in separate experiments where only one type of methylated lysine was added to the sample and the expected peak increased accordingly. In contrast to the profile with Madrid 1 SPA, ε–N-di- and trimethyllysine were not detected in the hydrolysate of SPA from the avirulent strain Madrid E^{r} (Fig. 1D). Comparable results were obtained with Madrid 2 and Madrid E, respectively (Table 1)

Since ε-N-monomethyl lysine could not be separated from unmodified lysine on the ion exchange column in the post column derivatization system, we separated their phenylisothiocyanate (PITC) adducts on reversed-phase columns using pre-column derivatization systems (Figure 2). The PITC adduct of ε-N-monomethyllysine elutes after that of lysine and is well separated from other extraneous peaks by Waters Pico-Tag. The same system could not be used for the analysis of ε-N-di- and trimethyllysine in the mixture of amino acids from a protein lysate since the PITC adduct of ε-N-trimethyllysine eluted with that of arginine, and the

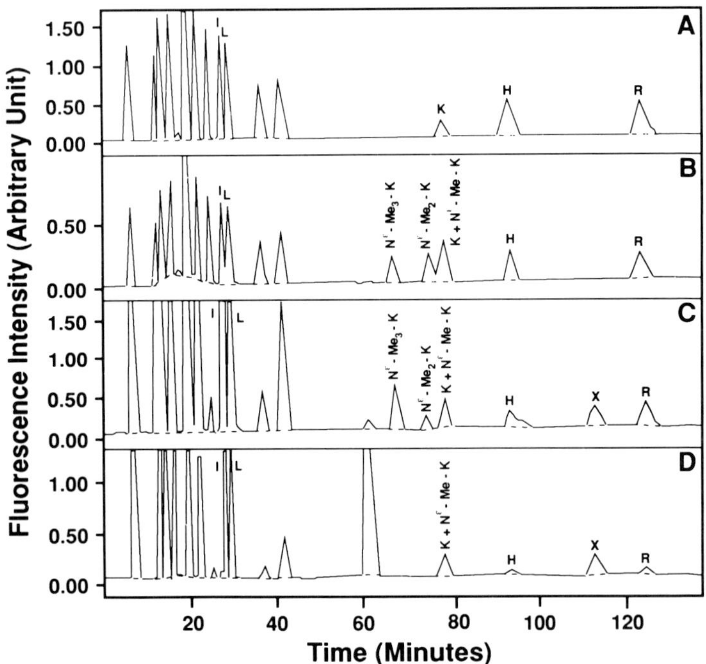

Figure 1. Analysis of multiply-methylated lysines by Waters post-column derivatization system. Samples: A. 500 pmole of amino acid standard. B. 500 pmole of amino acid standard plus 500 pmole of ε-N-methyllysine (N^{ε}-Me-K), ε-N-dimethyllysine (N^{ε}-Me$_2$-K) and ε-N-trimethyllysine(N^{ε}-Me$_3$-K). C. Hydrolysate of SPA from strain Madrid 1 (32 pmol). D. Hydrolysate of SPA from strain Madrid E^r (9 pmol).

Figure 2. Analysis of lysine and ε-N-monomethyllysine by Waters Pico-tag System. Samples: A. 500 pmole of amino acid standard. B. 500 pmole of amino acid standard plus 500 pmole of ε-N-methyllysine (N^{ε}-Me-K). C. Hydrolysate of SPA from strain Madrid E^r (8.8 pmol).

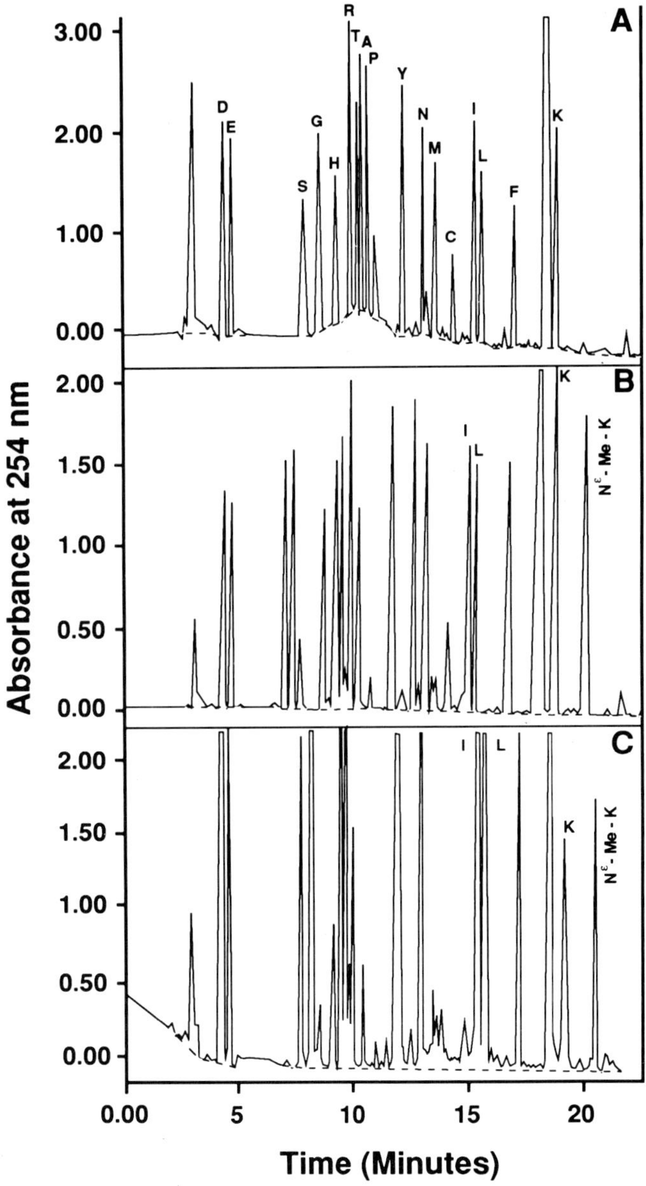
A
B
C
R
T
A
P
D
E
G
H
S
Y
N
M
C
I
L
F
K
N^{ε} - Me - K
Absorbance at 254 nm
3.00
2.00
1.50
1.00
0.50
0.00
5
10
15
20
Time (Minutes)

adduct of ε-N-dimethyllysine eluted between the arginine and threonine adducts. Neither could be separated adequately from the adjacent amino acid derivatives despite numerous attempts. Results on this column with the SPA of Madrid E^r are shown in Fig. 2C. The amount of ε-N-monomethyllysine relative to lysine can be very precisely determined. Similar results were obtained on the ABI Model 420H amino acid analyzer (not shown).

Table 1. Amino Acid Composition of SPAs from Strains of *Rickettsia prowazeki*

Amino Acid	Number of Residues per Molecule[1]					
	Madrid Strains				Breinl	
	1	2	E	E^r	Protein	DNA[2]
Asp/Asn	222	234	227	234	242	239
Thr	145	148	130	140	144	147
Ser	81	81	85	76	81	91
Glu/Gln	76	70	81	75	72	69
Gly	159	157	176	156	170	152
Ala	139	137	126	134	132	130
Val	93	92	85	92	101	98
Met	13	15	6	12	7	9
Ile	105	106	105	106	107	119
Leu	105	104	103	101	107	103
Tyr	17	17	16	12	13	18
Phe	50	47	49	45	49	47
Pro	30	31	30	23	23	25
His	9	10	14	11	11	13
Cys	n.d.	n.d.	n.d.	n.d.	3	3
Arg	18	16	12	14	13	15
Trp	n.d.	n.d.	n.d.	n.d.	n.d.	1
ε-N-Me_3-Lys	14	13	0	0	13	-
ε-N-Me_2-Lys	3	3	0	0	4	-
ε-N-Me-Lys	16	16	47	39	14	-
Lys	13	11	22	36	16	42

[1]Calculated based on the DNA-deduced molecular weight of 1321 amino acids of *R. prowazeki* Breinl SPA (135 kDa). The data represent the average value of 2 determinations. Most determinations differ by 1-2%. n. d., not determined.

[2]Deduced from DNA sequence (14).

Except for lysine, the variation among the amino acid compositions of SPA from each strain of *R. prowazeki* examined was similar to that between different preparations of the same SPA and comparable to that predicted from the nucleic acid sequence of Breinl strain SPA (Table 1). In the three virulent strains, the amount of lysine, ε-N-mono- and trimethyllysine was about the same, 11-16 residues per molecule, while only a small amount of ε-N-dimethyllysine, about 3-4 residues per molecule, could be found. In contrast in the two avirulent strains, neither ε-N-dimethyllysine nor ε-N-trimethyllysine was detected, and the amount of ε-N-monomethyllysine and lysine was more than double the amount found in the virulent strains. Surprisingly, the total number of lysine and ε-N-monomethyllysine residues in the avirulent strains, 69-75, is at least 50% greater than the total number of all types of lysines, 43-47, respectively, that are present in the virulent strains. The latter values are in good agreement with the 42 lysines that are predicted from the DNA sequence of the Breinl strain.

IV. Discussion

Post-column derivatization analysis permitted quantitation of ε-N-di- and trimethyllysine and the integrated sum of lysine and ε-N-monomethyl-lysine. Pre-column derivatization analysis provided quantitation of ε-N-monomethyllysine and lysine. When results from both systems were combined, the content of all three types of ε-N-methylated lysines could be determined. More acurate quantitative results for standard amino acid analyses have generally been obtained with the post column system than the pre-column system. Therefore, we used the ratio of ε-N-monomethyllysine to lysine from pre-column derivatization analyses to calculate their relative amounts in their common peak obtained in the post-column system. The isoleucine peak served as a common reference point to normalize the quantitation between different pre- and post-column gradient runs.

Two possible explanations may be offered for the increased content of lysine in the Madrid E strains relative to that found in the virulent strains. During the attenuation of strain Madrid 1 to Madrid E, not only the post-translational methylation of lysine changed, but mutations increasing the number of lysine residues present in the SPA molecule occurred. This seems unlikely since the content of other amino acids appears unchanged and the reversion of the Madrid E to Madrid EVir appears to occur readily under the proper selective pressure (10, 13). Rather a single mutation, possibly in a methylase or enzyme affecting methylation substrates rather than SPA, might account for the observed changes in methylation. Alternatively, some other material may be coeluting in the lysine/ε-N-monomethyllysine peak in the post column system and increase its relative content. It is also possible that the SPAs of Madrid E strains undergo post-translational β-protein cleavage at a site closer to the carboxy terminus than occurs in virulent strains of *R. prowazeki* (3); indeed, the amino terminus of the β-protein of *R. prowazeki* Breinl is relatively lysine-rich (14). DNA and protein sequence analyses on the Madrid 1 and Madrid E strains will be required to distinguish these different possibilities.

The precise function of rickettsial SPA is not known. Although Madrid E strain grows readily in fibroblast cell lines, it is unable to replicate efficiently in macrophages (15). The addition of methyl groups to lysines increases the hydrophobicity of proteins. Although a specific role for lysine methylation in rickettsial growth has not been demonstrated, protein methylation may be in-volved in the targeting of proteins for transport to specific organelles and protection of proteins from proteolytic degradation. If the SPA serves primarily as a protective barrier, as has been proposed for other S-layer proteins, without full methylation it may be an ineffective barrier to the active host cell degradative processes found in professional phagocytes. The concordance between loss and restoration of multiple methylation capacity and virulence in the Madrid 1-Madrid E-Madrid EVir (7-8) succession of strains is indeed striking and offers the best present explanation of the biochemical basis for the attenuation of the Madrid E strain of *R. prowazeki*.

Acknowledgments

This investigation was supported by the Naval Medical Research and Development Command, Research Task No. 62787A 3M162787A870.AN.121. We thank W. G. Sewell for expert technical assistance.

The opinions and assertions herein are not to be construed as official or as reflecting the views of the Navy Department or the naval service at large.

References

1. Ching, W.-M., Dasch, G. A., Carl M., and Dobson, M. E. (1990) *Ann. N. Y. Acad. Sci.* **590**, 334-351.
2. Vishwanath,S. (1991) *FEMS Microbiol. Lett.* **81**, 341-344.
3. Hackstadt, T., Messer, R., Cieplak, W., and Peacock, M. G. (1992) *Infect. Immun.* **60**, 159-165.
4. Carl, M., and Dasch, G. A. (1989) *J. Autoimmun.* **2** (Suppl.), 81-91.
5. Ching, W.-M., Carl, M., and Dasch, G. A. (1992) *Mol. Immunol.* **29**, 95-105.
6. Ching, W.-M., Dasch, G. A., Falk, M., Weaver, J., and Williams, R. W. (1992) Manuscript in preparation.
7. Paik, W. K., and Kim, S. (1990) In *Protein Methylation* (Paik, W. K., and Kim, S., eds.) pp 1-22, CRC Press.
8. Rodionov, A. V. (1990) *Bioorgan. Khim.* **16**, 1687-1688
9. Rodionov, A. V., Eremeeva, M. E., and Balayeva, N. M. (1991). In *Rickettsiae and Rickettsial Diseases* (Kazar, J., and Raoult, D., eds.) pp 95-101, Publishing House of The Slovak Academy of Sciences.
10. Balayeva, N. M., and Nikolskaya, V. N. (1972) *Acta. Virol.* **16**, 80-82.
11. Perez Gallardo, F., and Fox, J. P. (1948) *Am. J. Hyg.* **48**, 6-21.
12. Weiss, E., and Dressler, H. R. (1960) *J. Hygiene* **71**, 292-298.
13 Kazar, J., Brezina, R., and Urvölgyi, J. (1973) *Bull. WHO* **49**, 257-265.
14. Carl, M., Dobson, M. E., Ching, W-M., and Dasch, G. A. (1990) *Proc. Natl. Acad. Sci. USA* **87**, 8237-8241.
15. Turco, J., and Winkler, H. H. (1982) *Infect. Immun.* **35**, 783-791.

CHIRALITY DETERMINATIONS OF D- AND L-AMINO ACIDS OF SYNTHETIC AND NATURALLY OCCURRING PEPTIDES

Brenda L. Purcell and Muriel S. Doleman

Sterling Winthrop Pharmaceuticals Research Division,
Great Valley, PA 19355

I. Introduction

Biologically active peptides are often required to be stereochemically pure. If synthetically produced, racemization may occur as a result of the synthesis. This then requires the separation and identification of the optically pure bioactive product. Stereochemical configuration may also dictate the activity of small naturally occurring peptides. This needs to be investigated as part of the total characterization of the peptide.

The alpha carbon of the amino acid is the chiral center. Glycine is the only amino acid which does not have an asymmetric alpha carbon. The D and L designations refer to absolute configuration about the chiral carbon, based on glyceraldehyde as a standard of reference.

Several methods for stereochemical separations have been devised: RP-HPLC with chiral stationary phases (1-3), chiral metal mobile phases (4-6), and reactions with chiral derivatives (7-13).

We chose to use RP-HPLC separation of the hydrolyzed peptide derivatized with the chiral reagent: 1-fluoro-2,4 dinitrophenyl-5-L-alanine amide, Marfey's reagent (FDAA). The reaction is shown in Figure 1. Covalently bound diastereomeric derivatives produced can be resolved with ordinary stationary phases and eluants (14,15). The major problem which we addressed was to unambiguously resolve the mixture of DL-amino acids produced by hydrolysis of the peptides.

TECHNIQUES IN PROTEIN CHEMISTRY IV

Figure 2 shows the unusual amino acids in the synthetic peptide. They are abbreviated in the text as:

TIC Tetra isoquinoline-3-carboxylic acid
OIC Octahydro-indole-2-carboxylic acid

II. Materials and Methods

A. Peptide Sample

Synthetic Peptide

D-Arg L-Arg L-Pro L-Hypr Gly Thienyl-L-Ala L-Ser D-TIC L-OIC L-Arg

is a bradykinin antagonist.

B. Chemicals and Equipment

Hewlett-Packard 1050 Series quaternary gradient pump, injector and variable wavelength UV detector.
Beckman Ultrasphere ODS 5 μm column, 4.6 x 15 cm
Marfey's Reagent (FDAA)(Pierce Chemical Co.)
Silli-vap Argon Manifold (Pierce Chemical Co.)
Methane-Oxygen Flame Torch (Air Products, Inc.)
Pyrex Hydrolysis Tubes (15 x 75 mm)

C. Sample Preparation

A 5 mL ampule of 6N HCl is opened, 20 μL of 2-mercaptoethanol is added and the mixture immediately blanketed with argon for 20 minutes. 100 μL of this mixture is added to the sample in a pyrex hydrolysis tube. The tube is blanketed with argon and quickly sealed with the methane-oxygen flame torch. Samples are heated in an oven at 110°C for 20 - 24 hours, then dried to completion and dissolved in 0.5 M $NaHCO_3$. An aliquot of sample is placed in a microfuge tube and reacted with 1% Marfey's reagent in acetone, such that the μmoles FDAA : μmoles amino acid is approximately 1.5 : 1.0. The reactants are heated at 40°C for 90 minutes and cooled to room temperature. Then 20 μL of 2 M HCl is added and the sample allowed to degas.

Marfey's Reagent

1-fluoro-2,4-dinitrophenyl-5-L-alanine amide D-,L- amino acid

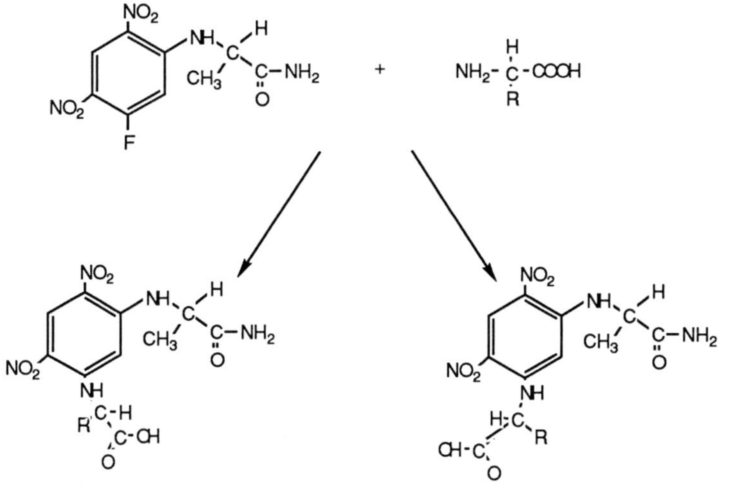

NO_2 NH C H CH_3 C−NH_2 O NO_2 F + H NH_2- C -COOH R

NO_2 NH C H CH_3 C−NH_2 O NO_2 NH R C-H C-OH O

NO_2 NH C H CH_3 C−NH_2 O NO_2 NH H C R OH-C O

L-L- Marfey's derivative **L-D- Marfey's derivative**

Figure 1

Unusual Amino Acids

L-OIC

Octahydro-indole-2-carboxylic acid

D-TIC

Tetra isoquinoline-3-carboxylic acid

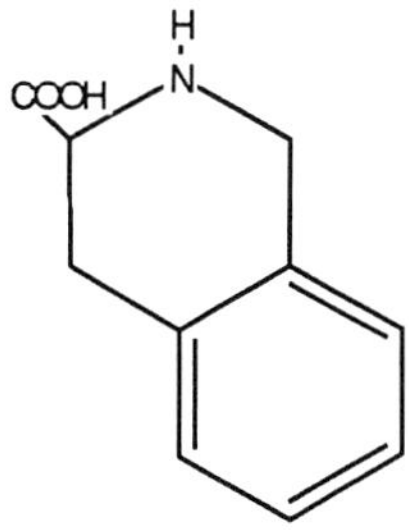

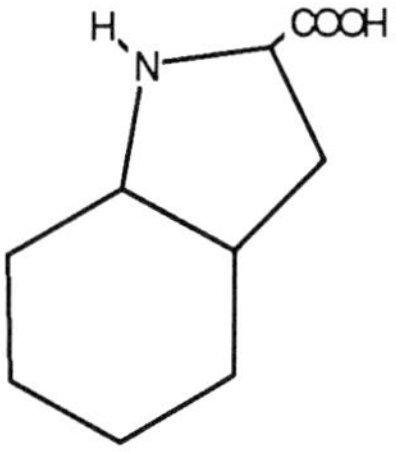

L-THIENYL-ALANINE

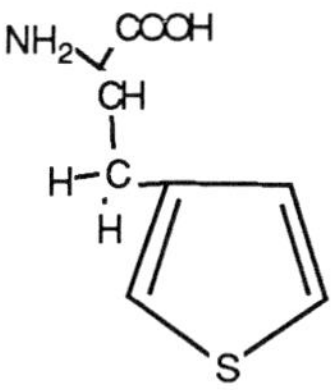

H N COOH

H N COOH

NH_2 COOH CH H−C H S

Figure 2

A
Mixed Standard
L-Ser
L-HyPro
D-Ser & L-Arg
D-HyPro
Gly
L-Pro
D-Arg
Marfey's Rgt
D-Pro
Thi-L-ala
L-Oic
D-Tic
Thi-D-ala
L-Tic
0
20
40
60
80
B
Woe 1114-108A
L-Ser
L-Arg
Gly
L-pro
D-Arg
Marfey's Rgt
Thi-L-ala
L-Oic
D-Tic
0
20
40
60
80
Time(min)
C
Woe 1114-108B
L-Ser
L-Arg
Gly
L-Pro
D-Arg
Marfey's Rgt
Thi-L-ala
L-Oic
L-Tic
0
20
40
60
80

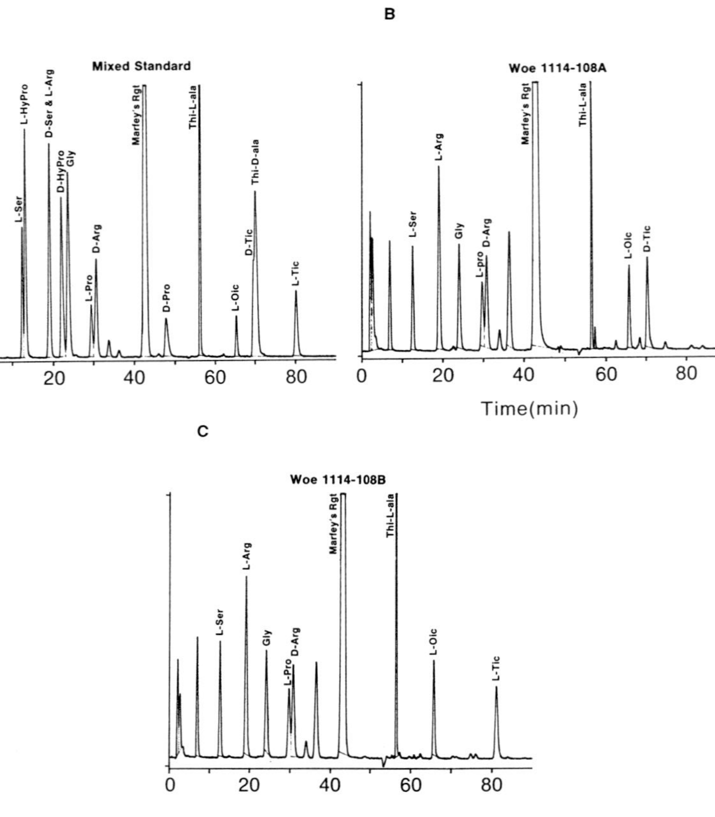

Figure 3

D. Chromatographic Conditions

The resolution of mixtures of amino acids in the peptide hydrolysate required selection of differing mixtures of eluants depending on the amino acids present. D- and L- standards were run first separately then as a mixture to find unambiguous retention times for each component. Peaks were monitored at 340nm.

Mobile Phases

Eluant A	Methanol
Eluant B	Acetonitrile
Eluant C	0.02 M Sodium Acetate pH 4
Flow Rate	1 ml/min
Wavelength	340 nm

Gradient Elution

	A		B		C
0 min	15	:	5	:	80
50 min	25	:	5	:	70
51 min	47	:	0	:	53
90 min	53	:	0	:	47

III. Results and Discussion

During final purification of the synthetic peptide Woe-1114-108, two major RP-HPLC fractions were collected, A and B. Racemization of the product was expected to have occurred, but with no indication of which amino acid(s) may be involved. This provided a challenging separation problem and the chromatography is shown in Figure 3 . A gradient elution system (see section IID) was devised which could resolve almost all the possible DL combinations of amino acids present in the peptide (Fig. 3A). Some coelution of D-serine and L-arginine, D-TIC and thienyl-D-alanine did occur. These coelutions did not compromise the results.

Figure 3B and C show the chromatography of the gradient applied to the samples Woe-1114-108 A and B, respectively.

One serine residue detected by amino acid analysis was found as the L enantiomer in the stereochemical separation. L-arginine and D-arginine were both found, in the expected ratios. Amino acid analysis had detected three residues of arginine. Thienyl-D-alanine was not detected in either sample. The chromatography shown in Fig. 3B and C clearly shows the only difference between the two samples: Woe-1114-108A contains D-TIC and Woe-1114-108B contains L-TIC. Sample Woe-1114-108A was found to be biologically active. Sample Woe-1114-108B was inactive. The stereochemistry of TIC was the only difference between the two peptides.

References

1 Yu, L.W. and Hartwick, R.A. J. Chromatographic Science,27, 176-185 (1989).
2. Yamashita J. et al J. Chromatography, 464 (1989), 411-415.
3. Yamashita J. et al J. Chromatography, 403 (1987), 275-279.
4. Van Der Haar et al J. Chromatography, 445 (1988), 219-224.
5. Marchelli R. et al J. Chromatography, 441 (1988), 287-298.
6. Weinstein S. et al J. Chromatography, 303 (1984), 244-250.
7. Bruckner, H. et al J. Chromatography, 476 (1989),73-82
8. Krull, I.S. et al Anal Chem. 61 (1989), 1548-1558.
9. Euerby, M.R. et al J. Chromatography, 483 (1989), 239-252.
10. Jerorov, A. et al J. Chromatography, 434 (1988), 417-422.
11. Einarsson, S. et al J. Liquid Chromatography, 10 (1987), 1589-1601.
12. Nimura, N., et al J. Chromatography, 352 (1986),169-177
13. Buck, R.H.,et al J. Chromatography, 315 (1984), 279-285.
14. Szokan, G. et al J. liquid Chromatography, 12 (1989), 2855-2875.
15. Szokan, G. et al J. Chromatography, 444 (1988), 115-122.

SECTION VI

Protein Separations

Development of Separation Strategies for Proteins by Capillary Electrophoresis

Judith A. Nolan and Richard Palmieri

Beckman Instruments, Fullerton CA 92634

I. Introduction

Classical electrophoretic techniques have been employed in a variety of protein analyses; e.g. sample purification, molecular weight estimation, binding interactions, subunit structure, refolding capabilities, and immune recognition (1). The introduction of capillary electrophoresis (CE) addressed several limitations in the traditional methods, including sample availability, analysis time, purity assessment and resolution.

The use of small diameter capillaries (20-75 μm inner diameter) requires only nanoliter sample amounts. The remainder of the sample is available for other analytical techniques. Small column volumes and high field strengths reduce analysis times from hours to minutes. Separations on the basis of charge-to-mass ratio provide complementary information to SDS-PAGE, HPLC and alternative techniques for assessing sample purity. Finally, the ability to dramatically alter separation schemes by simple buffer replacement promotes rapid methods development (2).

While all these improvements were fairly easily demonstrated for peptide analysis (3), proteins proved to be more challenging to adapt to free solution CE. Interestingly, theory predicts that larger molecules will have decreased diffusion in solution and therefore higher efficiency peaks should be obtained during electrophoresis(4). However, in most cases, a decrease in peak efficiency for proteins has been found. This is attributed to protein adsorption to negatively charged silanol groups on the fused silica capillary wall with resultant band broadening. A variety of methods have been developed to control this phenomenon. Modeled on gas and liquid chromatography chemistries (5,6,7), coatings for capillaries have been developed with some success. While the advancement of these chemistries will most definitely benefit CE, problems still exist with reproducibility, stability, efficiency and adsorption (8). Considering the diversity of protein applications, it is apparent that there will not be a singular approach to resolve protein adsorption issues. This paper will present several additional methodologies which have proven useful in our laboratory for developing separation strategies for proteins in CE.

II. Experimental

All experiments were performed on a P/ACE™ 2100 capillary electrophoresis system from Beckman Instruments, Fullerton CA. Reagents were purchased from Sigma Chemical, St. Louis MO and Pierce Chemical, Rockford IL. Sample buffers, run buffers and run conditions are listed in results and

TECHNIQUES IN PROTEIN CHEMISTRY IV

discussions. Sample were injected at 0.5 psi for 3-50 sec. UV absorbance was monitored at 214 nm for zwitterion containing buffer and eCAP™ SDS 200 and at 200 nm for all other buffer systems. Capillary environment temperature was maintained at 20°C +/- 0.1°C with a recirculating liquid. Proteins, 1 mg/ml in 25% phosphate buffered saline, were diluted 1:3 or 1:10 for use. Instrument control and data acquisition at a rate of 2 Hz was performed using Gold software on an IBM PS/2™ model 70.

III. Results and Discussion

A. Competitive binding

Early work by Lauer and McManigill (9) and recently Chen (10) shows significant reduction of wall effects with accompanying improvement in peak shape by the addition of high concentrations of salt to the buffer. However, these components increase buffer conductance so that the choice of capillary inner diameter that may be used is limited to sizes which prevent excessive heating. Alternatively, Jorgensen (11) demonstrated that zwitterions could be added to a buffer in high concentrations without the concomitant increase in current. However, unusually high concentrations of zwitterion (2M) which contain a free carboxyl group prevent low UV detection. Replacement of a phosphoryl for a carboxyl group provides a UV transparent alternative. An example of the use of this zwitterion is seen in Fig. 1. Five proteins ranging in isoelectric point (pI) from 5.2 to 11.0 are poorly resolved in a sodium phosphate buffer pH 7.3 (1A). The addition of O-phosphorylethanolamine at pH 6.5 is seen in (1B). The zwitterion was added to a low ionic strength buffer to examine the effect contributed most specifically by the zwitterion. While there is a dramatic improvement in resolution of the proteins, the use of 100 mM ionic strength buffer in conjunction with the zwitterion at relatively low concentrations (250 mM) provides optimal resolution and peak efficiency (1C). It is apparent that the phosphoryl group of the zwitterion could not replace free phosphate in the silica wall-phosphate buffer equilibrium. Loss of the carbonic anhydrase peak may be due to an effective change in pI upon binding of the zwitterion that results in precipitation. A peak was clearly visible for carbonic anhydrase at pH 5.8 with the same zwitterion (12). Although the precise mechanism by which O-phophorylethanolamine enhances the separation is unknown, we postulate that it derives its efficacy from both the interaction with the silica wall as well as the protein surface. At the same time, a minimum of 100 mM phosphate buffer appears to be necessary for optimal resolution. Clarification of this model is currently under investigation.

B. Voltage Ramping

McCormick, using coated capillaries with a 1.45 pH run buffer, has demonstrated improved resolution for CE protein separations by ramping the voltage during the initial minutes of the separation (13). The effectiveness of this technique was investigated with uncoated capillaries at a basic pH for a

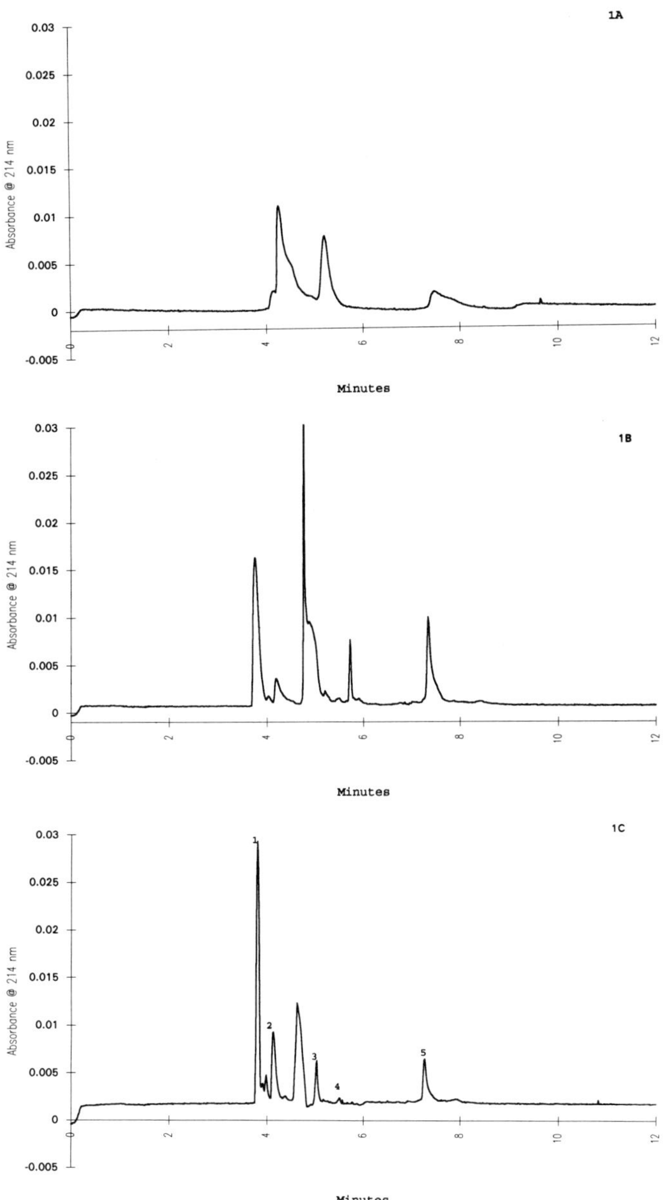

Fig. 1. Effect of O-phophorylethanolamine on Protein Separation Efficiency. A mixture of five proteins (1) lysozyme,pI 11.0; (2) cytochrome c,pI 10.7; (3) myoglobin,pI 7.0; (4) carbonic anhydrase,pI 5.9; (5)beta-lactoglobulin,pI 5.2 at 200 µg/ml were prepared in 25%PBS. Samples were injected for 4 sec. onto a 50 µm i.d. x 27 cm column (25°C) and run at 8 kV in several different buffers: (A) 100 mM sodium phosphate pH 7.3; (B) 10 mM sodium phosphate, 250 mM o-phosphorylethanolamine pH 6.5; (C) 100 mM sodium phosphate, 250 mM o-phosphorylethanolamine pH 6.5.

mixture of bovine serum albumin, rabbit immunoglobulin g, transferrin and alpha-2-macroglobulin (Fig. 2). Since the capillary used was only 27cm in total length, the ramp was applied over the entire run. The improved separation in Fig. 2B for the ramped voltage method may be explained by non-uniform electroosmotic flow (EOF) (14). Differences in ionic strength between the sample and run buffer are used to improve peak efficiencies by stacking. The initial high field strength experienced across the low ionic strength sample plug during a constant voltage separation may result in non-uniform EOF and disruption of resolution. Excessive joule heating in the sample zone may also have a negative effect (15). Fig. 2C illustrates the effect of using a step voltage in place of a linear voltage gradient. Identical resolution is obtained while decreasing analysis time. All the parameters affected by initial low voltage application are not fully understood. It is postulated that slowly electrophoresing the sample into the run buffer prevents the resolution disruption experienced by a rapid application of the maximum run voltage.

This voltage ramping technique was also instrumental in resolving multiple forms of a proprietary protein, SFP, known to dimerize in solution. Isoelectric focusing had previously been used to separate the monomer, dimer and a five amino acid carboxy terminal truncate for purposes of monitoring compound stability in solution. Initial attempts at resolving these components in free solution at constant voltage were poor (Fig. 3A). The charge-to-mass ratios for the monomer and dimer should be almost identical and may not be resolved by free solution techniques. However, Fig. 3B shows two well resolved forms of SFP, based upon the voltage ramping technique. A shoulder is visible on the first peak, which may represent the truncated form. To verify the identity of the SFP peaks, a new CE separation technique was employed. The eCAP™ SDS 200 system is a polymer filled capillary which permits protein separation on the basis of molecular weight (16), similar to traditional SDS-PAGE(17). Two samples of SFP were prepared, with and without the presence of 2-mercaptoethanol (2-ME). Fig. 3C overlays the two analyses, demonstrating reduction of the disulfide bond which dissociates the dimer to the monomer in the 2-ME treated sample. Molecular weights for the two peaks were estimated using the molecular weight option for Gold software based upon a standard curve and were found to be 21.7 kDa and 51kDa for the monomer and dimer, in good agreement with the predicted values of 23 kDa and 46 kDa.

C. SDS Differential Binding

The higher than expected molecular weight observed for the dimer in the non-reduced sample of SFP is not unexpected. In the classical model, SDS binds to proteins in a constant ratio of 1.4g detergent per gram of protein to give a rod-shaped SDS-protein micelle complex (18). However, this ratio is altered for some proteins by disulfide bond formation (19), amino acid composition (20) and post-translational modifications(21). Anomalous molecular weights are then obtained by SDS-PAGE. This differential SDS binding may be used

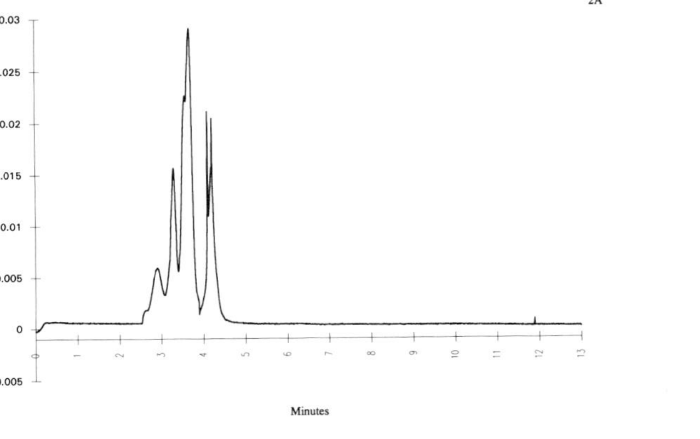

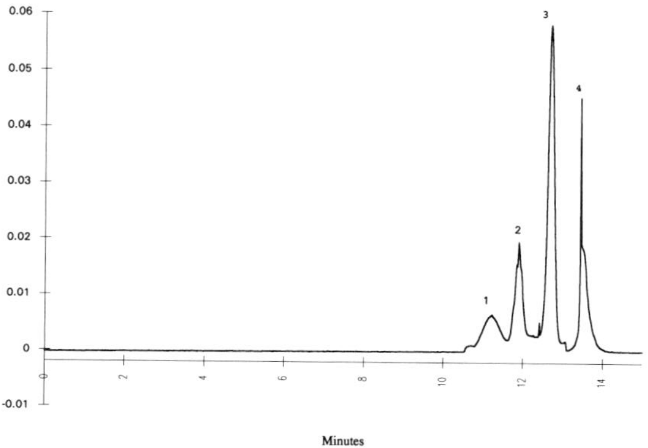

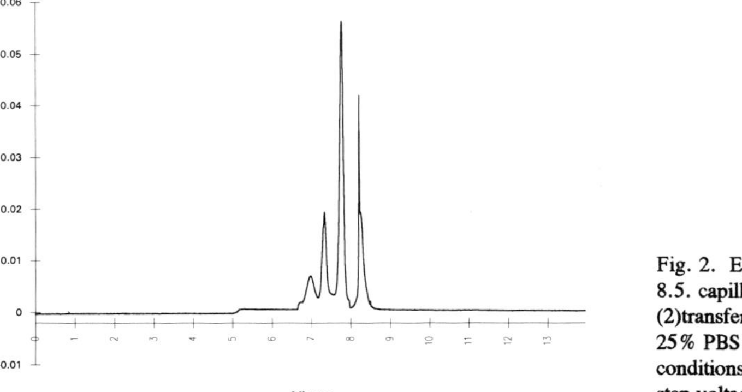

Fig. 2. Effects of Voltage Ramping on Resolution. Run buffer: 500 mM sodium borate pH 8.5. capillary dimensions: 50μmx27cm . samples A-C: (1)horse immunoglobulin, (2)transferrin, (3)alpha-2-macroglobulin, (4)bovine serum albumin. Samples were dissolved in 25% PBS at 1 mg/ml, then diluted 1:5 prior to injection. Injection: 5 sec. pressure. Voltage conditions: (A)constant voltage, 10 kV; (B) linear voltage gradient 0-10kV over 20 min.; (C) step voltage gradient, 2kV for 5 min., step to 10 kV for 10 min.

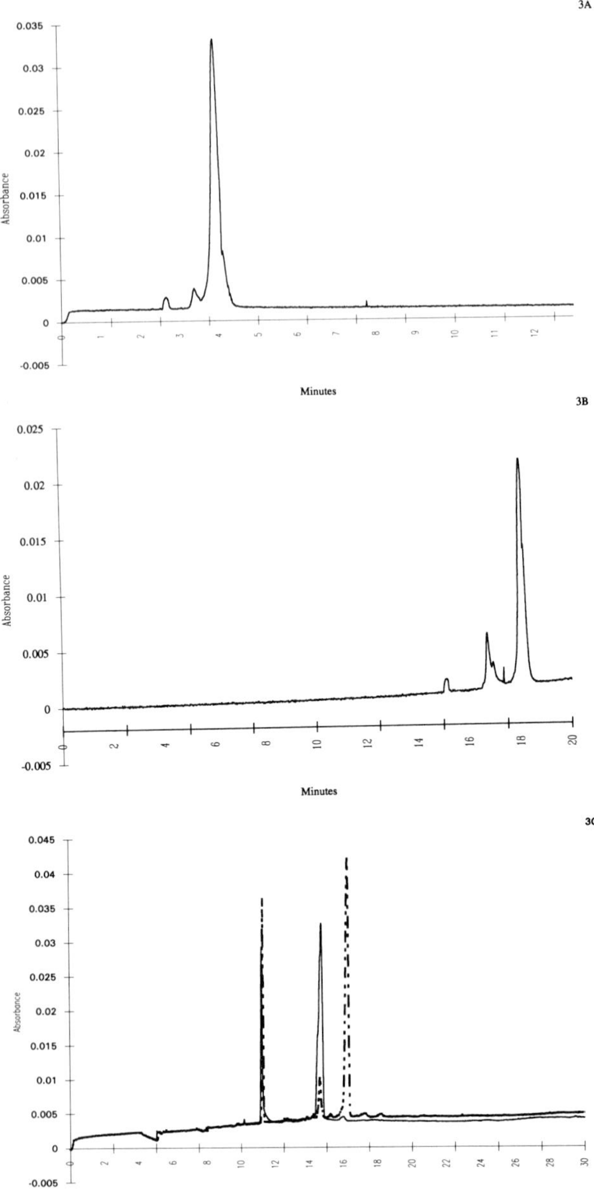

Fig. 3. **Characterization of SFP Forms using a Combination of Free Solution CE and eCAP™ SDS 200 gel. (A) and (B): sample and run buffer conditions as in Fig. 2; voltage (A) 10 kV constant voltage; (B) linear voltage gradient 0-10 kV over 20 min. (C): sample diluted and boiled in eCAP™ SDS 200 sample buffer at a concentration of 1.0 mg/ml. Run buffer: eCAP™ SDS 200 run buffer. Capillary: 100µm x 47cm coated eCAP™ SDS capillary. Voltage: 14.1 kV constant voltage. Solid trace: sample treated in presence of 2-ME; dotted trace: sample treated in absence of 2-ME. peaks (1) orange G internal standard, (2)monomer,(3)dimer.**

in CE to separate proteins by increasing differences in charge-to-mass ratio. To test this hypothesis, proteins which have been used as electrophoresis standards were evaluated(22). These proteins were chosen since it was believed that, as well characterized standards, that they would behave most closely in agreement with the SDS-protein binding model. If SDS bound in the same ratio to all the proteins, it would be predicted that no separation would occur in a free solution mode, given equal charge-to-mass ratios. However, as demonstrated in Fig. 4, it is possible to separate certain pairs of proteins using this technique. Although the resolution window is small, the high resolving power of CE permits separation to be achieved. This system would be useful for resolving proteins that vary in post-translational modifications, such as glycosylation. Differences that affect SDS binding will induce mobility changes in the analytes. The examples in Fig. 4 demonstrate this phenomenon, with proteins migrating in the order of least to most glycosylated. Fig. 5 is further illustration of the separation of a glycosylated (avidin) and a non-glycosylated (streptavidin) protein. Initial attempts to resolve this pair in borate buffer were unsuccessful, even with the voltage ramping technique applied. This technique may be further developed to determine variability in protein migration from predicted values due to incomplete SDS binding in gel systems.

IV. Conclusions

Protein analysis by CE has proven to be a challenge due to protein aggregation and protein-capillary wall interactions, as well as the small differences in mobility that are exhibited in free solution. Development of buffer systems with a new zwitterion additive has proven effective at significantly lower concentrations in reducing protein adsorption phenomena that decrease resolution and peak efficiencies. Differences in ionic strength between sample and run buffers improve peak efficiency through stacking. Non-uniform EOF over the high field strength experienced by the sample plug can decrease resolution, but this may be avoided by gradually applying voltage as a linear or step gradient. Finally, it is now possible to resolve SDS-protein complexes which have variations in glycosylation by differential detergent binding that alters the intrinsic charge-to-mass ratio. These techniques provide new tools which extend the capability of CE for the analysis of proteins.

Acknowledgements
The authors wish to thank Dr. Albert Chen, Dr. Ron Brown and Dr. Joe Olechno of Beckman Instruments for helpful discussions.

References

1. Hames, B. and Rickwood, D. (1981) In "Gel Electrophoresis of Proteins: A Practical Approach"(D.Hames and D. Rickwood. eds), IRL Press, Oxford, Washington.
2. Wallingford, R, and Ewing, A. (1989) Adv. Chromgr. **29**, 1.
3. McLaughlin, G., Palmieri, R., and Anderson, K. (1991) In "Techniques in Protein Chemistry II"(J. Villafranca, ed.), Chap. 1, Academic Press, New York.

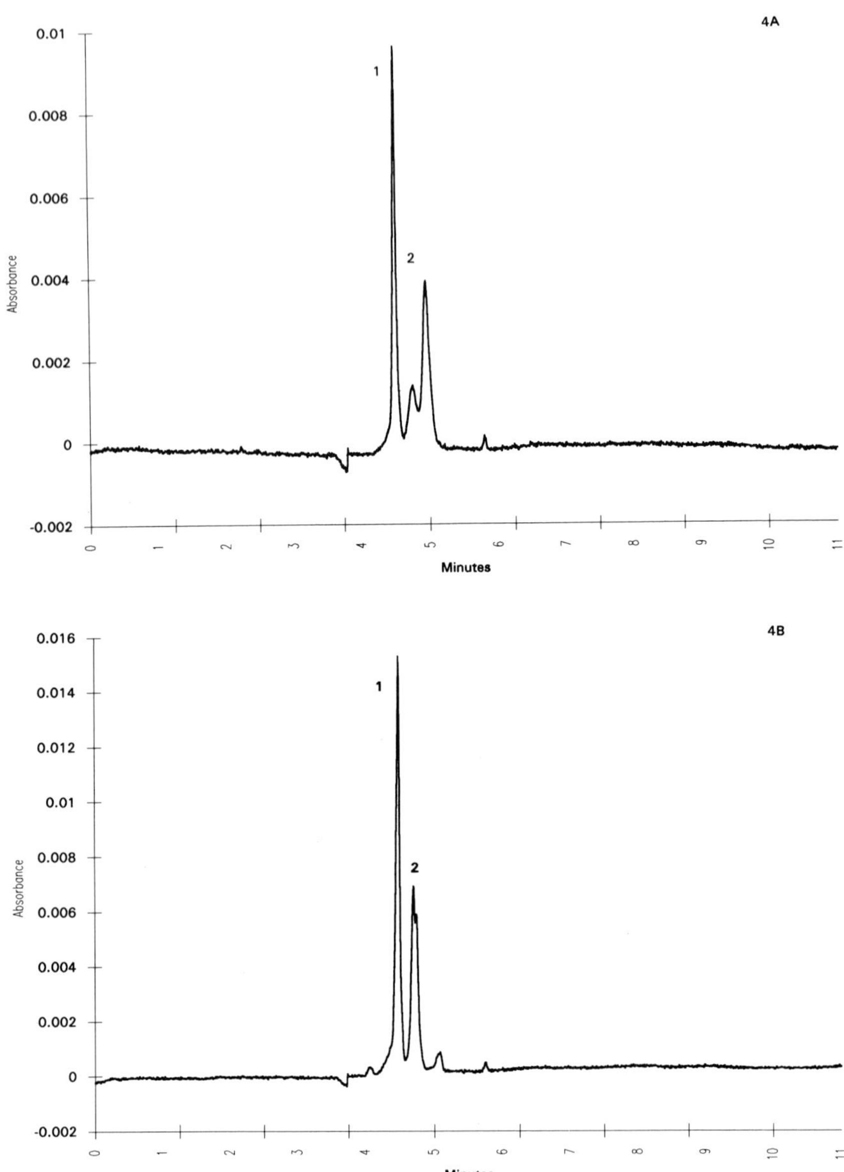

Fig. 4. Separation of SDS-protein Complexes at Low pH. Samples of initial concentration 1mg/ml were diluted 1:3 in sample buffer and boiled for 5 min. Sample buffer: 2.5 mM Tris-glycine pH 8.9,10% glycerol,1% SDS,5% 2-ME. Capillary dimensions: 20µmx27cm. Run conditions: 10kV constant voltage, reverse polarity mode (towards anode). Run buffer: 500 mM sodium phosphate pH 2.5, 0.1% SDS. Total carbohydrate content is listed after each protein: (A)(1)BSA,0% (2)bovine IgG, heavy and light chains represented by two, non-baseline resolved peaks, 2.6% (B) (1)BSA, (2)transferrin, 4%.

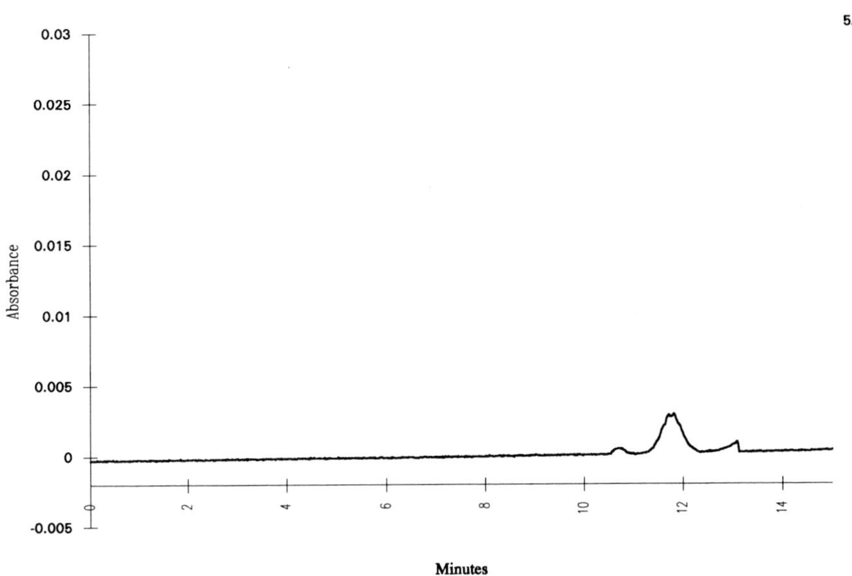

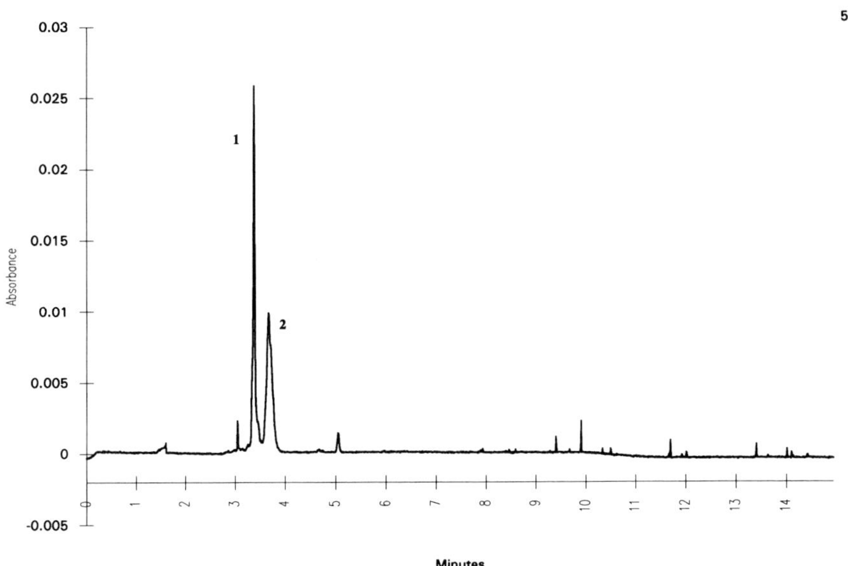

Fig. 5. Separation of Glycosylated and Non-glycosylated forms of Avidin. Avidin and streptavidin solutions were made 1mg/ml in 25% PBS, diluted 1:3 for injection. Capillary dimensions: 50μmx27cm. (A) Run buffer: 500 mM sodium borate pH 8.5; 0-10kV linear voltage ramp for 20 min. Sample diluted 1:3 with water. (B) Run buffer: 100 mM sodium phosphate pH 2.0,0.1% SDS; 10kV constant voltage, reverse polarity. Sample diluted 1:3 with sample buffer (see Fig. 4) and boiled. peaks (1)streptavidin, (2)avidin.

4. Jorgenson, J., and Lukacs, K. (1981) Anal. Chem. **53**, 1298.
5. Hjerten, S. (1985) J. Chromatogr. **347**, 191.
6. Towns, J., and Regnier, F. (1991) Anal. Chem. **63**, 1126.
7. Bruin, G., Huischen, R., Kraak, J., and Poppe, H. (1989) J. Chromatogr. **480**, 339.
8. Deyl, Z., and Struzinsky, R. (1991) J. Chromatogr.,**569**, 63.
9. Lauer, H., and McManigill, D. (1986) Anal. Chem. **58**, 166.
10. Chen, F-T. (1991) J. Chromatogr. **559**, 445.
11. Bushey, M., and Jorgenson, J. (1989) J. Chromatogr.**480**, 301.
12. Chen, F-T., Kelly, L., Palmieri, R., Biehler, R., and Schwartz, H. (1992) J. Liq. Chromatogr. **15**, 1143.
13. McCormick, R. (1988) Anal. Chem. **60**, 2322.
14. Chien, R-L., and Helmer, J. (1991) Anal. Chem. **63**, 1354.
15. Vinther A., and Soeberg H. (1991) J. Chromatogr. **559**,27.
16. Guttman, A., and Nolan, J. (1992) J. Chromatogr.,In press.
17. Laemmli, U. (1970) Nature (London) **277**, 680.
18. Reynolds, J., and Tanford, C. (1970) J. Biol. Chem. **245**, 5161.
19. Pitt-Rivers, R., and Impiombato, F. (1968) Biochem. J. **109**, 825.
20. Thomas, J., and Kornberg, R. (1975) Proc. Nat. Acad. Sci. **72**, 2626.
21. Hames, B.(1981) In "Gel Electrophoresis of Proteins: A Practical Approach", pp. 15-17, IRL Press, Oxford,Washington.
22. Andrews, A. (1986) In "Electrophoresis" (A. Peacocke and W. Harrington, eds.), pp. 118-122, Clarendon Press,Oxford.

High Resolution Full-Range (pI = 2.5 to 10.0) Isoelectric Focusing of Proteins and Peptides in Capillary Electrophoresis

Shiaw-Min Chen
John E. Wiktorowicz

Applied Biosystems, Inc.
Foster City, CA. 94404

I. Introduction

Due to the zwitterionic nature of their amino acid constituents, proteins exhibit a unique sensitivity to the pH of their environment. This property, which manifests itself as a net charge at a given pH, results from the aggregate of charges of the amino acid side chains, and their interplay in the higher order structure of the protein. The pH at which the net charge of a protein equals zero is known as its isoelectric point (pI). Isoelectric focusing (IEF) is a method that separates proteins on the basis of their isoelectric points (pI). It is accomplished by the electrophoresis of proteins or peptides through a stable pH gradient until their net charge is zero. At this pH, mobility is zero, and the isoelectric point is achieved. This method has been widely used in research (1), pharmaceutical (2) and clinical (3) areas to separate, isolate and analyze proteins from various biological products.

Initially, IEF was performed in solution in a vertical column of ampholytes (4). These ampholytes are small zwitterions which exhibited sufficient buffering capacity in order to maintain a stable pH gradient. Convective currents, generated by high field strengths, were minimized by the formation of a density gradient. These conditions were less than ideal, because convective currents generated by Joule heating limited the resolution achievable by this system. As a consequence, anticonvective media, such as cross-linked polyacrylamide, became popular. With the popularity of polyacrylamide, standard pre- and post-analysis techniques (polymerization, staining, destaining, densitometry, etc.) were applied to IEF of proteins. Nevertheless, the use of a gel-based separation and its reliance on semi-quantitative staining strategies for detection (e.g., loss of resolution due to Joule heating, inability to accurately quantify the amount of analyte in bands, long analysis time, labor intensive preparation and post-separation handling, and large sample mass requirements) limit the potential of IEF.

TECHNIQUES IN PROTEIN CHEMISTRY IV

The recent development of capillary electrophoresis (CE) led to the development of IEF by using CE (CE-IEF) as the separation strategy (5, 6, 7, 8, 9, 10, 11). By performing separations in a 50 μm diameter fused-silica capillary, the detrimental effects of convection created by Joule heating are circumvented. Shorter analysis times result from the use of high field strengths (volts per unit length). The narrow diameter capillary permits analysis with minute mass requirements, while the UV transparency of the fused silica permits on-line detection with precise and accurate quantification. Finally the free solution format eliminates the need for polymerization steps in preparation for electrophoresis and permits complete control over the composition of the separation buffer resulting in gradient linearity over the entire range of ampholytes (pH 3-10).

The CE-IEF was first developed by Hjerten et al. (5, 7) and was modified by Zhu et al. (6) Later, Mazzeo et al. (10) and Thormann et al. (11) independently reported using an uncoated capillary for IEF. However, due to various reasons (12), these CE-IEF methods failed to generate a linear pH gradient over the entire pH range of the ampholytes. Recently, Chen and Wiktorowicz (12) reported a CE-IEF method in which reproducibility, quantification, and linearity over the entire pH range of the ampholyte can be obtained. In this paper, effects of sample buffer matrix, surfactants, denaturant, and narrow pH range ampholyte using the same CE-IEF method will be discussed.

II. Materials and Methods

In this study Servalyt™ 3-10 and Servalyt™ 5-8 was purchased from Crescent Chemical Co. (Hauppauge, N.Y.), and methyl cellulose (1500 cps @ 2%) from Sigma (Sigma Chemicals, St. Louis, MO.). Carbonic anhydrase I (pI 5.4 and pI 5.9), carbonic anhydrase I (pI 6.6), ß-lactoglobulin A (pI 5.1), myoglobin (pI 7.2) and ribonuclease A (RNase A, pI 9.45) were obtained of the highest purity from Sigma. Unsulfated cholecystikinin (CCK) flanking peptide (pI 2.75) was obtained from Peninsula Labs (Belmont, CA.). Reduced Triton X-100 were purchased from Calbiochem (San Diego, CA) and Brij 35 from Pierce (Rockford, Il.) and human hemoglobin variants from Isolab (Akron, Oh.). Urea was the highest purity from Bio-Rad (Richmond, CA). All other reagents used were of analytical grade.

Capillary electrophoresis was performed on the Applied Biosystems Model 270A-HT Capillary Electrophoresis System (Foster City, CA.) using a DB-1 coated capillary with 50 μm internal diameter and 0.05 μm coating thickness (J&W Scientific, Folsom, CA.). The experiments were performed as described earlier (12) except in the event of narrow pH range ampholyte experiments Servalyt™ 5-8 was used and all other experimental variables were remained the same. In the studies of the effects of surfactants (reduced Triton X-100 and Brij 35) appropriate amount of additives were dissolved in both anode and cathode electrolytes and the ampholyte solutions. In the experiments on the effects of urea, the urea solution was deionized by passage through a mixed bed amberlyte and mixed with sample and markers to the same concentration as in all other electrode buffers and ampholyte solutions.

III. Results

Effect of surfactants. Reduced Triton X-100 and Brij 35 are two non-ionic surfactants that are commonly used to solubilize proteins and peptides. At a concentration of 0.2 %, both reduced Triton X-100 and Brij 35 showed no effect on the linearity of the pH gradient over the entire range of the Servalyt™ 3-10 when these surfactants were incorporated in all electrode buffers and ampholyte solution as shown in Figure 1.

Effect of Urea. Urea is a commonly used denaturation agent for proteins. In general commercially available urea contains some level of ionic species. To avoid any ionic species from contaminating the IEF system, the urea was de-ionized prior to use by passing through amberlyte mixed bed ion-exchanger. Three different urea concentrations, 0 M, 2 M, and 4 M were used in this study. Figure 2, shows the effect of urea on the linearity of the pH gradient. In this Figure the electropherogram of the markers, RNase A, myoglobin, ß-lactoglobulin A, and CCK flanking peptide with 4 M urea in both electrode buffers, ampholyte and markers is shown. The linearity demonstrates that urea had no effect up to a concentration 4 M.

Narrow range pH gradient. The linearity of narrow range pH gradient (pH 5 to 8) was determined by running myoglobin (pI 7.2), carbonic anhydrase I (pI 6.6), carbonic anhydrase II (pI 5.9), and carbonic anhydrase II (pI 5.4) as markers. The proteins were focused and mobilized as described in the Methods section. Figure 3 shows the electropherogram and pH gradient linearity plot. This Figure shows this CE-IEF method can be used with narrow range ampholyte and still provide linearity over the range of the ampholyte used. Human hemoglobin (hHb) variants A, S, F, and C were studied using the narrow pH range ampholyte and the separation is shown in Figure 4. The pI's of the variants A, F, S, and C were determined by CE-IEF as 6.99, 7.05, 7.17, and 7.40, respectively. The published pI's of the A and C variants were 6.95 and 7.40, respectively(13). Total protein loaded in this experiment was 50 ng.

Effect of salt, A monoclonal antibody 2.5 mg/ml in PBS buffer was injected as a discrete plug as described earlier (12) or was dissolved in ampholyte solutions (pH 5-8 and 3-10) to a final concentration of 50 µg/ml. When the sample was not dissolved in ampholyte , the buffer salt was not distributed evenly throughout the ampholyte solution and remained at the point of injection, resulting in a base line disturbance (data not shown). However, when the sample containing ampholyte solutions was loaded into the capillary and IEF were conducted as described in the Methods section, the electropherogram of narrow range ampholyte is comparable to that obtained from slab gel IEF (Figure 5). These results indicate that the sample salt effect, PBS buffer, can be resolved by using the modified method described in this paper without desalting the sample.

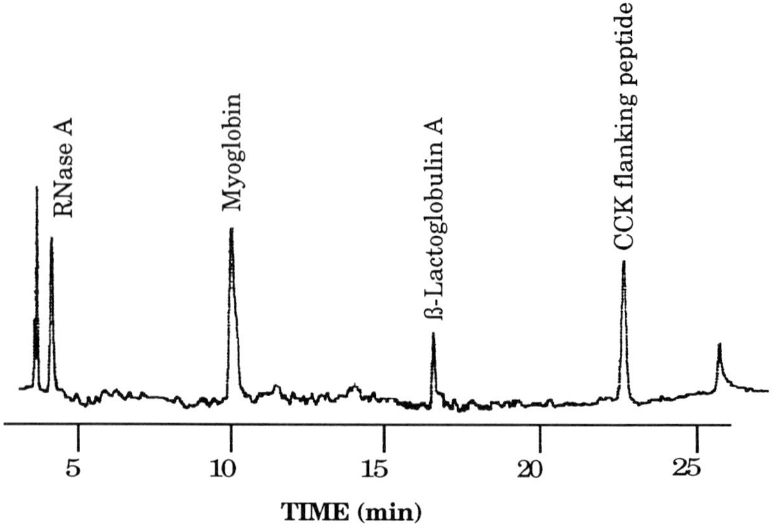

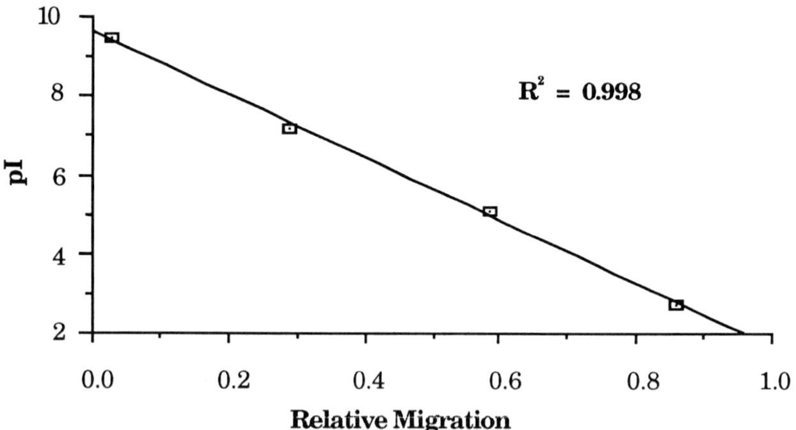

Figure 1. Reduced Triton X-100 was added to both electrode buffers and ampholyte solution at the concentration of 0.2 %. The IEF was performed as described in the Methods section. The data suggest that addition of reduced Triton X-100 has no effect on the pH gradient of CE-IEF. Identical results were also obtained with Brij 35.

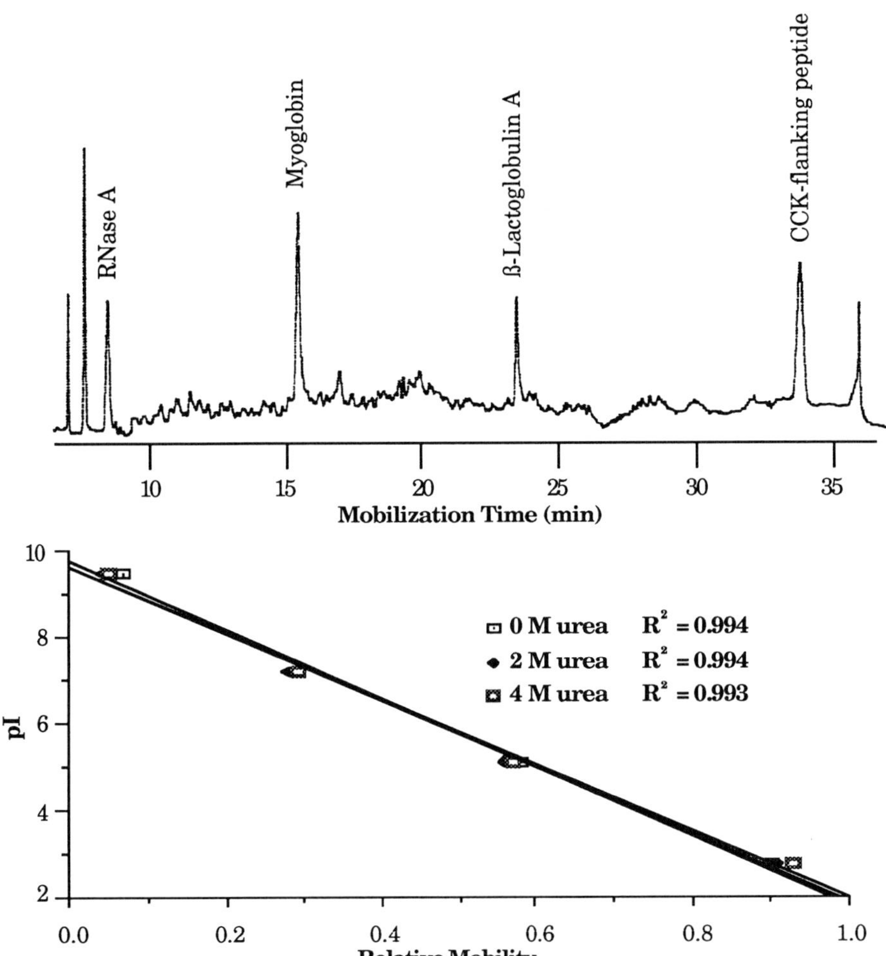

Figure 2. Deionized urea was added to both electrode buffers and ampholyte solution at concentration of 0, 2, and 4 M. The same concentration of urea was added to the markers. The IEF was performed as described in the Methods section. A electropherogram contains 4 M urea is shown. Linearity of the pH gradients with the three urea concentrations demonstrates that deionized urea has no effect on the pH gradient formation in CE-IEF.

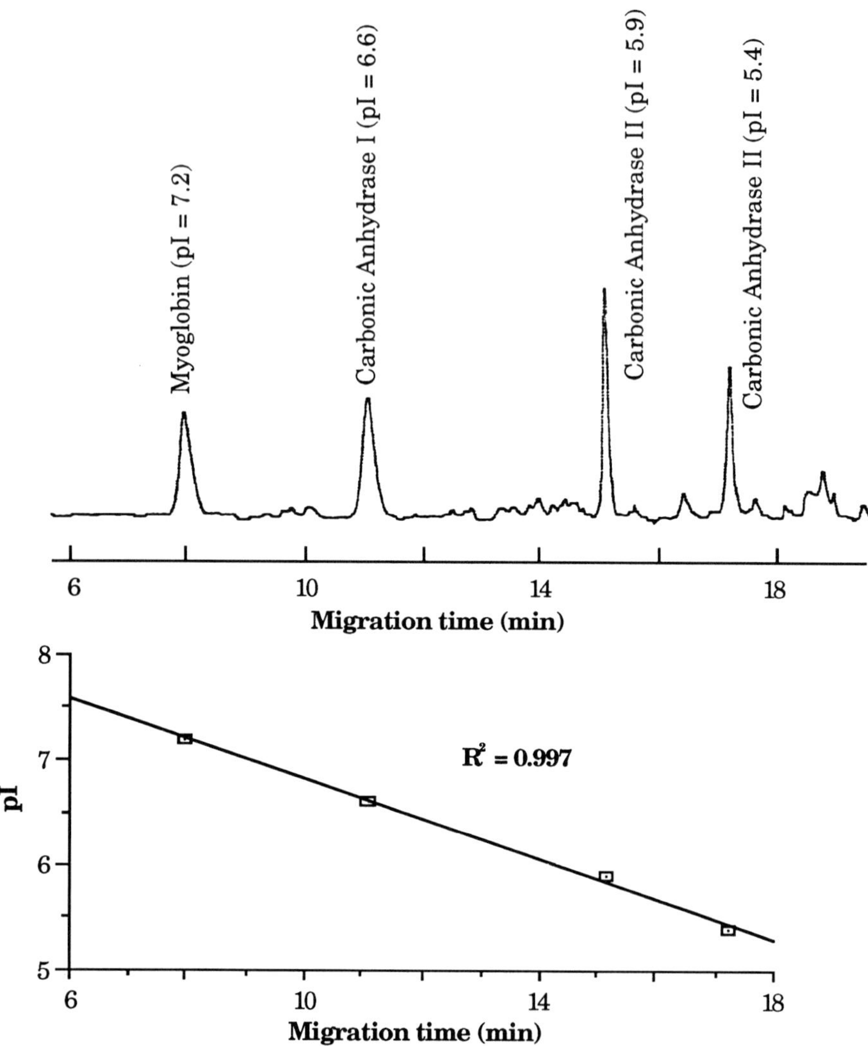

Figure 3. Linearity of a narrow range ampholyte (pH 5–8) is shown in the pI vs. migration plot. Myoglobin (pI 7.2), carbonic anhydrase I (pI 6.6), carbonic anhydrase II (pI 5.9), and carbonic anhydrase II (pI 5.4) were used under the same experimental conditions as described in the Methods section. This result indicates this experimental procedure can be used to study isoelectric focusing with narrow range ampholytes.

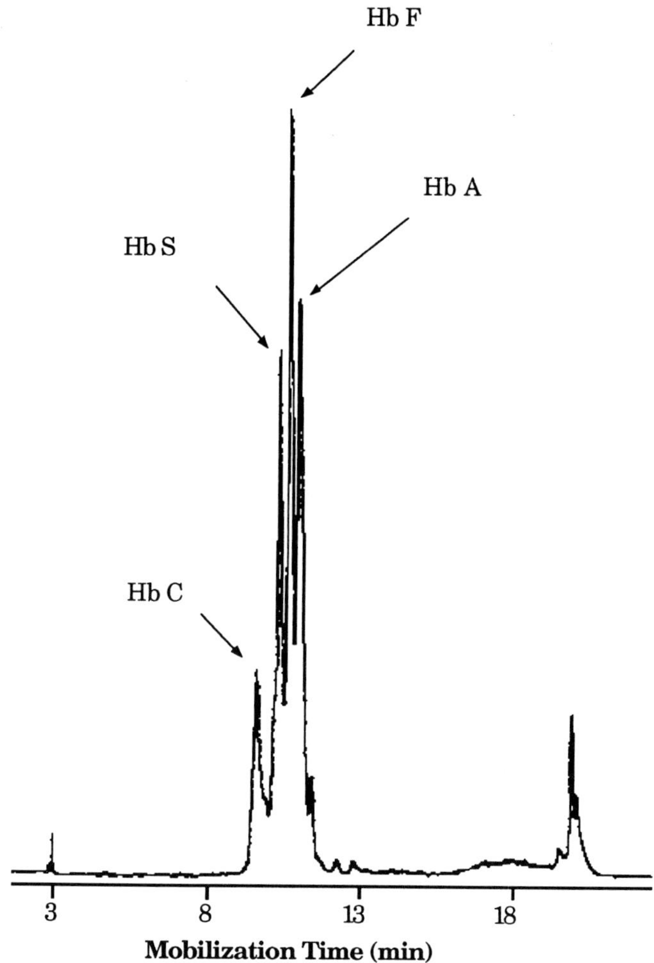

Figure 4. Human hemoglobin variants were separated by IEF as described in method section with ampholyte pH range 5-8. The major types, A, F, S, and C were well separated and the pI's were calculated to be 6.99 (6.95), 7.05, 7.17, and 7.40 (7.4), respectively (Data in the parentheses are published values (13)) .

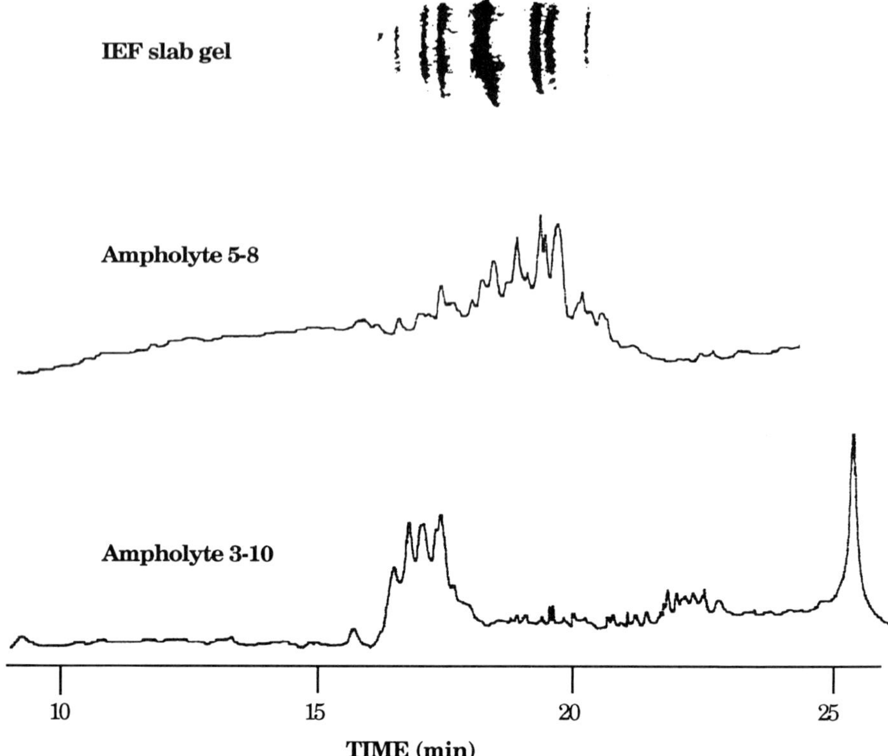

Figure 5. Recombinant monoclonal antibody in a PBS buffer was diluted in ampholyte solutions (5-8 or 3-10) to a final concentration of 50 μg/ml. This solution was loaded and focused. The electropherogram from ampholyte 5-8 is comparable to that obtained from slab gel IEF.

IV. Discussion

Chen and Wiktorowicz (12) demonstrated earlier that CE-IEF can produce a high resolution, reproducible wide range linear pH gradient that covers the entire range of the ampholyte. In this study, without any modifications, a linear pH gradient over the range of a narrow pH range ampholyte can be produced as shown in Figure 3. A direct comparison of narrow range and wide range ampholyte experiments (Figure 5) indicated better resolution was achieved with narrow range ampholyte. A simple and reliable method for hemoglobin analysis showed better than 0.1 pH unit resolution was reached by using Servalyt™ 5-8 (Figure 4).

For proteins or peptides only soluble in surfactants and denaturants, this method provides a way to analyze these samples without suffering any lost of linearity of the pH gradient. By dissolving samples, such as recombinant monoclonal antibodies, containing salts directly in the ampholyte solution, these samples can be analyzed by the CE-IEF without a desalting step.

In conclusion, this study indicated that the CE-IEF method developed by Chen and Wiktorowicz (12) has proven to be very flexible. It is clear that this CE-IEF method used on a free solution separation can be used to perform rapid high resolution analysis of proteins and peptides with various sample buffer matrices and additives.

REFERENCES

1. Burghes, A.H.M., Dunn, M.J., and Dubowitz, V., (1982).*Electrophoresis* 3, 354-363.
2. Bischoff, R., Roecklin, D., and Roitsch, C., (1992).*Electrophoresis* 13, 214-219.
3. Lessard, F. and Dion, R.,*Clin. Chem.* 35, 2116-2118.
4. Vesterberg, O. (1969) *Acta Chem. Scand.* 23, 2653-2666.
5. Hjerten, S. (1985) *J. Chromatogr.* 347, 191-198.
6. Hjerten, S., and Zhu, M. (1985) *J. Chromatogr.* 346, 265-270.
7. Hjerten, S., Liao, J., and Yao, K. (1987) *J. Chromatogr.* 387, 127-138.
8. Zhu, M., Rodriguez, R., and Wehr, T. (1991) *J. Chromatogr.* 559, 479-488.
9. Hjerten, S., Elenbring, K., Kilar, F., Liao, J., Chen, A., Siebert, C., and Zhu, M.D., *J. Chromatogr.* 403, 47-61.
10. Mazzeo, J. R., and Krull, I. S. (1991) *Anal. Chem.* 63, 2852-2857.
11. Thormann, W., Caslavska, S., and Chmelik, J., (1992) *J. Chromatogr.* 589, 321-327.
12. Chen, S.M. and Wiktorowicz, J.E., (in press) *Anal. Biochem.*
13. Righetti, P.G., and Caravaggio, T., (1976) *J. Chromatogr.* 127, 1-28.

Purification of Different Biologically Active Forms of Mouse β-Nerve Growth Factor by Immobilized Metal Ion Affinity Chromatography

Giuseppe Corona[1], Steven D. Skaper[a] and Lanfranco Callegaro

Advanced Technology Division, and [a]Department of Cell Biology, Fidia Research Laboratories, Abano Terme (PD), Italy

I. Introduction

Immobilized metal affinity chromatography (IMAC) is a protein separation method based on the interaction between proteins in solution and transition metal ions fixed to a solid support [1]. The specificity of the separation is determined by the occurrence and position of metal-coordinating residues on the protein surface. The exposed histidines and cysteines, when present, are the most important metal-coordinating ligands at neutral pH. When such ligands are in a sufficient number and are accessible, the protein can be efficiently purified from a complex mixture. The proteins produced through recombinant DNA technology can be engineered to have high affinity for the metal ion by adding a sequence of poly-histidines at the N-terminal position, and thus can be easily purified [2]. The interactions of histidines with the metal ions are sensitive to variations of polypeptide chain conformation. In fact, two histidines separated by three residues show a high degree of interaction with the metal only when they are in an α-helix, not when they are in a random coil or a β-sheet [3]. This high affinity is due to the imidazole

[1] Present anddess: Dr. Giuseppe Corona
Laboratory of CNS Studies, NINDS
National Institutes of Health
Bldg. 36/4A15
9000 Rockville Pike
Bethesda, Maryland 20892, USA

TECHNIQUES IN PROTEIN CHEMISTRY IV

groups of histidine, whose geometry is such that two groups can simultaneously bind a single metal ion in a mutidentate complex. The relationship between the protein interaction with the metal ions and its primary and secondary structures has made IMAC an attractive tool for exploring the topography of specific residues of a protein molecule [4].

Nerve growth factor (NGF) is a protein that plays a key role in the development and maintenance of neuronal functions in the peripheral and central nervous systems [5, 6]. The biologically active β subunit of the NGF complex is a 26.5 KDa basic protein composed of two identical polypeptide chains linked by strong non-covalent forces [7]. Each chain contains four surface histidines which show a high inter species conservation [8], making NGF a good candidate for IMAC chromatography. During the purification process from mouse submaxillary gland, NGF is partially modified. One modification involves cleavage of the C-terminal arginine by a B-like carboxypeptidase (des^{Arg} NGF), and another consists of the removal of the NH_2-terminal octapeptide ($bisdes^{1\text{-}8}$NGF) [9, 10]. In the latter, a specific endopeptidase present in the submaxillary gland cleaves the protein at a histidine/methionine bond, releasing the first eight aminoacids of βmNGF as an octapeptide. The preparation of mouse NGF usually contains variable amounts of this cleaved species, each present in a dimeric form but differeing in their isoelectric point and histidine content. This makes IMAC a potential technique to separate these different forms of mNGF without using any denaturing condition, in order to better understand the significance of this post-translational modification. Here we have evaluated the interaction of two different species of NGF with copper, cobalt, nickel and zinc. This procedure was applied to develop fast purification methods for different N-terminal cleaved forms of mNGF. The role of histidines in the interaction of NGF with metals also has been evaluated by diethyl pyrocarbonate modification of protein histidine residues.

II. Materials and methods

All metal chlorides and imidazole were purchased from Merck (Darmstadt, Germany). Diethyl pyrocarbonate (DEP) and hydroxylamine were from Fluka (Buchs, Switzerland). Double-distilled water was MilliQ grade (Millipore, Bedford, MA). Purified mouse NGF was obtained by the method of Mobley et al. 1971, [9] with a few modifications. Recombinant human NGF was obtained as previously described [11]. IMAC analytical runs were made using a chelating Superose column (10 x 5 mm), while a 6B IDA-Sepharose column (100 x 10 mm) was used for large-scale purifications (Pharmacia-LKB, Uppsala, Sweden). Both columns were matched to a

Pharmacia FPLC system equipped with a LCC-500 gradient controller. Reverse phase HPLC of NGF was performed using a Vydac C18 column (250 x4.6 mm) with a MilliporeWaters 600 system. Automated sequence analysis was performed on an Applied Biosystem Gas Phase sequencer, model 470A.

A. IMAC column preparation and elution

Metal ions were immobilized by loading 5 volumes of a 0.05 M solution of metal chlorides in water at a flow rate of 1 ml/min. Excess Ni(II), Zn(II) and Co(II) were washed out with 5 volumes of water, while 5 volumes of 0.1 M acetate buffer, 0.5 M NaCl, pH 4.0, were used to remove the excess Cu(II) [12]. Columns were equilibrated in buffer A. The column was subsequently loaded with 25 μg of purified rhNGF (or mNGF), washed with 5 ml of buffer A, then eluted with a 30 min gradient from 0 to 50 mM imidazole.

B. Chemical modification of histidines

Chemical modification of histidines was performed by treating rhNGF with a 25-fold molar excess of DEP in buffer A. The reaction was monitored by taking the difference spectra every 5 min between 230-350 nm using a Lambda 2 spectrophotometer (Perkin-Elmer, Norwalk, CT). The concentration of the modified histidine residues was calculated by recording absorbance at 242 nm (ε=3200, l x mol^{-1} x cm^{-1}) [13]. The reaction was complete in 30 min, and could be reversed by treatment with 20 mM hydroxylamine at room temperature for 2 hr. This was confirmed by the decrease in absorbance at 242 nm.

C. Assay of biological activity

NGF activity was evaluated by the ability to maintain neurons cultured from 8-10 -day old chicken embryos (E8-E10) dorsal root ganglion (DRG) [14]. Briefly, DRGs were dissected out, dissociated, enriched by a pre-plating step, and added to a 96-well microtiter tissue culture plate (2000 neurons/well). Wells were coated sequentially with polyornithine and laminin prior to cell seeding. Medium consisted of Dulbecco's modified Eagle's medium supplemented with 100 U/ml penicillin, 2 mM L-glutamine and 10% (v/v) heat-inactivated fetal calf serum. Surviving neurons were counted after 48 hr of culture.

III. Results and discussion

The three-dimensional structure of mouse NGF consists of a predominantly β–sheet domain with histidines located at positions 4, 8, 35 and 75. The structure of the N-terminal sequence is not available from X-ray data, but the internal histidines are all exposed on the polypeptide surface and are solvent-accessible [15]. Previous investigations concluded that DEP modification of histidines, in particular His35 and His75, prevents the binding of NGF to its receptor [16]. This indicates that histidine residues of NGF are potentially available for interaction with metal ions. In the present study, rhNGF interacted with all metal ions tested, with a Ni(II) column yielding the strongest affinity. Copper and cobalt displayed intermediate degrees of binding, while zinc had the lowest affinity, as 20 mM of imidazole was sufficient to elute the protein from this last column (Fig. 1). The interaction order, Ni(II)>Cu(II)>Co(II)>Zn(II), has already been observed for most proteins that contain exposed histidines [17]. A key role of histidines for interacting with metal interactions was demonstrated by blocking the imidazole groups with ethoxyformyl groups. The spectrophotometric analysis of the reaction of rhNGF with DEP indicated that a total of 8 histidines (4 residues per monomer) was modified, i.e., all residues in the molecule. When DEP-modified NGF was applied to a Zn(II) column, the protein eluted in the void volume (Fig. 2), indicating that histidine plays a key role in the interaction of rhNGF with Zn(II) ions. The interaction was restored after treatment with hydroxylamine.

Previous investigations have established that specific submaxillary gland proteases can produce different proteolyzed forms of mouse NGF. When purified mNGF was analyzed by IMAC, three forms were obtained after elution from a Zn-IMAC column. These species eluted between 15 and 25 mM imidazole (Fig. 3), suggesting the existence of three forms of mNGF having different affinities for zinc and thus different histidine contents. Peaks A and C eluted, respectively, at lower and higher imidazole concentrations (Fig. 3.), but both eluted as single peaks when analyzed by RP-HPLC as show in Fig. 4 (4-1 and 4-3). Conversely, the peak recovered at an intermediate imidazole concentration (Fig. 3, peak B) showed two distinct peaks of similar absorbance after RP-HPLC analysis (Fig. 4-2). N-terminal amino acid sequencing indicated that peak A (Fig. 3) was homogeneous in composition, with its N-terminal sequence starting from methionine 9 of mNGF. This N-terminal proteolyzed form has already been described as bisdes(1-8)NGF [9]. Peak 4-3 was also homogeneous and had the same N-terminal sequence as native mNGF.

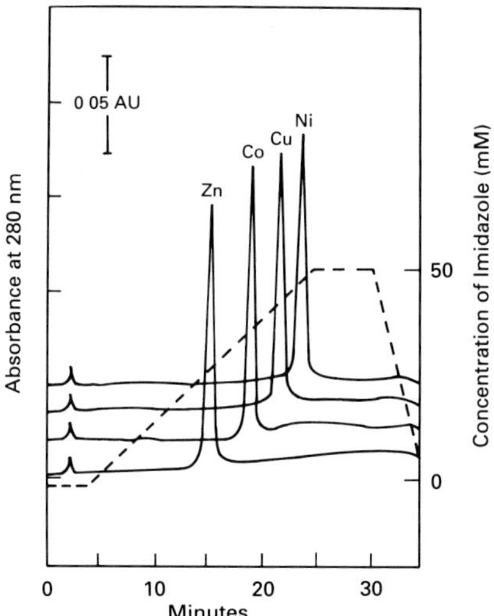

Fig. 1 Elution of rhNGF from different metal ion columns. The column used was a chelating IDA-Superose (50x5 mm) loaded with the indicated ions. All elutions were performed with the same gradient from 0 to 50 mM imidazole in buffer A in 20 min (dashed line). Flow-rate: 1 ml/min.

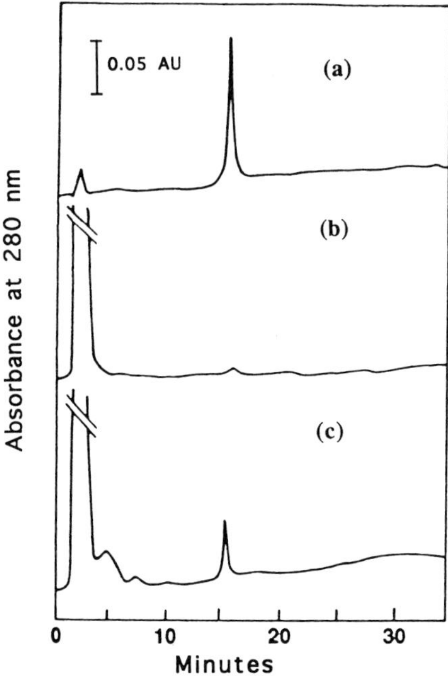

Fig. 2 IDA-Zn(II) chromatograms of modified rhNGF: (a) unmodified rhNGF, (b) DEP-modified rhNGF, (c) DEP-rhNGF after reemoval of carbethoxy groups by treatment with hydroxylamine.

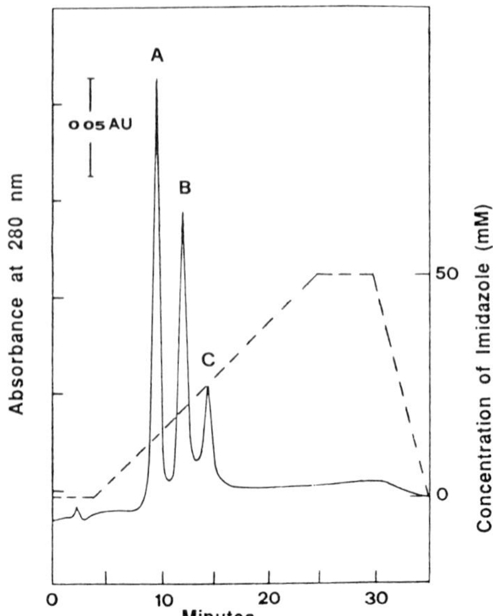

Fig. 3. Analysis by IMAC of purified mNGF. 50 ug of mNGF were loaded on a Zn(II)-Superose column (50 X5 mm). Elution was performed with a linear gradient of 0 to 50 mM imidazole over 20 min. Flow rate = 1 ml/min. The fractions indicated by A, B and C were collected.

Peaks 4-2 coincided with the retention times of peaks 4-1 and 4-3 and is most likely to be a heterodimer formed by one chain of the native monomer and one chain of the truncated N-terminal sequence (monodes$^{1-8}$NGF).

In vitro biologic assays indicated the specific activity of native NGF to be 20-fold higher than that of bisdes$^{1-8}$NGF, while intermediate values were obtained for monodes$^{1-8}$NGF (Fig. 5). Sequence data and RP-HPLC analysis indicated that the different forms of mNGF eluted according to their histidine content. Bisdes$^{1-8}$NGF, lacking 4 histidines, eluted first, while increasing retention was found for amonodes$^{1-8}$NGF and native mNGF. Mouse NGF forms cleaved at C-terminal arginine residues have been described, but no evidence for their presence was obtained here. The behavior of mNGF with the metal ion may indicate that histidines at positions 75 and 84 contribute similarly to the interaction as those at positions 4 and 8. It appears that no stabilizing interaction due to specific structural arrangements of histidines, i.e., a high-affinity metal binding site, is likely to be present [3]. Further confirmation for the existence of different dimers was also obtained by an *in vitro* biologic assay. It is important to note that the neuronal biologic activity of NGF has been associated with the dimeric form of the β subunit. These experiments

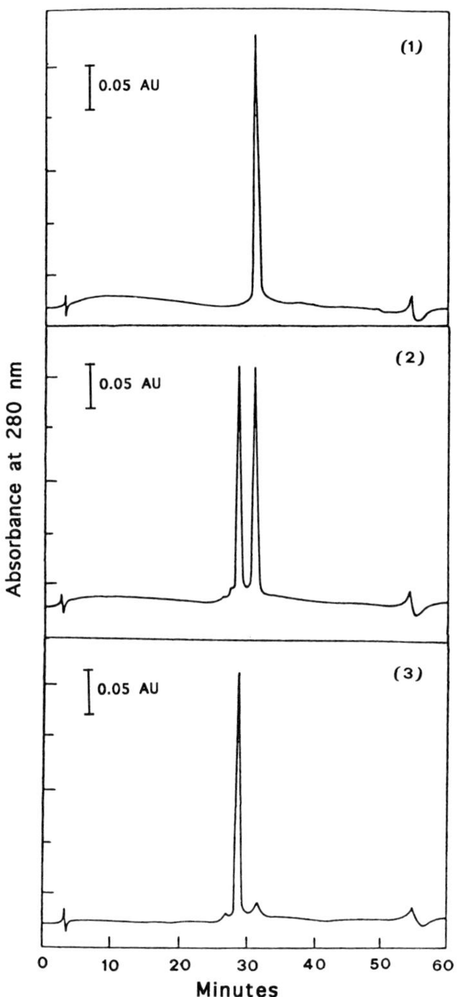

Fig. 4 RP-HPLC analysis of mNGF fractions from IMAC chromatography. The chromatographic traces reported as (1), (2) and (3) refer to fractions A, B and C of fig. 3. Conditions: reverse-phase HPLC column was Vydac C18, 7 µm 300A (250x4.6 mm), eluents were water-0.05% TFA (A) and acetonitrile-O.O5% TFA (B). The gradient was 0% B 5min., from 0 to 30% B 10 min., from 30 to 60% B 30 min. then to 90% B in 10 min. remain at 90% 5 min. and equilibrated for 10 min at 0%B. Flow rate = 1.2 ml/min.

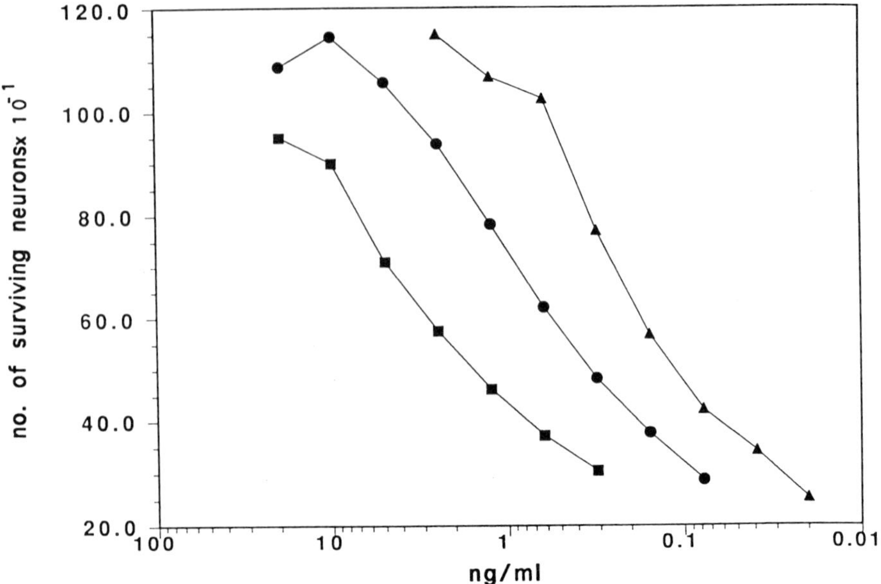

Fig. 5. Biological assay of fractions A, B and C of mNGF collected from IMAC column: ■ desocta mNGF, ● heterodimer mNGF, ▲ native mNGF.

demonstrate that the three forms express significantly different activities. Bisdes$^{1-8}$NGF was less active than both the monodes$^{1-8}$ and native NGF. These results agree with previous observations where the activity of des$^{1-8}$NGF was also assayed by measuring C-fos kinase activation in intact PC-12 cells [17]. These results contrast with previous data which indicated that removal of the N-terminal octapeptide had no effect on NGF activity (9). The N-terminal octapeptide can contribute to receptor binding or influence the bond through allosteric effects, but the reason remains unexplained.

IV. Conclusion

In conclusion, IMAC represents a useful tool for either preparative and analytic studies of NGF, or future applications for other similar proteins, e.g., neurotrophins. IMAC has been used to easily separate three different heterologous dimers of mNGF without the use of denaturing conditions, thus permitting further characterization of these cleaved species. This will lead to future knowledge of the physiological role of the highly conserved aminoterminal octapeptide. This chromatographic technique can also be applied in purification of the rhNGF from eucaryotic cell lines, or its capacity to separate isomer conformation may be used for *E. coli* NGF purification.

References

1. Porath J., Carlsson, J., Olsson, I., and Belfrage, G. (1975)*Nature* (London), 258, 598-599.
2. Hochuli, E., Bannwarth, W., Bobeli, H., Gentz, R., and Stuber, D. (1988). *Biotechnology* 80, 1321-1325.
3. Arnold, F., and Haymore, B. L., (1991). *Science,* 252, 1796-1797.
4. Hemdan, E. S., Zhao, Y-J., Sulkowski, E., and Porath, J. (1989). *Proc. Natl. Acad. Sci. USA* 86, 1811-1815.
5. Levi-Montalcini, R., (1987). *Science* 237, 1154-1162.
6. Angeletti, R. H., Bradshaw, R. H., and Wade, R. D. (1971). *Biochemistry* 10, 463-469.
7. Greene, L. A., Varon, S., Piltch, A. and Shooter, E. M., *Neurobiology*, 971, 1 , 37-48.
8. Thoenen, H., Bantlow, C., andHeumann, R. (1987). *Rev. Physiol. Biochem. Pharmacol.,* 109, 145-178.
9. Mobley, W. C., Schenkel, A., and Shooter, E. M. (1976). *Biochemistry* 15, 5543-5552.
10. Wilson, W. H., and Shooter, E. M. (1979). *J. Biol. Chem.* 254, 6002-6009.
11. Soranzo, C., Martini, I., Bigon, E., Callegaro, L., Cazzola, F., Corona, G., Skaper, S.D., and Negro, A. (1992). *In* "Growth factors of the vascular and nervous system", (Karger , ed.), p.71-79.
12. Belew, M., Yip, T. T., Anderson, L., and Ehrnstöm, R. (1987). *Anal. Biochem.* 164, 457-465.
13. Miles, E. W., (1977). *Methods Enzymol.* 43, 431-442.
14. Skaper, S. D., Facci, L., Milani, D., Leon, A., and Toffano, G., (1990), *in* "Methods in Neurosciences", (P. M. Conn, ed.) Vol. 2, pp. 17-33.
15. McDonald, N. G., Lapatto, R., Marvay-Rast, J., Gumning, J., Wlodawer, A., and Blundell, T. L. (1991). *Nature* (London) 354, 411-414.
16. Dunbar, J. C., Tregear, G. W., and Bradshaw, R. A. (1984). *J. Protein Chem.* 3, 349-356.
17. Taylor, R., Kerrigan, J. F., Longo, F. M., DeBoinsblanc, M., and Mobley, W. C. *Soc. Neurosci. Abstr.* 17, 236.7

Increasing the Antigen Binding Capacity of Immobilized Antibodies

Alvin S. Stern and Frank J. Podlaski

Department of Protein Biochemistry
Roche Research Center
Hoffmann-La Roche Inc.
Nutley, NJ 07110-1199

I. Introduction

Immunoaffinity chromatography has been used extensively for the purification of proteins via specific and reversible interactions of antibodies with specific protein antigens (1). Much research has been devoted to the preservation of the antibodies' antigen-binding activity upon immobilization to a solid support (2). An antibody, whose structure may be represented schematically by the letter "Y", consists of two F_{ab} fragments joined to a F_c region with the antigen binding sites at the ends of the F_{ab} units (3). With solid supports which couple antibodies through their primary amine groups (e.g. N-hydroxysuccinimide esters of derivatized, crosslinked agarose), antibodies may attach to the support at or near the antigen binding site, resulting in the loss of active sites on the antibody which bind antigen (Fig. 1A). This is because the four N-terminal amino groups on the antibody have a lower pK_a and are thus, more reactive than the ϵ-amino groups of lysine residues present in the antibody. In fact, the effects of attachment through these former groups would be even greater were it not for the overwhelming excess of ϵ-amino groups. This, however, leads to multi-site attachment and multiple orientations (Fig. 1A), and as a result, the antibody loses its antigen-binding activity (4,5).

A method for site-directed immobilization is the attachment of the antibody through the carbohydrate moiety in the F_c domain leading to an orientation which would potentially result in increased antibody binding capacity (6-9). Immobilization is initiated with the generation of aldehydes by the oxidation of the carbohydrate side chains of the antibody using periodate followed by the formation of stable hydrazone bonds with the hydrazide groups on a solid support. Although some studies have shown that the immunosorbents prepared by oriented coupling using hydrazide supports have greater binding capacities than those prepared by random coupling (6), others have suggested that the

TECHNIQUES IN PROTEIN CHEMISTRY IV

purported preferential F_c binding may not always occur (10). Although hydrazide chemistries work through interaction with carbohydrate moieties present predominantly on the F_c portion of the antibody molecule, recent work has shown that these moieties can be found on the F_{ab} region of antibodies as well (11). In addition, the distribution and prevalence of these carbohydrate moieties are highly antibody specific leading to variability with the hydrazide immobilization technique.

In order to improve the antigen binding capacity of antibodies bound to column matrices and to circumvent the above stated problems, Gersten and Marchalonis (12) linked antibody to *Staphylococcus* protein A (Protein A)-Sepharose. Protein A binds the F_c portion of the antibody molecule leaving the antigen specific sites free (Fig. 1B). This procedure avoids the problem of random antibody-matrix coupling inherent in binding via primary amine groups and the unpredictable antibody immobilization by the hydrazide-activated coupling chemistries. Modifying Gersten and Marchalonis' approach, by generating affinity crosslinked matrices with dimethylpimelimidate, led to immunoaffinity columns which did not leak antibody under low pH buffer elution conditions (13,14). Although Protein A-Sepharose is an efficient immunoaffinity isolation agent for human and rabbit antibodies and some monoclonal antibodies of murine and rat origin, many monoclonal antibodies of murine and rat origin bind very weakly, if at all (15). Therefore, it may not be possible to couple a high concentration of every antibody to Protein A-Sepharose. However, Protein G has a different spectrum of binding affinities than Protein A (2), and recombinant Protein G (with its serum albumin binding

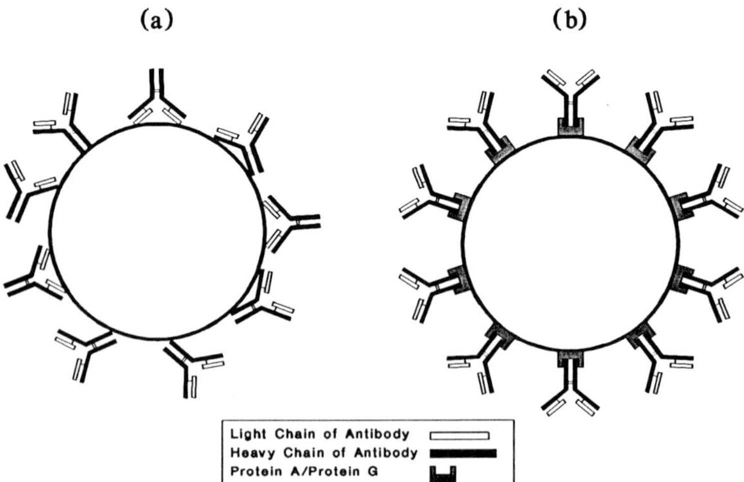

Figure 1. Orientation of immobilized antibodies on solid supports. Antigen binding sites are lost due to random orientation by direct coupling to beads (*panel a*) whereas site-directed immobilization with crosslinking to Protein A- or Protein G-Sepharose increases available antigen binding sites (*panel b*).

domain deleted) immobilized to agarose is now commercially available. The present study reports a procedure that yields complete and stable crosslinking of antibodies to Protein G-Sepharose while retaining maximal antigen-binding activity of the antibody when compared to antibodies immobilized via activated agarose.

II. Materials and Methods

A. *Antibody Generation and Purification*

A rat immunized with partially purified Interleukin 12 (IL-12) produced serum antibodies that neutralized IL-12 bioactivity (16). Hybridoma cells secreting monoclonal antibodies to IL-12 were produced by fusing NSO cells with splenocytes recovered from this rat (17). Antibodies were purified from ascites fluid by affinity chromatography on a GammaBind G-agarose column (Genex, Gaithersburg, MD).

B. *Generation of Antibody-Coupled Columns*

1. Preparation of anti-IL-12 antibody linked to Protein G-Sepharose

Recombinant Protein G-Sepharose 4 Fast Flow (Protein G-Sepharose, Pharmacia LKB Biotechnology Inc.) was washed with cold water and equilibrated in Dulbecco's phosphate-buffered saline without calcium chloride or magnesium chloride (PBS). Anti-IL-12 monoclonal antibody 20C2 (1 mg/ml) was added to an equal volume of Protein G-Sepharose and mixed for 30 minutes at room temperature. The slurry was centrifuged and unbound antibody was quantified by UV absorbance at 280 nm, assuming an extinction coefficient of 1.40 for 1 mg/ml of IgG antibody in a 1 cm path length (6). The amounts of antibody bound to the column was determined by subtracting the mg of unbound antibody from the mg of antibody added to the gel. The concentration of antibody coupled to the gel was 1 mg/ml. Bound antibody was crosslinked to the Protein G-Sepharose by collecting the gel on a sintered glass filter and washing it with 0.2 M triethanolamine in 0.1 M borate buffer, pH 9.0. The gel was then transferred from the sintered glass filter to 5 vol of 0.2 M triethanolamine in 0.1 M borate buffer, pH 9.0. Dimethylpimelimidate was added to a final concentration of 0.2 - 50 mM (as described under Results). The pH of this crosslinking solution was readjusted to 8.2 with concentrated NaOH. The gel crosslinking solution was continuously agitated for 45 min at which time the reaction was stopped by first washing the gel on a sintered glass filter with 20 vol of 50 mM ethanolamine, pH 8.2, and subsequently mixing the gel with 10 vol of 50 mM ethanolamine, pH 8.2, for 5 min. Finally, the gel was washed extensively with PBS and stored in that buffer containing 0.02% sodium azide.

The crosslinking efficiency was monitored by boiling samples of beads taken before and after coupling, and separating non-crosslinked antibody by SDS-PAGE (18).

2. Coupling to hydrazide activated supports

20C2 was coupled to Affi-Gel Hz gel (Bio-Rad Laboratories) following the procedure described by the manufacturer (19). Unbound antibody was quantified, after dialysis against 1000 vol of PBS, by UV absorbance at 280 nm as described for Protein-G-Sepharose coupling. The concentration of antibody coupled to the gel was 1 mg/ml. The column was stored in PBS containing 0.02% sodium azide.

3. Coupling to N-hydroxysuccinimide esters on derivatized supports

20C2 was coupled to Affi-Gel 10 (Bio-Rad Laboratories) following the procedure described by the manufacturer (20). The amount of bound and unbound antibody was calculated in the same manner as described for Protein G-Sepharose. The concentration of antibody coupled to the gel was 1.1 mg/ml. The column was stored in PBS containing 0.02% sodium azide.

C. IL-12 Immunoaffinity Chromatography

Columns containing immobilized antibody (0.5-1.0 ml) were gravity packed and equilibrated with 10 vol of PBS. All resins were mock eluted with 4 vol 0.2 N acetic acid, 0.15 M NaCl, pH 2.8, after incubation of the resin for 5 min with an equal volume of this buffer. After re-equilibration into PBS, the resins were incubated for 5 min with an equal volume of BSA (3 mg/ml) to block nonspecific binding sites. The resins were washed with 10 vol of PBS followed by 5 vol of 0.2 N acetic acid, 0.15 M NaCl, pH 2.8, until the absorbance at 280 nm was <0.01. After re-equilibration of the resin into PBS, recombinant IL-12, at a concentration equivalent to the theoretical capacity of the column, was applied to the resins. After washing the columns with PBS until the optical density of the column eluant stabilized, IL-12 was eluted with 0.2 N acetic acid, 0.15 M NaCl, pH 2.8. Protein concentration in individual samples was quantified by injecting microliter volumes into a fluorescamine protein assay system (21). IL-12, whose absolute concentration was determined by amino acid analysis, was used as the calibration standard.

III. Results

IL-12 is a 75-kDa heterodimeric cytokine which has been shown to cause the proliferation of activated T and NK cells, to enhance the lytic activity of NK cells, and to induce interferon-γ production by resting and activated T and NK cells (16,22). In an effort to develop an immunoaffinity process for the large-scale production of recombinant IL-12, a variety of chemistries for coupling an anti-IL-12 monoclonal antibody, 20C2, to a solid support, was investigated. The goal in these studies was the construction of an immunosorbent which utilized a coupling mechanism that immobilized the antibody to the solid-phase support without adversely affecting the antibodies' function to capture antigen. Since coupling antibodies to hydrazide-derivatized solid supports has been reported to not always successfully accomplish F_c-site-specific immobilization of monoclonal antibodies (10), a Protein A-Sepharose coupling was proposed for site-directed immobilization. However, because 20C2 is a rat IgG, its affinity for Protein A was low and Protein G-Sepharose, which binds 20C2, was substituted. Once the antibodies were bound to the Protein G-Sepharose, 20C2 was crosslinked to the Protein G via a bifunctional coupling reagent, dimethylpimelimidate. Previously it had been shown that the conditions for the crosslinking reaction were critical for the coupling of antibodies to Protein A-Sepharose (14). In addition, a potential problem with dimethylpimelimidate, whose binding groups couple to free amino groups, may be that the high concentration of reagent could result in crosslinking intrachain lysine residues that are in close proximity to the antigen-binding site. In an effort to minimize this reaction, the critical amount of dimethylpimelimidate was titrated down to a sufficient concentration to provide an antibody-Protein G-Sepharose complex stable to repeated acid elution. After 20C2 was bound to Protein G-Sepharose

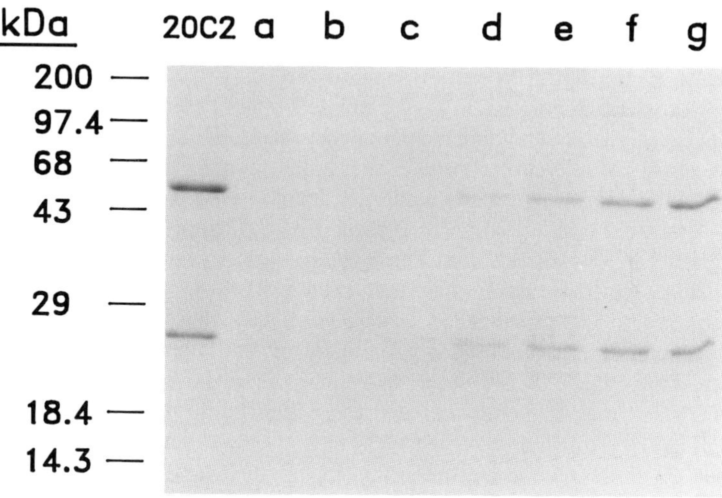

Figure 2. Crosslinking of 20C2 to Protein G-Sepharose. 20C2 (*lane 20C2, 3 μg*) was bound to Protein G-Sepharose (as described under Materials and Methods), crosslinked with dimethylpimelimidate at reagent concentrations of 20 (*lane a*), 10 (*lane b*), 5 (*lane c*), 2 (*lane d*), 1 (*lane e*), 0.5 (*lane f*), and 0.2 (*lane g*) mM. The crosslinking efficiency was monitored by boiling samples of beads in Laemmli sample buffer (18) and separating non-crosslinked antibody under reducing conditions by SDS-PAGE on a 12% slab gel which was stained with Coomassie Blue R-250. The molecular weights indicated in the margins were estimated from standards electrophoresed in a parallel lane.

Table I. Binding and Recovery of IL-12

Method	% IL-12 bound[a]	%IL-12 recovery[b]
Protein G-Sepharose	79	100
Hydrazide chemistry	75	60
N-Hydroxysuccinimide coupling	58	48

[a]The percentage of IL-12 bound to a resin was calculated by obtaining the difference between the amount of IL-12 loaded on a column and the amount of IL-12 which was not bound by that column divided by the amount of IL-12 loaded. This number was then multiplied by 100 to obtain the percentage of IL-12 bound.

[b]The percentage of IL-12 recovered after elution from a column was calculated by the amount of IL-12 recovered from that column divided by the amount of IL-12 which was bound. This number was then multiplied by 100 to obtain the percentage of IL-12 recovered.

(as described under Materials and Methods), aliquots of the resin were allowed to react with varying concentrations of dimelthylpimelimidate. Samples were boiled under denaturing and reducing conditions and monitored for released antibody by SDS-PAGE (Fig. 2). At concentrations of 5 to 50 mM dimethylpimelimidate, the bound antibody was completely crosslinked to Protein G. Therefore, since 5 mM dimethylpimelimidate was found to be the minimal concentration of reagent needed, this optimal amount was used for the preparation of the 20C2-Protein G-Sepharose resin.

20C2 crosslinked to Protein G-Sepharose bound 79% of the applied IL-12 and consistently bound a higher percentage of IL-12 than 20C2 coupled to agarose via N-hydroxysuccinimide esters and equivalent amounts compared to 20C2 immobilized on hydrazide-activated agarose (Table I). The recovery of protein after acid elution from 20C2-Protein G-Sepharose was greater than 2-fold higher than that eluted from the N-hydroxysuccinimide coupled resin and greater than 1.6-fold higher than that recovered from the 20C2-hydrazide-activated resin. While it is not clear why the hydrazide and N-hydroxysuccinimide resins bind a percentage of IL-12 irreversibly, the higher recovery from the 20C2-Protein G-Sepharose suggests that this type of immunosorbent may be better for generating resins with the ability to bind and release IL-12.

Based on these recoveries, antibody efficiency was determined (Table II) since in functional terms recovery of IL-12 following column elution is of more pertinence than mere antibody binding capacity. Assuming that 20C2 is divalent (3), a 100% efficient antibody will bind two IL-12 molecules, which is equivalent to 1 mg IL-12 per mg 20C2 (16). The 20C2-Protein G-Sepharose was consistently more efficient corresponding to a resin with 79% functional antibody which is 192% and 76% higher than the N-hydroxysuccinimide derivatized and hydrazide-activated resins, respectively.

Table II. Antigen-Binding and Efficiency of Immobilized Antibody

Method	Antigen-binding capacity[a]	Antibody efficiency[b]
Protein G-Sepharose	.790	79%
Hydrazide chemistry	.448	45%
N-Hydroxysuccinimide coupling	.269	27%

[a]The antigen-binding capacity of a resin was calculated by the amount (mg) of IL-12 recovered from a 1 ml column divided by the amount (mg) of anti-IL-12 antibody (20C2) immobilized to the solid support.

[b]Antigen-binding capacity was multiplied by 100 to obtain antibody efficiency. Antibody efficiency is the percentage of active antibody where 100% active antibody is 2 mol of IL-12 bound per mol of antibody or 1 mg IL-12/mg antibody.

IV. Conclusions

In terms of site-directed immobilization of proteins, the hydrazide chemistry used for coupling the 20C2 antibody increased antibody efficiency 1.6-fold over resin prepared by N-hydroxysuccinimide-activated beads. However, pre-coupling oxidation of the antibody with the strong oxidant, sodium periodate, can reduce or eliminate the antigen-binding activity of the molecules and cause precipitation of the protein. In addition, the periodate reaction is light sensitive and excess reagent must be removed by desalting before the reaction with hydrazide-activated supports. For these reasons, coupling by means of carbohydrate moieties found on the antibody can be inconvenient and inefficacious.

The 20C2-Protein G-Sepharose resin was almost twice as efficient as the hydrazide chemistry resin and 3-fold more efficient than the 20C2-N-hydroxysuccinimide coupled beads. An additional advantage of the Protein G-Sepharose coupling methodology is the ability to directly bind antibodies from cell culture supernatant solutions or ascites to produce an immunoaffinity column. In so doing, the antibody would not be subjected to denaturing elution conditions in purification of the antibody and it would consequently retain more antigen-binding activity thereby producing a more efficient immunoaffinity column.

Although the results described here are specific for the anti-IL-12 antibody, 20C2, we have also utilized this methodology for an IL-2-specific monoclonal antibody. This mouse IgG molecule did not bind its antigen when coupled to N-hydroxysuccinimide-activated resins, but did bind IL-2 when the antibody was bound and crosslinked to Protein G-Sepharose. Therefore, site-directed immobilization of antibodies on Protein G-Sepharose may be a useful, general technique for the preparation of immunoaffinity resins.

References

1. Chase, H.A. (1983) *Chem. Eng. Sci.* **39**, 1099-1150.
2. Harlow, E., and Lane, D. (1988). *In* "Antibodies: A Laboratory Manual," pp. 511-522.
3. Valentine, R.C., and Green, N.M. (1967) *J. Mol. Biol.* **27**, 615-617.
4. Cuatrecasas, P. (1970) *J. Biol. Chem.* **245**, 3059-3065.
5. Cress, M.C., and Ngo, T.T. (1989) *Am. Biotechnol. Lab.* **7**, 16-19.
6. Hoffman, W.L., and O'Shannessy, D.J. (1988) *J. Immunol. Methods* **112**, 113-120.
7. Matson, R.S., and Little, M.C. (1988) *J. Chromatogr.* **458**, 67-77.
8. Turková, J., Petkov, L., Sajdok, J., Kás, J., and Benes, M.J. (1990) *J. Chromatogr.* **500**, 585-593.
9. Domen, P.L., Nevens, J.R., Mallia, A.K., Hermanson, G.T., and Klenk, D.C. (1990) *J. Chromatogr.* **510**, 293-302.
10. Highsmith, F., Regan, T., Clark, D., Drohan, W., and Tharakan, J. (1992) *BioTechniques* **12**, 418-423.
11. Madurawe, R.D., Orthner, C.L., Tharakan, J., Highsmith, F.A., Drohan, W.N., Velander, W.H. (1991) *J. Chromatogr.* **558**, 55-70.
12. Gersten, D.M., and Marchalonis, J.J. (1978) *J. Immunol. Methods* **24**, 303-309.
13. Schneider, C., Newman, R.A., Sutherland, D.R., Asser, U., and Greaves, M.F. (1982) *J. Biol. Chem.* **257**, 10766-10769.
14. Sisson, T.H., and Castor, C.W. (1990) *J. Immunol. Methods* **127**, 215-220.
15. Langone, J.J. (1982) *J. Immunol. Methods* **55**, 277-296.
16. Stern, A.S., Podlaski, F.J., Hulmes, J.D., Pan, Y.-C.E., Quinn, P.M., Wolitzky, A.G., Familletti, P.C., Stremlo, D.L., Truitt, T., Chizzonite, R., and Gately, M.K. (1990) *Proc. Natl. Acad. Sci. USA* **87**, 6808-6812.
17. Chizzonite, R., Truitt, T., Podlaski, F.J., Wolitzky, A.G., Quinn, P.M., Nunes, P., Stern, A.S., and Gately, M.K. (1991) *J. Immunol.* **147**, 1548-1556.
18. Laemmli, U.K. (1970) *Nature* (London) **227**, 680-685.
19. Bio-Rad Laboratories. Affi-Gel Hz Immunoaffinity Kit Instruction Manual.
20. Bio-Rad Laboratories (1982) Activated affinity supports: Affi-Gel 10 and Affi-Gel 15. Bulletin 1085.
21. Stern, A.S., and Lewis, R.V. (1985) *In* "Research Methods in Neurochemistry, Vol. 6" Marks, N., and Rodnight, R., eds.) pp. 153-193.
22. Kobayashi, M., Fitz, L., Ryan, M., Hewick, R.M., Clark, S.C., Chan, S., Loudon, R., Sherman, F., Perussia, B., and Trinchieri, G. (1989) *J. Exp. Med.* **170**, 827-845.

SECTION VII

Peptide Mapping

MICROPREPARATIVE CAPILLARY ELECTROPHORESIS (MPCE) AND MICROPREPARATIVE HPLC OF PROTEIN DIGESTS

James W. Kenny, J.I. Ohms, and Alan J. Smith

Beckman Center, Stanford University Medical Center
Stanford, CA 94305

I. Introduction

Capillary electrophoresis (CE) has a number of strengths that make it an attractive analytical technology for a variety of applications. Those strengths are sensitivity, speed of analysis, high resolution, flexibility of pH and the ability to utilize a variety of separation matrices. In the case of peptide mixtures, the high resolution capability of CE has recently been extended from a purely analytical technique to a micropreparative technique (1-3). Two approaches have been utilized to achieve this goal: a single collection from a large diameter capillary (1) or multiple collections from an analytical capillary (2,3). Pure peptides have been isolated in sufficient quantity for characterization by N-terminal sequence analysis, mass spectrometry and amino acid analysis. However, most of these studies have been performed on relatively simple peptide mixtures (2) or on specifically labeled peptides from a mixture of unlabeled peptides (3).

The routine use of micropreparative capillary electrophoresis (MPCE) is potentially limited by several factors: i) the requirement for relatively high sample concentrations, ii) selective losses of individual peptides from a mixture due to irreversible adsorption to the capillary wall, iii) background contamination at the high sensitivity levels, iv) adsorption losses of all peptides at high sensitivity levels, and v) losses due to handling small amounts of the isolated peptides.

To a certain extent these problems are shared by the most frequently used preparative separation technique for peptides, reverse phase HPLC, where recoveries are severely limited at low

TECHNIQUES IN PROTEIN CHEMISTRY IV

digest load levels by these latter two constraints. We find that successful sequence characterization of peptides recovered from separations carried out on narrow-bore (2.1 mm) HPLC columns is limited to load levels above 50 pmols. At load levels below this amount we are rarely able to obtain usable protein sequence information.

Unfortunately the availability of greater than 50-100 pmol of protein for protein sequence characterization is becoming an increasingly rare event. When such material is available, it is frequently bound to an immobilized support such as PVDF or nitrocellulose after gel electrophoresis of a partially purified preparation. The utilization of *in situ* protease digestion of these samples (4,5) is frequently incomplete, resulting in the recovery of peptides at much lower than the expected levels. Consequently, we are finding an increasing need to separate and characterize digests at or below the 50 pmol level. Due to our lack of success in recovering peptide sequences from reverse phase HPLC at these levels, we have continued to investigate MPCE as a possible solution to these problems.

In our earlier studies (1) we demonstrated the ability to separate approximately 100 pmol of a β-lactoglobulin digest on a 200 μ capillary, thereby successfully characterizing individual peptides. However, our recoveries were variable (from 20-70%), as judged by N-terminal sequence analysis. This report will show that we have successfully stabilized peptide recoveries and have obtained sequences from low picomole load levels of complex mixtures using both large diameter 150 μ single capillary separations and by multiple separations on a single 50 μ analytical capillary.

II. Methods

A tryptic digest of β-lactoglobulin was prepared following the method of Stone *et al* (6). Protein was dissolved in 8 M urea then diluted 4-fold with 100 mM ammonium bicarbonate, pH 8.5. Trypsin was added to 4% followed by a 24 hr incubation at 37°C. The reaction was stopped with addition of acid, and stored frozen.

Peptides were loaded on to an Applied Biosystems, Inc. C_{18} reverse phase HPLC narrow-bore column (Aquapore OD-300, 7 μ, 2.1 x 220 mm) that was equilibrated in 5% Buffer B. Peptides were separated with a linear 45 min gradient extending from 5% B to 55% B (Buffer A contained 0.1% TFA in water; Buffer B, 70% acetonitrile, 0.085% TFA). The flow rate was 200 μl/min. Peaks were monitored at 210 nm. Fractions were hand collected to assure optimal peak pooling, then stored frozen.

For amino acid analysis, samples were acid hydrolyzed (in 6 N HCl, 0.1% phenol, 24 hrs, 110° C, *in vacuo*) and analyzed on a Beckman 6300 amino acid analyzer.

Peptides were sequenced on an Applied Biosystems, Inc. 475A or 473 protein sequencer in the presence of 3 mg Polybrene™ and NaCl.

Micropreparative CE on the Applied Biosystems, Inc. 270A-HT was performed according to the manufacturer's specifications using 50 mM phosphate, pH 2.7 in place of 20 mM citrate, pH 2.5 (3). A tryptic digest was diluted to 4 nmol/100 μl with water and vacuum loaded for 15 secs. Fractions (0.5 mins) were collected into 10 μl of 50 mM phosphate, pH 2.7 and ethylene glycol (7:3).

The analytical and micropreparative separations on the P/ACE 2000 CE (Beckman Instruments) were as described previously (1) with the exception that the inner diameter of the preparative capillary was 150 μ in this study.

III. Results and Discussion

In order to quantitate recoveries from the CE it was necessary to calibrate the pressure loading capabilities of the P/ACE unit. This was achieved by pressure injection of the sample for a fixed period followed by low pressure delivery of the column contents into a collection vial for subsequent amino acid analysis. The data (Table I) showed that there was good agreement between the calculated sample load based on injection time and the amount of recovered digest. This also suggested that little or no nonspecific irreversible column adsorption occurred at or above the 8.5 pmol level, which was the practical limit of the amino acid analyzer detection system.

Table I. Digest recoveries from CE by amino acid analysis

Initial load (pmol)	Recovery (by AAA) (pmol)	*N* (separations)
50	57 ± 13	4
25	25 ± 8.0	6
8.5	8.1 ±1.9	6
4.3	n.d.	3

n.d. not determined

After calibration, a series of MPCE separations were performed over decreasing load levels to ascertain the lowest level at which sequencable amounts of peptides could be recovered. Since it was impractical to sequence each peak from these separations, we selected 3 peptides from the mixture (Fig. 1). Representative recoveries are shown in Table IIA. For the MPCE of later eluting peaks, a pre-electrophoresis was carried out before the initiation of collection (1). This made it difficult to predict the exact migration times of specific peptides prior to sequencing. To compare MPCE-recoveries with HPLC, specific peaks were selected at each load level for yield calculations. Representative sequence recoveries from HPLC at differing load levels are shown in Table IIB.

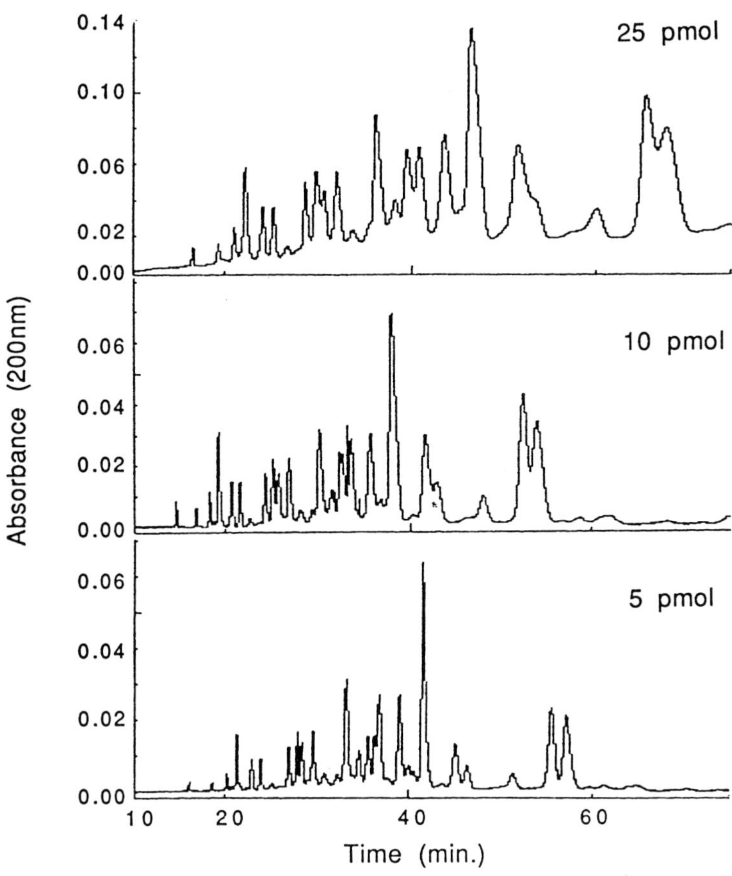

Figure 1. MPCE-P/ACE separation of peptide digest. MPCE chromatographs of β-lactoglobulin tryptic peptide digest load amounts from 25 pmol to 5 pmol are illustrated. MPCE separation conditions are summarized in Methods.

Table II. Sequence recovery from MPCE and HPLC separations

(A) MPCE			(B) HPLC		
peak	N-terminal sequence	% yield	peak	N-terminal sequence	% yield
25 pmol load			250 pmol load		
1	IIAEK	75	1	IDALN	36
2	GLDIQ	66	2	TPEVD	24
3	LIVTQ	60	3	LIVTQ	30
10 pmol load			100 pmol load		
1	IDALN	80	1	IDALN	16
2	TPEVD	44	2	ALPMH	15
3	LIVTQ	70	3	LIVTQ	10
5 pmol load			50 pmol load		
1	IDALN	66, 70	1	IDALN	7
2	xxxxxx	0	2	ALPMH	5
3	xxxxxx	0	3	VYVEE	4

The results clearly demonstrate that the recovery of sequencable peptides from MPCE is superior to HPLC at load levels below 50 pmol. The yield of recovered sequence is greater than 66% at load levels above 10 pmol. At load levels of 5 pmol, we have experienced variable recoveries (0-70%). The presence of ethylene glycol and phosphate in the collection medium (1) contributes glycine and serine at the 5 pmol level and can interfere with sequence calling at the low pmol level. The amino acid analysis recovery data suggests that the variable recovery of sequencable peptides at less than 10 pmol is not due to capillary adsorption.

Further support for this conclusion was demonstrated by MPCE separations on the ABI 270A-HT. This protocol calls for the collection of 0.5 minute fractions across the complete electrophoretic separation on a 50 μ analytical capillary. Since capillaries of this size can only hold low to subpicomole amounts of digest per injection, the collection regimen must be repeated multiple times to obtain sequencable amounts of peptides. We collected fractions from 30 separate injections of 1.3 and 0.7 picomoles to give combined separation levels of 40 and 21 pmol respectively. Peptide sequence recoveries from these fractions were in the 20-40% range. The fact that sequencable amounts of peptides were recovered from multiple loads of subpicomole amounts of digests strongly implies that recoveries were quantitative at the individual load levels. The less than quantitative recoveries of the pooled fractions were probably the result of losses associated with sample handling at these low levels. Alternatively, since recovery was determined by N-terminal sequence analysis, low recoveries could be the result of N-terminal blockage of the separated peptides.

One of the strengths of MPCE is that the resulting fractions can be screened for peptide content prior to selection for sequencing, i.e., fractions containing single components. The 270A-HT CE can be programmed automatically to screen fractions since the separation is also performed on the analytical column. The preparative column cartridge on the P/ACE CE has to be replaced with an analytical column before screening. The lower limit for screening fractions from an MPCE run appears to be about 5 pmol of loaded digest when monitored at 200 nm due to sample dilution during collection.

IV. Conclusion

Previously we had demonstrated that sequencable quantities of peptides could be recovered from a single MPCE separation of 100 pmol of digest. We have extended these findings and demonstrated the successful recovery of peptides at the low picomole level. These recoveries, as characterized by N-terminal sequence analysis, are in the 60-70% range at the 5-10 pmol load level. These studies demonstrate that MPCE is an effective technique for isolating sequencable quantities of peptides at much lower levels than are currently successful by HPLC

Furthermore, the successful recovery of peptides from multiple separations of subpicomole loadings would strongly suggest that recovery from MPCE is quantitative at these levels. However, our ability to characterize isolated peptides from these load levels is severely restricted. We are currently unable to screen fractions by analytical CE prior to sequencing due to the limited sensitivity of UV detection. The recent introduction of fluorescent detection offers a potential alternative in this area. In addition, we are restricted both by our limited ability to handle small quantities of isolated peptides and by the current N-terminal sequencing sensitivity of approximately 1 picomole.

REFERENCES

1. Smith, A. J., Kenny, J. W., and Ohms, J.I. (1991) *Techniques in Protein Chemistry III* (Ruth H. Angeletti, ed.) pp.113-120, Academic Press, San Diego.
2. Bergman, T. and Jornvall, H. (1991). *Techniques in Protein Chemistry III* (Ruth H. Angeletti, ed.) pp. 129-134, Academic Press, San Diego.
3. *Biosystems Reporte*r (Mar. 1992). Issue #15, Applied Biosystems Inc., Foster City, CA.
4. Aebersold, R., Leavitt, Saavedra, R.A., and Hood, L.E. (1987) *Proc. Natl. Acad. Sci., USA*, **84**, 6970-6974.
5. Tempst, P., Link, A.J., Rivere, L.R., Flemming, M., and Elicone, C. (1990). *Electrophoresis*, **11**, 537-553.
6. Stone, K., Lo Presti, M., Crawford, J.M., DeAngelis, R., and Williams, K.R. (1989) *A Practical Guide to Protein and Peptide Purification for Microsequencing* (Paul T. Matsudaira, ed.) pp 31-47, Academic Press, San Diego.

Peptide Mapping of 2-D Gel Proteins by Capillary HPLC

Susan C. Wong, Christopher Grimley, Allan Padua,
James H. Bourell, and William J. Henzel

Department of Protein Chemistry, Genentech, Inc.,
South San Francisco, CA 94080-4990

I. Introduction

Two-dimensional gel electrophoresis is a high resolution technique for resolving complex protein mixtures. This technique has been used to follow cellular events, protein purification, and recombinant protein expression. Proteins resolved by 2-D gel electrophoresis can be electroblotted onto inert supports such as PVDF membranes for automated sequence analysis (1). These supports also allow long term storage of 2-D gel proteins.

A significant number of intracellular proteins are shown to be N-terminally blocked, which necessitates removal of the blocking group or fragmentation of the protein to obtain protein sequence information. Although PVDF membranes have significant advantages as a support for sequence analysis of 2-D gel proteins, they have not been the membrane of choice for performing *in situ* enzymatic digestions. Electroblotting onto nitrocellulose, electroelution, or in-gel digestion (2-4) methods have produced better yields of peptide fragments than *in situ* digestion on PVDF membranes (5-8). However, these methods all have disadvantages. A 2-D gel can resolve hundreds of proteins, so the spot of interest must be known and excision of a small gel spot for in-gel digestion and electroelution is difficult. Electroblotting onto nitrocellulose also requires immediate identification of the spots of interest since nitrocellulose becomes brittle on storage.

One way to improve the yield from digestions on PVDF is to modify the protein. Several studies have been published that describe alkylation of cysteine on proteins electroblotted on PVDF (9-10). Recently, Iwamatsu combined carboxymethylation with *in situ* digestion of proteins electroblotted onto PVDF. This resulted in complete digestion of several proteins at the 50 pmol level (11). We have developed a micro method of *in situ* digestion on PVDF membranes for 2-D gel proteins that extends this methodology to the 5 pmol level. Peptides generated by the enzymatic cleavage are separated on reversed-phase capillary HPLC and characterized by protein sequencing and on-line HPLC electrospray mass spectrometry (LCMS) analysis.

TECHNIQUES IN PROTEIN CHEMISTRY IV

II. Materials and Methods

Two-Dimensional Electrophoresis

E. coli were lysed by sonication in lysis buffer containing 8 M urea, 2% Triton X-100, 2% 2-mercaptoethanol, and 8 mM phenylmethylsulfonylfluoride (PMSF). The solubilized proteins were centrifuged at 12,000 *g* in an Eppendorf microfuge and the supernatant was diluted 1:1 with the sample buffer recommended by Pharmacia (8 M urea, 2% 2-mercaptoethanol, 2% Pharmolyte 3-10, 2% Triton X-100 and 0.003% bromophenol blue. Samples were centrifuged prior to application to minimize streaking. Sample volume was kept between 40-60 µl containing 100-200 µg of total protein.

Two dimensional gel electrophoresis was performed on a Pharmacia Multiphor II electrophoresis apparatus using precast pH 4-6 immobilized gradient (IPG) strips for the first dimension and gradient 8-18% SDS gels for the second dimension. The gels were run following the Pharmacia protocol. Proteins were electroblotted onto ProBlott (Applied Biosystems) or Immobilon-PSQ (Millipore) membranes using 10 mM 3-(cyclohexylamino)-1-propane sulfonic acid (CAPS), pH 11.0, containing 20% methanol for 45 min. at 250 mA constant current in a BioRad trans-blot transfer cell (1). The PVDF membrane was stained with 0.1% Coomassie blue in 50% methanol for 1 min. and destained for 2-3 min. with 10% acetic acid in 50% methanol. The membrane was thoroughly washed with water and air dried prior to storage at –5°C.

In Situ Alkylation and Digestion

PVDF membranes containing individual electroblotted two-dimensional gel spots were wetted with 1-3 µl of methanol and transferred to a microcentrifuge tube. The electroblotted proteins were reduced with 100 µl of 0.5 M Tris-HCl (pH 8.5), 6 M guanidine-HCl, 5 mM EDTA, 10% acetonitrile and 7 mM DTT at 45°C for 1 hr. The solution was cooled to room temperature and 10 µl of 200 mM iodoacetic acid in 0.5 M NaOH was added. The solution was kept in the dark for 20 min. at room temperature. The blots were washed three times with 500 µl of 10% acetonitrile and then incubated in 200 µl of 0.25% PVP-40 in 0.1% acetic acid for 20 min. at room temperature on a shaker. PVP-40 prevents adsorption of the protease to the PVDF membrane. The blots were again washed three times with 10% acetonitrile. PVP-40 must be completely removed prior to digestion to prevent interference with peak detection on the HPLC. The electroblotted proteins were then digested in 100 µl of 0.1M Tris-HCl, pH 8.0 containing 10% acetonitrile with either 0.2 µg of Promega modified trypsin, or Lys-C (Wako) at 37°C for 24 hr. The supernatant was concentrated to 10-20 µl on a Savant SpeedVac and then injected directly on a capillary HPLC column. Large peptides generated by Lys-C were extracted from membrane, after removal of the supernatant, by incubating the blot in 20 µl of DMSO on a shaker for 10 min. (12). The DMSO was dried and combined with the supernatant prior to injection on the HPLC.

The Coomassie stain should be removed if several blots are to be combined or a large intensely stained blot is to be alkylated. This is easily performed by a modified chloroform-methanol precipitation (13). The blots are wetted with methanol and the following solvents added: 100 µl of water, vortexed; 400 µl of methanol, vortexed 2-3 min.; 100 µl of chloroform, vortexed 2-3 min. The solvents are removed and the electroblotted protein can be reduced and alkylated as described above.

Capillary HPLC Peptide Mapping

Peptides generated from *in situ* digestion were separated on a C18 0.32 x 100 mm capillary column (LC Packing, Inc.). The HPLC consisted of an Applied Biosystems, Inc. 140A microgradient pump system, and a model 783 UV detector equipped with a Z-shaped flow cell (LC Packings, Inc.). The pump was operated at 50 µl/min. and connected through a Valco tee which was used to reduce the flow to 6 µl/min. (14). Solvent A was 0.05% TFA and B was acetonitrile containing 0.05% TFA. Peptides were eluted using a linear gradient of 0 to 70% B in 50 min. and detected at 195 nm.

Protein Sequencing

Automated protein sequencing was performed on models 477A and 470A Applied Biosystems sequencers equipped with on-line PTH analyzers. Sequencers were modified to inject 80-90% of the sample. Peaks were integrated with Justice Innovation software using Nelson Analytical 760 interfaces. Sequence interpretation was performed on a VAX 8750 (15).

Mass Spectrometry

The capillary HPLC was connected on-line to a SCIEX API III mass spectrometer. The electrospray mass spectrometer was operated with articulated ionspray, interface plate, and orifice potentials set to 4600, 600 and 120 volts respectively. Spectra were acquired with a 11 second scan rate, with a mass range of 300-2200 Da and a step size of 0.2 Da.

III. Results and Discussion

Two-dimensional electrophoresis on the Pharmacia Multiphor unit is an efficient, reproducible method that provides high sensitivity and high resolution. High resolution is achieved by the incorporation of an immobilized pH gradient in the first dimension which replaces the use of ampholytes in conventional isoelectric focusing (IEF). This allows the use of higher voltages, resulting in higher resolution of closely migrating proteins (16-17). The IEF gel is bound to a plastic backing which prevents shrinkage or stretching when applied to the horizontal second dimension SDS gradient gel. The SDS gel is also bound to a plastic backing which is removed prior to electroblotting. The 2-D gel process focuses the protein into a small surface area which is ideal for digesting low picomole levels of protein. A two-dimensional gel of an *E. coli* extract electroblotted to a ProBlott membrane is shown in Figure 1. Spot #1 was

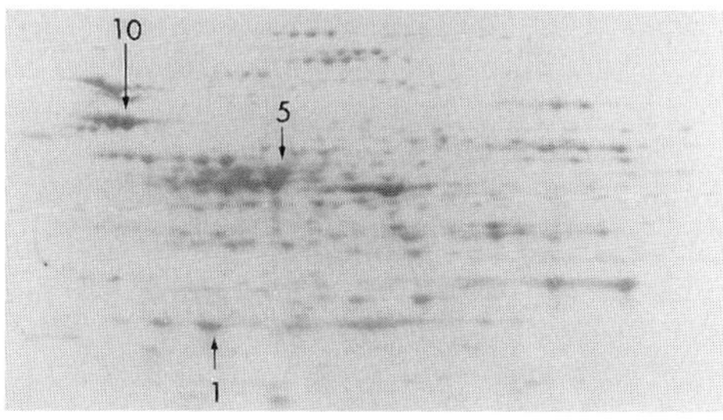

Figure 1. Two-dimensional gel of *E. coli* proteins electroblotted onto an Immobilon-PSQ membrane and stained with Coomassie blue.

Table I. Sequence analysis of 2-D gel spot #1

	1				5					10	
A.A.	S	L	I	N	T	K	I	K	P	F	
Yield (pmol)	2.3	2.3	1.9	1.9	1.3	0.9	1.6	1.0	1.3	1.3	
					15					20	
A.A.	K	N	Q	A	F	K	N	G	E	F	
Yield	1.0	1.2	1.2	1.2	1.0	0.5	1.1	0.6	0.5	0.7	
					25					30	
A.A.	I	E	I	T	K	K	D	T	E	G	R
Yield	0.9	0.4	0.8	0.6	0.3	0.2	0.4	0.4	0.1	0.2	0.1

sequenced in a Applied Biosystem 470A sequencer utilizing a blot cartridge. The initial yield indicated a total of 2-4 pmol of protein. The sequence yield combined with the Coomassie blue staining intensity of the sequenced spot can be used as a convenient estimate of the amount of protein in other spots on the electroblot. A computer search of the protein database identified spot #1 as alkyl hydroperoxide reductase (Table I).

Proteolysis of the electroblotted proteins is significantly enhanced if the protein is alkylated prior to digestion (11). We have examined the importance of protein reduction immediately prior to alkylation. Proteins are reduced prior to loading on the IEF first-dimension, but they may become oxidized during electrophoresis and electroblotting. A comparison of *in situ* tryptic digestion of electroblotted 2-D gel protein spot #5 with and without reduction prior to alkylation is shown in Fig. 2. Higher yields of peptides were obtained when the electroblotted protein was reduced prior to alkylation. We also examined proteins reduced in SDS sample buffer and separated on an SDS Laemmli gel. Amino acid analysis of bovine serum albumin and human growth hormone alkylated with iodoacetic acid on PVDF membranes without prior reduction on the blot resulted in lower yields of carboxymethyl cysteine (data not shown).

We have also noticed that Coomassie blue can affect the yield of alkylation as reported by Ploug et al. (9). This has not been a problem with single 2-D gel spots but the Coomassie should be removed if several blots are to be alkylated in the same tube. PVDF blots can be treated with a modified chloroform-methanol precipitation (13) which efficiently removes Coomassie without extracting protein. We have observed significant protein losses for some proteins if the blots are extracted with methanol alone.

Spot #10 was excised from the ProBlott membrane and alkylated with iodoacetic acid. The carboxymethylated protein was digested with Promega modified trypsin *in situ* on the ProBlott membrane in 100 µl of 0.1M Tris-HCl pH 8.5 containing 10% acetonitrile. The supernatant was removed from the blot, and concentrated in a SpeedVac to 20 µl, and the sample directly injected onto a reversed-phase capillary HPLC column (Fig. 3A). The same spot was excised from an identical blot and digested with Lys-C (Fig. 3B). We have found that *in situ* digestion with Promega modified trypsin consistently results in higher peptide yields than digestion with other endoproteases. The efficiency of a number of endoproteases can be evaluated from single protein spots isolated from replicate 2-D gels. The enzyme that produces the best peptide map can then be used to generate peptides for protein sequencing and mass spectrometry.

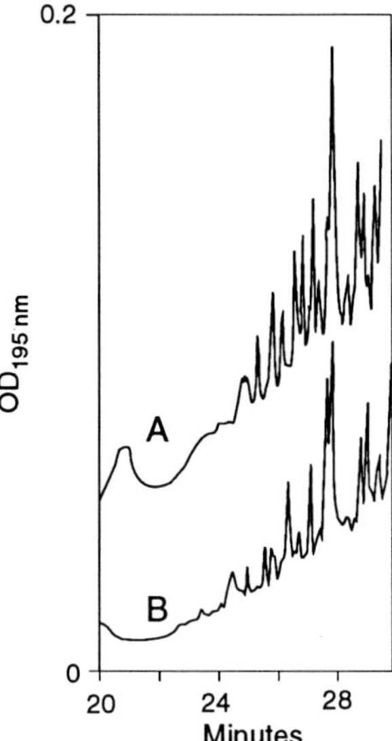

Figure 2. Comparison of *in situ* digestions of 2-D gel spot #5 with reduction (A) and without reduction (B) prior to alkylation with iodoacetic acid.

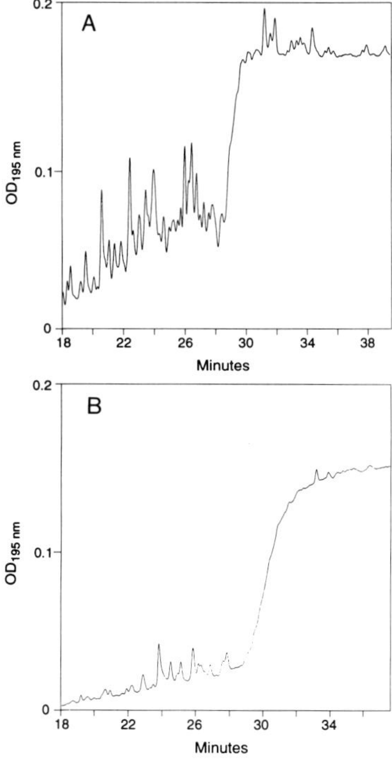

Figure 3. *In situ* digestion of 2-D gel spot #10 with trypsin (A) and with Lys-C (B) after alkylation on the blot with iodoacetic acid.

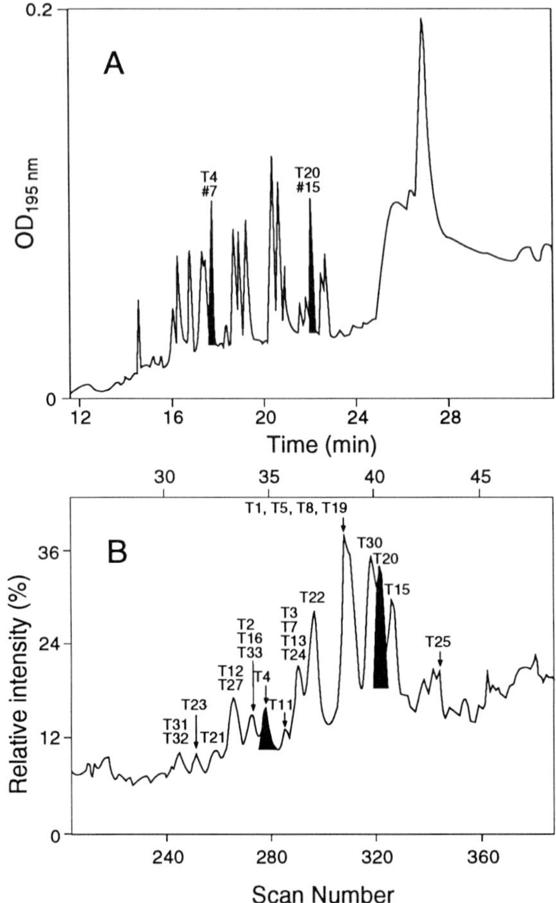

Figure 4. (A) Reversed-phase capillary HPLC separation of peptides from a tryptic digest of 2D gel spot #15 (phosphate binding protein). (B) Ion current trace of spot #15 from electrospray LCMS analysis. All peptides identified by mass spectrometry analysis are labeled as to their location in the ion current trace.

A protein spot #15 from a different 2-D gel of *E. coli* proteins (not shown) was carboxymethylated and digested with Promega trypsin. The supernatant was removed from the blot and concentrated in a SpeedVac to 20 μl and 50% of the sample was injected onto a capillary column. The two shaded peaks were sequenced and were found to match tryptic fragments from a known *E. coli* protein, phosphate-binding protein (Fig. 4A). The other half of the sample was injected onto the same column which was connected to a Sciex electrospray mass spectrometer. The ion current trace from the HPLC profile is shown in Fig. 4B. Mass spectra of all of the major peaks in the ion current provided identification of most of the tryptic peptides found in *E. coli* phosphate binding protein.

LCMS prior to sequence analysis provides an ideal measurement of peak purity. A typical LCMS trace may provide information on posttranslational modifications. This information may be extracted from LCMS analysis for novel proteins once the DNA sequence is determined. Mass data from the ion current trace can then be examined retrospectively for masses that do not correspond to the sequence. Peaks that were collected earlier that correspond to unidentified masses can then be characterized and the type of posttranslation modification determined.

Two typical spectra obtained from the on-line LCMS analysis are shown in Fig. 5. The experimental measurements are in close agreement (0.2 AMU) with the calculated masses obtained from protein sequence analysis.

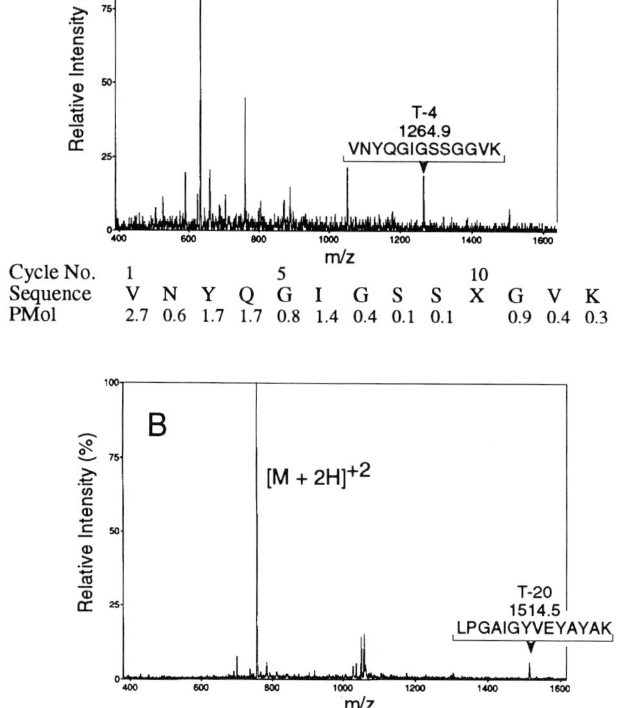

Figure 5. (A) Sequence analysis and electrospray mass spectra of peptide T4. (B) Sequence analysis and mass spectra of peptide T20.

IV. Conclusion

We have demonstrated that single 2-D gel protein spots at the 5 pmol level can be digested *in situ* on PVDF membranes. The resulting peptides can be separated by capillary HPLC and analyzed by on-line LCMS and microsequencing. In general, higher peptide and sequencing yields are obtained if the proteolytic digestions are performed on reduced and alkylated protein spots from the 2-D gels. Peptides generated by trypsin are generally small and most are solubilized by the digestion buffer which contains 10% acetonitrile. Digestion with Lys-C and Asp-N generally results in larger peptides which may require extraction of the PVDF membrane with DMSO prior to analysis by capillary HPLC or mass spectrometry. The use of a capillary HPLC column and low flow rates enables collection of peptide peaks in a volume of 1-6 μl. This minimizes the losses from adsorption to the collection tube observed when low level samples are collected and stored in larger volumes (200 μl) which are typical of the collection volumes for conventional columns. Although collecting peptides by hand from a capillary column can be tedious, this inconvenience is more than outweighed by the improved recovery that is obtained at the low picomole level.

References

1. Matsudaira, P. (1987). *J. Biol. Chem.* **262**, 10035.
2. Abersold, R.H., Teplow, D.B., Hood, L.E., and Kent, S.B.H. (1986). *J. Biol. Chem.* **261**, 4229.
3. Eckershorn, C., Mews, W., Goretzki, H., and Lottspeich, F. (1988). *Eur. J. Biochem.* **176**, 509.
4. Rosenfield,J., Capdevielle,J., Cluade Guillemont,J., and Ferrara,P. (1992). *Anal. Biochem.* **203**, 173-179.
5. Bauw, G.W., Van Damme,S., Puype, M., Vandekerchove, B.G., Ratz, G.P., Lauridsen, J.B., and Celis, J.E. (1989). *Proc. Natl. Acad. Sci.* USA **86,** 7701.
6. Simpson, R.J., Ward, L.O., Reid, G., Battenham, M.P., and Moritz, R.L. (1989). *J. Chromatogr.* **476**, 345.
7. Yuen, S.W, Chui, A.H., Wilson, K.J., and Yuan, P.M. (1989). *Biotechniques* **7,** 74.
8. Eckerson,C., and Lottspeich,F. (1989). *Chromatographia* **28**, 92.
9. Ploug, M., Stoffer, B., and Jensen, A.L. (1992). *Electrophoresis* **13**, 148-153.
10. Nokihara, K., Morita, N., Yamaguchi, M., and Watanabe, T. (1992). *Analytical Letters* **25,** 513-533.
11. Iwamatsu, A. (1992). *Electrophoresis* **13**, 142-147.
12. Wong, S., Padua, A. and Henzel, W.J. (1992). *In* "Techniques in Protein Chemistry III" (Angeletti, R.H. ed.) 3-9.
13. Wessel, D., and Flugge, U.I. (1984). *Anal. Biochem.* **138**, 141.
14. Henzel, W.J., Bourell, J.H., and Stults, J .T. (1990). *Anal. Biochem.* **187**, 228.
15. Henzel, W.J., Rodriguez, H., Watanabe, C. (1987). *J. Chromatogr.* **404**, 41-52.
16. Harrington, M.G., Gudeman, D., Zewert,T., Yun, M., and Hood, L. (1991). *Methods: Companion Methods in Enzymol.* **3**, 98-108.
17. Strahler, J.R., and Hanash, S. (1991). *Methods: Companion Methods in Enzymol.* **3**, 109-114.

Internal Peptide Sequence of Proteins Digested In-Gel after One- or Two- Dimensional Gel Electrophoresis

Pascual Ferrara, Jorge Rosenfeld, Jean Claude Guillemot, and Joel Capdevielle

Laboratoire de Biochimie des Protéines,
Sanofi Elf-BioRecherches, Labege Innopole, BP 137, 31676, Labege CEDEX, France

I. Introduction

The two most employed electrophoretic techniques, SDS-PAGE[1] and 2D-GE, when used in micropreparative scale, are powerful tools for the isolation of microgram quantities of proteins. Different techniques have been developed for obtaining either N-terminal (1,2) or internal sequences (3-6) from the electrophoretically isolated proteins. Among these techniques, the direct in-gel digestion of CBB stained proteins, followed by elution and HPLC separation of the resulting peptides, is an attractive option. However, there is only a limited number of reports employing this technique. This may be because the existing methods involve very long and laborious sample treatments before peptide sequencing (7-9), during which losses of material can not be excluded.

We recently developed a modified CBB staining protocol followed by a simple in-gel digestion technique which allowed us to obtain internal peptide sequences from 1-2 µg of protein separated by SDS-PAGE and 2D-GE (10). The method is rapid and allows the use of SDS-sensitive as well as SDS-resistant proteases. The in-gel digestion overcomes the electrotransfer or electroelution step and the need for the elution of peptides from PVDF membranes or glass-fiber filters, reducing considerably material losses and the risk of chemical amino acid damage.

In the present report we extend our study of the in-gel digestion technique, and we describe the identification, through

---

[1] Abbreviations used: 1D, one-dimensional; 2D, two-dimensional; PA, polyacrylamide; GE, gel electrophoresis; SDS, sodium dodecyl sulfate; DTT, dithiothreitol; ACN, acetonitrile; TFA, trifluoroacetic acid; CBB, Coomassie brillant blue R-250.

TECHNIQUES IN PROTEIN CHEMISTRY IV

internal sequence analysis, of E. coli proteins separated on SDS-PAGE, and of rat heart mitochondrial membrane proteins, separated on giant 2D gels.

II. Materials and Methods

A. Reagents and Protein Samples

Acrylamide, SDS and other reagents for SDS-PAGE and 2D-GE are from Bio-Rad. CBB R-250 and porcine trypsin are from SIGMA. Glycine and Tris base are from Riedel-deHaen. Urea is from Schwartz-Mann. All organic solvents are from SDS (Peypin-France).

Albumin is from Pierce. E. coli periplasmic proteins were prepared as described (11). The rat heart mitochondrial membrane proteins were prepared as previously described (10).

B. Gel Electrophoresis, CBB Staining, and In-Gel Enzymatic Digestion

SDS-PAGE was done essentially as described by Laemmli (12). Gels of 160 x 150 x 0.75 mm of 12.5% (w/v) acrylamide were left for polymerization overnight. The giant 2D-GE technique described by Levenson et al.(13) was adapted to larger 2D gels (38 x 45 x 0.1 cm) as described elsewhere (14). Up to 3 mg of protein were loaded and efficiently resolved in these gels.

After electrophoresis, 1D or 2D gels were stained with one of two procedures: the standard procedure, fixing with 50% methanol, 10% acetic acid, staining with 0.2% CBB in 50% methanol, 10% acetic acid and destaining in 40% methanol, 7.5% acetic acid; or the modified procedure, no fixing, staining with 0.2% CBB in 20% methanol, 0.5% acetic acid, and destaining in 30% methanol until protein bands became visible.

In-gel digestion was performed as described (10) with some modifications. Briefly, the stained protein bands were excised and washed two times with 150 µl of 50% ACN in 200 mM ammonium carbonate, pH 8.9, for 20 min at 30 °C in an Eppendorf-thermomixer. The gel slices were left to semi-dry at room temperature, on a Parafilm sheet covered with the top of a Petri dish (≈10 min depending on the fragment size). Then they were partially rehydrated with 5 µl of 200 mM ammonium carbonate pH 8.9, 0.02% Tween 20, and 2 µl of porcine trypsin solution (200 µg/ml dissolved in 200 mM ammonium carbonate, pH 8.9) was added.

After absorption of the protease solution, aliquots of 5 µl of rehydration buffer were added until the slices returned to their original size. Gels slices were then cut into 1 mm^3 pieces, transferred to an Eppendorf tube, and immersed in a minimum of rehydration buffer (30-50 µl). The digestions were carried out at 30°C for 4 hours. If necessary after the first hour, 10 µl of rehydration buffer was added. The reaction was stopped by the addition of 1.5 µl of TFA.

The resulting peptides were recovered by two 20 min extractions, with 100 µl of 60% ACN, 0.1% TFA by shaking in an Eppendorf-thermomixer at 30°C. The extracts were combined and concentrated to ≈20 µl in a Speedvac (Savant).

C. Peptide Purification and Characterization.

The peptides eluted from the polyacrylamide matrix were purified on a Brownlee Labs microgradient system equipped with a C18 Ultrasphere column (250 mm x 2.1 mm). A linear gradient from 5 to 70% ACN, 0.1% TFA was run for 60 or 90 min at a flow rate of 300 µl/min. Elution was monitored at 218 nm, and peaks were manually collected.

Sequence analysis was performed on a gas phase sequencer (470A, Applied Biosystems).

III. Results

We observed that peptide recoveries from in-gel digested CBB stained proteins were improved with a reduction in the methanol and the acetic acid concentrations employed during the staining-destaining procedure. An optimization of the staining-destaining solutions, based on the digestion yield of known proteins, resulted in the following: in the CBB staining solution, the acetic acid was reduced from 10 to 0.5% and the methanol from 50 to 20%, and in the destaining solution, the acetic acid was eliminated and the methanol was increased from 20 to 30%. These modifications did not affected the staining sensitivity, and 1-2 µg of standard proteins were easily visualized. Furthermore, no material was lost when the gels were stained with the modified procedure (not shown).

To get cleaner and more reproducible peptide maps of a CBB-stained protein, an exhaustive wash of the gel before the digestion is necessary. After evaluation of several solvents, a solution of 50% ACN in ammonium carbonate was selected. Two washes with this solution proved to be effective in reducing the artifactual peaks in the fingerprints. The dehydration resulting from the ACN wash

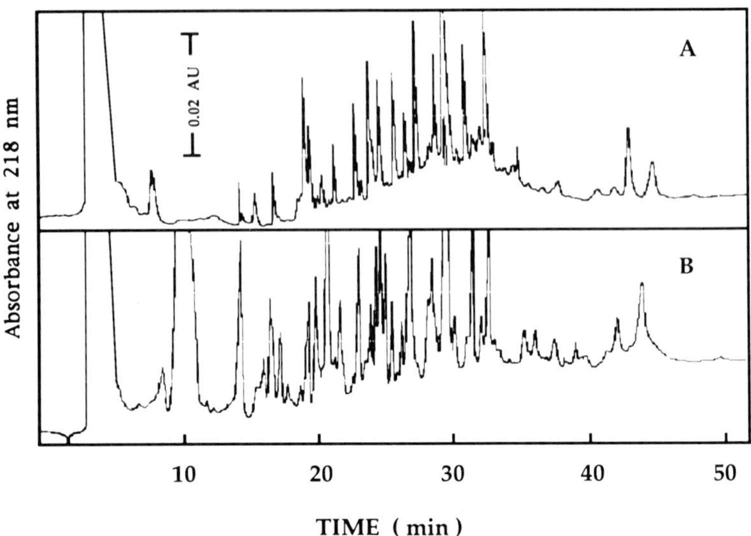

Figure 1. Peptide map of 50 pmol of albumin digested A) in-gel; and B) in-solution.

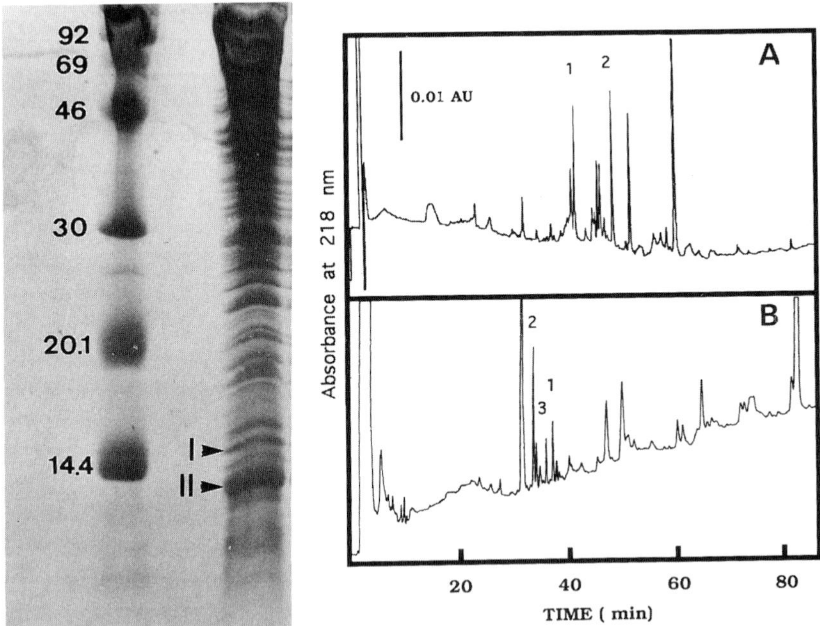

Figure 2. SDS-PAGE of periplasmic proteins from E. coli, and HPLC peptide maps resulting from in-gel digestion of A) all the 14.5 kDa band, and B) half of the 12 kDa band. The indicated peptides were sequenced.

step reduced considerably the size of the gel pieces. Moreover, the subsequent rapid evaporation of the organic solvent produced a second gel shrinkage that improved the absorption of the trypsin into the gel matrix when the gel slices were rehydrated with the protease solution. The improved absorption of the enzyme gave better digestion yields.

The in-gel digestion of BSA (50 pmol), performed as described above, gave a peptide map similar to the one obtained in solution (Figure 1, A and B). The recovery of the peptides after in-gel digestion was, on the average, 60% when compared, by amino acid analysis, to the peptides recovered when the protein was digested in solution (not shown).

The general utility of the described method is illustrated by the identification, through internal sequence analysis, of soluble E. coli derived proteins separated on SDS-PAGE, and of rat mitochondrial membrane proteins separated on 2D-GE.

The periplasmic proteins of an E. coli transfected with a plasmid containing the gene of human thioredoxin, were extracted by osmotic shock. 50 µg were directly separated on a 12.5% SDS-PAGE (Figure 2). The indicated bands were cut out and in-gel digested with trypsin. The band I (14.5 kDa) gave the fingerprint shown in the Figure 2A. The sequence of selected peptides are shown in Table I. A database search of these sequences show that this protein belongs to the Phe-Tyr operon. The digestion of half of the band II (12 kDa) gave the fingerprint shown in Figure 2B. Sequencing of several peptides (Table I) resulted in the identification of this protein as the recombinant thioredoxin.

Figure 3 shows a representative CBB-stained giant 2D gel. The total protein loaded in this gel was 2.5 mg, 124 protein spots were detected, and more than 50% of these spots could be directly in-gel digested and sequenced. As an example, we show the peptide maps resulting from the in-gel digestion of some of those spots. One third of the spot A (50 kDa, pI 5.3) was subjected to in-gel digestion, and the resulting peptide map is shown in Figure 4A. The indicated peptides were sequenced. The sequences obtained were correlated with the chain B of the ATP synthase, a protein present in the database (Table I). Half of the spot B (21.8 kDa, pI 5.4) gave the fingerprint shown in Figure 4B. The sequences of three peptides (Table I) allowed us to associate this spot, by homology with the bovine protein, with a subunit of the mitochondrial complex I. The molecular weight calculated from the sequence is in agreement with the value determined from the gel. Sequence information of the in-gel digested spot C (Figure 4C) with a molecular mass of 15 kDa and a pI of 6.1 allowed its identification as the ATPase chain D, by homology with the bovine protein. The molecular weight of this protein, 15065 Da, agrees with the experimental value .

IV. Discussion

Producing sequence information from proteins isolated by GE is a straight-forward approach to their chemical characterization. Moreover if one can obtain internal peptides by an in-gel digestion technique, the problem of natural or artifactual N-terminal protein

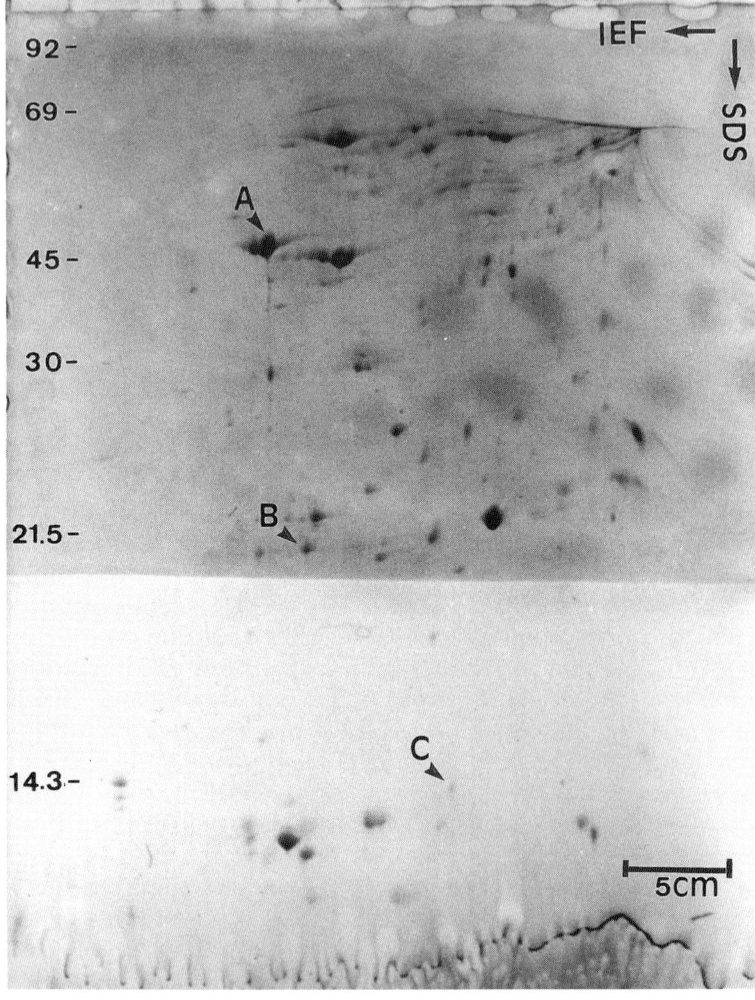

Figure 3. CBB stained preparative two-dimensional gel electrophoresis of the mitochondrial membrane fraction from rat heart. The indicated spots were cut out and digested in-gel as described in Materials and Methods.

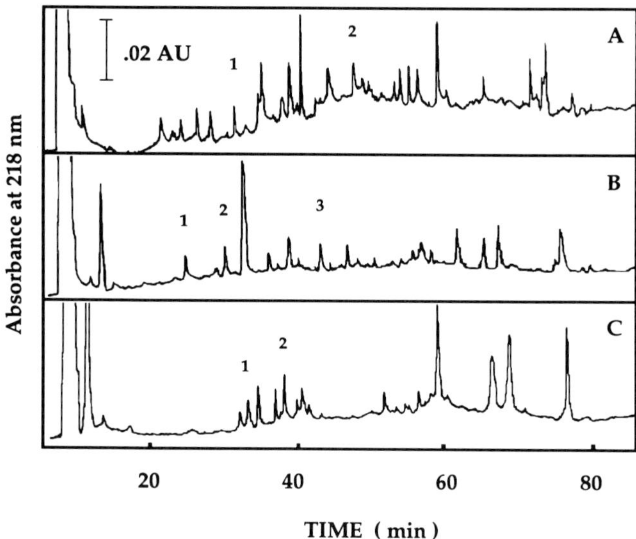

Figure 4. Peptide maps of the selected spots in Figure 3. The numbered peptides were sequenced.

blockage is circumvented. In spite of these advantages, only a limited number of reports using in-gel digestions have been published, probably due to the fact that the existing methods involve long, multistep procedures (7-9).

We have reinvestigated the in-gel digestion of proteins. As a result, we developed an improved method where a modified CBB staining procedure, and a rapid wash and equilibration of the gel slices in the protease solution, resulted in the successful production of peptides from 1-2 μg of in-gel digested proteins. The new staining conditions do not affect the detection limits, and improve the overall yield of the in-gel digestion probably avoiding irreversible aggregation of the protein in the gel. Washing the gel slices with 50% ACN in ammonium carbonate is sufficient to remove the dye and gel contaminants without loss of protein. This step improved dramatically the quality of the peptide maps. During the washing step, the water in the gel is reduced, thus the lyophilization step previously described (8,15) is eliminated. Since we (10) and others (15) observed that peptide yields are severely reduced if gel slices

Table I. Sequence of peptides generated by in-gel digestion of proteins isolated from one- or two-dimensional gels

Sample	Initial Yield[a] (pmol)	Sequence	Position in the sequence[b]
SDS-PAGE			
Band I. 14.5 kDa : identified as a protein of the Phe-Tyr operon of E. coli			
peptide 1	13	TMNITSK	2-8
peptide 2	15	WQTHLINPHIILSK	30-43
Band II. 12 kDa : identified as thioredoxin			
peptide 1	11	XTPTFQFFK	72-80
peptide 2	16	MIXPFFHSLSEK	36-47
peptide 3	9	EKLEATINEL	94-104
2D-GE			
Spot A (50 kDa, pI 5.3): identified as ATP synthase chain B			
peptide 1	8	VLDSGAPIK	79-87
peptide 2	15	GFQQILAGDYDHLPEQAFY	397-415
Spot B (21.8 kDa, pI 5.4): identified as a mitochondrial complex I subunit			
peptide 1	15	EPATINYPFEK	80-90
peptide 2	8	XQELDVDMK	43-51
peptide 3	13	YDIDMTK	143-151
Spot C (15 kDa, pI 6.1): identified as ATPase chain D			
peptide 1	12	ANLEK	132-136
peptide 2	6	IEANEALVK	157-165

[a] Calculated by extrapolation of the PTH-amino acid values to cycle 1.
[b] The proteins were identified by comparison with the sequences present in the SwissProt (release 92.22) and NBRF (release 92.33) databases.

are dried before digestion, the semi-dehydration that prevents complete dryness of the gel slices improved yields.

It should be noted that, as most of the SDS is washed out from the gel matrix, proteases other than those capable of working in the presence of SDS can be used with this technique.

The described method has been used to identify proteins isolated from SDS-PAGE and 2D gels. The initial yields of the sequenced peptides were sufficiently high, in general 10 pmol or more, allowing the identification of long stretches of amino acids with very few ambiguities. The correct identification of labile amino acids, such as methionine and tryptophan, probably is a consequence of the elimination of the electrotransfer and the short time needed to obtain the peptides

In conclusion, we describe in this report an improved method for generating peptide fragments for sequence analysis of proteins separated by 1D- or 2D-GE. The method is a straightforward approach to the rapid identification of proteins in complex cell extracts. Furthermore, as shown in the examples here described, since we succeeded in sequencing membrane proteins of the rat mitochondria isolated in a single preparative 2D-gel, the method should be of particular utility for sequencing other hydrophobic insoluble membrane proteins that are difficult to isolate by other means.

References

1. Matsudaira P. (1987) J. Biol. Chem. 262, 10035-10038.

2. Moos M., Nguyen N. Y., and Liu. T-Y. (1988) J. Biol. Chem. 263, 6005-6008.

3. Aebersold R.H., Leavitt J.R., Saavedra A., Hood L.E., and Kent S.B.H. (1987) Proc. Natl. Acad. Sci. U.S.A. 84, 6970-6974.

4. Tempst P., Link A.J., Riviere L.R., Fleming M., and Elicone C. (1990) Electrophoresis 11, 537-553.

5. Simpson R.J., Ward L.D., Reid G.E., Batterham M.P., and Moritz R.L. (1989) J. Chromatogr. 476, 345-361.

6. Szewczyk B., and Summers D. (1988) Anal. Biochem. 168, 48-53.

7.Ward L.D., Reid G.E., Moritz R.L., and Simpson R.J. (1990) J. Chrom. 519, 199-216.

8. Eckerskorn C., and Lottspeich F. (1989) Chromatographia 28, 92-94.

9. Kawasaki H., Emori Y., and Suzuki K. (1990) Anal. Biochem. 191, 332-336.

10. Rosenfeld, J., Capdevielle, J., Guillemot, J.C., and Ferrara, P. (1992) Anal. Biochem. 203, 173-179.

11. Joseph-Liauzun, E., Leplatois, P., Legoux, R., Guerveno, V., Marchese, E., and Ferrara, P. (1990) Gene-AMST 86, 291-296.

12. Laemmli U.K. (1970) Nature (London) 227, 680-685.

13. Levenson R.M., Anderson G.M., Cohn J.A., and Blackshear P.J. (1990) Electrophoresis 11, 269-279.

14. Rosenfeld, J., Capdevielle, J., Guillemot, J.C., and Ferrara, P. (1992) manuscript in preparation.

15. Simpson R.J., Moritz R.L., Nice E.C., and Grego B. (1987) Eur. J. Biochem. 165, 21-29.

IN SITU TRYPTIC DIGESTION OF PROTEINS SEPARATED BY SDS–PAGE: IMPROVED PROCEDURES FOR EXTRACTION OF PEPTIDES PRIOR TO MICROSEQUENCING

Tim Tetaza[a], Esther Bozas[b], Jerry Kanellos[c], Ian Walker[b] and Ian Smith[a]

Baker Medical Research Institute, Melbourne[a]; Department of Veterinary Science, University of Melbourne[b]; Millipore–Waters, Melbourne, Australia[c]

I. INTRODUCTION

Perhaps the most powerful technique for small–scale (1–10 μg) isolation of proteins is SDS–polyacrylamide gel electrophoresis (SDS–PAGE), with protein detection using stains such as Coomassie Brilliant Blue (CBB). However, the severe losses frequently associated with recovery of proteins from gels (by techniques such as electroelution or passive elution) and the subsequent need to remove gel–related contaminants prior to enzymatic fragmentation have previously limited the utility of SDS–PAGE as a procedure for purification of proteins for the purpose of obtaining internal amino acid sequence information. The recent development by Eckerskorn and colleagues (1) of a procedure for *in situ* tryptic digestion of proteins separated by SDS–PAGE thus represents a significant advance in microsequencing, since there is no need for removal of proteins from the gel matrix: proteins are digested simply by lyophilization of gel slices followed by rehydration with buffer containing trypsin. However, the published method for extraction of peptides from the gel matrix is time–consuming (2 days), hazardous (high concentrations of trifluoracetic acid) and costly.

In this report, we describe two alternative extraction procedures which are both quicker (30 min) and more economical than the standard TFA/acetonitrile extractions. The first modification involves fine crushing of the gel slices prior to digestion, and extraction of the peptides by sequential washing with 0.1 M ammonia. The second

TECHNIQUES IN PROTEIN CHEMISTRY IV

modification involves digestion of gel slices inside an empty HPLC precolumn and direct recovery of peptides on a downstream reversed–phase (RP–)HPLC column. We have evaluated these modifications by comparing the 3 extraction procedures in their ability to recover peptides from tryptic digests of low levels (100 pmol) of bovine serum albumin (70 kD) and human apolipoprotein A–I (28 kD). Peptides were separated by RP–HPLC on a microbore column (RP300, 1.0 mm I.D. x 50 mm) and peaks collected manually for direct application to the sequencer. Sequenceable peptides were obtained by all 3 extraction procedures, with initial yields ranging from 2 to 16 pmol, indicating that useful sequence information can be obtained from as little as 100 pmol of protein separated on a SDS–polyacrylamide gel.

II. METHODS

A. Materials

Apolipoprotein A–I was purified from human lymph chylomicrons by reversed–phase HPLC on a TSK Phenyl–5PW column (2). Bovine serum albumin (A–7638) and trypsin were purchased from Sigma and Worthington, respectively. Sequencing grade trifluoroacetic acid (TFA) was obtained from Applied Biosystems, HPLC grade water and acetonitrile from Mallinkrodt, and analytical grade ammonia solution from Ajax Chemicals, Sydney, Australia. Sodium dodecyl sulphate (SDS) and Coomassie Brilliant Blue R–250 (CBB) were purchased from Sigma, acrylamide from National Diagnostics and acetic acid from Merck. Reversed–phase RP300 C8 columns (1.0 mm I.D. x 50 mm) and TSK Phenyl–5PW columns (7.5 mm I.D. x 75 mm) were obtained from Applied Biosystems and Toyo Soda (Japan), respectively. Glass–lined, stainless steel HPLC precolumns (4.6 mm I.D. x 75 mm, 2.0 mm I.D. x 50 mm) were purchased from Scientific Glass Engineering Pty. Ltd., Australia.

B. SDS Polyacrylamide Gel Electrophoresis (SDS–PAGE)

SDS–PAGE was carried out by the procedure of Laemmli (3). Bovine serum albumin (bSA) and human apo A–I (hA–I) were dissolved in 0.1M NH_4HCO_3 at a concentration of 1 mg/ml, and dilutions of these stock solutions were quantitated by amino acid analysis. Samples containing 100 pmol of each protein were then prepared in 50 μl of 2% SDS, 5% 2–mercaptoethanol, 10% glycerol and 0.002% bromophenol blue, boiled for 30 seconds and loaded into the wells of a 10% polyacrylamide gel (width 12 cm, length 10 cm, thickness 0.75 mm). Gels were prerun for 1h at 50 mA, loaded and then run at 50 mA until the dye front reached the bottom of the gel. Gels were stained with CBB (0.1% in 10% methanol, 5% acetic acid) for one hour at room temperature, and then destained overnight in 10% methanol, 5% acetic acid with frequent changes of destain.

C. In Situ Tryptic Digestion

Three methods were used for *in situ* tryptic digestion of electrophoretically purified proteins and subsequent extraction of peptides from the polyacrylamide gel matrix.

1. TFA extraction

The method was essentially as described by Eckerskorn *et al.*(1). Briefly, gel slices containing protein bands were washed extensively (Milli Q water for 24h) then sliced into small cubes (approx 1 mm) and dried for 1–2 h in a SpeedVac (Savant). Dried gel pieces were then rehydrated in 0.1 M ammonium bicarbonate (NH_4HCO_3, pH 8.0) containing trypsin at an enzyme:substrate ratio of 1:10 (w/w), and incubated overnight at 37°C. Supernatants from the digests were removed and gel pieces were extracted with an equal volume of 75% TFA for 4 h, followed by 50% TFA/50% acetonitrile for 4 h. These two extractions were then repeated. Digest supernatants and extractions were pooled and dried in a SpeedVac prior to resuspension in 0.1% TFA for HPLC.

2. Ammonia extraction

Gel slices containing protein bands were crushed between siliconized microscope slides and transferred into siliconized glass test tubes. The crushed gels were washed 3 times (Milli Q water) and then lyophilized. The dried gels were then slowly rehydrated with 100 μl aliquots of digest buffer (0.1 M NH_4HCO_3) containing trypsin at a concentration of 50 μg/ml of swollen gel. Digest buffer was added until the gel was completely submerged. After overnight incubation at 37°C, protein extraction was carried out by the addition of sequential 500 μl aliquots of 0.1 M ammonia. The suspension was then extensively vortexed, spun and the supernatant removed. This was repeated until no more blue colour was visible in the gel. Supernatants were pooled, dried in a SpeedVac and then resuspended in 0.1% TFA for HPLC.

3. HPLC extraction

The method was based on the observation by Pearson and co–workers (4) that low Mr (up to 10–20 kD) proteins could be recovered efficiently from polyacrylamide gels by packing excised gel pieces into an empty HPLC column and using the high pressure of HPLC to force proteins out of the gel matrix. Tryptic digests were performed essentially as described above for the TFA extraction method. However, after rehydrating the dried gel pieces with buffer containing trypsin the gel pieces were immediately transferred into an empty HPLC precolumn. The column ends were sealed and the gel allowed to swell to fill the column. After overnight incubation at 37°C, the column was connected upstream from a microbore RP–HPLC column previously equilibrated with 0.1% TFA. Peptides were then removed from the gel matrix by pumping 0.1% TFA through the precolumn at high pressure (100 bar for 15–20 min) and recovered on the downstream RP–HPLC column. The precolumn was then switched out of line and peptides were purified as described below.

D. Purification and Sequencing of Peptides

Peptides were chromatographed on microbore RP300 C8 columns (1.0 mm I.D. x 50 mm) using a Hewlett Packard HP1090 HPLC system equipped with a diode array detector (DAD) and 2 ml Rheodyne injection loop. Column temperature was maintained at 40°C. Peptides extracted from gel digests by either TFA or ammonia were dissolved in 0.1% TFA and applied to the column. Alternatively, peptides were extracted from the gel by HPLC as described above. After the non–retained material had passed through the DAD, the loop was switched out of line and the flow rate decreased to 100 μl/min. Peptides were then eluted by a linear gradient (90 min) from Solvent A to Solvent B, where Solvent A was 0.1% TFA and Solvent B was 0.09% TFA in 60% acetonitrile. The column effluent was monitored at 215, 254 and 280 nm. Peptides were collected manually into Eppendorf tubes and loaded onto polybrene–coated glass fibre discs for automated sequencing. The tubes were rinsed with 2 successive 50 μl aliquots of TFA which were also applied to the glass fibre disc. Sequencing was performed using an Applied Biosystems Model 470A Protein Sequencer with an on–line Model 120A PTH Analyzer. To maximize recovery of PTH amino acids, the PTH analyzer was equipped with a 100 μl injection loop and the standard sequencing program was modified as described by Tempst and Riviere (5). PTH amino acids were chromatographed on an Applied Biosystems PTH–C18, 5 micron column (2.1 mm I.D. x 220 mm) at a temperature of 49°C and flow rate of 200 μl/min. The column effluent was monitored at 269 nm (0.005 AUFS) and PTH amino acids were eluted by a linear gradient (26 min) from 10 to 47% B where Solvent A was 5% tetrahydrofuran containing (per litre) 13.2 ml 3M sodium acetate pH 3.8, 1.4 ml 3M sodium acetate pH 4.6, 100 μl triethylamine and 100 μg L–tryptophan, while Solvent B was acetonitrile containing 500 nmol DMPTU per litre.

III. RESULTS AND DISCUSSION

Samples (100 pmol) of bovine serum albumin (bSA, 7.0 μg) and human apoA–I (hA–I, 2.8 μg) were electrophoresed on 10% polyacrylamide gels and stained with CBB. Protein bands were excised, dried and digested *in situ* with trypsin, and then peptides were extracted by one of 3 procedures (TFA extraction, ammonia extraction or HPLC extraction) described in detail in the Methods section. As a control, gel slices were excised from a region of the gel which did not contain detectable protein and these slices were treated as for the bSA and hA–I samples. Following extraction, tryptic peptides were separated by microbore RP–HPLC (Fig. 1). Since the very low amounts precluded further purification of peptide peaks by multidimensional RP–HPLC, selected peaks from the first dimension RP–HPLC run were applied directly to the sequencer. Table I shows the sequences obtained, with yields of PTH–amino acids in pmol for each cycle.

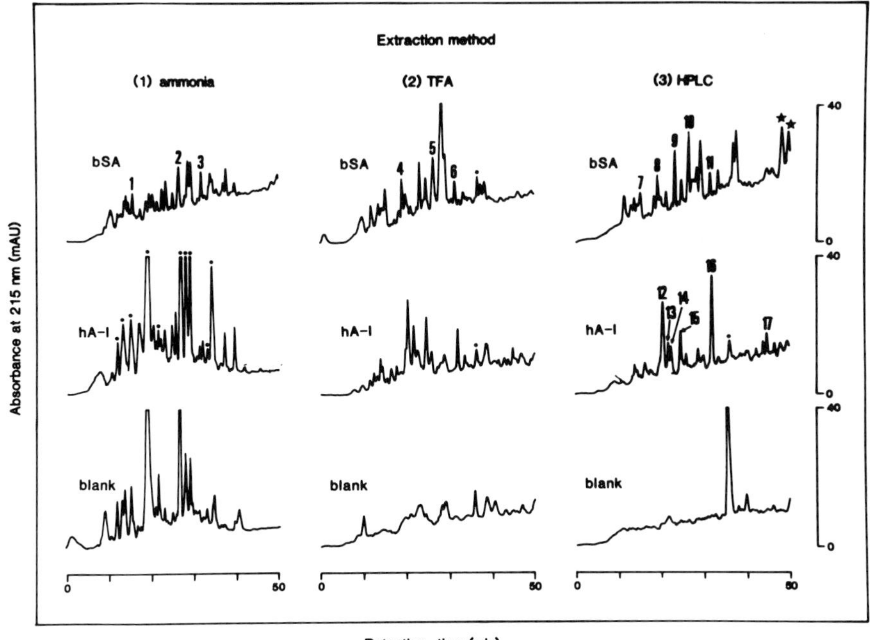

Fig. 1

A total of 17 peaks were sequenced and all but one (no.12) produced useful sequence information, with initial yields between 2 and 16 pmol (average 7.7 pmol). Although many of the peaks yielded more than one peptide sequence, in each case a dominant sequence (molar ratio ≥ 2:1) was read correctly without consultation to the known sequences (with the exception of peak no.12 which contained 4 peptide sequences at approx equimolar levels). A typical sequence analysis is shown in Fig.2. From 16 peptides sequenced, 124 residues were correctly assigned (68% of the 183 residues theoretically present). Residues which were not detected included tryptophan plus certain threonine residues, which would be expected when sequencing at such low pmol levels. A glycine residue was not assigned in 2 peptides sequenced (position 2 in sequence nos. 3 and 6) indicating that sample contamination may interfere with detection of certain hydrophilic residues close to the N–termini of peptides. Other residues not assigned were a lysine at position 2 in sequence no.4 and a histidine at position 8 in sequence no.8. The reasons for this are not known.

The modifications to the PTH analyzer and protein sequencer described by Tempst and Riviere (see Methods section) were crucial to obtaining sequence information at these low pmol levels. Further improvement might be achievable by covalent attachment of peptides to PVDF disks prior to sequencing, thus enabling longer sequences to be obtained by eliminating peptide "washout" during the extraction of ATZ amino acids. Another means by which to obtain extended peptide sequence lengths is by the use of endoproteinases of different specificities in order to generate overlapping peptide sequences. Such an approach should be possible using the *in situ* digestion and extraction techniques described, since we have shown (unpublished data) that endoproteinase Asp–N (and to a lesser extent Lys–C and Glu–C) works well with this technique. Similar observations have been made by Tempst *et al.*(6) with *in situ* digests of bSA and other proteins electroblotted from SDS–polyacrylamide gels to nitrocellulose.

Analysis of two sets of bSA sequences (peak nos. 2, 5, 10 and 3, 6, 11) suggests that peptides extracted by any one of the 3 described extraction procedures sequence equally well (10 of 15 residues and 9/10 of 13 residues, respectively, were assigned), although yields were

---

Fig.1 RP–HPLC profiles of tryptic peptides from *in situ* digests of 100 pmol bSA, 100 pmol hA–I and of blank gel slices. Peptides were extracted by TFA, ammonia or HPLC. Coomassie Blue–related peaks (✶) and trypsin peaks (✱) are shown. Peaks (numbered 1–17) were collected manually and sequenced without further purification. Digests extracted by ammonia were performed using trypsin at a concentration of 50 μg/ml (see Methods section) so that levels of tryptic peptides of trypsin vary according to the size of gel piece excised. Digests of bSA were performed in a separate experiment to those of hA–I and blank gel slices.

Table I. Sequences of tryptic peptides from in situ digests of 100pmol each of bSA and hA-I

Peak no.[a]	Amino acid sequence[b] (yield per cycle in pmol)															Residues sequenced[c]
	1	2	3	4	5	6	7	8	9	10	11	12	13	14	15	
1.	A(10.5)	E(7.7)	F(2.4)	V(1.7)	E(1.7)	V(0.4)	T(0.1)	K(0.4)								bSA 224-231
2.	K(11.6)	V(10.7)	P(7.4)	Q(4.4)	V(3.9)	S(0.6)	T(1.1)	P(2.7)	T(0.5)	L(2.3)	V(1.0)	E(1.3)	V(1.0)	S(-)	R(-)	bSA 412-426
3.	L(4.7)	G(3.6)	E(2.7)	Y(1.0)	G(3.1)	F(0.3)	Q(0.8)	N(2.5)	A(0.6)	L(1.0)	I(0.5)	V(0.2)	R(-)			bSA 396-408
4.	F(4.6)	K(0.9)	D(6.0)	L(4.7)	G(6.3)	E(1.9)	E(1.9)	H(1.2)	F(0.7)	K(0.2)						bSA 11-20
5.	K(16.1)	V(8.3)	P(8.1)	Q(5.6)	V(5.6)	S(1.6)	T(1.4)	P(2.7)	T(0.5)	L(2.1)	V(1.5)	E(1.5)	V(1.0)	S(-)	R(-)	bSA 412-426
6.	L(8.9)	G(4.8)	E(4.0)	Y(3.8)	G(5.0)	F(3.9)	Q(2.9)	N(3.7)	A(1.7)	L(0.8)	I(0.3)	V(0.3)	R(0.3)			bSA 396-408
7.	A(11.0)	E(7.0)	F(5.9)	V(2.8)	E(1.8)	V(0.8)	T(0.1)	K(0.4)								bSA 224-231
8.	F(5.8)	K(10.1)	D(7.4)	L(7.1)	G(6.4)	E(3.8)	E(4.3)	H(-)	F(0.9)	K(0.6)						bSA 11-20
9.	H(2.6)	L(4.3)	V(4.3)	D(2.3)	E(1.5)	P(0.9)	Q(0.7)	N(0.5)	L(0.3)	I(0.3)	K(-)					bSA 377-387
10.	K(15.5)	V(11.7)	P(12.8)	Q(7.7)	V(5.9)	S(1.9)	T(1.6)	P(4.0)	T(0.8)	L(2.1)	V(2.0)	E(1.3)	V(1.5)	S(-)	R(-)	bSA 412-426
11.	L(9.4)	G(7.8)	E(7.7)	Y(4.1)	G(6.7)	F(4.4)	Q(3.4)	N(4.6)	A(5.0)	L(4.7)	I(1.1)	V(1.0)	R(1.0)			bSA 396-408
13.	V(2.9)	Q(2.4)	P(3.4)	Y(3.1)	L(3.2)	D(1.5)	D(1.9)	F(0.3)	Q(0.4)	K(0.3)						hA-I 97-106
14.	Q(1.9)	K(2.6)	V(2.9)	E(2.8)	P(2.9)	L(3.0)	R(2.9)	A(2.0)								hA-I 117-124
15.	W(-)	Q(2.4)	E(3.0)	E(3.6)	M(1.4)	E(1.5)	L(1.4)	Y(1.1)	R(-)							hA-I 108-116
16.	L(5.3)	L(6.4)	D94.9)	N(2.6)	W(-)	D(3.4)	S(0.7)	V(1.7)	T(0.1)	S(0.7)	T(0.1)	F(0.5)	S(0.5)	K(0.3)		hA-I 46-59
17.	D(4.3)	L(1.9)	A(2.0)	T(0.3)	V(0.8)	Y(0.8)	V(0.5)	D(1.9)	V(0.4)	L(0.4)	K(0.1)					hA-I 13-23

a Peak no. 12 contained 4 peptides which sequenced at approx. equimolar levels.

b Underlined sequences were read without consultation to the known sequences. Residues which are not underlined were not readable above the background "noise" but are assumed to be present based on the known sequences.

c Amino acid sequences for bSA and hA-I were taken from the Swissprot data base. In each case, residue no.1 corresponds to the first residue of the mature polypeptide (i.e. the translated sequence minus signal and pro-sequences).

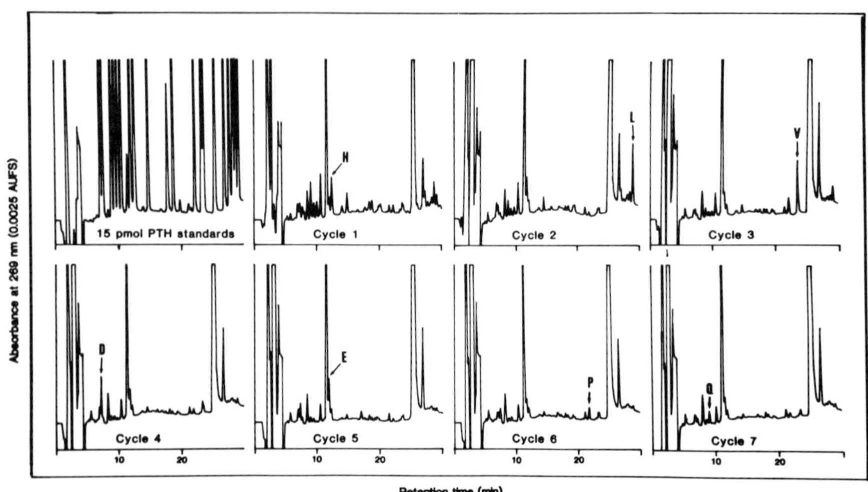

Fig.2 Sequence analysis of peak no. 9 (Fig.1). Yields of PTH amino acids obtained in each cycle are shown in Table I.

somewhat lower for the ammonia–extracted peptides. The 2 new extraction procedures described offer the advantage of being quicker (30 min compared to 2 days), safer and more economical than the standard TFA/acetonitrile extractions described by Eckerskorn *et al.* (1). Having demonstrated the utility of these approaches to obtaining internal sequence information from 100 pmol amounts of 2 known proteins, we are currently adapting this technology to the sequencing of several novel, low abundance hydrophobic membrane–bound proteases, kinases and receptors.

ACKNOWLEDGEMENTS

The authors wish to thank Rebecca Lew and Robert Andrews for critical review of the manuscript, Neil Potter for artwork and Julie Simpson and Catherine Wallace for careful preparation of the manuscript.

REFERENCES

1. C. Eckerskorn, J. Strahler, S. Hanash and F. Lottspeich (1989) in A.T. Endler and S. Hanash (Editors), Two Dimensional Electrophoresis, VCH Publishers, New York, pp 153–157.
2. T. Tetaz, E. Kecorius, B. Grego and N. Fidge (1990) J. Chromatogr. **511**:147–153.
3. U.K. Laemmli (1970) Nature **227**:680–685.
4. J.D. Pearson, D.B. De Wald, H.A. Zurcher–Neely, R.L. Heinrikson and R.A. Poorman (1986) in K.A. Walsh (Editor), Methods in Protein Sequence Analysis, Humana Press, New Jersey, pp.295–302.
5. P. Tempst and L. Riviere (1989) Anal. Biochem. **183**:290–300.
6. P. Tempst, A.J. Link, L.R. Riviere, M. Fleming and C. Elicone (1990) Electrophoresis **11**:537–553.

In Situ Proteolytic Digestions Performed on Proteins Bound to the Hewlett-Packard Hydrophobic Sequencing Column

William Burkhart

Dept. of Structural and Biophysical Chemistry
Glaxo Research Institute
Research Triangle Park, North Carolina 27709

I. Introduction

Proteins submitted for Edman sequence analysis arrive in the protein chemistry lab in various conditions. The protein is often at low concentration in a solution containing buffers and salts, frequently including denaturing and stabilizing components such as detergents, chaotropic agents, and glycerol. Techniques commonly used to concentrate and desalt proteins are susceptible to large sample losses. This is especially problematic when handling limited amounts of sample or when the protein is highly hydrophobic.

Reverse-phase HPLC has been an ideal method for preparing a protein for enzymatic digestion, if the protein can be recovered in sufficient yield. Various conditions have been evaluated to aid in greater recoveries such as the use of elevated temperature, polymeric supports, and organic modifiers other than acetonitrile (e.g. propanol and isopropanol) (1). All of these variations have had limited success. Hewlett-Packard has recently introduced the new G1000S Protein Sequencing System which enables proteins in complex matrices to be sequenced directly. This system employs a miniature hydrophobic column to which samples are applied and subsequently washed to remove any substances which will interfere with sequence analysis (2). The column containing bound protein is subsequently taken to the sequencer for on-column automated Edman degradation. Bente et al. (3) have shown that proteins can be loaded onto the sequencing column in various matrices and subsequently washed without suffering losses.

In addition to on-column chemical modification(4), it would be desirable to be able to manipulate the bound protein enzymatically. I have investigated methods for *in situ* proteolytic digestion using trypsin, and the endoproteinases Lys-C and Asp-N to generate peptides from the bound proteins. *In situ* reduction and alkylation of cysteines can be performed before digestion, if desired. Coupling the sequencing column

TECHNIQUES IN PROTEIN CHEMISTRY IV

to another HPLC column via the column adapter allows separation of peptides generated by enzymatic or chemical means without the need for transfer of the peptides or lyophilization. Thus, losses are minimized and potential problems such as adsorption to surfaces and sample insolubility are eliminated. Moreover, the advantages of reverse-phase chromatography are fully utilized while many problems, such as the inability to recover protein from the column, are avoided. A method for *in situ* digestion on a reverse-phase support provides a simple procedure for handling low amounts of problematic proteins. In this communication, I present data which shows that *in situ* enzymatic digestion of proteins can be accomplished by this technique.

II. Materials and Methods

Recombinant FKBP12 was produced and purified at Glaxo Inc. by Rebecca Gooding. Each digestion contained 5.7 nanomoles, as determined by amino acid analysis using the ABI 420 Analyzer. Myosin and ovotransferrin were purchased from Sigma. For digestions, 120 picomoles of myosin and 300 picomoles of ovotransferrin were used. Samples were sequenced on either the ABI 477A Protein Sequencer with the 120A PTH Analyzer or the Hewlett-Packard G1000S Protein Sequencing System with on-line PTH analysis. The miniature sequencing columns (1 X 20mm) packed with a proprietary ODS-like support, the G1001A Sample Prep Station, and the sequencing column adapter were supplied by Hewlett-Packard. Proteins were loaded onto the column according to the manufacturer's protocol. The column was pre-wetted using 1 ml of methanol, washed with 1 ml of 2% aqueous trifluoroacetic acid followed by 0.5 ml of water. The protein solution was then applied directly. It is important to note that proteins will not bind to the column if loaded in 1% SDS or Triton. If detergent is present, the sample should be diluted to about 0.1% detergent before application to the column. The bound protein samples were washed to remove any buffers, salts, etc. using 2 times 1 ml of deionized water. A final wash with 0.5 ml of 20% aqueous acetonitrile was made if detergent was present in the initial protein solution. If pyridylethylation of cysteine was desired, 0.4 ml of 20 mM DTT and 50 mM 4-vinylpyridine in 6 M guanidine hydrochloride containing 0.5 M Tris-HCl, pH 8.6 and 5 mM EDTA was dripped through the column leaving a small amount in the reservoir. After incubation for 15 min at room temperature, the column was washed with 0.5 ml of water and 1 ml of 20% acetonitrile to remove buffer and excess reagents.

Sequencing grade trypsin and endoproteinase Asp-N were from Boehringer-Mannheim (Indianapolis, IN). Endoproteinase Lys-C was purchased from Wako Chemicals (Dallas, TX). For all digestions, 0.4 ml of 0.1 M Tris-HCl, pH 8.5 containing 20% acetonitrile was used. To this buffer was added 5 μg of trypsin or Lys-C or 2 μg of Asp-N for the respective digestions. The digestion buffer containing protease was

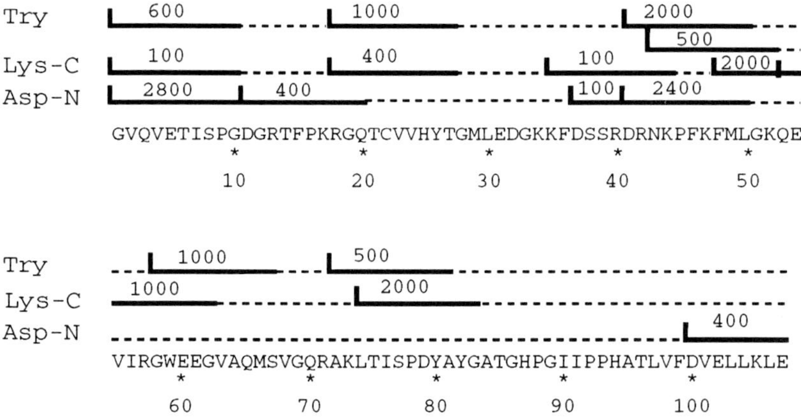

Figure 1. Sequence results of FKBP12 *in situ* digestions. After each digestion, the sequencing column was taken to the sequencer for 10 cycles of Edman degradation. Sequencing initial yields are given in picomoles and are rounded to the nearest hundred.

dripped through the column leaving a small volume in the reservoir. The column with reservoir attached was removed from the sample prep station and placed at 37 °C in a closed container for 20 hr. After digestion, the peptides were either batch eluted using 0.4 ml of 80% acetonitrile or the column was placed in-line on the HPLC using the column adapter. Peptides were separated using a Hypersil ODS (2.1 X 100mm; Hewlett-Packard) column employing a linear gradient of 8 to 48% acetonitrile in 0.1% TFA over 80 min with a flow rate of 100µl/min. After 80 min, the gradient was stepped up to 100% acetonitrile over the next 20 min. For smaller proteins (e.g. FKBP12), it was possible to take the digest directly to the sequencer after washing with water to remove digestion buffer. This allowed sequence determination of the peptide mixture directly without separation.

III. Results

As a model protein, FKBP12, a rotamase enzyme and receptor of FK506 and rapamycin, was chosen since it was available in large amounts and its structure had been determined recently by nuclear magnetic resonance and X-ray crystallography (5, 6). Figure 1 shows the resulting mixed sequence patterns obtained from *in situ* digestions with trypsin, Lys-C, and Asp-N. For all three proteases, the *in situ* digestion is restricted, in that certain cleavage sites are preferred over others. This contrasts to results obtained if the digestion is done in solution. Due to

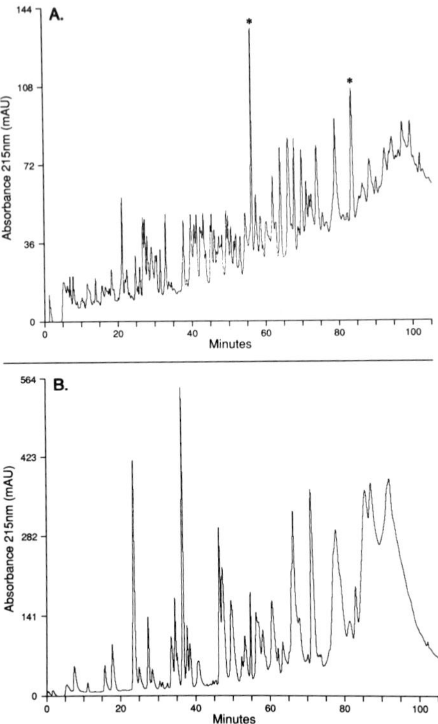

Figure 2. A, Chromatogram of peptides generated by *in situ* Lys-C digestion of rabbit myosin. Peaks marked with a (*) are autoproteolysis products of the Lys-C. B, Chromatogram of peptides generated by *in situ* Lys-C digestion of ovotransferrin .

the experimental design, short, hydrophilic peptides were lost by washing the column after digestion. In practice, this problem can be eliminated by either collecting any wash following digestion or by using the in-line column adapter.

In the tryptic digest, no cleavage was observed at Lys 73, although Arg 71 was cleaved. Cleavage at Arg 40 was about four times faster than at Arg 42, as determined by sequencing initial yields. No detectable cleavage was observed at Lys 47. Lys 44 is followed by a Pro residue and thus no cleavage is observed in the tryptic and Lys-C digests. The occurrence of restricted digestion has been observed with all of the proteins I have worked with using these new procedures. The conformation of the protein has an influence on the accessibility of cleavage sites. The effect on protein conformation when bound to the hydrophobic support is difficult to assess and would be protein dependent.

All three proteases readily cleave FKBP12 in the region around residues 40-50. This region is in a well exposed loop on the surface of the protein according to the NMR solution structure. With the Asp-N digestion, it is interesting to see that no cleavage occurred at Asp 79 *in situ*, but this site is readily cleaved in solution. The region around Asp 79 is in a beta sheet, which is most likely blocking access to the protease during *in situ* digestion. Using FKBP12, it has been demonstrated that limited peptide maps can be obtained for small, model proteins.

It remained to determine the utility of this method for the characterization of proteins of large molecular weight and of unknown structure. *In situ* Lys-C peptide maps have been obtained for the model proteins myosin (200kD) and ovotransferrin (78kD). Figure 2(A&B) shows the HPLC chromatograms obtained using the column adapter to place the sequencing column in-line with a narrowbore reverse-phase column. A novel protein, rat liver 3-phosphatase (50kD), was characterized by the *in situ* method. The protein arrived in a dilute solution containing buffer, a high concentration of salt, and CHAPS detergent. Previous attempts to concentrate and desalt the sample resulted in complete loss of the protein. *In situ* digestion after *in situ* pyridylethylation produced a set of peptides (see Table 1) which yielded sufficient sequence information to construct probes for cloning the protein. Figure 3 shows the HPLC chromatogram of the Lys-C digestion of the phosphatase using the column adapter.

IV. Discussion

A method for *in situ* proteolytic digestion on a reverse-phase column has been developed and demonstrated for proteins of both known and unknown structure, ranging from low picomole to

Table 1. Sequence Analysis Results for Peptides Isolated after *In Situ* Lys-C Digestion of Rat Liver 3-Phosphatase

Retention time (min)	Initial yield (pmol)	Cycles analyzed	Retention time (min)	Initial yield (pmol)	Cycles analyzed
27.3	5	4/4	58.9	9	9/9
31.8	11	9/9	75.6	14	21/26
35.7	5	12/12	76.7	13	20/25
41.6	17	9/9	78.6	5	18/20
54.3	11	10/10	81.7	3	22/24
55.7	12	20/20	90.0	20	27/30
57.4	13	20/20	93.7	6, 3	ND[a]

[a]Two peptides were present and the sequences were not determined.

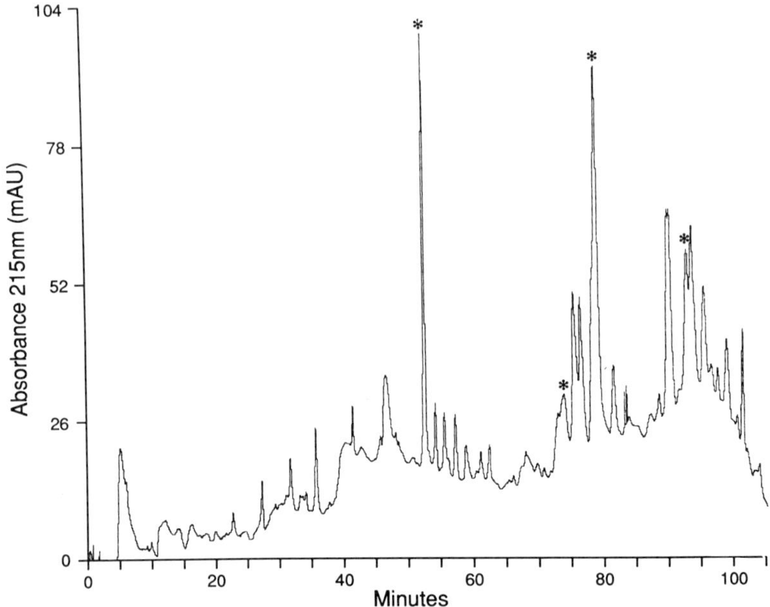

Figure 3. Chromatogram of peptides generated by *in situ* Lys-C digestion of rat liver 3-phosphatase. Peaks marked with a (*) are autoproteolysis products of the Lys-C.

nanomole amounts. The described method is in routine use and has already been an aid in handling a variety of proteins submitted for sequence analysis. These proteins have ranged from 8kD to 190kD in molecular weight and several have been available in very limited amounts. This procedure eliminates the need for many concentration and desalting steps that can cause high losses of protein and solubility problems. Since many proteins are resistant to proteolytic digestion, it will be useful to see if other less hydrophobic stationary phases, such as C-4 and phenyl, can enhance the efficiency of *in situ* digestion. Riviere et al. (7) have described the use of detergents and chaotropic agents for aiding in proteolytic digestion and it will be useful to evaluate if any *in situ* pretreatments with these agents can enhance digestion. However, denaturing, and thus exposing hydrophobic regions of a protein, could result in tighter binding to the column and actually reduce digestion. With some proteins, performing reduction and alkylation before loading the protein could render it less amenable to *in situ* proteolytic digestion.

A simpler approach to handling proteins that are resistant to proteolytic digestion could be to perform an *in situ* CNBr cleavage followed by an *in situ* proteolytic digestion. Stone et al. (8) have found such an approach useful in dealing with PVDF-blotted proteins. Riviere's findings also indicate other proteases such as chymotrypsin, endoproteinase Glu-C, and subtilisin are stable in the presence of acetonitrile. Preliminary trials using Glu-C and chymotrypsin have been successful in performing *in situ* digestions. Further experimentation will be to examine the feasibility of using additional proteases *in situ*. Lower amounts of protease have not been tested, but amounts could probably be reduced when working with small quantities of protein. Interference from autoproteolysis has not been any more of a problem than when performing digestions in solution. The presence of the 20% acetonitrile was found to be critical in order to obtain digestion. With trypsin and Lys-C, for example, no digestion occurred in the absence of acetonitrile.

The availability of an adapter to couple the sequencing column in-line to an HPLC, eliminates the need for batch elution before running the peptides on microbore LC. Although the lower level of detection of PTH-amino acids is in the mid-femtomole range and FABMS can sequence peptides at the low picomole range (9), investigators normally begin working with at least ten picomoles when attempting to obtain amino acid sequence on an unknown sample. The handling of a few picomoles of protein presents the protein chemist with formidable challenges. The method described in this paper has already been found very useful in obtaining sequence from small amounts of protein by allowing for the manipulation of a sample with the elimination of many steps which result in dramatic losses. As the detection limits of PTH-amino acids and FABMS techniques are pushed to lower and lower levels, it will be imperative that sample handling techniques be improved to adequately handle lower starting amounts of protein. The capability of performing *in situ* proteolytic digestions on the hydrophobic sequencing column promises to be a great aid in handling low amounts of protein for Edman sequencing and mass spectrometry applications.

References

1. Wehr, T., Lundgard, R., and Nugent (1989) *LC-GC* **7**, 32-37.
2. Miller, C., Bente, H., Fischer, S., Myerson, J., Wagner, G., Widmayer, R., and Horn, M. (1990) Poster T-143, Fourth Symposium of the Protein Society.
3. Bente, H., Fischer, S., Horn, M., Miller, C., Myerson, J., and Wagner, G. (1990) Poster T-144, Fourth Symposium of the Protein Society.
4. Wagner, G., Fischer, S., Myerson, J., Miller, C., Bente, H., and Horn, M. (1990) Poster T-141, Fourth Symposium of the Protein Society.
5. Michnick, S., Rosen, M., Wandless, T., Karplus, M., and Schreiber, S. (1991) *Science* **252**, 836-839.
6. Van Duyne, G., Standaert, R., Karplus, P., Schreiber, S., and Clardy, J. (1991) *Science* **252**, 839-842.
7. Riviere, L., Fleming, M., Elicone, C., and Tempst, P. (1991). *In* "Techniques In Protein Chemistry II"(Villafranca, J., ed.) 171-179.
8. Stone, K., McNulty, D., LoPresti, M., Crawford, T., DeAngelis, R., and Williams, K. (1992). *In* "Techniques In Protein Chemistry III" (Angeletti, R., ed.) 23-34.

9. Kassel, D., Williams, K., Musselman, B., and Smith, J. (1991) *Anal. Chem.* **63**, 1978-1983.

Acknowledgments

The author wishes to thank the investigators who submitted proteins for sequence analysis. I wish to thank Nora G. Geddie (Glaxo) for performing amino acid analysis. I am grateful to Mary Moyer and Drs. John Graham, Daniel Kassel, and Robert Anderegg for critical review of this manuscript. I also wish to thank the Protein Chemistry Systems Group at Hewlett-Packard for their helpful discussions.

SECTION VIII

Protein Sequencing

ROUTINE PROTEIN SEQUENCE ANALYSIS BELOW TEN PICOMOLES: ONE SEQUENCING FACILITY'S APPROACH

Donna Atherton, Joseph Fernandez, Michael DeMott, Lori Andrews, and Sheenah M. Mische

The Rockefeller University Protein Sequencing / Howard Hughes Medical Institute Biopolymer Facility, New York, N.Y. 10021

INTRODUCTION

The isolation of low picomolar amounts of purified proteins and peptides has become a common occurrence, due largely to substantial advances in the techniques available for sample purification. These include the advent of SDS-PAGE and electrophoretic transfer to polyvinylidene difluoride (PVDF) membrane , the introduction of microbore HPLC, and improved protocols for the enzymatic digestion of membrane bound proteins (1-4). As a result, there is an increased need for routine protein sequence analysis at low picomole levels. Over the last three years, The Rockefeller University Protein Sequencing / Howard Hughes Medical Institute Biopolymer Facility has handled 1294 samples, with 58% of the 947 successful runs having an initial yield of ten picomoles or less, 42% with five picomoles or less. To consistently perform at this sensitivity, both the sequencer and the on-line HPLC must be operated and maintained at optimal performance conditions beyond those stated by the manufacturer. Currently, Applied Biosystems, Inc. (ABI) provides service and guarantees performance of their sequencing systems based on 100 picomoles of β-lactoglobulin in solution spotted to a preconditioned polybrene filter (PCPB) with a minimum 92% repetitive yield. In our facility, three Applied Biosytems sequencers, two Model 470A and one Model 477A, are maintained so that ten picomoles of β-lactoglobulin spotted to PCPB, or alternatively, twenty-five picomoles of β-lactoglobulin analyzed by SDS-PAGE and transferred to PVDF, sequence with a minimum initial yield of 50%, a repetitive yield of greater than 93%, and positive identification of each of the first 15 amino acids. Here we describe our procedures for consistently operating at these specifications, including instrument modifications, scheduled preventive maintenance protocols, a daily instrument optimization program, a protocol for loading PVDF-bound samples, and modified sequencer control cycles.

TECHNIQUES IN PROTEIN CHEMISTRY IV

Table Ia. Preventive Maintenance for Applied Biosystems 470A and 477A

Six Week	Three Month	Six Month	Yearly	Procedure	Purpose
Clean Conversion Flask	Clean Conversion Flask	Replace Conversion Flask	Replace Conversion Flask	Clean w/ 2M KOH for 30 min (ABI user bulletin #24). Replace flask assembly every 6 months.	Eliminates problems due to precipitate buildup, including some 120A transfer line clogs.
Check R3 Valves	Check R3 Valves	Check R3 Valves	Replace R3 Vent Valve	Look for corrosion, including TFA resistant valves. Replace as necessary.	Eliminates source of background artifact peaks.
S1 and S2 Flow Rates	S1 and S2 Flow Rates	S1 and S2 Flow Rates	S1 and S2 Flow Rates	With graduated cyclinder, measure from outlet line; argon dry before reading. 470A rates: S1: (1.8-2.4 ml/300 sec) & S2: (1.5-1.8 ml/240 sec). 477A rate: S2 (1.0-1.5 ml/240 sec).	Flow rates should be consistent to maximize removal of by-products while minimizing washout of protein.
Flush Exhaust Line	Flush Exhaust Line	Flush Exhaust Line	Flush Exhaust Line	Flush exhaust line with methanol.	Assures clear path for waste gases, and prevents liquid slugs from causing back pressure buildup.
Change pump oil and vacuum test	Change pump oil and vacuum test	Change pump oil and vacuum test	Change pump oil and vacuum test	Drop < 1 inch Hg/10 minfor vacuum assist test. Not applicable to 477A.	Vacuum assist system critical to proper valve block operation.
Run ß-Lactoglobulin	Run ß-Lactoglobulin	Run ß-Lactoglobulin	Run ß-Lactoglobulin	In solution: 10 pmol on PCPB. On PVDF: load 25 pmol on SDS-PAGE.	Verifies integrity of instrument.
	Check acid trap	Check acid trap	Check acid trap	pH ~ 2-3.	Assure effectiveness of trap.
	Wash "E" and "A" Blocks	Wash "E" and "A" Blocks	Wash "E" and "A" Blocks	With ACN (5 min), then 25% TFA (30 min), then ACN (5 min). 470A from S3 position, 477A from S2.	Helps reduce buildup of salts and other contaminants which may affect baseline.
	Clean Fan Filters	Clean Fan Filters	Clean Fan Filters	Rinse under tap and dry.	Reduces heat buildup around electrical components, thereby reducing computer failures.
	Clean Touchsreen	Clean Touchsreen	Clean Touchsreen	Remove front panel to clean diodes.	Improves touchscreen response.
	Inspect R3 cap assembly (470A) Inspect R1-R4 cap assemblies (477A).	Inspect R1-R4 cap assemblies (470A) Inspect all cap assemblies (477A).	Inspect all cap assemblies	Replace worn or damaged parts.	Critical to proper solvent/reagent delivery, to exclusion of oxygen, and to elimination of corrosion byproducts which generate background artifacts.
	Leak Test relevant Cap assemblies.	Leak Test relevant Cap assemblies.	Leak Test relevant Cap assemblies.	Drop < 0.05 psi/5 min.	To avoid oxygen contamination and to ensure proper solvent/reagent delivery.
		Replace inlet and outlet lines.	Replace inlet and outlet lines.	Replace Inlet and Outlet lines to cartridge every six months.	Helps reduce buildup of salts and other contaminants which may affect baseline and hinder flow rates.

Table Ib. Preventive Maintenance for Applied Biosystems 120A

Six Week	Three Month	Six Month	Yearly	Procedure	Purpose
Check for Leaks	Check for Leaks	Check for Leaks	Check for Leaks		Any leak (liquid or dried salt) should be considered serious.
Replace Static Mixer	Replace Static Mixer	Replace Static Mixer	Replace Static Mixer		Routine replacement is a cost effective means of reducing many problems, including baseline fluctuations and overpressuring.
Sonicate Buffer Frits	Sonicate Buffer Frits	Sonicate Buffer Frits	Replace Buffer Frits	In methanol for 20 min.	Often neglected source of contaminants.
Check Lamp Status	Lamp Performance Test	Lamp Performance Test	Lamp Performance Test	Follow procedure in 120A user manual.	May signal imminent lamp failure. Also allows monitoring of electronic noise level as lamp ages.
	Replace Pump Seals	Replace Pump Seals	Replace Pump Seals	Front, Rear, and Cylinder Head.	Routine replacement is a cost effective means of minimizing baseline fluctuations, noise and especially shifting retention times.
	Acid Wash Pumps	Acid Wash Pumps	Acid Wash Pumps	Purge Pumps: 3x Milli-Q water, 3x 50% ACN/0.125% TFA, 3x ACN, 3x 50% ACN/0.125% TFA, 3x Milli-Q water.	Helps reduce buildup of salts and other contaminants which may affect baseline. Helps to seat new pump seals.
	Pressure Test	Pressure Test	Pressure Test	Drop < 75 psi/15 min (to purge port #3). Drop < 150 psi/15 min (to injector port #2).	To help detect invisible yet destructive leaks.
	Clean Photodiodes	Clean Photodiodes	Clean Photodiodes	With methanol and a cotton swab.	ABI recommended. Eliminates dust which may intefere with proper absorbance.
	Wash Column	Wash Column	Wash Column	90% B, 210 μl/min, 30 min.	ABI recommended. Organic solvent wash may reduce buildup of some contaminants.
			Replace Transfer Line		Yearly replacement helps reduce incidence of clogs.
			Replace Rheodyne Valve Seals	Replace seal in each valve type (purge, refill, injector). Follow installation guide. Purge valve seal is available only from Rheodyne.	Worn seals lead to leaks and therefore to baseline fluctuations.

MATERIALS AND METHODS

Chemicals and reagents: All sequencer reagents (sequencing grade) and related sequencer supplies were purchased from Applied Biosystems, except the acetonitrile (ACN) (HPLC grade) which is used for Solvent B on the Model 120A PTH analyzer and was from Burdick and Jackson. S4B was diluted 1:1 with Milli-Q water to achieve a final 10% acetonitrile solution. Polybrene was from Pierce; ß-lactoglobulin from Applied Biosystems or Sigma; and PVDF was either supplied by individual investigators or purchased from Millipore (Immobilon-p, 0.45 µm). All electrophoretic chemicals, as well as the preparation of Milli-Q water, were as previously described (4,6). Prepurified argon and nitrogen (99.998%) were supplied from Matheson Gas Products, Inc. A one inch basin strainer, No. 1505, was from Melard Mfg. Corp., Passaic, N.J.

SDS-PAGE and electroblotting of ß-lactoglobulin: ß-lactoglobulin was electrophoresed according to Laemmli, using a Hoefer Mighty Small II slab gel apparatus (7). Protein bands were then electrophoretically transferred to PVDF, using a Hoefer TE 22 Mini Transphor apparatus (1). Finally, membranes were stained with 0.1% amido black/10% acetic acid, destained using 5-10% acetic acid, and then rinsed twice in Milli-Q water. The air-dried PVDF was then stored at 4°C.

N-terminal sequence analysis: Sequence analysis was performed on either an ABI 470A gas-phase sequencer equipped with a 900A data-acquisition system or an ABI 477A pulsed-liquid sequencer. Argon gas was supplied to all three sequencers from main and backup tanks through a manifold system. The 60 µL load loop in each of the 470A sequencers was replaced by a 75 µL loop. In addition, the conversion flask temperatures were increased to 64°C, to insure proper drying of R4A in the flask. All sequencers were equipped with on-line Model 120A PTH analyzers (Applied Biosystems) and dual pen strip chart recorders (Kipp and Zonen type BD 41). Nitrogen gas was supplied from one tank through a manifold to all three PTH analyzers. PTH-C18 column temperatures were adjusted within a range of 51°C - 56°C to effect optimal separation and resolution of PTH-amino acids. All PTH analyzers were fitted with a 100 µL injector loop, and sequencer control cycles for all sequencer models were modified to deliver 150 µL of dilute S4B to the conversion flask (5). The linear gradient utilized in all instances was both a modification of the gradient recommended by Tempst and by ABI (5,11). Initial and final percentages of Solvent B were varied in order to achieve optimal separation of PTH-amino acids on a given instrument. PCPB filters were generated by spotting 3.6 mg of polybrene onto a TFA-treated cartridge filter (GF/C) and precycling with 3 ABI-recommended FIL-3 cycles. Samples were loaded according to the protocol presented below. A Best Co. 5 KVA uninterruptible power supply with 50 min backup batteries was interfaced with all sequencer components listed.

PREVENTIVE MAINTENANCE

The goal of the preventive maintenance protocol (see Tables Ia and Ib) is to avoid all predictable, and therefore, preventable instrument failures and contaminants. When operating a sequencer at high sensitivity, the degenerative by-products due to the aging of instrument components and the buildup of salts from the sequencer chemistry become visible as background noise on the HPLC chromatogram much sooner than when less sensitivity is required. Therefore, more frequent cleaning as well as replacement of some sequencer components is necessary to eliminate this baseline noise. The elimination of this noise enables us to load less than ten picomoles into the sequencer and identify, on HPLC, PTH-amino acids at the femtomole level. Figure 1 represents the chromatographic baseline from an instrument before and after scheduled maintenance. In Figure 1a-1c the baseline is erratic, making the assignment of PTH-His, Arg, and Ser difficult. While the identification of most PTH-amino acids is possible at five picomoles, it is not possible at the femtomole level. Figure 1d-1f represents a stable baseline from the same instrument following scheduled maintenance.

Any preventive maintenance protocol must address a diversity of potential sequencer malfunctions. Major mechanical or electronic malfunctions such as a

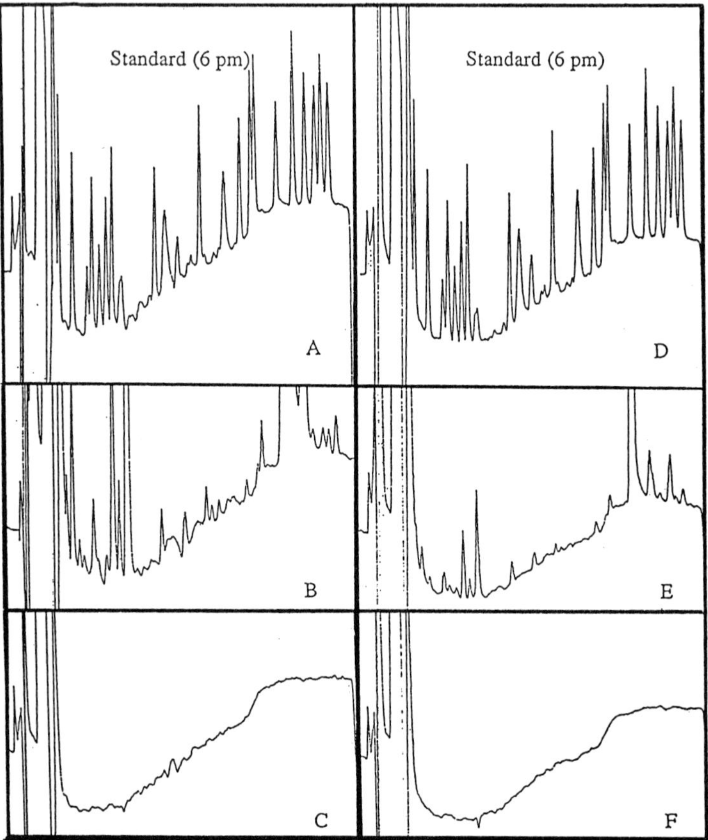

Figure 1. Representative PTH chromatograms from a sample (A-C) sequenced on a 470 A immediately before yearly preventive maintenance (see Tables Ia and Ib), and PVDF-bound ß-lactoglobulin (D-F) sequenced on the same instrument immediately after yearly preventive maintenance. The small baseline fluctuations in B and C are indicative of either aging components or a build up of contaminants in the sequencing system. No similar baseline disturbances were observed after preventive maintenance was performed. These baseline fluctuations are generally intermittent, making identification of less than 1 pmol of amino acid difficult and unreliable.

valve failure are adequately addressed by the manufacturer's recommended protocols. The deterioration of performance at high sensitivity due to baseline noise and loss of sample is minimized by our maintenance schedules. Background noise on the HPLC chromatogram can be caused by external sources such as reagent and solvent contaminants, and environmental electrical noise (power surges, brownouts). Or, it can be from internal sources such as electronic noise generated by improperly adjusted or faulty sequencer components, degenerative by-products from aging sequencer components, erratic flows, and buildup of salts throughout the sequencer. Sample loss can be caused by mechanical problems such as partial or complete blockage of lines, adsorption of sample to salt buildup in the conversion flask, transfer lines, or valve blocks, or adsorption of sample to an HPLC column. Electronic problems such as software bugs or static electricity can also result in loss of sample or data. Operator error is also a source of problems.

The preventive maintenance protocols used in our facility were empirically developed over the course of eight years to address the above sources of malfunctions. Their use as outlined in Tables Ia and Ib, as well as the use of an uninterruptible power supply, has helped to reduce the above sequencer malfunctions to less than one percent. This compares favorably to the 9.7% average of sequencer malfunctions reported over three years by the Association of Biomolecular Resource Facilities (ABRF) (12-14).

In addition to the preventive maintenance protocols in Table I, daily maintenance is performed prior to the sequence analysis of each new sample. First, a sequencer self test is performed, a PTH standard is run to optimize the chromatographic separation of all PTH-amino acids, and data from the previous sample is checked for any instrument problems, especially intermittent ones. Then the operating parameters of the instrument are efficiently checked and adjusted using the optimization program listed in Table II. Because sequencer flow rates will change with different types of filters, filters appropriate for the sample to be analyzed are used to optimize the instrument (PCPB for samples in solution; blank GF/C for PVDF samples). To eliminate any back pressure, the waste bottle is vented for two minutes to remove any accumulation of liquid in the back waste line. The log book for each instrument is completed, noting lot numbers of all replenished reagents, solvents, buffers, etc., as well as sample information and programs used. Separate logs and sample chromatograms are also kept of all sequencer software bugs and malfunctions encountered .

After the sample is loaded into the sequencer, lab personnel routinely use a checklist in order to minimize the most common sources of operator error. They check for adequate supplies of solvents, reagents, HPLC buffers, and fraction collector vials. Because identifications of PTH-amino acids are made from the strip chart recorder and not from the Model 900, an adequate supply of strip chart paper, its proper alignment in the recorder, and properly functioning pens are crucial. Also, the system will pause without sufficient paper in the printer. Pausing a sequencer run at 10 pmol sensitivity can adversely affect the analysis. So, if the printer is set to print, the paper supply should be checked. The flask cavity cover must be in place in order to maintain a constant conversion flask temperature (470A only). The cartridge is leak tested (drop <0.05 psi/ 5 min), the argon regulator to the cartridge is reset, the fraction collector tray is reset, and the status log is checked 100 seconds after the start of the run (some warnings take up to 100 sec to register).

The cost of this additional maintenance can be significant; however, when compared to the financial and time investment of investigators, the costs are far outweighed by the benefits. The ability to routinely operate at low picomole levels has actually increased income to the facility.

Table II. Representative Optimization Cycle for Model 470A Sequencer[a]

Reaction Cycle Listing		Conversion Cycle Listing	
Cycle Name	: RU-OPTIMIZE	Cycle Name	: RU-OPTIMIZE
Gradient	: NONE	Gradient	: NONE
Run Time	: 24 min 30 sec	Run Time	: 24 min 30 sec

STEP	FUNCTION	*b*	TIME	STEP	FUNCTION	TIME	
1	Argon Dry		70	1	Block Flush	10	
2	Prep Transfer		30	2	Deliver S4	25	
3	Transfer w/ S3		12	3	Argon Dry	4	
4	Pause		10	4	Empty	25	
5	Transfer w/ S3	*d*	25	5	Load R4	6	*c*
6	Pause		10	6	Ready to Recieve	194	
7	Transfer w/ Argon		3	7	Block Flush	10	
8	Transfer w/ S3		25	8	Argon Dry	150	
9	Pause		10	9	Load R4	6	*c*
10	Transfer w/ Argon		3	10	Argon Dry	400	*h*
11	Transfer w/ S3		25	11	Block Flush	20	
12	Pause		10	12	Clear Inj To Waste	120	
13	Transfer w/ Argon		30	13	Load S4	6	*c*
14	End Transfer		1	14	Argon Dry	4	
15	Argon Dry		60	15	Load S4	6	*c*
16	Prep R1	*e*	15	16	Argon Dry	4	
17	Deliver R1		3	17	Pause	10	
18	Argon Dry	*f*	20	18	Load Injector	50	*i*
19	Deliver R2	*e*	50	19	Inject	1	
20	Argon Dry		10	20	Pause	10	*j*
21	Deliver R2	*e*	50	21	Load S4	6	*c*
22	Argon Dry		10	22	Argon Dry	4	
23	Deliver R2	*e*	50	23	Load S4	6	*c*
24	Argon Dry		60	24	Argon Dry	4	
25	Deliver S1	*e,f*	100	25	Clear Inj. To FC	50	*k*
26	Deliver S2	*e,f,g*	140	26	FC Advance	1	
27	Argon Dry		120	27	Empty	20	
28	Deliver R3	*e*	50	28	Pause	110	
29	Argon Dry		10	29	Argon Dry	118	
30	Deliver R3	*e*	50	30	Empty	30	
31	Argon Dry		10	31	Prep R5	10	
32	Deliver R3	*e*	50	32	Load R5	6	*c*
33	Argon Dry		130	33	Block Flush	10	
34	Prep S3		10				
35	Transfer w/ Argon		30				
36	Transfer w/ S3		12				
37	Pause		10				
38	Transfer w/ S3	*d*	25				
39	Transfer w/ Argon		30				
40	Pause		1				
41	Argon Dry		100				

a : Optimization Cycle for 477A is similar. Some procedures as per ABI user manuals (8-10).
b : Swing injector into load postion.
c : Verify that sequencer loop fully loads (adjust regulator accordingly).
d : Verify time remaining in step when S3 first drips into flask (3 to 8 sec).
e : Adjust appropiate regulator to the proper setting.
f : Verify that solvent/reagent is delivered to reaction cartridge/flsk.
g : Verify that S2 drips into waste (30-80 sec remaining in step).
h : Flask should be completely dry with 50-80 sec remaining in step.
i : Check transfer time to 120A as per ABI User Manual procedure.
j : Swing injector to inject position.
k : Verify that S4 in flask is transferred to fraction collector.

PROTOCOL FOR SAMPLE TREATMENT

Samples on PVDF: First, the sequencer is optimized with a standard blank GF/C filter mounted in the sequencer upper cartridge block (see Table II). Then using a small glass plate, forceps, and a new razor blade (all cleaned thoroughly with methanol), narrow strips (~3 mm wide) of varying lengths are cut from the PVDF sample such that the strips will fit side by side, without overlap, within the 11 mm diameter of the upper cartridge block. To facilitate flow through the strips and the sequencer cartridge, each strip is slit at 1 mm intervals. Twenty-five µl of a 90 mg/ml stock solution of polybrene is diluted with 25 µl ACN. The slit strips are then arranged on the blank GF/C filter on the upper cartridge block and wetted (strips only) with the dilute polybrene solution. The addition of polybrene insures against the loss of any sample. After covering with a Melard basin strainer, the strips can be dried without danger of blowing about in the oven compartment. The strips are rewetted with polybrene and dried again. The sample is then ready to be loaded into the sequencer.

Samples in solution: The sequencer is optimized using a PCPB filter. If the sample is dry, it is dissolved in an appropriate solvent (usually 50% ACN/0.05% TFA) and allowed to set 20 min. All samples are sonicated prior to spotting onto a PCPB filter which is mounted in the upper cartridge block and then dried as above. The sample vial is washed with ACN, vortexed, added to the sample on the PCPB filter, and dried. Separate washes are repeated with 50% ACN/0.05%TFA and then neat TFA. The sample is ready to be loaded into the sequencer.

Table III. Sequencer 470A "Wash-100" Reaction Cycle Listing[a]

STEP	FUNCTION	TIME (sec)
1	Argon Dry	180
2	Deliver S3	10
3	Pause	20
4	Deliver S3	120
5	Argon Dry	120
6	Deliver S1	180
7	Deliver S2	240
8	Argon Dry	120
9	Deliver S1	120
10	Deliver S2	240
11	Argon Dry	120
12[b]	Deliver R3	870
13	Argon Dry	120
14	Deliver S3	180
15	Argon Dry	120
16	Deliver S1	120
17	Deliver S2	240
18	Argon Dry	180

[a] Solvent and acid wash run routinely on every sample prior to Begin cycle on sequencer. **Warning: if using a 900A, never use Wash-100 cycles when the 120A is set to remain inactive, since a scheduling error developes.**

[b] Step 12 on 470A Gas Phase Sequencer is replaced with the following 5 steps on the 477A Pulsed Liquid Sequencer: Load R3, 5 sec; Argon Dry, 4 sec; Pause, 400 sec; Load S2, 6 sec; Block Flush, 6 sec.

Table IV. Modified Sequencer Edman Cycles for Model 470A[a]

REACTION CYCLE			CONVERSION CYCLE		
Cycle Name	: SPIKE-100		Cycle Name	: SPIKE-100	
Run Time	: 51 mins 28 secs		Run Time	: 48 mins 6 secs	
STEP	**FUNCTION**	**TIME (sec)**	**STEP**	**FUNCTION**	**TIME (sec)**
1	Prep R2	6	1	Block Flush	10
2	Deliver R2	20	2	Load R4	6
3	Prep R1	6	3	Ready To Recieve	197
4	Deliver R1	3	4	Block Flush	10
5	Argon Dry	50	5	Argon Dry	150
6	Deliver R2	400	6	Load R4	6
7	Deliver R1	3	7	Argon Dry	4
8	Argon Dry	50	8	Pause	1200
9	Deliver R2	400	9	Argon Dry	400
10	Deliver R1	3	10	Block Flush	60
11	Argon Dry	50	11	Clear Inj. To Waste	120
12	Deliver R2	400	12	Load S4	6
13	Argon Dry	60	13	Argon Dry	4
14	Deliver S1	100	14	Load S4	6
15	Deliver S2	240	15	Argon Dry	4
16	Argon Dry	120	16	Pause	540
17[b]	Deliver R3	800	17	Load Injector	22
18	Argon Dry	30	18	Inject	1
19	Prep Transfer	30	19	Load S4	6
20	Transfer w/ S3	12	20	Argon Dry	4
21	Pause	10	21	Load S4	6
22	Transfer w/ Argon	3	22	Argon Dry	4
23	Transfer w/ S3	25	23	Clear Inj To FC	60
24	Pause	10	24	Deliver S4	25
25	Transfer w/ Argon	3	25	Argon Dry	4
26	Transfer w/ S3	25	26	FC Advance	1
27	Pause	10	27	Empty	30
28	Transfer w/ Argon	3			
29	Transfer w/ S3	25			
30	Pause	10			
31	Tansfer w/ Argon	30			
32	End Transfer	1			
33	Deliver S3	30			
34	Argon Dry	120			

[a]Reaction steps 20-31 adopt the S3 extraction outlined by Speicher, whereas reaction step 15 retains the 240 sec S2 delivery of the standard ABI program (11). The multiple, short deliveries of S2 outlined by Speicher were not adopted, as artifact levels were unacceptably high for routine sequence analysis below 10 pmol.

[b]Reaction step 17 in the 470A program is replaced with the following 5 steps in the 477A program: Load R3, 5 sec; Argon Dry, 4 sec; Pause, 300 sec; Load S2, 6 sec; Block Flush, 6 sec.

SEQUENCE ANALYSIS

A typical sequence analysis is composed of a WASH-100 (Table III) cycle, one BEGIN cycle, and as many SPIKE-100 (Table IV) cycles as needed. This program sequence generates baseline profiles of an HPLC blank, an S4b blank from the conversion flask, and a calibration table, followed by the profiles of the sequence of PTH amino acids from the sample.

Table III represents a wash program used to "clean up" experimental samples prior to the start of the Edman degradation. This preliminary wash increases the chances of obtaining successful sequence data from an experimental sample. Table IV represents the modified reaction cycle routinely used in the Facility.

DATA INTERPRETATION

Data interpretation at these sensitivities is affected by background noise displayed on the chromatogram (see Figure 1) and by non-linear elution of some of the amino acids on the HPLC column. Familiarity with the current baseline fluctuations is imperative in order to confidently make amino acid assignments. Also, since the column adsorption of certain amino acids increases with the age of the column, it is important to be familiar with the current relative responses of each amino acid. This can be assessed from the amino acid standard profile at the beginning of each run. This is particularly important when multiple sequences are recorded at each cycle. Two or three trained individuals assign the amino acids independently in order to increase the confidence level of the assignments.

Table V. Summary for Three Years of Samples that were Submitted for N-terminal Sequence Analysis

	1989		1990		1991		Average
	470A[a]	477A	470A[a]	477A	470A[a]	477A	per year
Samples							
Total number of samples submitted :	269	89	340	139	304	153	431
% of samples submitted in solution :	37.2	53.9	47.4	37.4	62.8	32.7	46.5
% of samples submitted on PVDF :	62.8	46.1	52.6	62.6	37.2	67.3	53.5
Results							
% of samples that successfully seq[b]:	65.8	73.0	68.5	71.2	78.6	86.3	73.2
% of samples with no data obtained[c] :	33.1	27.0	30.3	25.9	21.4	13.7	26.1
% of instruments that malfunctioned :	1.1	0.0	1.2	1.4	0.3	0.7	0.9
Sensitivity[d]							
% of successful samples ≤ 10 pm :	59.9	36.9	63.1	71.3	52.3	53.0	57.6
% of successful samples ≤ 5 pm :	44.1	21.5	48.1	54.4	36.8	36.4	41.7

[a]Data is the total from two 470A sequencers.

[b]Successful sequence required 5 positively identified amino acids.

[c]No sequence data could be obtained due to a blocked N-terminus, protein not present or present at a level below the detection limit of the instrument, or no PTH analysis by HPLC was requested.

[d] Sensitivity is based on the background corrected initial yield of the first identified amino acid.

SUMMARY

Sequencing is not an art, but a science and a developed skill. High throughput, high sensitivity, and low failure rate are all routinely achieved by adhering to rigorous standards of operation and maintenance and by using an improved protocol for sample treatment. Currently with three sequencers, an average of 431 experimental samples are analyzed each year. 58% of successful runs began with ten pmol or less of amino acid in the first cycle and 42% of successful runs with five pmol or less (see Table V). The average length of the successfully sequenced samples was 16 ±8, with a range from five to sixty residues, indicating that many samples required two nights of run time to obain data. In addition, β-lactoglobulin results from a full year of preventive maintenance show an average repetitive yield of 94.0% ±1.4 for 470A #1, 94.6% ±1.4 for 470A #2, and 94.4% ± 0.9 for the 477A. Our protocols, as outlined here, have reduced our sequencer malfunctions to less than one percent, and allow us to routinely and confidently analyze samples of low picomole quantities in a cost effective and timely manner.

ACKNOWLEDGEMENTS

Supported in part by NIH Biomedical Shared Instrumentation Grants and by funds provided by the U.S. Army and Naval Office for purchase of equipment.

REFERENCES

1. Matsudaira, P. (1987) *J. Biol. Chem.* **21**, 10,035-10,038.
2. Stone, K.L., LoPresti, M.B., Williams, N.D., Crawford, J.W., DeAngelis, R., and Williams, K.R. (1989) *in* Techniques in Protein Chemistry (Hugli, T.E., Ed.), pp.377-390, Academic Press, New York.
3. Bauw, G., Van den Bulcke, M., Van Damme, J., Puype, M., Van Montagu, M., and Vandekerckhove, J. (1989) *in* Methods in Protein Sequence Analysis (Wittmann-Liebold, B., Ed.), pp. 220-233.
4. Fernandez, J., DeMott, M., Atherton, D., and Mische, S.M. (1992) *Anal. Biochem.* **201**, 255-264.
5. Tempst, P., and Riviere, L. (1989) *Anal. Biochem.* **183**, 290-300.
6. Atherton, D. (1989) *in* Techniques in Protein Chemistry (Hugli, T.E., Ed.), pp. 273-283, Academic Press, New York.
7. Laemmli, U. K. (1970) *Nature* **227**, 680-685.
8. Applied Biosystems Model 470A/900A Protein Sequencer User's Manual (1989), pp 11.1 - 11.18.
9. Applied Biosystems Model 477A/900A Protein Sequencer User's Manual (1989), pp 11.1 - 11.16.
10. Applied Biosystems Model 120A PTH Analyzer User's Manual (1988), p.3-11.
11. Speicher, D. W. (1989) *in* Techniques in Protein Chemistry (Hugli, T. E., Ed.), pp. 24-35, Academic Press, New York.
12. Niece, R.L., Williams, K.R., Wadsworth, C.L., Elliot, J., Stone, K.L., McMurray, W.J., Fowler, A., Atherton, D., Kutny, R., and Smith, A.J. (1989) *in* Techniques in Protein Chemistry (Hugli, T.E., Ed.), pp. 89-101, Academic Press, New York.
13. Yüksel, K.Ü., Grant, G.A., Mende-Mueller, L.M., Niece, R.L., Williams, K.R., and Speicher, D.W. (1991) *in* Techniques in Protein Chemistry II (Villafranca, J.J., Ed.), pp. 151-162, Academic Press, New York.
14. Crimmins, D.L., Grant, G.A., Mende-Mueller, L.M., Niece, R.L., Slaughter, C., Speicher, D.W., and Yüksel, K.Ü. (1992) *in* Techniques in Protein Chemistry III (Angeletti, R.H., Ed.), pp. 35-43, Academic Press, New York.

Successful Peptide Sequencing With Femtomole Level PTH-Analysis: A Commentary

Hediye Erdjument-Bromage, Scott Geromanos, Amy Chodera and Paul Tempst

Molecular Biology Program and Microchemistry Group
Memorial Sloan-Kettering Cancer Center, New York, NY 10021

I. Introduction

Some exciting variations on the automated chemical sequencing methods, using novel Edman-type reagents (e.g. fluorophoric or quaternary amine PITC-derivatives) and detection techniques (e.g. laser-induced fluorescence or, respectively, mass spectrometry) have recently been proposed [3,4]. While these methods await testing in the field, an innovative new sequencing approach, using triple-quadrupole MS with electrospray ionization, has, in a few cases, resulted in spectacular peptide sequencing data [5]. However, many protein chemists still feel that the classical, applications-proofed sequencing techniques should not be abandoned, at least not until the new developments have widely and successfully entered the biological laboratory. Moreover, high performance of the new instruments / technologies, in the hands of a few selected, highly skilled individuals, is usually compared to the average performance of the thousand or so 'classical' sequencing machines in the field. It has always been our belief that, with only a moderate effort, those 'older' instruments could be significantly upgraded in terms of sensitivity. In this paper, we will comment on some practical aspects of successful chemical sequencing with femtomole level PTH-analysis. The reader should know that, since well over 90% of sequencing in our laboratory is on peptides (<30AA), many of arguments and suggestions we make here should be taken in that context.

II. Methods

Peptide preparations from electroblotted, digested proteins and the RP-HPLC equipment and running conditions used in this process were essentially as described [1,2]. Separations on 1mm columns (1x100mm, Inertsil 100GL1-ODS-I-10/5; SGE, Ringwood, Australia) were done, using a similar solvent system and modular instrument as in [2], except that a LC-Packings (San Francisco, CA) Kratos-compatible, capillary flow cell was directly connected to the column outlet; the gradient slope was 1% B/min at a flow of 25 µl/min.

TECHNIQUES IN PROTEIN CHEMISTRY IV

Peptide sequencing was done using a 477A/120A automated system (ABI, Foster City, CA), optimized exactly as described [6]; other changes were as indicated in text and figure legends.

Test peptide 'PEPEP II' was synthesized using t-Boc chemistry, cleaved from the resin with HF, purified by two successive rounds of RP-HPLC, derivatized with 4-vinyl pyridine and repurified by HPLC; additional handling is described in the text. Quantitation of stock solutions and working solutions was done by amino acid analysis. We are indebted to C. Elicone for preparing this peptide.

III. Results and Discussion

A. *Documented Sequencer Modifications*

Three years ago, we reported on several simple modifications of the ABI 477A-120A sequencing system to improve femtomole PTH-analysis and to make the process more routine [6]. Our suggestions included injecting a larger portion of the PTH's for analysis and changing the HPLC conditions to allow real time monitoring (stripchart recordings) at 0.001 AUFS; the 1989 modifications are summarized in table I. Since then, we have sequenced nearly 1,000 samples, combined, on several instruments thus modified. The amount of standard that we routinely use is 1.6 picomoles of each PTH-amino acid in the flask.

Table I. Modifications for efficient injection of PTH's and real time OD_{269} monitoring/recording at 0.001 AUFS

SEQUENCER / HPLC MODIFICATIONS (1989; FROM REF 6)

1. 100 µl injector loop
2. S4 = 10% acetonitrile
3. PTH-NLE delivered to flask from X2 (prior to ATZ's)
4. Bubble Argon (5 x 3 sec) in flask while dissolving PTH's
5. DMPTU (0.5 µM) in solvent B
   TRP (1 µM) in solvent A (+ adjust as necessary)
6. Modified gradient: MOD-1 (Flow = 200 µl/min)

Time	%B
Time 0 :	12%B
26:	47%B
27:	90%B

B. *Practical Problems With Low Picomole Level Sequencing*

Several investigators are now able to obtain extensive sequencing runs (over 25 residues) with 2-3 picomoles initial yield. However, over the past three years, we have identified several problems specifically associated with 'high-sensitivity' sequencing, problems that were totally unperceived until femtomole level PTH-analysis was routinely done.

Relative small baseline disturbances at 0.01 AUFS (OD_{269}) become giant 'junk-peaks' when monitoring PTH-analysis at 0.001 or 0.0005 AUFS. At these levels of sensitivity we also see, in addition to the familiar DMPTU, DPTU and DPU peaks, two consistent (A,B) and one occasional (C) background peaks, positioned in the chromatograms between Asp/Asn (A), between Asn/Ser (B) and on top of Ile (C). Although the precise molecular identity of the latter three contaminants is, to our knowledge, unknown at present, we have ample evidence that they are 'chemistry' related.

An old rule in sequencing holds that samples must be homogeneous and essentially free of amino acids and solvent components that interfere with the chemistry. Even so, whereas typical amino acid backgrounds of one to a few picomoles (e.g. Gly, Ser, Ala, etc.) do not cause much of a problem with analyte signals of, say, 5-10 picomoles, it represents unacceptable noise when looking at 1-2 picomoles, or less, of PTH's.

Thus, the most important challenge in low picomole sequencing is keeping down chemical background and reducing levels of free amino acids in the sample. In addition, collection, storage and handling of low picomole amounts of peptides, using instrument components and laboratory supplies that were designed in the "nanomole' era of protein chemistry, are becoming very difficult and will require some serious rethinking.

C. Preparation and Handling of Peptides.

The heavy emphasis on peptide (<30AA) sequencing in our laboratory led to the synthesis of a peptide standard, PEPEP II (25 AA), for use in sequencer performance evaluation and improvements:

ISC$_p$WAQIGKEPITFEHINYERVSDR

All common amino acids, except Met and Leu, are represented; Cys is modified to its pyridyl ethyl derivative (Cp). Four Ile and three Glu residues are evenly spaced for the purpose of repetitive yield calculations; Tyr (19) and His (16) can be iodinated to allow easy tracking of the peptide.

During subsequent sequencing experiments, using 5-10 picomoles PEPEP II, drastic variation of the initial yields was observed. This was quickly traced to sample losses during storage and handling. The legitimate worry that similar losses might occur with equally small amounts of unknown peptides, collected after micro-separation of proteolytic digests, prompted an investigation.

The results shown in figure 1 clearly indicate that, in the volume range tested (30-100 µl), about 50% of the peptide is not recovered from storage in 0.1% TFA (1 min to 1 week). When supplemented with TFA (33% final concentration) recoveries were on the average over 80%, regardless of storage time. The data presented in figure 1 are from a single peptide; clearly, more work with a variety of different peptides is needed to confirm these findings. Until then, we have decided to store HPLC-collected peptides at -70°C and add neat TFA in a 1 to 4 ratio (TFA/sample; vol/vol) after storage, just before loading on the sequencer disc.

D. Maintenance and Optimization Routines

HPLC baseline disturbances, such as noise, long range drift, dips and bumps, are highlighted at 0.001 AUFS (OD_{269}) and require frequent correction. The effects of a number of routine adjustments are illustrated in figure 2. The baseline is balanced by adding extra aliquots of Trp to solvent A (panel A) or DMPTU to solvent B and by changing to fresh solvent B at least once every week (panel B); detector lamps must be carefully prescreened (many lamps are unacceptable) and more frequently replaced (panel C). In addition, acid washing of the HPLC pumps, following the manufacturers procedure, is regularly done (result not shown).

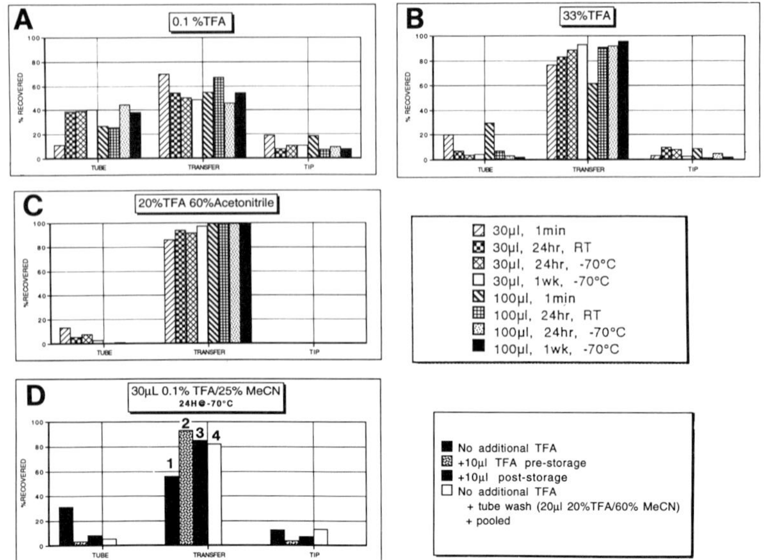

Figure 1. Peptide losses during storage in polypropylene tubes and handling. Ten picomoles PEPEP II (sequence: see text), spiked with 500 femtomoles ^{125}I-PEPEP (100,000 cpm) were stored in solution at room temperature (1', 24 h, 1 week) or at -70°C (24 h, 1 wk); volumes and solvent compositions were as indicated. Solutions were then pipetted to a second tube and counted; the emptied tube and pipette tip were also counted. Combined counts of tube A (empty), tube B (transferred solution) and tip were considered 100%. Panel D: peptide in 30 µl volume for 24 h at -70°C, as indicated. No additional TFA was added (1) or 10 µl TFA, either before (2) or after (3) storage (but before transfer); alternatively, the tube was rinsed and washing solution combined with the first transfer (4).

Artifactual peaks, due to byproducts of the Edman chemistry (see section IIIB), are kept to a minimum by replacing reagents (R1,2,3,4) once a week (result not shown) and by using a special, biweekly cleaning procedure as listed in table II; the result of the cleaning routine is shown in fig. 2, panel D. As a rule, discs with fresh polybrene are extensively precycled, i.e. at least two FIL cycles and then NORMAL cycles (usually 2 to 6) until the background is acceptable. Chemical background will continue to decrease with cycle number until it stagnates at manageable levels, usually after 20-40 cycles. Since peptides are always sequenced all the way to the C-terminus, or until no more signals are observed, several can be analyzed successively without reapplying polybrene. We routinely do that for five consecutive days (about 125 cycles); after that, peptide washout becomes unacceptably high (see figure 3). Empirically, we determined that, for the speediest background reduction, reagents should have been installed on the instrument, and Argon-conditioned, at least 48 h prior to precycling the filter and starting the first sequencing run. In practice, new polybrene is loaded every monday and reagents are installed every tuesday. Background can be further reduced by omitting DTT from solvents S1,2,3; if a tube must be rinsed after applying the sample (see section IIIC), do not use ABI's reagent R4 (25% TFA) because it also contains DTT. If despite all these precautions, sizable peaks are still obscuring certain PTH's (e.g. Asp, Ser,

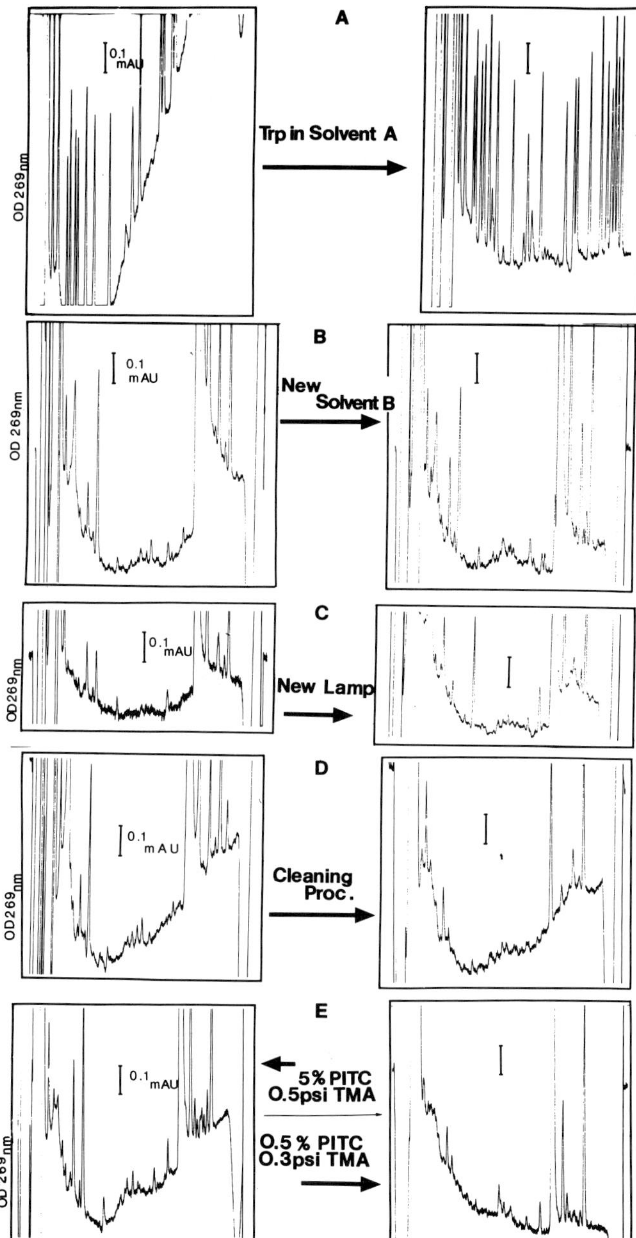

Figure 2. Baseline corections and reduction of 'chemical' noise on the ABI 477A/120A. Chromatograms shown are taken from runs just before or after a change was made or procedure carried out: A, addition of Trp to solvent A; B, fresh solvent B; C, new lamp; D, cleaning procedure (see table 2); E, reduction from 5% PITC / 0.5 psi (R2) TMA (left) to 0.5% PITC / 0.3 psi TMA (right). All chromatograms are real time recordings of OD_{269} at 0.001 AUFS; the indicator bar is 0.0001 AU.

Table II. Cleaning of reaction cartridge, conversion flask, transfer line and injector valve assembly

SEQUENCER CLEANING PROCEDURE

1. Remove filter and install new cartridge seal.
2. Replace S2 with neat acetonitrile.
   Open valves 8,10,15,37,38 until flask is 3/4's full.
   CNV function 22 (30") + CNV 15 (until last drop exits restrictor).
   Repeat 2 times.
3. Replace S2 with neat TFA and repeat step #2.
4. Replace S2 with neat acetonitrile and repeat step #2; then
   CNV function 13 & RXN function 24 (15") + CNV function 20 (30").
5. Loop loading :
   CNV function 12 (20") + CNV 22 (4").
   Repeat 2 times.
   CNV 15 (optimize load injector time).
6. Remove S2 bottle.
   Open valves 9,8 (30").
7. Install new ethyl acetate on S2.
   RXN function 17 (120") + RXN function 29 (240").
8. Install new cartridge seal and filter.

Ile), then some can be positioned away (e.g. between Asp/Asn or between Asn/Ser) by manipulating pH and ionic strength of solvent A. Although we haven't been able to position an occasionallyoccurring contaminant peak away from PTH-Ile, one can easily observe peak broadening, when the 'real' analyte is present, by inspecting the relevant section of the chromatogram after computer-assisted magnification. In general, the 'Ile'-junk peak is correlated with aged PITC.

E. *Preliminary Studies for Future Improvements*

In view of the recurring background problems, we explored the feasibility of a new series of simple changes for automated chemical sequencing. Because the amounts have remained unchanged since the days (1981) when nanomole sequencing was still common, Edman-reagents are now delivered in vast excess over the 5 picomoles or so of peptide in the cartridge. Although suggestions have been made to reduce chemical background noise by lowering PITC concentration and by employing a different coupling base (7), no thorough, systematic and quantitative evaluation of these changes on initial yields, repetitive yields, lag and noise reduction was attempted, especially not for peptides. Here we describe some preliminary data of a study on the effects of reduced PITC concentration, TMA (R2) pressure and polybrene amounts, all evaluated with 10 picomoles peptide on the disc (figure 3).

Although reducing PITC concentration tenfold (5 to 0.5%) did not affect the sequencing process, except for a slight decrease in RY (92 to 88-89%), the DPU peak was cut in half, from a peak height corresponding to 3.4 picomoles PTH-Trp to a value of 1.7. By itself, delivering TMA vapor at a lower pressure (0.3 psi instead of 0.5) did not result in DPU reduction. Surprisingly then, the DPU reduction at 0.5% PITC was more pronounced when the TMA pressure was again lowered to 0.3 psi; the combined effect did cut the background peak to a third (equivalent of 1.1 pmole of PTH-Trp). At 0.5% PITC, junk peaks obscuring Asp and Ser were practically gone (see fig.2E). Further reductions of

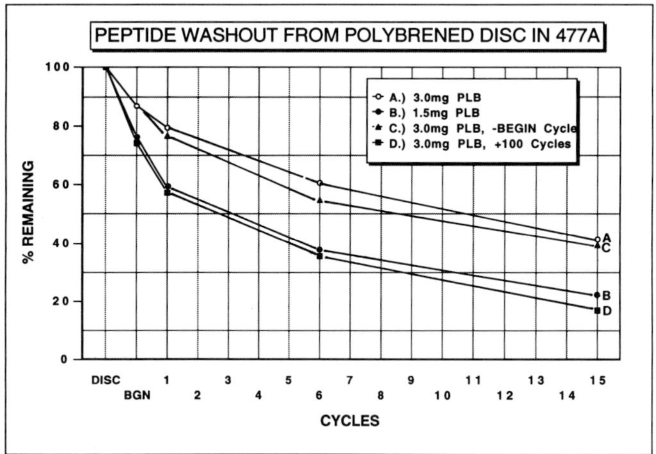

Figure 3. Peptide washout from polybrened discs in the ABI 477A. Ten picomoles PEPEP II (sequence in text), spiked with 500 fmoles ^{125}I-PEPEP II (100,000 cpm), were loaded on the filter disc after the fresh polybrene had been pretreated with 1 FIL cycle (A-C), except for experiment D where 100 NORMAL cycles had been run first. The filter was then removed from the cartridge, counted, repositioned and the sequencing run started. The filter was taken out again for counting after the BEGIN cycle and after amino acid cycles (NORMAL) 1, 6 and 15; no BEGIN cycle was run in experiment C. Data were plotted as % of counts that were on the filter before sequencing was started. Since no ^{125}I label is present in the first 15 amino acids of PEPEP II, reduction of counts reflects washout.

PITC concentration (0.1% and less) have a serious negative impact on IY and RY and cannot be used. When half the amount of polybrene on the disc was used (down from 3 to 1.5 mg), DPU levels were clearly reduced (equivalent of 1.4 pmole of PTH-Trp) but there was a substantially increased peptide washout problem (figure 3); after 15 cycles, only 22% remained on the disc versus 44% when 3 mg polybrene was used. Similar peptide washout kinetics were observed when using a polybrened (3 mg) filter that first had been subjected to 100 Edman cycles (figure 3). Finally, cutting the BEGIN cycle did not result in reduced peptide washout throughout the rest of the sequencing run (figure 3). Although more work is needed to further optimize automated chemical sequencing with high-sensitivity PTH-analysis, we concluded that, at this point, a combined regimen of 0.5% PITC / 0.3 psi TMA (R2) with 3 mg polybrene on the disc is preferential for peptide sequencing. The associated reduction in RY and higher lags are not a real disadvantage for analysis of peptides smaller than 25 residues.

F. *Conclusions*

During the past year, from a total of 605 peptide sequencing runs in our laboratory, about 87% were with initial yields (IY) under 10 picomoles: 17% IY between 5-10 pmoles, 38% between 2-5pmoles and 32% under 2 pmoles. In general, peptide peaks from the mid-part of the chromatograms (2 and 1 mm columns), with absorptions >1mAU (at 214 nm), were considered candidates for reliable sequencing. Routine operation of the PTH-amino acid analyzer, during these runs, was at a sensitivity of 0.001 AUFS; this is sometimes changed to 0.0005 AUFS in the course of an experiment, when judged neccessary. Amino acid background usually prevented unambiguous calls during the first few cycles.

Based on all these experiences and on some recent optimization studies, we have adopted the following rules and guidelines for use in our laboratory:

1. All instruments have been modified as described [6].
2. Baselines of HPLC analysis (OD_{269}) must be flat (titrate!) and noise free (lamp!) at 0.001 AUFS; 1 picomole (in flask) PTH-Ala must correspond to 0.2 mAU.
3. Replace Edman chemicals, polybrene and HPLC solvents once a week (100-125 cycles); column is replaced every 2-3 months.
4. Biweekly cleaning procedure (table II) + optimization.
5. Use 0.5% PITC / 0.5 psi TMA (R2) and 3 mg polybrene.
6. Precycle polybrene with chemicals that were 48 h on the instrument; lowest background after 20-40 cycles and beyond (max. 100 cycles).
7. Biweekly instrument performance test with 5-10 picomole PEPEP II: yield (@ cycle 5) > 35%; RY (I/E) > 88%; total lag (@cycle 10) < 18%.
8. HPLC-collected peptides are stored at -70°C; neat TFA is added after storage, just before loading, to a sample/TFA ratio of 4/1 (vol/vol).
9. Sample prep and sequencing are done in a 'semi-clean room' environment; use dedicated chemicals, consumables, etc.

With these procedures as standard practice, we feel that the limits to sensitivity of Edman sequencing have not yet been reached. However, this will require impoved micropreparations of polypeptides in even cleaner environments, higher purity reagents and solvents, instrument and HPLC miniaturization and solid-phase techniques.

Acknowledgements

The authors are indebted to Mary Lui and Lynne Lacomis for excellent technical assistance; P.T. thanks Lise Riviere for her valuable input during the early stages of this work.

References

1. Aebersold, R.H., Laevitt, J., Saavedra, R.A., Hood, L.E., and Kent, S.B.H. (1987) Proc. Natl. Acad. Sci. USA 84: 6970-6974.
2. Tempst, P., Link, A.J., Riviere, L.R., Fleming, M., and Elicone, C. (1990) Electrophoresis 11: 537-553.
3. Cheng, Y.-F., and Dovichi, N.J. (1988) Science 242: 562-564.
4. Aebersold, R., Bures, E.J., Namchuk, M., Goghari, M.H., Shushan, B., and Covey, T.C. (1992) Protein Science 1: 494-503.
5. Hunt, D.F., Henderson, R.A., Shabanowitz, J., Sakaguchi, K., Michel, H., Sevilir, N., Cox, A.L., Appella, E., and Engelhardt, V.H. (1992) Science 255: 1261-1263.
6. Tempst, P., and Riviere, L. (1989) Anal. Biochem. 183: 290-300.
7. Farnsworth, V., Carson, W., and Krutzsch, H. (1991) Peptide Research 4: 245-251.

ON-LINE MICROBORE HPLC DETECTION OF FEMTOMOLE QUANTITIES OF PTH-AMINO ACIDS

Russell W. Blacher
Michrom BioResources, Inc., Pleasanton, CA 94588

John H. Wieser
Column Engineering, Inc., Ontario, CA 91761

I. Introduction

PTH-Amino Acid identification by HPLC (1,2) is the standard analytical tool used to determine protein and peptide sequence from automated Edman sequencers (3,4). Currently, this HPLC identification (employing 2.1mm columns) is the chief limiting step toward increasing sequencer sensitivity. This study evaluates the next level in on-line PTH-AA identification utilizing recent advances in microbore HPLC and 1mm column technologies.

The routine application of 1mm microbore HPLC to on-line sequencing had to overcome several boundaries previously associated with this technology. These limits included adequate resolving capacity of the column, linearity over a wide range (several orders of magnitude), the ability to repeatedly inject large volumes (100µl, almost a full column volume on a 1 x 150mm) while maintaining satisfactory resolution, baseline drift during a gradient run at high sensitivity (characterized by an unacceptable baseline drop, rise or "smile" on some systems), retention time reproducibility, the ability to consistently pack small particles into 1mm columns and limited column lifetimes.

II. Materials

All of the analyses were performed on an Ultrafast Microprotein Analyzer (UMA) (5,6) from Michrom BioResources, Inc. (Pleasanton, CA). This instrument consists of a dual range, binary, microbore solvent delivery system, a 10 port automated injector, a variable wavelength UV/Vis detector with custom micro flow cell, a thermally controlled column and flow cell, a built-in helium degas and blanketing system and a PC/AT-based controller/data system. The column used for these analyses is a Reliasil-PTH (1 x 150mm), a custom designed reversed-phase material packed into the Michrom column cartridge. All chemicals and reagents used were the highest quality available.

TECHNIQUES IN PROTEIN CHEMISTRY IV

III. Results

The first goal of a PTH-AA analysis system is to achieve near baseline resolution of all common PTH's along with a stable modified cysteine derivative (i.e. PEC, S-ß-[4-pyridylethyl] cysteine). Also, the chemistry by-products common to commercial sequencers have to be well resolved to avoid gaps in sequence assignment, all in an appropriate time frame. PTH-AA standards were manually injected in the presence of 0.01% TFA/3% CH_3CN/H_2O to simulate the conversion flask environment after acidic conversion (figure 1a). In addition, a protective precolumn used to trap both mobile phase and sample introduced particulates cannot significantly diminish this separation (figure 1b).

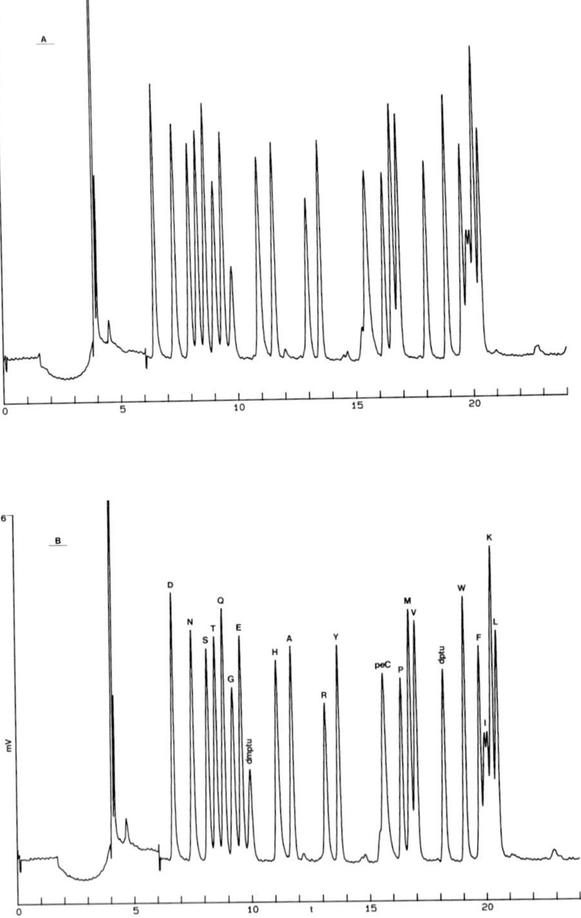

Figure 1. 10pm PTH-AA standards (Applied Biosystems)/100μl, flow rate = 50μl/min, without pre-column (A) and with pre-column (B).

To attain femtomolar sensitivity in this analysis, the baseline has to be steady throughout the area of PTH-AA elution. The major sources of baseline irregularities or drift (i.e. the rise, drop or "smile" effect) are related to the buffer/solvent system and the optics. Specific problems associated with high sensitivity detection include quality of line input power, thermal stability of the photodiode detector, the purity, optical compatibility and effective reproducible mixing of elution buffer and solvent, micro-bubble formation from insufficiently degassed eluants during gradient formation, and relative peak height equality.

Relative peak height equality among all the PTH-AA's can generally be achieved through gradient slope manipulation. The greatest challenge in most systems is to overcome the broad peak shape of the basic residues histidine and arginine. The sharpness of these derivatives in this system, is due to their interaction with the bonded phase and the alkylsulfonate ion pairing agent.

The components of the elution buffer and solvents were chosen both for their ability to resolve the PTH's and their optical compatibility during gradient formation. The A buffer consists of 25mM ammonium acetate pH 4.1/0.02% heptane sulfonic acid/5% acetonitrile (v/w/v), B solvent is composed of 70% acetonitrile/25% n-propanol/5% water (v/v/v). The A buffer also contains a proprietary stabilizing agent to prevent losses associated with low levels of labile derivatives. The thermally stable photodiode, uniform mixing of the buffer and solvent at low flow rates, and a helium degassing/blanketing system are all built-in features of the UMA.

The performance of all electrical equipment is dependent on the quality of the line input power. When operating analytical instruments at maximum sensitivities, power surges, spikes and sags, voltage/frequency deviations, and brief power interruptions can all affect ultimate sensitivities (not to mention the potential loss of precious samples). Power conditioning to handle all of the above issues was handled by a Ferrups uninterruptable power supply from Best Power Technology (Necedah, WI). Figure 2 illustrates the results of addressing all of these issues.

Retention time reproducibility, over the course of a protein sequence run, is dependent on several of the above items. In addition, the buffer/solvent system has to be free of readily oxidizable components, and the 1mm column has to be evenly, densely packed in order to repeatedly resolve large (100μl) injection volumes. A relatively low column temperature (40°C) and disposable precolumn also contribute to the reproducibility and extended column life. Figure 3 displays two cycles from a sequence run on an Applied Biosystems model 470 protein sequencer on-line to the UMA. The relative retention time standard deviation (based on 6 representative PTH's through the gradient) over the course of a 10pm, 25 cycle run of bovine serum albumin was less than 2.4%.

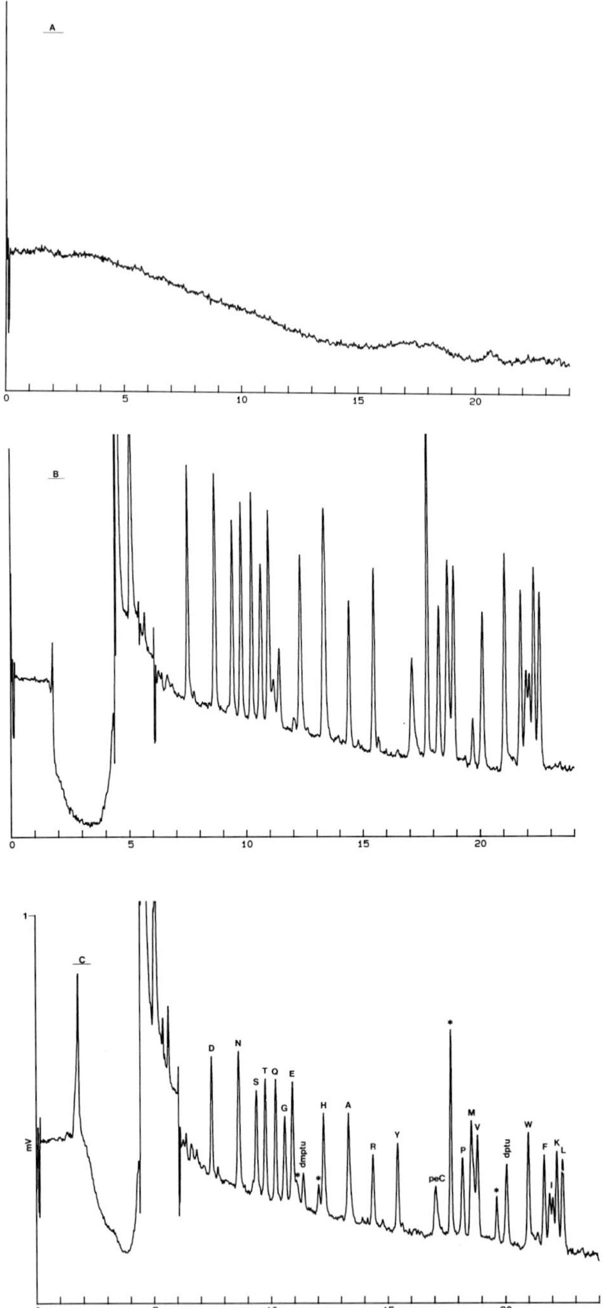

Figure 2. Blank gradient baseline at maximum detector sensitivity, 1mV full scale (A), 1.0pm (B) and 500fm (C) of PTH-AA standards (ABI). Peaks labeled with * are associated with the manual injection solvent.

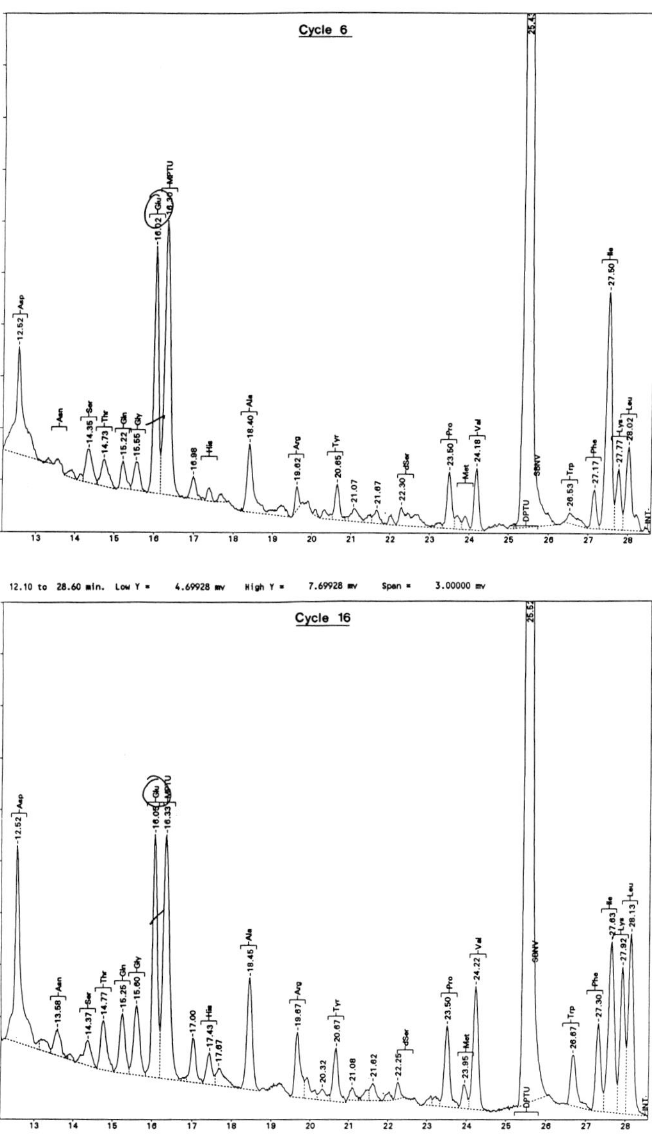

Figure 3. Repeating glutamic acid residues at cycles 6 and 16 from a 10pm sequence of BSA on an Applied Biosystems model 470, illustrate the retention time reproducibility.

Linearity of response and the ability of a microbore PTH-AA system to span a wide range of sample detection are imperative. Protein sequence sample quantities obtained from methods other than amino acid analysis often deviate by more than a factor of 10 from the determined amount. Using data obtained over three orders of magnitude (from 500fm, 5pm, and 50pm of PTH-AA standards), a linear response was achieved for all PTH-AA's (figure 4).

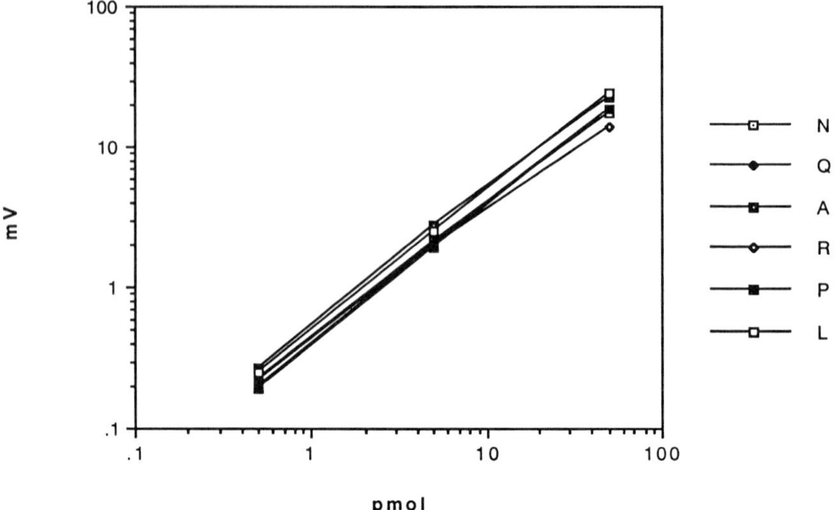

Figure 4. Plot illustrating the linearity of several representative PTH's at 500fm, 5pm, and 50pm.

IV. Discussion

The development of this true microbore PTH-AA system was the next logical step in obtaining higher sensitivities from existing automated sequencers. This technique has demonstrated a 2.5-10 fold increase in sensitivity over existing narrowbore (2.1mm) technology. The ability of this system to handle repeated large (100µl) injection volumes offers compatibility to modern protein sequencers requiring no sequencer chemistry, transfer volume, analysis time, software or hardware modifications. Another benefit of microbore technology to a system that frequently runs 24 hours a day is the four-fold reduction in buffer and solvent consumption.

High resolution on a relatively short (150mm) PTH column is attributed to having full control of the packing and materials. Starting with 3µ base silica, both the bonding chemistries and batch sizes were rigidly controlled in addition, newly developed column packing procedures that utilize constant high pressure/velocity packing have been incorporated. Also, a protective precolumn has been developed that enhances column lifetime without reducing efficiency. The elimination of column-to-detector system volume in the UMA also contributes to peak resolution at these low levels by eliminating band broadening.

Due to the high sensitivity of this method, UV absorbing impurities and breakdown products from the injection solvent (0.01% TFA/3% CH_3CN/H_2O) were observed and found to increase with time. Freshly made up injection solvent (from neat TFA) was found to minimize the number and size of these peaks. Also of note is the PTH-Ile doublet associated with the ABI PTH standards. This phenomenon was not observed with PTH standards obtained from Pierce (figure 5).

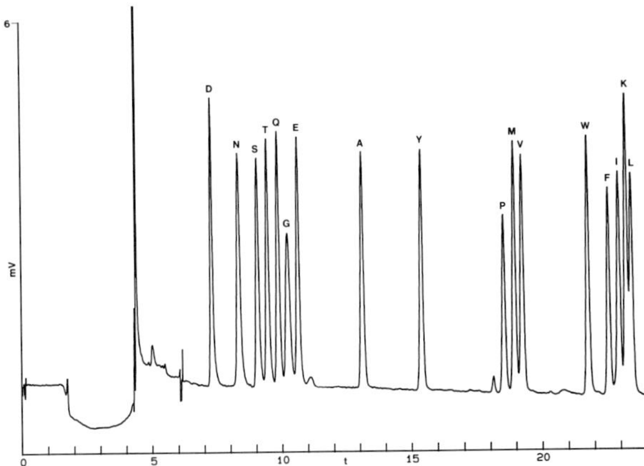

Figure 5. 10pm PTH-AA standards (Pierce)/100µl.

Attempts thus far to perform PTH analyses routinely using capillary liquid chromatography have been less than satisfactory. The gain in sensitivity has been more than offset by it's current disadvantages, such as severely limited sample injection volumes, inadequate resolution of the all the PTH's and sequencer chemistry by-products, and the inability to handle large numbers of injections per column. Capillary electrophoresis, which has to overcome many of the same limits (i.e. nanoliter sample loading volume), has yet to show if it will contribute to these analyses without major sequencing instrument modifications.

Acknowledgements

I would like to thank Drs. William Henzel and Christopher Grimley of Genentech, Inc. for their valuable discussions and generation of on-line sequence data.

References

1. Zimmerman, C.L., Apella, E., and Pisano, J.J. (1977) Anal. Biochem. **77** 569-573
2. Machleidt, W., and Hofner, H. (1980) *In* "Methods in Peptide and Protein Sequence Analysis" (C. Birr, ed.) pp. 35-47, Elsevier Press
3. Edman, P., and Begg, G. (1967) Eur. J. Biochem. **1,** 80-91
4. Hewick, R.M., Hunkapiller, M.W., Hood, L.E., and Dreyer, W.J. (1981) J. Biol. Chem. **256**, 7990-7997
5. Nugent, K., and Olsen, K. (1990) BioChromatography Vol.5, No.2, 101-105
6. Nugent, K.D., and Nugent, P.W. (1990) BioChromatography Vol.5, No.3, 142-148

Single Syringe-Pump Solid-Phase Protein Sequencer

Mark Stolowitz and Leroy Hood

Center For Molecular Biotechnology, Division of Biology 139-74
California Institute of Technology, Pasadena, CA 91125

I. Introduction

Solid-phase protein sequencing, as first proposed by Laursen over 20 years ago (1,2), exploits the principle that proteins immobilized on an insoluble support can readily be separated, without losses, from reagents and reaction by-products. Interest in solid-phase sequencing declined with the advent of the gas-phase protein sequencer (3) because of the complexity of the requisite immobilization chemistries. Renewed interest in the solid-phase approach has recently resulted from commercial availability of functionalized polyvinylidene difluoride (PVDF) membrane sequencing supports (4) and the development of covalent entrapment protocols for immobilization of proteins electroblotted onto PVDF membranes. As the potential of solid-phase methods has proven increasingly important when attempting to develop approaches to sub-picomole sequencing (6), we have developed a new simplified instrument to aide us in our research.

II. Materials and Methods

Kel-F solenoid valves of unique design (7,8) were purchased from Biomedical Instrumentation Services at the City of Hope (Duarte, CA). Gas-tight syringes were purchased from Hamilton (Reno, NV). Solid-phase sequencing reagents were purchased from the MilliGen/Biosearch Division of Millipore.

III. Results and Discussion

We describe the construction of a unique solid-phase protein sequencer having the capability to deliver reagents at precise flow rates with a single syringe-pump employed as a pneumatic displacement device. The instrument utilizes solenoid valves which were developed at the City of Hope and have been described previously (7,8). The following description will be limited to novel aspects of the instrument which have not been previously reported.

TECHNIQUES IN PROTEIN CHEMISTRY IV

Single Syringe-Pump Concept

The principles of the single syringe-pump concept are illustrated in Figure 1. Initially, the syringe is filled with argon gas which is metered by a pressure regulator through a port on a Rusnak valve. The designs of the Rusnak (7) and Hexagonal (8) valves have been described previously. Any one of the six reagents which access the Hexagonal valve may be metered into a 500 µl (0.030" I.D.) volumetric loop situated between the Rusnak and Hexagonal valves. The volumetric loop may be partially filled or completely filled to waste via the second port on a Rusnak valve. Reagent is delivered to the reaction cartridge in a flow-rate dependent manner by displacing the contents of the volumetric loop with argon gas delivered from the syringe. The software driver allows flow-rates to be corrected for solvent compressibility. Although compressibility correction factors can be calculated for pure solvents, they must be determined experimentally for solvent mixtures (Table 1). We have determined that reproducible flow rates can be obtained by this approach in the range 100-1500 µl/min. This range encompasses all flow rates previously utilized on Solid-phase sequencers, and the use of a single syringe-pump greatly reduces the complexity of instrument design. The syringe-pump mechanism need not be utilized in every instance. Reagents may be delivered directly to the reaction cartridge via the Hexagonal valve as either liquids or vapors.

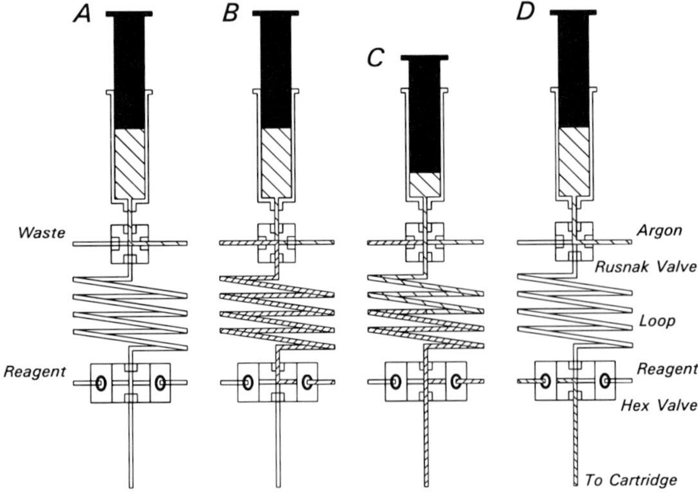

Figure 1. Single syringe-pump method of reagent delivery. **A**: Syringe barrel is first filled with argon gas. **B**: Volumetric loop is filled with reagent metered from the Hexagonal valve to the waste port on the Rusnak valve. **C**: Reagent is delivered to reaction cartridge by displacing the contents of the volumetric loop with argon gas delivered from the syringe. **D**: Reagent may be delivered directly to the reaction cartridge without utilizing the syringe-pump mechanism.

Table 1. Flow-rate Variation Resulting From Solvent Compressibility

Reagent/Solvent	Percent Delivery[a]
5% *N*-Methylmorpholine in 70:30 Methanol/Water[b]	77
Methanol[b]	45
Heptane/Ethyl Acetate	88
10 % Phenylisothiocyanate in Acetonitrile	57
Trifluoroacetic Acid	100

[a]Expressed as percent of programmed flow rate obtained on delivery of a 500 µl volume.
[b]Reagent/Solvent contains 0.1% cyclohexylamine.

Syringe-Pump Design

Compact syringe-pumps utilized in previous solid-phase sequencers have the axis of the lead screw offset from the axis of the syringe plunger. Such an offset drive may cause flexing of links that connect the plunger and lead screw, warping of the plunger and/or warping of the syringe barrel, resulting in inaccurate deliveries and leakage. These designs also exhibit substantial hysteresis, making them unsuitable for precision infusion applications. To overcome these limitations, a coaxially driven syringe-pump with all forces aligned with the plunger's axis of motion is utilized (Figure 2). Developed specifically for use on bioanalytical instrumentation, the pump design includes kinematic (three-point) mounts and a precision lead-screw which has a linear displacement of 1.5000 ± 0.0002 mm/revolution. The syringe barrel, syringe plunger, lead-screw, lead-screw housing and stepper-motor are all axially aligned. The syringe-pump accepts interchangeable syringe mountings which, in turn, accept 1.0, 5.0, 10.0 and 25.0 ml Hamilton gas-tight syringes. With a 10.0 ml syringe installed, the syringe pump is driven at a resolution of 0.625 µl/step. Additional details of the syringe-pump mechanism will be described elsewhere.

Figure 2. Syringe-pump utilized in the sequencer. The coaxial alignment of the syringe barrel, syringe plunger, lead-screw, lead-screw housing and stepper-motor (pictured from left to right, respectively) requires that the pump be of considerable length (53.5 cm).

The syringe-pump uses a four phase stepper-motor which provides a resolution of 1.8°/step. The stepper-motor drive state machine, which is implemented in a PAL (Programmable Logic Array) provides eight states for half step or 0.9° resolution. High step rates are obtained by powering the bridge from 24 volts and then chopper modulating to limit the motor winding currents to 0.5 amps. The stepper-motor drive state machine is located on the microcontroller printed circuit board.

Custom Microcontroller

The protein sequencer is designed to operate as an intelligent peripheral. A host computer downloads firmware and control sequences using a standard RS-232 serial communication port. Once initialized by the host system, the sequencer control board is capable of independent operation. The control board is based on the Intel 8032 microcontroller. This single chip microcomputer can address up to 64K of program memory and 64K of data memory. On the control board, the first part of program memory is in an EPROM (Erasable Programmable Read Only Memory) while the balance shares the SRAM (Static RAM) with the data memory. Having the program memory in SRAM allows new or revised firmware to be downloaded from the host. The Intel 8032 also features three timers plus an asynchronous communications port. The third timer is used to derive a baud rate clock from a 12 MHz microcontroller oscillator.

The first five channels of an 8-bit analog/digital converter are used for piezoresistive pressure sensors located on the control board. These devices, which feature a full scale output of 40 mv at a gauge pressure of 15 psi, have been calibrated and temperature compensated on chip for an offset of less than 2 mv and a linearity of 0.1%. The pressure sensors are connected directly to the gauge ports on five pressure regulators rated at either 0-2.5 or 0-10 psi.

The entire control board is powered from a single 24 volt, 140 watt switching power supply. The 24 volts is used by the stepper-motor drive, external solenoid drivers and relays. A three-terminal regulator provides 5 volts for all of the logic on the board. Another three-terminal regulator provides +10 volts for all of the analog circuitry, pressure sensor excitation, and RS-232 drivers. Thirty-six solenoid operated valves are powered by a third three-terminal regulator which can be programmed for either +24 or +12 volts. In operation, the solenoid valves are powered at +24 volts for 100 msec and then the power is dropped to +12 volts. The compact control board is a laminate comprised of four layers and measures only 25 X 21.5 cm (Figure 3).

Figure 3. Microcontroller printed circuit board. Connectors include (clockwise from the upper left-hand corner): Heater 1, Heater 2, Fan, Syringe-pump, Power Supply, General Interface, Front Panel Connector, TTL Interface, RS-232, Pressure Sensors, Conversion Flask Backlight, Reaction Cartridge Backlight, Conversion Flask Heaters, Reaction Cartridge Heaters, Solenoid Valve Drivers.

Sequencer Enclosure

The prototype single syringe-pump solid-phase protein sequencer is pictured in Figure 4. The compact design features a footprint only 1/3 that of commercial instruments, and supports both gas-phase and solid-phase operational modes. The waste bottle is located at the back left-hand corner of the instrument in proximity to the heat sinks on the power supply. This warms the bottle and facilitates the evaporation of volatile solvent waste. Five pressure regulator knobs are located immediately forward of the waste bottle. Reagent reservoirs are mounted at a 45° angle to maximize the surface area of reagent available for vapor-phase delivery. The reagent reservoirs are mounted on a hinged bulkhead allowing access to the pressurization solenoid valves without removal of the instrument cover. Similarly, an access panel located adjacent to the U-shaped enclosure which houses the reaction cartridge and conversion flask allows access to all additional liquid connections.

Engineering behind us, we have begun the process of optimizing the performance of this instrument with a focus toward the development of rapid and sub-picomole sequencing methods. In this regard, we are investigating a variation on a chemical degradation based on thiobenzoylation of the *N*-terminal amino acid (9). We have recently demonstrated that amino acid derivatives afforded by this method (4-substituted-5-acetoxy-2-phenylthiazoles) are suitable for detection by gas chromatography/mass spectrometry at low femtomole sensitivity utilizing an inexpensive single quadruple mass spectrometer. To facilitate the further development of this method, a prototype protein sequencer to gas chromatograph interface has been integrated into the design of the new instrument.

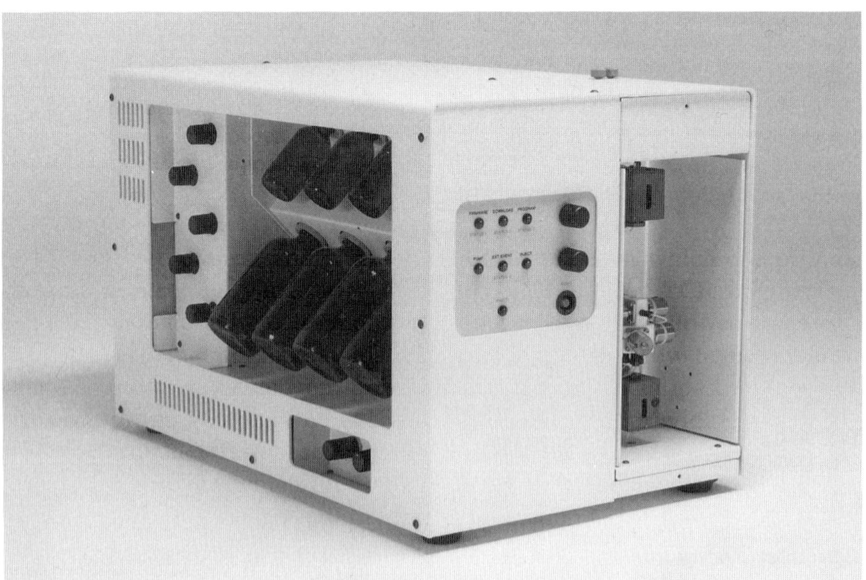

Figure 4. Single Syringe-pump Solid-phase Protein Sequencer. The enclosure measures only 36 X 36 X 56 cm. A panel with window which ordinarily encloses the reaction cartridge and conversion flask has been removed.

References

1. Laursen, R. A. (1966). *J. Am. Chem. Soc.* **88**, 5344-5346.

2. Laursen, R. A. (1971). *Eur. J. Biochem.* **20**, 89-102.

3. Hewick, R. M. , Hunkapiller, M. W., Hood, L. E., and Dryer, W. J. (1981). *J. Biol. Chem.* **256**, 7990-7997.

4. Pappin, D. J. C., Coull, J. M., and Koster, H (1990). *Anal. Biochem.* **187**, 10-19.

5. Coull, J. M., Pappin, D. J. C., Mark, J., Abersold, R., and Koster, H. (1991). *Anal. Biochem.* **194**, 110-120.

6. Abersold, R., Bures, E. J., Namchuk, M., Goghari, M. H., Shushan, B., and Covey, T. C. (1992). *Protein Science* **1**, 494-503.

7. Hawke, D. H., Harris, D. C., and Shively, J. E. (1985). *Anal. Biochem.* **147**, 315-330.

8. Calaycay, J., Rusnak, M., and Shively, J. E. (1991). *Anal. Biochem.* **192**, 23-31.

9. Barrett, G. C. (1967). *Chemical Commun.*, 487-488.

Acknowledgments

Advice and assistance provided by John Shively and Miro Rusnak at the City of Hope is gratefully acknowledged. The syringe-pump described wherein was designed and assembled by Steven Clark and Charles Spence. We thank Steven Marsh for assembly of the prototype instruments.

ANALYSIS OF COMPLEX PROTEIN MIXTURES ON THE HP-G1000A SEQUENCER

T.K. Slattery and R. N. Harkins
Department of Protein Chemistry, Berlex Biosciences, Alameda CA 94501

I. Introduction

Samples for gas phase protein sequencing are generally loaded in low volumes of a volatile liquid, such as dilute trifluoroacetic acid (TFA), in order to avoid undesirable effects of solvents with the sequencing chemistry. Except for HPLC purified samples, proteins and peptides are usually fractionated in aqueous buffers containing free amines, salts and often, denaturants. Therefore, additional manipulations such as desalting by HPLC, precipitation, or dialysis, are required to render the sample matrix compatible with sequencing. Each of these additional steps involves potential losses, especially when only microgram amounts of protein are present at low concentrations.

There would be a great advantage in loading a protein or peptide directly onto a sequencing support in any volume or buffer, washing away the interfering components, and leaving the protein bound to the support and ready for analysis. Hewlett-Packard Analytical Instruments has introduced a sequencer (HP-G1000A) which uses the standard Edman chemistry with a unique sample support system (1,2). A bi-phasic column is used to retain the sample during loading and sequencing. Proteins or peptides in solution are loaded onto the hydrophobic (reverse phase) half of the two part column by placing the sample solution in a 5 ml funnel attached to the column and applying nitrogen pressure (Figure 1). It is possible to do multiple sample additions for larger volumes or to use a second solution to wash the sample. The hydrophilic half of the column is assembled onto the hydrophobic half and placed in the sequencer. Sequencing takes place using standard automated Edman chemistry. The sequencer can deliver solvents and reagents in an upward or downward direction through the column, so that the protein will remain focused at the interface between the two halves of the column.

TECHNIQUES IN PROTEIN CHEMISTRY IV

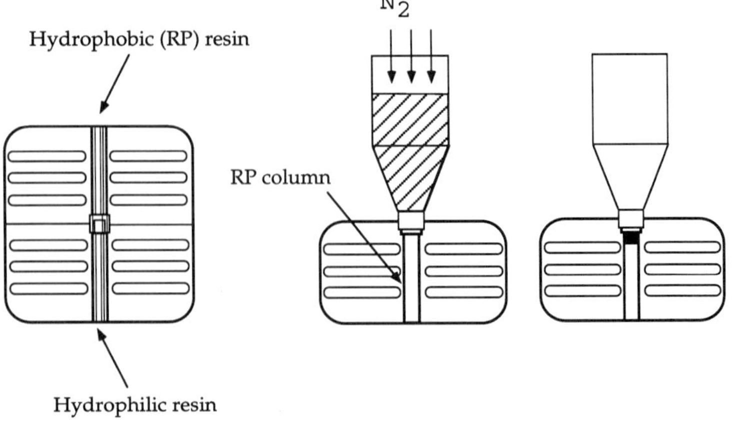

Figure 1. A. Assembled bi-phasic column, consisting of an upper hydrophobic half and a lower hydrophilic half.
B. Protein solutions are loaded onto the hydrophobic half of the column from a 5 ml funnel under nitrogen pressure.

To demonstrate the efficiency of this novel sequencing system and its ability to sequence proteins in the presence of complex solvent matrices, we examined the compatibility of several widely used methods for protein and peptide purification with direct sample application and sequence analysis using the HP sequencer. These methods included sequencing cyanogen bromide (CNBr) peptides purified off a sizing column in the presence of 6M guanidine HCl (G-HCl), proteins eluted directly out of polyacrylamide gel slices, proteins purified by isoelectric focusing in the presence of ampholytes, and final lot analysis of a purified formulated protein.

II. Experimental

A. *Materials*

All sequencing was performed on an HP-G1000A sequencer from Hewlett-Packard Analytical Instruments (Palo Alto, CA). All reagents and supplies for sequencing were from Hewlett-Packard. The protein standards produced at Berlex included recombinant human transforming growth factor-alpha (TGF-α), recombinant human fibroblast interferon Cys17Ser (Betaseron) and BACH 250, a humanized mouse monoclonal antibody raised against the erbB-2 proto-oncogene receptor. Beta-lactoglobulin was purchased from Applied Biosystems.

Comparison of Loading Solvents

The following three solvents were used to evaluate the effect of different components on sample loading and sequencing: solvent A was 2% TFA in water; solvent B was 6M G-HCl, 0.1M Tris pH 9.0; solvent C was 5M G-HCl, 5M urea, 0.2% TFA, 10% acetonitrile, 1% Zwittergent 3-08 (CALBIOCHEM). The test proteins were human serum albumin (HSA) (Sigma) (1.2 pmol μl^{-1} in 0.1% TFA) and TGF-α (10 pmol μl^{-1} in 0.1% TFA). One hundred pmol of the test protein was diluted to 1 ml in each solvent (A, B or C). Each sample was vortexed and allowed to sit for approximately fifteen minutes at room temperature. Sample loading onto the column was done according to the directions provided by HP. After loading was completed the samples were washed with 5 ml of 0.1% TFA, and sequenced 10 cycles.

B. Loading of β-Lactoglobulin in the Presence of SDS

Lyopholized β-lactoglobulin was dissolved in 0.1% TFA to make a 5 pmol μl^{-1} solution. Two hundred fifty pmol (50 µl) of this solution was diluted in an appropriate volume of 0.1% SDS, 0.025M Tris and 0.2M glycine, pH 8.5, to give a range of total SDS from 62.5 µg to 4.0 mg. Following loading, columns were washed with 5 ml of 0.1% TFA and sequenced 10 cycles.

C. Size Exclusion Chromatography

Two and one half nmol of β-lactoglobulin was dissolved in 50 µl of 6M G-HCl, 0.1M Tris HCl, pH 8.5, 10 mM EDTA. The sample was reduced by adding dithiothreitol (DTT) (CALBIOCHEM) to 20mM and incubating at 37° C for 30 min. Iodoacetamide (Sigma) was added to 70 mM and incubated at room temperature for 20 minutes. The solution was acidified by adding 200 µl of 0.2N HCl containing CNBr at 100 mg ml^{-1}. The solution was then incubated overnight at room temperature. The entire sample was then loaded onto two tandem Superdex 75 (Pharmacia) analytical columns (1 x 30 cm). The columns were equilibrated in 6M G-HCl, 50 mM Tris HCl, pH 8.0 (flow rate was 0.4 ml min^{-1}), and the effluent was monitored with a diode array detector (HP 1090M). Fractions were collected at 1 minute intervals, and pools of 5.0 to 8.0 ml were made across the profile (Figure 2). Half of each pool was applied to a sequencing column and run 15 cycles.

Selective NH_2-terminal blocking of non-proline residues during sequencing was accomplished with o-pthalaldehyde (OPA) (3). A solution of 0.2% (wt/vol) OPA and 0.6% (vol/vol) mercaptoethanol in acetonitrile was placed in the spare bottle position in the top bank of bottles on the sequencer. A method was written to deliver the OPA at a predetermined cycle during sequence analysis where proline had been observed in the mixed sequence during the previous series of runs (4).

D. Gel Elution

Three proteins, HSA, β-lactoglobulin and Betaseron, were run on 10-20% Tricine gels (Novex), and stained briefly with Coomassie blue R-250 for the minimum time to visualize the bands of interest. After destaining in 7% acetic acid 40% methanol the gel was thoroughly washed with water. The gel bands were cut out and the protein passively eluted using a method similar to Hoger and Burgess (1980) (6). Briefly, the excised bands were suspended in 300 µl of 0.1M Tris HCl, pH 9.0, 0.1% SDS. Using a small plastic pestle the gel pieces were thoroughly crushed and placed on a rotating platform for two hours at room temperature. The crushed gel and buffer mixture was transferred to a .45 µm centrifugal filter (Scientific Resources Inc.) and spun at 14,000 rpm in a microfuge until all possible liquid had been removed from the gel pieces. The nearly dry gel paste was resuspended in 300 µl of 0.1M Tris HCl, pH 9.0, and filtered two more times. Filtrate from the extraction and the two washes was combined and loaded onto a sequencing column, followed by a wash with 5 ml of 0.1% TFA. Sequencing was carried out for 10 cycles.

E. Free-flow Isoelectric Focusing

A sample of HSA was separated on a Rainin RF3 free-flow isoelectric focusing apparatus in a buffer of 1% CHAPS and 1% Rainin 3-10 ampholytes. The HSA sample focused in a fraction within the 4 to 5 pH range (pI = 4.9). A sample containing 1 nmol of HSA in 75 µl was submitted for sequencing. The sample was loaded directly onto the sequencing column and washed with 2 ml of 0.1% TFA. Sequencing was carried out for 15 cycles.

F. Sequencing a Formulated Protein

A sample of a liquid formulated protein (the monoclonal antibody, Bach 250) was submitted for final lot analysis. The protein concentration was 410 µg ml^{-1} in 20mM $NaPO_4$, 200 mM NaCl, pH 7.0. One hundred µl (270 pmol) was diluted to 1 ml with Milli-Q water before loading onto the sequencing column. The column was then washed with 4 ml of Milli-Q water and sequenced for 15 cycles.

G. Yield Calculations

Initial and repetitive yield were calculated by linear regression using the method of least squares. The slope and y intercept were calculated from the best fit line of cycle number (x) versus log(pm_i) (y), where pm_i is the pmol of the amino acid identified at cycle i. Initial yield (IY) is the antilog of the y intercept. Repetitive yield (RY) is the antilog of the slope. All cycles collected were used for these calculations and no exceptions were made for low yielding amino acids.

III. Results and Discussion

A. Effect of Buffers on Sequence Analysis

Tris, guanidine, and urea are commonly used in purifying proteins and each can interfere with sequencing if they are present during Edman degradation. To examine the effects of these components on sequencing by the HP sequencer, 100 pmol of a standard solution of either HSA or TGF-α was diluted to 1ml in each of the three test solvents A (TFA), B (G-HCl/Tris) or C (G-HCl/urea/TFA/acetonitrile /Zwittergent), to give a final concentration of 100 pmol ml^{-1} of the protein standard. For each run the initial and repetitive yields were calculated (Table I). Solvent A (2% TFA) is the standard solvent recommended by Hewlett-Packard and is the reference for comparison with B and C, each of which contained various denaturants. The IY was 41 pmol (41%) for solvents A and B, and increased slightly to 44 pmol (44%) for solvent C. The RY's for HSA were all comparable (96%-97%). TGF-α showed a little greater variation between the solvents, displaying an IY for solvent A of 55 pmol and increasing for solvents B and C to 66 and 85 pmol respectively. The RY 's for TGF-α were lower than those for HSA. The results indicated that the solvent had essentially no effect on the sequencing of HSA. TGF-α showed no significant difference when loaded in solvent A or B, and a slight improvement in yield when

Table I. Solvent Effects on Protein Recovery

Protein	Solvent	IY (pmol)	RY (%)
HSA	A	41	97
	B	41	97
	C	44	96
TGF-α	A	55	77
	B	66	70
	C	85	83

Figure 2. Sequencing of CNBr fragments of β-lactoglobulin separated by size exclusion

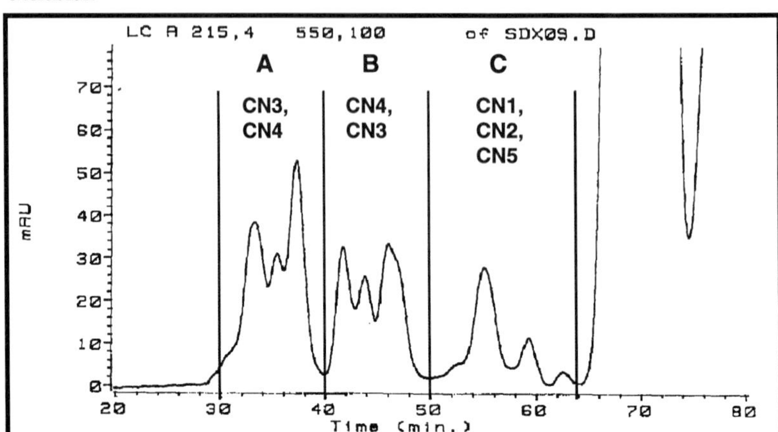

Figure 2A. Fractionation of β-lactoglobulin CNBr fragments by size exclusion chromatography. The pools used in sequence analysis are marked A, B and C. The CNBr fragments identified are shown within each pool.

	CN1	CN2	CN3	CN4	CN5
1	L	K	A	E	H
2	I	G	A	N	I
3	V	L	S	S	R
4	T	D	D	A	L
5	Q	I	I	E	S
6	T	Q	S	P	F
7	HSe	K	L	E	N
8		V	L	Q	P
9		A	D	S	T
10		G	A	L	Q
11		T	Q	A/V	L
12		W	S	C	E
12		Y	A	Q	E
14		S	P	C	Q
15		L	L	L	C
Mol. Wt.	805	1881	9382	4220	2065

Figure 2B. β-Lactoglobulin CNBr fragments. The first 15 residues in each of the 5 CNBr fragments of β-lactoglobulin. The boxed regions of CN4 and CN5 denote the sequences observed after OPA blocking.

loaded in C. The lower RY for TGF-α may be the result of the short (10 cycle) runs which for TGF-α tend to exaggerate the poor recoveries of serine at cycle 3 and histidine at cycle 4. The overall results show that these proteins at low concentration, in the presence of salts and denaturants, can be sequenced without additional manipulation or loss in efficiency. There may be a slight improvement in sequencing samples which were exposed to various denaturants before application to the sequencing column, a phenomenon noted previously for HPLC samples (3).

B. Mixture Sequencing of CNBr Fragments

Beta-lactoglobulin (2.5 nmols) was reduced and alkylated, cleaved with CNBr, and fractionated over a sizing column (Superdex 75) in 6M G-HCl (Figure 2). Fractions were combined into large pools, and half of each pool (approximately 3 ml) was loaded onto a sequencing column and sequenced 15 cycles. The results of sequencing the pooled fractions across the Superdex 75 profile are shown in Figure 2. Most of the pools contained at least two CNBr fragments, except pool C which contained fragments CN1, CN2 and CN5. Usually it was possible to uniquely identify the signal corresponding to each fragment because of differences in concentration of the two fragments present. All of the expected five fragments of β-lactoglobulin were found. The total protein recovery was 52%, based on the NH_2-terminal sequence of β-lactoglobulin found in pools A-C, nearly all of which was in pool C. The recovery of the other fragments ranged from 16% to 47%. Each pool contained mixtures of fragments, so o-pthalaldehyde (OPA) was used to block the primary amines at cycles where proline was observed during the mixture runs. The remaining half of pools B and C were each resequenced with OPA deliveries at cycle 6 (pool B run) and cycle 8 (pool C run). In each case a single sequence for CN4 and CN5 was obtained from the mixture of fragments. These results show that it is possible to obtain sequence directly from peptide mixtures in the presence of 6M G-HCl, and using OPA blocking, single sequences can be derived from the peptide mixtures.

C. SDS Tolerance

SDS is often a necessary sample component to maintain protein solubility. It is also present in proteins eluted from SDS-PAGE gels, which are commonly used in analytical and preparative protein purification. In the past, direct sequence analysis of SDS-PAGE separated proteins required either electroblotting onto a PVDF type membrane or removal of SDS by precipitation prior to analysis. Since it would be advantageous to directly sequence proteins eluted from

Table II. The results of sequencing 250 pmol of β-lactoglobulin in the presence of SDS

SDS (μg)	Initial Yield (pmol)	Repetitive Yield (%)
0	74	91.4
62.5	69	94.1
125	91	95.1
250	108	95.4
375	109	95.2
500	75	95.7
625	74	94.4
750	111	93.3
825	113	93.1
1125	76	94.1
1250	76	93.3
1500	67	90.8
2000	50	93.3
4000	38	87.1

gels, a series of experiments were performed to examine the effects of increasing amounts of SDS on sequencing. The results of loading 250 pmols of β-lactoglobulin in varying amounts of SDS ranging from 62.5 μg to 4.0 mg (Table II). The initial yields are some what variable, but remain fairly stable between 69 and 113 pmol over the range of 0 to 1.5 mg SDS. Starting at 2.0 mg of SDS the IY drops to 50 pmol, finally dropping to 38 pmol at 4.0 mg of SDS. Without SDS present the RY was 91.4%. In the presence of SDS, RY values range from 87.1 to 95.7%, the values rise steadily from 94.1% to 95.7% at 0.5 mg , after which they decline to 87.1% over the remainder of the range of SDS amounts. The data indicates that up to about 1.5 mg of SDS has almost no adverse effect on the sequencing of β-lactoglobulin, except for a slight decline in RY above 500 μg. The slight beneficial effect of low amounts of SDS (62.5 to 500 μg) may be caused by denaturing the protein before loading, similar to the effect seen when loading TGF-α in solvent C earlier. Above 4.0 mg (14 μmole), SDS appears to bind most available sites on the reverse phase resin, reducing the ability to bind β-lactoglobulin.

D. Passive Elution From Polyacrylamide Gels

To test the ability to sequence proteins extracted from a gel, Coomassie stained bands were cut from 10-20% Tris glycine gels and eluted using the technique described earlier. The results of the gel elution and sequencing are shown in Table III. Recoveries are based upon the signal observed on the sequencer relative to the amount of protein applied to the gel. No attempt has been made to correct for initial

Table III. Recovery of proteins by passive elution from 10-20% Tricine gels.

	Pmol of Protein Loaded onto Gel	Pmol IY From Sequencer	Recovery (%)
β-lactoglobulin	274	122	45
Betaseron	200	62	31
HSA	200	29	15

yield or other losses. Beta-lactoglobulin is an 18.3 kD protein that is readily soluble, and generally well behaved during most manipulations. Directly sequencing passive eluates of gel purified β-lactoglobulin gave a recovery of 45%, the highest of the three proteins tested. Betaseron, at 19.9 kD, is about the same size as β-lactoglobulin, but it is very hydrophobic, and gave a recovery of 31%. HSA, 66.47 kD gave a recovery of 15%. There were faint bands above HSA in the gel that may have been dimer and trimer, accounting in part for the low recovery of HSA. Our laboratory experiences typical recoveries of 10 to 30% for proteins electroblotted from gels onto polyvinylidene difluoride (PVDF) membranes. The results with the passive elution/direct sequencing give equal or better results, with the additional advantage of leaving the protein in solution for further characterization.

E. *Isoelectric Focusing*

An HSA standard was run in a Rainin RF3 free-flow isoelectric focusing instrument in a buffer containing CHAPS and ampholytes. After focusing, fractions were collected across the pH gradient. The pH range 4 to 5 contained the HSA and 1 nmol was loaded onto the sequencer. Two sequences were found, HSA at an initial yield of 350 pmol and HSA missing the first two NH_2-terminal residues at 100 pmol. Together this represented a yield 45% of the loaded protein. No interference was seen from the ampholytes. Previously IEF could not be used for preparing proteins for sequencing without additional steps to remove the ampholytes.

F. *Formulated Protein*

A sample of 270 pmol of Bach 250 in 20mM $NaPO_4$, 200mM NaCl pH 7.0 was loaded onto the sequencer and run. The only manipulation was to dilute the sample to 1 ml in Milli-Q water before loading. Two sequences were observed that matched the heavy and light chains of Bach 250, with IY's of 80 and 135 pmol and an average RY of 95%. From the sequence data a purity of 90% was determined.

With the salt content of this sample, it would normally require an additional manipulation to remove the salts. The problem with dialysis, HPLC desalting or any of the other methods is the possibility of altering the proteins in the sample. Loading the sample directly onto the sequencer support avoids this problem.

References

1. Miller, C.G., Bente, H.B., Fischer, S., Myerson, J., Wagner, G., Widmayer, R., Horn, M.J., Fourth Symposium of the Protein Society, San Diego, CA (1990) Abstract T143

2. Myerson, J., Bente, H.B., Fischer, S., Horn, M.J., Miller, C.G., Wagner, G., Fourth Symposium of the Protein Society, San Diego, CA (1990) Abstract T144

3. Brauer, A.W., Oman, C.L., Margolies, M.N. (1984) *Anal.Biochem.* **137**,134-142.

4. Wank, S.A., Harkins, R., Jensen, R.T., Shapira, H., de Weerth, A., Slattery, T.K. (1992) *Proc.Natl.Acad.Sci.USA* **89**,3125-3129.

5. Nugent, K.D., Burton, W.G., Slattery, T.K., Johnson, B.F., Snyder, L.R., (1988) J.*Chrom.* **443**, 381-397.

6. Hager, D.A., Burgess, R.R. (1980) *Anal.Biochem.* **109**, 76-86.

PROTEIN SEQUENCING OF POST-TRANSLATIONALLY MODIFIED PEPTIDES and PROTEINS: DESIGN,CHARACTERIZATION AND RESULTS OF ABRF-92SEQ

Sheenah M. Mische[1], K. Ümit Yüksel[2], Liane M. Mende-Mueller[3], Paul Matsudaira[4], Dan L. Crimmins[5], and Philip C. Andrews[6]

[1] Protein Sequencing/HHMI Biopolymer Facility, The Rockefeller University, New York, NY 10021

[2] Biopolymer Analysis Laboratory, Texas College of Osteopathic Medicine, Fort Worth, TX 76107

[3] Protein/Nucleic Acid Shared Facility, Medical College of Wisconsin, Milwaukee, WI 53226

[4] Whitehead Institute, M.I.T., Cambridge, MA 02142

[5] Howard Hughes Medical Institute Core Protein/Peptide Facility, Washington University School of Medicine, St. Louis, MO 63110

[6] Protein Structure Facility, University of Michigan Medical School, Ann Arbor, MI 48019

I. INTRODUCTION

Protein sequencing is a highly sensitive technique that has been invaluable for providing critical primary structural data on isolated proteins, quality control for both recombinant proteins and synthetic peptides, and internal amino acid sequence from peptides derived from proteins that are used to design oligonucleotide DNA probes in recombinant DNA studies. One of the major problems facing protein chemistry core facility personnel, in addition to the ever diminishing sample quantity submitted for analysis, is the identification of post-translationally modified amino acids. The complete covalent structure of a protein entails knowledge of the primary structure as well as the chemical nature and positions of all the modifications to the protein. There are over 150 known modifications (1,2), and the definitive identification of covalent modification

TECHNIQUES IN PROTEIN CHEMISTRY IV

sites has become increasingly important.

In order to begin to address this aspect of sequence analysis, the Association of Biomolecular Resource Facilities (ABRF) Sequencing Committee designed a peptide that would test the ability of core facilities to identify modified amino acids. The modified amino acids included in the test peptide would not be so unusual to preclude any identification, and would be common enough that the majority of facilities would be expected to identify one if not both of them.

Two hundred and twenty-five samples were distributed with detailed instructions for completion of a brief sequencing worksheet and limited questionnaire. The facilities were instructed to use any analytical technique that was available to them for correct assignment of the sequence. Previous samples designed by the ABRF have provided significant findings regarding the performance of the average facility. But, more importantly, these studies have allowed the participating facilities to objectively evaluate their strengths (3-6). This report summarizes the results from the analysis of 74 data sets returned by July 1, 1992.

II. MATERIALS AND METHODS

A. *Design of the ABRF-92SEQ Test Sample*

The test sample, designated ABRF-92SEQ, was designed in order to evaluate the ability of core facilities to correctly identify two relatively common post-translational modifications, as well as other sequencing identification problems. The peptide was 37 amino acids long and contained a phosphoserine at residue 4 and hydroxyproline at residue 11. The peptide was phosphorylated at residue 4 utilizing the cAMP dependent consensus site (residues 2-5). In addition, some standard problem regions were incorporated, with two sets of doublets (G_6,G_7 and Q_{12},Q_{13}), a triplet hydrophobic region (A_{18} V_{19} L_{20}), and E_{27} Q_{28} E_{29} to evaluate the effects of deamidation on sequence assignment. Alanine at residues 1,5,9,and 18 was included for repetitive yield calculations (A_{18} after Pro to determine lag). Sites for trypsin (K_{31}-I_{32}, and R_{15}-H_{16}), cyanogen bromide (M_{24}-G_{25}), endoproteinase V-8 (D_{27}-Q_{28}, E_{29}-A_{30}) and endoproteinase Asp-N (Q_{13}-D_{14}) cleavage were also included (Figure 1).

NH_2-A-R-G-U-A-G-G-S-A-K-J-Q-Q-D-R-H-P-A-V-L-T-N-G-M-G-S-E-Q-E-A-K-I-H-N-F-A-K-COOH
10 20 30

Figure 1: ABRF-92SEQ peptide. Single letter codes used are standard except for phosphoserine (U) and hydroxyproline (J).

B. Peptide Synthesis and Characterization

The 37 residue peptide was synthesized on a Milligen/Biosearch synthesizer, Model 9050, by the Protein Facility of the Medical College of Wisconsin. FMOC-activated esters of amino acids were used for the synthesis with the exception of hydroxyproline. In this case, 0.8 mmol FMOC-t-butyl hydroxy-L-proline (Aminotech, Nepean, Ontario, Canada) was coupled with 0.8 mmol benzotriazole-1-yl-oxy-tris (dimethylamino) phosphonium hexafluorophosphate (BOP) in N-methyl morpholine/dimethylformamide in the presence of 0.8 mmol hydroxy-benzo-tetrazolehydrate (HOBt). The reaction was performed in an automated mode with a standard coupling cycle (Milligen/Biosearch Operator Manual p. 124). The peptide was removed from the resin by TFA cleavage and deprotected for 8 hr at room temperature in TFA (90%) in the presence of mixed scavengers (thioanisole (5%) ethanedithiol (3%) and anisole (2%)).

The crude product (2 μmol) was purified by preparative HPLC using a Vydac C18 column (5 x 25 cm). The purified peptide (1.324 μmol) was phosphorylated using a cAMP dependent kinase catalytic subunit from bovine heart (Sigma) using the conditions described by Beavo (7). The modified protein (approximately 80%) was separated from the nonphosphorylated peptide using a Vydac C18 preparative column and separation was confirmed by co-elution with ^{32}P-labelled peptide, and laser desorption time-of-flight mass spectrometry (LDTOFMS). The purified phosphopeptide was distributed to several of the authors for N-terminal sequence analysis, amino acid analysis, mass spec analysis (both LDTOF and ES), and cleavage studies (chemical and enzymatic digestion, followed by RPHPLC and sequence analysis). The peptide was judged to be pure by sequence analysis of 100-500 pmol with initial yield (%) of 55.61% ±28.4% (average ±sample SD, 5 independent determinations) with a range of 28.4-91.04%. Phosphoserine at residue 4 was confirmed by sequence analysis after modification to yield S-ethyl cysteine (8). The two independent mass spectral analyses yielded a m/z of 3946 and 3945 (calculated m/z 3946) as the only species present.

C. Preparation and Distribution of the test peptide

Five hundred picomoles of the purified peptide was aliquotted into pre-washed polypropylene microfuge tubes (1.5 ml), and then vacuum dried (3,4). ABRF-92SEQ peptide was mailed to 225 core facility members of the ABRF along with a detailed set of instructions for sample solubilization and data reporting. Data were returned to an independent third party who removed any identifying marks. These anonymous data were then forwarded to the authors for analysis.

III. Results and Discussion

A. *Survey Results*

The number of facilities that participated in this year's study was lower than the previous year (75 vs 90; one participant did not send data), but more significantly, the response rate was lower, with only 32.9% (74/225) responding, as compared to the previous years response rate of 45.5% (90/198). The reason for the low return rate is unknown, but may be indicative of the work load of ABRF member core facilities.

Based on the brief data sheet returned by the participating facilities, the average age of the sequencers increased from 4.2 years (0.8-9) to 6.7 years (1-10). As in previous years most of the facilities have Applied Biosystems sequencers (470 = 18, 473 = 4, 475 = 18, 477 = 23). The remaining facilities have Porton 2090 (5), 2020 (2), and Milligen 6625 (2), 6600 (1) sequencing systems.

Seventy-three percent of facilities used glass fiber filters as a solid support for sequencing. The remainder used Porton peptide or GF membrane filters (13.5%), Sequelon arylamine (4.0%), or PVDF, Sequelon DITC membrane or CD matrix (all less than 2%). Sixty-seven percent of the facilities used polybrene as a pretreatment of their membranes, with 50% of these using 1-4 ABI FIL cycles prior to sequence analysis. Seventy percent of the facilities who responded used standard procedures for sequence analysis.

The average reconstitution volume was 144.4 ±136.7 µl (20-1000 µl), with an average of 46.7% ± 31.2% (10-100%) of the sample used for sequence analysis. This represents an average of 234 ± 156 pmols analyzed by Edman degradation. Of the seventy-four returns, sixteen correctly identified the peptide length of 37 residues, and five over-called the sequence, with 38 residues or longer (Figure 2). Twenty-six facilities (34.6%) were able to provide a molecular weight estimate, and three to within 3 mass units, using mass spectrometry, amino acid analysis, and/or protein sequencing.,

The extent of additional studies that were reported with the 500 picomoles of peptide was quite low; 34.6%, 26.6%, and 5.6% used one, two or three additional techniques, respectively, for the generation of peptides to confirm and/or complete sequence analysis. Of these analyses, 41% used amino acid analysis, and 21% used enzymatic or chemical cleavage for the generation of peptides. Seven facilities used cyanogen bromide cleavage, four used trypsin cleavage, one used endoproteinase Lys-C, two used endoproteinase Asp-N, and one facility each used endoproteinase Glu-C or ammonium hydroxide cleavage. Only 22.6% of the facilities used either RPHPLC (15/74) or CZE (2/74) for separation (10/74, 1/74) or analysis (5/74, 1/74, respectively) of ABRF-92SEQ. The availability and use of mass spectrometry increased, with 29.3% of the facilities using this technique for the determination of modified amino acids.

B. Sequence Results

A total of 2445 sequencing cycles were returned for this study. Of the positive assignments, 81.2% (1986) were correct, and 4.2% of the tentative assignments were correct, with an accuracy of 93.6% and 64.1%, respectively (Figure 2). The percent accuracy is comparable to those of previous years in which the amount of sample was high, for example 1988 (STD-1) and 1989 (ABRF-89SEQ) (3,4). The number of unassigned residues (161) accounted for 6.5% of the total. Instrument malfunctions were very low, (1.33% of the total). This compares very favorably to 10% for ABRF-91SEQ.

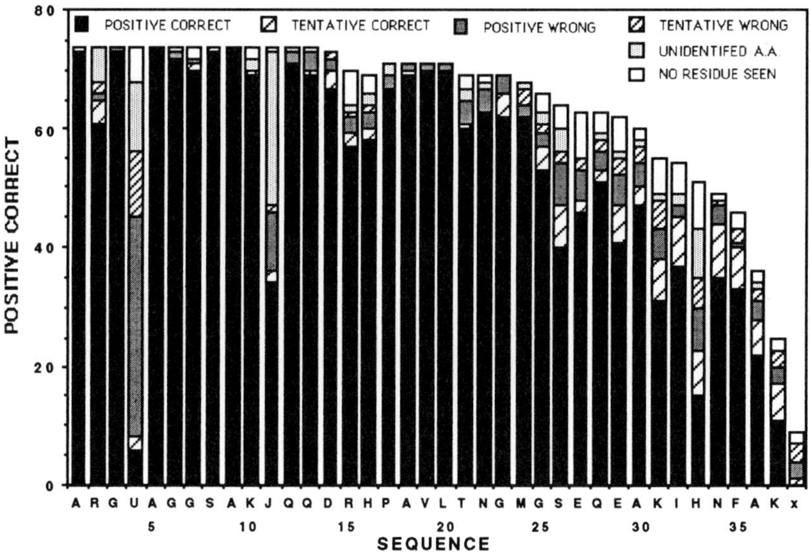

Figure 2: Results of Sequence Assignments for ABRF-92SEQ. The sequence of the peptide is listed across the bottom, with the amino terminus at the left. Standard amino acid abbreviations are used except for phosphoserine (U) and hydroxyproline (J). When more than one residue was listed in a cycle, it was plotted as wrong. Cycles for which no residue was called or unidentified residues were labelled as such.

The most correctly identified amino acid was Ala at residue 5 and 9 (100%) and the most difficult amino acid to identify was phosphoserine at position 4, with an accuracy of 13.9% (6/43) (Figure 3). It was most often identified as Ser, followed by Cys (Figure 4A). Hydroxyproline was correctly identified by 46% (34/74) of the participating facilities with 77% accuracy (34/44). It was most often misidentified as an X (35.1% (26/74)) or H (10.8%

(34/44). It was most often misidentified as an X (35.1% (26/74)) or H (10.8% (8/74)) (Figure 4B). The C-terminal residue, Lys_{37} was correctly identified by 13.5% (10/74) of the facilities (see Figure 3).

With regards to the identification of phosphoserine at residue 4, three facilities of the nine (including three tentative assignments) provided definitive identification by sequence analysis using the procedure of Meyers et al (8). Two facilities used the ratio of serine to dehydroalanine to determine the identity of phosphoserine (8). The remaining four facilities gave no method of analysis. Based on these results, it appears that O-phosphorylated residues remain a major obstacle to most facilities.

Of the seventy-four sequences returned, there were no perfect sequences reported, although two facilities had all residues correct, except for residue 4, with one correct tentative identification of phosphoserine (#178), and one misidentification as serine (#361). Two other facilities had all correct except for residue 4, but they misidentified it as either serine (#206), or dehydroalanine (#186), and each overcalled the peptide by one residue. These analyses are summarized in Table I. The most complete sequence analysis was reported by facility #178, with both sequence analysis and mass spectrometry used for identification of the modified amino acids, and reduction/alkylation with 4-vinyl pyridine to rule out cysteine at residue 4.

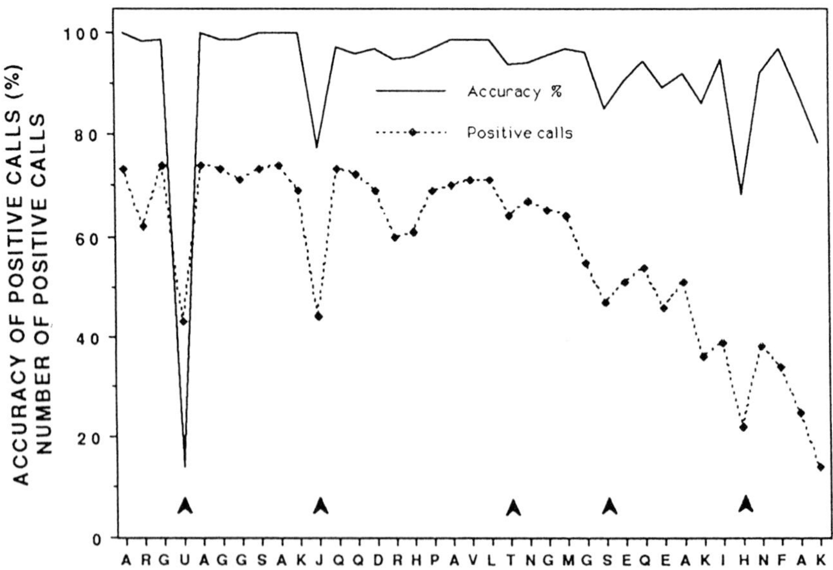

Figure 3: Accuracy of Positive Calls for ABRF-92SEQ. The accuracy of each call was determined using the number of positive calls for each residue.

Response Number	Sequence Call Categories: Positive Correct	Tentative Correct	Positive Wrong	Tentative Wrong	Overcall
178	36	1 (u_4)			no
186	36			1 ($∂_4$)	yes
361	36		1 (S_4)		no
206	36		1 (S_4)		yes
291	35	2 (u_4,k_{37})			no
225	35			1 (s_{34})	yes
262	35			2 (u_{33}, n_{37})	no
261	35		2 (S_4, J_{33})		no

Table I: Best calls for ABRF-92SEQ. The single letter code for the amino acids are standard except for phosphoserine (U), hydroxyproline (J) and dehydroalanine (∂). The subscript refers to the residue numbering of ABRF-92SEQ (see Figure 1).

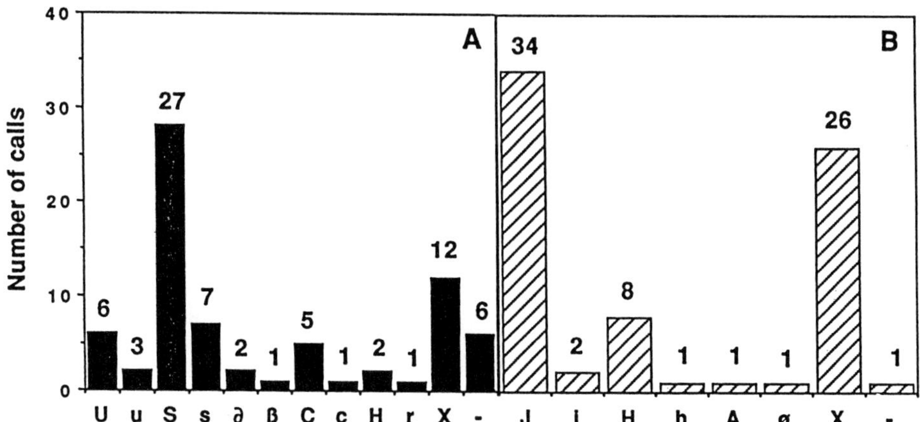

Figure 4: Distribution of calls. Sequence assignments for position 4 (A) and position 11 (B) are shown. Standard nomenclature for single letter code assignments are used except for phosphoserine (U), hydroxyproline (J), dehydroalanine (∂), unidentified serine derivative (ß) and ornithine (ø). Upper and lower case denote positive and tentative calls, respectively.

IV. Conclusions

ABRF-92SEQ was designed to address different problems of sequence analysis and was more challenging than previous ABRF sequencing samples in that respect. The decision to provide 500 pmols of ABRF-92SEQ was based on feedback from ABRF members who expressed interest in having a more challenging "real-world" sample for sequence analysis and wanted to try several analytical techniques. ABRF-92SEQ encompassed several of the standard types of sequence problems that facilities are faced with, but included two amino acids that were post-translationally modified. The cover letter requested the most complete sequence possible, and it was hoped that the more than adequate sample size (500 pmols) would be used to perform further analyses. In addition, the sequencing committee was interested how the participating facilities identified and reported these modified amino acids residues.

Overall, the accuracy of the sequence calls was very high (93.6%) which reflects the adequate sample size, and confirms previous ABRF sequencing studies that the accuracy of the sequence analysis is directly related to the amount of sample (3,4). In this study an average of 46% of the sample was used for sequence analysis alone. Of the additional analyses that were performed, amino acid analysis, was used most frequently, followed by mass spectral analysis. Very few facilities performed cleavage and peptide isolation by RPHPLC or CZE. The ABRF-92SEQ sample size was believed to be ample for additional analyses, but the reported results reflect otherwise. Several factors may have contributed to this, including the unavailability of these services in the majority of facilities (as several of the respondents reported).

Based on a response rate of 32.9%, the results presented here suggest that the majority of core facilities were not able to correctly identify the sequence of a 37 residue peptide that contained two modified amino acids. Thirty-three percent of the facilities used sequencing for the identification of modified amino acids. Eighteen percent used amino acid and sequence analysis, and eight percent used mass spectrometry in addition to sequencing. The ability to correctly identify phosphoserine was limited to only six facilities (13.9%). These results suggest that the average facility is limited in their ability to provide sequence data for phosphorylated amino acids. Determination of modified amino acids takes additional time and planning, and this may not be feasable for the majority of core facilities to undertake. However, it also highlights the inadequacy of the current sequencing instrumentation and methodology available from the major sequence instrumentation companies regarding phosphorylated amino acids. In contrast, the identification of hydroxyproline was easily identified by over 45% of the facilities and most used only Edman degradation to identify this residue. This too, is a reflection of what is currently available, as documentation for the chromatographic identification of this PTH amino acid is available (9).

In summary, the identification of post-translationally modified amino acids represents a challenge in sequence analysis. These analyses require additional time, and presently, standardized and reproducible methods for modified amino acids are limited. However, the demand for these identifications increases, and standardized methods will need to be developed for these applications. The ABRF is currently compiling a compendium of chromatographically identified modified amino acids (ABRF News (1992) 3:2).

It would benefit all ABRF core facilities to contribute to this project so that sequence identification of modified amino acids becomes the norm, rather than the exception, for protein sequencing core facilities.

Acknowledgements

This work was partially supported by NSF grant DIR 9003100 to John Crabb (W. Alton Jones Cell Science Center) on behalf of the ABRF. The cooperation of all the anonymous ABRF member facilities that participated in this survey by analyzing the sample and providing the data are gratefully acknowledged. The assistance of Ms. Pamela Baud in providing the anonymity of the respondents is appreciated. The authors are especially grateful to Steven Lewis (Medical College of Wisconsin) for the synthesis of the peptide, Brady Stoner (Medical College of Wisconsin) for the preparation of the final samples; Anne-Marie Michon (Whitehead Institute) for performing the phosphorylation of ABRF-92SEQ peptide; Richard Thoma (Washington University, St. Louis) for peptide cleavage and sample reconstitution/recovery studies; and Joseph Fernandez, Michael DeMott and Lori Andrews (Rockefeller University) for sequence analysis, peptide cleavage/HPLC studies, and phosphoserine identification.

References

1. Wold, F. (1981) Ann. Rev. Biochem. 50:783-814
2. Yan, C.B., Grinnell, B.W., and Wold, F. (1989) Trends Biochem Sci. 14:264-268
3. Niece R.L., Williams, K.R., Wadsworth, C.L., Elliot, J., Stone, K.L., McMurray, W.J., Fowler, A., Atherton, D., Kutny, R., and Smith A.J. (1989) *in* Techniques in Protein Chemistry, T.E. Hugli, ed., pp 89-101 Academic Press, San Diego,CA
4. Speicher, D.W., Grant, G.A., Niece, R.L., Blacher, R.W., Fowler, A.V., and Williams, K.R. (1990) *in* Current Research in Protein Chemistry II, J.J. Villafranca, ed., pp 151-162 Academic Press, San Diego, CA.
5. Yüksel, K.Ü., Grant, G.A., Mende-Mueller, L.M., Niece, R.L., Williams, K.R., and Speicher, D.W. (1991) *in* Techniques in Protein Chemistry II, J.J. Villafranca, ed., pp151-162 Academic Press, San Diego
6. Crimmins, D.L., Grant, G.A., Mende-Mueller, L.M., Niece, R., Slaughter, C., Speicher, D.W., and Yüksel, K.Ü. (1992) in Techniques in Protein Chemistry III, R.H. Angeletti, Ed.. pp 35-43. Academic Press, San Diego, CA.
7. Beavo, J.A., Bechtel, P.J., and Krebs, E.G. (1974) Methods in Enzymology vol 38 pp 299-307 Academic Press, N.Y.
8. Meyers, H.E., Hoffmann-Porsorske, E., Korte, H., and Heilmeyer, L.M.G., Jr. (1986) FEBS Lett. 204:61-66
9. Hunkapillar, M. W. (1985) ABI Users Bulletin 14:25

Carboxy-terminal Protein Sequence Analysis using Carboxypeptidase P and Electrospray Mass Spectrometry

Christine E. Smith and Kevin L. Duffin

Monsanto Corporate Research, Monsanto Company, St. Louis, MO 63198

I. Introduction

The determination of the carboxy-terminal sequence of a protein is desirable for characterizing the primary structure of a protein. Carboxy-terminal sequence information can assist in establishing the integrity of recombinant proteins. Additionally, it is desirable to determine the protein termination site derived from a cloned DNA sequence. Although this is usually accomplished through translation of the gene sequence and location of stop codons, post-translational processing, protease cleavage during purification, or stop codon suppression may lead to an unexpected C-terminal sequence. Also, for cloning purposes, C-terminal sequence information can be used in the generation of an oligonucleotide probe.

Sequential chemical degradation of the carboxy-terminus of a protein is an evolving technology, and therefore, classical methods of C-terminal sequencing have entailed the digestion of single amino acids from the C-terminus of a protein using carboxypeptidases. Analyses of the released amino acids are usually performed at several time intervals using ion-exchange chromatography or RP-HPLC. When monitoring the released amino acids, problems in data interpretation can arise. The presence of double residues is often missed or misinterpreted. Additionally, free amino acid background can contribute to the complexity of interpretation. Since the kinetics of the digestion reaction are dependent on the protein sequence, variation in the temporal release of amino acids commonly occurs. Therefore, the enzymatic reaction must be monitored at frequent time intervals so the true sequence of amino acids can be determined. These problems associated with the analysis of released amino acids from a carboxypeptidase digestion can be circumvented if the analysis is performed directly on the digested protein.

TECHNIQUES IN PROTEIN CHEMISTRY IV

Several groups have reported success at the analysis of carboxypeptidase-digested peptides using Fast Atom Bombardment-Mass Spectrometry (FAB-MS) and Thermospray LC/MS (1,2,3). Accurate mass analysis of digested peptides has been demonstrated with a level of sensitivity at 0.5-5 nmoles. Additionally, Plasma Desorption Mass Spectrometry (PDMS) has been used to monitor carboxypeptidase digestions (4,5). The recent development of Electrospray Mass Spectrometry (ESMS) allows mass spectrometric analysis of large molecular weight molecules (e.g. proteins) by producing multiple charge states of the compound during ionization. This decreases the effective mass-to-charge ratio of the ionized protein so it can be measured by conventional quadrupole mass analyzers. Electrospray mass analysis yields accurate protein molecular weights (+/- 0.01%) with enough resolution to distinguish mass differences of a single amino acid. This study involves the applicability of electrospray mass spectrometric analysis to monitor carboxypeptidase digestions of peptides and proteins.

II. Materials and Methods

Neurotensin and superoxide dismutase (SOD) were purchased from Sigma-Aldrich (St. Louis, MO). The recombinant human Interleukin 3 (IL-3) was purified by Monsanto Corporate Research (St. Louis, MO). The reduction and alkylation of superoxidase dismutase was performed in the following manner. The protein was solubilized in 6M Guanidine-HCl, 0.5M Tris pH8.5, reduced in the presence of 10mM DTT for 2 hours at 37C, and subsequently carboxymethylated in 20mM iodoacetic acid (Pierce, Rockford, IL) for 30 minutes at room temperature. The protein was then purified on a Sephadex G25 gel filtration column (Pharmacia, Piscataway, NJ).

Carboxypeptidase P (CPP) was purchased from Takara Biochemicals (Berkeley, CA) and Boehringer-Mannheim (Indianapolis, IN). Enzyme preparations which contained sodium citrate were desalted by gel filtration using a Sephadex G25 column.

The peptide or protein (typically 1-2 nmoles) to be digested was suspended in a buffer of 15 mM ammonium acetate and the pH was adjusted to 4.0 with formic acid. CPP was added at an enzyme-to-substrate ratio of 1:100 or 1:50 as detailed below. At several time points, aliquots were removed and acidified with methanol: water:formic acid (50:50:2) for mass spectral analysis.

Solutions of the CPP-digested peptide or protein were introduced to an electrospray mass spectrometer at 4 μL/min with a syringe pump. A Sciex API III triple-quadrupole mass spectrometer (Sciex, Inc., Toronto, Canada) equipped with an atmospheric pressure ion source was used to sample positive ions produced from a pneumatically assisted electrospray interface (6). Sample ions that were generated by the electrospray process were mass-analyzed by scanning the first quadrupole in 0.1 amu increments over the mass range of the mass spectrometer (2400 amu) in approximately 10 s. Approximately twenty scans were averaged together to obtain the mass spectra in this study. Mass calibration of the mass spectrometer was accomplished just prior to mass-analysis of the protein samples by matching ions of polypropylene glycol to known reference masses stored in a mass calibration table of the mass spectrometer. The average molecular weights of the protein and CPP-digested protein were calculated from the multiply-protonated ions in the mass spectrum using software based upon published equations (6).

III. Results and Discussion

Electrospray mass spectrometric (ESMS) analysis of proteins yields optimal data under low ionic strength and volatile buffer conditions, therefore, a carboxypeptidase that functions under these conditions was required. Carboxypeptidase P (CPP) has been reported to be a non-specific carboxypeptidase, capable of cleaving all amino acids from the C-terminus, although slow release of serine and glycine residues has been observed. Lu et al. have demonstrated activity of CPP at pH 4.0 in a low molarity of sodium acetate (7). In this study, 15mM ammonium acetate (pH 4.0) was used. This buffer meets the requirements of ESMS analysis. At various timepoints, aliquots from the digestion reaction were acidified and mass-analyzed.

Carboxy-terminal digestion with Carboxypeptidase P was first investigated with a model peptide, neurotensin. The digestion of neurotensin (10 nmoles) was carried out at an enzyme-to-substrate ratio of 1:100 in 100µl of digestion buffer at room temperature. Aliquots of 10µl were removed at 1 min, 15 min, 30 min, 35 min, and 60 min, diluted 1:10 with methanol:water:formic acid and analyzed by ESMS. The approximate quantity analyzed at each time point was 200 pmoles. The results suggested that CPP digests proline residues much more slowly than other amino acids present in neurotensin. Compilation of mass spectra obtained at several timepoints yielded the C-terminal sequence of R-R-P-Y-I/L-I/L-OH, corresponding to the known sequence of neurotensin (data not shown). The experiment was terminated after 60 minutes.

In order to test the applicability of the ESMS analysis method for the CPP digestion of proteins, recombinant *E. coli* derived human IL-3 was chosen as a model. IL-3 has a C-terminal sequence of ...T-T-L-S-L-A-I-F-OH. The digestion conditions detailed above were used to digest 2.7 nmoles of IL-3. Aliquots of 500 pmoles were removed at several time points, 10 min, 20 min, and 50 min, and analyzed by ESMS. At the 10 minute time point, the appearance of multiply-charged ions representing consecutive losses of five C-terminal amino acids was observed in the mass spectrum (Figure 1A and B). The average molecular weight of each truncated species can be determined through deconvolution of the mass spectrum (6). Figure 2A shows a closeup view of the $(M+9H)^{+9}$ charge state of CPP-digested IL-3. The mass differences

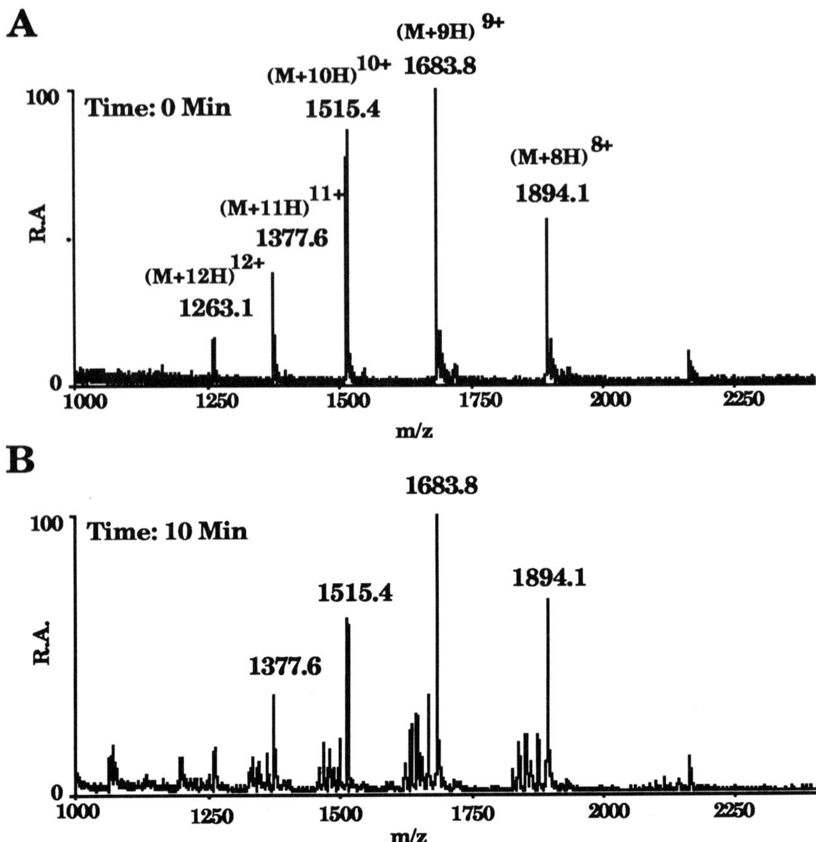

Figure 1. Positive ion electrospray mass spectra of recombinant human IL3. (A) IL3 time = 0, (B) IL3 + CPP (1:100) time = 10 minutes. (R.A.=Relative Abundance)

of these ions reflect the sequential loss of amino acids from full-length IL-3. The mass spectrum obtained from the 10 minute time point alone provided enough information to establish the C-terminal sequence as ...S-I/L-A-I/L-F-OH. Although serine residues have been reported to release slowly, under these experimental conditions the serine residue was effectively removed in approximately the same time frame as other amino acids. Additional time points yielded another mass ion representing the loss of the next residue, either an isoleucine or leucine (Figure 2B). Further degradation was not observed under these conditions. The sequence data obtained was consistent with the expected C-terminal sequence of IL-3.

Further application of this study was extended to include an acetylated protein, superoxide dismutase, from which N-terminal sequence would be unattainable by Edman degradation. Superoxide

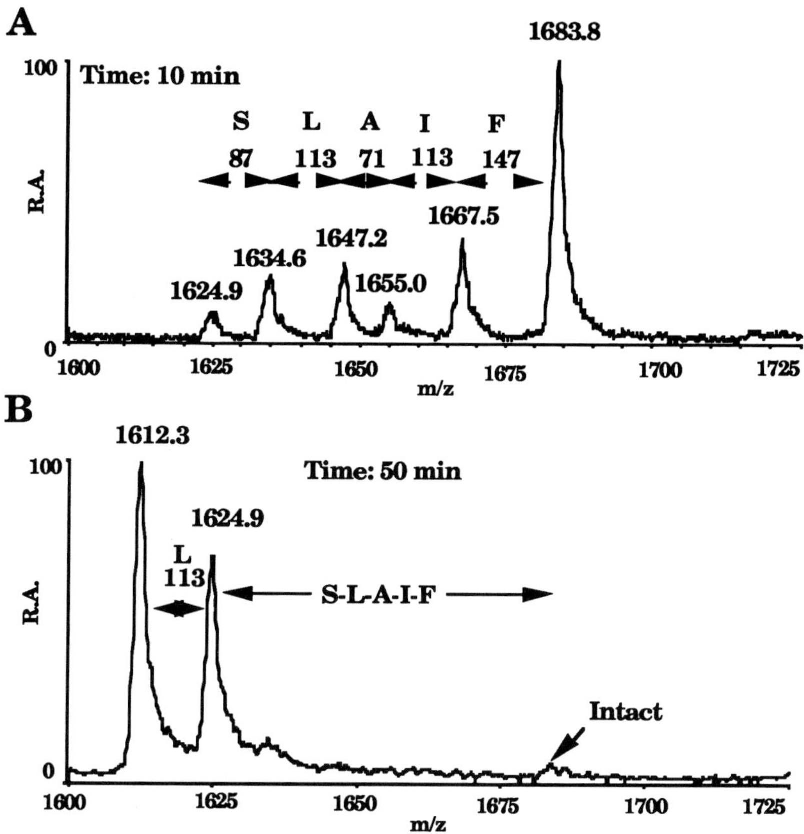

Figure 2. Expanded view from positive electrospray mass spectra of the +9 charge state from IL3 + CPP. (A) time = 10 minutes, (B) time = 50 minutes. (R.A.=Relative Abundance)

dismutase has a C-terminal sequence of ...C-G-V-I-G-I-A-K-OH. Initially, minimal digestion of the protein was observed in the mass spectrum, even at an enzyme-to-substrate ratio of 1:50 (Figure 3A &B). Only a low abundance truncation of lysine, the expected C-terminal residue, was observed. Subsequent reduction and carboxymethylation of superoxide dismutase improved the degree of C-terminal cleavage by CPP (Figure 3C & D). The C-terminal lysine residue was observed to cleave extremely fast after reduction and carboxymethylation and further sequencing yielded a C-terminal sequence which corresponded to the reported sequence of ...I-A-K at 20 minutes. The reaction did not appear to proceed through the glycine residue.

Analytical characterization of an *E. coli* produced truncated version of human LACI (LACI recK2) yielded data that was not consistent with the expected primary structure as predicted by the

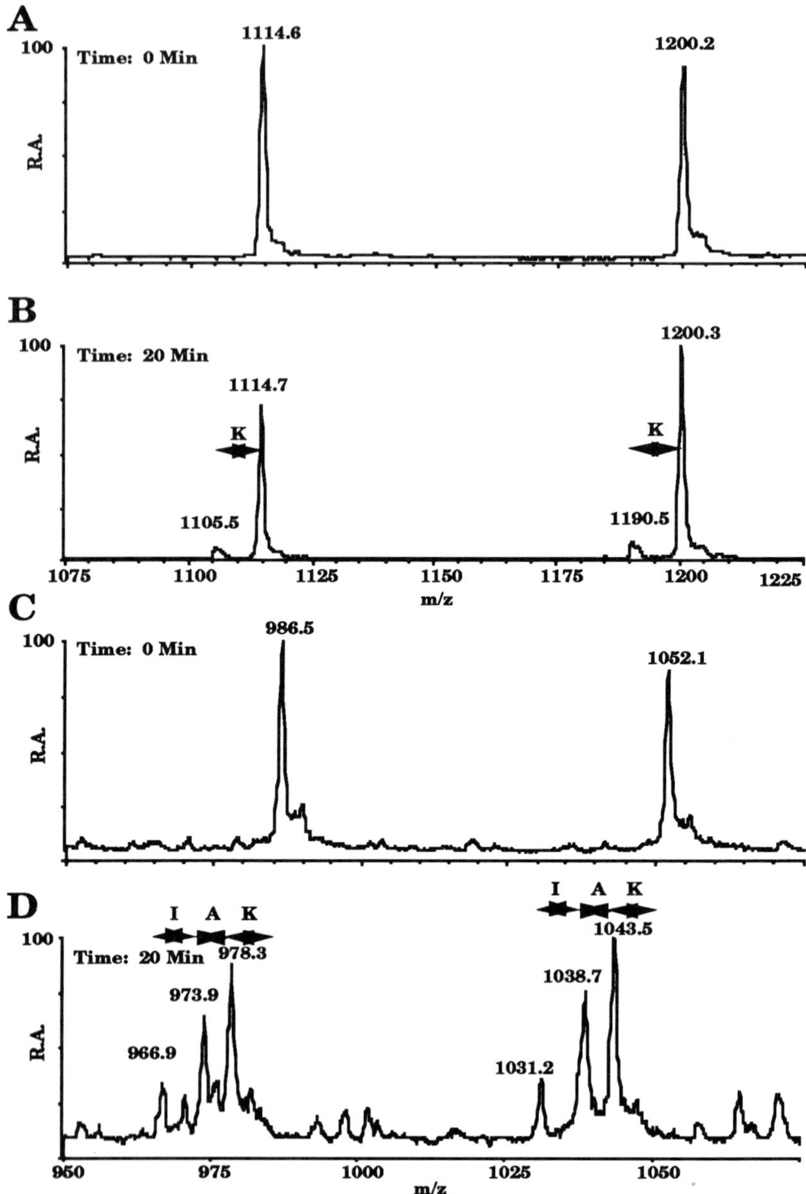

Figure 3. Positive ion electrospray mass spectra of superoxide dismutase. Comparison of the +15 and +16 charge states from superoxide dismutase before (A) and after (B) digestion with CPP. Comparison of the +15 and +16 charge states from reduced and carboxymethylated superoxide dismutase before (C) and after (D) digestion with CPP. The shift in mass observed before and after carboxymethylation reflects the addition of three carboxymethyl groups to the protein. (R.A.=Relative Abundance)

DNA sequence. Compositional analysis indicated the presence of additional amino acids and ESMS analysis indicated a molecular weight 1241 amu higher than expected. In an effort to evaluate these inconsistencies, CPP digestion on the intact protein followed by ESMS was used to establish the C-terminal residue (data not shown). The results, in conjunction with close inspection of the cloning vector, indicated that the desired stop codon had been suppressed and a second stop codon 11 residues downstream had been read as the actual protein termination site.

IV. Conclusions

The results obtained to date indicate that ESMS is an effective tool for the analysis of carboxypeptidase-degraded proteins and peptides. It compliments the analytical techniques currently used for obtaining C-terminal sequence information. However, the method is not capable of differentiating isobaric amino acids (i.e. leucine/isoleucine, lysine/glutamine) and careful mass analysis is necessary to differentiate amino acids that differ by 1 amu. For potential protein samples, several factors must be considered in evaluating the applicability of this method. As with any carboxypeptidase digestion method, the rate of digestion is strictly dependent upon the sequence itself. Additionally, steric hindrance of a protein's C-terminus to the active site of CPP can prevent C-terminal digestion. In the course of this study several proteins in their native form were not digested by CPP, and steric hindrance is suspected as a probable cause for this lack of activity. In the case of superoxide dismutase, this steric hindrance problem was circumvented by reducing and alkylating the protein prior to CPP digestion. For unambiguous sequence determination, proteins must be fairly homogeneous to eliminate spectral overlap and complicated data interpretation.

Further experimentation with other carboxypeptidases (or combinations thereof), in particular carboxypeptidase Y, will be useful in generalizing this methodology. Additionally, to aid in sequence interpretation involving amino acids of similar mass, simultaneous ESMS monitoring for the released amino acids and the degraded protein in a carboxypeptidase digestion mixture may be helpful.

References

1. Self, R. and Parente, A., (1983) *Biomedical Mass Spectrometry* **10**, 78-82.

2. Caprioli, R.M. and Fan, T., (1985) *Anal. Biochem.* **154**, 596-603.

3. Kim, H.Y., Pilosof, D., Dyckes, D.F., and Vestal, M.L., (1984) *J. Am. Chem. Soc.* **106**, 7304-7309.

4. Klarskov, K., Breddam, K., and Roepstorff, P. (1989) *Anal. Biochem.* ***180***, 38-27.

5. Wang, R., Cotter, R.J., Meschia, J.F., and Sisodia, S.S., (1992) in *Techniques in Protein Chemistry III*, (Ed) Angeletti, R.H., 505-513.

6. Bruins, A.P.; Covey, T.R.; Henion, J.D., (1987) *Anal. Chem.* **59**, 2642.

7. Lu, H.S., Klein, M.L., and Lai, P-H., (1988) *J. Chrom.* **447**, 351-364.

Protein Ladder Sequencing: A Conceptually Novel Approach To Protein Sequencing Using Cycling Chemical Degradation and One-Step Readout by Matrix-Assisted Laser Desorption Mass Spectrometry

Rong Wang, Brian T. Chait
The Rockefeller University, New York, NY 10021

Stephen B.H. Kent
The Scripps Research Institute, La Jolla, CA 92037

I. Introduction

Stepwise degradation from the amino terminus and analysis of the released derivatives has been used as a long-standing technique to determine the sequence of amino acids in a polypeptide chain. Almost all highly sensitive protein and peptide sequencing is currently done by automated Edman degradation (1,2) whereby practical analysis of 10-100 picomole amounts of protein has become routine. There is, however, a great current need for more rapid, highly sensitive protein sequencing methods, which give data that are easy to interpret. The most efficient and accurate techniques for sequencing of biopolymers are those that involve the readout of the entire sequence-defining data set in one operation. Thus, optimally, all members of a sequence-defining set of fragments of the biopolymer, each differing by one monomer unit, are simultaneously examined by a high resolution read-out method. The sequence of the parent polymer is deduced from the set of fragmentation products. Such a data set contains mutually interdependent information that determines the identity and order of monomer units in the parent molecule.

TECHNIQUES IN PROTEIN CHEMISTRY IV

The advent of matrix-assisted laser desorption mass spectrometry (3) together with the development of new matrix materials (4,5) have provided a powerful new tool for the accurate measurement of the mass of intact polypeptide chains containing up to hundreds of amino acids. Strong, clean mass spectrometric data are obtained in the order of 1-3 minutes from picomole or subpicomole amounts of peptide or protein samples (5).

In this paper, we describe a completely new principle in protein sequencing: controlled chemical generation of a family of sequence-defining fragments from a polypeptide chain, followed by read-out of the complete data set as a *protein sequencing ladder* in a single operation, using matrix-assisted laser desorption mass spectrometry. The sequence-defining fragments are generated by chemical degradation of the peptide chain in the presence of a terminating agent which blocks a small fraction of the peptide chain at each residue. In one variation of the approach, a peptide was subjected to repeated cycles of the Edman degradation using 5% phenylisocyanate as a terminating agent, without separation and analysis of the low molecular weight reaction products. The final unfractionated product mixture was subjected to mass spectral analysis, to give a "protein sequencing ladder" data set. Mass differences between adjacent peaks identify each amino acid, and the position in the data set defines the sequence of the original peptide chain.

II. Experimental

A. Peptide and Chemicals

[Glu^1]-fibrinopeptide B was purchased from Sigma Chemical Co. and used with no further purification. It has an amino acid sequence of $E^1GVNDNEEGFFSAR^{14}$. Phenylisothiocyanate (PITC), pyridine, and trifluoroacetic acid (TFA) were purchased from Pierce. Phenylisocyanate (PIC), trimethylamine (TMA), ethyl acetate, and 1,1,1,3,3,3-hexafluoro-2-propanol (HFIP), α-cyano-4-hydroxy-cinnamic acid (4HCCA) were purchased from Aldrich. Heptane was purchased from Burdick & Jackson.

B. Chemical Ladder Reaction

A modified manual Edman degradation procedure was employed to generate the desired peptide ladder. Instead of using PITC as coupling agent alone, a mixture of PITC and PIC (20:1, v/v) was used in the stepwise degradation reaction. Here, PITC was used as a coupling agent and PIC was used as a terminating agent. All of the reaction cycles were carried out in a single 1.5 ml Eppendorf microcentrifuge vial under a stream of dry N_2. Peptide (200 pmoles to 10 nmoles) was dissolved in 20 μl of 25% TMA:Pyridine (1:1). 20 μl of the coupling agent containing PITC:PIC:Pyridine:HFIP (20:1:76:4, v/v) was added to the reaction vial. The coupling reaction was allowed to proceed at 50 °C for 3 minutes. The coupling agent, and side-reaction products were extracted using heptane and ethyl acetate solvents. 400 μl of heptane:ethyl acetate (10:1, v/v) was added to the reaction vial, gently vortexed and centrifuged to clear phases. The upper phase was aspirated and discarded. The above washing procedure was repeated, followed by washing twice with heptane:ethyl acetate (2:1, v/v). The remaining solution in the reaction vial was dried in a Speed-Vac centrifuge. The cleavage step was carried out by adding 20 μl of anhydrous TFA to the dry residue in the reaction vial and allowing the reaction to proceed at 50 °C for 5 minutes, followed by centrifuge-vacuum drying. The above-described coupling-cleavage cycle was repeated the desired number of times. The low molecular weight ATZ/PTH derivatives released at each cycle were not separated/analyzed. Washes/extractions were performed after each coupling step because preliminary experiments showed that otherwise an intractable polymeric material was formed. Analysis was performed only once after the completion of all cycles of the degradation chemistry.

C. Amino Acid Sequence Read-Out by Laser Desorption Spectrometry

After the designated number of cycles, the peptide mixture was analyzed by matrix-assisted laser desorption mass spectrometry. An aliquot, 1 μl (~ 50 pmol total peptide), was mixed with 9 μl of a solution of α-cyano-4-hydroxy-cinnamic acid as matrix (5 g/l in 0.1% TFA:acetonitrile, 2:1, v/v) and 0.5 μl of this mixture of total peptides and matrix was applied to the probe tip and dried in a stream of air at room temperature. Mass spectra were acquired in positive ion mode using a laser desorption time-of-flight instrument constructed at The Rockefeller University (6). Spectra obtained from 200 laser pulses were acquired over 80 seconds and added to give a mass spectrum of the sequencing ladder. No peaks from the α-cyano-4-hydroxy-cinnamic acid matrix material used were detected above 500 Da.

III. Results and Discussion

Sequence-defining sets of fragments can potentially be generated in a controlled fashion, from a polypeptide chain in a number of ways. One straightforward way is to carry out stepwise degradation in the presence of a terminating agent. Thus, a protein sequencing ladder can be generated by carrying out rapid stepwise Edman degradation in the presence of a small-amount of terminating agent such as phenylisocyanate (Figure 1). In this way, a small proportion of N-terminally blocked peptide chain is generated at each cycle of the

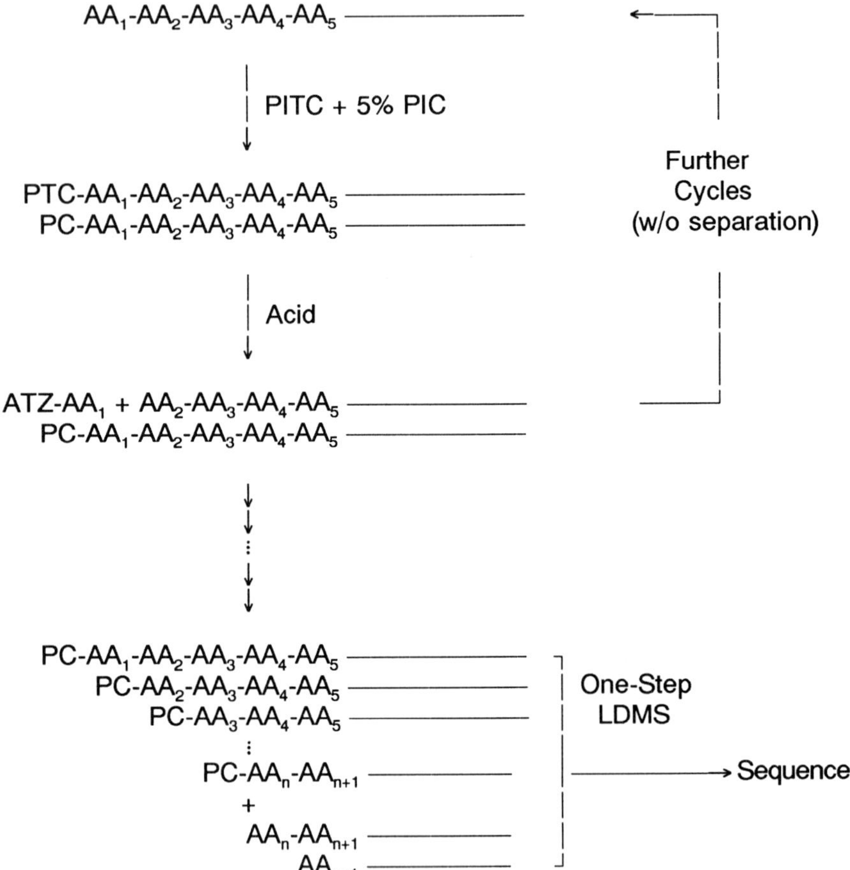

Figure 1. Schematic of the generation of a sequence-defining set of fragments by cycling Edman degradation in the presence of a terminating agent, phenylisocyanate. A small fraction of the peptide is blocked by phenylisocyanate in each cycle. These blocked peptides will not undergo further degradation in the following cycles and will accumulate forming the protein sequencing ladder.

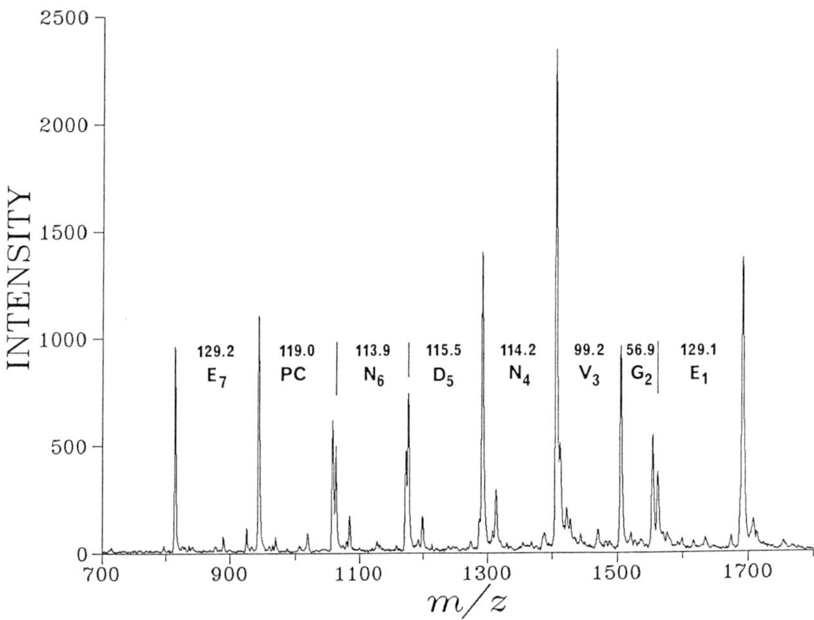

Figure 2. Protein ladder sequencing of [Glu[1]]-fibrinopeptide B. Positive ion matrix-assisted laser desorption mass spectrum was acquired in a single operation after seven cycles of the protein ladder sequencing reaction. Each amino acid is identified by the mass difference between neighboring peaks and the sequence of parent peptide is defined by the position of each amino acid in the same spectrum.

degradation. After the desired number of cycles has been performed, without intermediate separation and analysis of the released amino acid derivatives, the resulting mixture of terminated peptide chains is read out in a single operation by matrix-assisted laser desorption mass spectrometry, to yield a *protein sequencing ladder* (*i.e.*, a mass spectrum consisting of protonated molecules with masses corresponding to each terminated polypeptide species present). The mass differences between consecutive peaks each correspond uniquely to an amino acid (with the exception of Leu/Ile), and their order of occurrence in the data set defines the sequence of amino acids in the original peptide chain.

A sequence-defining set of fragments was generated from [Glu[1]]-fibrinopeptide B by seven cycles of the Edman degradation carried out in the presence of 5% v/v phenylisocyanate, as described in Methods, without separation or analysis of the low molecular weight ATZ/PTH derivatives resulting from each cyclization/cleavage step. The mass spectrum of the sequencing ladder is shown in Figure 2. All peaks present in the mass spectrum were identified (7). The mass differences between each adjacent peak were calculated and compared with the known mass of each amino acid residue. The interpretation of the data as amino acid sequence is illustrated in Figure 2.

In this example, the ratio of degradation/terminating reagent (PITC/PIC) was arbitrarily selected. The spectrum yields a simple and useful sequencing ladder (Figure 2). No effort was made to optimize the coupling or cleavage yields in the chemical degradation because incomplete reactions of coupling or cleavage in a given cycle, will be continued in following cycles. Thus, the accuracy of protein ladder sequencing is unaffected by yields of individual steps, over a wide range. Success depends only on the ratio of reactions, terminating-to-degradation. Consequently, reactions used in this approach can be simple and fast. This should be contrasted with the stringent yield requirements of the standard stepwise Edman degradation sequencing procedure (8).

The amino acid sequence is simply read off from the protein sequencing ladder. Mass differences between neighboring peaks identify each amino acid, and their position in the data set defines the sequence in the parent peptide. Because the degradation was started from the N-terminus, the amino acid sequence is read out from the high mass end to lower mass. For example, the mass difference between peaks at m/z 1690.9 and 1561.8 is 129.1 Da which corresponds to the residue mass of glutamic acid (calculated residue mass 129.1 Da), at the first position. The next pair of peaks at m/z 1561.8 and 1504.9 gives a mass difference of 56.9 Da which represents the amino acid residue glycine (calculated residue mass 57.1 Da), at the second position. The measured mass differences observed for [Glu^1]-fibrinopeptide B are given in Figure 2. The mass accuracy obtained was sufficient to unambiguously distinguish all of the amino acids. Because the amino acid identification relies on mass differences, less stringent demands are placed on absolute mass measurement. The laser desorption time-of-flight instrument used in these studies routinely gives mass accuracies of 0.01% (1 part in 10,000) (9). Furthermore, because the sequence-defining data set is read-out in one operation, the data is mutually interdependent. Thus, the likelihood of errors in the determination of the identity of an amino acid is reduced.

Because the sequence-defining set of fragments is read-out in one operation, protein ladder sequencing can be very fast. The relatively slow (typically 25-30 min) reverse-phase HPLC analysis of each released PTH amino acid derivative (used in conventional Edman sequencing) is not required. The ladder-generating chemistry lends itself to being performed simultaneously on a large number of samples. Because the laser desorption mass spectrometric read-out is fast (~ 1-3 minutes per sample), by multiplexing the ladder-generating chemistry, this technique can potentially give a throughput of 10-20 residues per minute (~ 1,000 residues per hour), at very low cost per amino acid residue. Such rapid, inexpensive sequencing could vastly expand the applications and use of protein sequence analysis in biological research.

The protein ladder sequencing is, in principle, applicable to the analysis of peptides and proteins with post-translational modifications. All modified residues that are stable to the chemical conditions used to generate the sequence-defining set of terminated peptides will show up as an additional mass difference in their residue masses. Partly stable modifications may actually yield even more useful data, in that two series of peaks will be seen. The mass difference between the two series will define the mass of the modification. Both the site and identity of the modified amino acid can be determined, and the nature of the modification can in many cases be deduced.

The principle of protein ladder sequencing is also directly applicable to C-terminal sequence analysis using simple variations of existing chemistries, and also can be used for sequencing of mixtures of peptides.

IV. Conclusions

The protein ladder sequencing approach involves the use of controlled degradation chemistry to generate a sequence-defining set of terminated peptides. The protein ladder is read-out in a single operation by matrix-assisted laser desorption mass spectrometry. The protein sequencing ladder data set is generated as a whole, and is straightforward to interpret. The mass difference between neighboring peaks defines the identity of an amino acid residue (with the exception of Leu/Ile since they have the same residue mass), while the position of each mass difference in the data set establishes the sequence of amino acids in the parent peptide chain. The overall process is extremely rapid and has a very low cost per amino acid residue sequenced.

This robust, practical new approach to the rapid determination of protein amino acid sequences has considerable potential for increasing the application and use of protein sequencing in biological research.

Acknowledgements

This work was supported by grants from the National Institutes of Health (RR00862, GM38274), and by funds from the Markey Foundation.

Reference

1. Edman, P. (1950), *Acta Chem. Scand.* **4**, 283-293.
2. Edman, P. and Begg, C. (1967), *Eur. J. Biochem.* **1**, 80-91.
3. Karas, M., Bachmann, D., Bahr, U. & Hillenkamp, F. (1987) *Int. J. Mass Spectrom. Ion Proc.* **78**, 53-68.
4. Beavis, R.C. & Chait, B.T. (1989) *Rapid Commun. Mass Spectrom.* **3**, 432-435.
5. Beavis, R.C., Chaudhary, T. & Chait, B.T. (1992) *Org. Mass Spectrom.* **27**, 156-158.
6. Beavis, R.C. & Chait, B.T. (1989), *Rapid Commun. Mass Spectrom.* **3**, 233-237.
7. The peak at m/z 1553.8 (third peak from the right) corresponds to the protonated [Pyr^1]-fibrinopeptide B, which is the N-terminal dehydration product of [Glu^1]-fibrinopeptide B.
8. Laursen, R.A., Lee, T.T., Dixon, J.D. & Liang, S.P. (1991) *Methods Protein Sequence Anal. [Proc. Int. Conf.] 8th.*
9. Beavis, R.C. & Chait, B.T. (1990) *Anal. Chem.* **62**, 1836-1840.

SECTION IX

Protein Conformation, Folding, and Modeling

The Use of Molecular Modelling to Delineate B-Cell and T-Cell Epitopes of Human Sperm-Specific LDH-C_4

Patricia A. O'Hern and Erwin Goldberg
Department of Biochemistry, Molecular Biology and Cell Biology
Northwestern University, Evanston, IL 60208

I. Introduction

The sperm-specific isozyme of lactate dehydrogenase (LDH-C_4) has been purified from mouse testes and has been shown to suppress fertility in female mice, rabbits and baboons (1-4). These results have suggested that LDH-C_4 might form the basis of an immunocontraceptive vaccine that would not rely on disrupting the endocrine milieu of the recipient. However, it has not been feasible to produce sufficient amounts of the natural protein for vaccine studies; a synthetic peptide-based vaccine is required. A combination of immunoaffinity purification of tryptic peptides and epitope mapping using synthetic peptides has identified several peptides which can elicit LDH-C_4 reactive antibodies (5-8). Molecular modelling has been used to localize these peptides within the crystal structure of the protein (9). The immunogenic peptides were grouped into three clusters, or epitopes, on the surface of the mouse LDH-C_4 tetramer (Table I). Because of the 222 axis of symmetry in the molecule, each of the three epitopes is actually dimerized in the tetramer.

Recently, the cDNA coding for human LDH-C_4 has been cloned and sequenced and engineered for expression in *E. coli* (10,11). The sequences of mouse and human LDH-C_4 are very homologous (approximately 70%). However, there are some significant differences between the two, especially in the putative B-cell epitope regions.

Table I. B-cell Epitopes of Mouse LDH-C_4[1]

Epitope	Residues
1 (R axis)	5-16[2]
2 (Q axis)	42-58
	101-111
	231-243[2]
3 (P axis)	211-220
	304-316[2]

[1]data from Hogrefe *et al.* (9).
[2]highly antigenic residues.

TECHNIQUES IN PROTEIN CHEMISTRY IV

Indeed, rabbit antisera to mouse LDH-C_4 have cross-reacted only weakly with purified human LDH-C_4. Since the goal of this work is to develop an immunocontraceptive vaccine for human use, it has therefore been necessary to identify the immunodominant epitopes of human LDH-C_4 instead of attempting to use the known epitopes of the mouse protein.

In the current study, we used computer-assisted secondary structure analyses together with our experience with mouse LDH-C_4 to predict an immunodominant B-cell epitope of human LDH-C_4. This peptide, when conjugated to diphtheria toxoid (DT-hC9-20), was highly immunogenic in rabbits. In addition, similar analyses enabled us to identify a T-cell epitope of human LDH-C_4 which could substitute for the normally required carrier molecule in eliciting an immune response to hC9-20. A completely synthetic peptide vaccine will greatly reduce the expense and batch-to-batch variability inherent in hapten/carrier conjugates, as well as reduce the risk of adverse reactions in vaccinated individuals.

The salient features of the protein epitopes identified in these studies, as well as our analysis of the reliability of the predictive algorithms that we used may be of interest to other investigators attempting to produce synthetic peptide antigens either for vaccine development of for general immunological applications.

II. Methods and Results

A. *Delineation of the B-Cell Epitopes of Human LDH-C_4*

The structural characteristics of B-cell epitopes have been well studied, and several recurring patterns have been observed (12-17). Antigenicity has been shown to be associated with surface accessibility, temperature factors, hydrophilicity, secondary structures such as alpha helices and beta turns, amino acid sequence variability and even nucleotide mutation rates. Several computer programs have been developed to predict the physical properties of a protein from the primary structure alone, in an attempt to aid in the prediction of antigenic determinants (18-20).

In addition to the minimum size requirements for antigenic peptides [$\geq$10 amino acids (21)], the data of Hogrefe *et al* (8) have indicated that surface accessibility and amino acid sequence variability (degree of "foreignness") are the parameters most closely correlated with the occurrence of B-cell epitopes in mouse LDH-C_4. Secondary structure features such as alpha helices or loop structures are also associated with immunogenicity, but not as reliably. Therefore, these three parameters (surface accessibility, secondary structure features and sequence variability) were used to predict the B-cell epitopes of human LDH-C_4.

1. Surface accessibility

The "PEPTIDESTRUCTURE" program of the Genetics Computer Group Sequence Analysis software package (22) contains a subroutine which predicts the surface accessibility of individual amino acids within a 6-residue "window",

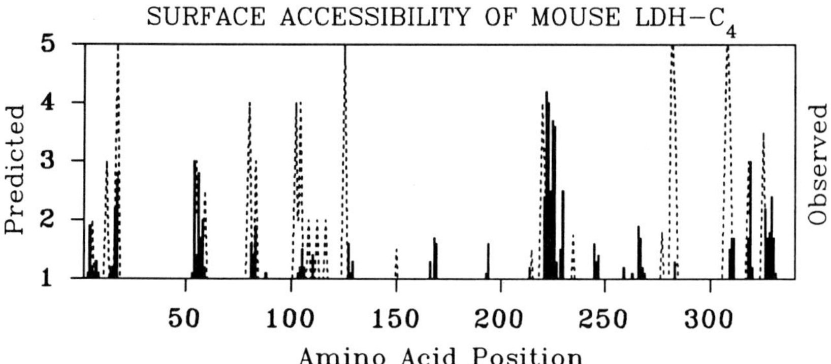

Figure 1. Localization of surface accessible residues of mouse lactate dehydrogenase. The predicted occurrence of surface accessible residues in mouse LDH-C_4 (solid bars) was compared to the observed surface accessibility determined crystallographically (dashed line). When the Threshold of Probability (vertical axis) was set to 1.0, the curves were essentially coincident.

or region, of a protein [Emini *et al.* (12)]. To determine the accuracy of this algorithm, the predicted surface accessibility of the amino acids of mouse LDH-C_4 was compared with the known surface accessibility as determined from the crystal structure of the protein (9). The surface accessibility of the residues of mouse LDH-C_4 to an antibody-sized probe (10Å radius sphere) was accurately predicted by the Emini subroutine when the threshold of detection was set to 1.0 (THReshold=1 command) (Figure 1). Therefore, this threshold was used to predict the surface probability of the amino acids of human LDH-C_4. According to this analysis, 5 regions of the protein were predicted to be particularly accessible to antibody binding: 1-20; 75-85; 95-105; 215-230 and 310-330.

2. Secondary structure predictions

The GCG software contains two subroutines designed to predict the occurrence of alpha helices, beta sheets and beta turns from primary sequence data (18,19). As with the surface accessibility studies, the predicted secondary structure of mouse LDH-C_4 was compared with the known secondary structures that had been determined crystallographically (9,23). Mouse LDH-C_4 contains 11 highly conserved alpha helices, designated αA through αH. The length and position of these helices are shown in Table II, along with the number of residues accurately predicted by the algorithms of Chou-Fasman (CF) (18) and Garnier *et al* (GOR) (19). Of the 122 residues of mouse LDH known to be involved in helices, 52 (43%) were correctly assigned by the CF method, and 56 (46%) by the GOR method, giving an overall success rate of less than 50% for the algorithms. Similar results were obtained for β-sheet structures (data not shown). This was considered to be an unacceptable degree of accuracy, and the program was not used to predict the secondary structure of human LDH-C_4.

Table II. Comparison of Actual and Predicted Alpha Helices in Mouse LDH-C_4

Alpha Helices	Residues	C-F Prediction	GOR Prediction
αA	3V-6Q (4)	4	4
αB	28N-38L (11)	0	8
αC	54N-64L (11)	6	8
αD	107L-117A (11)	11	8
αE	119V-127P (9)	1	0
α1F	138V-148I (11)	0	0
α2F	162N-175K (14)	0	0
α1G	226K-233V (8)	7	3
α2G	234E-243K (10)	7	0
α3G	248W-263K (16)	0	8
αH	308A-324M (17)	16	17
Total	(122)	52 (43%)	56 (46%)

3. Sequence variability

Antigenic regions of proteins are usually those whose amino acid sequence differs most from that of the host animal. The amino acid sequences of rabbit LDH-A (complete) and rabbit LDH-B (partial) have recently been published (24). Therefore, the sequences of these rabbit somatic isozymes were compared with the sequences of mouse and human LDH-C_4 to identify the most divergent, and therefore the most probably antigenic, regions of the testicular isozymes. Such comparisons are usually presented in terms of per cent homology between proteins or peptides. However, we decided to compute the number of amino acid differences within a window of 7 residues, as is now commonly done with computer analyses of other physical properties of proteins (*e.g.* Figure 1). Since no program was available in the GCG software for this purpose, the comparisons were performed manually and presented graphically in Figure 2.

As with the surface accessibility analyses described above, almost the entire length of the polypeptide chain contained one or more substitutions within each window of 7 residues. However, there were 5 regions which contained long stretches ($\geq$ 6 residues) of 3 or more sequence differences per window between mouse testicular and rabbit somatic LDH's (Figure 2). These regions were therefore defined as "hypervariable" for the purposes of epitope prediction: mC1-24; 54-67; 101-111; 120-130 and 309-315. Two of these segments (mC5-16 and mC304-316) have already been shown to be highly antigenic in rabbits (6,8), and a monoclonal antibody to mouse LDH-C_4 has been shown to recognize the coenzyme binding loop (amino acids 101-115), particularly in the region of Thr_{108} and Asp_{111} (5). Peptides 54-67 and 120-130 have not been tested.

Similar analyses to define amino acid sequence differences between human LDH-C_4 and the rabbit somatic isozymes identified three hypervariable regions: hC1-24; 74-81; and 90-97 (Figure 2). There were also short regions of sequence variability centering around residues 120, 230, 260 and 310, but they consisted only of short stretches of 3 to 4 amino acids each and so were eliminated from further consideration.

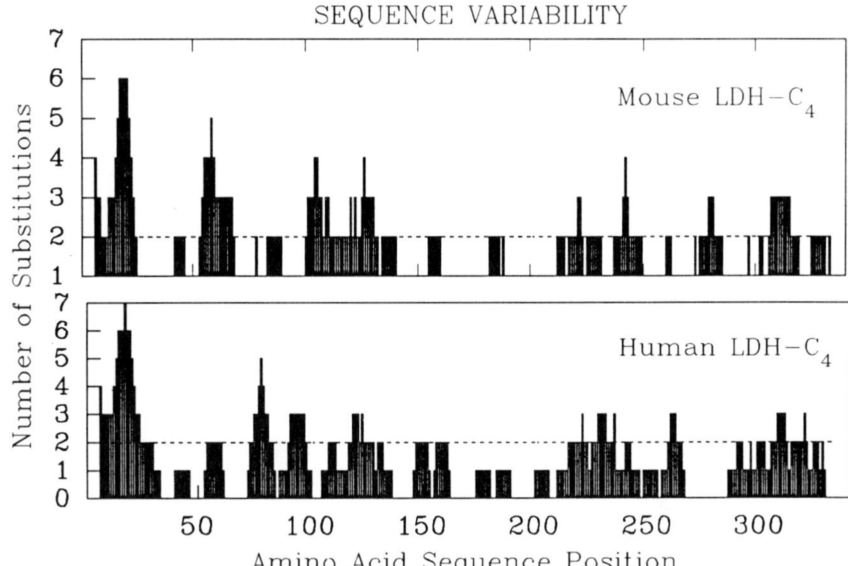

Figure 2. Amino acid sequence differences between rabbit somatic LDHs and mouse LDH-C_4 (top panel) and human LDH-C_4 (bottom panel). The total number of sequence differences within each "window" of 7 residues was calculated along the entire length of the molecule, similar to the analysis of surface accessibility presented in Figure 1.

From these data, we identified two regions of human LDH-C_4 which fulfilled three of the four criteria we established for antigenic peptides: equal to or greater than 10 residues in length, surface accessible, and hypervariable. (It was not possible to accurately predict secondary structure elements with the computer programs available to us). These regions were hC1-20 and hC74-85. The final choice of peptide vaccinogen was predicated on our intent to use the peptide as an immunocontraceptive in non-human primates and, eventually, in humans. To maximize immunogenicity in primates and to minimize any possible cross-reactivity with the somatic forms of lactate dehydrogenase (LDH-A_4, LDH-B_4 and the A/B heterotetramers) the amino acid sequence of the human C subunit was compared to the sequences of human A (25) and human B (26) subunits in the regions of interest. Residues 9-20 differed most from the human somatic isozymes, containing only one residue in common with LDH-A_4 and LDH-B_4. In addition, the naturally occurring cysteine at position 20 allowed the peptide to be conjugated to a carrier molecule while maintaining the same (free N-terminal) orientation which the peptide exhibits in the native molecule.

4. Immunogenicity in rabbits

Three overlapping peptides based on the amino terminal "arm" region (residues 1-20) of human LDH-C_4 were synthesized and tested for their ability to elicit

LDH-C_4 reactive antibodies in female rabbits (Figure 3). Peptide hC1-16 was not immunogenic in rabbits. Peptide hC5-16 was moderately immunogenic, eliciting antibody titers in the range of 10^3 to 10^4 in both rabbits, similar to the ELISA titers reported previously for this peptide (8). Peptide hC9-20 was clearly a superior immunogen, eliciting antibody titers 1 to 2 orders of magnitude greater than hC5-16; and approximately the same as those obtained with the native molecule (Figure 3B).

From these data, we conclude that hC1-15 does not contain a B-cell epitope; hC5-16 probably contains a partial epitope; and hC9-20 contains a complete epitope. Since the antibody titers obtained with this peptide were approximately the same as those obtained with the native molecule, it is probable that the 9-20 region contains the immunodominant B-cell epitope of human LDH-C_4.

B. Delineating a T-Cell Epitope of Human LDH-C_4

From the foregoing experiments, we concluded that peptide hC9-20 contained a B-cell epitope, but hC1-15 and hC5-16 did not. We next attempted to identify a T-cell epitope of human LDH-C_4 that would be able to substitute for diphtheria toxoid as a carrier molecule. Not only does carrier conjugation increase the expense and batch-to-batch variability of the vaccine, but the immune response to the carrier can overwhelm and even eliminate the antigen-specific response (27,28). Analysis of T-cell epitopes is a relatively new undertaking compared to the analysis of B-cell epitopes. As recently as 10 years ago, very little was known about what constitutes a T-cell epitope, or even whether antigen-specific T-cells recognize native or processed antigen (29).

Although progress in the field has been astonishingly rapid, the exact nature of peptides capable of stimulating helper T-cells (Th) or cytotoxic T-cells (Tc) remains unclear. It is known that T-cell epitopes are short, linear peptides (8-12 amino acids long)

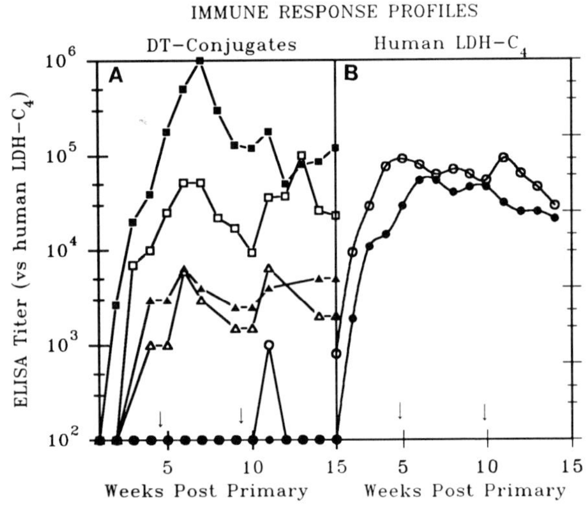

Figure 3. Identification of a B-cell epitope of human LDH-C_4. Female rabbits were immunized with one of three peptides conjugated to diphtheria toxoid (DT) emulsified in Freund's Complete Adjuvant (left). Circles: DT hC1-15; triangles: DT hC5-16; squares: DT hC9-20. Two additional females were similarly immunized with human LDH-C_4 (right). Booster injection (arrows) were administered in PBS. ELISA titers were determined on microtiter plates coated with purified human LDH-C_4.

A.A.	S	T	V	K	E	Q	L	I	E	K	L	I	E	D	D	E	N	S	Q	C
#	1	2	3	4	5	6	7	8	9	10	11	12	13	14	15	16	17	18	19	20

Figure 4. Putative T-cell epitopes in peptide hC1-20. The tetrapeptide T-cell epitope motif suggested by Rothbard and Taylor (34) (charged or glycine: hydrophobic: hydrophobic: charged) occurs three times within the first 13 residues of human LDH-C_4 (shaded residues). The known B-cell epitope (hC9-20) is underlined.

rather than the complex topographic determinants recognized by B-cells. By studying the structure and amino acid sequences of known T-cell epitopes, two different models had been proposed to predict antigenic sites in proteins. The first (30) suggested an association between amphipathic alpha-helices and T-cell reactivity. To test this hypothesis, a panel of synthetic peptides based on the carboxy-terminal alpha-helix of mouse LDH-C_4 (amino acids 310-327) was synthesized (31) and tested for immunogenicity in mice and rabbits (32). The results indicated that an 18-residue amphipathic alpha-helix did not stimulate T-cells *in vivo*, although a 36-residue alpha-alpha fold peptide did.

More recently, Sette, *et al.*(33) and Rothbard and Taylor (34) proposed that T-cell epitopes have an "extended" conformation which was subsequently found to be consistent with the crystal structure of MHC class I proteins (35,36). The sequence of human LDH-C_4 was examined for the presence of the tetrapeptide motif suggested by Rothbard and Taylor (34): charged or glycine, hydrophobic, hydrophobic, polar or glycine. Since the amino acid sequence of T-cell epitopes, as for B-cell epitopes, must differ from the host's protein, the analysis was confined to the "hypervariable" regions of thehuman LDH-C_4 molecule described above. Three examples of the Rothbard motif were detected in the first 13 residues of human LDH-C_4 (Fig. 4). To test whether one or more T-cell epitopes were contained within this region, 2 female rabbits were immunized with either hC9-20 (a known B-cell epitope) or with hC1-20 (the B-cell epitope synthesized in tandem with the putative T-cell epitope(s)). Neither peptide was conjugated to a carrier molecule. The results (Fig. 5) show that one rabbit did not respond to peptide hC9-20 and the other produced a modest primary response which was not effected by a booster

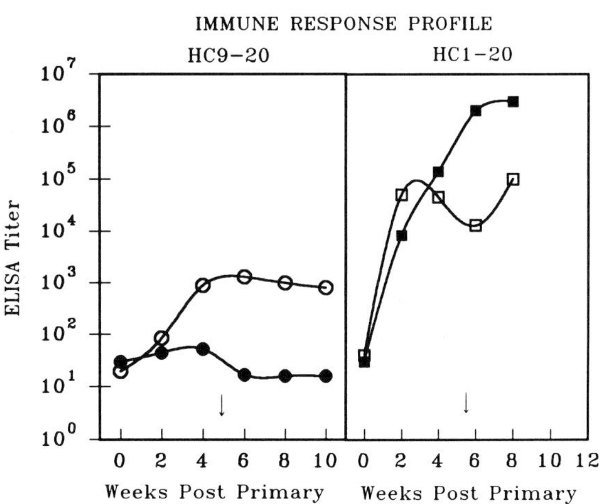

Figure 5. Immune response of rabbits to non-conjugated peptides. Two female rabbits were immunized with either hC9-20 (left) or hC1-20 (right) in Freund's Complete Adjuvant. They were boosted (arrows) with the same immunogen in PBS. ELISA titers were determined on microtiter plates coated with human LDH-C_4.

immunization. Both rabbits immunized with peptide hC1-20 produced high antibody titers which were sustained for more than 6 weeks and which increased after the secondary immunization.

III. Discussion

There has been increasing evidence over recent years that most antibody combining sites of protein antigens are topographic determinants composed of amino acids which may be far apart in the primary structure, but closely juxtaposed through secondary and tertiary interactions (37). Small, linear peptides do not adequately mimic such complex epitopes, and antibodies to such peptides often do not recognize the native molecule, or do so with very low affinity (38). There have been and continue to be efforts in several laboratories to produce synthetic peptides [*e.g.* "surface simulation" peptides (39) or "super-secondary structure" peptides (31)] which can mimic the topographic determinants of proteins. The design and synthesis of such antigens is still a very complex undertaking, however, and producing high-affinity anti-peptide antibodies is difficult.

Meanwhile, rapid advances in DNA cloning and sequencing, particularly the introduction of reverse transcriptase/polymerase chain reaction techniques to directly amplify and sequence diverse mRNAs, has resulted in a growing body of cDNA and deduced amino acid sequences available for analysis and experimentation. In many cases, researchers would like to use the B-cell epitopes of a deduced amino acid sequence to generate protein-reactive antibodies. The antibodies could then be used to study tissue localization or biological effects, or to aid in protein purification and characterization. In such cases, the synthesis of simple, linear peptides may be more feasible than the design and synthesis of complex topographic determinants.

Based on our experience, the most important criterion for predicting the antigenic sites of a protein is amino acid sequence variability. Although this concept is not new, it is sometimes overlooked because most reports of peptide antigenicity concern viral or bacterial proteins which are, by definition, very dissimilar to the host's proteins. However, when attempting to elicit antibodies to relatively non-immunogenic mammalian proteins, the identification of unique sequences, or of "hypervariable" regions within a family of related proteins (*e.g.* immunoglobulins or heat shock proteins) is extremely valuable.

The "moving window" analysis of sequence variability that we developed for these studies appears to be superior to a simple *per cent* homology analysis. A seven-residue window was chosen because it approximates the minimum size of the antigen binding cleft of antibodies and MHC molecules (35,36). Therefore, the investigator is able to "see" what the paratope "sees", particularly when the data are combined with similar analyses of the structural features of the segments.

The second most important criterion, at least for B-cell epitopes, is surface accessibility. We found that Emini's subroutine accurately predicts the surface accessibility of the residues of mouse LDH-C_4 when the threshold of probability is set to 1.0. This is the same threshold recommended by the original authors

(12). Surface probability is not necessarily more important than other predictors of antigenic sites mentioned above (*e.g.* hydropathy, segment mobility, temperature factors, *etc.*). Rather, the Emini method represents a statistical analysis of the observed frequency of a given amino acid appearing on the surface of proteins, while the other methods attempt to understand why this occurs. The two types of analyses are not in conflict but, for the purposes of epitope prediction, surface probability appears to be more reliable. Inaccuracies in the theoretical analyses probably underlie the errors we observed in attempting to predict the secondary structure elements of LDH-C_4.

Much less is known about the nature of T-cell epitopes than is known about B-cell epitopes. Because T-cell epitopes are the result of antigen processing and presentation, they do not necessarily occur on the surface of a protein. From the information currently available, it appears that T-cell epitopes, like B-cell epitopes, differ from the amino acid sequence of the host's proteins, but the requirements are less stringent. Charge interactions, as defined by Rothbard's tetrapeptide motif, are probably very important, which is consistent both with the structure of empirically determined epitopes and with the observed structure of the MHC binding cleft. By using these principles (size, sequence variability and tetrapeptide motif) we were able to identify a peptide (hC1-20) that contains both a B-cell epitope and a T-cell epitope(s). The general applicability of these principles remains to be determined.

We succeeded in our original goal of using a rational, knowledge-based approach to identifying both a B-cell and a T-cell epitope of human LDH-C_4, starting only with the cDNA sequence of the protein. Future experiments will focus on more precisely defining the critical residues required for T-cell stimulation, and on using the peptide as an immunocontraceptive vaccine in non-human primates.

Acknowledgments

Support for this project was provided by NIH HD23771 and by the Contraceptive Research and Development Program (CSA-92-099) under a Cooperative Agreement with the U.S. Agency for International Development (U.S.A.I.D.) (DPE-3044-A-00-6063-00) which in turn receives funds for AIDS research from an interagency agreement with the National Institute of Child Health and Human Development. The views expressed by the authors do not necessarily reflect the views of U.S.A.I.D. and CONRAD.

References

1. Goldberg, E. (1963). *Science* **139**, 602-603.
2. Goldberg, E. (1973). *Science.* **181**, 458-459.
3. Goldberg, E., Wheat, T.E., Powers, J.E. and Stevens, V.C. (1981). *Fertil. Steril.* **35**, 214-217.
4. Goldberg, E. and Shelton, J. (1986). *In* "Immunological Approaches to Contraception and Promotion of Fertility" (G.P. Talwar, ed.), p.219. Plenum Publishing Corp., New York.
5. Goldman-Leikin, R.E., and Goldberg, E. (1983). *Proc. Natl. Acad. Sci. USA* **80**, 3774-3778.

6. Gonzales-Prevatt, V., Wheat, T.E., and Goldberg, E. (1982). *Mol. Immunol.* **19**, 1579-1585.
7. Goldberg, E. (1987). *In* "Isozymes: Current Topics in Biological and Medical Research" (M.S. Rattazzi, J.G. Scandelios and G.S. Whitt, eds.), Vol. 14, pp. 103-122. Alan R. Liss, Inc., New York.
8. Hogrefe, H.H., Kaumaya, P.T.P. and Goldberg, E. (1989). *J. Biol. Chem.* **264**, 10513-10519.
9. Hogrefe, H.H., Griffith, J.P., Rossmann, M.G., and Goldberg, E. (1987). *J. Biol. Chem.* **262**, 13155-13162.
10. Millan, J.L., Driscoll, C.E., LeVan, K.M. and Goldberg, E. (1987). *Proc. Natl Acad. Sci. USA* **84**, 5311-5315.
11. LeVan, K.M. and Goldberg, E. (1991). *Biochem. J.* **273**, 587-592.
12. Emini, E.A., Hughes, J.V., Perlow, D.S., and Boger, J. (1985). *J. Virol.* **55**, 836-839.
13. Feiser, T.M., Tainer, J.A., Geysen, H.M., Houghten, R.A. and Lerner, R.A. (1987). *Proc. Natl. Acad. Sci. USA.* **84**, 8568-8572.
14. Hopp, T.P. and Woods, K.R. (1981). *Proc. Natl. Acad. Sci. USA* **78**, 3824-3828.
15. Schulze-Gahmen, U., Klenk, H.-D. and Beyreuther, K. (1986). *Eur. J. Biochem.* **159**, 283-289.
16. Urbanski, G. and Margoliash, E. (1977). *In* "Immunochemistry of Enzymes and Their Antibodies" (MRJ Galton, ed.) p. 203, Wiley, New York.
17. Frömmel, C. (1988). *J. theor. Biol.* **132**, 171-177.
18. Chou, K.C. and Fasman, G. (1978). *Adv. Enz.* **47**, 145-147.
19. Garnier, J., Osguthorpe, D.J. and Robson, B. (1978) *J. Mol. Biol.* **120**:97-120.
20. Modrow, S., Hahn, B.H., Shaw, G.M., Gallo, R.C., Wong-Staal, F. and Wolf, H. (1987). *J. Virol.* **61**, 570-578.
21. Tanaka, T., Slamon, D.J. and Cline, M.J. (1985). *Proc. Natl. Acad. Sci. USA* **82**, 3400-3404.
22. Devereux, Haeberli and Smithies (1984). *Nucleic Acids Research* **12**, 387-395.
23. Musick, W.D.L. and Rossman, M.G. (1979) *J. Biol. Chem.* **254.** 7611-7620.
24. Sass, C., Briand, M., Benslimane, S., Renaud, M. and Briand, Y. (1989). *J. Biol. Chem.* **264**, 4076-4081.
25. Tsujibo, H., Tiano, H.F. and Li, S.S.-L. (1985). *Eur. J. Biochem.* **147**, 9-15.
26. Sakai, I., Sharief, F.S., Pan, Y.C.E. and Li, S.S.-L. (1987). *Biochem. J.* **248**, 933-936.
27. Herzenberg, L.A., Tokuhisa, T. and Herzenberg, L.A. (1980). *Nature.* **285**, 664-667.
28. Schutze, M.-P., Leclerc, C., Jolivet, M., Audibert, F. and Chedid, L. (1985). *J. Immunol.* **135**, 2319-2322.
29. Infante, A.J., Kimoto, M. and Fathman, C.G. (1982). *In* "Isolation, Characterization, and Utilization of T Lymphocyte Clones", pp. 367-374, Academic Press, Inc.
30. DeLisi, C. and Berzofsky, J. (1985). *Proc. Natl. Acad. Sci. USA.* **82**, 7048-7052.
31. Kaumaya, P.T.P., Berndt, K.D., Heindorn, D.B., Trewhella, J., Kezdy, F.J. and Goldberg, E. (1990). *Biochemistry* **29**, 13-23.
32. O'Hern, P.A. (1991). *Molec. Immunol.* **28**, 1047-1053.
33. Sette, A., Buus, S., Colon, S., Smith, J.A., Miles, C. and Grey, H.M. (1987). *Nature* **328**, 395-399.
34. Rothbard, J.B. and Taylor, W.R. (1988). *EMBO J.* **7**, 93-100.
35. Saper, M.A., Bjorkman, P.J. and Wiley, O.C. (1991). *J. Mol. Biol.* **219**, 277-319.
36. Madden, D.R., Gorga, J.A., Strominger, J.L. and Wiley, D.C. (1991). *Nature* **353**, 321-325.
37. Jemmerson, R. (1987). *Proc. Nat. Acad. Sci.* **84**, 9180-9184.
38. Jemmerson, R. and Blankfeld, R. (1989). *Molec. Immunol.* **26**, 301-307.
39. Atassi, M.Z. (1978). *Immunochemistry* **15**, 909.

Hydrophobic Contact Density Distribution Functions

Wayne J. Becktel and Lorraine M. Rellick

Department of Biochemistry, College of Biological Sciences
The Ohio State University, Columbus, OH 43210

I. Introduction

The hydrophobic effect, as it concerns the stability of globular proteins, remains actively investigated and discussed (1-3). One particular group of models of hydrophobic interactions counts the number of apolar atoms, separated by distances of less than an arbitrary value (generally 4Å). The number of such "contacts" may then be related to some experimentally observed quantity. Often, a phenomenological relationship may then be derived to allow the structural information to predict various thermodynamic properties. When the latter are difficult to obtain experimentally, reliable theoretical predictions are of considerable utility.

Given the relatively large number of globular proteins having known structures, it should be possible to directly examine the distances between apolar carbons and determine the distribution of the distance between any two atoms. This article describes such a comparison. It is based on a group of 17 globular proteins for which structural and, in most cases, thermodynamic results are available. It is shown below that the main result is that a distribution of contact distances exists and that this distribution is a sum of nearest, next nearest, etc.

TECHNIQUES IN PROTEIN CHEMISTRY IV

interactions in what amounts to a regular lattice arrangement of the apolar atoms in globular proteins.

II. Materials and Methods

A. Choice of amino acid apolar atoms

The choice of which atoms, in which amino acids, are regarded as being apolar is, by necessity, an arbitrary one. The choices made included atoms which were (1) carbon atoms with hydrogen atoms attached, (2) not bonded to charged groups and (3) not bonded to polar groups. The atoms which were selected by these criteria are given in the following Table.

Table 1. Apolar carbons

Residue	Atoms Included
Ala, Cys, Gln, Glu	C_β
Arg	C_β, C_γ
Val	C_β, $C_{\gamma 1}$, $C_{\gamma 2}$
Pro, Lys	C_β, C_γ, C_δ
Met	C_β, C_γ, C_ε
Leu	C_β, C_γ, $C_{\delta 1}$, $C_{\delta 2}$
Ile	C_β, $C_{\gamma 1}$, $C_{\gamma 2}$, $C_{\delta 1}$
Trp	C_β, $C_{\varepsilon 3}$, $C_{\zeta 2}$, $C_{\zeta 3}$, $C_{H 2}$,
Phe	C_β, $C_{\delta 1}$, $C_{\delta 2}$, $C_{\varepsilon 1}$, $C_{\varepsilon 2}$, C_ζ
Tyr	C_β, $C_{\delta 1}$, $C_{\delta 2}$, $C_{\varepsilon 1}$, $C_{\varepsilon 2}$

The number of atoms included, as compared to the total number of atoms in the protein, is limited. For T4 lysozyme there were 298 apolar carbon atoms selected out of a total of 1257 atoms (not including solvent and hydrogens).

B. Choice of Proteins

The initial group of proteins selected consisted of those used by Privalov and Gill (3). To these were added proteins which had two or more of the following attributes: (1) a well resolved xray structure, (2) well known thermodynamic properties, and (3) globular nature. The final group of proteins consisted of the 17 listed in the following Table.

Table 2 Protein Data Set

Protein	MW	Protein	MW
Crambin	5200	Myoglobin	17900
Trypsin inhibitor	6500	T4 lysozyme	18700
Parvalbumin	11500	Papain	23400
T1 Nuclease	11800	β-Trypsin	23800
Cytochrome c	13300	α-chymotrypsin	25200
α-lactalbumin	13800	Concanavalin A	26800
RNase A	14000	Carbonic anhydrase	29000
Hen lysozyme	14300	Pepsinogen	40000
Staph. Nuclease	16800		

Xray structures for these proteins were obtained from the Brookhaven data base, and specifically were downloaded from the National Research Council of Canada "Cryspro" file server.

C. *Interval Sampling*

For a given distance interval, using the xray coordinates and our choice of apolar atoms, all contacts were summed to ±0.5Å. This means that, for an interval centered at 3.0 Å, all contacts between 2.5 and 3.5 Å would be counted. This is an intentional over sampling of the data which was removed at a later point by means of a sliding Fourier smoothing procedure. The sampling was carried out from 2.0 to 32.0 Å, at every 0.1 Å, for all 17 proteins. This yielded all possible combinations of 1 Å intervals and the number of apolar contacts contained within those intervals. Calculations were also performed with non-overlapping intervals of ±0.1Å. No significant differences were observed between the two types of calculations. All calculations were carried out on a Cray Y-MP8/864.

III. Results and Discussion

The total number of apolar contacts, over each interval for T4 lysozyme, is shown in the Figure below. The absolute number of contacts continues to increase to approximately 14 Å. After this point, the number of contacts first plateaus and then falls off rapidly as the interval in question begins to exceed the maximum distance between apolar atoms in this protein. The height and starting interval for the plateau vary with the size, composition and structure of the protein.

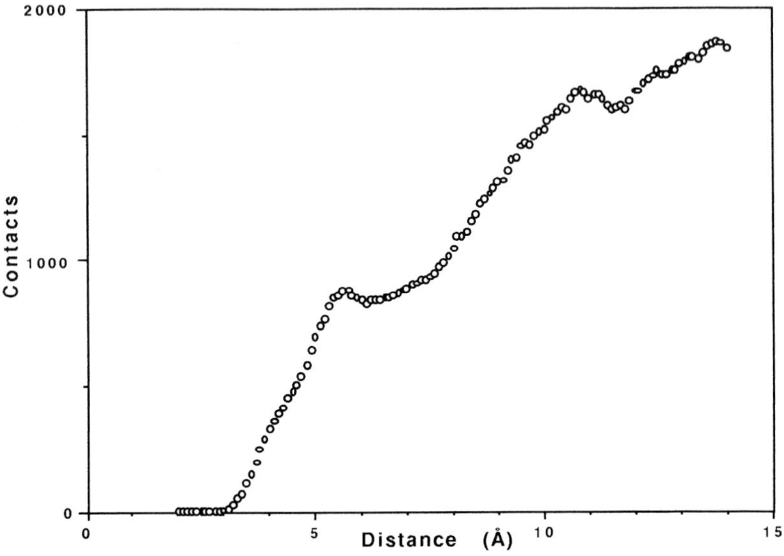

Figure 1: Variation of the number of apolar contacts in T4 lysozyme as a function of the distance between atoms. The intervals were ±0.5 Å from 2.0 to 14.0 Å in 0.1 Å increments.

It is unreasonable to expect that a hydrophobic contact is independent of the distance between atoms. One would expect that the interaction falls off as some function of the distance. The nature of that distance function is open to question. Figure 2 indicates the dependence of the distribution of contacts with distance as a function of R^{-x}, where "x" is an exponent ranging from 1 to 7. In this figure, the maximum of each curve has been normalized to unity to allow comparison of all seven resulting distributions. As may be seen a maximum quickly appears and the position of that maximum changes asymptotically to approximately 3.9 Å. The actual change in position of the maximum ranges from 5.4 Å to 3.8 Å in this figure. An exponent of 6 is attractive as it is seen in Leonard

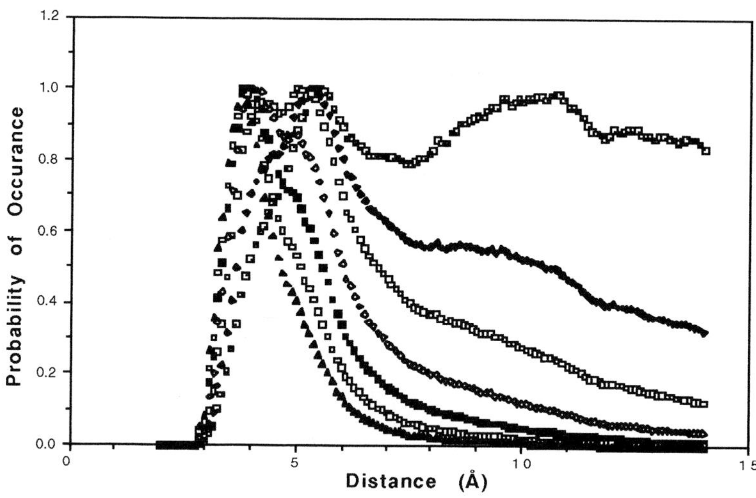

Figure 2: The variation of the probability of occurrence of an apolar contact in T4 lysozyme with the coefficient of the distance. All the plots represent the number of contacts seen in Figure 1 divided by the interval distance to a power ranging from 1 to 7. At 7 Å, scaling factors of R^1 through R^7 yield progressively smaller probabilities of occurrence.

Jones potentials and other non-bonded interactions which may be associated with apolar contacts. This value of the exponent was chosen and a transformation of the number of contacts on a given interval in T4 lysozyme was made. Finally, recall that we have over sampled the intervals and that they are not, strictly speaking, completely independent of one another. The transformed data was, therefore, smoothed with a sliding Fourier transformation where the width of the transform was equal to 10 points, or 1Å. The resulting data sets were then normalized to set the maximum value equal to unity.

The same R^{-6} transformation was carried out on the sum of all the contacts, over each interval, for all 17 proteins. The resulting set of data was again Fourier smoothed. The final spectrum appears in the third Figure as the smooth line. The maximum in this Figure is again near 3.9 Å and there is a shoulder near 5 Å.

This Figure bears a strong similarity to skewed Gaussian line shapes seen in probability distributions and in some types of chromatography. There is a maximum at approximately 3.9 Å. There is some probability of contacts both a greater and lesser distances. These decrease rapidly at distances of less than 3.9 Å, while the change is less rapid at distances greater than 3.9 Å. Since this function represents the probability distribution of contacts of apolar or hydrophobic groups, we suggest that such functions be called "hydrophobic contact density distribution functions".

The line shape of the smooth curve in Figure 3 suggests that there may be more than one probable distance at which contacts take place. This may be addressed in a number of ways. The simplest is to assume that there is a sum of normally distributed interactions having different maximal probabilities and centered at different distances. This assumption amounts to a sum of Gaussian curves. Such a deconvolution was carried out and also appears in Figure 3. In this figure, the integrated intensity of the sum of the four Gaussian curves was set to unity and the integrated intensity of each band adjusted to reflect this. There are maxima at 3.9, 5.3, 6.5 and 7.5 Å. The peak maxima decrease in intensity from left to right.

This deconvolution suggests a number of things about apolar contacts in globular proteins. First of all, representation of hydrophobic contacts by either a delta function or a cutoff function is likely a poor approximation. This is not a particularly surprising result. The dispersion of the first band is approximately 0.64 $Å^2$. This suggests a possible range of distances for interactions between nearest neighbor apolar atoms. The dispersions of the next three bands are similar.

The actual positions of the next three bands suggests that there are layers of interactions between apolar atoms in globular proteins. The first, at a distance of 3.9 Å is very similar to twice the Van der Waals radius of a carbon atom. A simple model for the next three Gaussian band positions may be approached by lattice models. For a perfect cubic lattice of equally sized spheres with nearest neighbors 3.9 Å apart, the distance to the next nearest neighbor would be expected to be 5.5 Å. If, instead, hexagonal packing is assumed the next nearest neighbor would be found at a distance of 6.7 Å. Finally, the next, next nearest neighbor would simply be 7.7 Å.

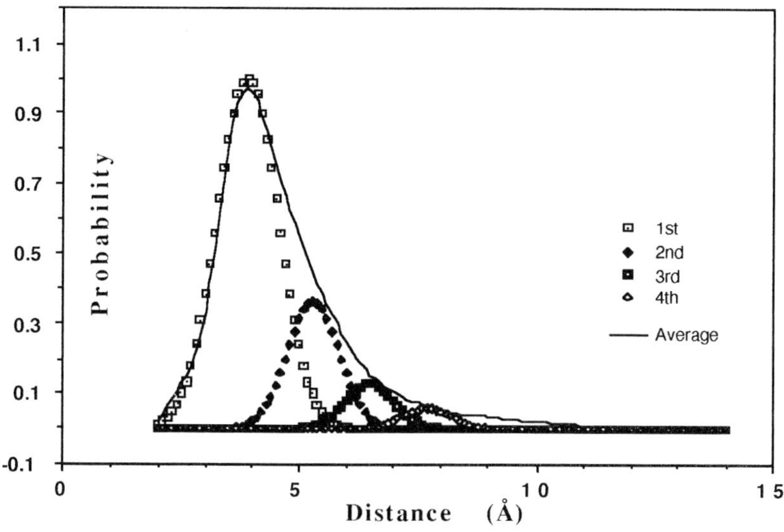

Figure 3: Average hydrophobic contact distribution for the 17 proteins used in this study (smooth line), and a set of Gaussian bands into which it has been resolved. The curve has been Fourier filtered as described in the accompanying text.

Compare these values to 3.9, 5.3, 6.5 and 7.5 Å for the band positions derived in the Gaussian deconvolution of the average distribution function. It would appear that the derived values correspond to neighboring apolar atoms which are closely packed in the protein core. This is consistent with packing theories of the hydrophobic core of proteins and for the packing of small, hydrophobic residues (2).

At this point it is necessary to interject a word of caution about the derived results. Recall that the distance geometry and values were derived from xray structures. Part of the derivation of these structures involves a refinement which takes into account bond angles, distances and Van der Waals radii. If the assumed radius of carbon is ~2 Å in the refinement, and if the refinement were directed by those values, it is possible to argue that the observed 3.9 Å result merely recapitulates the choice of Van der Waals radius in the refinement. The distribution represents the error in the protein data set refinement rather than a fundamental property of globular proteins. Now, if that were true, is this compatible with (1) the distribution of distances and (2) the higher order packing observed?

The dispersion of the average nearest neighbor model seems rather large for an error in the simulation. If one examines the individual maxima for the 17 proteins, the values range from 3.7 to 4.2 Å, implying a Van der Waals radius of an apolar carbon of between 1.85 and 2.10 Å. If the assumed 2 Å value absolutely had to be reproduced, then the error in individual positions in the proteins is ±0.35 Å. This seems somewhat large, but is not outside the realm of possibility.

This does not, however, invalidate the observations concerning next nearest and higher interactions. Either way, the nature of long range packing remains the same. This, in turn, has an important consequence to the prediction of protein folding. Whether or not the 3.9 Å value is artificial, the long range packing offers a constraint on how proteins have to fold. Specifically, any model predicting the folding of a protein must include a distribution of apolar atoms which roughly mimics the hydrophobic contact distribution function derived for globular proteins in general. Specifically, the distribution should have components corresponding to a sum of nearest and higher order neighboring carbons.

Acknowledgements

This research was supported by the National Science Foundation (DMB-9004902) and by the Ohio Supercomputer Center (PAS733-2).

References

1. P.L. Privalov *Adv. Protein Chem.* (1979) 33, 167.
2. K.P. Murphy and S.J. Gill, *Thermochem. Acta* (1990) 172, 11.
3. P.L. Privalov and S.J. Gill *Adv. Protein Chem.* (1988) 39, 191.

Construction and Functional Selection of a T4 Lysozyme Gene Library Randomly Mutagenized at Five Specific Sites

Enoch Baldwin, Jian Xu, and Omid Hajiseyedjavadi
Howard Hughes Medical Institute
Institute of Molecular Biology
University of Oregon
Eugene, OR 97403

I. Introduction

Random mutagenesis followed by selection or screening for a specific phenotype have been long used to identify functionally important genetic elements and interactions, from the cellular to the amino acid level. The advantages of such an approach are a) little prior understanding of the system under investigation is required, other than the means and limitations of the particular mutagenesis and selection employed, and b) a large spectrum of mutants can be scanned. In contrast, specific questions concerning the roles of particular amino acids in a protein can be addressed using oligonucleotide-directed mutagenesis. The two approaches can be combined by using degenerate oligonucleotides to create diverse mutant populations at specific sites followed by selection procedures that facilitate recovery of relatively rare individuals from these large pools or "libraries". Selections usually involve complementation of a cellular (1) or viral function (2), resistance to an antibiotic (3) or bacteriophage (4), or direct interaction with solid matrix (5). Screens for *in situ* enzymatic activity have also been utilized (6). By these methods, the roles of several amino acids and the potential interactions between them can be probed with minimal bias.

A major advantage of using T4 lysozyme as a model for mutational analysis of protein function and stability is the ease of high resolution crystallographic analysis and the large amount of existing thermodynamic and structural data. Here we report a method using the polymerase chain reaction (PCR) for generation of mutant genes that are randomized at sites far apart in the DNA sequence, but are proximal in the three-dimensional structure. A library of mutants at five sites was constructed to study interactions in the hydrophobic core of T4 lysozyme. The library was cloned into a phage λ vector which permitted selection, activity screening, and sequencing of functional genes and subsequent production of the encoded proteins.

TECHNIQUES IN PROTEIN CHEMISTRY IV

The selection relied on complementation of a defect in the λ lysis gene R by the cloned T4 lysozyme gene (7). A total of 134 DNA sequences of active T4 lysozyme genes were determined from a total pool of ~2.5 x 10^4 phage yielding 110 unique codon combinations and 106 unique mutant amino acid combinations. The relative stabilities and activities of each candidate were crudely estimated using an *in situ* activity assay (8) at 20°C and 37°C. Seven variants containing two to five replacements were purified and the free energies of folding (9) were determined. The order of stabilities indicated by the *in situ* assay corresponded to the melting transitions of the tested mutant enzymes. Six proteins crystallized isomorphous with wildtype T4 lysozyme (10) yielding data quality crystals.

II. Methods and Materials

A. *Bacterial Strains, Phages, and DNA*

Phages λZAP (11), λ (cI857, ind1, Sam7), VCSM13 and R408 M13 helper phages (12), and E. coli strains XL1-blue (thi^-, F'(tet^r)) (13), and VCS257 (tonA53, dapD8, lacY1, supE44, del(gal-uvrB)47, supD, supF58, gyrA29, del(thyA57), hsdS3) were obtained from Stratagene. E. coli A585 (C600: recD, supF) (14) was kindly supplied by Frank Stahl. Phage DNAs were packaged using Gigapack extracts (Stratagene). The F' derivatives of VCS257 and RR1 (15), VCS(F') and RR1(F'), were made by conjugation with XL1-blue followed by selection for the episomal tetracycline resistance and chromosomal thi^+ or streptomycin resistance, respectively. The tac-promoter driven expression vector pHN1403 derived from pHSe5 (16), and M13mp18 containing the cysteine-free wildtype (WT^*) T4 lysozyme gene (17) ligated into the BamHI and HindIII sites (M13mp18T4e*), were previously described (8). The plasmid named here pT4e is pHN1403 containing the WT^* gene. A phagemid version of pHN1403 containing the M13 origin of replication and the cheW gene was kindly supplied by A. Roth (pCW, ref. 18). DNA manipulations were carried out as previously described (19) or as to manufacturers instructions. Oligonucleotides for PCR amplification from standard polylinker flanking sequences (A: 5' side, 27-mer, GAGCGGATAA CAATTTCACACAGGAAC; and B: 3' side , 33-mer, contains an XhoI site, CCCCTCGAGTCACGACGTTGTAAAACGACGGCC) and saturation mutagenesis (C: site 121, 33-mer ATCCCAGCGTTTTTGTTG(A,C)NNCAT ACGTAAAGA; D: sites 129 and 133, 42-mer, CAAAAACGCTGGGAT GAANN(G,T)GCAGTTAACNN(G,T)GCTAAAAGT; E: site 149, 33-mer, ACACCTAATCGCGCAAAACGANN(G,T)ATTACAACG, F: site 153, 30-mer, GTCCCAAGTGCCAGTTCT(A,C)NNCGTTCTAAT) were synthesized (Applied Biosystems Model 380B DNA synthesizer) and purified using an OPC cartridge (Applied Biosystems). The degenerate sites were obtained using the mixed site feature of the synthesizer. HPLC analysis of model mixed site products showed essentially equimolar base compositions (P. Lindorfer, unpublished data). Protein preparations, crystallizations (8,10), and thermal denaturations (9) were performed as previously described.

The λ selection vector phage was constructed in the following way. The T4 lysozyme gene was amplified from M13mp18T4e* using PCR with primers A and B. The crude fragment was ligated into BamHI/HindIII cleaved and blunted pCW to give pT4oriSX, containing the T4 lysozyme gene flanked by 5' SacI and 3' XhoI sites. The R gene of phage λ (cI857ts, ind1, Sam7) was inactivated by insertion of a fragment derived from the mp8 polylinker (EcoRI/AccI) into the unique NarI site. The right arm generated by PvuII digestion was isolated. Phage λZAP(R$^-$), containing a similarly interrupted R gene, was cleaved with NaeI and the left arm isolated. Plasmid pT4oriSX was cleaved with NaeI and ligated with the two lambda arms described above. The mixture was packaged and the appropriate phage was identified by the ability to form plaques on VCS(F') and to produce ampicillin-resistance conferring particles after coinfection with M13 helper phage, and by restriction mapping. The T4 lysozyme gene of the phage was inactivated by generation of an internal deletion using SnaBI. The resultant phage was designated λ1401oriR$^-$ (cI857, nin5, ind1, Sam7, Δ(orf206a-Ea59)::pT4(ΔSnaBI)oriSX, R$^-$ (NarI::linker), Sam7). DNA was prepared from ampicillin resistant XL1-blue lysogens by temperature induction, yielding one hundred micrograms of purified phage DNA from 2 liters of induced lysogen.

B. Mutant T4 Lysozyme Gene Library

M13 templates for amplification of the T4 lysozyme gene fragments, lacking the 5' or 3' sequences corresponding to the annealing sites of oligonucleotides A or B, were produced from M13mp18T4e* by deletion with AvaII and BamHI, or HindIII and PvuI, to give M13T4Δ5' and M13T4Δ3' respectively. Dimeric E + F (EF) was produced by mixing equimolar E and F (100 pmoles each) and cycling as described below. Primer pairs A + C and D + EF (M13T4Δ3') or EF + B (M13T4Δ5') were used to amplify fragments from the appropriate single stranded templates (Figure 1). Initial amplifications were performed with Taq polymerase (2 mM $MgCl_2$, 1x Cetus buffer, 2.5 units Taq/100 µl reaction, 50 µM dNTPs, 250 ng template (100 fmole), 20 pmoles each primer, program: 95°C, 30 sec; 28°C, 45 sec; 72°C, 90 sec, 30 cycles). The resulting fragments (10-50% of a 100 µl reaction) were gel purified (native acrylamide, 6%, Tris-Borate-EDTA), eluted and ethanol precipitated. Linkage reactions and amplification of the linked products were performed using Vent polymerase (New England Biolabs) and the terminal primers (6 mM $MgSO_4$, 1x NEB Vent buffer, 100 µg/ml

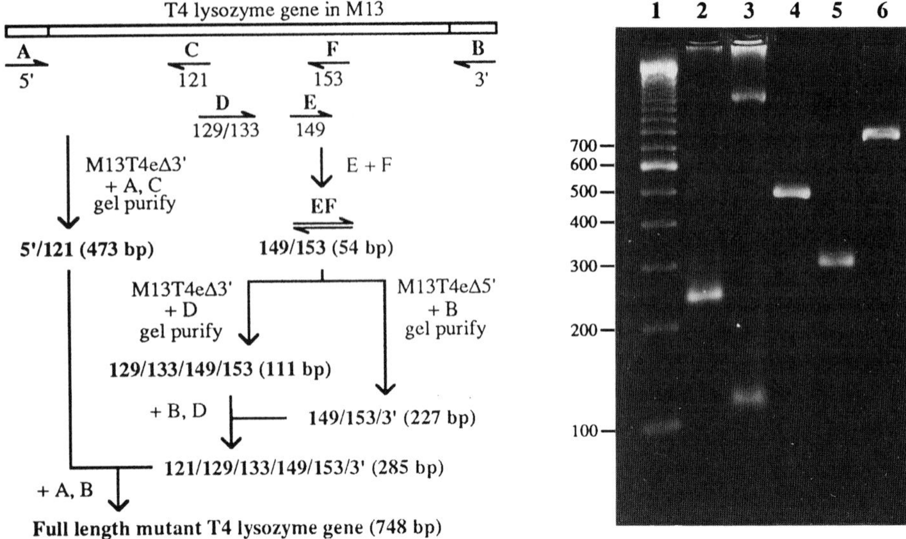

Figure 1. Assembly of the mutant T4 lysozyme gene library; (left) PCR scheme for assembly; (right) electrophoretic analysis of the PCR products (4% Nusieve agarose, Tris-Borate-EDTA), lanes: 1) 100 bp ladder; 2), 149/153/5'; 3) 129/133/149/154; 4) 5'/121; 5) 129/133/149/153/3'; 6) full length mutant T4 lysozyme gene.

acetylated BSA, 2.5 units Vent/100 µl reaction, 250 µM dNTPs, 10-30% of the gel purified fragments, 20 pmoles each primer, program: 97°C, 60 sec; 25°C, 30 sec; slow ramp (1 degree/sec) to 35°C, 15 sec; 75°C, 90 sec, 30 cycles). Linkage of the two final fragments was performed without gel purification. The final gene library (approximately 100 ng) was cleaved with SacI/XhoI, desalted, and concentrated with a Centricon-100 filter (Amicon). Ten micrograms of SacI/XhoI cleaved λ1401oriR$^-$ (above) was precipitated with library DNA using 6.5% PEG6000/0.8M NaCl and ligated in a volume of 10 ul with 1 unit of T4 ligase. One to three microliters were then packaged and stored at 4°C.

C. Selection and Screening of Lytic Phage

Total phage titers were determined on VCS257(pT4e). Approximately 2000 pfu/plate were plated on lawns of A585 at 37°C yielding 10-30 plaques. The plaques were grown 6 hours at 37°C and 12-16 hours at 20°C and then treated with chloroform to visualize halos indicative of T4 lysozyme activity (20). The plaques were sorted by halo size and packaged phagemid was obtained in two ways: by coinfecting VCS(F') with a plaque of the desired lambda clone and M13 helper phage (11); or by preparing an ampicillin-resistant lysogen in VCS(F') and subsequently infecting with helper phage. Culture supernatants (10-100 cfu/μl) were used to infect strain RR1(F') and plasmid bearing transfectants were selected on ampicillin plates. Single stranded DNA (ssDNA) for sequencing was produced by infection with M13 helper phage and preparation by the usual methods.

Activity screening of transfectants was performed at two different temperatures using a previously described plate assay (8). Four freshly transfected colonies of RR1(F') from each clone were toothpicked onto two separate indicator plates and grown at 37°C for 3 hours. One set was placed at room temperature for one hour. The plates were treated with chloroform vapor and incubated at room temperature or 37°C for 15 hours. The radius of each halo was measured using a circular template, averaged, and rounded to the nearest half-millimeter. The standard deviation for the set of four averaged 0.5 mm.

Phage-plasmid crosses were used to generate test phage expressing previously characterized T4 lysozymes. Five microliters of stocks of λ1401oriR$^-$ (10^{10} pfu/ml) were spotted onto lawns of RR1 harboring the desired mutant lysozyme gene in pHN1403. After overnight growth at 37°C, the spotted area was cored and resuspended in SM buffer (18), treated with chloroform, and titered on A585 and VCS257(pT4e). The proportion of R$^+$ recombinants was typically 0.01-2%.

III. Results and Discussion

Residues Leu121, Ala129, Leu133, Val149, and Phe153 pack together in an uninterrupted volume in the C-terminal domain of T4 lysozyme, bounded by five helices. Because of the distance between the target codons, existing cassette or primer mutagenesis methods were impractical. Higuchi *et al.* (21) showed that two overlapping double-stranded DNA segments could mutually prime during PCR resulting in a spliced product. We used PCR with degenerate oligonucleotide primers to generate 3 segments (5'/121 (**A** + **C**), 129/133/149/153 (**D** + **EF**), and 149/153/3' (**EF** + **B**)) and then splice them together (Figure 1). The degenerate codon NN(G,T) was used to specify all twenty amino acids and one stop (amber) codon for each site (strain A585 carries both Gln and Tyr suppressor mutations). In order to limit misincorporation errors by Taq polymerase, a large amount of template (100 fmoles) was used to minimize the number of logarithmic amplification cycles. Vent polymerase was used in the joining reactions because of its lower error rate (22). Only 4 clones containing single base errors were observed of approximately 65,000 bases sequenced. Contamination by the unmutagenized gene resulting from the products of **A** and **B** priming was a major problem using M13mp18T4e* as a template. We reduced the fraction of unmutagenized contaminant to approximately 0.05-0.1% of the total library

Table I. Distributions of mutations

a. number of unique sequences with *n* changes per gene

n	codons	redundant	amino acids
0	9	9	9
1	11	4	12
2	6 (0)[a]	1	30
3	18 (5)	5	38
4	39 (28)	3	21
5	35 (65)	3	5

b. number of changes and codons and amino acid types by position[b]

site	number with non-wt		number of types		variability[c]	
	codon	amino acid	codon	amino acid	codon	amino acid
121	93	73	21	14	140	44
129	94	76	23	14	110	50
133	92	53	16	10	94	20
149	70	23	10	6	28	8
153	59	56	17	11	37	23

[a] Numbers in parenthesis assuming that all codons at 121 and 153 are mutated.

[b] The population of unique sequences was used including one wildtype sequence.

[c] Variability = number of amino acids or codons represented/frequency of the most common amino acid or codon (25). At all sites, the wildtype amino acid and codon were most frequent except for position 133 (amino acid only).

(9/133 selected clones, Table Ia) by use of templates in which one priming site was deleted and purification of the template-directed PCR fragments using acrylamide gel electrophoresis. Primer tests of gel purified fragments showed that remaining contamination likely arose from low level amounts of single-stranded products from primers B and EF extended past the cognate upstream priming site on the M13mp18T4e* template, providing a source of wildtype sequence for subsequent reactions. A similar methodology was recently reported in which the flanking primers incorporated sequences absent in the template to prevent wild type contamination (23).

Vector λ1401oriR⁻ was constructed to facilitate selection, recovery, and expression of variant lysozymes. Complementation of the linker-inactivated lambda R gene by a functional T4 lysozyme gene allows for the formation of plaques on a non-lysozyme expressing host, as had been previously demonstrated with λ and phage P22 (7). No R^+ revertants were obtained from 10^{10} phage, the largest number tested. The vector incorporates a feature of λZAP in which single-stranded copies of an imbedded plasmid can be excised by coinfection with a filamentous phage (11), and likewise directs the excision of expression plasmid pT4oriSX for production of protein and ssDNA for sequencing. A version of λZAP in which the R gene had been similarly disrupted was initially used but frequent deletions in the T4 lysozyme coding sequence were encountered, due to the particular strand (-) that is packaged (D. Muchmore, A. Roth, unpublished observations).

The stability threshold required for complementation was investigated using the high rate of plasmid-phage recombination to prepare phage bearing lysozymes of different stabilities (-1.0 to -8.0 kcal/mole). All of the phage made plaques of normal size at 37°C. At 42°C, the least stable mutant made somewhat smaller plaques. A phage was constructed containing a lysozyme deletion mutant with 5% of WT* activity in a turbidity assay (R. DuBose, unpublished data) which made small plaques at 37°C. Thus, little activity is required for phage viability. In general, plaque sizes of the library-selected mutants were highly variable and not well correlated with the halo size.

A library of approximately 2.5 x 10^4 clones was screened in three different experiments for the ability to form plaques on non-lysozyme expressing hosts. The recD strain A585 was used because plaque titers were 10-100 fold higher than with VCS257. Approximately one-half of the plaques formed gave detectable halos when treated with chloroform at room temperature and the remainder were presumably of very low stability or activity. Packaged expression phagemids (pT4oriSX) carrying mutant lysozyme genes were prepared from the halo-forming clones and used to infect RR1(F') hosts, and the lysozyme activity of each was assessed using a plate assay (8). Using this plasmid, a strain expressing a mutant lysozyme that is destabilized by 5 kcal/mole (L99A, ref. 24) forms no halos at 37°C but forms substantial halos at 20°C, while a strain expressing a lysozyme destabilized by 8 kcal/mole (L99A/F153A, ref. 24) fails to form halos at this temperature. It should be noted that halo size is dependent on both activity and stability of the expressed lysozyme and only provides a rough estimate of fitness. Nonetheless, of the variants for which the stabilities are already known, halo size correlated reasonably with stability (Table II).

Codons of the wild-type sequence for 129, 133, and 149 contain A or C in the third position which are not encoded by the mutagenesis primers. Thus, the efficiency of the PCR can be estimated from the occurrences of non-mutant codons in the selected population at positions 129, 133, and 149, while at positions 121 and 153 it is possible to distinguish only when a wildtype codon is not present. The distributions of the number of mutated positions at the DNA and amino acid level, and the distribution of the numbers of non-wildtype amino acids, amino acid and codon diversity and variability (25) are shown in Table I. Codon conversions occurred at five sites in at least 35/107 of the unique sequences, with an upper limit of 65 using the assumption that mutagenesis was complete at sites 121 and 153. The least randomized positions were 149 and 153 (67 and 56 mutated sequences), with single codon changes occurring only at 121 (11/107

Table II. Properties of Purified Selected Mutant Lysozymes[a]

Mutation(s)	T_M	ΔT_M	halo radius (mm) 20°C	37°C	crystals
WT* (none)	51.8	-	6	10	-
129L	48.7	-2.9	6	8	-
121A/129M/153L	48.0	-3.7	4.5	6	yes
121A/129M/149I	47.8	-3.9	6.5	5.5	yes
121I/129L/133M/153W	47.3	-4.5	5.5	6	yes
121A/129L/133M/149M	45.5	-6.3	6.5	5.5	yes
121M/129L/133M/149I/153W	45.4	-6.4	5.0	5.5	yes
121C/129V/133M	43.7	-8.1	5.5	2	yes
121A	42.5	-9.1	4.5	1.5	-
A129M/153A	36.5	-15.0	5.5	0	yes

[a]Melting transitions (in pH 3.05, 20 mM KPO_4, 25 mM KCl) and halo sizes of selected mutants. Mutants 121A (25) and 129L were selected but had been characterized previously.

sequences). Mutants with wildtype codons exclusively at positions 149 or 153 (10 and 19 sequences), or both (11 sequences) accounted for a large fraction of the triple and quadruple mutants. The relatively low mutagenesis at these sites and the preponderance of single changes at 121 was likely due to contamination of the 149/153/3' fragment with unmutagenized runoffs from priming by **B**, which are then primed directly by 5'/121 or **D**, with **B** in the joining reactions. The use of **EF** dimers contaminated with monomers probably caused the exclusive mutations at positions 149 and 153.

Several unusual occurrences were noted. First, several sets of identical DNA sequences with multiple changes were recovered more than once (Table Ia). This occurrence is statistically unlikely for a randomized population, and may be due in part to phage replication during the infection period before plating. Nonetheless some identical mutants were recovered in different experiments suggesting that there may also be preferential primer annealing or PCR amplification. Secondly, two clones with mutations at position 129 had the wild-type codon at position 133, although these positions were encoded by the same primer. The source of such mutants could be dA contamination of the mixed site or partial repair of a heteroduplex with an unmutagenized strand, again likely arising from template priming by **B**.

Seven mutant proteins were produced (8) from the excized expression plasmids in sufficient amounts for thermodynamic measurements and crystallization experiments. All seven variants crystallized isomorphous with the wildtype enzyme, six of which gave data quality crystals (Table II).

In summary, we have demonstrated the ability to select functional but highly mutated T4 lysozyme molecules from a large pool with reasonable efficiency. A more critical analysis of the biases of the mutagenesis procedure and structural analyses of the crystallized variants is underway.

Acknowledgements

The authors would like to thank B.W. Matthews, A. Poteete, A. Roth, F. Stahl, and R. Dubose for necessary materials and helpful consultations. Thermal denaturations were generously performed by W.A. Baase. This work was supported in part by grants from the NIH (GM21967 and GM12989-03) and the Lucille B. Markey Charitable Trust.

References

1. Hermes, J., Blacklow, S., and Knowles, J. (1990). *Proc. Natl. Acad. Sci.* **87**, 696- 700.
2. Rennel, D., Bouvier, S., Hardy, L., and Poteete, A. (1991). *J. Mol. Biol.* **222**, 67-88.
3. Oliphant, A., and Struhl, K. (1989). *Proc. Natl. Acad. Sci.* **86**, 9094-9098.
4. Lim, W., and Sauer, R. (1989). *Nature* **339**, 31-36.
5. Scott, J., and Smith, G. (1990). *Science* **249**, 386-390.
6. Lehtovaara, P., Koivula, A., Bamford, J., and Knowles, J. (1988). *Prot. Engineer.* **2**, 63-68.
7. Rennel, D., and Poteete, A. (1985). *Virology* **143**, 280-289.
8. Poteete, A., Dao-Pin, S., Nicholson, H., and Matthews, B.W. (1991). *Biochemistry* **30**, 1425-1432.
9. Dao-Pin, S., Baase, W., and Matthews, B.W. (1990). *Proteins 7*, 198-204.
10. Weaver, L., and Matthews, B.W. (1987). *J. Mol. Biol.* **193**, 189-199 ; Remington, S., Anderson, W., and Owen, J., Ten-Eyck, L., Grainger, C., and Matthews, B.W. (1978). *Biochemistry* **118**, 81-98.
11. Short, J., Fernandez, J., Sorge, J., and Huse, W. (1988). *Nucl. Acids Res.* **16**, 7583-7600.
12. Russel, M., Kidd, S., and Kelly, M. (1986). *Gene* **45**, 333-338.
13. Bullock, W., Fernandez, J., and Short, J. (1987). *Biotechniques* **5**, 376-380.
14. Thaler, D., and Stahl, F. *unpublished work.*
15. Bolivar, F., Rodriguez, R., Greene, P., Betlach, M., Heyneker, H., Boyer, H., Crosa, J., and Falkow, S. (1977). *Gene* **2**, 95-99.
16. Muchmore, D., McIntosh, L., Russel, C., Anderson, D., and Dahlquist, F. (1989). *Methods Enzymol.* **177**, 44-73.
17. Matsumura, M., Signor, G., and Matthews, B.W. (1989). *Nature* **342**, 291-293; Matsumura, M., and Matthews, B.W. (1989). *Science* **243**, 792-794.
18. Roth, A., and Dahlquist, F. *unpublished work.*
19. Maniatis, T., Fritsch, E., and Sambrook, J. (1989). *In* "Molecular Cloning, A Laboratory Manual, Second Edition". Cold Spring Harbor Press. Cold Spring Harbor, NY.
20. Alber, T, and Wozniak, J. (1985). *Proc. Natl. Acad. Sci.* **82**, 747-750.
21. Higuchi, R., Krummel, B., and Saiki, R. (1988). *Nucl. Acids. Res.* **16**, 7351-7367.
22. Mattila, P., Korpela, J., Tenkanen, T., and Pitkanen, P. (1991). *Nucl. Acids. Res.* **19**, 4967-4973.
23. Merino, E., Osuna, J., Bolivar, F., and Soberon, X. (1992). *BioTechniques* **12**, 508-510.
24. Eriksson, A., Baase, W., Zhang, X.-J., Heinz, D., Blaber, M., Baldwin, E., and Matthews, B.W. (1992). *Science* **255**, 178-183.
25. Wu, T., and Kabat, E. (1970). *J. Exp. Med.* **132**, 211.

Horse Heart Ferricytochrome c: Anions, Buffers and Stability

Yash P. Myer
Department of Chemistry, Institute of Hemoproteins,
State University of New York at Albany,
Albany, NY 12222

The stability of horse heart ferricytochrome c in water, in the presence of phosphate, cacodylate, ClO_4^-, Cl^-, and a mixture of cacodylate + $NaClO_4$ has been investigated through thermal unfolding studies. The 222-nm ellipticity profiles are typically bi-phasic, with a high-temperature unfolding step and a low-temperature heme crevice perturbation step. The profiles are analyzed as a combination of two independent processes, each conforming to a two-state model with constant but finite ΔC_p. The T_m of the low-temperature component varies from 25°C in water to 43°C in cacodylate buffer, ΔH_m from 15 to 32 kcal/m and ΔC_p nearly zero except in perchlorate, 0.8 ± 0.3 kcal/m. The unfolding T_m is insensitive to the presence or the absence of buffers or anions, 77 ± 2°C, but ΔH_m increases from a value of 39 to 94 kcal/m, and similarly, ΔC_p from 0.8 to 2.4 kcal/m. The composite free energy of unfolding at 20°C in water is about 2.7 kcal/m, 4.0 kcal/m in the presence of anions and about 5 kcal in the two buffers. Both anions and buffers stabilize the protein in the following order:- water < Cl^- ≈ ClO_4^- < cacodylate < phosphate. The stabilization appears to be a reflection of an increasing degree of coalescence of the hydrophobic groups.

1. Introduction

Anions such as chloride, perchlorate and phosphate bind to ferricytochrome c and affect the conformational state of the protein (a review, Moore et al., 1984; Goto et al., 1990; Fen and Englander, 1990; Gopal et al., 1988; Arean et al., 1988). Three molecular domains have been identified as possible binding-perturbation sites, the top left of the heme crevice, the front of the crevice, and the front left of the molecule. Salt-dependent changes have also been reported in the region of the tryptophan-59 residue (Liu et al., 1988), which is buried deep in the heme crevice. It is generally thought that the effect is because of binding and/or the binding-linked shift of equilibrium between the open and the closed form (Trewhella et al., 1988).

Recently we showed that the addition of anions at neutral pH results in a slight tightening of the crevice, possibly because of the shielding of the electrostatic repulsive interactions of the charged groups surrounding the heme crevice opening (Myer and Saturno, 1991). Here we report findings from studies of thermal unfolding under various solution conditions. These studies show that both anions and buffers cause stabilization of the protein, possibly through enhancement of the degree of hydrophobicity of the molecule.

TECHNIQUES IN PROTEIN CHEMISTRY IV

2. Experimental

A. Materials. Monomeric horse heart ferricytochrome c was chromatographically prepared from Sigma type III according to the procedure of Brautigan et al. (1978). All other materials were of analytical grade.

B. Measurements. Circular dichroism measurements were conducted using an updated JASCO instrument, i.e. with a photo-elastic modulator, an ITHACO 391A lock-in amplifier, solid state AC/DC amplifiers and servos systems, an electrically thermostated cell enclosure, and online to an IBM computer with a 12-bit A/D converter for signal output. The temperature of the samples was controlled to $\pm$ 0.1^{o}, and the rate of heating/cooling was 0.6 $\pm$ 0.1^{o}C/min unless otherwise specified. The concentration of the protein stock solution, in water, pH 7.00 $\pm$ 0.05, was determined using an absorption extinction of 10.4 mM at 528 nm, which was transferred to the required system with appropriate dilution. To ensure a complete Fe^{3+} state of the protein, a minimal amount of potassium ferricyanide, $<$10% of protein equivalent, was added to the water solutions. Only freshly prepared solutions, i.e. within 2 hrs, were used throughout these studies.

3. Results and Discussion

A. Analysis of Thermal Profiles and Thermodynamic Parameters. The estimation of ΔG, the free energy of unfolding, at a defined temperature, usually 20 or 25°C, is the premise for determining the relative stability of a protein. Most commonly, the estimation is based upon the analysis of the thermal or denaturant unfolding profiles (Schellman, 1988; Becktel and Schellman, 1987), or a combination of the two (Pace and Laurents, 1989), or calorimeteric studies (Sturtevant, 1987; Privalov and Potekin, 1986; Privalov and Khechinashvili, 1974). The analysis of the thermal unfolding profiles is based upon the use of the Gibbs-Helmholtz equation in the form:

$$\Delta G(T) = \Delta H_m(1 - T/T_m) + \Delta C_p(T - T_m - T \ln(T/T_m)) \quad (1)$$

where $\Delta G(T)$ is the free energy difference at temperature T; T_m, the melting, mid-transition temperature; ΔH_m, the enthalpy change at the melting temperature; and ΔC_p, the heat capacity difference between the unfolded and the folded forms. Equation 1 is based on the model: (i) the unfolding process is a two-state reaction; (ii) the protein is stable at some temperature, i.e., G(T) is positive; and (iii) the ΔC_p is constant and finite (Schellman and Hawkes, 1980; Becktel and Schellman, 1987).

The procedure commonly used for the analysis of the thermal unfolding data is, first, the analysis of the profiles in terms of ln $K_{eq}(T)$ vs 1/T, where $K_{eq}(T)$ is the equilibrium constant of the folded <=> unfolded process at temperature T, to determine ΔH at a number of temperatures in the transition zone $(d(\ln K_{eq}(T)/d(1/T)) = -\Delta H(T)/R)$, and T_m the temperature at which $\ln K_{eq} = 0$, followed by the determination of ΔC_p through analysis of $\Delta H(T)$ vs temperature profile, $(d(\Delta H)/d(T) = \Delta C_p)$. $K_{eq}(T)$ is calculated using the relationship $K_{eq}(T) = (Y_f - Y(T))/(Y(T) - Y_u)$ where $Y(T)$, Y_f and Y_u are the values of a probe, linearly dependent on concentration, at temperature T, of the folded form, and of the unfolded form, respectively. Thus, in order to perform an

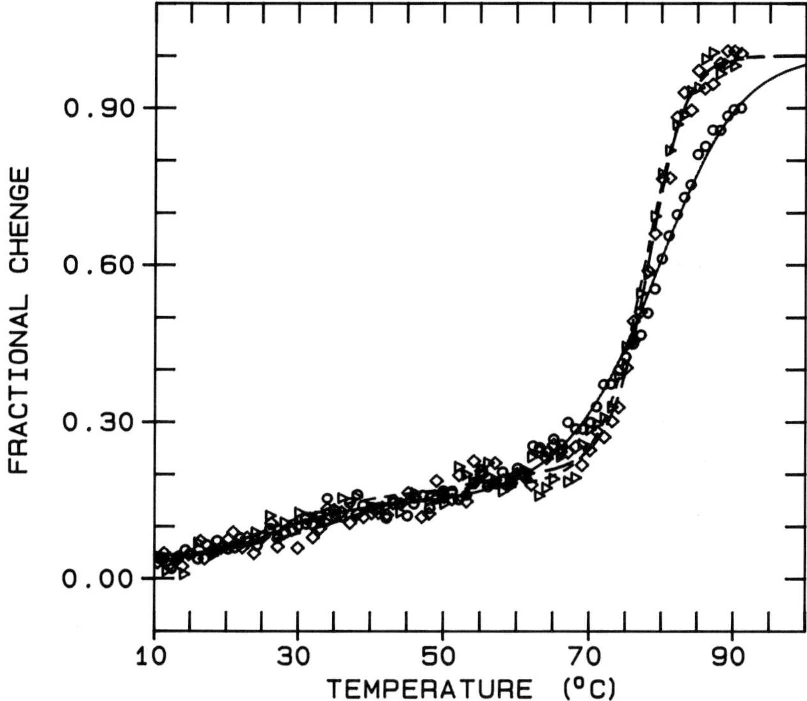

Fig. 1. Representative observed and calculated thermal transition profiles of HH ferricytochrome c in the absence and the presence of anions/buffers. Protein concentrations, 14 ± 2 μM; pH 7.0 ± 0.1. Data shown for: Water,.—o—; 0.2 M KCl in water, ---◁---; 0.1 M phosphate buffer, — ◇—. Symbols, normalized observed data, 1/2 the observed density. Lines, calculated curves using parameters listed in Table I.

analysis of the transition profiles, one requires accurate determination of values of probes, not only in the transition zone, but also of the folded and the unfolded forms. In most instances, i.e., where the melting temperatures are in the 45-55°C range and the ΔH_m's are larger than 30 kcal/mole, this approach has been quite successful (Pace and Laurents, 1989; Schellman 1987; Feng and Slinger 1991).

The analysis of the thermal transition profile of HH ferricytochrome c based on ellipticity changes at 222 nm (Figure 1 - representative only) cannot be adequately accomplished using the above approach, because (i) the profiles are bi-phasic in nature, thus requiring resolution of components, and (ii) the high melting temperature, >75°C, precludes determination of Y_u with any degree of accuracy. This difficulty is further enhanced by the observation that at temperatures higher than 88°C, the protein undergoes irreversible denaturation, which further restricts the high temperature limit of the measurements.

We have employed a direct approach to analyze these profiles, i.e., fitting of the profiles directly using a non-linear, least-square procedure, the Marquardt algorithm, based on the following relationship:

$$Y(T) = Y_1 + \sum_{i=1}^{n} (Y_{i+1} - Y_i)(1/(1+ \exp(\Delta G_i(T)/RT))) \qquad (2)$$

where n is the number of two-state processes, i is the number of molecular species (n+1), Y(T) is the value of the probe at temperature T, Y_1 is the value of the probe of pure form 1 (the folded form), and Y_i and Y_{i+1} are the values of the probe for the pure initial and final forms of the ith reaction, respectively. $\Delta G_i(T)$ is the Gibbs free energy for the ith reaction at temperature T as defined by Equation 1. Equation 2 in terms of fractional change becomes:

$$F(T) = \sum_{i=1}^{n} (F_{i+1} - F_i)(1/(1 + \exp(\Delta G_i(T)/RT))) \qquad (2a)$$

where $F(T) = (Y(T) - Y_1)/(Y_{n+1} - Y_1) = (Y(T) - Y_f) / (Y_u - Y_f)$ and the quantity, $F_{i+1} - F_i$ is the maximum fractional contribution of the ith process to the total fractional change, 0 to 1, of the overall reaction.

The resulting expression, Eq. 2 + Eq. 1, is relatively complex with multiple roots. In the case of a two-component transition, there is a total of nine variables (three limiting values of the probe, Y_A, Y_B and Y_C, and three thermodynamic parameters for each of the components, ΔH_m, T_m and ΔC_p) requiring adjustment for the minimization of error. This complexity introduces a multiplicity of minimum error states, which makes it difficult to discern the real minima from the secondary ones.

We adopted the strategy of gradual step-wise convergence, i.e., repeated fitting with an increasing number of variables, one at a time. The following is a typical sequence of analysis. (i) Error minimization is performed with all Y_i's, whether observed or guessed, and the ΔC_p's are treated as constants, allowing concurrent adjustment of both the ΔH_m's and the T_m's. (ii) Fitting is repeated n+1 times, each with an additional Y as a variable, selected in the reverse order of their degree of uncertainty, the best known first followed by unknowns/guesses. (iii) Finally, the ΔC_p's are then added to the adjusting variables, one at a time, starting with that of the component dominating the profile.

The reliability of the procedure was tested using a simulated unfolding profile with varying degree of error, data density and transition range, and the results from the analysis of a single-component profile are shown in Table I.

It is apparent from the data in Table I that reliable estimations of the thermodynamic parameters can be made from measurements with as high as ±5% error, provided there is substantial density of observations in all three regions of the transition, >80 data points in the transition zone (5-95%) region and ≥20 data points in each of the pre- and the post-transition regions. In the case of measurements with a maximum random error of 2.5% (the uncertainty of the measurements in the intrinsic region of the instrument at our disposal), the reliable estimation of both ΔH_m and T_m is feasible with only 1/2 the density of observations, 98% or better, but the reliability of ΔC_p decreases significantly, 40-50% only. Further lowering

Table 1.
Effect of Instrumental Error, Data Density and Profile Limits on the Thermodynamic Parameters

Error[1] (± %)	Trans. Limits[2] (%)	Data Points[3]	ΔH_m (kcal/m)	T_m (^oC)	ΔC_p (kcal/m)
0.0	0-100	160	50.0 ± 0.0	50.00 ± 0.00	1.00 ± 0.0[4]
1.25	0-100	160	50.0 ± 0.2	50.00 ± 0.02	1.04 ± 0.06
2.50	0-100	160	49.5 ± 0.9	50.00 ± 0.07	1.0 ± 0.2
5.00	0-100	160	49.5 ± 3.3	49.8 ± 0.2	1.1 ± 0.3
6.25	0-100	160	53.2 ± 6.0	49.8 ± 0.9	1.4 ± 0.7
2.50	0-100	80	50.1 ± 1.2	49.9 ± 0.1	1.1 ± 0.5
2.50	0-100	53	49.8 ± 3.2	49.7 ± 1.5	0.9 ± 0.8
2.50	0-95/ 5-100	130	49.3 ± 1.4	50.1 ± 0.2	1.0 ± 0.3
2.5	0-90/ 10-100	121	49.3 ± 1.9	49.9 ± 0.4	1.1 ± 0.4
2.5	0-90/ 10-100	242	49.7 ± 0.9	50.0 ± 0.2	1.0 ± 0.2
2.5	0-85/ 15-100	115	49.3 ± 2.3	49.9 ± 0.5	1.1 ± 0.8
2.5	0-80/ 20-100	110	47.8 ± 3.0	50.2 ± 0.7	0.9 ± 0.9
2.5	5-95	90	49.5 ± 1.8	50.1 ± 0.3	0.8 ± 0.4
2.5	5-95	180	49.7 ± 1.0	49.9 ± 0.2	0.9 ± 0.2
2.5	10-90	71	41.5 ± 9.0	48.7 ± 1.6	2.0 ± 0.9

[1]Maximum, randomly adjusted error added to the simulated data.
[2]Limits of the transition used for analysis.
[3]Total number of data points in the profile subjected to analysis.
[4]Parameters of the simulated profiles. Data was generated at an increment of 0.25^oC, giving a maximum of 180-200 data points in the 5-95% transition zone and 30-40 data points in the pre-, <1%, and the post-, >99%, transition limits.

the density of observations makes the outcome even worse, with the reliability of the ΔC_p's ≤ 10%. For situations where measurements are lacking, either in the post- or the pre-transition region, a reliable estimation is feasible only for systems with 90% or more of the profile observed, and for cases with observations missing in both the pre- and the post-transition regions, reasonable analytical results are possible only if the observations cover, at the very least, the 5-95% range of the transition profile. It should be noted that, by increasing the density of observations for any one of the situations, the reliability of the calculated parameters can be improved.

Based on the above considerations, and since the transition profiles of ferricytochrome c are limited to only the pre- and the transition zone (Fig 1), we opted to collect data at an increment of 0.25^oC, providing at the minimum a total of 320 observations in each profile. In addition, we

Table II.
Thermal Unfolding Parameters and the Stabilizing Free Energy at 20°C Based on Ellipticity Changes at the 222-nm Minimum of Horse Heart Ferricytochrome c in the Absence and the Presence of Buffers and Anions.

Parameter[1]	System[2]					
	H_2O	KCl	$NaClO_4$	Caco.	Phos.	Mixt.
Low-Temperature Component						
T_m (±<2)(°C)	25	25	29	43	34	23
ΔH_m(±<4)(kcal/m)	32	22	32	15	16	25
ΔC_p(±<0.3)(kcal/m)	0.4	0.0	0.8	0.1	0.4	0.0
ΔG_{20C}(±0.3)(kcal/m)	0.5	0.4	0.8	1.1	0.6	0.2
High-Temperature Component						
T_m (±<2)(°C)	79	78	75	77	78	77
ΔH_m(±<5)(kcal/m)	39	81	85	94	92	86
ΔC_p(±<0.2)(kcal/m)	0.8	1.9	2.2	2.3	2.4	2.1
ΔG_{20C}(±0.5)(kcal/m)	2.2	3.6	3.2	3.8	4.5	3.9
$\Delta G_{20C,TOTAL}$ (kcal/m)	2.7	4.0	4.1	4.9	5.1	4.1

[1]The values and the degrees of uncertainty are based upon 4-5 independent measurements of freshly prepared solutions, i.e. within 2 hrs of preparation.
[2]pH, 7.0 ± 0.2; KCl, 0.2 M KCl in water; $NaClO_4$, 0.1 M $NaClO_4$ in water; Caco., 0.1 M Cacodylate + NaOH buffer; Phos., 0.1 M NaH_2PO_4 + NaOH buffer; Mixt., 0.1 M Cacodylate + 0.1 M $NaClO_4$.

conducted, at the minimum, five independent measurements of each system. The results in Table II are the average of the findings from independent analyses of each of such measurements.

B. Horse Heart Ferricytochrome c, Buffers and Anions

The addition of anions, chloride, perchlorate, and/or buffers such as phosphate has been shown to cause a number of distinct perturbations in the state of ferri-cytochrome c at neutral pH's, ranging from perturbations of surface group topology, lysine residues at the opening of the crevice (Feng and Englander, 1990), to changes in the environments of the groups buried deep in the heme crevice (Liu et al., 1989) and of the heme crevice, including the heme moiety itself (Myer and Saturno, 1991). The intrinsic CD spectra of ferricytochrome c at neutral pH's in the absence or the presence of anions or buffers and at room temperature are essentially the same, a negative minimum at about 222 nm and a second inflection in the region 208-210 nm, except for small but definite variations of the amplitude of the 222-nm minimum and different levels of resolution of the 208-nm peak (data not shown; see Myer and Saturno (1991) for some details). The CD spectrum at high temperatures, 87 ± 2°C, is typical of the unfolded protein, an inflection-free spectrum in the region 210-240 nm with a tendency to a negative minimum below 200 nm (data not shown). Here again, there are differences, but in terms of the amplitude of the plateau only. These findings are consistent with the idea that anions and/or buffers do alter the conformational state of the protein at room temperature, but only subtly. The spectral differences at high temperatures could simply be a reflection of different degrees of unfolding, i.e., differing degrees of stability under different solution conditions.

The thermal unfolding profile of ferricytochrome c, based upon ellipticity changes at 222 nm and irrespective of the solution conditions, is bi-phasic (Fig. 1 - representative members only), with a minor low-temperature component in the region 10-60°C and the major transition at higher temperatures. The bi-phasic transition profile in water is consistent with the report made long ago, which also showed that the low-temperature component is primarily due to perturbations of heme-protein interactions, and the high-temperature component, the unfolding (Myer 1968).

The melting temperature of the unfolding transition is essentially the same for all the solutions systems, 77 ± 2°C (Table II). Both ΔC_p and ΔH_m, on the other hand, are larger in the presence of anions as well as buffers. The ΔG_{20^oC}, a direct measure of protein stability, also becomes more positive, in the following order:- water < NaCl ≈ $NaClO_4$ ≈ cacodylate + $NaClO_4$ < cacodylate < phosphate (Table II). The effect of anions and buffers on the state of heme-protein interactions, the low-temperature component, appears to be distinct for the two. The presence of buffers results in higher mid-transition temperatures, smaller ΔH_m's without any significant variation of the ΔC_p's. In contrast, the presence of $NaClO_4$ results in a higher T_m, a slightly larger ΔC_p, but a nearly unchanged ΔH_m, and the addition of KCl imparts a transition with a smaller ΔH_m and an unchanged T_m and ΔC_p. The contribution of protein-heme interactions to protein stability, ΔG_{20^oC}, remains essentially unchanged upon addition of anions and phosphate buffer, but does increase to double its magnitude upon addition of cacodylate buffer, 0.5 to 1.1 kcal/m (Table II). In terms of the overall stability of the protein, $\Delta G_{20^oC,TOTAL}$, the order remains the same as reflected by the unfolding process, water < NaCl ≈ $NaClO_4$ ≈ cacodylate + $NaClO_4$ < cacodylate < phosphate.

The ΔC_p of unfolding is shown to arise primarily from two sources, large positive contributions from the solvent exposure of the apolar groups and a significant negative contribution from the exposure of the polar groups (Murphy and Gill, 1991). The 2-3 fold increase of ΔC_p in the presence of anions/buffers could be interpreted in a number of ways. It could come from an increase of the degree of hydrophobicity in the folded protein, i.e., the solvent exposure of a larger number of apolar groups on unfolding, or it may simply be a reflection of the exposure of a larger number of apolar groups or the exposure of a smaller number of polar groups. Among all these possibilities, we tend to favor the first, because recent resonance Raman studies have shown that the addition of anions does cause tightening of the heme crevice (Myer and Saturno, 1991), which would result in a higher degree of hydrophobicity. A similar inference regarding the state of the heme crevice also comes from the H-exchange studies reported by Liu et al. (1989). The finding that the radius of gyration of ferricytochrome c decreases in high ionic strength solutions (Trewhella et al., 1988) reflects the greater compactness of the globular form, i.e., the stabilization of forces, both apolar interaction as well as hydrogen bonding.

References

Andersson, T., Thulin, E. and Forsen, S. (1979) *Biochemistry* **18**, 2487-2493.

Arean, C.O., Moore, G.R., Williams, G. and Williams, R. (1988) *Eur. J. Biochem.* **173**, 607-615.

Becktel, W.J. and Schellman, J.A. (1987) *Biopolymers* **26**, 1859-1877.

Brautigan, D.L., Ferguson-Miller, S. and Margoliash, E. (1978) *Meth. Enzymol.* **53**, 128-164.

Feng, Y. and Englander, W. (1990) *Biochemistry* **29**, 3505-3509.

Feng, Y. and Sligar, S.G. (1991) *Biochemistry* **30**, 10150-10155.

Gopal, D., Wilson, G.S., Earl, R.A. and Cusanovich, M.A. (1988) *J. Biol. Chem.* **263**, 11652-11656.

Goto, Y., Takahashi, N., and Fink, A.L. (1990) *Biochemistry* **29**, 3480-3488.

Liu, G., Grygon, C.A. and Spiro, T.G. (1989) *Biochemistry* **28**, 5046-5050.

Margalit, R. and Schejter, A. (1973) *Eur. Bio. Chem.* **32**, 500-505.

Moore, G.R., Eley, C.G.S., and Williams, G. (1984) *In* "Advances in Inorganic and Bioinorganic Mechanisms" (Skyes, A.G., ed.) Academic Press, New York, Vol. 3, pp. 1-96.

Murphy, K.P. and Gill, S.J. (1991) *J. Mol. Biol.* **222**, 699-709.

Myer, Y.P. (1968) *Biochemistry* **7**, 765-776.

Myer, Y.P and Saturno, A.F. (1991) *J. Protein Chem.* **10**, 481-494.

Pace, C.N. and Laurents, D.V. (1989) *Biochemistry* **28**, 2520-2525.

Privalov, P.L. and Khechinashvili, N.N. (1974) *J. Mol. Biol.* **86**, 665-684.

Privalov, P.L. and Potekhin, S.A. (1986) *Meth. Enzymol.* **131**, 4-51.

Schellman, J.A. (1987) *Ann. Rev. Biophys. Biophys. Chem.* **16**, 115-137.

Schellman, J.A. and Hawkes, R.B. (1980) *In* "Protein Folding" (Jaenicke, R., ed.) Elsevier/North-Holland Biomedical Press, pp. 331-341.

Sturtevant, J.M. (1987) *Anu. Rev. Phys. Chem.* **38**, 463-488.

Trewhella, J., Carlson, V.A.P., Curtis, E.H. and Heidorn, D.B. (1988) *Biochemistry* **27**, 1121-1125.

Measurement of Amide Hydrogen D/H Fractionation Factors in Proteins by NMR Spectroscopy

Stewart N. Loh and John L. Markley
Department of Biochemistry, University of Wisconsin-Madison, Madison, WI 53706

I. Introduction

The equilibrium constant K_f for isotope exchange at a site in a molecule, for example, a backbone amide group within a protein that exchanges its proton with a solvent deuteron,

$$NH + D^+ \rightleftharpoons ND + H^+ \qquad (1)$$

is known as the fractionation factor. A K_f value of unity reflects an equal distribution of protons and deuterons between amide groups and solvent: a value greater than unity indicates a preference for D over H, and a value less than unity indicates a preference for H over D on the nitrogen. The position of the equilibrium is thought to depend primarily on the relative vibrational energies of the N-H and N-D bonds (1). Deuterium fractionation factors for exchangeable carboxy, hydroxyl, imidazole and amide hydrogens have been found to be generally close to unity (2,3), however, it must be stressed that these measurements historically have been made with small molecules often in non-aqueous solution. One may expect significantly different fractionation factors from hydrogens buried inside a protein, where they may be subjected to unusual steric constraints, hydrophobic environments, or strong internal hydrogen bonds. Hydrogen bonding, in particular, has been shown

TECHNIQUES IN PROTEIN CHEMISTRY IV

to produce very large isotope effects in certain situations. In the case of single-well and "low-barrier" hydrogen bonds, where a proton is shared between two groups of similar pK_a, the hydrogen bond is unusually strong, and the fractionation factor drops to around 0.5 (4). Examples of this have been recently found for the hydrogen involved in a hydrogen bond between Glu-168 and Glu-211 in enolase and that between Glu-217 and N-1 of adenosine in adenosine deaminase (5).

In addition to hydrogen bonding, other factors may contribute toward raising or lowering of the fractionation factor beyond unity. One can envision interactions involving the carbonyl oxygen that would modulate its electron withdrawing effects on the amide group and alter the N-H bond order. Close van der Waals packing around the amide group might inhibit N-H stretching or out-of-plane bending modes and affect the K_f value as well. Thus, the fractionation factor potentially contains valuable structural information.

Here we report a new method, which employs two-dimensional heteronuclear NMR spectroscopy of ^{15}N labeled proteins, for measuring D/H fractionation factors at individual backbone amides. In developing this technique, we have used as a model system the enzyme staphylococcal nuclease H124L (16.8 kD) labeled uniformly with ^{15}N to 99 atom %. Nuclease H124L (here abbreviated Nase) is the recombinant protein produced from *Escherichia coli* whose sequence is identical to that of the nuclease produced by the V8 strain of *Staphylococcus aureus*. Nase differs from the variant produced by the Foggi strain by having leucine at position 124 in place of histidine. The enzyme, which requires Ca^{++} for activity, catalyzes the degradation of DNA and RNA to mono- and oligonucleotides. Most importantly for this study, nearly complete ^{15}N NMR peak assignments are available for Nase (6,7), and three high resolution crystal structures have been solved for the Foggi strain enzyme (8-10).

II. Materials and Methods

A. *Sample Preparation*

^{15}N-labeled Nase was produced biosynthetically as previously described (6). Several samples of ^{15}N labeled protein were prepared that differed only in the ratio of ^1H_2O to ^2H_2O in the solvent. After equilibrium was reached, the proton occupancy at each amide site was investigated by ^1H-^{15}N two-dimensional NMR. When making measurements of this type, precision in sample preparation is very important. Care must be taken to insure that protein concentrations are uniform, ^1H_2O and ^2H_2O ratios are precise, and full exchange with solvent has taken place. These points are addressed separately below.

1. Uniformity of protein concentrations

To insure that all samples contained the same amounts of protein, identical volumes of a protein solution consisting of 5 mM Nase buffered with 50 mM succinate-d_4 (Aldrich Chemical Co.), pH 5.1 were pipetted into eppendorf tubes, whereupon the samples were lyophilized to dryness. Alternately, if the protein is soluble to a concentration of 15 mM or greater, one may prepare each sample by aliquoting a small volume of the protein dissolved in (^1H_2O or ^2H_2O) water and simply diluting with ^1H_2O and/or ^2H_2O as described below.

2. Solvent composition

These measurements require that the *ratios* of ^1H_2O and ^2H_2O, rather than their absolute concentrations, be accurate. To this end, a single pipettor adjusted to a single volume was used to add identical aliquots of (natural abundance) H_2O and 100% ^2H_2O (Cambridge Isotopes, Inc.) to the dry protein. Seven samples were made ranging from 14% to 100% ^1H_2O content. Since the dry protein contains several hundred exchangeable protons as well as an appreciable number of bound waters per molecule, the samples were exchanged once in their new solvents, lyophilized, and re-dissolved as before. We elected not to adjust the pH of each sample, so as to avoid introducing uncertainty into the isotope ratios; therefore a small deuterium-induced pH variation exists. However, this should not exceed 0.3 pH units between the samples with highest and lowest mole fraction ^1H_2O, and should be considerably less among the intermediate samples.

3. Equilibration of exchange

It must be emphasized that the fractionation factor measures an *equilibrium* isotope effect. To insure that exchange had gone to completion, we dissolved a Nase sample in 100% ^2H_2O under conditions of the experiment and observed the ^1H signal as it decayed to zero as a function of time (see NMR section below). This took approximately one day at 45°C for unliganded Nase and ten days for the Nase-Ca^{++}-pdTp ternary complex. Samples studied by NMR were allowed to equilibrate for times at least as long as these. Note that equilibration must be reached under the conditions of the experiment, i.e., one cannot unfold and refold the protein to hasten exchange. One can, however, test for complete equilibration by using two samples: one that approaches equilibrium from the fully N-D state and the other from the fully N-H state. We describe here results for the Nase-Ca^{++}-3',5' thymidine bisphosphate (pdTp) ternary complex. NMR samples contained 3 mM protein, 9 mM pdTp (Pharmacia LKB Biotechnology, Inc.), 18 mM ultrapure $CaCl_2$ (EM Industries, Inc.) in 50 mM succinate-d_4, pH 5.1.

B. NMR Spectroscopy

^1H-^{15}N spectra were acquired at 45°C on a Bruker AM500 spectrometer by using the HSMQC experiment of Zuiderweg (11). The 512 t_1 increments of 2048 complex t_2 points that constituted each experiment were zero filled to yield final digital resolutions of 2 point/Hz on the ^1H axis and 1 point/Hz on the ^{15}N axis. Solvent supression was by means of a 1.1 s DANTE pulse sequence, and ^{15}N decoupling was achieved by GARP phase modulation during aquisition. Proton occupancy was determined by measuring the peak volumes with the FELIX software package (version 2.05, Hare Research, Inc.). To correct for errors due to variation in spectrometer sensitivity, inconsistent phasing, field inhomogeneity, or other potential spectral artifacts, peak volumes were normalized to that of ^{15}N-acetyl glycine in 50% ^2H_2O, which was present in a capillary insert (in the 5mm NMR tube) that was transferred to each NMR sample prior to analysis. This same insert also provided the deuterium lock signal for samples with low ^2H_2O content.

C. Determination of K_f

From the data described above, one can plot the relative proton occupancy at each NH site along the protein backbone as a function of the solvent composition. Fractionation factors were obtained by nonlinear least-squares analysis of the function

$$y = \frac{x}{K_f(1 - x) + x} \quad (2)$$

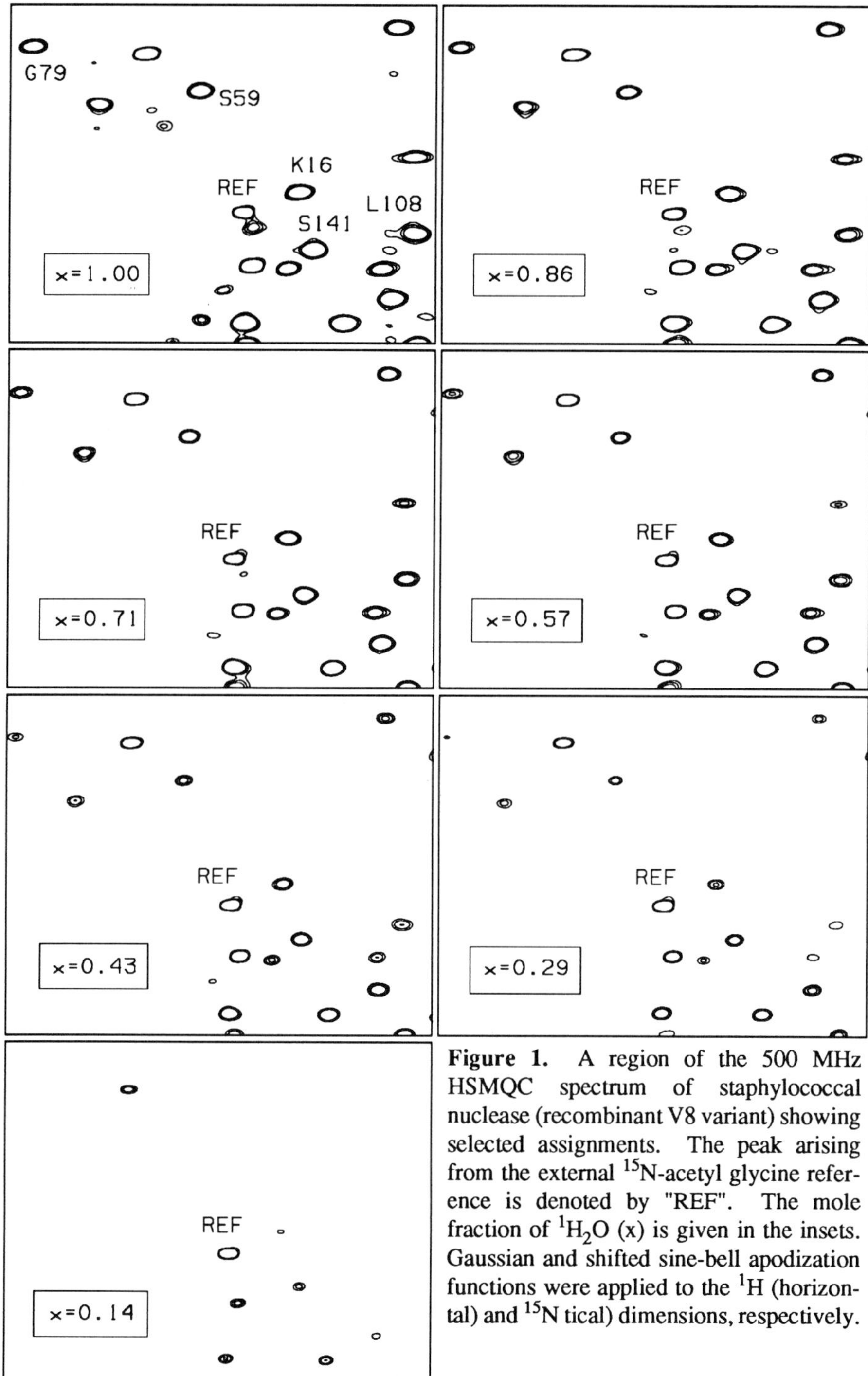

Figure 1. A region of the 500 MHz HSMQC spectrum of staphylococcal nuclease (recombinant V8 variant) showing selected assignments. The peak arising from the external ^{15}N-acetyl glycine reference is denoted by "REF". The mole fraction of $^{1}H_2O$ (x) is given in the insets. Gaussian and shifted sine-bell apodization functions were applied to the ^{1}H (horizontal) and ^{15}N tical) dimensions, respectively.

where x is the mole fraction ^1H_2O and y is the fraction of signal intensity. Alternately, one can extract K_f from a linear least-squares fit to the rearranged form of Eq. 2 shown below.

$$\frac{1}{y} - 1 = K_f\left(\frac{1 - x}{x}\right) \tag{3}$$

III. Results and Discussion

Although fractionation factors can be obtained in principle from a single spectrum of known intermediate solvent composition, for example 50% ^2H_2O, it is more accurate to estimate K_f from a series of samples containing different ^1H_2O / ^2H_2O ratios. ^1H-^{15}N HSMQC spectra of Nase were collected at seven different ^1H_2O mole fractions: 0.14, 0.29, 0.43, 0.57, 0.71, 0.86, and 1.0. Crosspeaks have been assigned to 120 of the 143 non-proline residues in Nase, and 80-90 of these peaks were sufficiently well resolved to permit the determination of K_f values (Fig. 1 shows sample results).

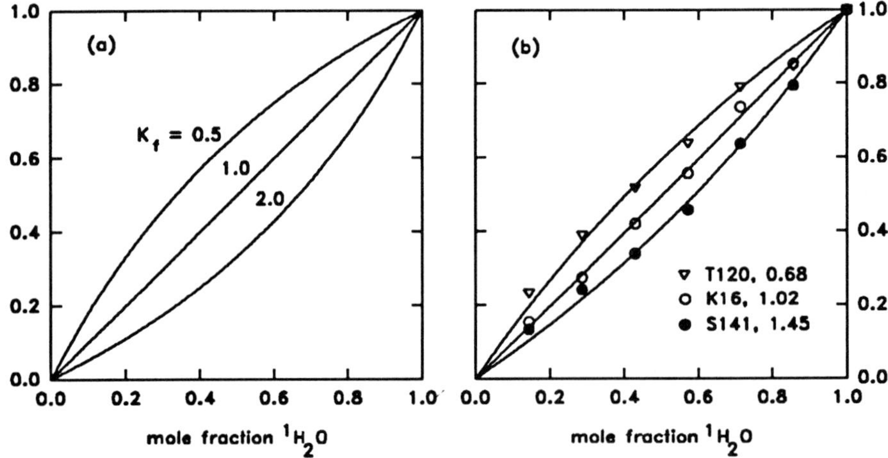

Figure 2. (a) Plot of Eq. 2 showing three theoretical values for K_f. (b) Plot of the relative peak volumes as a function of the mole fraction ^1H_2O in the solvent for some of the peaks in Fig. 1, with assignments and K_f values shown in the legend. Lines represent the best fit of Eq. 2 to the data.

Figure 2a illustrates the ideal behavior of a labile hydrogen in a mixed D/H protic solvent as described by Eq. 2. An upwards or downwards bowing of the straight line of unit slope (K_f = 1.0) indicates a respective enrichment of H or D at that site. As illustrated by Fig. 2b, Nase contains residues that exhibit fractionation factors greater than, less than, and equal to unity. Values below unity were uncommon: the lowest value found was K_f = 0.68 for Thr-120. The majority of K_f values were found to lie within 10% of unity, but many fell in the range 1.2 - 1.5.

We now turn to a discussion of the errors in the measurements. The reproducibility of the results was improved by phasing the spectra independently (rather than by implementing the same phasing parameters for all) and by reducing baseplane roll by correcting the first points of the free induction decays in both dimensions by complex linear prediction. An estimate of the maximum error can be obtained from the volumes of the peak arising from the ^{15}N-acetyl glycine standard placed into each sample prior to data acquisition. The relative standard deviation in the volume of this crosspeak, obtained from the seven spectra shown in Fig. 1, was 4.6%. Since the protein crosspeak volumes in individual spectra were normalized by these values, it is likely that the errors associated with the NH peak volumes are smaller.

Complete fractionation factor data have been collected for both unliganded Nase and the Nase-Ca^{++}-pdTp ternary complex at 45°C, pH 5.1. Preliminary analysis of these results show that the K_f values for many residues change upon ligand binding. It is not yet known to what extent the K_f value depends on solution conditions such as pH, salt concnetration, and temperature. Future studies will investigate these and other possible correlations between fractionation factors and hydrogen bonding, amide hydrogen exchange kinetics, or steric factors.

IV. Conclusions

NMR is the only method suited for obtaining complete site-specific fractionation factor information for biological macromolecules. We have demonstrated a technique for measuring the D/H fractionation factors of backbone NH hydrogens in proteins that makes use of ^{15}N labeling and two-dimenisonal ^{1}H-^{15}N NMR spectroscopy. Our results indicate that, although the majority of residues in staphylococcal nuclease exhibit fractionation factors that are near unity, a substantial number become enriched in deuterium relative to the aqueous solvent, and a few become depleted. These results are important insofar as fractionation factors of unity are assumed in most hydrogen exchange studies of proteins. We are currently attempting to interpret observed differences in K_f on the basis of studies of model compounds.

Acknowledgement

The authors thank Professor W. Wallace Cleland for many helpful suggestions and discussions.

References

1. For reviews, see Fersht, A. (1985). *Enzyme Structure and Mechanism,* W.H. Freeman and Co., Walsh, C. (1979). *Enzymatic Reaction Mechanisms,* W.H. Freeman and Co., and references therein.
2. Jansco, G. and Van Hook, W.A. (1974). *Chemical Reviews* **74,** 689-750.
3. Cleland, W.W. (1980). *Methods Enzymol.* **64,** 105-125.
4. Kreevoy, M.M. and Liang, T.M. (1980). *J. Am. Chem. Soc.* **102,** 3315-3322.
5. Cleland, W.W. (1992). *Biochemistry* **31,** 317-319.
6. Wang, J., Hinck, A.P., Loh, S.N. and Markley, J.L. (1990). *Biochemistry* **29,** 102-113.
7. Wang, J., Hinck, A.P., Loh, S.N., LeMaster, D.L. and Markley, J.L. (1992). *Biochemistry* **31,** 921-936.
8. Cotton, F.A., Hazen, E.E., Jr. and Legg, M.J. (1979). *Proc. Natl. Acad. Sci. USA* **76,** 2551-2555.
9. Loll, P.J. and Lattman, E.E. (1989). *Proteins* **5,** 183-201.
10. Hynes, T.R and Fox, R.O. (1991). *Proteins* **10,** 92-105.
11. Zuiderweg, E.R.P. (1990). *J. Magn. Reson.* **86,** 346-357.

A Model for the Molten Globule State of CTF Generated Using Molecular Dynamics

Valerie Daggett
Department of Cell Biology
Stanford University Medical School
Stanford, California 94305-5400

I. Introduction

A protein must assume a particular well-ordered conformation to be biologically active. Much is known about these conformations, however, little is known of how these native conformations are acquired, or the steps involved in protein folding. Experimentally, it has proved very difficult to elucidate the structural details of the folding process. It is the aim of this study to explore this process using theoretical means, namely to disrupt the structure of a protein and monitor its unfolding and the resulting partially unfolded state. The C-terminal fragment (CTF) of the L7/L12 ribosomal protein was chosen for study because it is small (68 residues) and has a high degree of regular secondary structure (approximately 84% of the protein). CTF is also very well-behaved in molecular dynamics simulations (Åqvist et al, 1985; Åqvist et al, 1989; Åqvist and Tapia, 1990; Daggett and Levitt, 1991).

We have performed a molecular dynamics (MD) simulation at high temperature (498 K) in an attempt to "denature" the molecule. The control simulation at room temperature (298 K) has already been described (Daggett and Levitt, 1991). The main focus here is characterization of the partially unfolded form reached with molecular dynamics. This same approach has been used for the bovine pancreatic trypsin inhibitor (BPTI) (Daggett and Levitt, 1992a), and the rationale for using molecular dynamics for studying unfolding is outlined in this earlier account.

TECHNIQUES IN PROTEIN CHEMISTRY IV

II. Methods

The starting conformation for the molecular dynamics simulation was the X-ray crystal structure of CTF with a sulfate ion (Leijonmarck and Liljas, 1987). Various procedures were performed to prepare the crystal structure for molecular dynamics and they have already been described for the room temperature simulation (Daggett and Levitt, 1991). These same steps were used for the high temperature simulation with the exception that the volume of the water box was adjusted to give the experimentally observed density at 498 K, 0.829 gm/ml (Kell, 1967). Classical molecular dynamics was then performed on the full systems at constant volume, employing periodic boundary conditions, for 200 ps at 298 K and 166.2 ps at 498K (the simulation was discontinued at this point as it became unstable). Structures were saved every 0.2 ps for analysis, yielding 1000 structures at 298 K and 830 at 498 K.

III. Results

A. *Size and Shape*

The radius of gyration of CTF was well-maintained at 298 K and increased 5% at 498 K(Table I, Figure 1). This increase at high temperature translates into a 16% increase in volume.

The solvent accessible surface area increased considerably both at 298 K and 498 K (Table I). Much of this increase at low temperature was due to charged surface residues adopting different conformations in the crystal structure, where crystal contacts dominate (or as a result of refinement) and in solvent where these residues tend to become more extended. At high temperature approximately 50% of the observed increase in the solvent accessible surface area resulted from the exposure of nonpolar groups.

Table I: Size and Shape

	Temperature (K)	
Property [a]	298	498
ΔRadius of Gyration	0	5
ΔProtein Volume	0	16
ΔAccessible Surface Area	9	18

[a]The percentage change from the crystal structure is given. The radius of gyration was calculated as described by Levitt and Sharon (1988). The protein volume was estimated from the radius of gyration ($V = 4/3\pi r^3$). The water accessible surface area was calculated using the method of Lee and Richards (1971).

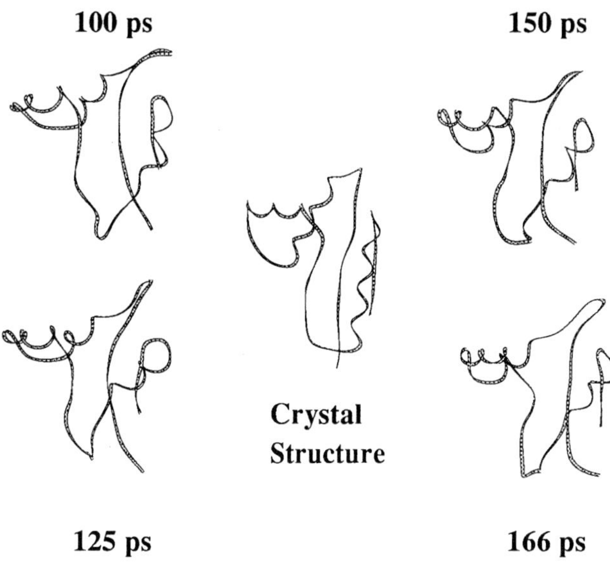

Figure 1: Snapshots from the simulation of CTF at 498 K, showing the heterogeneity of the partially unfolded state.

B. Internal Motion

Overall, the α-carbon backbone remained close to the crystal structure conformation at 298 K (Table II). And, in fact, the average conformation generated during MD only deviated by 0.80 Å. In contrast, the deviation at 498 K was considerably higher, and at some times was over 5 Å (Figure 2).

The fluctuation of the α-carbons about the mean structure at low temperature is in good agreement with experiment (Table II; Daggett and Levitt, 1991). The value was high at 498 K, indicating a large degree of conformational sampling (Figure 1). A two-fold increase in the mainchain and side chain dihedral angles was observed (Table I). Thus, the side chains were rotating rapidly in the partially unfolded conformations, however they were not freely rotating (~90° would be expected if that were the case).

Table II: Internal Motion

	Temperature (K)	
Average Property	298	498
C_α rmsd from X-ray Structure (Å) after 50 ps	1.1	3.8
C_α Fluc about Mean Structure (Å) last 50 ps	0.7	2.1
Fluctuation in ϕ (°)	14	28
Fluctuation in ψ (°)	13	29
Fluctuation in $\chi1$ (°)	22	44
Fluctuation in $\chi2$ (°)	27	45

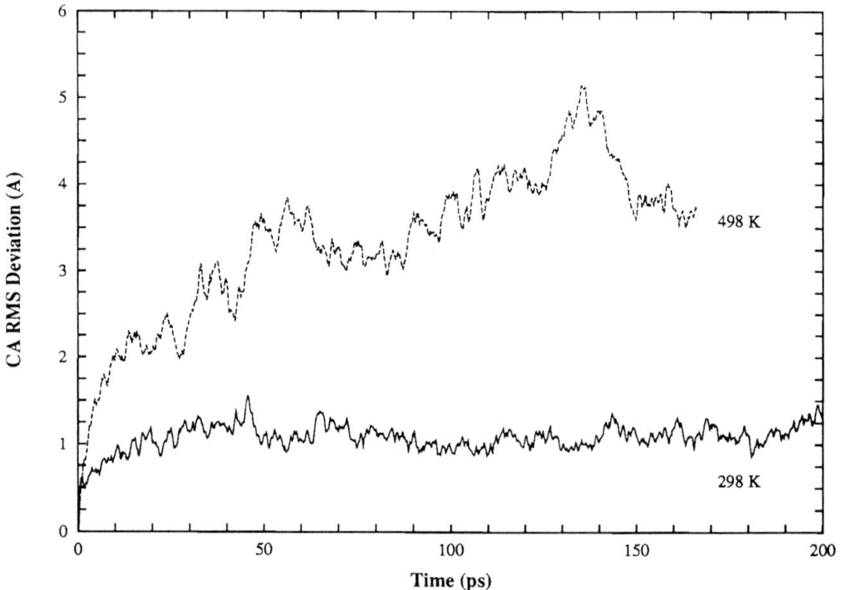

Figure 2: The α-carbon root-mean-square deviation from the crystal structure as a function of simulation time.

C. Secondary Structure

The amount of α-helical and β-structure in the crystal structure was well-reproduced in the 298 K simulation (Table III). At 498 K, however, the secondary structure contents dropped but show that the majority of the secondary structure was maintained .

Table III: Secondary Structure and Intramolecular Hydrogen Bonds

		Temperature (K)	
Average Property[a]	Crystal Structure	298	498
% α Structure	55	56	44
% β Structure	32	30	19

[a]Secondary structure assignments are based on dihedral angles as described by Daggett and Levitt (1992a, b).

D. Internal Packing

The percentage of the total number of tertiary contacts intact during MD was high at 298 K (Table IV). The majority of the crystal structure contacts were reproduced (89%). The drop in native contacts exaggerates the differences between the crystal structure and MD generated conformations, as only 4% of the total number of contacts were nonnative and all of these represent conservative interactions that preserve the packing observed in the crystal structure. At 498 K, however, the total number of contacts dropped greatly and many were nonnative (Table IV). Only 66% of the native contacts were maintained on average during MD.

Table IV: Internal Packing

	Temperature (K)	
Property[a]	298	498
% Total Tertiary Contacts Intact	92	77
% Native Tertiary Contacts Intact	89	66
% Total 3° Contacts that are Nonnative	4	14
No. of Ring Flips	0	5
ΔAccessible Surface Area of Phe 54 (Å^2)	25	52

[a]Intact tertiary interactions are defined as non-neighboring residue-residue contacts ≤ 4.5 Å. Ring flips are defined as 180°rotations about $\chi 2$. The change in water accessible surface areas given are relative to the crystal structure.

Due to the increase in the volume and internal motion at high temperature, the side chain of the lone aromatic group in the protein, Phe 54, experienced ring flips, while it did not flip at 298 K (Table IV). The solvent accessible surface area of Phe 54 also increased. Thus, the aromatic ring was much more exposed to solvent at high temperature.

IV. Discussion

The structural features of a partially unfolded form of CTF have been described. Overall, the conformations generated at high temperature have all of the properties of the molten globule state. Table V lists the generally agreed upon features of the molten globule state observed for a variety of proteins using many different experimental techniques. The corresponding properties from the high temperature simulation are given for comparison.

Table V: General Properties of the Molten Globule State and Comparison Between Simulation and Experiment

General Properties of the Molten Globule State	Experimental Technique or Observation[a]	Corresponding Properties from the Simulation
Dense, compact state but expanded compared to N (~10-15% of N)	Ultracentrifugation Electrophoresis Gel Chromatography X-ray Scattering Fluorescence	Radius of Gyration ↑ 5% Accessible Surface Area ↑ 9% Volume ↑ 16%
Increased motion of mainchain and side chains	H/D Exchange NMR	Cα rmsd ↑ HB ↓ $\delta\phi$, $\delta\psi$, $\delta\chi 1$, $\delta\chi 2$ ↑ Cα fluctuation ↑
Native-like Secondary Structure (~60-200 % of N)	CD ORD IR	79% of Native α Structure Maintained 63% of Native β
Lose unique and rigid packing of side chains	NMR Fluorescence Difference UV	Ring flips ↑ Protein Contacts ↓ 23 %
Increased exposure of hydrophobic residues	Binding of hydrophobic probes↑ Interactions with phospholipid vesicles ↑ Calorimetry---ΔC_p ↑ Solubility in H_2O ↓	~50% of ΔA ↑ due to exposure of hydrophobic residues Phe exposure ↑

[a]The relevant experimental studies have been reviewed Ptitsyn (1987) and Kuwajima (1989).

Experimental studies have shown that molten globules are compact but expanded compared to the native state. The 498 K simulation showed increases in size of the proper magnitude, such that the conformation remained compact. Molten globules, as the name suggests, are highly mobile. This increase in mobility compared to the native state is true for both the mainchain and side chain groups. An increase in the extent of conformational sampling of the mainchain and side chain dihedral angles was observed at 498 K. However, a portion of this increase is almost certainly due to just the temperature increase, as was observed for BPTI (Daggett and Levitt, 1992a).

A heterogeneous ensemble of conformations was generated at high temperature that was distinct from the crystal structure. Overall, the partially unfolded form of CTF remained fairly close, conformationally, to the crystal structure. In contrast, the modelled molten globule state of BPTI deviated more from the crystal structure (Daggett and Levitt, 1992a). It has been suggested that the molten globule state is near the native conformation and since the transition from the molten globule to unfolded state is gradual, there may be a continuum of conformations, displaying the properties of a molten globule on average, which would most likely depend on the environmental conditions.

The secondary structure contents of molten globules are native-like. Roughly native α-helical and β-structure contents were observed in the high temperature simulation. The structure retained in the partially unfolded form was mostly native-like in position along the sequence. The unfolding was localized to the turn and loop regions of the protein along with some fraying of the native secondary structure elements. This is consistent with experimental studies that have detected portions of native-like structure in partially unfolded intermediates while other regions of these intermediates appear to be unstructured (Baum et al., 1989; Dobson, 1992; Harding et al., 1991; Hughson et al, 1990).

Experimentally, the internal side chain packing is disrupted in the molten globule state. As a result of the increase in volume in the high temperature simulation, the residue-residue contacts decreased, which were short-lived compared to the 298 K simulation. Many nonnative interactions were made at 498 K, which also tended to be transient, in accord with having a mobile structure. The lone aromatic group in CTF became more exposed to solvent at high temperature. In fact, approximately 50% of the increase in accessible surface area was due to the exposure of nonpolar residues, which is consistent with the observation that molten globules are more hydrophobic than native proteins.

The high temperature simulation reproduces all of the known features of the molten globule state. Unfortunately, the molten globule state of CTF has not yet been characterized experimentally and therefore only qualitative comparisons to experiment can be made. In any case, models like the one presented here can be very helpful for visualizing and interpreting data for partially unfolded states and folding intermediates, as the nature of these conformations makes detailed experimental structure delineation difficult.

So, what is the importance of the molten globule state? It has been suggested that this state is a common early intermediate during protein folding (Ptitsyn et al., 1990). Due to its increased hydrophobicity, it has also been suggested that the molten globule state is important during protein translocation (Bychkova et al., 1988). Given the special features of this state, one could imagine that it would be advantageous to be in a molten globule conformation prior to proper localization of the protein, since this conformation probably confers some amount of protection to proteolytic processing. Once within the proper compartment, further folding could be triggered by the environment, for example, by the ion concentration, as is the case for α-lactalbumin (Ptitsyn, 1987), or by dissociation of a chaperone protein. In fact, the molten globule state may be the relevant unfolded state in vivo, and as such could accelerate folding by allowing folding to occur within a condensed volume.

Acknowledgements

The work was performed in the laboratory of Michael Levitt, whose support is much appreciated. This work was funded by the NIH (GM-41455 to M.L.) and The Jane Coffin Childs Memorial Fund for Medical Research. I am also thankful for the help of David Hinds.

References

Åqvist, J., van Gunsteren, W.F., Leijonmarck, M., & Tapia, O. (1985). J. Mol. Biol. **183**: 461.
Åqvist, J., Leijonmarck, M. & Tapia, O. (1989). Eur. Biophys. J. **16**: 327.
Åqvist, J. & Tapia, O. (1990). Biopolymers **30**: 205.
Baum, J., Dobson, C.M., Evans, P.A. & Hanley, C. (1989). Biochem. **28**: 7.
Bychkova, V.E., Pain, R.H. & Ptitsyn, O.B. (1988). FEBS Lett. **238**: 231.
Daggett, V. & Levitt, M. (1991). Chem. Phys. **158**: 501.
Daggett, V. & Levitt, M. (1992a). Proc. Natl. Acad. Sci. USA **89**: 5142.
Daggett, V. & Levitt, M. (1992b). J. Mol. Biol. **223**: 1121.
Dobson, C. (1992). Curr. Opin. in Struct. Biol. **2**: 6.
Harding, M.M., Williams, D.H. & Woolfson,D.N. (1991). Biochem. **30**: 3120.
Hughson, F. , Wright, P. & Baldwin, R.L. (1990). Science **249**: 1544.
Kell, G.S. (1967). J. Chem. Eng. Data **12**: 66.
Kuwajima, K. (1989). Proteins **6**: 87.
Lee, B. & Richards, R.M. (1971). J. Mol. Biol. **55**: 379.
Leijonmarck, M. & Liljas, A. (1987). J. Mol. Biol. **195**: 555.
Levitt, M. & Sharon, R. (1988). Proc. Natl. Acad. Sci. USA **85**: 7557.
Ptitsyn, O.B. (1987). J. Protein Chem. **6**: 273.
Ptitsyn, O.B., Pain, R.H., Semisotnov, G.V., Zerovnik, E. & Razgulyaev, O.I. (1990). FEBS Lett. **262**: 20.

Urea and Guanidine-HCl Yield Different Unfolding Free Energies for CheY: Which Denaturant Provides the Most Reliable Free Energy Values?

Gregory T. DeKoster
Andrew D. Robertson
Dept. of Biochemistry, University of Iowa, Iowa City, IA 52242

Ann M. Stock
Center for Advanced Biotechnology and Medicine, Piscataway, NJ 08854

Jeffry B. Stock
Dept. of Molecular Biology, Princeton University, Princeton, NJ 08544

I. Introduction

Much of the current effort in protein chemistry is aimed at establishing the link between the conformation of proteins and the energetics of protein folding. The free energy of unfolding, ΔG_u, is an often used measure of the thermal stability of proteins and is the basis for quantitative evaluation of perturbations to protein structure, such as mutations or changes in solvent conditions (1). Such perturbations are being used to derive correlations between conformation and energetics, to the extent that we are beginning to describe the atomic-level origins of the macroscopic thermodynamic parameter, ΔG_u (1-3).

A number of methods are available for measuring ΔG_u, the most direct of which is differential scanning calorimetry (DSC)(4). DSC is used to detect the excess heat capacity accompanying thermal denaturation of proteins which, when integrated over temperature, permits determination of the enthalpy of unfolding, ΔH_u. In addition, proteins show a characteristic difference in heat capacity, ΔC_p, between the native and denatured states which can be measured either 1) directly in a single calorimetric scan if the data are of sufficient quality or 2) by monitoring ΔH_u as a function of temperature, where the denaturational temperature is altered by changing the pH of the sample. In conjunction with ΔH_u and the temperature at the midpoint of the thermal transition, T_u, ΔC_p can be used to calculate ΔG_u at any other temperature, T, using a modified Gibbs-Helmholtz equation (4,5):

$$\Delta G_u = \Delta H_u(1 - T/T_u) - \Delta C_p[(T_u - T) + T*\ln(T/T_u)] \quad (1)$$

The limited availability of high-sensitivity calorimeters has led to the development of alternative and indirect methods for characterizing the thermodynamics of unfolding (6). One approach is to determine ΔH_u by a van't Hoff analysis of data obtained from thermal denaturation studies in which a

TECHNIQUES IN PROTEIN CHEMISTRY IV

spectroscopic probe is used to follow the progress of unfolding. ΔC_p can, in principle, be determined by varying T_u, as described above for the calorimetric experiments. Unlike the calorimetric approach, however, the spectroscopic methods, and indeed most indirect methods, typically rely on a two-state model for unfolding.

A more recent development is the use of chemical denaturants to determine ΔG_u (6). In these experiments $\Delta G_u(H_2O)$, ΔG_u in the absence of denaturant, is obtained by linear extrapolation of unfolding free energies measured in the unfolding transition, i.e. in the presence of denaturant, back to zero denaturant concentration. This simple and popular approach is summarized in Equation 2:

$$\Delta G_u(H_2O) = \Delta G_u + m[D] \quad (2)$$

where m is the dependence of ΔG_u on the concentration of denaturant, [D].

There are both experimental and theoretical arguments for the validity of the linear extrapolation method (LEM) (7-9). In the few instances where the results of LEM and calorimetric studies have been compared using matched solution conditions, the two methods give essentially identical results (8,9). The LEM, however, is still based largely on empirical observation and relies on assumptions that must be tested with each protein to be studied.

One such assumption is that the chemical denaturants affect only the intrinsic stability of the protein. It follows that if no additional effects are operant, then a variety of denaturants should yield the same extrapolated value for $\Delta G_u(H_2O)$. In fact, it has been demonstrated that GuHCl can act to both stabilize oxidized thioredoxin as a salt and destabilize it as a denaturant (9). Binding of GuHCl appears to stabilize β-lactoglobulin (10). And in the case of rat intestinal fatty acid binding protein, urea-induced denaturation provided a nearly two-fold higher $\Delta G_u(H_2O)$ than that observed with GuHCl (11).

We have obtained very similar results in our studies of CheY, a small (M_r 14,000) globular protein involved in bacterial chemotaxis (12,13). We believe, however, that the ambiguous results with CheY can be resolved in favor of those obtained using GuHCl through examination of 1) the effects of ligand binding and ionic strength and 2) the results from thermal denaturation experiments. This approach may prove to be generally useful in those instances where different chemical denaturants yield different values of $\Delta G_u(H_2O)$ for the same protein.

II. Materials and Methods

CheY from *Salmonella typhimurium* was purified as described previously (14). Protein was stored at -20°C in 50 mM sodium phosphate buffer at pH 7. Protein concentration was determined by measuring absorbance at 280 nm and using an extinction coefficient for the native protein of 6420 $M^{-1} \cdot cm^{-1}$, which was determined by the method of Edelhoch (15).

Spectral quality guanidine hydrochloride (GuHCl) was purchased from Heico Chemicals, Inc. Ultra-pure urea was purchased from Boehringer Mannheim. The concentration of GuHCl and urea was determined by refractive index measurements (16).

Denaturant-induced unfolding was followed using an AVIV 62DS circular dichroism (CD) spectropolarimeter equipped with a thermoelectric temperature controller equilibrated at 25 ± 0.1°C. Unfolding was achieved by addition of concentrated denaturant to the protein solution in a semi-micro cuvette (4 mm width x 10 mm path-length) equipped with a stir bar. During the experiment, the CheY concentration was diluted from 0.2 mg/ml to 0.03 mg/ml in 50 mM sodium phosphate, pH 7 (plus any added salts). The CD signals at 222 nm and 400 nm were recorded for 2.5 minutes; the mean value at 400 nm was then subtracted from that at 222 nm. An additional correction was made by subtracting the signal of a buffer solution lacking protein. All CD data are reported as mean residue ellipticity, [Θ] ($deg \cdot cm^{-2} \cdot dmol^{-1}$).

Linear extrapolations were determined by a least squares fit of the CD data to the following equation, which describes the transition as well as pre- and post-transitional baselines (17):

$$y_{obs} = \frac{[(y_n + m_n[D]) + (y_u + m_u[D])(\exp\{-\Delta G_u(H_2O)/RT + m[D]/RT\})]}{[1 + \exp\{-\Delta G_u(H_2O)/RT + m[D]/RT\}]} \quad (3)$$

where y_{obs} is the observed ellipticity, [D] is the concentration of denaturant, T is the absolute temperature, R is the gas constant [1.987 cal/(mol•deg)], m_n and m_u are the slopes of the native and denatured baselines, respectively, y_n and y_u are the intercepts at zero denaturant concentration for the native and denatured baselines, respectively, m is the slope of the line describing the dependence of ΔG_u on [D], and $\Delta G_u(H_2O)$ is the free energy of unfolding extrapolated to zero denaturant concentration. Nonlinear least squares fitting of the data was performed as described previously (17). The reported errors are a single standard deviation.

The enthalpy of unfolding (ΔH_u) and the midpoint of the thermal transition (T_u) were determined by van't Hoff analysis of the thermally-induced unfolding of CheY as followed by UV-CD. Protein (32 μg/ml) in 50 mM sodium phosphate, pH 7, was stirred continuously in a 10 mm path-length cuvette. Temperature was monitored with a probe inserted in the solution and positioned just above the beam; the cuvette was sealed to prevent evaporation. In an effort to minimize the time during which protein was exposed to high temperatures, the temperature was increased continuously from 0 to 80°C at rates varying from 5.0°C/min to 0.25°C/min; ΔH_u and T_u reached constant values at scan rates below 0.5°C/min. Thermal denaturation under these conditions was approximately 90% reversible. Thermodynamic parameters were determined by a least squares fit of the data to the following equation:

$$y_{obs} = \frac{[(y_n + m_nT)+(y_u + m_uT)(\exp\{-\Delta H_u/RT + \Delta S_u/R\})]}{[1 + \exp\{-\Delta H_u/RT + \Delta S_u/R\}]} \quad (4)$$

where the baseline parameters are as described for Eqn. 3. ΔH_u and ΔS_u are the enthalpy and entropy of unfolding, respectively.

III. Results

CD spectra for native and chemically denatured CheY are shown in Figure 1. The spectrum of native protein shows minima at 222 and 208-210 nm, characteristic of the α-helical peptide chromophore. CheY is an α/β protein consisting of approximately equal numbers of residues in α-helical and β-strand conformations (18,19), but the mean residue ellipticity observed for the helical conformation is much stronger than that seen for β sheets (20). The spectra of CheY in concentrated GuHCl and urea are similar, although urea has a slightly more negative signal below 235 nm. We chose to follow the progress of chemical denaturation at 222 nm, as this is the wavelength at which the maximum difference between native and chemically denatured CheY is observed.

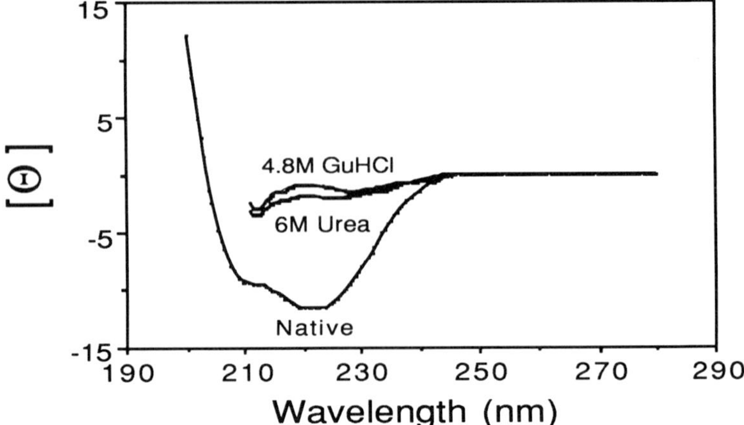

Figure 1. CD spectra of native and chemically denatured CheY. The protein concentration is 15 μm in 50 mM sodium phosphate, pH 7, at 25°C. The samples were in a 2 mm pathlength cuvette. The units for [Θ] are [deg/(cm^2•dmol^1) x 10^{-3}].

A plot of $[\Theta]_{222}$ versus denaturant concentration is shown in Figure 2, along with curves representing least squares fits to Eqn. 3. The parameters derived from the fitting procedure are presented in Table I. The observed agreement between the data and fitted curves is representative of all of the experiments discussed below. The most striking result from these experiments is that the values of $\Delta G_u(H_2O)$ obtained from GuHCl- and urea-induced denaturation of CheY are approximately 3 kcal/mol and 7 kcal/mol, respectively (Table I).

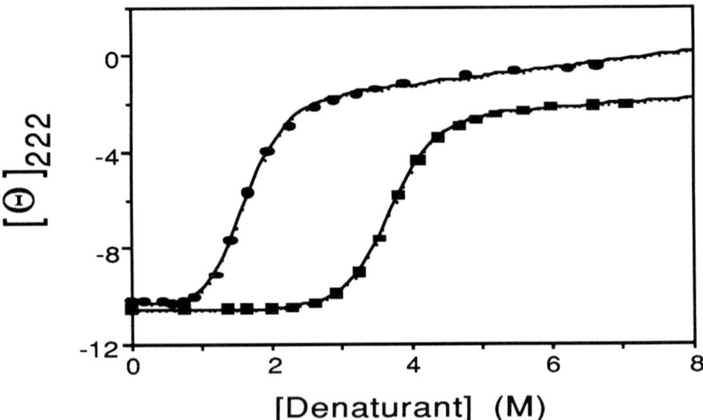

Figure 2. Chemical denaturation of CheY by GuHCl (circles) and urea (squares) in 1 mM EDTA, 50 mM sodium phosphate, pH 7, at 25°C. The thin line represents the least squares fit of the data to Equation 3. The units for $[\Theta]_{222}$ are [deg/(cm^2•dmol^1) x 10^{-3}].

Table I. Chemical Denaturation of CheY at pH 7.0, 25°C[a]

Additions	m (cal/mol/M)	$[D]_{1/2}$ (M)[b]	$\Delta G(H_2O)$ (kcal/mol)
GuHCl			
	1990 ±470	1.56 ±0.37	3.1 ±0.9
1mM EDTA	1950 ±190	1.50 ±0.17	2.9 ±0.4
1mM EDTA	1940 ±200	1.55 ±0.17	3.0 ±0.4
1M NaCl	1560 ±350	2.10 ±0.23	3.3 ±0.2
1M NaCl	1550 ±190	2.19 ±0.17	3.4 ±0.4
10mM $MgCl_2$	2590 ±320[c]	1.70 ±0.19[c]	4.4 ±0.6[c]
Urea			
	1790 ±480	3.73 ±0.38	6.7 ±1.8
	1930 ±680	3.72 ±0.49	7.2 ±2.5
1mM EDTA	1840 ±100	3.66 ±0.08	6.7 ±0.4
1mM EDTA	1810 ±60	3.71 ±0.04	6.7 ±0.2
1M NaCl	1430 ±150	6.05 ±0.15	8.7 ±0.9
1M NaCl	1440 ±180	5.80 ±0.17	8.3 ±1.0
10mM $MgCl_2$	1630 ±140	4.34 ±0.12	7.1 ±0.6
10mM $MgCl_2$	1720 ±180	4.22 ±0.14	7.3 ±0.7

[a] All samples contain 50 mM sodium phosphate.

[b] $[D]_{1/2} = \Delta G(H_2O)/m$

[c] These values are the mean and standard deviation for three independent measurements.

CheY contains a binding site for divalent cations (21), hence contaminating metals in the urea might explain an elevated value of $\Delta G_u(H_2O)$ in this denaturant (see below). This appears to be unlikely, as the addition of 1 mM EDTA leads to little or no change in $\Delta G_u(H_2O)$. Moreover, repeated recrystallization of urea resulted in no change in $\Delta G_u(H_2O)$ (data not shown). The results of additional experiments, described below, are consistent with the conclusion that contaminating metals cannot explain the differences in extrapolated free energies.

Another possible explanation resides in a fundamental physical difference between GuHCl and urea: the former is a salt and the latter is not. To be consistent with our results, high ionic strength (i.e. high GuHCl) would have to destabilize CheY. It was found that, in fact, adding 1 M NaCl to the experiments in which urea is the denaturant resulted in an apparent *increase* in the stability of CheY (Table I). Ionic strength effects thus do not explain the discrepancies in $\Delta G_u(H_2O)$.

In principle, one predictable property of CheY is an increase in stability in the presence of divalent cation (22). CheY has a single binding site for divalent cation, so the predicted increase in $\Delta G_u(H_2O)$ in the presence of ligand (L), $\Delta\Delta G_u(H_2O + L)$, can be calculated using Eqn. 5.

$$\Delta\Delta G_u(H_2O + L) = \Delta G_u(H_2O + L) - \Delta G_u(H_2O) = RT \ln(1 + [L]/K_D) \qquad (5)$$

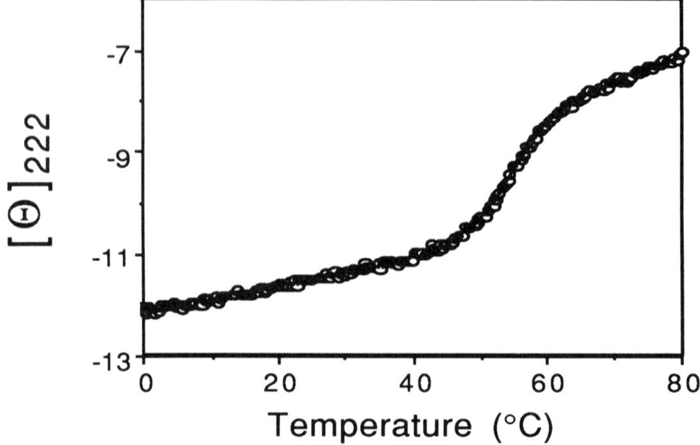

Figure 3. Thermal denaturation of CheY in 50 mM sodium phosphate, pH 7. The thin continuous line represents the least squares fit of the data to Equation 4. The units for $[\Theta]_{222}$ are $[\text{deg}/(\text{cm}^2 \bullet \text{dmol}^1)] \times 10^{-3}$.

Dissociation constants for a variety of metals have been determined under conditions similar to those employed in the present studies (21). For example, the K_D for magnesium at pH 7 and 20°C is 0.45 mM, so addition of 10 mM $MgCl_2$ to dilute CheY should result in an increase in $\Delta G_u(H_2O)$ of about 1.8 kcal/mol under these conditions. At pH 7 and 25°C, an increase in $\Delta G_u(H_2O)$ of about 1.4 kcal/mol is observed in GuHCl while no significant increase is seen when urea is the denaturant (Table I). There is clearly an increase in the $[D]_{1/2}$ for urea in the presence of magnesium, but m has undergone a compensatory decrease. These results suggest that denaturation by urea is not providing reliable values of $\Delta G_u(H_2O)$.

Thermal denaturation experiments were performed to provide an additional test of the validity of the $\Delta G_u(H_2O)$ values obtained with the two denaturants (Figure 3). A van't Hoff analysis of these data (Equation 4) yielded a ΔH_u of 59 kcal/mol and a T_u of 54°C. These parameters and $\Delta G_u(H_2O)$ can be used in Eqn. 1 to estimate ΔC_p for CheY: ΔC_p estimates from the GuHCl and urea experiments are 1.7 kcal/(mol•deg) and -1.3 kcal/(mol•deg), respectively. The latter value is clearly unrealistic and provides an additional argument against the use of urea in characterizing the thermodynamics of CheY unfolding.

Preliminary results of fluorescence studies, in which additional denaturants are being investigated, are consistent with the results presented above. Urea-induced unfolding yields a $\Delta G(H_2O)$ value (7.4 kcal/mol) similar to that observed in the analogous CD experiments, while the value for $\Delta G(H_2O)$ obtained with 1,3-dimethylurea as the denaturant, about 4 kcal/mol, is closer to that seen with GuHCl-induced unfolding as monitored by CD. Additional experiments are underway to determine if the difference in $\Delta G(H_2O)$ obtained with 1,3-dimethylurea and GuHCl is significant.

IV. Discussion

There is a significant difference in the value of $\Delta G_u(H_2O)$ for CheY when GuHCl and urea are used as denaturants: 3.0 kcal/mol and 6.7 kcal/mol, respectively

(Table I). Plausible explanations for this discrepancy include 1) destabilization of CheY by high ionic strength, 2) contamination of urea by divalent cation which, as a result of binding, is stabilizing CheY, and 3) preferential binding of urea by native CheY.

The first hypothesis was tested by adding 1 M NaCl to the experiments in which urea was used as the denaturant. The resulting $\Delta G_u(H_2O)$ value was, in fact, higher than that observed in the absence of salt, so high ionic strength appears to stabilize CheY. The presence of EDTA does not significantly alter the values of $\Delta G_u(H_2O)$, which suggests that the second hypothesis is likewise untrue. These results do not allow us to rule out completely the possibility of some contaminating ligand in the urea, but this ligand would have to either bind with great avidity or be present at very high concentrations to explain the approximately 4 kcal/mol difference in the extrapolated free energies.

Two additional observations lead us to conclude that denaturation by urea is not providing reliable estimates of $\Delta G_u(H_2O)$. First, $\Delta G_u(H_2O)$ does not show the expected increase in the presence of $MgCl_2$ when urea is the denaturant, whereas such an increase is observed with GuHCl as the denaturant. And second, when $\Delta G_u(H_2O)$, thermal denaturation data, and Equation 1 are used to estimate ΔC_p, the result is an unrealistically negative value when using values of $\Delta G_u(H_2O)$ obtained in the urea experiments. We feel that the simplest, and as yet unproven, explanation for the results with urea is the preferential binding of urea by native CheY.

The question remains as to whether or not the data obtained with GuHCl are providing better estimates of ΔG_u. The results with added $MgCl_2$ and preliminary fluorescence studies suggest that GuHCl is providing values of $\Delta G_u(H_2O)$ that are reasonable estimates of ΔG_u. The addition of 1 M NaCl to the urea denaturation experiments leads to an increase in $\Delta G_u(H_2O)$, so the extrapolated free energies obtained using GuHCl may be too high. We currently are attempting to control for possible ionic strength effects (9).

A van't Hoff analysis of data obtained from the thermal denaturation of CheY provides estimates for ΔH_u and T_u which, along with $\Delta G_u(H_2O)$, have been used to calculate a ΔC_p of about 1.7 kcal/(mol•deg). This value is very close to those obtained by a variety of purely calculational approaches (23,24). There are, moreover, a number of proteins similar in size to CheY for which ΔC_p values, ranging from 1 to 2 kcal/(mol•deg), have been determined (25). If we use these ΔC_p values and the estimates for ΔH_u and T_u for CheY determined in this study, then estimates for ΔG_u at 25°C obtained using Eqn. 1 range from 2.6 kcal/mol to 3.9 kcal/mol. Thermally- and GuHCl-induced unfolding of CheY are thus providing results that are consistent with the expected properties of a globular protein the size of CheY.

More generally, it appears that thermal denaturation experiments can provide valuable information in cases where the LEM provides ambiguous results. The utility of this approach will of course depend on the magnitude of the observed difference in extrapolated free energies and the errors in ΔH_u and T_u.

Acknowledgements

This work was supported in part by grants from the University of Iowa Cancer Center (A.D.R.) and from the Public Health Service (AI 20980 to J.B.S.). A.M.S. is a Lucille P. Markey Scholar and this work was supported in part by a grant from the Lucille P. Markey Trust. We thank Drs. D. Wayne Bolen (Southern Illinois University) and C. N. Pace (Texas A & M University) for helpful discussions and valuable advice.

References

1. Dill, K. A. (1990) *Biochemistry* **29**, 7133-7155.
2. Alber, T. (1989) *Annu. Rev. Biochem.* **58**, 765-798.
3. Serrano, L., Kellis, J. T., Jr., Cann, P., Matouschek, A., and Fersht, A. R. (1992) *J. Mol. Biol.* **224**, 783-804.
4. Privalov, P. L. (1992) in "Protein Folding" (T. E. Creighton, ed.) 83-126, W H. Freeman and Company, New York.
5. Becktel, W., and Schellman, J. A. (1987) *Biopolymers* **26**, 1859-1877.
6. Pace, C. N., and Laurents, D. V. (1989) *Biochemistry* **28**, 2520-2525.
7. Schellman, J. A. (1987) *Biopolymers* **26**, 549-559.
8. Hu, C.-Q., Sturtevant, J. M. , Thomson, J. A., Erickson, R. E., and Pace, C. N. (1992) *Biochemistry* **31**, 4876-4882.
9. Santoro, M. M., and Bolen, D. W. (1992) *Biochemistry* **31**, 4901-4907.
10. Pace, C. N., and Marshall, H. F. (1980) *Arch. Biochem. Biophys.* **199**, 270-276.
11. Ropson, I. J., Gordon, J. I., and Friedan, C. (1990) *Biochemistry* **29**, 9591-9599.
12. Stock, J. B., Lukat, G. S., and Stock, A. M. (1991) *Annu. Rev. Biophys. Biophys. Chem.* **20**, 109-136.
13. Bourret, R. B., Borkovich, K. A., and Simon, M. I. (1991) *Annu. Rev. Biochem.* **60**, 401-441.
14. Stock, A. M., Koshland, D. E., Jr., and Stock, J. B. (1985) *Proc. Natl. Acad. Sci. U.S.A.* **82**, 7989-7993.
15. Edelhoch, H. (1967) *Biochemistry* **6**, 1948-1954.
16. Pace, C. N. (1986) *Meth. Enzymol.* **131**, 266-280.
17. Santoro, M. M., and Bolen, D. W. (1988) *Biochemistry* **27**, 8063-8068.
18. Stock, A. M., Mottonen, J. M., Stock, J. B., and Schutt, C. E. (1989) *Nature* **337**, 745-749.
19. Volz, K., and Matsumura, P. (1991) *J. Biol. Chem.* **266**, 15511-15519.
20. Woody, R. W. (1985) *The Peptides* **7**, 15-114.
21. Lukat, G. S., Stock, A. M., and Stock, J. B. (1990) *Biochemistry* **29**, 5436-5442.
22. Pace, C. N. (1990) *TIBTECH* **8**, 93-98.
23. Spolar, R. S., Livingstone J. R., and Record, M. T., Jr. (1992) *Biochemistry* **31**, 3947-3955.
24. Murphy, K. P., and Gill, S. J. (1991) *J. Mol. Biol.* **222**, 699-709.
25. Pfeil, W. (1986) in "Thermodynamic Data for Biochemistry and Biotechnology" (H. -J. Hinz, ed.) 349-376, Springer-Verlag, New York.

Learning an Objective Alphabet of Amino Acid Conformations in Protein[1]

Robert T. Miller, Richard J. Douthart, and A. Keith Dunker(1)

Department of Biochemistry and Biophysics, Washington State University, Pullman, WA 99164-4660; Life Sciences Center, Pacific Northwest Laboratory, Richland, WA 99352, and (1) the Center for Visualization, Analysis, and Design in the Molecular Sciences, Washington State University, Pullman, WA 99164-1220.

ϕ, ψ, and ω backbone dihedral angles in 97 proteins of known structure have been analyzed using a simple cluster-seeking algorithm. With no *a priori* hypothesis as to the number of distinct conformations, the approach generates 125 distinct cluster centers from 19,121 residues. The twenty most populated centers capture 90% of the input set, while the least populated sixty centers capture only 0.2% and probably represent errors in the data or very rare conformations. To validate the approach, the total data set is divided into the protein structural classes all-α, all-β, α/β, and other, amongst which cluster centers are hierarchically established. For each class, centers are identified from the twenty amino acid subsets which partition it. These twenty sets are themselves clustered to generate centers for each of the four protein structural classes, which are then clustered to produce a set of 'generic' conformations. Based on the clustering algorithm and early results, two classes of initial conditions are identified; the entire process is repeated for each and the results combined heuristically to generate the final set of conformational cluster centers. All commonly recognized conformational classes are represented by at least one of these centers, and the beta sheet (extended) region of the Ramachandran plot is separated into several distinct sub-classes. The identified conformational cluster centers may be directly mapped to individual amino acids in proteins, and hence are amenable to the many sequence-analysis tools already in use.

[1] By acceptance of this article, the publisher and/or recipient acknowledges the U.S. Government's right to retain a nonexclusive, royalty-free license in and to any copyright covering this paper. This research was supported by the Northwest College and University Association for Science (Washington State University) under grant DE-FG06-89ER-75522 with the U.S. Department of Energy.

I. INTRODUCTION

The fundamental problem of predicting protein structure from sequence remains largely unsolved[1, 2, 3, 4]. A significant aspect of the prediction puzzle is the language used to describe the substructures comprising the final folded form (e.g. [5]). Many approaches endeavor to predict amino acid membership in the general classes helix, sheet (or extended), turn, and random coil (or other) [1, 2, 3, 6, 7]. The main features of these classes were originally predicted on theoretical grounds, and subsequently observed in crystal structures of globular proteins, confirming their overall significance as structural building blocks [8, 9, 10]. Unfortunately, attempts to objectively assign secondary structure class labels to proteins of known structure have revealed that these classes are actually both vague and subjective [11, 12, 13].

An objective approach to amino acid structure representation using dihedral angle measurements is presented based on the maxi-min distance algorithm presented in Tou and Gonzalez [14]. The approach differs from [15, 16, 17, 18] by having a clear termination criteria and by its application to individual residues rather than sequences, and from [1, 5, 19, 20] in that it objectively identifies centers which are not dependent on a 'rasterization' of dihedral angle space. The results indicate the amino acid variation among the various structures and reveal the existence of several rare, but amino acid specific conformations. Although the exact results of any cluster-seeking algorithm are unavoidably dependent on the distribution of the analyzed data set, validation efforts indicate that the centers presented here are reasonably invariant, enumerate significant amino acid conformational classes, and hence are valuable in describing and understanding protein structure.

II. MATERIALS AND METHODS

All software was initially developed on a Commodore Amiga system. Later versions incorporating a larger data set were executed on a DEC MicroVAX III at the Center for Visualization, Analysis, and Design in the Molecular Sciences (VADMS).

The data set used in this study is based on a union of the proteins listed in Kneller *et al* [4] and Unger *et al* [18], limited to the files located in the November, 1990 version of the Brookhaven Protein Data Bank [21]. This data set is currently available at the VADMS Center (e-mail: PRCADAMS@wsuvms1.csc.wsu.edu). Multiple chain proteins were reduced to single examples of each unique chain, and only non-terminal residues (those for which all dihedral angles are defined) were used in this study. The 97 protein data set generated a total of 19,121 non-terminal residues.

Backbone dihedral angles were extracted from protein coordinate data using software written in C. Cluster analysis routines were written in David Betz's XLisp on the Amiga and Common Lisp [22] on the microVAX III.

The maxi-min distance algorithm is described in Tou and Gonzalez [14]; it identifies a series of cluster centers from an input data set, without user intervention and without an *a priori* hypothesis of the number of centers to be identified. Each dimension of the pattern space corresponds to a physical measurement of amino acid conformation, e.g., the ϕ, ψ, and ω dihedral angles define the three dimensional pattern space used. Points in the pattern space are compared by Euclidean distance measures. The algorithm is as follows:

1. Select the first element and the element which is farthest from it from the data and set them as the first two cluster centers. Save the distance between these two centers in a list.

2. For each of the remaining data elements, identify the closest cluster center and the distance to it.

3. Select the element which is the farthest from its identified closest cluster center. If the distance is greater than *one-half* the average value in the list of distances between cluster centers, append this distance to the list, set this element as a new cluster center, and go to step two.

4. Classify all data elements to their nearest cluster center, as identified in steps one to three; for each such group, average the values in each dimension of the pattern space and set the class average as the cluster center in that dimension.

5. Reclassify all data elements to their nearest cluster centers.

To reduce computation time and highlight differences among amino acids, maxi-min is applied separately to the twenty residue subsets which partition the input set. The resulting centers are combined and subjected to an additional application of maxi-min. This second level of clustering groups the similar representatives identified in each of the twenty subsets, and generates a manageable set of conformational classes which assymptotically captures the input set. To reduce representational inaccuracies at this step, the final position averages are weighted by the number of example instances assigned to each cluster.

Two problems exist in the algorithm as described: selection of the first center is not specified, and final center positions are sensitive to the outlying points of each data set. The two most common centers identified in early runs captured examples in all data set partitions and exhibited α-helical and parallel β-strand dihedral angles. To justify comparison between final sets of centers, each data set is analyzed separately starting with the best example of each of these conformations, then the results are combined as described below. To reduce the data set dependence of the results, clustering is applied separately to four protein structural class partitions (all-α, all-β, α/β, and other [23], following Kneller *et al* [4]), then these four sets of centers are subjected to a final application of maxi-min.

Our final methodology is as follows:

1. Partition the data set by protein structural class, then partition each of these subsets by amino acid type.

2. Apply maxi-min to each amino acid subset, selecting the best alpha-helical example (based on ϕ and ψ angles identified in earlier runs) as the first center each time.

3. Combine the twenty sets of centers for each structural class separately using maxi-min, again selecting the best α-helical example as each first center.

4. Combine the four resulting sets of centers (one for each structural class) using maxi-min, as in step (3).

5. Repeat steps (1) through (4), this time selecting the best example of the parallel beta strand conformation for each initial center.

6. Combine the two resulting sets as follows: for each member of each set, find the nearest member in the other set. Average the positions of those centers which are closest to each other, then add to this list the remaining centers. This generates a list containing the centers which are essentially equivalent independent of starting with the best helical or best strand representatives, plus those centers which are dependent upon the initial conditions.

III. RESULTS

88 centers are paired between the results from the two runs started with different initial centers; however, 11 (α-first set) and 26 (β-first set) centers do not match. This suggests that the 88 are relatively invariant but, as three of the 37 were able to capture 1-15% of the examples, all 125 were included as distinct centers. Only the top 48 centers capture at least 0.2% of the data set each (totalling to 97.8%). The remaining 77 include all of the *cis*-proline conformations, but are otherwise sparsely populated and represent very rare conformations or possibly erroneous data (many have non-planar ω angles). Table 1 lists the 18 most populated centers and five *cis* centers, grouped to match the dihedral angle labels in [5, 19, 24]. Figure 1 plots the ϕ and ψ angles of these 23 centers.

Plots like figure 1 for the four protein structural class partitions (and for the unpartitioned data set) exhibit very similar patterns of centers, including a group of four bridging the $\psi = 0°$ line near $\phi = -100°$ (forming a diagonal ellipsoid anchored to the alpha-helical center near $-65°$, $-40°$, centers 1,2,4, and 5 in figure 1), a pair in a similar diagonal near $\phi = 100°$ (corresponding to the left-handed helix, centers 19 and 20), and a central beta sheet representative near $-85°$, $135°$ (center 9) surrounded by four to six other 'extended' centers (10-14, 16 and 17 in figure 1).

In the all-α and 'other' partitions, the alpha helix center ranked first, followed by the anti-parallel β strand conformation; the α/β partition also populated the α-helix position first, but the parallel β strand conformation ranked second in this set, reflecting the defining features of the structural class. For the all-β partition, the parallel strand center (9 in figure 1) was more populated than its anti-parallel partner by only two percent of the data set, but these were nine and seven percent (respectively) ahead of the third ranked α-helical representative.

IV. DISCUSSION

The goal of this work is to develop procedures that objectively identify a set of distinct representatives for modeling amino acid conformations in proteins, without prior knowledge of the cardinality of such a set. The selected maxi-min clustering procedure depends on the conformation of the first residue chosen (as it becomes the first center) and on the contents of the data set, but not on the order of elements in the data set, any arbitrary specification of the number of nearest neighbors, or any physical cutoff definition of 'difference' between conformations. It is possible that the use of the neural net based ART II clustering algorithm (see [25]) may result in more representative centers, but it too is subject to a variable controlling the cluster resolution (the 'vigilance') which is not as clearly specified as the 0.5 value for maxi-min. We further note that, given the continuums of dihedral angle values observed in proteins of known structure, some arbitrariness is likely to be unavoidable in any assignment of observed conformations to representative centers.

Finally, as Richardson and Richardson [26] noted in plotting ϕ and ψ angles of amino acids in non-repetitive conformations, the dihedral angles of repeating structures (helices and beta sheets) are also quite common in regions of unclassified structure. This emphasizes the point that, although the centers presented here are *representative* of various structural classes, classification to a given center does not necessarily imply membership in a particular secondary structure.

V. CONCLUSION

The conformational cluster centers presented here are sufficiently resolved to adequately represent alpha helices, 3-10 helices, and parallel and anti-parallel beta strands. Sequences of these centers are capable of differentiating each of the beta turns described in Wilmot and Thornton's new nomenclature [5], as well as capturing Leszczynski and Rose's Ω loops [27]. In future reports we will detail the representation and analysis of proteins using these conformational centers, their amino acid sequence dependence, the development of regular expressions to assign secondary structures from conformational center sequences, and the significance of conformational center adjacency in a structure prediction algorithm.

The authors wish to thank Dr. Michael Wick, Dr. Michael Gribskov and Dr. Scot Wherland for helpful discussion.

Table 1: The major conformational centers discussed in the text and their distribution among the twenty amino acids. The top row of numbers gives the total occurence of each amino acid in the data set, as well as the data set total in the far right column. The far left cloumn lists arbitrarily assigned identifiers for each center, selected to correspond with the work of Rooman *et al* [24], Wilmot and Thornton [5], and Efimov [19]; corresponding numbers are used in figure 1. The next three columns from the left specify the average backbone dihedral angles of each center. The remaining values indicate the fraction (normalized to 1000) of each column total captured by each center; for example, conformation 1 captures only 12.9% of the 1722 glycines, but 29.1% of the complete, 19,121 residue data set.

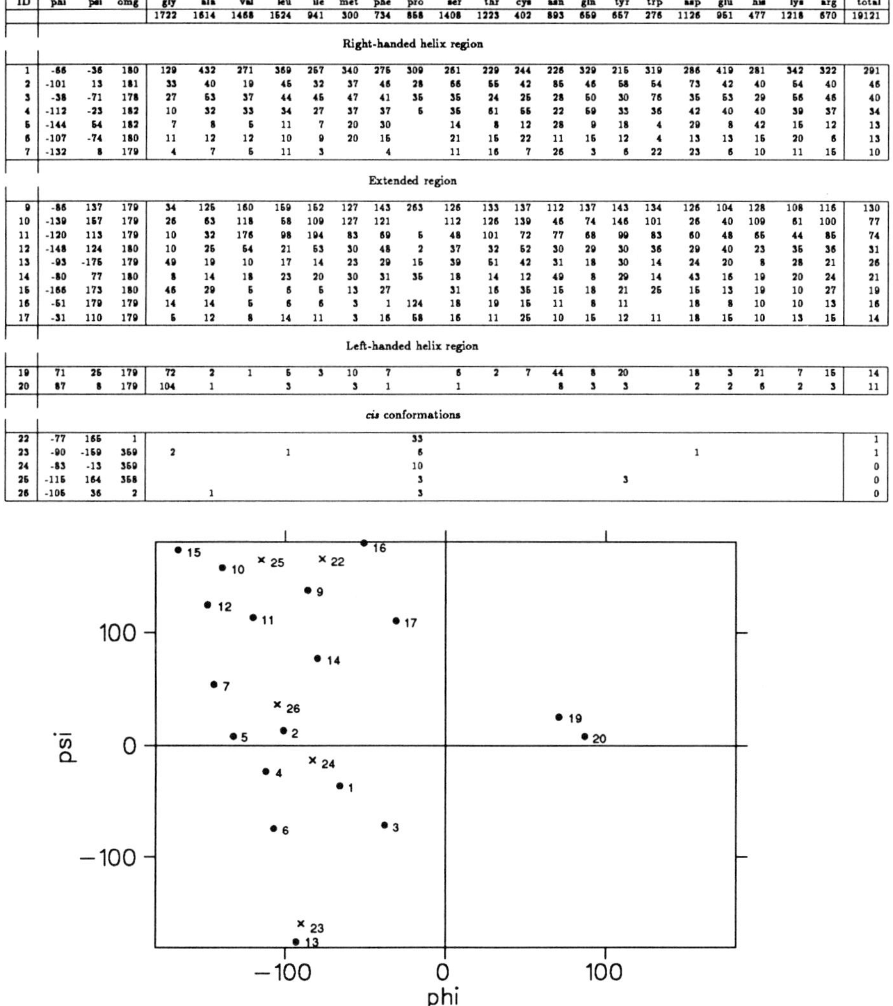

ID	phi	psi	omg	gly	ala	val	leu	ile	met	phe	pro	ser	thr	cys	asn	gln	tyr	trp	asp	glu	his	lys	arg	total
				1722	1614	1468	1524	941	300	734	858	1408	1223	402	893	659	657	276	1126	951	477	1218	670	19121
	Right-handed helix region																							
1	-66	-36	180	129	432	271	369	267	340	276	309	261	229	244	226	329	215	319	286	419	281	342	322	291
2	-101	13	181	33	40	19	45	32	37	46	28	66	55	42	85	46	58	54	73	42	40	54	40	46
3	-38	-71	178	27	53	37	44	45	47	41	35	35	24	25	28	50	30	76	35	53	29	56	46	40
4	-112	-23	182	10	32	33	34	27	37	37	5	35	61	55	22	59	33	36	42	40	40	39	37	34
5	-144	54	182	7	8	5	11	7	20	30		14	8	12	28	9	18	4	29	8	42	15	12	13
6	-107	-74	180	11	12	12	10	9	20	15		21	15	22	11	15	12	4	13	13	15	20	6	13
7	-132	8	179	4	7	5	11	3		4		11	16	7	26	3	6	22	23	6	10	11	15	10
	Extended region																							
9	-86	137	179	34	125	160	159	152	127	143	263	126	133	137	112	137	143	134	126	104	128	108	116	130
10	-139	157	179	26	63	118	58	109	127	121		112	126	139	46	74	146	101	26	40	109	61	100	77
11	-120	113	179	10	32	176	98	194	83	69	5	48	101	72	77	68	99	83	60	48	65	44	85	74
12	-148	124	180	10	25	54	21	53	30	48	2	37	32	52	30	29	30	36	29	40	23	35	36	31
13	-93	-175	179	49	19	10	17	14	23	29	15	39	51	42	31	18	30	14	24	20	8	28	21	26
14	-80	77	180	8	14	18	23	20	30	31	35	18	14	12	49	8	29	14	43	16	19	20	24	21
15	-166	173	180	46	29	5	6	5	13	27		31	16	35	15	18	21	25	15	13	19	10	27	19
16	-51	179	179	14	14	5	6	6	3	1	124	18	19	15	11	8	11		18	8	10	10	13	16
17	-31	110	179	5	12	8	14	11	3	16	58	16	11	25	10	15	12	11	18	15	10	13	15	14
	Left-handed helix region																							
19	71	25	179	72	2	1	5	3	10	7		6	2	7	44	8	20		18	3	21	7	15	14
20	87	8	179	104	1		3		3	1		1			8	3	3		2	2	6	2	3	11
	cis conformations																							
22	-77	165	1								33													1
23	-90	-159	359	2			1				6								1					1
24	-83	-13	359								10													0
25	-115	164	358								3						3							0
26	-105	36	2		1						3													0

Figure 1: A $\phi - \psi$ plot showing the locations of the 18 centers capturing at least one percent of the data set each (solid circles) and of the five *cis* conformation centers (x's).

REFERENCES

[1] Burgess, A. W., Ponnuswamy, P. K., and Scheraga, H. A., Is. J. Chem., vol. 12 (1974), pp. 239-286.
[2] Chou, P. Y. and Fasman, G. D., Biochem., vol. 13, number 2 (1974), pp. 211-222.
[3] Maxfield, F. R., and Scheraga, H. A., Biochem., vol. 15, no. 23 (1976), pp. 5138-5149.
[4] Kneller, D. G., Cohen, F. E., and Langridge, R., JMB, vol. 214 (1990), pp. 171-182.
[5] Wilmot, C.M., and Thornton, J.M., Prot. Engr., vol. 3, no. 6 (1990), pp. 479-493.
[6] Cohen, F. E., Abarbanel, R. M., Kuntz, I. D., Fletterick, R. J., Biochem., vol. 25(1) (1986), pp. 266-275.
[7] Wilmot, C.M., and Thornton, J.M., JMB, vol. 203 (1988), pp.221-232.
[8] Pauling, L., Corey, R. B., and Branson, H. R., PNAS USA, vol. 37 (1951), pp. 205-211.
[9] Pauling, L. and Corey, R. B., PNAS USA, vol. 37 (1951), pp. 729-740.
[10] Venkatachalam, C. M., Biopolymers, vol. 6, pp. 1425-1426.
[11] Kabsch, W., and Sander, C., Biopolymers, Vol. 22 (1983), pp. 2577-2637.
[12] Levitt, M., and Greer, J., JMB, vol. 114 (1977), pp. 181-293.
[13] Richards, F. and Kundrot, C., Proteins, vol. 3 (1988), pp. 71-84.
[14] Tou, J. T. and Gonzalez, R. C., *Pattern Recognition Principles*, Addison-Wesley, 1974. ISBN 0-201-07586-5
[15] Prestrelski, S. J., *Experimental and Computational Approaches to Protein Secondary Structure*, doctoral dissertation, City University of New York, 1990.
[16] Rooman, M., Rodriguez, J., and Wodak, S. J., JMB, vol. 213 (1990), pp. 327-336.
[17] Sippl, M. J., JMB, vol. 213 (1990), pp.859-853.
[18] Unger, R., Harel, D., Wherland, S., and Sussman, J. L., Proteins, vol. 5 (1989), pp. 355-373.
[19] Efimov, A. V., Mol. Bio. (Moscow), vol. 20, no. 1 (1985), pp. 250-260 (English pp. 208-216).
[20] Zimmerman, S. S., Pottle, M. S., Nemethy, G., and Scheraga, H. A., Macromol., vol. 10, no. 1 (1977), pp. 1-9.
[21] Protein Data Bank Atomic Coordinate Entry Format Description, July 1981.
(a) Bernstein, F.C., Koetzle, T. F., Williams, G. J. B., Meyer, E. F. Jr., Brice, M. D., Rodgers, J. R., Kennard, O., Shimanouchi, T., and Tasumi, M., JMB, vol. 112 (1977), pp. 535-542.
(b) Abola, E. E., Bernstein, F. C., Bryant, S. H., Koetzle, T. F., and Weng, J., in *Crystallographic Databases* F. H. Allen, G. Bergerhoff, and R. Sievers, eds., Data Commission of the International Union of Crystallography, Bonn/Cambridge/Chester, 1987, pp. 107-132.
[22] Steele, G. L. Jr., *Common Lisp The Language*, Digital Press, 1984. ISBN 0-932376-41-X
[23] Levitt, M., and Chothia, C., Nature, vol. 261 (1976), pp. 552-558.
[24] Rooman, M. J., Kocher, J. A., and Wodak, S. J., JMB, vol. 221 (1991), pp. 961-979.
[25] Hertz, J., Krough, A., and Palmer, R. G., *Introduction to the Theory of Neural Computation*, Addison-Wesley, 1991, ISBN 0-201-51560-1 (pbk).
[26] Richardson, J. S., and Richardson, D. C., in *Prediction of Protein Structure and the Principles of Protein Conformation*, Fasman, G. D, ed., Plenum Press, 1989, ISBN 0-306-43131-9, pp. 1-98.
[27] Leszczynski, J. F., and Rose, G. D., Science, vol. 234 (1986), pp. 849-855.

ΔΔG°(WT-mut) for BPTI Hydrophobic Core Mutants Measured by Hydrogen Isotope Exchange

Feng Tao, James Fuchs, and Clare Woodward*

Department of Biochemistry, University of Minnesota, St. Paul, MN 55108

* This work is supported by NIH grant GM26242.

I. Introduction

Tyr 21, Phe 22, Tyr 23, Phe 33 and Phe 45 in the hydrophobic core region of bovine pancreatic trypsin inhibitor (BPTI) (Figure 1) have been replaced by alanine, using site-directed mutagenesis. The high resolution X-ray crystal structures for F22A, Y23A and F45A have been determined (Danishefsky et al., 1992). They show little deviation from wild type (WT) except for the removal of side chain atoms. Replacement of the aromatic rings with methyl groups creates large crevices in the protein surface. The thermostabilities of these mutants are significantly decreased compared to WT. At pH 2, T_m for WT, F22A, Y23A and F45A is 87, 78, 54 and 49 °C, respectively (Kim et al., 1992); for Y21A and F33A, T_m is 65 and 57 °C (Tao, et al., unpublished results).

In proteins, nitrogen-bound protons are labile and they exchange isotope with solvent hydrogens. When a protein purified with ^1H aqueous solvents is dissolved ^2H_2O, peptide amide $-N^1H$ exchanges hydrogen isotope and becomes $-N^2H$. Those on the protein surface exchange rapidly, with a rate constant usually the same or within an

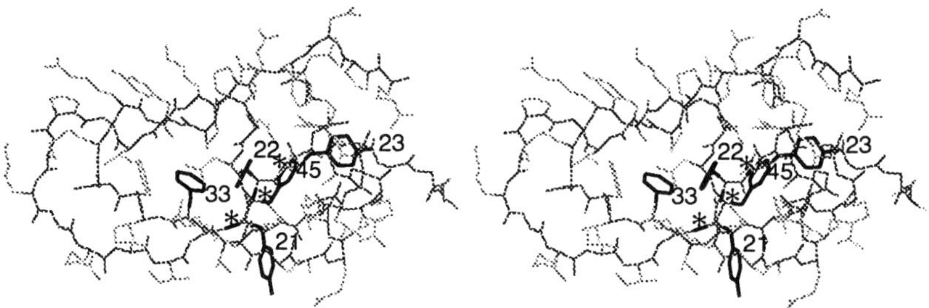

Figure 1. Stereo view of WT BPTI. The mutated side-chains of residue Y21, F22, Y23, F33 and F45 are drawn in thick lines, while other side-chains are in dotted lines. The three * symbols indicate the approximate positions of the three most slowly exchanging -NH in BPTI on residues 21, 22 and 23.

TECHNIQUES IN PROTEIN CHEMISTRY IV

order of magnitude smaller than k_{cx}, the rate constant of an oligopeptide -NH. Buried -NH protons exchange slower than free -NH, but with a broad range of rates. Hydrogen exchange of buried protons is traditionally used to assess the internal flexibility of the protein in the locale of the exchanging proton. However, hydrogen exchange can also be used to monitor the cooperative unfolding reaction of proteins, because each -NH in a protein may exchange either from the folded state (giving a measure of local flexibility) ***or*** by an unfolding mechanism (giving a measure of the folding/unfolding equilibrium) (Woodward et al., 1982).

The three slowest exchanging protons, the peptide NH's of residues 21, 22 and 23, which exchange *only* by the unfolding mechanism, define the hydrophobic core of BPTI (Figure 1, Woodward and Hilton, 1980). ^1H-^2H isotope exchange in WT, F22A, Y23A and F45A for 30 of the 53 exchangeable amide protons were measured by ^1H NMR. The increase in exchange rates of the peptide amide protons of residues 21, 22 and 23 in the mutants is due only to the shift in the folding/unfolding equilibrium. Assuming that the values of k_{cx} do not change by mutations, the ratio of the exchange rate constant of any of these protons in WT to the rate constant of the same proton in the mutant, is equal to the ratio of the equilibrium constants, K_{eq}, for the folding/unfolding reaction in WT versus mutant. From this, $\Delta\Delta G°_{(WT\text{-}mut)}$, the change in free energy of the folding/unfolding reaction for the WT versus the mutant, is obtained. Values of $\Delta\Delta G°_{(WT\text{-}mut)}$ measured by hydrogen exchange in this way at pH 3.5 agree reasonably well with those measured by differential scanning calorimetry at pH 2.0 (Kim et al., 1992). Using as examples with the exchange rate constants of protons in F22A, Y23A and F45A and WT BPTI, we compare here the exchange behavior of protons which can be used to measure changes in the folding/unfolding equilibrium, and those which cannot. The application of this method for determination of $\Delta\Delta G°_{(WT\text{-}mut)}$ in other protein systems is discussed.

II. Materials and Methods

Wild type BPTI, purchased from Novo Industries, Copenhagen (Aprotinin Novo, Ultrapure Grade, Batches 32-65), was further purified by gel filtration on Sephadex G-25 in 20 mM potassium phosphate buffer, pH 7.0, with 0.02% NaN_3, filtered through Millipore Millex GV (0.22μm pore size), lyophilized and stored at 4°C.

The BPTI mutants were produced from the expression system described by Goldenberg (1988), and Housset et al. (1991). Protein samples were dialyzed against distilled water at the same pH of the NMR experiment, filtered (0.22μm filter) and lyophilized. 4-5 mM of samples were adjusted to pH 3.50 ± 0.02 on ice (< 25 °C experiments) or at room temperature (≥ 25 °C experiments); no correction of isotope effect on pH is made. NMR was performed on a GE 500MHz spectrometer. Sequential assignment strategy and spin system identification (Wüthrich, 1986) were utilized for resonance assignments in the spectra of the mutants. Data were processed by

FELIX (Hare Research Inc.) software. Samples for hydrogen exchange are put in the magnet immediately after adjusting the pH. Exchange rates were measured over the temperature range 10-68 °C. After shimming for 10-20 minutes, 8k complex data points × 128 scans one-dimensional and 2k × 256 blocks × 4 or 8 scans each block COSY data were collected. Peak heights from 1D and areas of COSY cross peaks were selected and calculated using FELIX. Both 1D and 2D data were used to determine the exchange rate constants. Deviations of the rate constants from 1D and 2D are less than 10%.

III. Results

We have the complete chemical shift assignments for the mutant peptide -NH protons (Tao & Woodward, unpublished results). Table I lists the assignments for the NH and $C_\alpha H$ protons discussed herein.

The temperature dependence of the exchange rate constants for most of the protons in F22A, Y23A and WT have been measured at pH 3.5, and apparent activation energies have been estimated. The amide protons of residue Tyr 21, Phe 22, and Tyr 23 have the highest activation energies among all observed -NH protons. They are about 80, 69 and 52 kcal/mol for WT, F22A and Y23A, respectively. Arrhenius plots for these protons at pH 3.5 are shown in Figure 2. For F45A, data have been obtained only at 30 °C (Figure 2).

Within the same protein, the rate constants of 21, 22 and 23 -NH are about the same at all temperatures (Figure 2). In contrast, those for Glu 7 and Met 52 show a different pattern (Figure 3). For easy comparison, dotted lines in Figures 3 and 4 reproduce the curves in Figure 2. In WT and F22A, Glu 7 and Met 52 exchange slower than 21, 22 and 23 in the respective protein. In Y23A, the exchange rate constants, and the temperature dependence of Glu 7 and Met 52 are approximately the same as 21, 22 and 23. The rate constant for Glu 7 of F45A at 30 °C (one point in Figure 3) is within error of the exchange rates for 21, 22 and 23 in F45A.

In Figure 4, a third type of behavior is illustrated for Ile 19. The exchange of the -NH of Ile 19 is much faster than any proton in Figure 2 or 3. Further, the exchange rate constant, and its temperature dependence, is the same in WT and in all three mutants.

Table I. Chemical shifts[a] of NH and $C_\alpha H$ proton resonances in the ^1H NMR spectrum of three BPTI mutants at 25°, pH 3.5

	F22A		Y23A		F45A	
Protons	NH	$C_\alpha H$	NH	$C_\alpha H$	NH	$C_\alpha H$
Glu 7	7.36	4.65	7.57	4.58	7.50	4.61
Ile 19	8.77	4.30	8.76	4.32	8.76	4.29
Tyr 21	9.09	5.20	9.21	5.68	9.02	5.57
Phe 22	9.32	4.98	9.88	5.24	9.83	4.93
Tyr 23	10.58	4.47	10.44	4.32	9.89	4.23
Met 52	8.64	4.25	8.54	3.94	8.48	4.05

[a] ppm from TSP (sodium 2,2,3,3-tetradeutero-3-trimethyl-silypropionate, 0.00ppm)

IV. Discussion

The difference in the free energy change of the folding/unfolding equilibrium of WT and a mutant, ΔΔG°(WT-mut), is widely used in describing the protein stability changes associated with amino acid mutations. Conventional methods for obtaining ΔΔG°(WT-mut) are differential scanning calorimetry (DSC) and chemical denaturation monitored spectroscopically and extrapolated to zero denaturant concentration. These methods sometimes are limited by uncertainties in ΔC_p for DSC experiments, and in linear extrapolation in the chemical denaturation experiments. The use of hydrogen isotope

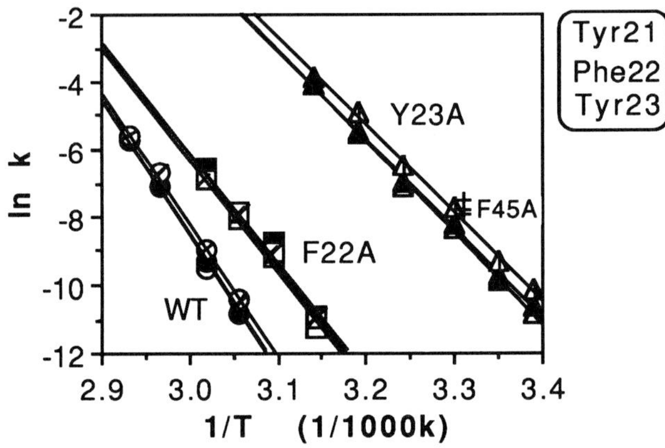

Figure 2. The temperature dependence of the slowest exchanging -NH protons in WT (- ○ -), F22A (- □ -) and Y23A (- Δ -). The exchange rate constants of the amide -NH of Tyr 21 (open symbols), Phe 22 (filled symbols) and Tyr 23 (half crossed symbols) are shown. For F45A, the exchange rate constants of 21, 22 and 23 at 30 °C are indicated by + signs.

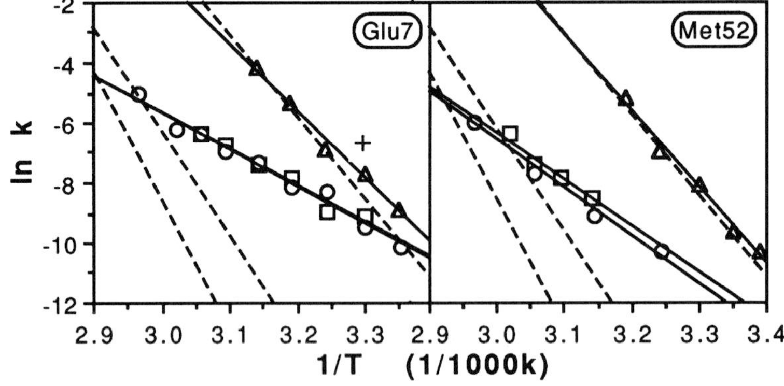

Figure 3. The temperature dependence of more rapidly exchanging -NH protons in WT (- ○ -), F22A (- □ -) and Y23A (- Δ -) (a) The exchange rate constants of amide -NH of Glu 7. For F45A, data at 30 °C is indicated by a + sign. (b) The exchange rate constants of the -NH of Met 52. For F45A, Met 52 -NH was not measured.

exchange, described here, may be more convenient for some proteins.

In the "two process" model of hydrogen isotope exchange in proteins, described in Woodward & Hilton (1980) and Woodward et al., (1982), each proton may exchange by either of two mechanisms. In one, exchange occurs from the native, or folded state, with the protein in a conformation approximated by the crystal structure. This is illustrated in Figure 5a.

The observed exchange rate constant for exchange from the folded state (Figure 5a) is

$$k_N = \beta \bullet k_{cx} \quad (1)$$

where k_{cx} is the rate constant for an equivalent -NH in a small oligopeptide (Molday et al. 1972). The factor, β, is the probability that local motions within the protein will transiently expose buried -NH groups to the solvent components necessary for exchange, namely, catalyst (hydrogen or hydroxyl ion) and water. This mechanism reflects the dynamic structure of proteins. The crystal structure represents a time average conformation; internal motions within the protein result in local excursions of atoms from their average positions. The internal motions responsible for exposure of buried NH groups has been the subject of literature debate, but is not further covered here since the exchange mechanism of concern in this context is the unfolding mechanism (Figure 5b). The local fluctuations supporting exchange of buried protons vary throughout the interior of the molecule. Some buried protons exchanging by the scheme in Figure 5a exchange fairly rapidly; their exchange rates are within several orders of magnitude of the rates of oligopeptide -NH exchange rates. Other buried protons exchanging by the scheme in Figure 5a exchange much more slowly, some on the month-year time scale. These are the protons most likely to be observed in the unfolding mechanism. In the unfolding mechanism, isotope exchange

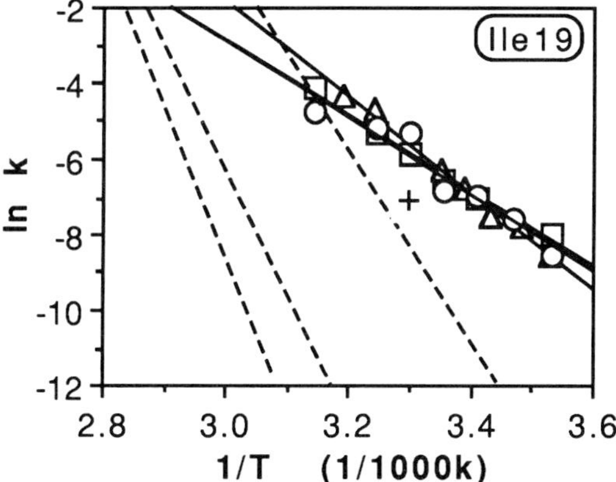

Figure 4. The temperature dependence of more rapidly exchanging -NH proton, Ile 19 in WT (- ○ -), F22A (- □ -) and Y23A (- Δ -). For F45A, only one temperature, 30°C, was measured; the rate constant of Ile 19 -NH is indicated by a + sign.

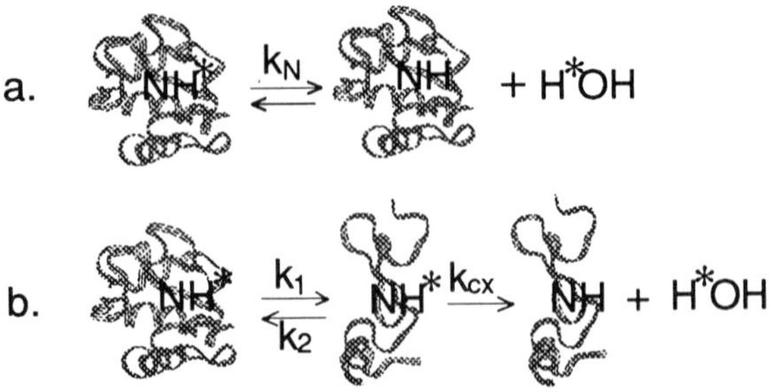

Figure 5. The two pathways for exchange of a peptide proton in proteins. (a) Exchange from the folded state, with no contribution from cooperative unfolding. (b) Exchange preceded by cooperative unfolding.

is preceded by cooperative denaturation of the protein, as illustrated in Figure 5b. The rate constants for unfolding and folding, are k_1 and k_2. The rate constant for exchange by the unfolding mechanism, k_U, depends on the relative values of k_{cx}, k_1 and k_2. Under native conditions, $k_1 << k_2$. When $k_2 >> k_{cx}$, then

$$k_U \approx k_1 / k_2 \bullet k_{cx} \approx K_{eq} \bullet k_{cx} \quad (2)$$

where K_{eq} is the equilibrium constant for the folding/unfolding. The exchange will be first order in catalyst concentration, reflecting the pH dependence of k_{cx}. When $k_2 << k_{cx}$, then

$$k_U \approx k_1 \quad (3)$$

and exchange will have the pH dependence of k_1. This is a rare occurrence, and has been observed only for a small range of pH at high temperatures for the slowest exchanging protons of BPTI. Under most conditions where the mechanism of Figure 5b operates, exchange is given by eq. 2. This can be verified by showing that observed exchange has first order catalysis, i.e. there is an order of magnitude increase in rate for each pH unit as the pH is raised (> 4).

For each exchanging proton at a given pH and temperature, the observed rate constant for exchange, k_{obs}, is the sum of two terms,

$$k_{obs} = k_U + k_N \quad (4)$$

At some pHs and temperatures, $k_U >> k_N$, and $k_{obs} = k_U$. Under other conditions, $k_{obs} = k_N$. For the slowest exchanging protons in a protein, exchange from the folded state may be so slow that exchange occurs more readily by unfolding, and for these protons $k_U >> k_N$ even under native conditions. These are the protons whose exchange rate constants give us the equilibrium constant of the folding/unfolding, assuming the value of k_{cx} is approximately same for the equivalent protons in WT and mutants. When the exchange rate of one of these protons is measured in WT and in a mutant, the ratio of the exchange rate constants in WT and mutant gives the ratio of equilibrium constants for cooperative denaturation in WT and mutant:

$$\frac{k_{obs}^{WT}}{k_{obs}^{mut}} \approx \frac{K_{eq}^{WT} \bullet k_{cx}}{K_{eq}^{mut} \bullet k_{cx}} \approx \frac{K_{eq}^{WT}}{K_{eq}^{mut}} \quad (5)$$

and $$\Delta\Delta G°(\text{WT-mut}) \approx -RT \bullet \ln \frac{K_{eq}^{WT}}{K_{eq}^{mut}} \quad (6)$$

An example of how $\Delta\Delta G°_{(WT\text{-}mut)}$ values are computed in this way for BPTI is given in Table II. For each of the three protons, the right side of eq. 6 is computed from the rate constants in the upper portion of the table, and averaged for each mutant to give $\Delta\Delta G°_{(WT\text{-}mut)}$ determined from hydrogen exchange, listed as $\Delta\Delta G_{HX}$ in Table II. These values are reported in Kim et al. (1992) along with the $\Delta\Delta G°_{(WT\text{-}mut)}$ values from DSC (pH 2.0), listed as $\Delta\Delta G_{DSC}$ in Table II.

To obtain $\Delta\Delta G°_{(WT\text{-}mut)}$ from hydrogen exchange rate constants for other proteins, the only requirement is that the proton(s) whose exchange rate is measured must exchange only by the unfolding mechanism (Figure 5b). Stated alternatively, for the proton being measured, k_U must be >> k_N. In BPTI, this is true for the -NH protons of 21, 22 and 23 (Figure 2), but <u>not</u> for 7 and 52 (Figure 3), nor for Ile 19 (Figure 4). In other proteins, the protons which may be used to determine $\Delta\Delta G°_{(WT\text{-}mut)}$ should be easy to identify if the following points are kept in mind.

1) If any protons exchange only by the unfolding mechanism, they will be the last several protons to exchange-out. These are the simplest to identify by ^{1}H NMR since their resonances are the most long lived when the protein is dissolved in ^2H_2O.

2) If more than one proton exchanges by the unfolding mechanism, all will have approximately the same observed rate constant, because k_1 and k_2 (eq. 2), will be the same for each of the protons, and the nearest neighbor effect on k_{cx} is small. This is illustrated by the data in Figure 2.

3) Exchange by the unfolding mechanism, k_U, will have an apparent activation energy approximately equal to the enthalpy of the folding/unfolding reaction plus the activation energy of k_{cx}, ~17 kcal/mol. Since the enthalpy of unfolding is usually large, the apparent activation energy of k_U is large. The activation energy for

Table II. Calculation of ΔΔG from the exchange rate constants at pH 3.5, 30°

	k (min^{-1}) ± s.d.							
Proton	F22A[a]		Y23A		F45A		WT[a]	
Tyr 21	5.5×10^{-8}	-	2.4×10^{-4}	±0.4	3.6×10^{-4}	±0.1	5.5×10^{-9}	-
Phe 22	10.0	-	2.7	±0.4	4.3	±0.1	3.1	-
Tyr 23	12.0	-	4.5	±0.5	5.4	±0.5	8.8	-

	ΔΔG (kcal/mol)		
	F22A	Y23A	F45A
Tyr 21	1.4	6.4	6.7
Phe 22	2.1	6.9	7.1
Tyr 23	1.6	6.5	6.6
$\Delta\Delta G_{HX}$[b]	1.7 ± 0.4	6.6 ± 0.3	6.8 ± 0.3
$\Delta\Delta G_{DSC}$[c]	1.2	5.9	6.9

[a] For F22A and WT, the exchange rate constant at 30° is linearly extrapolated from 45-58° and 54-68° data. Values of s.d. (standard deviation) of k_{obs} are given for Y23A and F45A. No correction for a nearest neighbor effect on k_{cx} is made.
[b] average ΔΔG ± s.d. calculated from 21, 22, 23 exchange rate constants listed above.
[c] values from differential scanning calorimetry (pH 2.0, 25°) (Kim et al. 1992)

k_N will be in the range 17 - 40 kcal/mol. The simplest way to demonstrate that a given proton exchanges by the unfolding mechanism is to determine the exchange rate constants over a wide range of temperatures. Typically, Arrhenius plot of $\ln(k_{obs})$ versus 1/T will be curved indicating a switch in mechanism when the relative magnitudes of k_U and k_N are reversed. Figure 2 does not show this curvature because the temperature is not varied over a sufficient range since the usually hydrogen exchange observation time window is 15 minutes to 3 weeks. If exchange rates for 21, 22 and 23 were measured at low temperatures and plotted onto Figure 2, the plot would be curved, indicating that the mechanism has shifted from that in Figure 5a to that in Figure 5b. (A curvature also arises from the temperature dependence of the enthalpy of denaturation.)

4) Some protons may exchange primarily from the folded state in the WT, but by the unfolding mechanism in a destabilized mutant. These will be the protons for which k_N is just larger than k_U in WT. Although these protons exchange far more slowly than surface protons, they still exchange more rapidly by mechanism in Figure 5a than the unfolding mechanism (Figure 5b). Arrhenius plots of the -NH protons, E7 and M52 from WT and mutant BPTI illustrate this (Figure 3). In the very destabilized Y23A and F45A, these protons exchange the same as 21, 22 and 23, as indicated by their coincidence with the dotted line in Figure 3. In other words, in Y23A and F45A, k_U is greatly increased, while k_N stays the same, and $k_U >> k_N$. In contrast, in WT and F22A, -NH protons of 7 and 52 exchange at the same rate, and faster than 21, 22 and 23 (dotted lines). In WT and F22A, k_U for Glu 7 and Met 52 is small, and $k_U < k_N$.

5) The faster exchanging protons are those that exchange from the folded state under most conditions, that is, $k_U << k_N$ under most conditions. These have the same exchange rate constants in mutants that are very different in stability. This phenomenon is further discussed elsewhere (Kim et al., 1992). An example of this is Ile 19 in Figure 4. Ile 19 has the same exchange rate constants, and the same temperature dependence, in WT and in all three mutants. These protons exchange only by the mechanism in Figure 5a.

References

Danishevisky, A., Housset, D., Kim, K. S., Tao, F., Wlodawer, A., Fuchs, J. & Woodward, C., (1992) *manuscript in preparation.*

Goldenberg, D.P. (1988) *Biochemistry* **27**, 2481-2489.

Housset, D., Kim, K.S., Fuchs, J., Woodward, C., & Wlodawer, A. (1991) *J. Mol. Biol.* **220**, 757-770.

Kim, K.S., Tao, F., Danishevisky, A., Housset, D., Wlodawer, A., Fuchs, J. & Woodward, C., (1992) *manuscript in preparation.*

Molday, R., Englander, S.W., & Kallen, R. (1972) *Biochemistry* **11**, 150-158.

Wüthrich, K. (1986) *NMR of Proteins and Nucleic Acids* Wiley, New York.

Woodward, C., & Hilton, B. (1980) *Biophys. J.* **32**, 561-575.

Woodward, C., Simon, I. & Tüchsen, E. (1982) *Mol. Cell. Biochem.* **48**, 135-160.

Thermodynamics of Side Chain Internal Rotations - Effects on Protein Structure and Stability

Andrew J. Doig[1]
Mark Gardner
Mark S. Searle
Dudley H. Williams
Cambridge Centre for Molecular Recognition, University Chemical Laboratory, Lensfield Road, Cambridge CB2 1EW, England

1. Introduction

One factor which will affect the stability of a protein is the unfavorable change in free energy which results from the "freezing" of internal rotations in side chains when a protein folds. Figure 1 shows the internal rotations present in the side chains of the 20 naturally occurring side chains in proteins, numbered from their terminal ends. Rotation is presumably possible about all of these bonds in the unfolded protein and may be restricted in the folded state by the close packing and specific interactions of the protein core. The free energies, enthalpies and entropies of each of the rotors shown in Figure 1 are calculated here, using the theory of Pitzer (1).

2. Theory

The partition function (Q_f) for a free rotation (i.e. an internal rotation with no energy barrier) is given by equation 1

$$Q_f = 2.7935 \frac{\sqrt{10^{38} I_r T}}{n} \qquad (1)$$

---

[1]Present address: Department of Biochemistry, Stanford University Medical Center, Stanford, CA 94305.

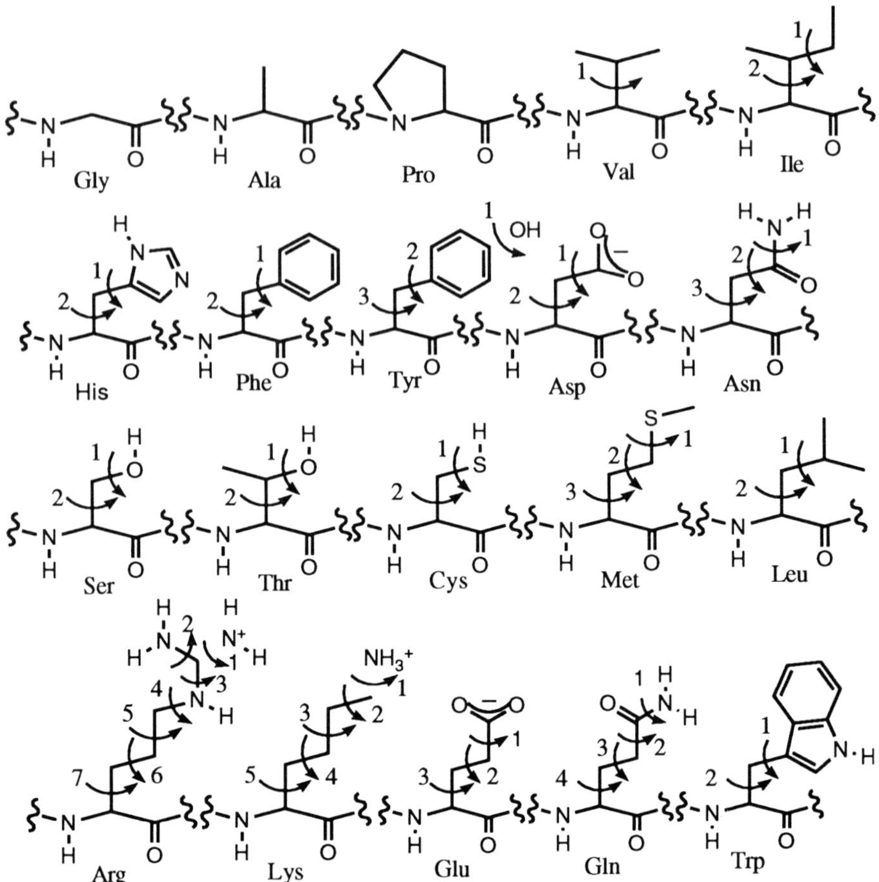

Figure 1. Internal Rotations Present in the Side Chains.

where I_r is the reduced moment of inertia about the bond axis in gcm^2, T is the absolute temperature (here 298K) and n is the symmetry number of the rotation (the number of identical structures per 360° rotation). I_r is calculated by measuring the moment of inertia of the side chain about the bond axis for the atoms attached (directly or indirectly) to the terminal end of the side chain only. The assumption of ignoring the remainder of the protein on the backbone side of the bond axis can be justified by considering the symmetric, coaxial top equation $I_r = I_1 I_2/(I_1+I_2)$ (where I_1 and I_2 are the moments of inertia of the two groups on either end of the bond), which is approximate for the asymmetric side chains and is a simplification of the general case. If the group on one end of the bond is very large (e.g. a polypeptide backbone) and the group on the other end is comparatively small (e.g. part of one side chain) then $I_1 >> I_2$ and $I_r \approx I_1 I_2/(I_1+I_2) \approx I_2$.

The more general, exact equation for the reduced moment of inertia can be simplified similarly. It is therefore not necessary to know the moment of inertia of the very large polypeptide backbone about the bond in question; only the moment of inertia of the smaller group is needed.

The free energy of a free internal rotation with no potential barrier is -RT ln Q_f. This is now corrected to take into account the potential energy barrier. The correction is a function of the barrier height and partition function and is found from a table in Pitzer (1). This correction factor assumes that the potential energy curve is a cosine. Preliminary work on CH_3-CONH-CHR-CONH-CH_3 peptides (where R is the side chain of the amino acid under investigation), using MacroModel (2) and the AMBER force field (3) suggests that a cosine function is reasonably accurate. For example, the conformations of lowest energy about the internal rotation differ by generally only about 10% of the barrier (results not shown). The entropy of an internal rotation is given by equation 2.

$$S = R(0.5 + \ln Q_f) - (S_f - S) \quad (2)$$

where $R(0.5 + \ln Q_f)$ is the entropy of the free rotation and $(S_f - S)$ is the correction taken from a table to take account of the energy barrier. The enthalpy is available from Pitzer's tables as a function of the partition function and the barrier.

When rotation takes place about the bonds shown in Figure 1, a number of water molecules which are solvating the polar groups will presumably also take part in the rotation. This will increase the effective moment of inertia of the side chain about the bond and hence the free energy. The following numbers of water molecules are assumed to rotate with the following functional groups: CO_2^- (5), NH_3^+ (3), $CONH_2$ (4), uncharged imidazole (2), charged imidazole (4), guanidine (5), SH (1), OH (1) and NH (1). The assignment of the number of waters around polar groups comes from molecular dynamics (CONH, $CONH_2$, CO; (4)), low temperature nmr (5) or chemical intuition (e.g. His^+ has more waters than His^0). In a survey of 15 crystal structures Baker & Hubbard (6) found the following numbers of waters per side chain: Asp (1.89), Glu (2.08), Lys (0.79), Arg (1.22), Ser (0.66), Thr (0.81), Asn (0.89), Gln (1.16), Tyr (0.74), His (1.02) and Trp (0.32). These are expected to be underestimates as disordered waters will not be observable and some side chains are buried. The uncertainty in the numbers and locations of these bound waters makes the results of the calculations on the polar side chains somewhat less reliable than the non-polar side chains.

3. Results

The potential energy barriers to the rotations were calculated using MacroModel (2), the AMBER force field (3) and a dielectric constant of 80, a value appropriate for rotations in water, at 298K. Each of the amino acids

Table I. Moments Of Inertia, Potential Barriers, Free Energies, Enthalpies and Entropies of Side Chain Internal Rotations at 298K

Residue & Bond No.	M. of I. (1040 gcm2)	Barrier (kJmol-1)	Free Energy (kJmol-1)	Enthalpy (kJmol-1)	Entropy (JK-1mol-1)	Total Free Energy (kJmol-1)
Ser 1	150	4.8	-8.4	2.5	37	-16.5
2	550	16.4	-8.1	2.6	36	
Cys 1	190	6.8	-8.2	2.7	36	-16.5
2	720	18.9	-8.3	2.5	36	
Thr 1	150	4.7	-8.4	2.5	37	-16.5
2	680	20.8	-8.1	2.6	36	
Arg 1	340	50.0	-6.1	2.4	29	-57.4
2	260	49.9	-5.8	2.4	28	
3	1100	49.0	-7.5	2.5	34	
4	1300	17.9	-9.1	2.6	39	
5	2400	17.6	-9.8	2.6	42	
6	1500	25.0	-8.8	2.5	38	
7	3700	18.0	-10.3	2.6	43	
Phe 1	90	13.4	-4.6	2.5	24	-13.0
2	720	16.5	-8.4	2.6	37	
Tyr 1	130	15.0	-6.5	2.6	30	-23.5
2	380	13.4	-8.0	2.7	36	
3	1200	16.5	-9.0	2.6	39	
Met 1	54	14.5	-5.5	2.5	27	-18.6
2	120	19.5	-6.0	2.5	29	
3	270	18.0	-7.1	2.5	32	
Val 1	72	22.5	-5.3	2.4	26	-5.3
Trp 1	510	12.0	-8.5	2.7	37	-18.3
2	2200	16.8	-9.8	2.6	42	
His^+ 1	870	5.3	-10.3	2.6	43	-19.3
2	1200	17.3	-9.0	2.6	39	
His 1	460	5.3	-9.5	2.6	41	-18.3
2	990	17.3	-8.8	2.6	38	
Ile 1	36	21.1	-4.6	2.4	23	-10.8
2	180	24.4	-6.2	2.5	29	
Leu 1	73	22.5	-5.3	2.4	26	-11.5
2	160	22.9	-6.2	2.5	29	
Asn 1	310	66.2	-5.6	2.4	27	-23.6
2	560	8.3	-9.1	2.7	40	
3	1300	19.1	-8.9	2.6	39	
Gln 1	310	66.2	-5.6	2.4	27	-32.6
2	560	5.6	-9.7	2.6	41	
3	1300	19.8	-8.9	2.6	38	
4	790	18.3	-8.4	2.6	37	
Asp 1	620	5.5	-8.2	2.6	36	-17.5
2	1600	18.5	-9.3	2.6	40	
Glu 1	620	2.5	-9.1	2.1	38	-26.8
2	1600	19.8	-9.2	2.6	39	
3	860	18.2	-8.5	2.6	37	

Lys 1	500	11.9	-5.8	2.6	28	-39.0
2	500	16.7	-8.0	2.6	35	
3	600	19.8	-8.0	2.5	35	
4	1100	26.0	-8.4	2.5	37	
5	1000	18.0	-8.8	2.6	38	
Mean	750	19.9	-7.9	2.5	35	-21.4

was placed in a CH_3-CONH-CHR-CONH-CH_3 molecule (where R is the side chain of the amino acid under investigation). The structure was energy minimized and the energy of the extended conformation calculated. The dihedral angle of the bond being considered was constrained into a high energy, eclipsed conformation while the remainder of the molecule was allowed to relax into a nearby low energy conformation to remove too-close van der Waals contacts etc. The energy of the molecule was calculated, after energy minimization and then the removal of the constraint. The potential energy barrier to the internal rotation is the difference between this energy and the energy of the energy minimized molecule. All of the possible eclipsed conformers were investigated to ensure that the highest energy conformer had been found. The results obtained are given in Table I. Potential barriers were also calculated in an identical manner except using a dielectric constant of 1, the dielectric of a vacuum. It was found that while the absolute energies varied greatly from those calculated with a dielectric of 80, the energy barriers, which are the differences between two absolute energies, agreed to within 10% in the few cases considered (results not shown). This is because the dielectric is not part of the dihedral angle energy term which dominates these energy changes. Individual water molecules bound to polar groups were not considered when estimating the potential barriers. A more accurate method for determining the internal energy as a function of dihedral angle has recently been described by Wade & McCammon (7).

The moments of inertia of the terminal ends of the side chains about the bonds were calculated using the energy minimized structure, which included any water molecules bound to polar groups, and a FORTRAN program which used a MacroModel data file. The required number of water molecules were placed near to the polar groups to which they are assumed to bond and energy minimized to a local low energy conformation. The results obtained are shown in Table I. Clearly, the side chains will not always be in their lowest energy conformations which introduces an error into the estimates of the moments of inertia.

It is notable that the internal rotations which are restricted by conjugation (such as those within the guanidinium group) have smaller, but still significant, free energies. It is likely that some cooperative crankshaft rotation will take place within some of the larger side chains. (Crankshaft motion occurs when a small central portion of a molecule rotates about two or more single bonds, thus avoiding large motions at the ends of the chain.

Its expected prevalence within the protein backbone is why the methods here cannot be used for backbone internal rotations.) If this is significant, the free energies of the internal rotations will be less certain, particularly for the internal rotations close to the backbone of the longer side chains. Bound waters discourage the burial of polar side chains slightly, as desolvation leads to a decrease in the moment of inertia about the internal rotations and hence a smaller (more positive) free energy. It is also possible that the values given here are overestimates for aqueous solution, as a fairly viscous solvent, such as water, is likely to decrease the rate of internal rotation and make it easier for such motions to be frozen out. Such effects cannot be quantified using the present theory. Hydrogen bonding of polar side chains back to the main chain may also lower rotational entropies in the unfolded state.

Although there are many approximations in these calculations (discussed above), the thermodynamic quantities obtained for rotations in the unfolded state (Table I) are relatively insensitive to the assumptions made and to the different moments of inertia or potential barriers. The range in moments of inertia calculated is 36×10^{-40} to $3700 \times 10^{-40} gcm^2$ and the range in potential barrier is 2.5 to $66.2 kJmol^{-1}$. These correspond to factors of 100 and 26 respectively, but the free energies range from -4.6 to $-10.3 kJmol^{-1}$, a factor of only 2.2. This is primarily because the free energy depends only on the logarithm of the reduced moment of inertia.

A major uncertainty in these results arises from potentially large variations in the solvation numbers of the polar side chains. It may even be the case that no waters actually rotate with the side chains at all. It is thus important to estimate how the free energies vary with differing numbers of solvating water molecules due to a change in the moment of inertia. The results for His^+ (4 waters) compared to neutral His (2 waters) show that each additional water molecule adds about $0.5 kJmol^{-1}$ to the free energy of the rotor in this case. The free energies of the rotors Ser 2, Cys 2, Thr 2 and Tyr 2 have been calculated with and without a single water attached to the OH or SH groups in the side chains. The moments of inertia increased approximately 8-fold with the addition of a single water giving increases in free energies of 2 to $3 kJmol^{-1}$. This increase is very large as the rotating group without the water is small. The effect of waters on the free energies can also be obtained approximately by considering similarly shaped side chains whose moments of inertia differ principally by solvation number. In this way, the rotors Val 1, Asp 1, Thr 2, Leu 1 and Glu 1 can be compared as well as Met 2 with Lys 3 and Met 3 with Lys 4. There will also be smaller changes due to different side chain masses, bond lengths, bond angles and potential energy barriers which are less important and ignored here. These comparisons suggest that each water increases the free energy of a rotor by $\approx 0.6 kJmol^{-1}$. The error associated with uncertainties in the number of waters bound to each polar group can thus be roughly quantified.

The enthalpies in Table I are positive thus showing that restriction of side chain motion is enthalpically favorable. This is because the hindered rotation will be on average closer to the bottom of the potential energy well.

The freezing of internal rotation is unfavorable in terms of free energy and entropy as a smaller range of dihedral angles can now be sampled.

4. Relevance to Protein Folding

The results listed in Table I should be reasonable estimates of the free energies, enthalpies and entropies of side chain internal rotations in the unfolded state relative to a stationary rotor at the bottom of its potential energy well (i.e. zero Kelvin). They are therefore Third Law (absolute) entropies. However, within a protein core, or any complex involving peptides, there will be some residual motion about the bonds, such as in the form of a torsional vibration, which will retain some of the free energy, entropy and enthalpy of the internal rotation in the unfolded or unbound state. Additionally, some groups will actually continue to rotate (e.g. the aromatic rings of tyrosine or phenylalanine about their C_1-C_4 axes) within the folded protein on occasion (for example, refs. 8, 9). It is therefore important to stress that the results listed in Table I can give *upper limits only* of the thermodynamic changes which occur in protein folding and other situations where internal rotations are hindered. Not all the calculated free energy will be lost.

An estimate of the entropic change resulting from hindrance of internal rotations when a protein folds can be obtained by considering the melting of hydrocarbons, a process which may be analogous to protein unfolding as protein cores have densities similar to organic crystals (10). The entropies of fusion for homologous series of organic compounds suggest that the entropic cost of restricting each internal rotation is 5 to $12JK^{-1}mol^{-1}$ at 300K (11, 12). If these results are compared to the mean absolute entropy of $34JK^{-1}mol^{-1}$ per rotor (Table I) it is seen that only about 15 to 35% of the absolute entropy of the internal rotation may be lost in protein folding. A larger fraction may be lost where the constraints are greater than within solid hydrocarbons such as during covalent bond or transition state formation.

5. Discussion

It is noteworthy that hydrophobicity and the effect of the freezing of internal rotations act in the opposite sense for non-polar side chains. As a non-polar side chain increases in length, the hydrophobic effect, as measured by solvent transfer experiments to a liquid non-polar environment, such as an organic solvent, becomes more favorable. However, a longer side chain will generally have a larger number of internal rotations, which will discourage the burial of such a group within a solid-like protein core. This will be offset by improved van der Waals bonding in a denser protein core (13) and opposed by the introduction of strain. It is thus not necessarily the case that it is more favorable to bury a larger non-polar side chain in a protein core, than a smaller side chain, even if the side chain packing is good in both cases. Note that the sum of the free energies of internal rotations for side chains will not be proportional to the solvent accessible surface area of the side chain since internal rotations depend upon shape as well as size. The contribution a non-polar residue makes to the free energy of protein folding is therefore not expected to be proportional to surface area.

There is a direct relationship between the sum of the free energies of the internal rotors in a side chain and the probability of that residue being buried in a folded protein. An "equilibrium constant" for partitioning between a protein surface and interior can be obtained from the fraction of residues buried. The logarithm of this equilibrium constant is equivalent to a free energy for transfer between the protein surface and interior (ΔG_{tr}). These transfer free energies have been calculated by Radzicka & Wolfenden (14), based upon a survey of protein crystal structures by Chothia (15), who defined a buried residue as one which has <5% of the surface area of the residue in a Gly-X-Gly tripeptide. If ΔG_{tr} is plotted against the cost in free energy of completely freezing out the rotors in a side chain (ΔG_{rot}), a good correlation is obtained (Figure 2). The actual free energy change resulting from restricting side chain motion in a folded protein will be smaller than ΔG_{rot}. Note that an equally good correlation will be observed as long as the free energy a residue has in the folded state is proportional to its free energy in the unfolded state.

In Figure 2 the polar and non-polar residues fall into two separate regions. The residues classed as non-polar are Ala, Val, Ile, Leu, Met and Phe. His is labeled as charged and uncharged as both forms are commonly found in proteins. The data for the fraction of His residues which are buried does not distinguish between charged and uncharged imidazole groups so ΔG_{tr} for His is plotted against the mean value of ΔG_{rot} for both neutral and charged species.

Non-polar residues are more likely to be buried in folded proteins than to be found on the protein surface; polar residues show the opposite trend. This observation has traditionally been explained by the fact that it is more favorable to transfer a non-polar residue to a protein core than a polar residue, as a result of the hydrophobic effect (16). However, while this is clearly a real and important effect, the evidence of Figure 2 suggests the partitioning of residues between the protein core and the protein surface is also affected by the energetic cost of freezing the internal rotations in a side chain. The folding of polar residues is discouraged to some extent as a result of polar groups generally being located on the ends of long, unbranched flexible chains. Non-polar residues are generally easier to bury as they are short and branched. Branching decreases the number of internal rotations in a side chain and therefore makes burial easier. It is not easy to separate the relative influences of hydrophobicity and internal rotations on the surface/core partitioning ratio as no residues have many internal rotations and are hydrophobic, or are very inflexible and polar.

Hydrophobicity is important in determining the fraction of residues buried as revealed by some deviations from the line of best fit in figure 2. In general the non-polar residues are more buried than expected solely on the basis of the free energy of their internal rotations and vice versa. Thus, non-polar residues tend to fall below the line of best fit in figure 2. These results

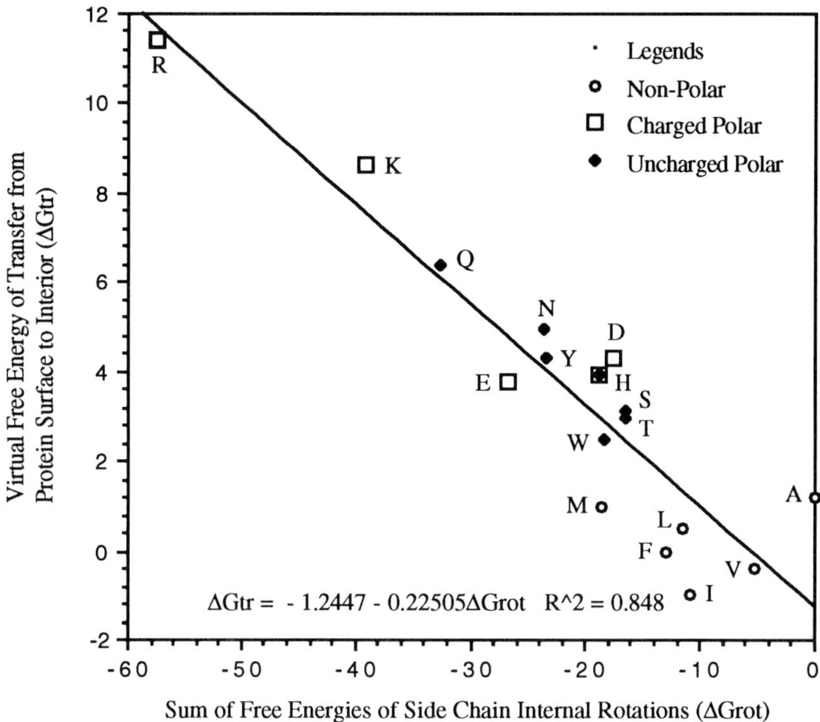

Figure 2. Correlation Between Free Energy of Side Chain Internal Rotations and Virtual Free Energy of Transfer from Protein Surface to Interior. In this graph the residues Cys, Pro and Gly are not included. Cys is not included as it is commonly found in disulfide bonds and the cost of freezing a disulfide unit is unknown and cannot be calculated by the method used here as the side chains form part of a loop. Gly and Pro are not included as the internal rotations in their backbones differ greatly from the other residues. Gly is more flexible than normal and Pro has only a single backbone rotor due to the 5-membered ring it possesses.

may provide some insight into why particular amino acids were selected by evolution. The arguments presented here are purely thermodynamic. Hence, the factors which are important in the kinetics of folding may be quite different.

Acknowledgments

We thank the S.E.R.C. (U.K.), The Upjohn Company (Kalamazoo), SmithKline Beecham and I.C.I. for financial support. A.J.D. wishes to thank I.C.I. for a Student Scholarship and the S.E.R.C. for a NATO Postdoctoral Fellowship We wish to thank Buzz Baldwin, Doug Laurents, Steve Mayo, Ken Dill, A.T. Urtle and an anonymous referee for helpful discussions.

References

1. Pitzer, K.S. (1953). "Quantum Chemistry", Constable, London, UK.
2. Mohamadi, F., Richards, N.G.J., Guida, W.C., Liskamp, R., Lipton, M., Caufield, C., Chang, G., Hendrickson, T., and Still, W.C. (1990). *J. Comp. Chem.* **11**, 440.
3. Weiner, S.J., Kollman, P.A., Case, D., Singh, U.C., Alagona, G., Profeta, S., and Weiner, P. (1984). *J. Am. Chem. Soc.* **106**, 765.
4. Rossky, P.J., and Karplus, M. (1979). *J. Am. Chem. Soc.* **101**, 1913.
5. Kuntz, I.D. (1971). *J. Am. Chem. Soc.* **92**, 514.
6. Baker, E.N., and Hubbard, R.E. (1984). *Prog. Biophys. Mol. Biol.* **44**, 97.
7. Wade, R.C., and McCammon, J.A. (1992). *J. Mol. Biol.* **225**, 679.
8. Williams, R.J.P. (1987). *Carlsberg Res. Commun.* 52, 1.
9. Williams, R.J.P. (1989). *Eur. J. Biochem.* **183**, 479.
10. Richards, F.M. (1977). *Ann. Rev. Biophys. Bioeng.* **6**, 151.
11. Nicholls, A., Sharp, K.A., and Honig, B. (1991). *Proteins: Struct., Funct., Genet.* **11**, 281.
12. Searle, M.S., and Williams, D.H., unpublished.
13. Bello, J. (1978). *Int J. Pept. Protein Res.* **12**, 38.
14. Radzicka, A., and Wolfenden, R. (1988). *Biochemistry* **27**, 1664.
15. Chothia, C. (1976). *J. Mol. Biol.* **105**, 1.
16. Kauzmann, W. (1959). *Adv. Protein Chem.* **14**, 1.

SECTION X

NMR Analysis of Protein and Peptide Structure

Evaluation of NMR Based Structure Determination of Flexible Peptides: Application to Desmopressin

Jianjun Wang, Frank D. Sönnichsen, Robert Boyko,
Robert S. Hodges, and Brian D. Sykes
MRC Group of Protein Structure and Function
Protein Engineering Network of Centres of Excellence
Department of Biochemistry, University of Alberta
Edmonton, Canada T6G 2S2

I. Introduction

Desmopressin is a nine-residue peptide hormone analog with strong antidiuretic and antibleeding activities (1,2). It consists of a six-residue disulfide-linked loop and a three-residue tail (see Figure 1). We have studied the conformation of *desmopressin* in aqueous and trifluoroethanol (TFE) containing solutions using ^{1}H-NMR spectroscopy. These studies are typical of NMR studies of flexible peptides where many of the measured parameters such as the NOEs and coupling constants reflect an average of many possible conformations. The presence of several conformations complicates the use of standard procedures for NMR structure determination (3,4). Current methods such as restrained molecular dynamics and distance geometry calculations use NOE-derived distance information, which implies that each NOE cross peak is associated with a single interproton distance. In the case of flexible molecules this assumption is not valid, since the observed NOE is an average with contributions from every conformation present. Nonetheless, the standard approach has been used for flexible molecules, predominantly in cases where only a part of the molecule is flexible. Frequently in these cases, many structures were obtained which are proposed to represent the flexibility in the molecule, but this has not been proven. Recently, Torda *et al.* introduced a time averaged NOE approach which accounts for flexibility in the calculation (5,6). For peptides in which flexibility extends over the complete molecule, the value of this approach is untested. On the more theoretical side, many investigators have used molecular dynamics to simulate peptide dynamics. Again the question is how valid this approach is.

TECHNIQUES IN PROTEIN CHEMISTRY IV

The focus of this paper is to develop a method to evaluate how well a family of structures generated by a given approach actually represents the flexibility of the peptide. We have developed a program to calculate observable NOE intensities for any set of input structures and then to statistically compare these calculated data with the experimental data. We have used *desmopressin* as a model system for our studies.

II. Theory and Methods

Over the past years, various laboratories including ours have developed a full relaxation matrix approach for structure refinement of macromolecules (7,8,9). This approach was recently expanded to include the transferred nuclear Overhauser enhancement experiment (TRNOE) and describes the NOE buildup rate of a ligand/protein system (10,11) in which the ligand undergoes fast chemical exchange between a free and a bound state. In this paper, we extend the approach to a system of multiple conformations which are in fast exchange. The exchange is assumed to be slow compared to the molecular tumbling correlation time τ_c ($\tau_e / \tau_c >>1$) where observed cross relaxation rates are population averaged. For small peptides τ_c is in the range of 0.1 - 1 ns, so that this approach is valid for exchange rates as fast as 10 ns. In this case the time evolution of the longitudinal magnetizations of n homonuclear spins takes the following generalized form as shown by Landy & Rao (12):

$$d(\Sigma_i \mathbf{m}_i) /d\tau_m = - (\Sigma_i f_i \mathbf{W}_i) (\Sigma_i \mathbf{m}_i) \quad (1)$$

where $\mathbf{W}_i$ is the n-dimensional relaxation matrix for the spins in conformer i, and where the relaxation matrix elements and spectral density functions have the standard forms (10).

The program calculates an individual relaxation matrix for each conformer and then averages these matrices weighted by the fractional population of each conformer. The resulting ensemble or average relaxation matrix is diagonalized to yield the eigenvalue matrix and used to calculate the matrix of normalized NOE intensities. The final statistical analysis involves proper scaling of the two data sets and the calculation of a residual error function. We currently define the deviation between experimental and calculated NOE data by

$$RMSD = \{ [\Sigma_i (NOE_i^{obs} - NOE_i^{calc})^2] / [\Sigma_i (NOE_i^{obs})^2] \}^{1/2} \quad (2)$$

^{1}H-NMR spectra were performed on a Varian UNITY 500 spectrometer using 2 mM samples of *desmopressin* in 20% TFE-d_3 and 80% H_2O at 5°C. NOESY spectra were acquired with several mixing times between 50 and 400 ms. The NOE buildup curves exhibit a linear buildup of the NOEs almost up to a mixing time of 300 ms. Therefore, distance restraints were obtained from cross peaks in the 300 ms NOESY spectra. Cross peaks were integrated and grouped into strong, medium and weak which corresponds to distance restraints of 1.8 - 2.7, 1.8 - 3.3 and 2.3 to 5Å. 157 restraints were used in simulated annealing calculations using Discover (Biosym, San Diego, CA). We used a protocol developed by Nilges *et al.* (13) which starts from a completely random array of atoms. It circumvents the folding problem generally associated with real space methods by setting the force constants of all terms in the target function to very low values during the early stages of the stimulation. As a result the atoms can move essentially independently of each other to satisfy the restraints. During these stages the NOE-force constant is scaled up rapidly, while the force constants for bonded and nonbonded terms are increased slowly. This is followed by the scaling up of all force constants except the NOE-force constant, which was kept at the low value of 10 kcal / mol $Å^2$. This ensured that during this stage proper covalent geometry and nonbonded contacts could be restored at the expense of some NOE violations thereby reducing the probability of obtaining a high energy, physically meaningless average structure.

III. Results and Discussion

Two-dimensional ^{1}H-NMR spectra were measured for *desmopressin* in aqueous solution containing 20% TFE at 5 °C and assigned by standard methods (3). Figure 1 and Table I summarize the NOE connectivities of *desmopressin* under

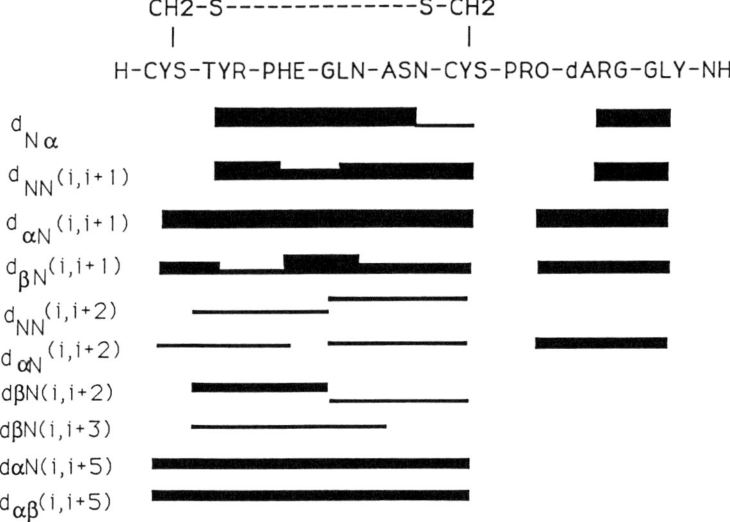

Figure 1. Summary of NOE connectivities of *desmopressin* in 20% TFE solution at 5°C.

Table I. Number and distribution of NOE connectivities of *desmopressin*.

	All		Loop		Tail		Loop-tail
	NOE[a]	NOE/res.	NOE	NOE/res.	NOE	NOE/res.	NOE
Total	157	35.0	102	34.0	43	28.6	12
interresidue NOE	78	17.4	51	17.0	15	10.0	12
sequential	52	11.6	30	10.0	14	9.4	8
medium range	22	4.8	17	5.6	1	0.6	4
long range	4	0.8	4	1.4	0	0	0

[a] NOE means the number of NOEs; NOE/res is the average number of NOEs per residue.

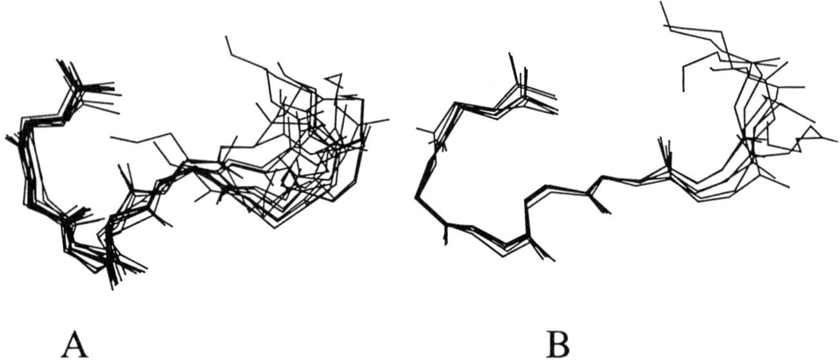

Figure 2. Backbone presentation of 18 superimposed simulated annealing structures. Two distinct groups of A) 13 and B) 5 structures were obtained. The structures were superimposed by best-fitting the backbone atoms of the loop (residue 1 - 6; C, CA, N, O) of each structure to the respective group average structure.

these conditions. Simulated annealing calculations were carried out for *desmopressin* using the distance restraints shown in Table I. Twenty structures were calculated and divided into two major conformational groups consisting of 13 and 5 structures (Figure 2) based on distinct backbone conformations of the loop region. All structures show reasonably low energies and fulfill almost the complete set of distance restraints. In each structure no more than 10 NOE violations are found (> 0.3 Å). In both families of structures the tail region exhibits widely varying conformations which presumably originates in the low number of NOEs per residue and indicates the flexibility in this region of *desmopressin*. The loop, however, is quite well defined with atomic rms differences for backbone atoms being less than 0.6 Å within each group, despite the presence of different backbone conformations for ASN 5 in group 1. These

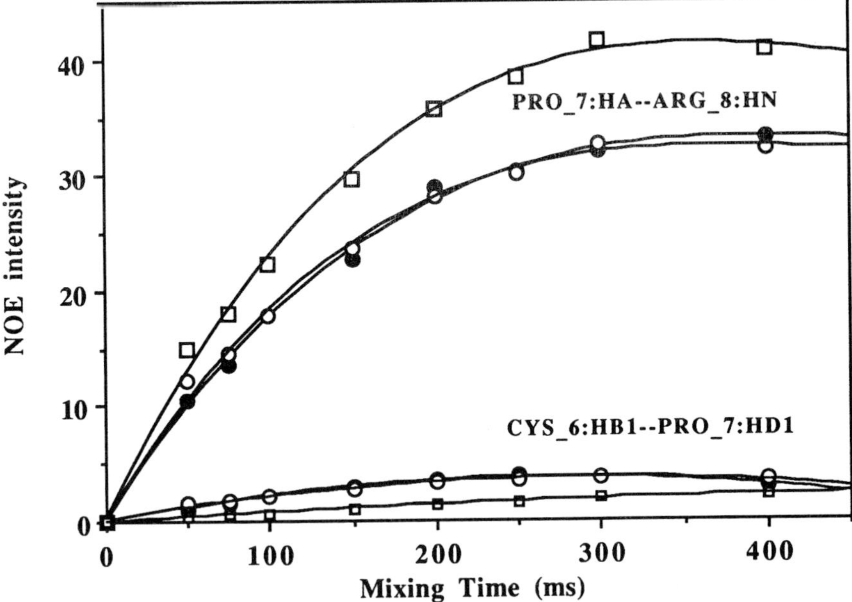

Figure 3. NOE intensities as a function of mixing time for two proton pairs. The black circles are the experimental NOEs at different mixing times, the open circles are the calculated NOEs based on the 18 SA structures shown in Fig. 2. The open boxes are the calculated NOEs based on one representative structure.

findings raised several questions, e. g. whether simulated annealing with low NOE-force constants can be used to obtain meaningful structures from average NOE data, whether these structures are representative of the conformational ensemble in solution, and whether variations between these structures reflect the real flexibility of the peptide. In order to answer these questions, we used our approach of calculating NOE cross peaks for an ensemble of structures by a full relaxation matrix analysis to evaluate the quality of the simulated annealing structures for *desmopressin.*

Figure 3 compares experimental and calculated NOE intensities as a function of mixing time for two NOEs in the tail region of *desmopressin.* It is clearly shown that the NOEs calculated from 18 simulated annealing structures compare much better to these experimental NOEs than those from only one structure. This implies that the experimental NOE data might reflect an ensemble of conformations rather than one single structure and indicates the flexibility in that region.

Table II. Comparison of RMS deviations between experimental and calculated NOE intensities of different ensembles of simulated annealing structures

		RMS deviations[a]					
No.[b]	group[c]	all NOE	loop	tail, loop-tail	b-b[d]	s-b	s-s
1	1 (A)	0.35 ± 0.02	0.31 ± 0.02	0.43 ± 0.03	0.57 ± 0.07	0.67 ± 0.03	0.21 ± 0.02
1	2 (B)	0.38 ± 0.03	o.34 ± 0.04	0.44 ± 0.03	0.70 ± 0.08	0.64 ± 0.03	0.22 ± 0.02
2	1 (A)	0.31 ±0.02	0.28 ± 0.03	0.37 ± 0.02	0.51 ± 0.11	0.61 ± 0.02	0.18 ± 0.01
5	2 (B)	0.35	0.34	0.37	0.71	0.56	0.19
13	1	0.28	0.25	0.34	0.43	0.58	0.16
18	1+2	0.29	0.25	0.34	0.52	0.56	0.17
20	(all)	0.30	0.27	0.34	0.56	0.56	0.17

[a] The root mean square deviations were calculated for a mixing time $\tau_m = 300$ ms and a correlation time $\tau_c = 0.8$ ns. [b] Number of structures used for NOE calculation using [c] structures from group 1 or 2 (Figure 2., A or B). [d] b-b = backbone to backbone NOEs, s-b = sidechain to backbone NOEs and s-s = sidechain to sidechain NOEs.

The overall statistical results are presented in Table II. RMSD values of about 0.3 based on all 157 NOEs indicate a reasonable agreement between experimental and calculated data. Generally, structures from group 1 fit the experimental data better than group 2 structures. The RMSD values decrease in proportion to the number of group 1 structures used in the ensemble calculation. The improvement originates mainly from a better fit of NOEs in the tail region, which indicates that several different conformations are required to simulate NOE data for this highly flexible region of the peptide. 18 structures, i.e. all structures from group 1 and group 2, yield a similar RMSD for all NOEs as compared to group 1 alone. In this case an improved fit for sidechain-backbone NOEs is compensated by a less good description of backbone-backbone restraints. The RMSD value is calculated to be 0.71 and indicates that group 2 structures do not reproduce the experimental NOEs. Therefore, these structures seem not to be relevant for the description of the ensemble of conformations in solution. Group 1 structures describe the data reasonably well with an RMSD of 0.25 for all NOEs in the loop region. However, the RMS deviations for backbone-backbone NOEs and especially backbone-sidechain NOEs are significantly higher than expected. Obviously, the obtained variations of the backbone conformation within group 1 are not sufficient to completely simulate the experimental data, i.e. additional backbone conformations of the loop are required to describe the conformational ensemble.

The program allowed us to evaluate the quality of structures. The results indicate that simulated annealing enabled us to obtain a group of reasonable structures (group1) as well as it yielded a group of less probable structures for *desmopressin* in solution. However, the range of structures does not reflect the real ensemble of conformations. Possibly the static distance restraints derived from average data restrict the range of calculated structures to those near to an average structure despite using a lowered NOE-force constant. In the future we will use this approach to investigate various other methods, which will include the approaches proposed by Torda *et al.* (5, 6) and Brüschweiler *et al.* (14).

References

1. Vávra, I., Machová, A. and Krejci, I. (1974). *J. Pharmacol. Exp. Ther.* **188**, 241-247.
2. Mannucci, P.M., and Rota, L., (1980). *Thromb. Res.* **20**, 69-76.
3. Wüthrich, K. (1986). *NMR of Proteins and Nucleic Acids*, Wiley, New York.
4. Clore, G.M., and Gronenborn, A.M. (1989). *Crit. Rev. Biochem. Mol. Biol.* **24**, 479-564.
5. Torda, A.E., Scheek, R.M., and Van Gunsteren, W.F. (1989). *Chem. Phys. Lett.* **157**, 289-294.
6. Torda, A.E., Scheek, R.M., and Van Gunsteren, W.F. (1990). *J. Mol. Biol.* **214**, 223-235.
7. Borgias, B.A., and James, T.L. (1990). *In "Methods in Enzymology, Vol 176", Part A,* (Oppenheimer, N.J., and James, T.L., eds.) 169-183.
8. Boelens, R., Koning, T.M.G., and Kaptein, R. (1988). *J. Mol. Struct.* **173**, 299-311.
9. Baleja, J.D., Moult, J., and Sykes, B.D. (1990). *J. Magn. Reson.* **87**, 375-384.
10. Campbell, A.P., and Sykes, B.D. (1991). *J. Magn. Reson.* **93**, 77-92.
11. Campbell, A.P., and Sykes, B.D. (1991). *J. Biomolec. NMR* **1**, 391-402.
12. Landy, S.B., and Rao, B.D.N. (1989). *J. Magn. Reson.* **81**, 371-377.
13. Nilges, M., Clore, G.M., and Gronenborn, A.M. (1988). *FEBS Lett.* **239**, 129-324.
14. Brüschweiler, R., Blackledge, M., and Ernst, R.R. (1991). *J. Biomolec. NMR* **1**, 3-11.

Acknowledgements

We are grateful for the financial support of the Protein Engineering Network of Centres of Excellence.

Paramagnetic Proton NMR Methods Used in Studying the Hemeprotein Subunit of *Escherichia coli* Sulfite Reductase[1]

Jeffrey Kaufman, Lewis M. Siegel and Leonard D. Spicer
Departments of Biochemistry and Radiology,
Duke University Medical Center, Durham, North Carolina 27710

I. Introduction

NMR spectroscopy of solutions containing paramagnetic metalloproteins is a rapidly developing, specialized area in the larger field of protein NMR. In the presence of a paramagnetic prosthetic group, hyperfine-shifted resonances may be observed outside the normal diamagnetic spectral envelope (0-10 ppm) and can often be used as a probe of the protein's electronic and molecular structure near the active site. A distinct advantage of this hyperfine shift is that meaningful spectra can be obtained with proteins which are otherwise too large to be examined by current diamagnetic NMR techniques. Examples of large proteins that have been studied using paramagnetic NMR (PMR) include horse radish peroxidase, M_r 42,000 (1); lactoperoxidase, M_r 78,000 (2); and human superoxide dismutase, M_r 32,000 (3). In addition to hemeproteins and iron-sulfur containing enzymes, the PMR technique has also been used quite successfully to study proteins containing non-heme iron, uteroferrin (4), cobalt (II); copper (II), Co_2Cu_2 superoxide dismutase (5); and nickel (II), Ni_2 substituted azurin (6), for example. The purpose of this paper is not to review the extensive body of literature devoted to paramagnetic NMR (for good reviews see 7,8,9,10) but to focus attention on those questions PMR can answer and on the specific techniques involved in acquiring meaningful spectra. Sulfite reductase hemeprotein subunit (SiR-HP), a large paramagnetic protein that is studied in our laboratory, will be used as our PMR model.

NADPH-sulfite reductase of *Escherichia coli* is a multimeric ($\alpha_8\beta_4$) hemoflavoprotein (M_r 685,00) which catalyzes the 6-electron reductions of SO_3^{2-} to S^{2-} and NO_2^- to NH_3. This enzyme's physiological role, as shown by genetic studies (11), is the provision of reduced sulfur for biosynthesis. In 3-4 M urea, the *E. coli* enzyme complex dissociates into an α_8 octamer and 4 β-monomers. The β-subunit, M_r 63,000, contains two prosthetic groups, a siroheme (an Fe - isobacteriochlorin; (12,13)) and an iron-sulfur cluster of the [4Fe-4S] type. The existence of a chemical linkage between the siroheme and the cluster is strikingly manifest in the form of magnetic exchange coupling between the two prosthetic groups, which on the basis of Mossbauer (14,15,16) and EPR spectroscopy

[1] This work was supported by grants from the NIH and VA. The Duke Magnetic Resonance Spectroscopy Center was established with grants from the NIH, NSF, and the North Carolina Biotechnology Center.

(17) has been shown in frozen samples to persist in various redox states of the enzyme, both in the presence and absence of exogenous ligands. A model of the active site has been proposed (Figure 1) in which one cluster Fe is covalently bridged to the siroheme Fe by the sulfur of a cysteinal cluster ligand (17). The X-ray crystal structure (18) is consistent with this model although it lacks sufficient resolution to identify the bridging ligand. Questions about SiR-HP that we can answer utilizing PMR include the following: What is the identity of the bridging ligand? Is the bridge maintained in solution in all three redox states? Are there any ionizable amino acid residues near the heme? What is the spin state of the heme on reduction? How does the siroheme cluster coupling change on heme reduction?

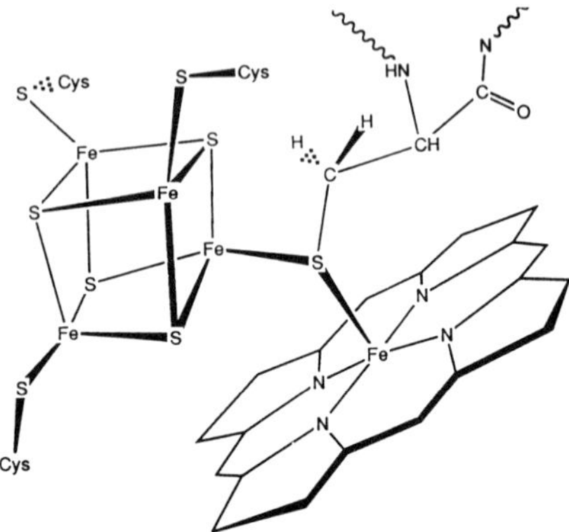

Figure 1. Schematic representation of the exchange coupled [4Fe-4S]-siroheme in the active site of sulfite reductase with cysteine acting as the bridging ligand.

II. Theoretical Considerations in Paramagnetic NMR

The observed chemical shift (δ_{obs}) in PMR can be expressed as the sum of three terms:

$$\delta_{obs} = \delta_{dia} + \delta_c + \delta_p \qquad (1)$$

where δ_{dia} is the diamagnetic chemical shift, i.e., the chemical shift in the absence of unpaired electrons. δ_c is the contact (or Fermi) shift and δ_p is the pseudocontact (or dipolar) shift. The sum of the contact and pseudocontact shifts ($\delta_c + \delta_p$) is called the hyperfine, paramagnetic (δ_{para}), or isotropic (δ_{iso}) shift. The contact shift arises from the transfer of unpaired spin density from the paramagnetic center to the nucleus in question via the σ or π bonding framework. The pseudocontact shift results from the through-space dipole coupling of the nuclear and electron magnetic moments in a highly distance dependent manner.

In addition to causing the isotropic shift, the presence of an unpaired electron can greatly enhance proton NMR relaxation, severely shortening the observed T_1 and T_2. The strength of this interaction can be appreciated if one considers that the dominant proton relaxation mechanism in diamagnetic compounds is a dipolar proton-proton interaction. A similar effect occurs in paramagnetic compounds (a dipolar electron-proton interaction) but the magnetic moment of an electron is 658 times that of a proton. The enhancement of relaxation rates sometimes dictates that a paramagnetic complex will not demonstrate resolvable resonances. The restriction on the limit of the unpaired electron's spin-lattice relaxation time (T_{1e}) which allows observation of discrete nuclear resonances is $T_{1e} < 10^{-11}$ sec. A general guideline is that if the species in question gives sharp EPR signals at room temperature, it will not yield a resolvable NMR spectrum, and if an EPR spectrum is unattainable at room temperature, a well resolved NMR spectrum can be expected.

The small difference in the Boltzmann population of the electron spin levels gives rise to a finite static electric moment that is always aligned along the external magnetic field. This can act to relax protons via a dipolar interaction modulated by molecular tumbling (7,8). This mechanism is usually called magnetic susceptibility or Curie spin relaxation to reflect its relationship with the magnetic susceptibility of a sample (Curie law). The influence of the Curie spin effect on T_1 is extremely small even in cases where it dominates T_2 (7,8). The dependence of the Curie relaxation mechanism on the square of the applied field and the rotational correlation time signifies that it is most important in high molecular weight molecules at high applied fields. For example unligated SiR-HP, which contains a high spin ferric siroheme coupled to a diamagnetic cluster, gives a well resolved spectra at 300 MHz (Figure 2), but at 500 MHz the peaks are broadened past our ability to resolve them. Hence, in PMR, particularly with large proteins, higher fields are not always better.

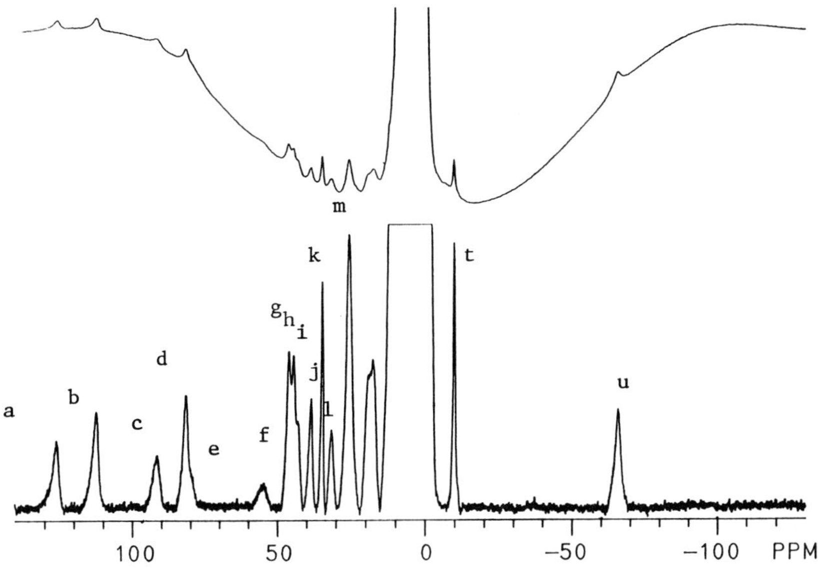

Figure 2. 300 MHz ^1H NMR spectrum of 1.5 mM unligated SiR-HP in 99.9% D_2O (pD=7.7) 50 mM potassium phosphate buffer. The top spectrum is the uncorrected spectrum while the lower spectrum is the same data after baseline correction utilizing the GE intercalate subroutine.

One of the most useful NMR experiments in protein assignment and structural studies is nuclear Overhauser effect (NOE) spectroscopy. The observed NOE between two protons is due to dipolar interaction of spins and indicates that the protons are within 5Å of one another. The nuclear Overhauser enhancement factor (η) is defined as the fractional change in intensity of spin i when spin j is saturated and is given by

$$\eta(t)_{j \rightarrow i} = \sigma_{ij}/\rho_i[1 - e^{-\rho_i t}] \quad (2)$$

where σ_{ij} is the cross relaxation rate defined as

$$\sigma_{ij} = -\gamma^4 h^2 \tau_c/(10 r_{ij})^6 \quad (3)$$

ρ_i is the spin-lattice relaxation rate ($1/T_{1i}$), and t is the time of irradiation of peak j. τ_c is the rotational correlation time. In the case of paramagnetic proteins, where relaxation times are short and ρ_i is large, the observable NOEs are decreased. For example, decreasing the T_1 of a peak from 50 msec (a typical value in low spin hemes) to 5 msec (common in high spin hemes) results in a decrease of the observed NOE value by a factor of ten, assuming a constant cross relaxation rate. Nevertheless, NOEs on the order of 5 - 10% can still be observed for peaks with T_1s as short as 2 msec (19). In SiR-HP, peak *f* has been assigned to a cysteine β-CH_2 by deuterium labelling. Despite its short T_1 (1.8 msec) and large linewidth (~1000 Hz), NOEs can be seen to protons arising from the heme ring as illustrated in the 1D spectrum of Figure 3. This provides evidence that cysteine acts as the ligand bridging between cluster and heme (Figure 1) and is an example of the usefulness of 1D-NOEs.

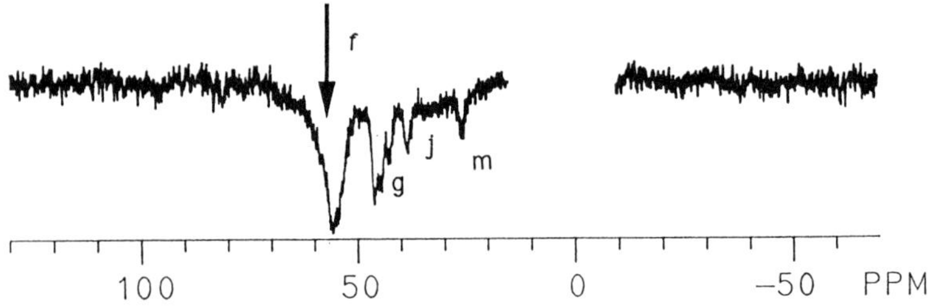

Figure 3. Difference spectrum showing 1D NOEs from a cysteine derived β-CH_2 peak to resonances from the heme periphery in SiR-HP. Irradiation is at the arrow.

III. Practical Considerations in Obtaining PMR Spectra

The characteristic features of a paramagnetic proton NMR spectrum that differ from a diamagnetic spectrum are broad peaks shifted well outside the 0 to 10 ppm range. The large dispersion in chemical shifts stipulates the use of wide sweep widths which may not be available on all spectrometers. The spectrum of SiR-HP in Figure 2 was run with a sweep width of 80,000 Hz. This large sweep width requires the use of short pulse times to obtain uniform excitation across the entire spectral window. A general guideline is that the pulse time necessary for uniform excitation is given by $t_p \leq 1/(4SW)$. A width of 80,000 hertz would therefore require the use of a 90° pulse of 3.1 μsec, however, the measured 90° pulse time is 12 μsec which is linear over only a range of 20,830 Hz. As a compromise in the SiR-HP system, a 7.5 μsec observe pulse was used in most experiments (56°, 0.83 intensity, linear over 33,330 Hz, nulls at 133,000 Hz). While this is adequate in most cases, several experiments require the use of true 90° or even 180° pulses. For example, the inversion-recovery method of measuring T_1s uses the pulse sequence 180°-τ-90°. A 180° pulse time of 24 μsec is linear over the range of only 10,415 Hz, therefore, the T_1s were determined by dividing the spectra into 10,000 Hz sections and repeating the experiment with the transmitter centered in each section.

The size of the sweep width also determines the rate of digitization i.e. the dwell time or delay between digitization points. With quadrature phase detection, the sampling rate is set to twice the sweep width. Since large sweep widths are used in PMR fast digitizers are necessary. The number of points sampled is set by the user and will in conjunction with the dwell time determine the acquisition time. For a diamagnetic spectrum on a 300 MHz spectrometer, a 10 ppm sweep width yields a dwell time of 167 μsec. If an 8K block size is used; this results in an acquisition time of 1.34 sec. The paramagnetic spectrum in Figure 2 has a 280 ppm width yielding a 6.0 μsec dwell time, therefore, 8K points would correspond to an acquisition time of only 48.7 msec. This short acquisition time is satisfactory since we are only interested in the hyperfine shifted peaks, all of which have T_1s of less than 10 msec. Use of a 20 msec decoupler presat pulse to suppress the residual water signal and no recycle delay permits 15 scans to be run every second. In PMR this rapid cycle time permits adequate signal to noise to be achieved in reasonable periods of time despite small broad peaks. The spectrum of unligated SiR-HP in Figure 2 consists of 32,000 transients and was obtained in approximately one hour.

Another advantage of rapid recycle times is the suppression of the large diamagnetic envelope. The large peaks generated from the diamagnetic envelope can fill the dynamic range of the digitizer making the weaker hyperfine shifted peaks undetectable. With rapid pulsing, these peaks never have the opportunity to relax; and so their intensity is diminished. Special pulse sequences have also been designed to minimize the diamagnetic signal such as the super WEFT sequence (20). Most modern spectrophotometers are now routinely equipped with a sixteen bit digitizer. The GN 300 spectrometer used in most of our studies is equipped with both a 12 and 16 bit digitizer. Unfortunately the 16 bit digitizer possesses a limited sweep width (42,000 Hz), so that we have used it only for studying the low spin complexes of SiR-HP or limited sections of the high spin spectrum.

It is necessary to slightly delay the start of data acquisition after the observe pulse to avoid pulse breakthrough in the FID. If this time is a significant fraction of the signal's T_2, a large portion of the FID can be lost. This delay results in the rolling baseline shown in Figure 2. While several routines such as the spline fit, intercalate and baseline fix subroutines can flatten the baseline, the results are less than optimal. They can lead to distortion in peaks away from true Lorentzian shape along with changes in relative peak area.

IV. Assignment Strategies and Data Analysis

Achieving unambiguous resonance assignment is a prerequisite to utilizing the full power of paramagnetic NMR spectroscopy. The problem of resonance assignments is complicated in paramagnetic systems by the loss of the usual shift/functional group correlations established for diamagnetic systems. Several techniques are commonly used to achieve assignments in paramagnetic systems. One of the most powerful involves isotopic substitution by either incorporating specifically labeled amino acids or reconstitution of heme proteins with isotopically labelled hemes. Unfortunately, reconstitution of SiR-HP can not be accomplished because extraction of siroheme invariably leads to cluster destruction.

Incorporation of deuterium labelled cystine has been used to successfully identify resonances arising from the cluster cysteines. Peaks *b*,*c*,*f* and *u* in Figure 2 are absent in the spectrum obtained on SiR-HP isolated from cells grown on β-CD_2-cystine, hence they can be assigned to cluster cysteine β-CH_2s. The size and presence of upfield hyperfine shifts for β-CH_2 of cysteine residues clearly indicates that through-bond coupling between heme and cluster exists in solution.

In proteins of less than 14,000 molecular weight with reasonably narrow lines, it is also possible to assign resonances of a coupled spin system using common homonuclear decoupling methods (8). Unfortunately, the high MW of SiR-HP makes this technique impossible to use. In the isolated siroheme-CN complex; however P-COSY spectra was successfully employed in making assignments and will be discussed below .

Another powerful method utilized for both making assignments and gleaning structural information is the 1D-NOE experiment (21, 22). Two types of 1D-NOE experiments are typically performed in paramagnetic NMR: the steady state (ssNOE) and the truncated (tNOE) experiment. In the ssNOE experiment, the peak of interest is saturated for a period of time much longer than its T_1 with the resulting NOE being equal to σ_{ij}/ρ_i in equation #2. The most important considerations in performing this experiment are: to obtain sufficient signal-to-noise (many scans), and to adequately discriminate true NOEs from off resonance effects. Due to the rapid relaxation and broad lines of hyperfine-shifted resonances, high pulse power is needed to saturate the resonance. As off resonance effects are proportional to the quadratic of the pulse power, increased effects are observed. Typically the NOE difference spectrum is generated by subtracting a control spectrum with the decoupler off resonance from one with the decoupler on resonance. Techniques involved in minimizing or discriminating off-resonance effects include; placing the off-resonance decoupler midway between the two peaks, varying the pulse power, varying the off-resonance decoupler $\pm$ δ from the saturated peak, and "walking" the decoupler across a peak. The last technique is exceedingly useful in helping to separate NOEs arising from two overlapping peaks. The decoupler frequency is moved in small increments across a peak. The intensity of the saturated peaks along with the peaks suspected of possessing NOEs are plotted as a function of decoupler position. As the decoupler moves closer to a a peak, the intensity of off-resonance effects should increase quadratically. However, true NOEs should show a maximum at the point at which the dipolar coupled peak is saturated. In conjunction with through-bond coupling data, 1D NOEs by themselves may be sufficient to completely assign all hyperfine shifted resonances (23).

The second type of NOE experiment, the tNOE experiment, is utilized much less frequently than the ssNOE experiment. In the truncated NOE experiment very short saturation times are used compared to the relaxation rate (T_1), and the size of the NOE becomes $\eta(t)_{j\rightarrow i} = \sigma_{ij}t$. Thus for short saturation times with respect to T_1 the NOE is independent of the nuclei's relaxation rate. However this technique can only be used on protons with relatively long T_1s. A plot of NOE size verse irradiation time can be used to calculate the cross relaxation rate.

Considerable attention has recently been focused on applying 2D NMR techniques to paramagnetic proteins. Examples of proteins with short T_1s and broad peaks that have been successfully examined by 2D methods include metmyoglobin (24) and *Anabaena* 7120 ferredoxin (25). Cross peaks from hyperfine shifted resonances are difficult to detect because of paramagnetically induced relaxation which shortens both spin-spin (T_2) and spin lattice (T_1) relaxation times, sometimes so severely that detection of cross peak intensity is prevented. For the COSY experiment, short T_2s decrease the time available for the buildup of coherence. A similar problem occurs in the NOESY experiment in that short mixing times are necessary because of rapid proton relaxation so only a fraction of the analogous diamagnetic cross peak intensity can be observed. Also, in phase sensitive COSY experiments, the cross peaks exhibit anti-phase properties, which can result in cancellation of detectable coherence. This limitation increases with increasing linewidth. For this reason the magnitude COSY experiment is better suited to samples with large linewidths and sweepwidths. The phase sensitive experiment, however, is useful for spectra with narrow lines and a section of the 2D P-COSY spectrum of the siroheme-CN complex is given in Figure 4. Using this spectrum along with T_1 and relative intensity measurements we were able to classify all the observed peaks. The choice of spectral parameters (mixing time, block size, spectral bandwidth, number of acquisitions) and processing conditions are crucial for good results.

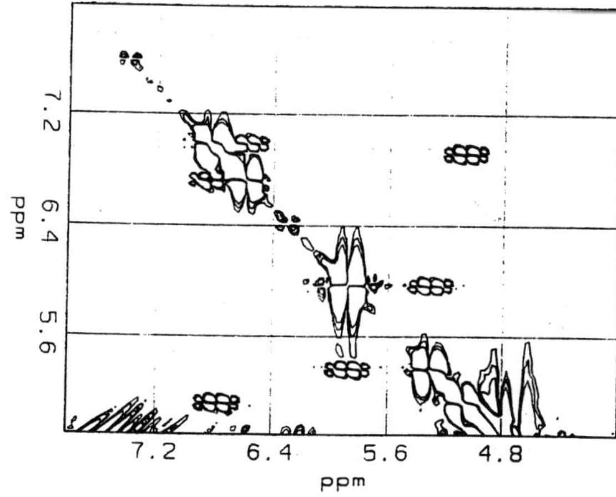

Figure 4. A section of the P-COSY spectrum of siroheme-CN in pD 6.0 50 mM potassium phosphate.

The remaining assignment methods are less definitive then those discussed above and should be used with caution. These techniques include comparison with model compounds and the use of temperature dependent studies and differential dipolar relaxation data. In SiR-HP comparison with model compounds is of limited utility because an iron isobacteriochlorin linked to a [4Fe-4S] cluster has not yet been synthesized . The use of variable temperature data depends on the fact that most hyperfine shifted peaks positions are temperature dependent. In uncoupled paramagnetic systems with orbitally non-degenerate ground states, plots of 1/(absolute temperature) verse peak position give high temperature intercepts at the proton's diamagnetic position. In systems with orbitally degenerate ground states, such as low spin Fe(III), the intercepts are not readily interpretable. In coupled metal systems, hyperfine shifts can be present even in nominally diamagnetic samples due to the population of excited states at room temperature. In this case, as the sample temperature is lowered, the excited state population drops and the net paramagnetic shift decreases. This is the reverse of the normal Curie law behavior in which larger hyperfine shifts are seen at lower temperature.

When dipolar relaxation dominates the nuclear relaxation mechanism, the ratio of relaxation rates for any two non-equivalent protons (a,b) yields a ratio of their respective distances from their paramagnetic center. This is given by

$$T_{1a}/T_{1b} = r_b^6/r_a^6 \quad (4)$$

where protons a and b are r_a and r_b from the metal ion. Therefore, determination of T_{1a} and T_{1b} under the condition that metal centered dipolar relaxation is the sole contribution to to the proton's spin relaxation time, can yield the relative distance of two protons (8).

V. Summary

We have attempted to present some of the principles and techniques necessary to successfully utilize paramagnetic NMR spectroscopy. Structural information about the active site can be obtained from the NMR spectra of paramagnetic metalloproteins, particularly in the case of coupled metal systems. Resonance assignments can be achieved by a variety of techniques including isotopic substitution, 1D-NOE measurements, 2D NOESY/COSY maps, variable temperature intercepts, differential dipolar relaxation, and comparison with model compounds. The size and magnitude of hyperfine shifted resonances along with their temperature dependence and relaxation properties are also valuable in understanding the electronic and magnetic properties of paramagnetic proteins.

References

1. La Mar, G.N., de Ropp, J.S., Smith, K.M., and Langry, K.C. (1980) **J. Biol. Chem. 255**, 6646-6652.
2. Goff, H.M., Gonzalez-Vergara E., and Ales, D.C.(1985) **Biochem. and Biophysical Res. Comm. 133**, 794-799.
3. Bertini, I., Capozzi, F., Luchinat, C., Piccioli, M., and Viezzoli, M.S. (1991) **Eur. J. Biochem. 197**, 691-697.
4. Holz, R.C., Que, L., and Ming, L.J. (1992) **J. Am. Chem. Soc. 114**, 4434-4436.
5. Banci, L., Bencini, A., Bertini, I., Luchinat, C., and Piccioli, M. (1990) **Inorg. Chem. 29**, 4867-4873.
6. Blaszak, J.A., Ulrich, E.L., Markley, J.L., and McMillin, D.R. (1982) **Biochemistry 21**, 6253-6259.
7. Bertini, I., and Luchinat, C. (1986) **NMR of Paramagnetic Molecules in Biological Systems** Benjamin Cummings Pub, Menlo Park, CA.
8. Satterlee, J.D. (1986) **Annual Reports on NMR Spectroscopy 17**, 79-177.
9. Que, L., and Maroney, M.J. (1987) **Metal Ions in Biological Systems 21**, 87-120
10. La Mar, G.N., and Walker, F.A. (1979) in **The Porphyrins Vol. IV** (Dolphin, D.ed.) pp 61-157, Academic Press, New York.
11. Kemp, J.D., Atkinson, D.E., Ehret, A., and Lazzarini, R.A. (1963) **J. Biol. Chem. 238**, 3466-3475.
12. Murphy, J.M., Siegel, L.M., Kamin, H., and Rosenthal, D. (1973) **J. Biol. Chem. 248**, 2801-2814.
13. Scott, IA., Irwin, A.J., Siegel, L.M., and Shoolery, J.N. (1978) **J. Am. Chem. Soc.** 100, 7987-7994.
14. Christner, J.A., Janick, P.A., Siegel, L.M., and Munck, E. (1983a) **J. Biol. Chem. 258**, 11157-11164.
15. Christner, J.A., Munck, E., Janick, P.A., and Siegel, L.M. (1983b) **J. Biol. Chem. 258**, 11147-11156.
16. Christner, J.A., Munck, E., Kent, T.A., Janick, P.A. Salerno, J.C., and Siegel, L.M. (1984) **J. Am. Chem. Soc. 106**, 6786-6794.
17. Janick, P.A., and Siegel, L.M. (1982) **Biochemistry 21**, 3538-3547.
18. McRee, D.E., Richardson, J.S., and Siegel, L.M.(1986) **J. Biol. Chem. 261**, 10277-10281.
19. Dugad, L.B., La Mar, G.N., Banci, L., and Bertini, I. (1990) **Biochemistry 29**, 2263-2271.
20. Inubishi,T. & Becker, E.D. (1983) **J. Magn. Reson. 51** 128-136.
21. Banci, L., Bertini, I., Luchinat, C., and Piccioli, M. (1991) in **NMR and Biomolecular Structure** (Bertini I.., Molinari, H., and Niccolai, N., Eds.) pp 31 - 60, VCH Publishers, New York.
22. Lecomte, J.L., Unger, S.W., & La Mar, G.N. (1991) **J. Magn. Reson. 94**, 112-122.
23. Licoccia, S., Chatfield, M.J., La Mar, G.N., Smith, K.M., Mansfield, K.E., & Anderson, R.R. (1989) **J. Am. Chem. Soc. 111**, 6087-6093.
24. de Ropp, J.S., and La Mar, G.N. (1991) **J. Am. Chem. Soc. 113**, 4348-4350.
25. Skjeldal, L., Westler, W.M., Oh, B.H., Krezel, A.M., Holden, H.M., Jacobson, B.L., Rayment, I., & Markley, J.L. (1991) **Biochemistry 30**, 7363-7368.

One- and Two-Dimensional ^{19}F-NMR Methods for Proteins and Nucleic Acids

Fraydoon Rastinejad[1,2,†], Pam Artz [2], and Ponzy Lu [2,*]

[1]Departments of Biochemistry & Biophysics, and [2]Chemistry
University of Pennsylvania, Philadelphia, PA. 19104

I. Introduction

There are some 300 protein and nucleic acid structures determined by x-ray crystallography to date. While the interactions observed in these structures are quite informative for the chemist, they are constrained by required crystal environment. It is not always clear to the biologist just how these static structures are to be altered to carry out their dynamic cellular roles. The major advantage of using NMR for structural studies has been the ability to study biomolecules in solution environments more akin to *in-vivo* conditions. This has been constrained as well by narrow chemical shift range of ^{1}H-NMR and the large number of protons in biological macromolecules and their solvent environments.

^{1}H-NMR experiments involving two-, three-, and four-dimension have not to date been applied to protein structures larger than 153 residues (Clore *et al.*, 1991), whereas the median size protein contains some 250 amino acids. Moreover, one would ideally like to obtain structural data for

---

* Author to whom all correspondence should be addressed.

† Present Address: Howard Hughes Medical Institute, Yale University School of Medicine, New Haven CT, 06536-0812

TECHNIQUES IN PROTEIN CHEMISTRY IV

proteins in the presence of their bound ligands or in association with other proteins. Studies of protein-nucleic acid interactions by ^{1}H-NMR have been limited due to those problems. To date, there are only two detailed ^{1}H-NMR studies of protein-DNA interaction, both of which are based on small protein fragments rather than intact proteins (Boelens *et al.*, 1987; Otting *et al.*, 1990).

As an alternative to ^{1}H, other NMR detectable nuclei may be used to obtain structural details for large macromolecular complexes. Fluorine-19 offers a sensitivity of detection only 17% lower than proton. ^{13}C, ^{15}N, ^{17}O, and ^{31}P have much lower natural abundance or gyromagnetic ratios, causing their absolute sensitivity of detection to be four to five orders of magnitude lower than ^{19}F. Detection of these nuclei must be improved by (expensive) isotopic enrichment, increased sample concentrations, or use of indirect detection methods. We have extended the applications of fluorine-19 NMR originally demonstrated by Sykes (1974), Gerig (1976), and Ho (1975).

^{19}F chemical shifts are extended over a 70-fold wider ppm range than that of ^{1}H. The intrinsic sensitivity of a given nucleus to its environment is determined by the density, distribution, and symmetry of the orbital electrons surrounding the nucleus. In hydrogen, only the s-electron exerts an orbital magnetic field at the nucleus. However, in covalently bound fluorine, with a valency shell containing unequal populations in its three 2 p-orbitals, the fluorine chemical shift tensor is quite large and orientation sensitive (Feeney, 1967; Gerig, 1989). As a result, the ^{19}F-NMR observed peak positions are highly dependent on the micro-environment of the fluorine nucleus.

Fluorine atoms on aromatic rings of amino acids and nucleotides can significantly alter pKs of ring protons, leading to altered structures and reactivity. The inductive effects of the fluorine substitution on the thymine or uracil base are strong enough to significantly lower the pKa of the N3 (imino) proton, and increase its acidity. Because the thymidine N3 proton participates in Watson-Crick base-pairing with the adenine N1 atom, its increased acidity is expected to strengthen the hydrogen bond and stabilize the DNA helix. Nevertheless, limited fluoro-substitutions often produce only negligible changes in local structures. For example, in 5-FdU substituted hexamer DNA, the helix to coil transition had a midpoint transition temperature 0.7 °C higher than that of the wild-type

Table I. Fluorinated analogs of nucleosides and nucleoside triphophates

Molecule	Name	Use	Reference(s)
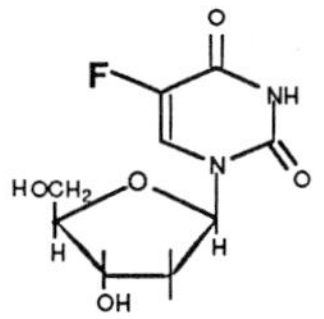	5-fluoro-2'-deoxyurdine (5F-dU)	^{19}F-NMR observations of DNA binding by λ cro repressor; T7 RNA polymerase; *lac* repressor.	Metzler & Lu (1989); Rastinejad (1992)
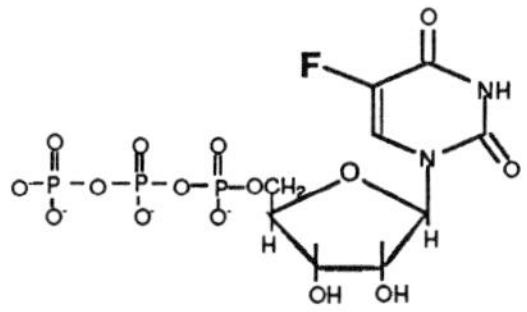	5-fluorouridine triphosphate (5F-UTP)	^{19}F-NMR studies of tRNA (val) structure and interactions with synthetase; 5-S RNA structure; T7 RNA polymerase transcription.	Gollnick *et al.* (1987); Hardin et al. (1987); Chu & Horowitz (1991); Marshall & Smith (1977); Rastinejad (1992)
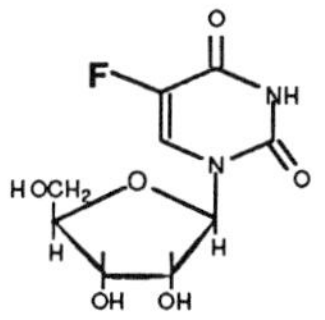	5-fluorouridine (5-FU)	tRNA studies	Gollnick *et al.* (1987); Hardin et al. (1987); Chu & Horowitz (1991); Marshall & Smith (1977)
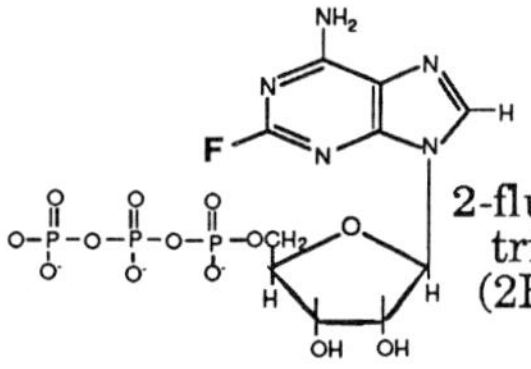	2-fluoroadenosine triphosphate (2F-ATP)	^{19}F-NMR observations of binding by myosin subgfragment-1	Baldo *et al.* (1983)
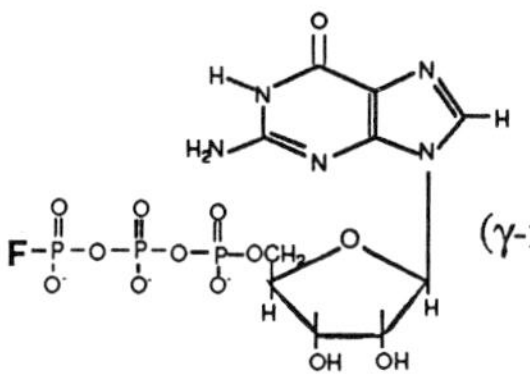	guanosine 5'-(γ-fluorotriphosphate) (GTP-γF)	^{19}F-NMR spectroscopy of tubulin binding	Monasterio (1987).

hexamer (Metzler, 1988), and the essential helical parameters remain unaltered (Metzler & Lu, 1989).

Fluoro-amino acids that can be used in NMR studies of proteins include 4-F-Phe (Lu, 1990; Richmond, 1963; Pliska & Marbach, 1978), 3-F-Tyr (Sixl, 1990; Kimber, 1977, Lu, 1976; Anderson, 1975; Sykes, 1974), and 4-, 5-, and 6-F-Trp (Pratt,

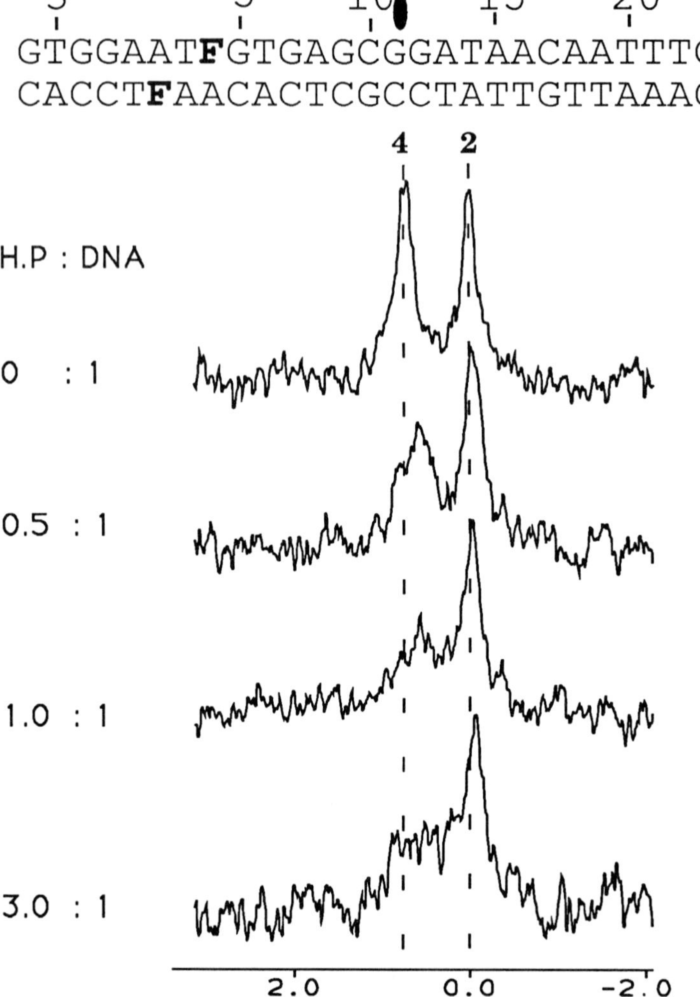

Figure 1. 19F-NMR spectra showing the addition of *lac* repressor headpiece (H.P) to 5-FdU substituted *lac* operator DNA. The stoichiometries are indicated at left. Bold Fs mark the 5-FdU positions in the DNA, and a dark oval marks the pseudo dyad axis. Assignments were made by comparison with the singly substituted operator DNAs.

1990; Sixl, 1990; Pratt & Ho, 1975; Brown & Otvos, 1976; Robertson, 1977; Kimber, 1977). In addition, there are several fluorinated analogs of nucleosides (shown in Table 1). Fluoro-substitution allows ^{19}F-NMR observation of a small number of

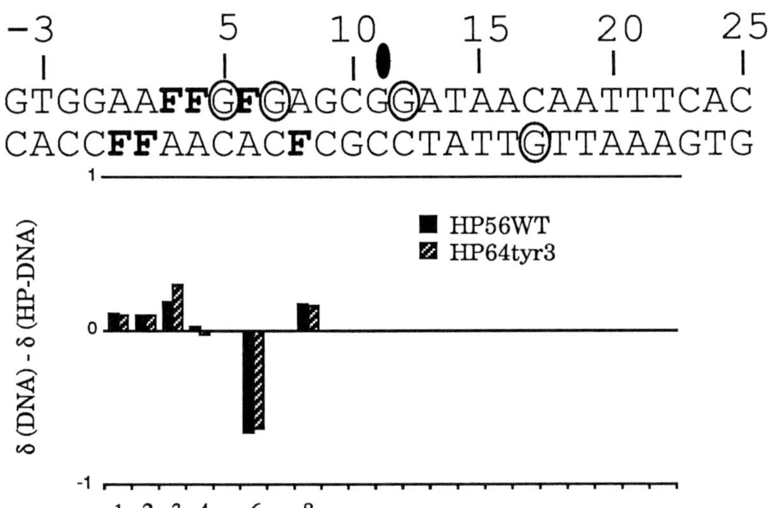

Figure 2. ^{19}F-NMR resonance shifts caused by binding of *lac* headpieces to 5-FdU containing operator DNA. HP56WT and HP64tyr3 are 56 and 64 residue polypeptides respectively, the latter contains a pro3 to tyr mutation. HP-DNA refers to 2.5 : 1 protein : DNA complexes. Numbers along the bottom indicate operator DNA positions (sequence is shown above). Circles indicate guanines protected from methylation in presence of *lac* repressor.

well resolved ^{19}F-NMR resonances from pre-selected sites in proteins and nucleic acids. Fluorine NMR spectra from a fluoro-substituted macromolecule complexed with non-fluorinated species allows characterization of the specific interaction without increased spectral complexity and loss of resolution. Moreover, complete freedom can be exercised in choosing an appropriate buffer or solvent system since these components are generally invisible to ^{19}F-NMR.

Fluorine-19 spin lattice (T_1) relaxation is mainly dipolar for proteins and nucleic acids at both low and high magnetic fields. This means that the ^{19}F-{^{1}H}-NOE dependence on rotational correlation time displays its maximal intensity of -1 for macromolecules with correlation time $> 5 \times 10^{-8}$ sec. As a result, two dimensional ^{19}F-{^{1}H}HOESY experiments are ideally suited for polypeptides and nucleic acids (Metzler *et al.*, 1988). This is in contrast to many other proton-heteronuclear NOE experiments where the NOE intensity nulls with increasing τ_c (Wuthrich, 1986), making them useful only for small molecules. The spin-spin relaxation (T_2) for fluorine-19 is dominated by chemical shift anisotropy for large macromolecules (> 20 kDa) and for magnetic fields > 300 MHz (Hull & Sykes, 1975). Superior resolution is obtained at lower fields (Hull & Sykes, 1975; Kimber, 1977).

II. One-dimensional ^{19}F-NMR spectroscopy

Figures 1 and 2 show that 5-FdU substitutions in the *lac* operator DNA give ^{19}F-NMR signals that map specific interaction sites of the *lac* repressor DNA binding domain (headpiece). 5-FdU differs from deoxythymidine only at the 5 position, where the methyl is replaced by fluorine. As a result, ^{19}F-NMR signals monitor the local major groove electronic and magnetic environments of the DNA. These local environments are changed appreciably when the *lac* headpiece amino acid side chains interact at sites proximal to fluorines. The operator DNA base 6, a site adjacent to a pair of guanine N-7 atoms implicated by methylation protection (Ogata & Gilbert, 1979) shows the largest chemical shift perturbation (Figure 2).

There is no quantitative method for interpreting fluorine chemical shifts unambiguously in terms of structural details, although much progress has been made in identifying shielding contributions that affect fluorine chemical shifts (Gregory & Gerig). The use of two-dimensional heteronuclear Overhauser spectroscopy (HOESY) described below provides the advantages and sensitivity of fluorine detection and allows determination of real distances separating ^{19}F and ^{1}H nuclei.

III. Two-Dimensional ^{19}F-{^{1}H} Heteronuclear NOE Spectroscopy (HOESY)

Figure 3 shows the ^{19}F-{^{1}H}HOESY spectrum of *lac* repressor headpiece that contains biosynthetically incorporated 3-fluorotyrosines (Lu *et al.*, 1990). The pulse sequence of Yu & Levy was used (1983, 1984), with a mixing time of 300 msec. The measurement parameters have been described previously (Metzler *et al.*, 1988). Assignments of the fluorine resonances have been published previously (Arndt *et al.*, 1982). The crosspeaks connect the fluorines with dipolar coupled protons at their respective chemical shifts. Each fluorine resonance should contain a cross-peak in the 6-7 ppm (^{1}H) region, corresponding to position 2 hydrogens adjacent to the fluorine on the tyrosine ring. Other cross-peaks in the 6-7 ppm region

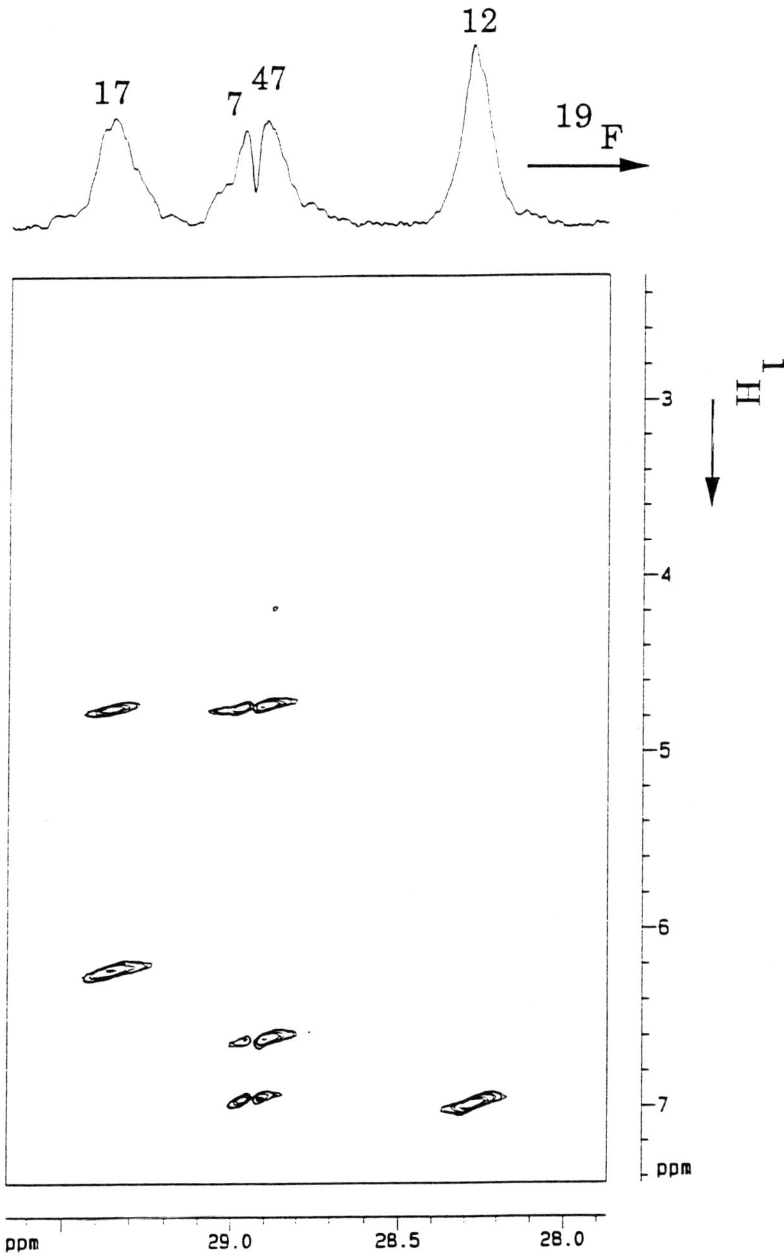

Figure 3. HOESY spectrum of *lac* headpiece HP56WT containing 3-F-tyr at sites 7, 12, 17, and 47. ^{19}F and ^{1}H dimensions are indicated. Crosspeaks at 6-7 ppm of ^{1}H are to the H2 hydrogens of the tyrosine rings. Crosspeaks at 4-5 ppm are inter-residue connections.

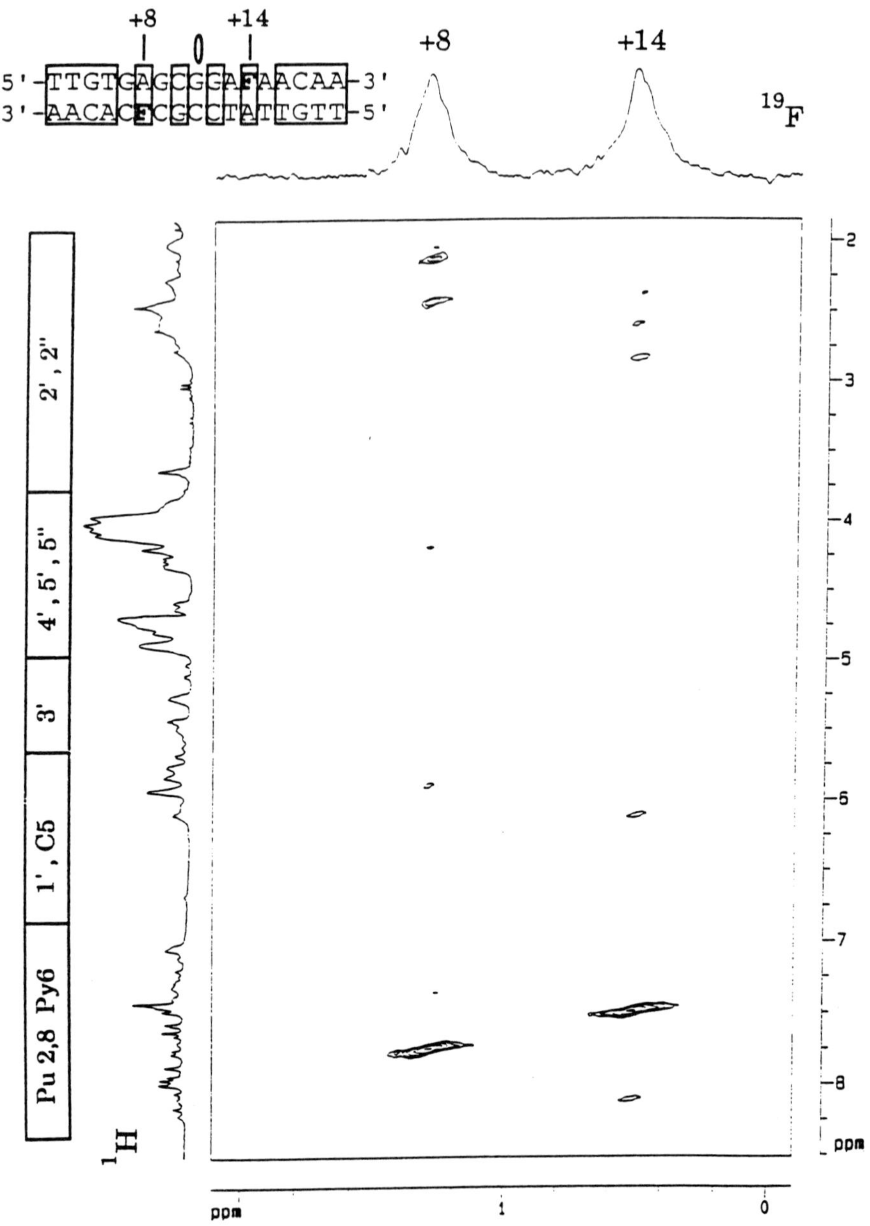

Figure 4. HOESY spectrum of *lac* operator DNA (sequence shown) with 5-FdU substitutions at pseudo-dyad related sites +8 and +14. Symmetric sequences are boxed. Spectrum was obtained using 500 ms mixing time.

may correspond to 2, 4, or 5 protons belonging to other tyrosines in the headpiece. The cross-peaks to the 4-5 ppm region likely represent short distances to C-alpha protons of residues adjacent to the fluorines in the headpiece structure.

In 5-FdU substituted B-DNA, one expects a large number of protons within 5 Å of the fluorine atom. These range from 2.6 Å to the intranucleotide H6, to the internucleotide H8 or H5 at 4.7 Å. In addition, one expects crosspeaks to the 1', 2', and 2" sugars of the 5' nucleoside. Figure 4 shows crosspeaks corresponding to many of these fluorine-proton distances in the ^{19}F-{^{1}H}HOESY spectrum of fluoro-substituted *lac* operator DNA. While the HOESY spectra in Figures 3 and 4 can provide direct intramolecular structural information, the HOESY spectrum of a protein-DNA complex has the potential of providing intermolecular connections. We have found intermolecular contacts between *lac* repressor headpiece and 5-FdU substituted lac operator DNA using this method (Rastinejad, 1992; Artz, Rastinejad, & Lu, manuscript in preparation).

References

Anderson, R. A., Nakashima, Y., & Coleman, J. E. (1975). *Biochemistry* **14**, 907-917.

Arndt, K., Nick, Ha., Boschelli, F., Lu, P., & Sadler, J. (1982). *J. Mol. Biol.* **161**, 439-57.

Baldo, J. H., Hansen, Poul E., Shriver, John W., Sykes, Brian D. (1983). *Can. J. Biochem. Cell Biol* **61**, 115-19.

Boelens, R., Scheek, R. M., Van Boom, J. H., & Kaptein, R. (1987). *J. Mol. Biol.* **193**, 213-16.

Bolin, J. T., Filman, D. J., Matthews, D. A., Hamlin, R. C., & Kraut, J. (1982). *J. Biol. Chem.* **257**, 13650-13662.

Boschelli, F., Jarema, M. C., & Lu, P. (1981). *J. Biol. Chem.* **256**, 11595-9.

Brown, K. D., & Otvos, D. (1976). *Biochem. Biophys. Res. Commun.* **68**, 907-913.

Chu, W. C. and J. Horowitz (1991). *Biochemistry* **30**, 1655-63.

Clore, G. M., Kay, L. E., Bax, A., & Gronenborn, A. M. (1991). *Biochemistry* **30**, 12-18.

Feeney, J. (1967). In Nuclear Magnetic Resonance for Organic Chemists. D. W. Mathieson, ed. New York, Academic Press. 151-158.

Gerig, J. T., & McLeod, R. S. (1976). *J. Am. Chem. Soc.* **98**, 3970-5.

Gerig, J. T. (1978). In Biol. Magn. Reson. L. Berliner & Reuben, J., ed. New York, Plenum Press. 139-203.

Gerig, J. T. (1989).*Methods Enzymol* **177B**, 3-23.

Gollnick, P., Hardin, C. C., & Horowitz, J. (1987). *J. Mol. Biol.* **197**, 571-84.

Gregory, D. H. and J. T. Gerig (1991). Prediction of fluorine chemical shifts in proteins. *Biopolymers* **31**, 845-58.

Hardin, C. C., Gollnick, P., Kallenbach, N. R., Cohn, M., & Horowitz, J. (1986). *Biochemistry* **25**, 5699-709.

Hardin, C. C. and J. Horowitz (1987). *J. Mol. Biol.* **197**, 555-69.

Holland, D. P., Klein, R. L., & Briggs, A. H., (1964). Introduction to Molecular Pharmacology. New York, MacMillan.

Horowitz, J., Ofengand, J., Daniel, W. E., & Cohn, M. (1977). *J. Biol. Chem.* **252**, 4418-20.

Horowitz, J., Cotten, M. L., Hardin, C. C., & Gollnick, P. (1983). *Biochim. Biophys. Acta* **741**, 70-6.

Hull, W. E., & Sykes, B. D. (1974). *Biochemistry* **13**, 3431-7.

Hull, W. E., Sykes, B. D. (1975). *J. Chem. Phys.* **63**, 867-80.

Hull, W. E., & Sykes, B. (1975). *J. Mol. Biol.* **98**, 121-153.

Jardetzky, O., & Roberts, G. C. K. (1981). NMR in Molecular Biology. New York, Academic Press.

Jarema, M. C., Lu, P., & Miller, J. H. (1981). *Proc. Natl. Acad. Sci. U. S. A.* **78**, 2707-11.

Jarema, M. C., Arndt, K. T., Savage, M., Lu, P., & Miller, J. H. (1981). *J. Biol. Chem.* **256**, 6544-7.

Kimber, B. J., Griffiths, D. V., Birdsall, B., King, R. W., Scudder, P., Feeney, J., Roberts, G. C. K., & Burgen, A. S. V. (1977).*Biochemistry* **16**, 3492-3500.

Lu, P., Jarema, M., Mosser, K., & Daniel, W. E. (1976). *Proc. Natl. Acad. Sci. U. S. A.* **73**, 3471-5.

Lu, P., Metzler, W. J., Rastinejad, F., & Wasilewski, J. (1990). In Struct. Funct. Nucleic Acids Proteins , F.H. Y. Wu, and K. Wu, eds. 19-35.

Marshall, A. G. and J. L. Smith (1977). *J. Am. Chem. Soc.* **99**, 635-6.

Metzler, W. J., Leighton, P., & Lu, P. (1988). *J. Mag. Res.* **76**, 534-539.

Metzler, W. J. (1988). Thesis: University of Pennsylvania.

Metzler, W. J., & Lu, P. (1989). *J. Mol. Biol.* **205**, 149-64.

Monasterio, O. (1987). *Biochemistry* **26**, 6099-6106.

Ogata R. T., & Gilbert, W. (1979). *J. Mol. Biol.* **132**, 709-728

Otting, G., Qian, Y. Q., Billeter, M., Mueller, M., Markus, A., Gehring, W. J., & Wuthrich, K. (1990). *EMBO J.* **9**, 3085-92.

Pliska, V., & Marbach, P. (1978). *Eur. J. Pharmacol.* **49**, 213-22.

Pratt, E. A., & Ho, C. (1975). *Biochemistry* **14**, 3035-3040.

Rastinejad, F. (1992). Thesis: Univeristy of Pennsylvania, 172-198.

Richmond, M. H. (1963). *J. Mol. Biol.* **6**, 284.

Sixl, F., King, R. W., Bracken., M., & Feeney, J. (1990). *Biochem. J.* **266**, 545-52.

Sykes, B. D., Weingarten, H. I., & Schlesinger, M. J. (1974). *Proc. Natl. Acad. Sci.* **71**, 469-473.

Sykes, B. D., & Weiner, J. H. (1980). Manetic Resonance in Biology. J. S. Cohen, ed. New York, John Wiley & Sons. 171-196.

Taylor, H. C., Richardson, D. C., Richardson, J. S., Wlodower, A., Komoriya, A., & Chaiken, I. M. (1981). *J. Mol. Biol.* **149**, 313-317.

Wuthrich, K. (1986). NMR of Proteins and Nucleic Acids. New York, John Wiley & Sons.

Yu, C., & Levy, G. C (1983). *J. Am. Chem. Soc.* **105**, 6994-6996.

Yu, C., & Levy, G. C (1984). *J. Am. Chem. Soc.* **106**, 6533-6537.

"Active Site" Dynamics: A Comparison of the Wild Type and Single Tryptophan Variants of *E. Coli* Thioredoxin

Marvin D. Kemple[a], Kenneth E. Nollet[b], Peng Yuan[a], and Franklyn G. Prendergast[b]

[a]Department of Physics, Indiana University-Purdue University at Indianapolis, Indianapolis, IN 46205-2810

and

[b]Department of Biochemistry and Molecular Biology, Mayo Foundation, Rochester, MN 55905

I. Introduction

Escherichia coli thioredoxin (TRX) is a low molecular weight (11,700), almost spherical redox protein. TRX functions *in vivo* as part of a system including NADPH and a flavoprotein, thioredoxin reductase. The "active site" of *E. coli* TRX consists of cysteine residues at positions 32 and 35. These thiol side chains participate in electron transfer reactions, and, by forming a disulfide bond, reduce other disulfides, such as those of ribonucleotide reductase, methionine sulfoxide reductase, and adenosine 3'-phosphate 5'- phosphosulfate reductase. Reduced thioredoxin also plays a structural role in the life cycles of phages M13, T4, and T7 (1,2). TRX has two tryptophan residues, one at position 28 and the other at position 31. X-ray (3,4) and NMR (5) structural studies have shown that trp-31 lies on the surface of the protein as part of a hydrophobic patch implicated in substrate binding, and that trp-28 is partially buried and spatially proximate to the pair of cysteines. Because of their close relationship to the active site of TRX, the tryptophan residues are sensitive spectroscopic probes of that region. TRX is very suitable for spectroscopic studies since it is quite stable and either of its oxidation states is easily maintained over the time required for making the measurements.

Fluorescence spectral and lifetime studies of TRX have been reported (6-8). In this paper, we describe some of the results of our ^{13}C NMR relaxation and fluorescence anisotropy inquiry into the nature and extent of subnanosecond internal motions of the tryptophan residues of the wild type (WT) and of two single tryptophan mutants of *E. coli* TRX, W28F and W31F, in which phenylalanine replaces trp-28 and trp-31 respectively. We have incorporated two differently, isotopically labeled tryptophans into WT-TRX and the mutants, one with ^{13}C at the δ_1 (or 2) position of the indole ring and one with ^{13}C at the ε_3 (or 4) position of the indole ring. In solution the NMR relaxation rates of these ^{13}C nuclei are determined primarily by the modulation due to molecular tumbling of (i) the magnetic dipolar interaction between the ^{13}C and its attached proton and of (ii) the anisotropic part of the chemical shielding interaction. Details of the rotational motion of the labeled tryptophans in the proteins can then be deduced from

TECHNIQUES IN PROTEIN CHEMISTRY IV

measurements of ^{13}C relaxation rates. Tryptophan fluorescence anisotropy is also sensitive to the rotational motion of the indole ring. Thus we can take advantage of two different sets of spectroscopic measurements on the same samples to learn about the dynamics of the proteins (9,10).

In general, in all of the native proteins examined, the internal motion of the tryptophan rings was found to be quite restricted as reflected in high order parameters. A slight increase in the amplitude of the motion of trp-31 was observed in the reduced form of the protein in contrast to the oxidized form. The mutant proteins gave indications of more motional freedom of their tryptophan ring than was apparent in WT-TRX. We also subjected the reduced proteins to urea denaturation. The internal motion of the tryptophan rings is significantly less restricted in the denatured proteins as would be expected.

II. Materials and Methods

A tryptophan-auxotrophic strain of W3110 *E. coli* was transformed to express WT-TRX, W28F, or W31F in the laboratory of Jim Fuchs (Department of Biochemistry, University of Minnesota). Cultures were grown in rich media which included acid hydrolyzed (tryptophan-free) casein and, initially, yeast extract. The final tenfold dilution of the cell broth was made into the same medium without yeast extract. In this protocol, natural tryptophan exhaustion occurred at a cell density ($A_{660} \approx 0.8$) suitable for plasmid induction. Subsequent induction with IPTG was accompanied by the addition of either $^{13}C\delta_1$-trp, $^{13}C\varepsilon_3$-trp (Cambridge Isotope Laboratories), or an equimolar mixture of the two. Whole cells were packed by centrifugation, blended in a threefold excess of pH 7.5, 20 mM TRIS, 1 mM EDTA, and sonicated. Cell debris was packed by centrifugation, and the supernatant was stirred with streptomycin sulfate to precipitate nucleic acids. After another centrifugation step, the supernatant was filtered and loaded onto a 7 liter Sephadex S-100 column. The active fractions were identified with a colorimetric DTNB assay (11), pooled, and loaded onto a 2 liter DEAE ion exchange column. Thioredoxin eluted at a KCl concentration of about 0.3 M. Protein purity was verified using silver-stained SDS PAGE PhastGel media according to manufacturer's instructions (Pharmacia LKB). The pooled fractions were concentrated by ultrafiltration, dialyzed into potassium phosphate buffer, and lyophilized. For the NMR experiments, the protein was dissolved in 99.98 % D_2O to effect a sample concentration of approximately 5 mM in 50 mM pH 6.5 potassium phosphate buffer. The protein samples for NMR were subsequently reduced by adding a 3-10 times molar excess of DTT. There were no indications of interconversion of the proteins between the oxidized and reduced forms during the experiments. Fluorescence measurements were made on 50 mM potassium phosphate solutions with protein concentrations of 100 μM; a 20 fold molar excess of DTT was used to reduce the proteins in those samples.

^{13}C NMR measurements were performed at 75.4 MHz on a Nicolet NT-300 FT spectrometer and on a Varian Unity 300 MHz FT spectrometer. The broadband probe used with the Unity spectrometer was obtained from Cryomagnet Systems, Inc. Proton decoupled ^{13}C NMR spectra were obtained using either broadband noise or WALTZ-16 decoupling during signal acquisition. Proton-coupled spectra were acquired as well in order to measure differential broadening of the ^{13}C doublets. ^{13}C spin lattice relaxation time (T_1) measurements were made using the inversion recovery method with proton decoupling during acquisition. The extracted T_1 values, which were derived from three parameter fits of the recoveries to A {1- (1+W) exp (-t/T_1)} where t is the recovery time, were the same within experimental uncertainties whether or not the proton magnetization was saturated during the recovery period. Steady state ^{13}C -^{1}H nuclear Overhauser effect (NOE) measurements were made by collecting alternate scans (with and without NOE) of the ^{13}C signal into different memory blocks. Spin-spin relaxation (T_2) measurements were made with a CPMG pulse sequence with proton decoupling during acquisition only. Unless otherwise specified the NMR probe temperature was maintained at 20 °C.

Steady state trp fluorescence anisotropy measurements were made on an SLM-4800 fluorometer set to an excitation wavelength of 300 nm with a 1 nm bandpass. Fluorescence emission was detected through a Schott WG-345 cut-on filter. Fluorescence lifetimes were measured with a laser-based time correlated single photon counting instrument. The fluorescence intensity decays generally were fit to a sum of up to four exponentials. Only the average fluorescence lifetime was required in the analysis presented below (12). The sample temperature for these measurements was also 20 °C.

III. Theory

Since the primary ^{13}C relaxation mechanisms in this case consist of the ^{13}C-^{1}H magnetic dipolar interaction (DD) and the ^{13}C chemical shift anisotropy (CSA), T_1, T_2, and the NOE are given in terms of the following equations (13, 14).

$$T_1^{-1} = R_{DD}^{(1)} + R_{CSA}^{(1)} \qquad (1)$$

$$T_2^{-1} = R_{DD}^{(2)} + R_{CSA}^{(2)} \quad \text{and} \qquad (2)$$

$$NOE = 1 + \frac{1}{4}\alpha^2 \left(\frac{\gamma_H}{\gamma_C}\right) \left[\frac{6J(\omega_H + \omega_C) - J(\omega_H - \omega_C)}{R_{DD}^{(1)} + R_{CSA}^{(1)}}\right] \qquad (3)$$

where

$$R_{DD}^{(1)} = \frac{1}{4}\alpha^2 \left[J(\omega_H - \omega_C) + 3J(\omega_C) + 6J(\omega_H + \omega_C)\right] \qquad (4)$$

$$R_{CSA}^{(1)} = \frac{1}{3}\beta^2 J(\omega_C) \qquad (5)$$

$$R_{DD}^{(2)} = \frac{1}{8}\alpha^2 \left[4J(0) + 3J(\omega_H - \omega_C) + 3J(\omega_C) + 3J(\omega_H) + 6J(\omega_H + \omega_C)\right] \qquad (6)$$

$$R_{CSA}^{(2)} = \frac{1}{18}\beta^2 \left[4J(0) + 3J(\omega_C)\right] \qquad (7)$$

with $\alpha = \frac{\gamma_C \gamma_H \hbar}{r_{CH}^3}$ and $\beta = \frac{3}{2}\omega_C \delta_{ZZ}\left(1+\frac{\eta^2}{3}\right)^{1/2}$

and ω_C and ω_H are the carbon and proton resonance frequencies, γ_C and γ_H are their respective gyromagnetic ratios, $\hbar$ is Planck's constant over 2π, r_{CH} is the carbon-proton distance, δ_{ZZ} is the largest magnitude principal element of the carbon CSA tensor (in ppm), and η is the asymmetry parameter of the CSA tensor.

The information on molecular motion is contained in the spectral density, $J(\omega)$. In the analysis presented here, the Lipari & Szabo approach (14) is used to express the spectral density as

$$J(\omega) = \frac{2}{5}\left\{ \frac{S^2\tau_m}{1+\omega^2\tau_m^2} + \frac{(1-S^2)\tau}{1+\omega^2\tau^2} \right\} \tag{8}$$

where S^2 is a generalized, model-free order parameter related to the amplitude of the internal motion of the C-H vector ($0 \leq S^2 \leq 1$) and $\tau^{-1} = \tau_m^{-1} + \tau_e^{-1}$ with τ_e an effective correlation time for the internal motion, which has meaning when $S^2 < 1$, and τ_m a correlation time for the overall rotational motion of the molecule assumed to be isotropic which is reasonable for TRX. Values for these three motional parameters, τ_m, S^2, and τ_e, can be found from the measured relaxation rates by substituting Eq (8) for the spectral density into Eqs (1)-(7).

When both the CSA and the ^{13}C-^1H dipolar interactions make significant contributions to the relaxation rates, the interference of those interactions, called cross correlation, may become apparent as a differential broadening of the components of a scalar-coupled multiplet and therefore can be observed in ^1H-coupled ^{13}C spectra (15,16). For the case of a single coupled proton, the difference in width between the two ^{13}C lines, $\Delta(LW)$ in Hz, can be written

$$\Delta(LW) = \frac{1}{6\pi}\alpha\beta'\left[4J(0) + 3J(\omega_C)\right] \tag{9}$$

where

$$\beta' = \frac{3}{2}\omega_C \delta_{ZZ}\left[3\cos^2\theta - 1 - \eta\sin^2\theta\cos 2\phi\right] \tag{10}$$

and θ and ϕ are spherical coordinates of the C-H bond in the principal axis frame of the CSA tensor. The other symbols were defined above. Differential broadening thus gives another measure of the motional parameters.

Motional information can be obtained from the steady state fluorescence anisotropy, $\bar{r}$, from the equation (10)

$$\bar{r}/r_o = \frac{(1-S^2)}{1+\tau_f(\tau_e^{-1}+\tau_m^{-1})} + \frac{S^2}{1+(\tau_f/\tau_m)} \tag{11}$$

where τ_f is the fluorescence lifetime. $S^2 = r_\infty/r_0$ is an orientational order parameter. r_0 is the fundamental anisotropy which is a function of the relative orientation of the fluorescence emission and absorption dipoles at a given excitation wavelength and r_∞ is the limiting hindered anisotropy of the fluorophore. The NMR-derived order parameter in Eq (8) can be expected to be the same as the fluorescence order parameter only when the emission dipole is colinear with the relevant C-H vector. That condition is approximately met for the $^{13}C\delta_1$-trp label but not for $^{13}C\varepsilon_3$-trp. In Eq (9), it is sufficient to use the mean fluorescence lifetime. Inclusion of a multi-exponential form for the fluorescence intensity decay in the derivation of Eq (9) does not have a significant effect on the motional parameters obtained. Our approaches for finding values of the motional parameters from the data are described below. In the analysis, the following values for the CSA tensor were used for $^{13}C\delta_1$-trp and $^{13}C\varepsilon_3$-trp respectively (17 and Separovic unpublished results); δ_{ZZ} = -78 ppm, η = 0.936, θ = 26 °, ϕ = 90 °; δ_{ZZ} = -115 ppm, η = 0.639, θ = 0 °. Also $r_0 = 0.3$ at λ_{ex} = 300 nm (18) was used in Eq (11) above.

IV. Results and Discussion

The ^{13}C chemical shifts for WT-TRX and the single tryptophan mutants are summarized in Table I along with representative T_1, T_2, steady state NOE, and $\bar{r}$ values. Measurement uncertainties are on the order of 10 %. The assignment of the resonances to trp-28 and trp-31 in WT-TRX were made from a 2D heteronuclear COSY spectrum of the protein on the basis of published ^1H assignments for WT-TRX (19). Subsequently the assignments were corroborated from spectra of the single tryptophan mutants. As apparent from Table I, changes in the chemical shifts of both the δ_1 and ε_3 carbons of trp-28 and trp-31 were observed upon reduction of the protein.

Motional parameters obtained from least squares fits of the data to Eqs (1)-(8) & (11) are given in Table II. For WT-TRX, T_1, T_2, and the NOE were included in the analysis for each data set, but $\bar{r}$ was not because it was not possible to separate the contributions to $\bar{r}$ from the two tryptophans. $\bar{r}$ was included in the analysis of the $^{13}C\delta_1$ data for W31F and W28F however, and it was assumed that the NMR and fluorescence order parameters were the same in those cases. T_2 was not included for the mutant proteins and a common τ_m was solved for each protein.

The differential line broadening data, representative values of which are listed in Table III, were analyzed separately. The uncertainties in these data are fairly large (~ 2-4 Hz for the WT and 5-10 Hz for the mutants). Linewidths were more difficult to measure accurately in the latter case because of the interference of natural abundance ^{13}C signals with the signals of interest. This interference was more prevalent in the mutant proteins because the incorporation of the isotopic labels was not as good in the mutant samples as it was in the WT samples due to variations in sample preparation. For WT-TRX, only data for $^{13}C\varepsilon_3$ are listed because the $^{13}C\delta_1$ peaks of trp-28 and trp-31 overlapped in the coupled spectrum. In the limit of restricted internal motion, *i. e.* S^2 close to 1, which applies here (see Table II), and very rapid internal motion $\tau_e << \tau_m$, it can be shown from Eqs (8) and (9) that Δ(LW) depends mainly on the product $S^2\tau_m$. Values of $S^2\tau_m$ calculated in this manner from Eq (9) are also included in Table III. These values for $S^2\tau_m$ are in reasonable agreement with those given in Table II taking account of the experimental uncertainties.

Table I. Examples of chemical shifts, T1, T2, NOE, steady state anisotropy, and fluorescence lifetime data for thioredoxin

Protein	δ (ppm)	T_1 (s)	T_2 (s)	NOE	$\bar{r}$	τ_f (ns)
WT-TRX (S-S), W28, $^{13}C\delta_1$	124.6	0.320	0.015	1.12		
WT-TRX (S-S), W31, $^{13}C\delta_1$	127.3	0.320	0.019	1.12		
WT-TRX (S-S), W28, $^{13}C\varepsilon_3$	118.0	0.340	0.015	1.11		
WT-TRX (S-S), W31, $^{13}C\varepsilon_3$	119.2	0.280	0.018	1.11		
WT-TRX $(SH)_2$, W28, $^{13}C\delta_1$	124.7	0.260	0.017	1.19		
WT-TRX $(SH)_2$, W31, $^{13}C\delta_1$	127.2	0.280	0.018	1.36		
WT-TRX $(SH)_2$, W28, $^{13}C\varepsilon_3$	118.0	0.280	0.016	1.12		
WT-TRX $(SH)_2$, W31, $^{13}C\varepsilon_3$	119.5	0.270	0.018	1.23		
W28F (S-S), W31, $^{13}C\delta_1$	126.9	0.446	0.006	1.28	0.137	2.6
W28F (S-S), W31, $^{13}C\varepsilon_3$	119.2	0.399	0.007	1.22		
W28F $(SH)_2$, W31, $^{13}C\delta_1$	126.6	0.440	0.005	1.24	0.154	2.1
W28F $(SH)_2$, W31, $^{13}C\varepsilon_3$	119.4	0.419	0.010	1.22		
W31F (S-S), W28, $^{13}C\delta_1$	124.8	0.349	0.006	1.15	0.209	3.2
W31F (S-S), W28, $^{13}C\varepsilon_3$	118.9	0.386	0.008	1.18		
W31F $(SH)_2$, W28, $^{13}C\delta_1$	124.7	0.406	0.006	1.19	0.175	5.1
W31F $(SH)_2$, W28, $^{13}C\varepsilon_3$	118.6	0.390	0.009	1.17		

Table II. Motional parameters obtained from the NMR relaxation data and fluorescence anisotropy data for thioredoxin

Protein	τ_m (ns)	τ_e (ps)	S^2
WT-TRX (S-S), W28, $^{13}C\delta_1$	12.2		1
WT-TRX (S-S), W31, $^{13}C\delta_1$	10.7		1
WT-TRX (S-S), W28, $^{13}C\varepsilon_3$	12.5		1
WT-TRX (S-S), W31, $^{13}C\varepsilon_3$	10.2		1
WT-TRX $(SH)_2$, W28,$^{13}C\delta_1$	15.8	3260	0.73
WT-TRX $(SH)_2$, W31,$^{13}C\delta_1$	14.1	1050	0.80
WT-TRX $(SH)_2$, W28,$^{13}C\varepsilon_3$	10.9		1
WT-TRX $(SH)_2$, W31,$^{13}C\varepsilon_3$	12.1	1370	0.88
W28F(S-S), W31, $^{13}C\delta_1$	9.2	18	0.58
W28F(S-S), W31, $^{13}C\varepsilon_3$	9.2	13	0.63
W28F$(SH)_2$, W31, $^{13}C\delta_1$	9.5	15	0.62
W28F$(SH)_2$, W31, $^{13}C\varepsilon_3$	9.5	13	0.62
W31F(S-S), W28, $^{13}C\delta_1$	10.5	1	0.91
W31F(S-S), W28, $^{13}C\varepsilon_3$	10.5	12	0.80
W31F$(SH)_2$, W28, $^{13}C\delta_1$	11.8	16	0.84
W31F$(SH)_2$, W28, $^{13}C\varepsilon_3$	11.8	11	0.82

TABLE III. Differential broadening data (obtained at 25 °C) and results of their analysis

Protein	Δ(LW) (Hz)	$S^2\tau_m$ (ns)
WT-TRX (S-S), W28, $^{13}C\varepsilon_3$	13.9	6.5
WT-TRX (S-S), W31, $^{13}C\varepsilon_3$	10.3	4.8
WT-TRX $(SH)_2$, W28, $^{13}C\varepsilon_3$	10.1	4.7
WT-TRX $(SH)_2$, W31, $^{13}C\varepsilon_3$	10.8	5.1
W28F $(SH)_2$, W31, $^{13}C\delta_1$	9.1	7.8
W28F $(SH)_2$, W31, $^{13}C\varepsilon_3$	24.5	11.5
W31F $(SH)_2$, W28, $^{13}C\delta_1$	7.3	6.3
W31F $(SH)_2$, W28, $^{13}C\varepsilon_3$	23.2	10.9

The τ_m values in Table II are larger than values calculated from the Stokes-Einstein equation based on the molecular weight. For WT-TRX these may result from the inclusion of T_2 in the analysis, but also there is evidence that the large τ_m values are related to the temperature of the measurements (see below). Since the order parameters found were close to 1, the accuracy of the τ_e values is not high. Examination of the equations indicates that τ_e values less than ~ 10 ps are not likely to be significant. Also values of τ_e in the ns range may call for the use of a more complex spectral density. Clarification awaits the availability of more data, *viz.* T_1, T_2, NOE, and linewidth values at additional frequencies, and measurements of the time dependent fluorescence anisotropy as well as a detailed error analysis.

The trends in the motional information available for WT-TRX, independent of the specific values of the motional parameters, can be seen by a direct inspection of the data. For example, consider just the NOE values. In the limit of no internal motion, *i. e.*, $S^2 = 1$, it is found from Eqs. (3)-(8) that the steady state NOE approaches 1.14 (for τ_m ~ 5-10 ns). Similar considerations apply to the steady state anisotropy. Our results indicate that both tryptophans are rather rigidly held in TRX. Trp-31, the one nearer the surface, becomes slightly more mobile than trp-28 in the reduced form of the protein. An indication of how that could occur is shown in Fig. 1.

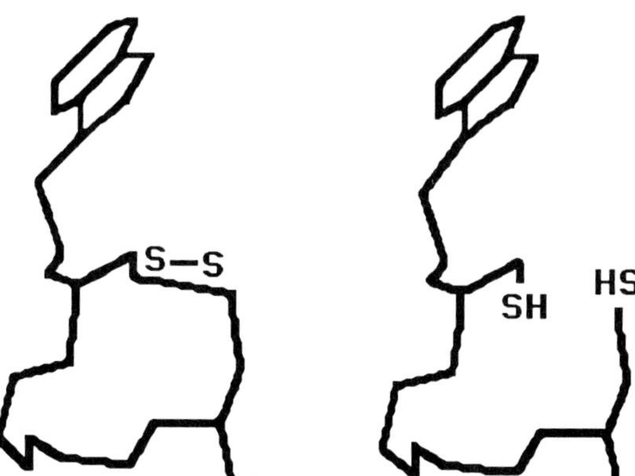

Figure 1. Backbone of oxidized (left) and reduced (right) WT-TRX from residues 31-35. Side chains are shown only for trp-31 and cys-32 and 35.

In the oxidized protein, the disulfide can be seen to form a sort of tether for trp-31, which, when broken upon reduction, could lead to somewhat greater mobility for trp-31. Trp-28 did not show a consistent difference in mobility in the reduced form from our measurements. In the single tryptophan variants, both tryptophans showed somewhat more mobility compared with WT-TRX. Again higher mobility was associated with trp-31. Little difference was found between the reduced and oxidized forms of the mutant proteins. These results are in general agreement with those of Elofsson *et al.* (8).

Values obtained here for τ_m (Table II) are somewhat larger than those found from ^{15}N relaxation measurements on WT-TRX at 35 °C (Stone *et al.* private communication). However, we have also made NMR relaxation measurements on TRX, W28F, and W31F at temperatures up to 50 °C. Preliminary analysis of the T_1 and NOE data indicated little change in the amplitude of the internal motion of the tryptophan residues. τ_m on the other hand decreased in line with the results of Stone *et al.* The lack of significant change in the internal motion of the trp residues at these temperatures is consistent with observations that WT-TRX is stable when exposed to temperatures as high as 80 °C (20). In contrast from NMR relaxation measurements on the mutant proteins, we have found that the amplitude of internal motion of each tryptophan is increased substantially when the proteins are denatured by exposure to high urea concentrations. In particular, smaller values of T_1, larger values of the NOE, and a sharp decrease in the differential broadening were observed for the denatured proteins. A preliminary analysis of the relaxation data yields S^2 values that are smaller (but distinctly not zero) in the denatured proteins as compared with the native proteins.

One of the principal objectives of this work is validation of the methodology for the detection and *quantitation* of picosecond dynamics in proteins. Molecular dynamics simulations of protein motion have been a strong driving force for these studies focused as they are on molecular motions occurring on similar time scales. The tack we have taken generally is to study the mobility of the trp side chain specifically enriched with ^{13}C thereby allowing fairly facile measurement of both ^{13}C NMR relaxation parameters and fluorescence anisotropy. In principle, use of these techniques should provide us with two independent measures of trp side chain motion which should yield similar rates and amplitudes.

As the results above indicate, accurate determination of the rate of trp side chain motion in proteins is not as straightforward as might be expected. In part this is the consequence of methodological difficulties, *e.g.* with the accurate determination of T_2. But it is also due to the high apparent order parameters we have found. The side chains of both tryptophan residues in WT-TRX are effectively nearly immobilized and show only the motion of the protein as a whole. Examination of the crystal structure of TRX suggests that this may be because trp-31 is in effect lying on the surface of the protein, seemingly adsorbed there by non-bonding interactions. In such a location, the indole moiety would be capable only of either collective motions of the entire protein matrix or whole protein rotation but not of local rotations (librations) of the ring. Trp-28 also appears to be largely immobilized but in this case tight packing of the protein matrix around the trp side chain is probably the cause of the apparent immobility. These results may be gainfully contrasted with the relatively free mobility of the trp side chain in the peptide melittin (10) and the partially restricted motion of this moiety in the melittin tetramer. Those results make us more confident that quantitatively meaningful data can be obtained by use of the fluorescence anisotropy and NMR relaxation techniques particularly when both sets of data are employed in a common analysis. The effectiveness of these experimental methods for detecting and quantifying picosecond side chain motions in proteins is also given credence by the results alluded to above with protein denaturation. Finally, even though the single trp TRX mutants were essentially isofunctional with WT-TRX, the NMR-derived results indicate increased mobility of both trp side chains. Given the relatively conservative nature of the structural substitutions in these mutant forms, it appears that the NMR relaxation parameters must be very sensitive to changes in the dynamics of the protein matrix as reflected in trp side chain mobility.

We are currently extending the use of the basic approach described here to other proteins. Additionally we are actively exploring how to measure actual picosecond motion accurately even when order parameters are fairly high. The NMR methodology is being buttressed by the use of picosecond time resolved fluorescence anisotropy techniques.

Acknowledgments

This work was supported in part by PHS Grant GM34847 to FGP and NSF Grant DMB-9105885 to MDK.

References

1. Holmgren, A. (1985) *Ann. Rev. Biochem.* **54**, 237-271.
2. Holmgren, A. (1989) *J. Biol. Chem.* **264**, 13963-13966.
3. Holmgren, A., Söderberg, B.-O., Eklund, H., & Bränden, C.-I. (1975) *Proc. Natl. Acad. Sci. U. S. A.* **72**, 2305-2309.
4. Katti, S. K., Lemaster, D. M., & Eklund, H. (1990) *J. Mol. Biol.* **212**, 167-184.
5. Dyson, H. J., Gippert, G. P., Case, D. A., Holmgren, A. and Wright, P. E. *Biochemistry* **29**, 4129-4136 (1990).
6. Holmgren, A. (1972) *J. Biol Chem.* **247**, 1992-1998.
7. Merola, F., Rigler, R., Holmgren, A., & Brochon, J.-C. (1989) *Biochemistry* **28**, 3383-3398.
8. Elofsson, A., Rigler, R., Nilsson, L., Roslund, J., Krause, G. and Holmgren, A. (1991) *Biochemistry* **30**, 9648-9656.
9. Weaver, A. J., Kemple, M. D. and Prendergast, F. G. (1988) *Biophys. J.* **54**, 1-15.
10. Weaver, A. J., Kemple, M. D. and Prendergast, F. G. (1989) *Biochemistry* **28**, 8624-8639.
11. Holmgren, A. & Sjöberg, B.-M. (1972) *J. Biol Chem.* **247**, 4160-4164.
12. Weaver, A. J., Kemple, M. D., Brauner, J. W., Mendelsohn, R. and Prendergast, F. G. (1992) *Biochemistry* **31**, 1301-1313.
13. Abragam, A. (1961) *Principles of Nuclear Magnetism*, Clarendon, Oxford.
14. Lipari, G. & Szabo, A. (1982) *J. Amer. Chem. Soc.* **104**, 4546-4559.
15. Goldman, M. (1984) *J. Magn. Reson.* **60**, 437-452.
16. Nageswara Rao, B. D. & Ray, B. D. (1992) *J. Amer. Chem. Soc.* **114**, 1566-1573.
17. Separovic, F., Hayamizu, K., Smith, R. and Cornell, B. A. (1991)*Chem. Phys. Lett.* **181**, 157-162.
18. Valeur, B. & Weber, G. (1977) *Photochem. Photobiol.* **25**, 441-444.
19. Dyson, H. J., Holmgren, A. and Wright, P. E. (1989) *Biochemistry* **28**, 7074-7087.
20. Reutimann, H. Straub, B., Luisi, P. L., & Holmgren, A. (1981) *J. Biol. Chem.* **256**, 6796-6803.

Proton Nuclear Magnetic Resonance Studies of Non-Covalent Complexes of Yeast Cytochrome *c* Peroxidase with Cytochromes *c*

Qian Yi, Steve Alam, Yihong Ge and James D. Satterlee
Department of ChemistryWashington State University, Pullman, WA 99164-4630

James E. Erman
Department of Chemistry, Northern Illinois University, DeKalb, IL 60115

I. Introduction

Cytochrome *c* peroxidase (CcP) from Baker's yeast catalyzes the hydrogen peroxide oxidation of ferrous cytochrome *c* (1). The enzyme mechanism includes intermediate steps involving a single electron transfer from ferrous cytochrome *c* to each oxidized CcP intermediate (CcP-I & CcP-II). Such electron transfers are each characterized by the net oxidation of ferrous cytochrome *c*, producing ferricytochrome *c* (2-7). A key development in understanding CcP behavior was formulation of a CcP:cytochrome *c* complex model by Poulos and Kraut (8) and this model has stimulated a great deal of work on CcP complexes. In addition to electron transfer measurements (2-7), modelling (9), equilibrium (10), kinetic (11, 12), spectroscopic (13-17) and biochemical (18, 19) studies have been carried out on noncovalent and covalent complexes of CcP with various cytochromes *c*.

There seem to be three primary reasons for interest in CcP:cytochrome *c* complexes. (A) Both proteins are water soluble, as is the noncovalent complex. This situation makes them ideal for solution spectroscopy and for handling. (B) The relatively small sizes of the two proteins (CcP: Mr~34 kD; cyt. *c*: Mr~13 kD) means that the 1:1 stoichiometric complexes are small enough (Mr~47 kD) to be easily handled and are amenable to many solution-state techniques of analysis. (C) Although CcP complexes with several cytochromes *c* have been studied, the (yeast) CcP:yeast cytochrome *c* complexes are particularly relevant, in the sense that both of these proteins are found in the mitochondrial intermembrane space as physiological partners. These CcP complexes as a group provide a paradigm for heme protein complexes, with the expectation that many details may be common to larger, more complicated protein-protein complexes, such as terminal oxidase with cytochrome *c* .

So far NMR has proven uniquely capable for detecting and characterizing CcP:cytochrome *c* complexes (14-17). Many other spectroscopic techniques have been employed in studies of these types of complexes, but have provided less specific information. Furthermore, the degree of sophistication in exploiting modern NMR methods for studying these complexes is now in a period of growth. In keeping with this trend, this article will be devoted to explaining the different kinds of CcP:cytochrome *c* complexes currently being investigated in this

TECHNIQUES IN PROTEIN CHEMISTRY IV

laboratory, as well as describing the scope and limitations of studying these complexes by proton NMR spectroscopy.

II. Methods and Materials

Cytochrome *c* peroxidase was isolated from Baker's yeast by methods that carefully avoid use of any acetate-containing compounds (20, 21). All cytochromes *c* that were used were purchased from Sigma and prepared for NMR studies as previously described (16). Solutions for NMR spectroscopy were created in 99.9% or 99.99% D_2O (Isotec) with appropriate salts and/or buffers as described in the text and figure captions. NMR spectroscopy was carried out on a Varian VXR500 spectrometer with a nominal proton frequency of 500 MHz, and with active temperature regulation. Where necessary residual water resonance intensity was suppressed using continuous wave decoupler irradiation during the relaxation delay period in all experiments. Standard 1D and 2D (NOESY, MCOSY) experiments that are part of the spectrometer program library were used. Phase cycling was carried out by the States-Haberkorn-Reuben method (22).

Table I. Summary of Possible 1:1 Complexes Between CcP and Cytochrome *c* Categorized by Heme Iron Ion Spin and Oxidation State

	CcP Heme				Cytochrome *c* Heme		
	Oxidation State	Spin State	M_S[1]	Example	Oxidation State	Spin State[2]	M_S
[3]Oxidized Intermediate Compound-I							
a.	+4	low	3/2	CcP-I	+3	low	1/2
b.	+4	low	3/2	CcP-I	+2	low	0
[4]Oxidized Intermediate Compound-II							
c.	+4	low	1	CcP-II	+3	low	1/2
d.	+4	low	1	CcP-II	+2	low	0
Resting State, Native Enzyme							
e.	+3	high	5/2	CcP	+3	low	1/2
f.	+3	high	5/2	CcP	+2	low	0
g.	+3	low	1/2	CcP-CN	+3	low	1/2
h.	+3	low	1/2	CcP-CN	+2	low	0
Reduced Enzyme							
i.	+2	high	2	CcP(Fe^{2+})	+3	low	1/2
j.	+2	high	2	CcP(Fe^{2+})	+2	low	0
k.	+2	low	0	CcP-CO	+3	low	1/2
l.	+2	low	0	CcP-CO	+2	low	0

[1]$M_S=\Sigma m_S$; the total electron spin-only angular momentum due to all unpaired electrons.

[2]Assumes maintenance of the cytochrome *c* six-fold heme iron coordination.

[3]The initial product formed after reaction with hydrogen peroxide is formulated as a ferryl heme iron ion plus a protein-centered free radical.

[4]Compound-II, the second oxidized intermediate, which is reduced one equivalent from Compound-I, is formulated as a heme ferryl iron ion.

III. Results and Discussion

A. CcP:cytochrome *c* Complexes

CcP has three accessible heme oxidation states, the most stable under ambient conditions being the resting enzyme ferric state (heme iron=+3), followed in order of decreasing ambient stability by the intermediate ferryl (iron=+4), and the ferrous (iron=+2) states. In addition to differing oxidation states differing ligation states are also encountered for the heme iron at the enzyme's active site. Combined with the two most common heme iron oxidation states found for cytochrome *c* (Fe^{+2}; Fe^{+3}) one readily sees that there are twelve possible 1:1 combinations of CcP:cytochrome *c* pairs for each species of cytochrome *c* (Table I). Only stoichiometric 1:1 complexes will be considered here since there is ample direct spectroscopic evidence, as well as kinetic evidence, that 1:1 complexes predominate in solutions of varying protein concentrations between ~100 micromolar and ~2 mM (10, 13, 14, 17).

Which are amenable to study by solution NMR methods? Complexes of oxidized CcP (CcP-I, CcP-II) and ferrous cytochromes *c* (Table I: b,d) are at best transiently stable since they spontaneously react to form products, obviating their study by NMR and other nontransient methods. The reduced enzyme, ferrous CcP, has not been implicated in the catalytic cycle, so its complexes are, at least initially, less interesting. It has also been reported that CcP catalytically turns over dithionite ion a common reducing agent, which raises unresolved questions about possible enzyme modification by dithionite decomposition products. Also, ferrous CcP, CO-ligated or not, is terribly oxygen sensitive, making handling of such complexes (Table I: i-l) difficult. This leaves the "half reactant complexes" (Table I: a,c,f), the "product" complex (Table I: e), the reactant complex "model" (Table I: h) and the "derivative" product complex (TableI: g) as appropriate systems for study by NMR. So far complexes a, e, g, and h have been studied in this laboratory by proton NMR spectroscopy and the remainder of this article describes experimental aspects of that work that have not yet been published.

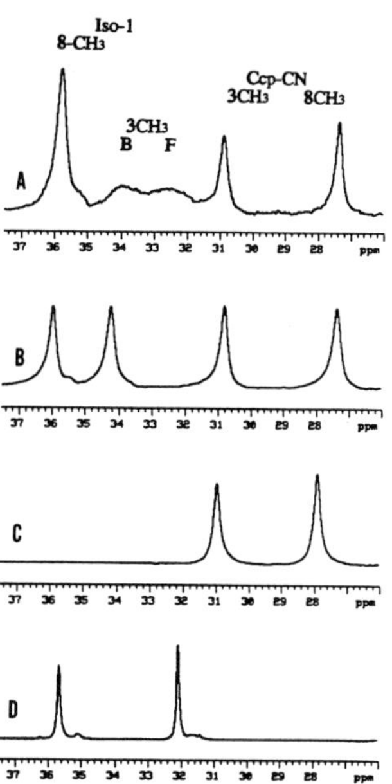

Figure 1. Downfield proton hyperfine region at 500 MHz of the following mixtures of CcPCN and yeast iso-1 ferricytochrome *c* : (A) 1:2 mole ratio; (B) 1:1; (C) CcPCN; (D) iso-1. In (A) letter B denotes resonance of iso-1 bound to CcPCN, F denotes free form. Solution conditions: D_2O, 10 mM KNO_3; pH'=6.5 19.7 °C, internal HDO ref=4.70 ppm.

B. Proton NMR Spectroscopy of the Complexes

Initially we have used complexes in which one or both of the proteins was in a paramagnetic state (Table I). The paramagnetic proteins have the advantage that their proton hyperfine shifts: (1) provide increased shift dispersion (23,24; Figs. 1,2), (2) are a result of the heme acting as an intrinsic shift (and relaxation) reagent, and (3) thereby transmit information about how the protein-protein association affects the active site electronic, secondary and tertiary structures. Consequently the "product" complex consisting of resting state CcP and ferric cytochromes *c* (Table I: e) were the first studied. Four ferricytochromes *c* have been used: horse, tuna, yeast iso-1 and yeast iso-2 (16). The choice of complex was due to anticipation that proton hyperfine shifted NMR resonances would be extremely sensitive to complex formation. It is no coincidence that the paramagnetic heme centers are the active sites of each protein. The hyperfine-shifted proton resonances provide highly sensitive, highly specific and highly selective information on the consequences of complex formation for each heme active site (14, 16, 23, 24).

More recent efforts have been directed to the CcP-CN:ferricytochrome *c* complex (Table I: g) in which horse, tuna and yeast iso-1 ferricytochromes *c* have been employed (17). Preliminary data on the CcP-CN:yeast iso-1 ferrocytochrome *c* complex (Table I: h) have also recently been obtained.

Turning to the CcP-CN:ferricytochrome *c* complexes, both proteins are low-spin and paramagnetic with the result that relatively narrow hyperfine shifted proton resonances occur. This particular complex is considered to be a "model" for the initial product complex (Table I: c) in the sense that CcP-CN is a closely related (ie. one electron different) low-spin model for the CcP-II low-spin ferryl compound and yeast iso-1 ferricytochrome *c* is identical to the initial reaction product. For the first time a complex-induced effect on the peroxidase spectrum has been observed (17) (Fig. 1). Figure 1 shows proton NMR specta of the heme methyl hyperfine shift region for CcP-CN and yeast iso-1 ferricytochrome *c* as an equimolar 1:1 mixture (CcP:iso-1) and as a 1:2 mixture. Several conclusions can be drawn from this figure. First, Fig. 1 B-D shows that complex-induced proton hyperfine resonance shifts occur for both CcPCN and yeast iso-1 ferricytochrome *c*. Studies designed to elucidate the origin of these shifts are in progress. However, in view of the buried nature of the CcP heme group, these shifts are very likely to be caused by a complex-induced structural rearrangement initiated at the protein-protein interaction surface and communicated to the heme active site. Second, comparing A and B it is clear that when the concentration of iso-1 ferricytochrome *c* exceeds that of CcP-CN the iso-1 ferricytochrome *c* heme 3-methyl group splits into two broadened peaks, representing iso-1 bound to the CcP-CN and iso-1 free in solution. This establishes the fact that at these concentrations the complex stoichiometry is 1:1, just as seen for the protein mixtures of native CcP and iso-1 ferricyt. *c* (16). Third, the simultaneous observation of heme 3-methyl resonances for the free and bound forms of iso-1 indicates that the exchange kinetics are similar to those found for the protein mixtures involving yeast iso-1 ferricytochrome *c* and native, resting state CcP (16).

There are also complications for NMR studies of these complexes. The simplest regards the tendency for protein precipitation in low salt solutions, especially in mixtures of CcP and yeast iso-1 cytochrome *c*. This complicates accurate construction and maintenance of solutions with precisely known protein concentrations and with precisely adjusted concentration ratios. Another complication is the rather large size of these types of complexes (M_r~47 kD), as regards NMR studies. For cases in which one or more of the partner proteins is diamagnetic virtually all of the proton resonances for the diamagnetic protein will lie in the rather crowded 12 ppm to -2 ppm region. Even when both proteins are paramagnetic the proton resonances of well over 400 amino acids lie in that chemical shift range, thereby restricting the discernible NMR information. This also explains our choice of basing initial NMR studies on complexes in which both CcP and the cytochromes *c* were paramagnetic, thereby taking advantage of the paramagnetic proton hyperfine shifts that range over ~120 ppm.

These advantages are also the source of limitations. Paramagnetism causes efficient nuclear relaxation (22, 23). It is not unusual for apparent proton nuclear spin-lattice relaxation times (T_1) to be as short as 10 msec (or even shorter) in

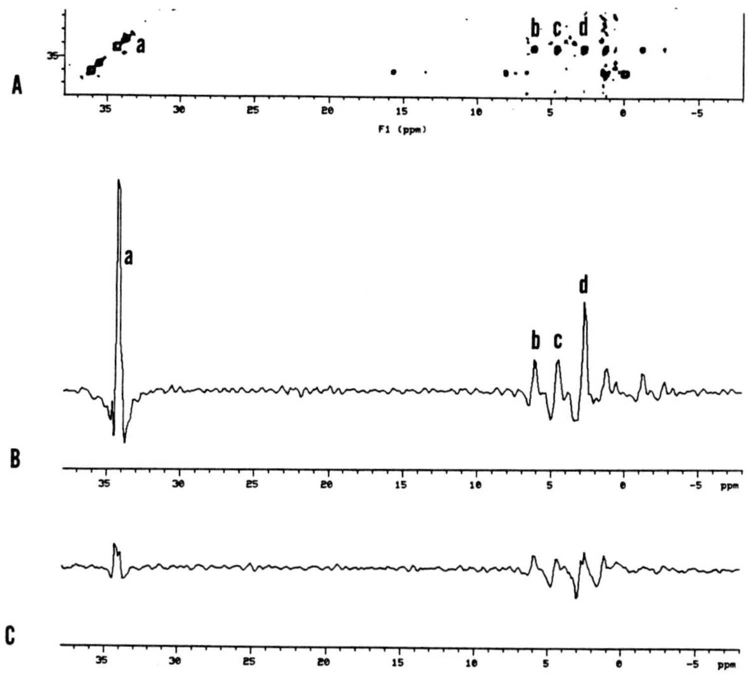

Figure 2. (A) Section of a 500 MHz proton homonuclear phase sensitive NOESY contour plot for the CcP-CN/yeast iso-1 ferricytochrome *c* complex, mixing time=30 msec, at 20 °C, pH 6.5 (diagonal phased positive and only positive contours plotted). (B) Slice through the iso-1 3-methyl resonance for the experiment with mixing time=30 msec. (C) Identical slice of identically processed data plotted at identical vertical scale to (B), but for the experiment with mixing time=75 msec. Peak labels: a=iso-1 heme 3-methyl; b=iso-1 phe-82; c=iso-1α-meso; d=iso-1 heme 4β-methyl (29).

these proteins. This obviously complicates application of many types of modern 2D pulse sequences. Short T_1s and accompanying short transverse relaxation times (T_2) mean that nonequilibrium or transverse magnetization does not persist long enough for large NOEs to develop or for maximally efficient coherence transfer. This point is emphasized by the data shown in Figure 2. This figure shows a section of a phase-sensitive proton homonuclear NOESY contour plot for the CcP-CN/yeast iso-1 ferricytochrome *c* 1:1 complex using a 30 msec mixing time, and slices through the yeast iso-1 3-methyl peak at mixing times of 30 msec and 75 msec. The contour plot shows that some types of 2D experiments can produce excellent data even for a protein complex of this size. The slices show that optimal mixing times may be quite short for complexes of this size, in which one or more of the proteins is in a paramagnetic state.

Another potential complication is how the normal concept of a T_1 translates into a fast-relaxing paramagnetic system that is also of comparatively large size. In these cases, as Figure 3 shows, recovery of longitudinal magnetization created via inversion-recovery (180°-tau-90°) pulse sequences (Fig. 3A) does not follow a single exponential equation as standard theory predicts (28). Instead, as shown in Figure 3B, semilog$_e$ plots of relaxation data are curved, indicating the presence of multiple exponential relaxation, most likely due to the presence of strong spin diffusion. The complication that these results reveal is that, whereas many formulations of standard experimental concepts in NMR are predicated on the

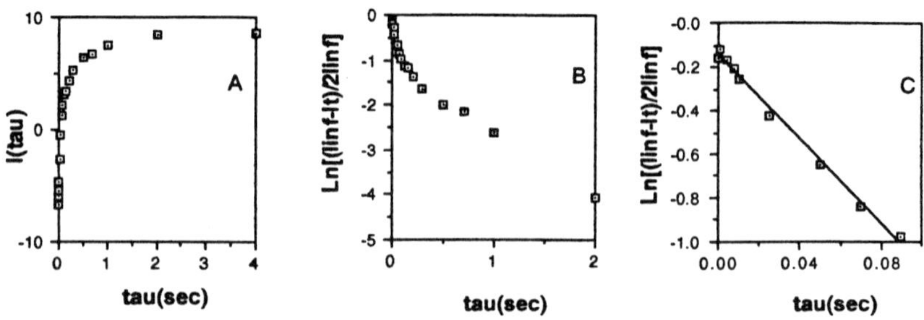

Figure 3. Inversion recovery data at 500 MHz for the heme 8-methyl hyperfine shifted proton resonance of the CcP/horse ferricytochrome *c* 1:1 complex, pH=6.6, 5°C. (A) Plot of observed intensity against delay time (τ); (B) Plot of natural logarithm of Intensity Ratio (Iinf=fully relaxed intensity; It=intensity at τ) vs τ for τ up to 4 sec; (C) Plot of initial data points from (B) using the maximum number of initial points that generate a linear least squares analysis corr. coefficient =1.00.

assumption that nuclear relaxation proceeds by a single exponential process, multiple-exponential relaxation common in these types of compounds may complicate simple interpretation of NMR data. Figure 3C shows that for the initial points in data like Figure 3B (ie. short tau values in the inversion-recovery pulse sequence) semilog_e plots of the data are linear and indicate that a single exponential time constant does govern relaxation in this delay-time regime. We call this time constant the "apparent T_1" (T_1^{app}).

C. Techniques for Studying Complex Exchange Dynamics

The observed splitting of the yeast iso-1 ferricytochrome *c* heme 3-methyl resonance into two resonances, attributable to free and bound states, when the ferricytochrome *c* concentration is in excess in mixtures with both CcP (16) and CcP-CN (Fig. 1A; 17), means that in principle the exchange dynamics of these complexes can be quantitated by NMR. The NMR approach in such a case can be illustrated with respect to Figure 1A. It requires selective perturbation of one of the two exchange-coupled peaks (either iso-1 free or iso-1 bound) and observation of the resulting effect on the unperturbed peak of the exchange coupled pair. For a pair of peaks as close together in the spectrum as are the iso-1 heme 3-methyl free and bound peaks selectivity is the key to success. A selective perturbation applied to one of the exchange-coupled peaks must miss the other.

Two NMR methods are common for applying a perturbation: (1) Saturation Transfer (27); and (2) Inversion Transfer (27, 28). Preliminary measurements via Saturation Transfer have indicated that dynamic exchange is characterized by a pre-exchange lifetime on the order of 0.003 sec for iso-1 ferricytochrome *c* in a mixture of CcP and iso-1 in mole ratios of 1:2. Subsequently, we are applying Inversion Transfer experiments to accurately determine the pre-exchange lifetime, which is the residence time for a ferricytochrome *c* molecule in either the free or CcP-bound states.

Originally the Inversion Transfer technique was implemented using a soft-pulse method for the inversion (perturbing) pulse (28), however, for these complexes the extremely short nuclear T_1s combined with their very rapid exchange rates (a pre-exchange lifetime on the order of a few msec) require (in our

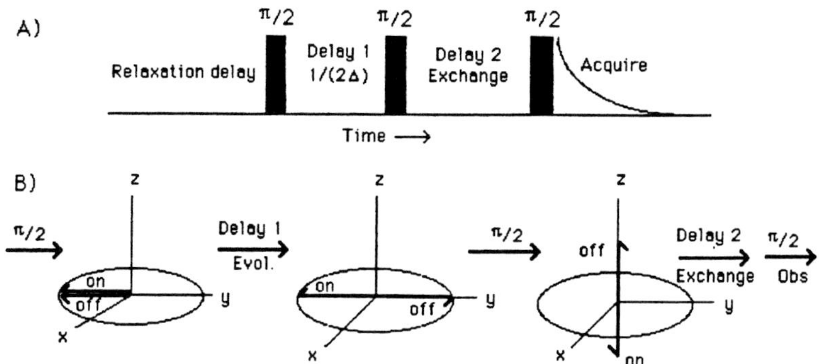

Figure 4. (A) On-resonance hard pulse Inversion-Transfer pulse timing diagram all 90° pulses have identical phases. Delta is the chemical shift difference (in Hz) between on- and off-resonance peaks. (B) Rotating frame vector diagram of the pulse sequence shown in (A), where "on" refers to the vector representing the on-resonance peak and "off" refers to the vector representing the off-resonance peak.

hands) higher power/shorter length pulses to achieve inversion on a useful time scale. The "useful time scale" means that inversion should be completed in a period of time short compared to the pre-exchange lifetime, which in these cases means that inversion should be completed on the order of tens of μsec. The practical problem this presents is that achieving short inversion times requires high pulse power level, which engenders a corresponding loss of selectivity that occurs with increasing pulse power (which is the advantage of the soft pulse method). To compensate for this requirement we first attempted achieving required selectivity on a useful time scale using a Dante pulse sequence, but were unsuccessful. Since we do not have access to a spectrometer with pulse shaping, or waveform generation capabilities, we have begun using a hard pulse method given by the sequence in Figure 4A. In this case the transmitter frequency is set to the frequency of the resonance to be inverted (ie. on-resonance) and three successive $\pi/2$ pulses are applied with appropriate delay times between them. A simple rotating frame vector diagram of the expected behavior of the exchange-coupled peaks is presented in Figure 4B. After the first $\pi/2$ pulse the delay time is chosen to allow the exchange-coupled resonances to become anti-parallel [delay $1=1/(2\Delta)$, where Δ=chemical shift difference in Hz between free and bound resonances]. The second $\pi/2$ pulse creates magnetization along the + and - z-axes corresponding to the off-resonance and on-resonance peaks, respectively. The subsequent second delay period allows exchange to occur between the two chemically inequivalent sites (in this case between free and bound environments). The second delay time is varied from experiment to experiment in order to change the extent of inversion transfer that occurs. The third $\pi/2$ pulse reads the state of the system.

The data, which are peak intensities for each delay-2 time interval, are then analysed using the Bloch equations modified for chemical exchange, as previously derived (28). Graphs of the predicted behavior for a hypothetical pair of exchange-coupled resonances with idealized spin-lattice relaxation times and pre-exchange lifetimes typical of what we expect to find in these complexes are shown in Figure 5. The presentation in Figure 5 is in the form of plots of resonance intensity vs the time interval of the second delay period. The "on-resonance" peak, the one that is inverted, behaves as if it were undergoing an inversion-recovery experiment (Trace A), while the "off-resonance" exchange-coupled peak (Trace B) exhibits more complex behavior. These curves are qualitatively similar to those found for a pair of more slowly relaxing, diamagnetic, exchanging proton resonances (28) and indicate that these experiments are feasible for the CcP/yeast iso-1 ferricytochrome *c* complexes.

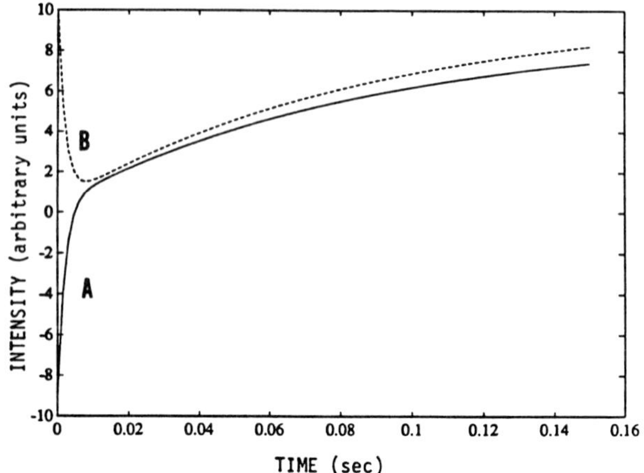

Figure 5. Graphs of expected behavior according to the exchange modified Block equations (28) for a pair of exchange coupled methyl resonances with the following characteristics: T_1s for both resonances = 90 msec, pre-exchange lifetime = 4 msec. These are values close to those expected for the CcP/yeast iso-1 ferricytochrome *c* complex based on preliminary work. The solid line (A) refers to the on-resonance peak and the dashed line (B) refers to the off-resonance peak of the exchange coupled pair.

Acknowledgements

We wish to thank the National Institutes of Health (HL01758 RCDA , GM45986) for direct support of this work. NMR spectroscopy at WSU has been supported by the NIH (RR0631401) and by the Battelle Pacific Northwest Laboratories.

References

1. Bosshard, H.R., Anni, H. and Yonetani, T. (1991) in *Peroxidases in Chemistry and Biology* (Everse, J., Everse, K.E., and Grisham, M.B., Eds.) Vol II, pp 51-83, CRC Press, Boca Raton, FL.
2. Kang, C.H., Ferguson-Miller, S., and Margoliash, E. (1977) *J. Biol. Chem.* 252, 919-926.
3. Geren, L., Hahn, S., Durham, B. and Millett, F. (1991) *Biochemistry* 30, 9450-9457.
4. Pan, L.P. Frame, M., Durham, B. Davis, D. and Millett, F. (199)) *Biochemistry* 29, 3231-3236
5. Hazzard, J.T., Poulos, T., and Tollin, G. (1987) *Biochemistry* 26, 2836-2848.
6. Hazzard, J.T., McLendon, G., Cusanovich, M.A., Das, G., Sherman, F. and Tollin, G. (1988) *Biochemistry* 27, 4445-4451.
7. Everest, A.M., Wallin, S.A., Stemp, E.D.A., Nocek, A.J.M., Mauk, A.G. and Hoffman, B.M. (1991) *J. Am. Chem. Soc.* 113, 4337-4338.
8. Poulos, T.L. and Kraut, J. (1980) *J. Biol. Chem.* 255, 8199-8205; *J. Biol. Chem.* 255, 10322-10330.
9. Northrup, S.H. Boles, J.O. and Reynolds, J.C.L. (1988) *Science* 241, 67-70.
10. Erman, J.E. and Vitello, L.B. (1980) *J. Biol. Chem.* 255, 6224-6227.

11. Erman, J.E., Kim, K.L., Vitello, L.B., Moench, S.J. and Satterlee, J.D. (1987) *Biochim. Biophys. Acta* 911, 1-10.
12. Kim, K.L., Kang, D.S., Vitello, L.B., and Erman, J.E. (1990) *Biochemistry* 29, 9150-9159.
13. Kang, D.S. and Erman, J.E. (1982) *J. Biol. Chem.* 257, 12775-12779.
14. Satterlee, J.D., Moench, S.J. and Erman, J.E. (1987) *Biochim. Biophys. Acta* 912, 87-97.
15. Moench, S.J., Satterlee, J.D., and Erman, J.E. (1987) *Biochemistry* 26, 3821-3826.
16. Moench, S.J., Erman, J.E. and Satterlee, J.D. (1992) *Biochemistry* 31, 3661-3670.
17. Yi, Q., Satterlee, J.D. and Erman, J.E. (1992) *J. Am. Chem. Soc.* 114, in press.
18. Bechtold, R. and Bosshard, H.R. (1985) *J. Biol. Chem.* 260, 5191-5200.
19. Kang, C.H., Brautigan, D.L., Osheroff, N. and Margoliash, E. (1978) *J. Biol. Chem.* 253, 6502-6510.
20. Vitello, L.B., Huang, J. and Erman, J.E. (1990) *Biochemistry* 29, 4283-4288.
21. Moench, S. J. (1984) *Doctoral Dissertation*, Colorado State University.
22. States, D.J., Haberkorn, R.A., and Ruben, D.J. (1982) *J. Magn. Reson.* 48, 286-292.
23. Satterlee, J.D. (1986) *Annual Reports on NMR Spectroscopy* (Webb, G.A., Ed.) Vol 17, pp 79-178, Academic Press, London.
24. Satterlee, J.D. (1987) *Metal Ions in Biological Systems* (Sigel, H., Ed.) Vol 21, pp 121-186, Marcel Dekker, New York.
25. Satterlee, J.D. and Erman, J.E. (1991) *Biochemistry* 30, 4390-4405.
26. Satterlee, J.D., Russell, D.J. and Erman, J.E. (1991) *Biochemistry* 30, 9072-9077.
27. Martin, M.L., Delpuech, J.J. and Martin, G.J. (1980) *Practical NMR Spectroscopy*, Heyden, London.
28. Alger, J.R. and Prestegard, J.H. (1977) *J. Magn. Reson.* 27, 137-141.
29. Moench, S.J. and Satterlee, J.D. (1989) *J. Biol. Chem.* 264, 9923-9931.

The Application of Chemical Shift Calculation to Protein Structure Determination by NMR

Timothy S. Harvey [1] and Wilfred F. van Gunsteren [2]

[1] Department of Biochemistry, South Parks Road, Oxford. OX1 3QU U.K.
[2] Laboratorium für Physikalische Chemie, ETH-Zentrum, CH-8092 Zürich, Switzerland

1 Introduction

The determination of the three-dimensional structure of small proteins ($M_r <$ 20 kDa) by two– and three– dimensional nuclear magnetic resonance (NMR) techniques is now commonplace (for reviews see Clore and Gronenborn 1991, Wüthrich 1989). This process relies on the derivation of distance and dihedral angle restraints from experimentally measured nuclear Overhauser effects (NOEs) and $^3J_{NH-C\alpha H}$ coupling constants. This information is subsequently used by computer programs to calculate a three-dimensional structure compatible with the input data. Molecular dynamics simulations, which solve Newton's equations of motion to determine the movements of a molecule under an applied forcefield, are often used both to create and refine these structures. Extra terms containing the distance and angle restraints are added to the forcefield which drive the system toward that observed experimentally (Brünger and Karplus 1991).

The increase in the speed of computers over recent years, together with a rise in the overall quality of structures calculated from NMR data, has resulted in an increase in the possible complexity of the methods used and a corresponding reduction in the amount of subjective data-interpretation involved. The need to assign upper limits based on NOE intensity may be removed by calculation of the full relaxation and NOE matrices, which may then be used to back-calculate spectra (Yip and Case, 1989, Nilges *et al.* 1991, Bonvin *et al.* 1992). Dihedral angle restraints rely on their interpretation in terms of the Karplus relationship before they may be used. A method has been developed whereby the structure may be driven to satisfy the measured coupling constant instead of the angle calculated from the Karplus relationship (Kim and Prestegard 1989). Time-averaging has been applied improving the accuracy of the treatment of the experimental data (Torda *et al.* 1990). The use of time-dependent NOE restraints has improved the modelling of the functional form of the NOE restraint, enabling the simulation of systems which are not well-represented by a single static conformation or are inherently flexible, and a similar method has been applied to dihedral restraints (Torda *et al.* 1992).

Thus, the degree of sophistication of the methods used to calculate and refine NMR-derived structures is steadily increasing, and the need for subjective interpretation is diminishing. There remain, however, many structures where the total number of conformationally significant distance restraints obtained is relatively low, or which contain regions which are poorly defined by the experimental data. Whatever the possible reasons (*e.g.* experimental conditions or conformational disorder) the consequences for the resulting structures are significant (Gao *et al.* 1991). Conversely, it is often a problem in larger proteins to assign NOEs between specific resonances because of the large number of NOEs and

TECHNIQUES IN PROTEIN CHEMISTRY IV

consequent spectral overlap. This crowding may render many structurally important restraints intractable, and is one advantage of using multi-dimensional heteronuclear NMR methods (Clore and Gronenborn 1991).

However, there remains one further parameter obtained from NMR spectra which contains structural information once a spectrum has been assigned; the chemical shift of each resonance. Of all the NMR observables, this is in principle the most valuable structurally, since it is the most accurately measured, uniformly distributed and conformationally sensitive. It also requires no interpretation in its measurement. Thus there is potentially a great deal of information in the measured chemical shifts if they can be usefully used in the structure calculation/refinement process. The use of chemical shift information is not a new idea; much of the early NMR work concentrated on the correlation of ^{1}H chemical shifts with known crystal structures, and was qualitatively successful (Perkins and Wüthrich 1979). Recently, there has been renewed interest, utilizing the increased number of known protein structures for which resonance assignments are known, and important correlations between ^{1}H chemical shifts and secondary structure were noted (Wishart *et al.* 1992a, 1992b). It was shown for the first time by Williamson *et al.* (1991, 1992) that differences between shifts calculated from crystal structures and those measured experimentally are due to differences between the solution and crystal structures, which although very small, are significant in terms of chemical shift. The use of chemical shift–derived penalty functions has also been discussed before (Gippert *et al.* 1990, Ösapay and Case 1992) and it is our aim to assess the usefulness of this approach in terms of the current approach to NMR-based protein structure determination, and how it can complement the existing methods.

2 Computational Methods

2.1 Chemical Shift Contributions

The observed chemical shifts (δ) of a magnetically susceptible nucleus is the sum of a number of individual terms, including local and non-local effects. Local effects, including diamagnetic and paramagnetic shielding are assumed to be independent of the conformation or environment of the molecule - the random coil shift. Other contributions, which produce the conformation–dependent or secondary shifts ($\Delta\delta$), are due to non-local effects, and have been suggested to arise from electric field, van der Waals contact, magnetic anisotropy, and ring current effects from nearby groups (Buckingham 1960, Zürcher 1967, Sternlicht and Wilson 1967):

$$\Delta\delta = \delta_{ef} + \delta_{vdW} + \delta_{anis} + \delta_{rc} \tag{1}$$

The term δ_{vdW} is thought to be negligible for ^{1}H nuclei, and was not included. Contributions from δ_{ef} were not included here for two reasons. Firstly, they will be extremely forcefield–dependent, and secondly, there will be significant contributions to this term from solvent, which is not present in the simulations presented here. The secondary shifts calculated here contain δ_{rc} and δ_{anis}. δ_{rc} was calculated using the classical ring-current model of Johnson and Bovey (1958), where the secondary shift at a given point (x, y, z), in parts per million (ppm) is described (in c.g.s/Gauss units) by

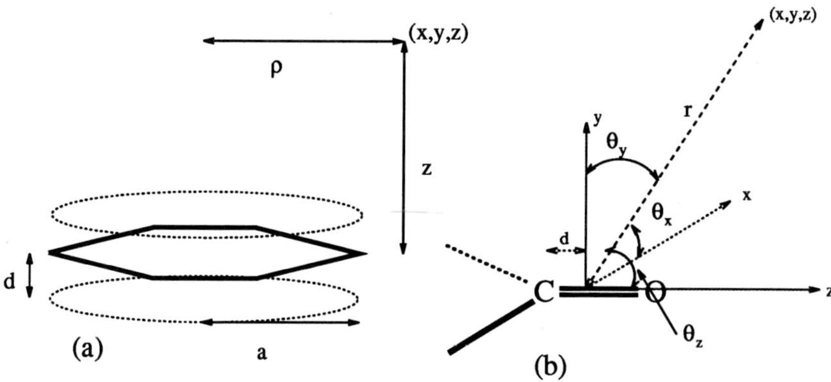

Figure 1: The geometrical parameters used in the calculation of ring-current and anisotropy contributions to secondary shift. (a) shows those used for ring currents, with dotted lines representing the two clouds of π-electrons at distance d from the plane of the ring. Anisotropy contributions are calculated using the quantities shown in (b), with the center of anisotropy indicated at distance d along the bond. The principle axes are re-oriented along the C-N bond vector to calculate these contributions.

$$\delta_{rc} = \frac{ne^2}{6\pi mc^2 a} \frac{a}{[(a^2+\rho)^2+z^2]^{\frac{1}{2}}}[K + \frac{a^2-\rho^2-z^2}{(a-\rho)^2+z^2}E] \tag{2}$$

where n is the number of electrons circulating in the π-orbital of the ring, e is the charge on an electron, mass m; c is the speed of light; a is the radius of the ring system; and ρ and z are cylindrical coordinates centered on the ring. K and E are complete elliptic integrals of the first and second kind, shown in Figure 2.1(a).

Effects due to anisotropic groups (δ_{anis}) were calculated using the equation of ApSimon *et al.* (1967):

$$\delta_{anis} = \frac{1}{3r^3}[\Delta\chi1(1-3cos^2\theta_y) + \Delta\chi2(1-3cos^2\theta_x)] \tag{3}$$

where δ_{anis} is again in ppm. r is the distance from the center of the anisotropy to the nucleus of interest, $\chi1$ and $\chi2$ are the magnetic anisotropies between the y and z, and between the x and y axes, respectively. θ_x and θ_y are the angles between the vector connecting the center of anisotropy and the nucleus and the x and y axes(Figure 2.1(b)).

2.2 Chemical Shift Potential

An extra term representing the chemical shift potential has been incorporated into a new GROMOS all-atom forcefield (van Gunsteren and Berendsen 1987, Harvey unpublished). The form of the potential has two components: an attractive part represented by

$$V_{\delta r}(\delta(t)) = 0 \quad if \;\; \delta(t) + ERR \leq \delta_0 \tag{4}$$

$$= \tfrac{1}{2}K_{\delta r}[|\delta(t) - \delta_0| - ERR]^2 \quad if \;\; \delta(t) > \delta_0 + ERR \tag{5}$$

and a repulsive part,

$$V_{\delta r}(\delta(t)) \quad = 0 \qquad if \;\; \delta(t) - ERR \geq \delta_0 \tag{6}$$
$$= \tfrac{1}{2}K_{\delta r}[|\delta(t) - \delta_0| - ERR]^2 \quad if \;\; \delta(t) < \delta_0 - ERR \tag{7}$$

where $\delta(t)$ represents the calculated secondary chemical shift, δ_0 represents the experimentally determined secondary shift, and $K_{\delta r}$ is the force constant for the chemical shifts restraints. Secondary shifts were calculated by subtracting the random coil values obtained given by Bundi and Wüthrich (1979) from those determined experimentally. ERR represents the estimated experimental error in the measurement of the ^{1}H chemical shifts. Typically, shifts are reported with an accuracy of $\pm$ 0.02 ppm, and since a similar error may be expected in the random coil values with no attempt to correct for experimental conditions/referencing, a value of 0.05 ppm for ERR was used. This is still rather small, given the absence of electric field contributions, but is partially compensated for by the use of a weak force constant (see later). Analytical derivatives of ring-current and anisotropy contributions were used in the calculation of the chemical shift restraint forces.

The recent work of Williamson and Asakura (1992) has focussed on the refinement of parameters used in the calculation of ring-current and anisotropy components, and these are used here. Thus, for the ring current intensity factors, values of 0.43, 1.05, 0.92, 0.90 and 1.04 for the aromatic groups of His, Phe, Tyr and the 5– and 6–membered rings of Trp, respectively. In the calculation of anisotropy effects, Cα-N contributions were omitted as indicated by Williamson *el al.* (1992), and values of -13.0 and -4.0 $\times$ 10^{-30}cm^3 for $\Delta\chi1$ and $\Delta\chi2$, with the center of anisotropy 1.1 Å along the bond vector were used for carbonyl bonds, whilst for C-N bonds, values of -11.0 and 1.4 $\times$ 10^{-30}cm^3 with the center of anisotropy 1.1305 Å along the bond were used. The contribution from the carboxyl groups of Asp and Glu sidechains were averaged in the cases where the sidechain was known to be unprotonated.

2.3 Computational Methods

Two systems were chosen for study. BPTI was chosen because this molecule has been used in much of the recent work of Williamson *et al.* (1992). The starting coordinates came from the Brookhaven entry 6PTI, and the chemical shift assignments were those reported by Wagner *et al.* (1987). The second protein, human epidermal growth factor (EGF), was chosen as an example of a recently-determined NMR structure which contains regions of well– and poorly defined structure (Hommel *et al.* 1992, Cooke *et al.* 1990). Only non-glycine CαH shifts were restrained. The structures were energy minimized for 500 steps without restraints. Two trajectories were then started from each structure, one unrestrained and one with chemical shift restraints. $K_{\delta r}$ was set to 50 kJ mol^{-1}ppm^{-2}. For each trajectory, the system was first equilibrated at 300K for 1 ps by coupling to a temperature bath at that temperature. Trajectories were then followed for 10ps. Throughout, the timestep used was 0.002 ps and bond-lengths were kept rigid using the SHAKE algorithm. Coordinate sets were

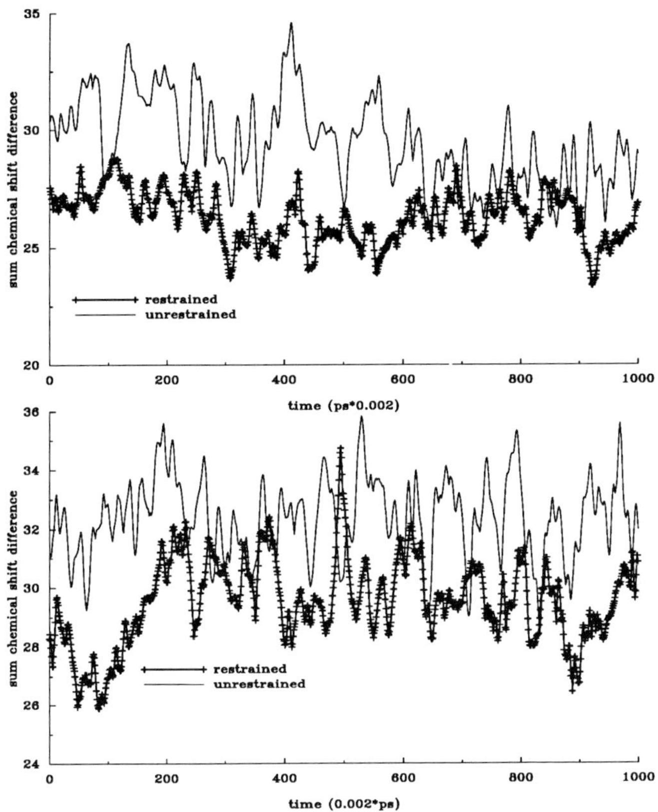

Figure 2: Timecourse of the sum of the total chemical shift difference for the EGF (above) and BPTI (below) simulations. The time period shown is 2 ps, whilst the sum of the chemical shift difference is in ppm.

saved every 0.1 ps for subsequent analysis. The non-bonded pair list, which was also used to determine groups contributing to the chemical shift term for a given nucleus, used a cut-off of 8 Å and was updated every 10 steps. The coordinates from the final 2 ps of the trajectory were used for averaging.

The trajectories were followed by calculating the sum of the absolute differences of the calculated and experimental secondary shifts. The effect of the inclusion of restraints is shown in Figure 2 where the chemical shift difference with and without restraints is shown from equivalent times in the restrained and unrestrained simulations. BPTI shows a slight improvement with respect to the calculated shifts; the average for the unrestrained run is 32.3 ± 1.3 ppm, whilst for the restrained run the average is lower, 29.5 ± 1.4 ppm. EGF shows a greater difference between the two, with that average summed difference being 29.4 ±1.8 ppm unrestrained, 26.3 ±1.1 ppm restrained. To evaluate the effects of the restraints, average structures and fluctuations about these were

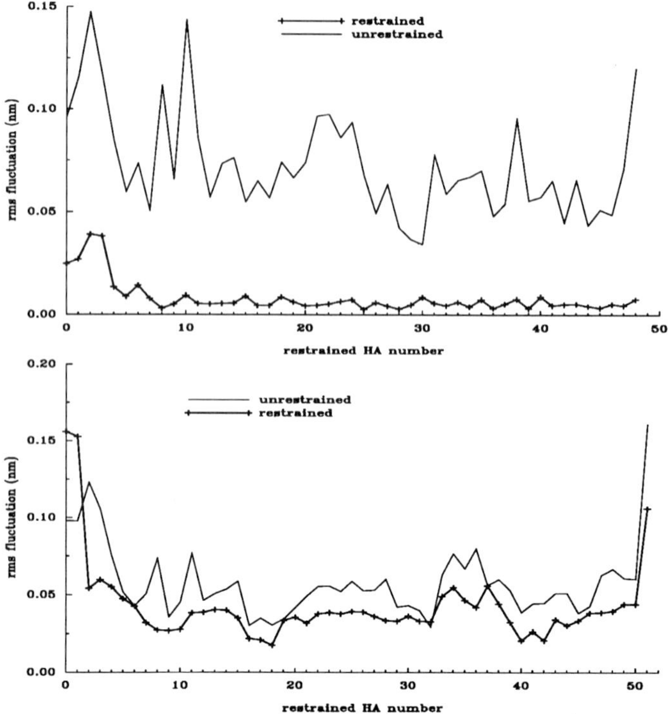

Figure 3: A comparison of the fluctuations about the mean positions of the restrained and unrestrained CαH proton positions with and without chemical shift restraints, following their sequence order, for EGF (above) and BPTI (below). Data for Gly residues are not included in this plot.

calculated for the final 2 ps of each simulation. The rms fluctuations for the restrained proton shifts are shown in Figure 3. Here, a notable difference is seen between the restrained and unrestrained simulations. The mobility of all the restrained atoms is significantly reduced, especially in the case of EGF.

3 Discussion

The results presented above indicate that there is a significant amount of structural information in chemical shifts. Whilst a perfect agreement between calculated and experimental data was not achieved, this could not be expected, given the omission of the electric field contribution. We are currently investigating the inclusion of this effect into simulations including explicit solvent molecules; a closer agreement may be possible.

The reduction of movement in the restrained atom positions will obviously be of use in structures where there are few restraints. However, the use of chemical shift restraints should be used in conjunction with other NMR data. The first 4 residues of EGF are known to be disordered from ^{15}N relaxation measurements; their secondary shifts are near to random coil, indicating that the measured shifts may be dynamically averaged. Their positions could be very well-defined by the use of chemical shift data, but the use of 'instantaneous' restraints would be inappropriate. The use of time-averaged restraints would be more appropriate, and is currently under investigation.

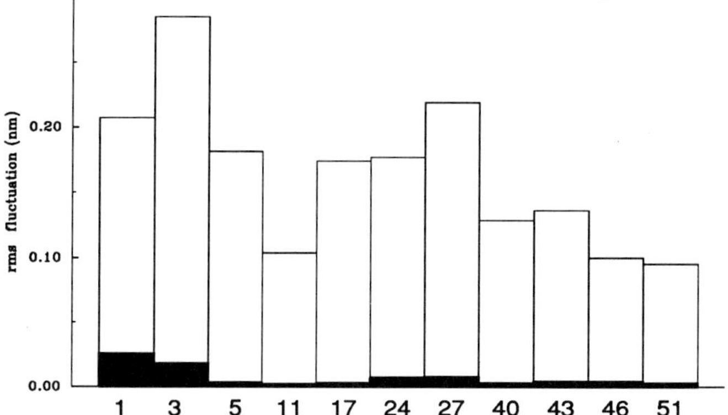

Figure 4: The rms fluctuations for sidechains containing carbonyl oxygens calculated from the restrained (filled bars) and unrestrained (open) simulations of EGF.

It is also important to note that chemical shift restraints not only affect the positions and dynamics of the restrained protons, but also the groups which contribute to the observed shifts. Whilst aromatic rings often give rise to many NOE-derived restraints and the positions consequently well-defined, little or no information can be obtained to define the position of sidechains such as Asp, Glu or Gln. A comparison of the rms variation of sidechain carbonyl oxygen in EGF atoms is shown in Figure 4. As can be seen, the effect on these groups is similar to that seen for the CαH positions. Again, whilst in certain conditions these sidechains would be expected to have well-defined conformations, here they are on the surface of the protein and on the whole would be expected to be disordered. This may be an artifact of the exclusion of electric field contributions, and explicit solvent.

The preliminary results presented here show the potential value of using chemical shift restraints. A natural progression would be to use the shifts of all protons bonded to carbon, given the good correlation of methyl shifts in earlier work (Perkins and Wüthrich 1979, Perkins and Dwek 1980). This will clearly add yet more significant information. There is as yet no good indication that amide chemical shifts can be calculated with any reliability, but already correlations between ^{13}C chemical shifts and protein secondary structure are observed (Spera and Bax 1992) and so it may be possible to use heteronuclear shifts in a similar manner.

Acknowledgements

We wish to thank EMBO for a Short Term Fellowship (TSH), ICI and Biomos for funding, Dr. A. E. Torda for many useful discussions, the Oxford Center for Molecular Sciences for use of computer facilities and Dr. M.P. Williamson for the communication of a manuscript prior to publication.

References

ApSimon, J.W. Craig, W.G. DeMarco, P.V. Mathieson, D.W and Saunders, L. (1967) *Tetrahedron* **23**, 2357-2373.

Bonvin, A.M.J.J. Boelens, R. and Kapein, R.M. (1992) *J. Biomol. NMR*

Brünger, A.T. and Karplus, M. (1991) *Acc. Chem. Res.*, **24**, 54-61.

Buckingham, A.D. (1960) *Can. J. Chem.* **38**, 300-307.

Bundi, A. and Wüthrich, K. (1979) *Biopolymers* **18**, 285-297.

Clore, G.M and Gronenborn, A.M. (1989) in *CRC Crit. Rev. Biochem.* **24** 479-563.

Clore, G.M and Gronenborn, A.M. (1991) *Annu. Rev. Biophys. Biophys. Chem.*, **20**, 29-63.

Cooke, R.M. Tappin, M.J. Campbell, I.D. Khoda, D. Miyake, T. Fuwa, T. Miyazawa, T. and Inagaki, F. (1990) *Eur. J. Biochem.* **193**, 807-817.

Gao, Y. Veitch, N.C. and Williams, R.J.P. (1991) *J. Biomol. NMR.* **1**, 457-471.

Gippert, G.P. Yip, P. Wright, P.E. and Case, D.A. (1990) *Biochem. Pharm.* **40**, 15-22.

Gregory, D.H and Gerig, J.T. (1991) *Biopolymers* **31** , 845-858.

Hommel, U. Harvey, T.S. Driscoll, P.C. and Campbell, I.D. (1992) *J. Mol. Biol* in Press.

Johnson, C.E. and Bovey, F.A. (1958) *J. Chem. Phys.* **29** 101-1014.

Kim, Y. and Prestegard, J.H. (1990) *Proteins* **8**, 377-385.

Nilges, M. Habazettl, J. Brünger, A.T. and Holak, T.A (1991) *J. Mol. Biol.* **219**, 499-510.

Ösapay, K. and Case, D.A. (1991) *J. Am. Chem. Soc.* **113**, 9436-9444.

Perkins, S.J and Wüthrich, K. (1979) *Biochim. Biophys. Acta* **576** 409-423.

Perkins, S.J and Dwek, R.A. (1980) *Biochemistry* **19**, 245-258.

Spera, S. and Bax, A. (1992) *J. Am. Chem. Soc.* **113** 5490-5492.

Sternlicht, H. and Wilson, D. (1967) *Biochemisty* 6, 2881-2892.

Torda, A.E. Scheek, R.M. and van Gunsteren, W.F. (1990) *J. Mol. Biol.* **214** 223-235.

Torda, A.E. Brunne, R.M., Huber, T., Kessler, H. and van Gunsteren, W.F. (1992) *J. Biomol. NMR*, submitted.

van Gunsteren, W.F. and Berendsen, H.J.C (1987) *Groningen Molecular Simulation (GROMOS) Library Manual.* Biomos, Nijenborgh 16, Groningen.

Wagner, G. Braun, W. Havel, T.F. Schaumann, T. Go, N. and Wüthrich, K. (1987) *J. Mol. Biol.* **196**, 611-639.

Williamson, M.P. and Asakura, T. (1991) *J. Magn. Reson.* **94**, 557-562.

Williamson, M.P. and Asakura, T. (1992) *J. Magn. Reson.*, submitted.

Williamson, M.P. Asakura, T. Nakamura, E. and Demura, M. (1992) *J. Biomol. NMR* **2** 83-98.

Wishart, D.S. Sykes, B.D. and Richards, F.M. (1992a) *Biochemistry* **31**, 1647-1651.

Wishart, D.S. Sykes, B.D. and Richards, F.M. (1992b) *J. Mol. Biol.* **222**, 311-333.

Wüthrich, K. (1989) *Science, (Wash.)* **243**, 45-50.

Yip, P. and Case, D.A. (1989) *J. Magn. Reson.* **83**, 643-648

Zürcher, R.F. (1967) *Prog. NMR Spectrosc.* **2**, 205-257.

Toward the Structure of Mosaic Proteins: Expression, Purification and Structural Analysis of a Pair of Fibronectin Type1 Modules

Michael J. Williams, Isabelle Phan, Robin T. Applin[a], Martin Baron and Iain D. Campbell

Department of Biochemistry, Oxford University, Oxford OX1 3QU, UK
and
[a]Dyson Perrins Laboratory, Oxford University, Oxford OX1 3QU, UK

I. Introduction

The rapid growth in protein sequence and structural data has confirmed that many proteins are related via common ancestors. Within families, their tertiary structures have been shown to be highly conserved even though their sequences may be considerably divergent. Thus it would appear that only a few residues are necessary to maintain a structure while others can drift to allow the evolution of new or varied functions. Smaller units of conserved sequence have been found repeated in many diverse proteins (1). They can be detected by a consensus sequence typically of 40-100 amino acids with conserved disulphides and hydrophobic residues which define their structures. They form compact structures, often encoded on single exons with the same phase. Thus their duplication, insertion and deletion can aid the development of mosaic proteins, without altering their reading frame (2-4). Mosaic proteins are involved in numerous diverse biological activities including embryogenesis, cell migration, blood coagulation and fibrinolysis, tissue integrity and cell signalling. In some cases certain module families have been implicated in specific interactions with other macromolecules such as the fibronectin type1 module which can bind to fibrin.

The use of X-ray crystallography to examine the structure of mosaic proteins has been hindered by the difficulties of crystallization, however nuclear magnetic resonance (NMR) has proved to be a powerful tool for studying the solution structures of individual modules. The list of those structures determined by NMR include epidermal growth factor-like domains (5), complement control protein repeat (6), kringle (7) and fibronectin type 1, 2 and 3

TECHNIQUES IN PROTEIN CHEMISTRY IV

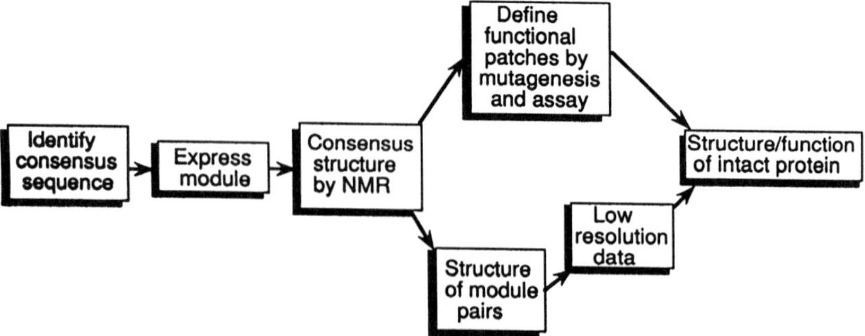

Figure 1: The strategy employed to study the structure and function of mosaic proteins.

domains (8-10). In a number of cases, several examples of a particular module structure are now known, confirming the conservation of their consensus structures. Thus it appears that once the structure of one module has been determined, all the other members of the family can be modelled with some confidence. To model larger regions or whole mosaic proteins it is important to confirm that the module structures are maintained and to find out how they fit together in pairs and triplets, before combining this with low resolution data such as electron microscopy.

This article describes our present strategy for the study of proteins with an aim to further understand the relationships between structure and function, by dissecting large proteins into manageable domains for NMR and rebuilding models for the intact protein, in conjunction with the identification of functional motifs (figure 1). The expression, purification and structural analaysis of the 4th and 5th type1 modules as a 93 amino acid pair from fibronectin ($F1^4.F1^5$) is discussed as an example. This is the first description of the structure of a module pair.

A. *Fibronectin, a Mosaic Protein*

Fibronectin is a large multidomain glycoprotein (11) found as a soluble dimer in plasma, where it is involved in blood clotting through it's affinity for both fibrin and platelets. It is also a major constituent of the extracellular matrix, providing a network of anchorage sites for cell surface receptors which assist the maintenance of tissue integrity and cell migration. Fibronectin's diverse functional properties are due to its ability to interact with many different macromolecules through its various domains (12). The protein consists of three types of modules called types 1,2 & 3 (13,14) which account for 90% of the sequence.

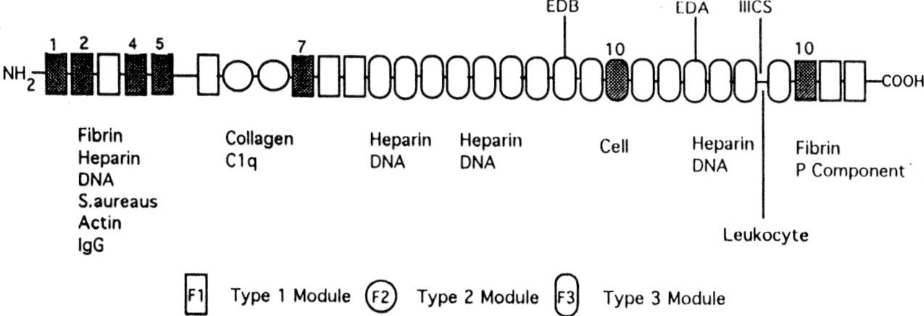

Figure 2: Modular structure of a fibronectin monomer showing the various domains involved in binding interactions. Shaded modules have been successfully expressed using the alpha factor system.

B. Identification of Type1 Consensus Sequences

The type1 module is repeated twelve times in fibronectin and single copies are also found in tPA and blood factor XII. It is typically of about 45 amino acids in length and is defined by a consensus amino acid sequence of conserved residues, encoded on a single phase1 exon. The consensus sequence is characterised by four cysteines which form two intra domain disulphide bonds and a number of hydrophobic residues and glycines. The sequences can be quite divergent. NMR structural determinations of the 7th type1 module from fibronectin and the N-terminal type1 module in tPA (8,15) confirm that only 8 conserved residues are sufficent to maintain the structural homology. The consensus sequence developed from these repeats can be used to search the protein sequence data base for further examples.

II. Protein Expression and Purification

NMR structural studies require tens of milligram quantities of folded, soluble protein with the possiblity of heteronuclear ($^{15}N/^{13}C$) labelling being important for peptides greater than about 10-15kD. Protein modules have been successfully produced by protein synthesis, proteolysis and recombinant expression. Protein synthesis is limited by the technical difficulty of making peptides larger than about 40-50 residues and by the need for refolding, while proteolysis is reliant on the availability of suitable proteolytic sites and a considerable supply of the intact protein. Recombinant expression in both prokaryotic and eukaryotic hosts has on the other hand, proved highly versatile and a wide range of systems are now available (16).

In addition to providing suitable quantities of the peptide of choice, efficient expression can allow heteronuclear labelling and the production of mutants for further functional studies. Potental problems include the difficulty of isolating the correct peptide in the absence of specific assays and refolding, especially for peptides containing disulphide bonds. However a number of studies have demonstrated that a yeast secretion system based on the α-factor leader sequence (17) can direct the secretion of authentically folded, biologically active heterologous proteins into the medium (16-19).

The cloned gene is ligated in phase with the leader sequence and expression of the hybrid is directed by a suitable yeast promoter. The authentic product is released by a peptidase that cuts at the C-terminal end of the dibasic pair Lys.Arg and can make up to 90% of the extracellular protein as is the case with human EGF, greatly simplifying purification (20).

A. Vector Construction

The polymerase chain reaction (PCR) was used to amplify a DNA fragment encoding the $F1^4.F1^5$ module pair from a cloned fragment of the human fibronectin gene. The 5' oligonucleotide primer had the sequence GCTGAGAAGTGTTTTGATCATG and the 3' primer had the sequence TAATGGATCCTTAAGAGGTGTGCCTCTCACACT. The former commencing with the codon for the first N-terminal residue and the latter with a sequence, complementary to the C-terminal coding region followed by 3' non-complementary sequence encoding a stop codon and *Bam* HI site.

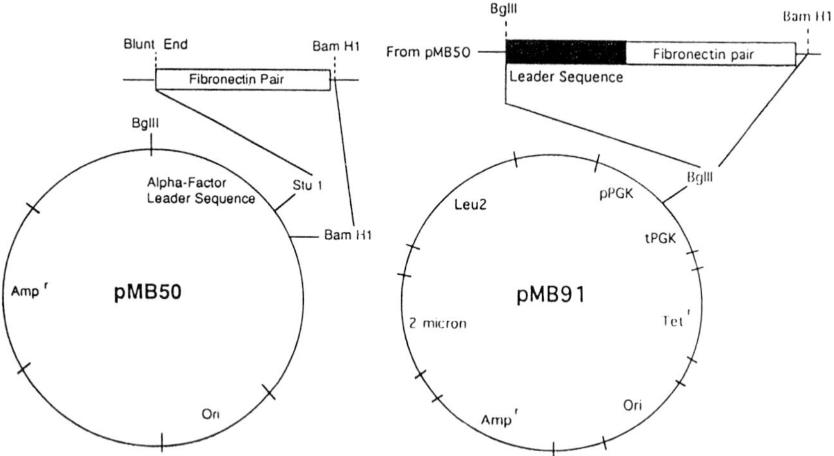

Figure 3: Construction of the pMA91 expression vector. Positive transformants were selected for by ampicillin resistance in E.Coli and ability to grow in leucine minus media for yeast.

The reaction was performed over 25 cycles, each one consisting of a 1min melting step at 94°C, a 1min annealing step at 56°C and a 3min extension step at 72°C . The PCR product was subsequently cut with *Bam* HI and ligated between the *Stu* I and *Bam* HI sites of pMB50, in phase with the yeast a-factor leader sequence.

After sequencing the cloned DNA the leader/module pair fusion was excised as a *Bgl* II/*Bam* HI fragment and ligated into the unique *Bgl* II site of the yeast shuttle vector pMA91 (21) to form the expression plasmid pMA91/$F1^4.F1^5$.

B. Expression and Purification of $F1^4.F1^5$

One litre cultures of yeast cells (*Saccharomyces cerevisae* MD50/a,a/leu2,3) transformed with pMA91/$F1^4.F1^5$ were grown for 60 hours in rotary shaking incubators at 30°C. The expressed peptide was batch absorbed to C-18 reverse phase silica beads present in the media, before being eluted off with 60% acetonitrile/0.1% trifluoroacetic acid (TFA). After lyophilizing and redissolving in a small volume of water, the crude sample was purified by reverse phase HPLC. An increasing gradient of 80% acetonitrile/0.1%TFA was run against 0.1%TFA on a C-8 reverse phase semi-prep column, at 280nm. The major protein peak was collected and rerun, resulting in 0.5-1mg of pure protein per litre of yeast culture.

C. Characterisation of $F1^4.F^5$

The authenticity of the peptide's primary structure was confirmed by SDS-page electrophoresis and mass spectroscopy.

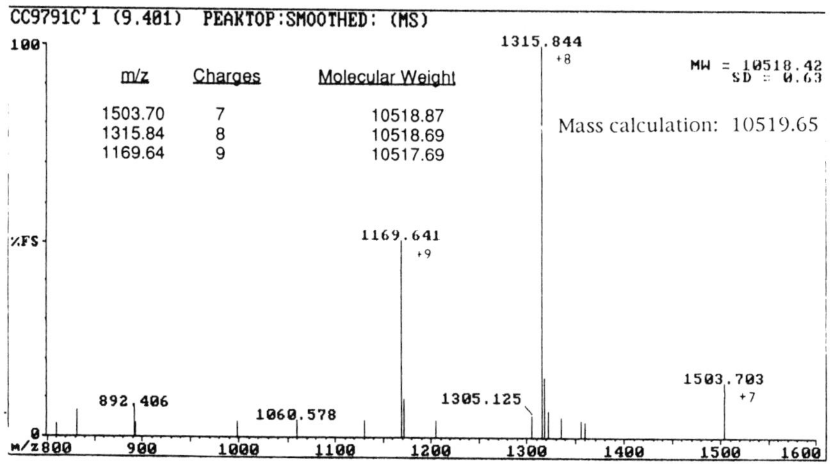

Figure 4: Mass spectrum of $F1^4.F1^5$ showing multiple charged species.

A VG Biotech BioQ atmospheric pressure quadrupole spectrometer equipped with an electrospray interface was used to obtain the low resolution spectrum shown in figure 4. The sample (c.a.20pmol/ml) was run from an acetonitrile/water (1:1) solution containing 1% formic acid. The derived molecular weight of 10,518.42+/-0.63Da, from a single measurement is sufficently accurate compared to the calculated average value of 10519.65Da, to confirm accurate N-terminal processing, formation of disulphide bonds and the lack of unexpected post transcriptional modifications. Multiple analysis on proteins below 20KDa using an internal standard affords mass measurements with a SD error of +/-0.2Da. The electrospray technique can also indicate the presence of impurites below the 2% level.

The presence of secondary structure in the peptide was confirmed from 1D ^{1}H NMR spectra. Alpha resonances were shifted downfield from their random coil positions towards the aromatic ring resonances, suggesting extensive beta sheet structure. While the sharpness of the peaks indicated the homogeneity of the peptide sample and its conformation.

III. NMR Structural Analysis of $F1^4.F1^5$

The basic methodology used to study protein structures by ^{1}H NMR has been previously described (15, 22). NMR experiments were recorded on a 3mM sample dissolved in either D_2O or $90\%H_20/10\%D_20$ at 37^oC and pH4.5 (uncorrected meter readings for D_20) on AM-500MHz and AM-600MHz Bruker spectrometers. To shift overlapping resonances, experiments were repeated at 30^oC and/or pH5.5. Nuclear Overhauser effect (NOE) spectra (23) and double quantum filtered (DQF) correlation spectra (24) were recorded in a phase sensitive manner by the time proportional phase increment method. Mixing times of 125msec and 200msec were used for NOESY experiments. Homonuclear Hartmann-Hahn (HOHAHA) spectra were recorded in reverse mode using the WALTZ-17 mixing sequence (25), with mixing times of approximately 45msec. Irradiation of the solvent resonance was carried out during the relaxation delay for spectra recorded in D_2O. For spectra recorded in $90\%H_20/10\%D_20$, a "jump-return" sequence (26) was used to suppress the solvent resonance. The receiver phase was adjusted to eliminate base-line distortions (27). Amide protons that exchanged relatively slowly with the solvent were identified by lyophilizing a sample from H_20 and redissolving in D_20 before recording an 8hour HOHAHA experiment. For $90\%H_20/10\%D_20$ spectra, the free induction decays were deconvoluted in the time domain with a Gaussian function prior to Fourier transformation in F2 to remove the residual water signal (28) and all spectra were processed using Felix software (Hare Research Inc, Washington).

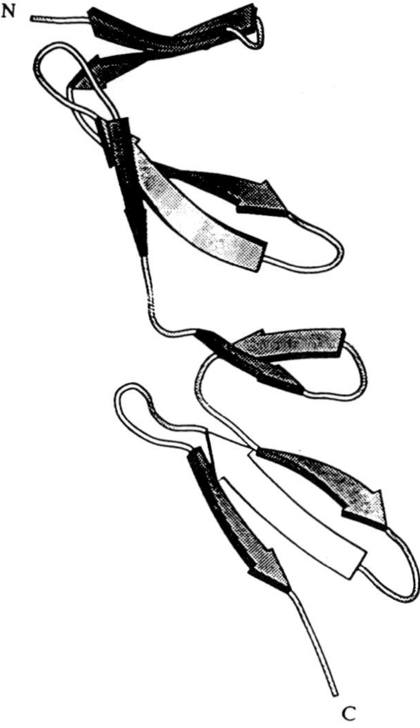

Figure 5: MOLSCRIPT diagram (30) of the $F1^4.F1^5$ module pair equivalent to residues 151-244 in the mature protein. A more detailed account of the structure will be published elsewhere.

The experimental restraints were used to compute three-dimensional structures using the program X-PLOR (29). Approximate folds were generated using a dynamical simulated annealing protocol starting from randomly generated backbone configurations with extended side-chains. These structures were further refined with successive rounds of an efficient simulated annealing protocol which relies on the SHAKE algorithm to maintain bond lengths (15). A MOLSCRIPT representation of the structures obtained is shown in figure 5.

The structures of two type1 modules have previously been described (8,15). The consensus structure is dominated by two regions of anti-parallel B-sheet, both of which have a right-handed twist and are stacked on top of each other to enclose a hydrophobic core of conserved residues. The structure is constrained by two disulphide bonds. The one between the first and third cysteines links the two B-sheets and the other links two strands of on of the B-sheets. Figure 5 shows the first module pair structure to be determined. It can be seen that both modules have maintained the consensus structure and have a well defined interface which restricts the movement of one relative to the other.

IV. Discussion

The combination of recombinant expression and NMR is a powerful technique for the exploration of protein module structure. Authentically folded protein domains can now be rapidly obtained in sufficent quantities for NMR and at expression levels that make heteronuclear labelling feasible. Heteronuclear NMR in 3D and 4D allow the determination of medium sized protein structures of up to about 25KD, but most intact mosaic proteins are too large for direct structural determination. However knowledge of the consensus structures of a mosaic protein and the possible interfaces between them can be used in conjuction with low resolution data of larger domains and the intact protein, to give a better indication of the structure of the intact protein.

The consensus structures are now known for all three fibronectin modules and electron microscopy studies have indicated that under certain conditions the intact fibronectin monomer exists as an elongated structure with a diameter of 2-2.5nm, a length of approximately 50nm and several flexible regions (31,32). The clusters of type1 modules shown in figure 2 can be modelled on the basis of the low resolution data available for fibronectin and the structure of the type1 module pair. It is now important to study other module interfaces in fibronectin so that a detailed tertiary model of the intact molecule can be built.

Further work is underway to map the biological functions associated with the fibronectin type1 family to the expressed peptides (figure 2), using various biological assays. The structure of functionally active modules can then be used to identify potentially important exposed residues for the design of site directed mutagenesis experiments, with the aim of localising functional patches using a similar procedure to that described by Handford *et al.* (33) to identify the calcium-binding motifs of EGF-like modules.

Acknowledgements

We thank Paul Driscoll for help with NMR, Tim Harvey for assistance with the structural computation. We also thank Dr Kingsman for the pMA91 plasmid, Helen Marden for supplying the cloned fibronectin gene and Tim Dudgeon of British Biotechnology for technological assistance with expression. This is a contribution from the Oxford Centre for Molecular Sciences which is supported by the SERC and the MRC.

References

1. Engel J., (1991). *Current Opinion in Structural Biology* **3**, 779-785.
2. Doolittle R.F., (1989). *TIBS* **14,** 244-245.
3. Patthy L., (1985). *Cell* **41**, 657-663.
4. Patthy L., (1987). *Febs. Lett.* **214**, 1-7.
5. Cooke, R.M., Wilkinson, A.J., Baron, M. Pasrore, M., Tappin, M.J., Campbell, I.D., Gregory, H. & Sheard, B., (1987). *Nature* **327**, 339-341.
6. Barlow, P., Norman, D.G., Baron, M. & Campbell, I.D., (1991). *Biochemistry* **30**, 997-1004.
7. Atkinson, A. & Williams, R.J.P., (1990). *J.Mol.Biol* **212**, 541-552.
8. Baron, M., Norman D., Willis A. & Campbell I.D., (1990). *Nature* **345** 642-646.
9. Constantine, K.L., Ramesh V., Banyai L., Trexler M., Patthy L., & Llinas M., (1991). *Biochemistry* **30**, 1663-1672.
10. Baron M., Main A.L., Driscoll R.C., Marden H.J., Boyd J., & Campbell I.D., (1992). *Biochemistry* **31**, 2068-2073.
11. Moser, D., (1988). *Fibronectin,* Academic Press Inc., San Diego.
12. Hynes R.O., (1990). *Fibronectins*, Springer Verlag, Berlin.
13. Petersen T.E., Thogersen H.C., Skorstengaard K., Vibe-Pedersen K., Sahl P., Sottrup-Jensen L., & Magnusson S., (1983). *Proc. Natl. Acad. Sci. USA* **80**, 137-141.
14. Ruoslahti, E., (1988). *Annu. Rev. Biochem.* **57**, 375-413.
15. Downing, A.K., Driscoll, P.C., Harvey, T.S., Dudgeon, T.S., Smith, B.O, Baron, M. & Campbell, I.D., (1992). *J. Mol. Biol.* **225**, 821-833.
16. Harris T.J.R., (1990). *Protein Production by Biotechnology*, Elsevier Applied Science.
17. Kuryan & Herskiwitz, (1982). *Cell* **30**, 933-943.
18. Bitter B.A., Chen K.K., Banks A.R. & Por-Hsiung L., (1984). *Proc. Nat. Acad. Sci. USA* **81**, 5330-5334.
19. Zsebo K.M., Lu H., Fieschko J.C., Goldstein L, Davis J., Duker K., Suggs S.V., Lai P. & Bitter G.A., (1986). *J. Biol. Chem.* 261, 5858-5865.
20. Brake A.J., Merryweather J.P., Coit D.G., Heberliein U.A., Masiarz F.R., Mullenbach G.T., Urdea M.S., Valenzuela P. & Barr P.J., (1984). *Proc. Nat. Acad. Sci. USA* **81,** 4642-4646.
21. Mellor J., Dobson M.J., Roberts N.A., Patel T., Kingsman A., & Kingsman S.M. (1983). *Gene* **24**, 1-14.
22. Wuthrich, K., (1986). *NMR of Protein and Nucleic acids*, John Wiley and Sons, New York.
23. Kumar A., Ernst R. & Wuthrich K., (1981). *Biochem. Biophys. Res. Commun,* **95**, 1-6.
24. Rance,P., Sorenson O.W., Bodenhausen G., Wagner G., Ernst R.R. & Wuthrich K., (1983). *Biochem. Biophys. Res. Commun.* **117**, 479-485.
25. Bax A., Sklenar V., Clore G.M. & Gronenborn A.M., (1987). *J. Am. Chem. Soc.* **109**, 6511-6513
26. Plateau, P. & Gueron, M. (1982). *J. Amer. Chem. Soc.* **104**, 7310-7311.
27. Marion D., & Bax A., (1988). *J.Magn.Reson.* **79**, 352-356.
28. Driscoll P.C., Clore G.M., Beress L., & Gronengurg A.M., (1989). *Biochemistry* **8**,2178-2187.
29. Brunger, A.T, (1988). *XPLOR Manual*, Yale University, New Haven , C.T.
30. Kraulis, P.J., MOLSCRIPT, (1991). *J. Appl. Cryst.* **24**, 946-950.
31. Odermat, E., Engel J., Richter H. & Hormann H., (1982). *Molec. Biol.* **156,** 109-123.
32. Erickson H.P., & Carrell N.A., (1983). *J. Biol. Chem.* **258**, 14539-14544.
33. Handford P.A., Mayhew M., Baron M., Winship P.R., Campbell I.D. & Brownlee G.G.,(1991). *Nature* **351**, 164-167.

Index

C

D

E

F

G

R

S

T

U

V